Birkhäuser

S. Ponnusamy

Foundations of Mathematical Analysis

S. Ponnusamy
Department of Mathematics
Indian Institute of Technology Madras
Chennai 600 036
India
samy@iitm.ac.in

ISBN 978-0-8176-8291-0 e-ISBN 978-0-8176-8292-7
DOI 10.1007/978-0-8176-8292-7
Springer New York Dordrecht Heidelberg London

Library of Congress Control Number: 2011941616

Mathematics Subject Classification (2010): 26-01, 26Axx

Printed on acid-free paper

Springer is part of Springer Science+Business Media
(www.birkhauser-science.com)

To my parents

Saminathan Pillai and Valliammai

Preface

Mathematical analysis is central to mathematics curricula not only because it is a stepping-stone to the study of advanced analysis, but also because of its applications to other branches of mathematics, physics, and engineering at both the undergraduate and graduate levels. Although there are many texts on this subject under various titles such as "Analysis," "Advanced Calculus," and "Real Analysis," there seems to be a need for a text that explains fundamental concepts with motivating examples and with a geometric flavor wherever it is appropriate. It is hoped that this book will serve that need. This book provides an introduction to mathematical analysis for students who have some familiarity with the real number system. Many ideas are explained in more than one way with accompanying figures in order to help students to think about concepts and ideas in several ways. It is hoped that through this book, both student and teacher will enjoy the beauty of some of the arguments that are often used to prove key theorems—regardless of whether the proofs are short or long.

The distinguishing features of the book are as follows. It gives a largely self-contained and rigorous introduction to mathematical analysis that prepares the student for more advanced courses by making the subject matter interesting and meaningful. The exposition of standard material has been done with extra care and abundant motivation. Unlike many standard texts, the emphasis in the present book is on teaching these topics rather than merely presenting the standard material. The book is developed through patient explanations, motivating examples, and pictorial illustrations conveying geometric intuition in a pleasant and informal style to help the reader grasp difficult concepts easily.

Each section ends with a carefully selected set of "Questions" and "Exercises." The questions are intended to stimulate the reader to think, for example, about the nature of a definition or the fate of a theorem without one or more of its hypotheses. The exercises cover a broad spectrum of difficulty and are intended not only for routine problem solving, but also to deepen understanding of concepts and techniques of proof. As a whole, the questions and exercises provide enough material for oral discussions and written assignments, and working through them should lead to a mature knowledge of the subject presented.

Some of the exercises are routine in nature, while others are interesting, instructive, and challenging. Hints are provided for selected questions and exercises. Students are strongly encouraged to work on these questions and exercises and to discuss them with fellow students and teachers. They are also urged to prepare short synopses of various proofs that they encounter.

Content and Organization: The book consists of eleven chapters, which are further divided into sections that have a number of subsections. Each section includes a careful selection of special topics covered in subsections that will serve to illustrate the scope and power of various methods in real analysis. Proofs of even the most elementary facts are detailed with a careful presentation. Some of the subsections may be ignored based on syllabus requirements, although keen readers may certainly browse through them to broaden their horizons and see how this material fits in the general scheme of things. The main thrust of the book is on convergence of sequences and series, continuity, differentiability, the Riemann integral, power series, uniform convergence of sequences and series of functions, Fourier series, and various important applications.

Chapter 1 provides a gentle introduction to the real number system, which should be more or less familiar to the reader. Chapter 2 begins with the concept of the limit of a sequence and examines various properties of convergent sequences. We demonstrate the bounded monotone convergence theorem and continue the discussion with Cauchy sequences. In Chapter 3, we define the concept of the limit of a function through sequences. We then continue to define continuity and differentiability of functions and establish properties of these classes of functions, and briefly explain the uniformly continuity of functions. In Chapter 4, we prove Rolle's theorem and the mean value theorem and apply continuity and differentiability in finding maxima and minima. In Chapter 5, we establish a number of tests for determining whether a given series is convergent or divergent. Here we introduce the base of the natural logarithm e and prove that it is irrational. We present Riemann's rearrangement theorem for conditionally convergent series. We end this chapter with applications of Dirichlet's test and summability of series. There are two well-known approaches to Riemann integration, namely Riemann's approach through the convergence of arbitrary Riemann sums, and Darboux's approach via upper and lower sums. In Chapter 6, we give both of these approaches and show their equivalence, along with a number of motivating examples. After presenting standard properties of Riemann integrals, we use them in evaluating the limits of certain sequences. In this chapter, we meet the fundamental theorem of calculus, which "connects the integral of a function and its antiderivative." In Chapter 7, we discuss the convergence and the divergence of improper integrals and give interesting examples of improper integrals, namely, the *gamma function* and the *beta function*. Our particular application emphasizes the integral test, the convergence of harmonic p-series, and the Abel–Pringsheim divergence test. We deal with a number of applications of the Riemann integral, e.g., in finding areas of regions bounded by curves and arc lengths of plane curves.

Chapter 8 begins with the theory of power series, their convergence properties, and Abel's theorem and its relation to the Cauchy product. Finally, we present some methods of computing the interval of convergence of a given power series. Chapter 9 contains a systematic discussion of pointwise and uniform convergence of sequence of functions. Students generally find it difficult to understand the difference between pointwise and uniform convergence. We illustrate this difference with numerous examples. We examine the close relationship between uniform convergence and integration—on the interchange of the order of integration and summation in the limit process—followed by a similar relationship between uniform convergence and differentiation. In Chapter 10, we introduce Fourier series with their convergence properties. In addition, we present a number of examples to demonstrate the use of Fourier series, such as how a given function can be represented in terms of a series of sine and cosine functions. The reader is encouraged to make use of computer packages such as *Mathematica*® and Maple™ where appropriate. Finally, in Chapter 11, we introduce a special class of functions, namely functions of bounded variation, and give a careful exposition of the Riemann–Stieltjes integral.

Numbering: The various theorems, corollaries, lemmas, propositions, remarks, examples, questions, and exercises are numbered consecutively within a chapter, without regard to label, and always carry the number of the chapter in which they reside. The end of the proof of a theorem, corollary, lemma, or proposition is indicated by a solid square ■ and the end of a worked-out example or remark by a bullet •.

Acknowledgments: Special thanks are due to my friend G.P. Youvaraj, who read the entire first draft of the manuscript with care and made many valuable suggestions. It is a great pleasure in offering my warmest thanks to Herb Silverman, who read the final manuscript and assisted me with numerous helpful suggestions. My Ph.D. students, especially, S.K. Sahoo, Allu Vasudevarao, and P. Vasundhra, helped me with the preparation of the LaTeX files on different occasions. Figures were created mainly by S.K. Sahoo. I thank them all for their help. I am grateful to my wife, Geetha, daughter, Abirami, and son, Ashwin, for their support and encouragement; their constant reminders helped me in completing this project on time.

The book was written with support from the *Golden Jubilee Book Writing Scheme* of the Indian Institute of Technology Madras, India. I thank IIT Madras for this support. It gives me immense pleasure in thanking the publisher and the editor, Tom Grasso, for his efficient responses during the preparation of the manuscript.

IIT Madras, India S. Ponnusamy

Contents

1

The Real Number System

This chapter consists of reference material with which the reader should be familiar. We present it here both to refresh the reader's memory and to have them available for reference. In Section 1.1, we begin by recalling elementary properties of sets, in particular the set of rational numbers and their decimal representations. Then we proceed to introduce the irrational numbers. In Section 1.2, we briefly discuss the notion of supremum and infimum and state the completeness axiom for the set of real numbers. We introduce the concept of one-to-one, onto, and bijective mappings, as well as that of equivalent sets.

1.1 Sets and Functions

1.1.1 Review of Sets

The notion of a set is one of the most basic concepts in all of mathematics. We begin our discussion with some set-theoretic terminology and a few facts from the algebra of sets. A set is a collection of well-defined objects (e.g., numbers, vectors, functions) and is usually designated by a capital letter $A, B, C, \ldots, X, Y, Z$. If A is a set, we write $a \in A$ to express "a is an element (or member) of A" or "a belongs to A." Likewise, the expression $a \notin A$ means "a is not an element of A" or "a does not belong to A." For instance, $A = \{a, b\}$ means that A consists of a and b, while the set $A = \{a\}$ consists of a alone. We use the symbol "$\emptyset$" to denote the empty set, that is, the set with no elements.

If B is also a set and every element of B is also an element of A, then we say that B is a subset of A or that B is contained in A, and we write $B \subset A$. We also say that A contains B and write $A \supset B$. That is,[1]

$$A \supset B \Longleftrightarrow B \subset A \Longleftrightarrow a \in B \quad \text{implies that } a \in A.$$

[1] The symbol $\Longleftrightarrow$ and the word "iff" both mean "if and only if."

S. Ponnusamy, *Foundations of Mathematical Analysis*,
DOI 10.1007/978-0-8176-8292-7_1,

Clearly, every set is a subset of itself, and therefore to distinguish subsets that do not coincide with the set in question, we say that A is a *proper subset* of B if $A \subset B$ and in addition, B also contains at least one element that does not belong to A. We express this by the symbol $A \subsetneq B$, a proper subset A of B. Since $A \subset A$, it follows that for any two sets A and B, we have

$$A = B \Longleftrightarrow B \subset A \quad \text{and} \quad A \subset B.$$

In this case, we say that the two sets A and B are equal. Thus, in order to prove that the sets A and B are equal, we may show that $A \subset B$ and $B \subset A$. When A is not a subset of B, then we indicate this by the notation

$$A \not\subset B,$$

meaning that there is at least one element $a \in A$ such that $a \notin B$. For every $A \subset X$, the *complement* of A, relative to X, is the set of all $x \in X$ such that $x \notin A$. We shall use the notation

$$A^c = X \backslash A = \{x : x \in X \text{ and } x \notin A\}.$$

The complement X^c of X itself is the empty set $\emptyset$. Also, $\emptyset^c = X$.

We often use the symbol $:=$ to mean that the symbol on the left is defined by the expression on the right. For instance,

$$\mathbb{N} := \{1, 2, \ldots\}, \quad \text{the set of natural numbers.}$$

A set can be defined by listing its elements or by specifying a property that determines the elements in the set. For instance,

$$A = \{2n : n \in \mathbb{N}\}.$$

That is, $A = \{x : P(x)\}$ represents the set A of all elements x such that "the property $P(x)$ is true." Also, $B = \{x \in A : Q(x)\}$ represents the subset of A for which the "property $Q(x)$" holds. For instance,

$$B = \{1, 3\} \quad \text{or} \quad A = \{x : x \in \mathbb{N}, 2x^3 - 9x^2 + 10x - 3 = 0\}.$$

For a given set A, the *power set* of A, denoted by $\mathcal{P}(A)$, is defined to be the set of all subsets of A:

$$\mathcal{P}(A) = \{B : B \subset A\}.$$

If A and B are sets, then their *union*, denoted by $A \cup B$, is the set of all elements that are elements of either A or B:

$$A \cup B = \{x : x \in A \text{ or } x \in B\}.$$

Clearly $A \cup B = B \cup A$. The *intersection* of the sets A and B, denoted by $A \cap B$, is the set consisting of elements that belong to both A and B:

$$A \cap B = \{x : x \in A \text{ and } x \in B\}.$$

Note that $A \backslash B$ is also used for $A \cap B^c$. Two sets are said to be *disjoint* if their intersection is the empty set. The notion of intersection and union can be extended to larger collections of sets. For instance, if Λ is an indexing set such as $\mathbb{N}$, then

$$\bigcup_{i\in\Lambda} A_i = \{x : x \in A_i \text{ for some } i \in \Lambda\}$$

and

$$\bigcap_{i\in\Lambda} A_i = \{x : x \in A_i \text{ for every } i \in \Lambda\}.$$

A collection of sets $\{A_i : i \in \Lambda\}$ is said to be *pairwise disjoint* if

$$A_i \cap A_j = \emptyset \quad \text{for } i, j \in \Lambda,\ i \neq j.$$

We do not include here basic set-theoretic properties, since these should be familiar from high-school mathematics.

We now list Giuseppe Peano's (1858–1932) *five axioms for* $\mathbb{N}$:

- $1 \in \mathbb{N}$.
- Each $n \in \mathbb{N}$ has a successor, namely $n+1$ (sometimes designated by n').
- 1 is not the successor of any $n \in \mathbb{N}$.
- If m and n in $\mathbb{N}$ have the same successor, then $m = n$, i.e., two distinct elements in $\mathbb{N}$ cannot have the same successor.
- Suppose $A \subset \mathbb{N}$. Then $A = \mathbb{N}$ if the following two conditions are satisfied:
  (i) $1 \in A$.
  (ii) If $n \in A$, then $n+1 \in A$.

The last axiom is the basis for the *principle of mathematical induction*, and so it is called the *induction axiom*.

The principle of mathematical induction reads as follows.

Theorem 1.1 (Principle of mathematical induction). *Suppose that $P(n)$ is a statement concerning $n \in \mathbb{N}$. If $P(1)$ is true and if $P(k+1)$ is true whenever $P(k)$ is true, then $P(n)$ is true for all $n \geq 1$.*

As an illustration of this theorem, the following can easily be proved:

(a) $\displaystyle\sum_{k=1}^{n} k = \frac{n(n+1)}{2}$ for all $n \geq 1$.

(b) $\displaystyle\sum_{k=1}^{n} k^2 = \frac{n(n+1)(2n+1)}{6}$.

(c) $\displaystyle\sum_{k=1}^{n} k^3 = \frac{n^2(n+1)^2}{4}$.

(d) $(n+2)! > 2^{n+1}$ for all $n \geq 1$.

(e) $(3+\sqrt{5})^n + (3-\sqrt{5})^n$ is an even integer for $n \geq 1$ (use the identity $a^{k+1} + b^{k+1} = (a^k + b^k)(a+b) - (a^{k-1} + b^{k-1})ab$).

We shall now begin to introduce the set $\mathbb{Q}$ of rational numbers and the set $\mathbb{R}$ of real numbers.

1.1.2 The Rational Numbers

A quotient of integers m/n ($n \neq 0$) is called a *rational* number. We assume that readers are familiar with the properties of the following basic sets:

$$\mathbb{Z} := \{\ldots, -2, -1, 0, 1, 2, \ldots\} \quad \text{the set of integers,}$$
$$\mathbb{Q} := \left\{\frac{m}{n} : m \in \mathbb{Z}, n \in \mathbb{N}\right\} \quad \text{the set of rational numbers.}$$

Clearly, $\mathbb{N} \subsetneq \mathbb{Z} \subsetneq \mathbb{Q}$. We remark that the representation of a rational number as a ratio of integers is not unique; for instance,

$$\frac{1}{2} = \frac{2}{4} = \frac{3}{6} = \cdots.$$

However, in the form m/n, if we assume that m and n have no common factor greater than 1, then the representation is unique. We frequently represent positive rational numbers in their decimal expansions. By a (positive) decimal fraction, we mean a number

$$0 \cdot a_1 a_2 a_3 \ldots,$$

where each a_k, $k \geq 1$, is an integer with $a_k \in \{0, 1, 2, \ldots, 9\}$. Here the ten integer values are called *digits*. When a decimal terminates, it means

$$0 \cdot a_1 a_2 a_3 \ldots a_n = \sum_{k=1}^{n} \frac{a_k}{10^k},$$

which is clearly a positive rational number. Thus, $\mathbb{Q}$ contains all terminating decimals such as

$$-0.123 = -\frac{123}{1000}, \quad 0.789 = \frac{789}{1000}, \ldots.$$

More generally, a decimal is an expression of the form

$$c_0 \cdot a_0 a_1 \ldots,$$

where $c_0 \in \mathbb{Z}$ and $a_k \in \{0, 1, 2, \ldots, 9\}$, $k = 1, 2, \ldots$.

Thus, there are two types of decimals, namely terminating (finite) and nonterminating (infinite). For instance, applying long division to $1/3$ and $3/7$ gives

$$\frac{1}{3} = 0.333\ldots \quad \text{and} \quad \frac{3}{7} = 0.428571428\ldots,$$

respectively. These nonterminating decimals are *repeating* and so may be abbreviated as

$$0.333\ldots = 0.\overline{3} \quad \text{and} \quad 0.428571428\ldots = 0.\overline{428571},$$

respectively. Thus, we formulate the following definition (omitting some technical details).

Definition 1.2. *A rational number is a number whose decimal expansion either* terminates after a finite number of places *or* repeats.

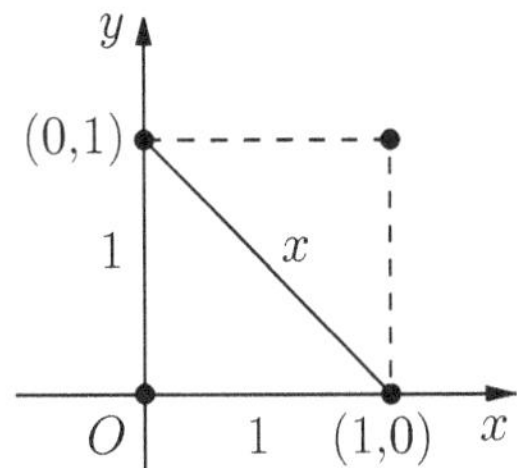

Fig. 1.1. The unit square.

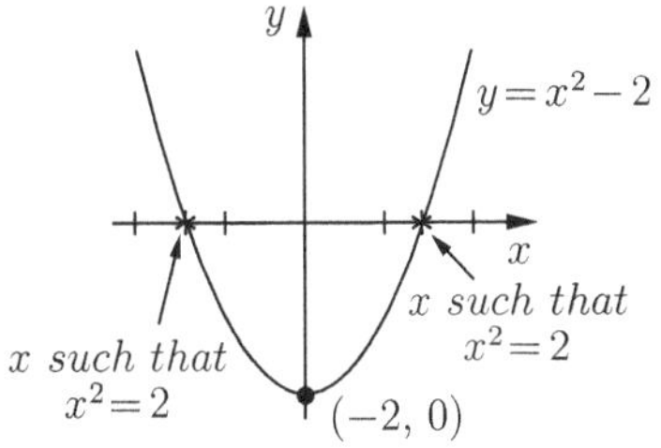

Fig. 1.2. Graph of $y = x^2 - 2$.

1.1.3 The Irrational Numbers

Although the set $\mathbb{Q}$ of rational numbers is a nice algebraic system, it is not adequate for describing many quantities such as lengths, areas, and volumes that occur in geometry. For example, what is the *length of the diagonal* in a square of unit length? (See Figure 1.1.)

What is the side length of a square with area 2? 3? 5? 7? In other words, is there a rational number x such that

$$x^2 = 2 \quad \text{or} \quad x^2 = 3 \quad \text{or} \quad x^2 = 5 \quad \text{or} \quad x^2 = 7?$$

What is the area of the closed unit disk $x^2 + y^2 \leq 1$?

Theorem 1.3. *There is no rational number x such that $x^2 = 2$.*

Proof. Suppose for a contradiction that $x = m/n$, where m and n have no common factors. Then

$$\left(\frac{m}{n}\right)^2 = 2, \quad \text{i.e., } m^2 = 2n^2,$$

where $m \in \mathbb{Z}$ and $n \in \mathbb{N}$ have no common factors other than 1. This shows that m^2 is even, and so is m (if m were odd, then m^2 would be odd). Hence, there exists $k \in \mathbb{Z}$ such that $m = 2k$. This gives

$$(2k)^2 = 2n^2 \quad \text{or} \quad 2k^2 = n^2,$$

and therefore n is also even. The last statement contradicts our assumption that m and n have no common factor other than 1. ■

It turns out, then, that the solution of $x^2 - 2 = 0$ is not a rational number. We denote it by $\sqrt{2}$ and call it an irrational number.

If we draw the graph of $y = x^2 - 2$ (see Figure 1.2), the value of x at which the graph crosses the y-axis is thus a "new type" of number x, which satisfies the equation $x^2 - 2 = 0$. It is called an *irrational number* (see Questions 1.11(7)).

1.1.4 Algebraic Numbers

A natural number is called a *prime number* (or a prime) if it is greater than one and has no divisors other than 1 and itself. For example, $2, 3, 5, 7$ are prime numbers. On the other hand, $4, 6$ are not prime (since $4 = 2 \times 2$ and $6 = 2 \times 3$). There are infinitely many primes, as demonstrated by Euclid around 300 BC, and there are various methods to determine whether a given number n is prime.

Definition 1.4. *A number x is called algebraic if there exists an $n \in \mathbb{N}$, $a_0, a_1, \dots, a_n \in \mathbb{Z}$ such that*

$$a_0 + a_1 x + \cdots + a_n x^n = 0 \quad (a_n \neq 0).$$

For instance, it follows that

- Every rational is algebraic ($x = m/n$ implies $m - nx = 0$).
- $7^{1/3}$, $3^{1/2}$, $2^{1/2}$ all represent algebraic numbers, since they are the solutions of
$$x^3 - 7 = 0, \quad x^2 - 3 = 0, \quad x^2 - 2 = 0,$$
respectively.

Our next basic result shows that a rational number has a special relationship to polynomial equations with integer coefficients.

Theorem 1.5 (Rational zeros theorem). *Suppose that a rational number $r = p/q$ (in lowest term) solves the polynomial equation*

$$a_0 + a_1 x + a_2 x^2 + \cdots + a_n x^n = 0 \quad (a_0 \neq 0, \ a_n \neq 0),$$

where $n \geq 1$ and $a_k \in \mathbb{Z}$ for $0 \leq k \leq n$. Then

(a) *p divides a_0, i.e., $a_0 = p \cdot k$ for some integer k.*
(b) *q divides a_n, i.e., $a_n = q \cdot m$ for some integer m.*

Proof. By hypothesis,

$$a_0 + a_1 \left(\frac{p}{q}\right) + \cdots + a_n \left(\frac{p}{q}\right)^n = 0.$$

Multiplying by q^n gives

$$a_0 q^n + a_1 p q^{n-1} + \cdots + a_{n-1} p^{n-1} q + a_n p^n = 0,$$

or

$$a_n p^n = -q[a_0 q^{n-1} + a_1 p q^{n-2} + \cdots + a_{n-1} p^{n-1}].$$

It follows that q divides $a_n p^n$. But since p and q have no common factors, q cannot divide p^n, and so q must divide a_n. Similarly, solving the equation for $a_0 q^n$ shows that p must divide a_0. ∎

Example 1.6. Consider $x^2-2=0$. Then $a_0=-2$, $a_1=0$, and $a_2=1$. Thus, the only possible rational solutions of $x^2-2=0$ are ± 1, ± 2 (if $x=p/q$, then integer values of p for which p divides 2 are ± 1, ± 2, and the natural number q for which q divides 1 is 1). Substituting these possible solutions shows that $\sqrt{2}$ cannot be rational. •

1.1.5 The Field of Real Numbers

We have the natural proper inclusions

$$\mathbb{N} \subsetneq \mathbb{Z} \subsetneq \mathbb{Q} \subsetneq \mathbb{R},$$

where the set $\mathbb{R}$ consists of rational numbers (terminating and repeating decimals) and all irrational numbers (nonrepeating decimals) such as the algebraic number $\sqrt{2}$. The mathematical system on which we are going to base our analysis is the set $\mathbb{R}$ of all real numbers (see Section 1.2).

A *field* is a set F that possesses two binary operations, namely *addition* $(+)$ and *multiplication* $(\cdot)$, such that F is closed with respect to these two operations (meaning that $a, b \in F$ implies $a+b \in F$ and $a \cdot b \in F$) and satisfies the familiar rules of arithmetic:

- Addition is *commutative*, i.e., $a+b=b+a$ for each $a, b \in F$.
- Addition is *associative*, i.e., $(a+b)+c=a+(b+c)$ for each $a, b, c \in F$.
- There exists an element $0 \in F$ such that $0+a=a$ for all $a \in F$ (0 is called the *additive identity*).
- To every $a \in F$ there corresponds an *additive inverse* $-a \in F$ such that $a+(-a)=0$.
- Multiplication is *commutative*, i.e., $a \cdot b=b \cdot a$ for each $a, b \in F$.
- Multiplication is *associative*, i.e., $(a \cdot b) \cdot c=a \cdot (b \cdot c)$ for each $a, b, c \in F$.
- There exists an element $1 \in F$, $1 \neq 0$, such that $1 \cdot a=a$ for all $a \in F$ (1 is called the *multiplicative identity*).
- To every $0 \neq a \in F$ there corresponds a *multiplicative inverse* $a^{-1} \in F$ such that $a \cdot a^{-1}=1$.
- Multiplication is *distributive* over addition:

$$a \cdot (b+c)=a \cdot b+a \cdot c \quad \text{for each } a, b \in F.$$

We state the following elementary properties, and we leave their proofs as simple exercises.

Theorem 1.7. *In a field F, the following are consequences of the field axioms:*

(a) *The additive identity and the multiplicative identity are unique.*
(b) *The additive inverse of an element and the multiplicative inverse of a nonzero element are unique.*
(c) $a \cdot 0=0$ *for every* $a \in F$.
(d) $a \cdot (-b)=(-a) \cdot b=-(a \cdot b)$ *for every* $a, b \in F$.

(e) $-(-a) = a$ *for every* $a \in F$.
(f) $(-a)(-b) = ab$ *for every* $a, b \in F$.
(g) $a + c = b + c$ *implies* $a = b$ *for each* $a, b, c \in F$.
(h) $ab = 0$ *implies either* $a = 0$ *or* $b = 0$ *for each* $a, b \in F$.
(i) $ac = bc$ *and* $c \neq 0$ *implies* $a = b$.

1.1.6 An Ordered Field

Definition 1.8. *A field* F *is said to be an* ordered field *if there is a nonempty subset* P *of* F, *called* positive, *satisfying the following additional axioms:*

(a) *if* $a, b \in P$, *then* $a + b \in P$ *and* $a \cdot b \in P$ *(closed under addition and multiplication, respectively),*
(b) *for every* $a \in F$, *exactly one of the following holds:*

$$a \in P, \quad -a \in P, \quad a = 0.$$

Thus, for elements $a, b \in F$, we say that

- $a < b$ (or $b > a$) if $b - a \in P$,
- $a \leq b$ if $a < b$ or $a = b$,
- $b \in P \iff b > 0$,
- $-a \in P \iff a < 0$, i.e., a is called *negative.*

Property **(a)** in Definition 1.8 may be read as follows:

$$a, b \in P, \quad \text{i.e., } a > 0 \text{ and } b > 0 \implies a + b > 0 \text{ and } ab > 0.$$

Property **(b)** implies that for any pair of elements $a, b \in F$, exactly one of the following holds:

$$a < b, \quad a = b, \quad a > b.$$

The most familiar examples of fields are the set $\mathbb{Q}$ of rational numbers and the set $\mathbb{R}$ of real numbers.

We may now write down some familiar properties concerning the ordered relation of $\mathbb{R}$.

- If $a, b \in \mathbb{R}$, then *exactly* one of the following holds:
  $a < b \quad \text{or} \quad a > b \quad \text{or} \quad a = b.$ **[Law of trichotomy]**
- If $a, b, c \in \mathbb{R}$, then we have
  (a) $a < b$ and $b < c$ implies $a < c$. **[Law of transitivity]**
  (b) $a < b$ implies $a + c < b + c$. **[Law of compatibility w.r.t. addition]**
  (c) $a < b$ and $c > 0$ implies $ac < bc$.
  (d) $a < b$ and $c < 0$ implies $bc < ac$.
  [Law of compatibility w.r.t. multiplication]
  (e) $a \neq 0$ implies $a^2 > 0$.
- If $a \in \mathbb{R}$, then there exists a positive integer n such that $n > a$.
  [Archimedean property]

The field axioms together with an ordered relation make both $\mathbb{Q}$ and $\mathbb{R}$ what are called ordered fields. We do not include the details, since that would defeat the purpose of this book. So, we accept the following.

Theorem 1.9. *$\mathbb{Q}$ and $\mathbb{R}$ are ordered fields.*

For $a \in \mathbb{R}$, its *modulus* $|a|$ is defined by

$$|a| = \begin{cases} a & \text{if } a \geq 0, \\ -a & \text{if } a < 0. \end{cases}$$

We call $|a|$ the *absolute value* of a. It is particularly useful in describing distances: we interpret $|a|$ as the distance along the real line between 0 and a. In the same way, for $a, b \in \mathbb{R}$, we let $|a - b|$ denote the distance between a and b, which is same as the distance along the real line from 0 to $a - b$. Here is a list of basic properties of the absolute value.

Theorem 1.10. *For $a, b, c \in \mathbb{R}$,*

(a) $|a| = 0$, *with equality iff* $a = 0$.
(b) $-|a| \leq a \leq |a|$.
(c) *For* $r > 0$, $|a| < r$ *if and only* $-r < a < r$.
(d) $|ab| = |a| \cdot |b|$.
(e) $|a + b| \leq |a| + |b|$.
(f) $|a - b| \geq \big| |a| - |b| \big|$.
(g) $|a - b| = |b - a|$.
(h) $|a - c| \leq |a - b| + |b - c|$.

1.1.7 Questions and Exercises

Questions 1.11.

1. For what values of x does $x^2 \in \mathbb{N}$ imply $x \in \mathbb{Q}$?
2. For each a, b, c real, does there always exist a real x such that $ax^2 + bx + c = 0$?
3. If a set A has n elements, how many elements does the power set $\mathcal{P}(A)$ have?
4. Why do we usually express a fraction in lowest terms (i.e., without common factors)?
5. Is there a rational number x such that $x^3 - x - 7 = 0$?
6. Is $\sqrt{3 + \sqrt{2}}$ rational or irrational?
7. If $p > 1$ is a prime number, can $\sqrt{p}$ be a rational number?
8. If a and b are two irrational numbers, what can you say about $a + b$, $a - b$, and ab? How about αa, where α is a rational number?

Exercises 1.12.

1. Show that if $m, n \in \mathbb{N}$ and $x^m = n$ has no integer solution, then $\sqrt[m]{n}$ is irrational.
2. Show that neither $\sqrt[3]{6}$ nor $\sqrt{2+\sqrt{2}}$ is a rational number.
3. Show that $\sqrt{n}$ is irrational for every natural number n that is not a perfect square.
4. Show that following numbers are irrational:
   (a) $\sqrt{12}$. (b) $\sqrt{n+1}+\sqrt{n-1}$.
5. For $a, b \in \mathbb{R}$, using the axioms of an ordered field, show that
   (a) $0 < a < b$ iff $1/a > 1/b$.
   (b) $a < b \implies a^p < b^p$ whenever $0 < a < b$ and $p > 0$.
6. Show that for all $a_1, a_2, \dots, a_n \in \mathbb{R}$,

$$|a_1 + a_2 + \cdots + a_n| \leq |a_1| + |a_2| + \cdots + |a_n|.$$

7. Prove by the method of induction that
   (a) $\displaystyle\sum_{k=1}^{n} k^3 = \left(\sum_{k=1}^{n} k\right)^2$. (b) $\displaystyle\sum_{k=1}^{n} k(k+1) = \frac{n(n+1)(n+2)}{3}$.

1.2 Supremum and Infimum

Definition 1.13. *Suppose that A is a nonempty subset of $\mathbb{R}$.*

(a) *If there exists an M such that $x \leq M$ for every $x \in A$, then we say that M is an* upper bound *for A. In this case, we say that A is bounded above by M. Geometrically, this means that no point of A lies to the right of M on the real line.*

(b) *If there exists an m such that $x \geq m$ for every $x \in A$, then m is called a* lower bound *for A. In this case, we say that A is bounded below by m. Geometrically, this means that no point of A lies to the left of m on the real line.*

(c) *The set A is said to be* bounded *if it is bounded above and bounded below.*

We remark that a set A is bounded iff there exist real numbers m and M such that $A \subset [m, M]$, or equivalently, if there exists a positive number a such that $A \subset [-a, a]$. A set that is not bounded is said to be *unbounded.* Thus, a set S is unbounded if for each $R > 0$ there is a point $x \in S$ such that $|x| > R$. For instance, the set $A_1 = \{1/n : n \in \mathbb{N}\}$ is a bounded set, but $A_2 = (-\infty, 2]$ and $A_3 = \{n : n \in \mathbb{N}\}$ are unbounded.

Any finite set of real numbers obviously has a *greatest* element and a *smallest* element, but this property does not necessarily hold for infinite sets. For instance, $(0, 1]$ has a greatest element, namely 1, but neither the set $\mathbb{N}$ nor the interval $[0, 1)$ has a greatest element. On the other hand, $[0, 1)$ is bounded above by 1, and $\mathbb{N}$ is not bounded above by any real number.

We have seen that not all sets are bounded above. However, if a nonempty set of real numbers is bounded, it has a *least* upper bound. What does this mean?

1.2.1 Least Upper Bounds and Greatest Lower Bounds

Definition 1.14 (Least upper bound). *Let A be a nonempty subset of $\mathbb{R}$. Then a real number M is said to be the* least upper bound *(lub) of A in $\mathbb{R}$ if:*

(a) *A is bounded above by M.*
(b) *For any $\epsilon > 0$, there exists a point $y \in A$ such that $y > M - \epsilon$. That is, M is the smallest among all the upper bounds of A.*

If A has a least upper bound M, we write $M = \operatorname{lub} A$.

The condition (b) is equivalent to saying that $\alpha < M$ implies that α is not an upper bound for A. Equivalently, it means that if M' is an upper bound for A, then $M' \geq M$. For instance, every $M' \geq 2$ is an upper bound for the set $A = [0, 2)$, whereas 1.99999 is not. The set $\mathbb{N}$ is not bounded above, because for each M, there is a positive integer n with $n > M$, by the archimedean property of $\mathbb{R}$.

Lemma 1.15. *If the least upper bound of a set exists, then it is unique.*

Proof. Let $A \subseteq \mathbb{R}$, where A is bounded above. Suppose that α and α' are both least upper bounds for A. Then both α and α' are upper bounds for A. Since both α and α' are least upper bounds, we must have

$$\alpha' \leq \alpha \quad \text{and} \quad \alpha \leq \alpha'.$$

Thus, $\alpha = \alpha'$, as required. ∎

Definition 1.16 (Greatest lower bound). *A real number m is said to be the* greatest lower bound (glb) *of a set $A \subset \mathbb{R}$ if:*

(a) *A is bounded below by m.*
(b) *For any $\epsilon > 0$, there exists a point $y \in A$ such that $y < m + \epsilon$. That is, m is the largest among all the lower bounds of A.*

If A has a glb m, we write $m = \operatorname{glb} A$.

The condition (b) means that if $m' > m$, then m' is not a lower bound for A.

Lemma 1.17. *If a set has a greatest lower bound, then it is unique.*

Proof. The proof follows from arguments similar to those of the proof of Lemma 1.15, and so we omit the details. ∎

The completeness properties of the real numbers can be expressed in the following forms.

Definition 1.18 (Least upper bound property). *Every nonempty subset A of real numbers that has an upper bound has a least upper bound,* $\operatorname{lub} A$.

Definition 1.19 (Greatest lower bound property). *Every nonempty subset A of real numbers that has a lower bound has a greatest lower bound* $\operatorname{glb} A$.

It is this property that distinguishes $\mathbb{R}$ from $\mathbb{Q}$. For example, the algebraic equation

$$x^2 - 2 = 0$$

is solvable in $\mathbb{R}$ but not in $\mathbb{Q}$. There are uncountably many numbers in $\mathbb{R}$, such as π, that are neither rational nor algebraic.

Both these facts are intuitively obvious. These two theorems—also called the *continuum properties*—are fundamental results of analysis. In conclusion, $\mathbb{R}$ is an ordered field that satisfies the continuum properties.

Now we extend the notions of $\operatorname{lub} A$ and $\operatorname{glb} A$ in a convenient form as follows. For a nonempty subset A of real numbers, we define the supremum of A (denoted by $\sup A$) and the infimum of A (denoted by $\inf A$) by

$$\sup A = \begin{cases} \infty & \text{if } A \text{ has no upper bound,} \\ \operatorname{lub} A & \text{if } A \text{ is bounded above,} \end{cases}$$

and

$$\inf A = \begin{cases} -\infty & \text{if } A \text{ has no lower bound,} \\ \operatorname{glb} A & \text{if } A \text{ is bounded below,} \end{cases}$$

respectively. We remark that the symbols $\sup A$ and $\inf A$ always make sense and that $\inf A \leq \sup A$.

For $a \leq b$, important subsets of $\mathbb{R}$ are intervals:

- $(a, b) = \{x \in \mathbb{R} : a < x < b\}$. **[open interval]**
- $[a, b] = \{x \in \mathbb{R} : a \leq x \leq b\}$. **[closed interval]**
- $(a, b] = \{x \in \mathbb{R} : a < x \leq b\}$. **[half-open interval]**
- $[a, b) = \{x \in \mathbb{R} : a \leq x < b\}$. **[half-open interval]**

The two endpoints a and b are points in $\mathbb{R}$. A set consisting of a single point is sometimes called a *degenerate interval.* It is sometimes convenient to allow the symbols $a = -\infty$ and $b = +\infty$, so that

$$\begin{aligned} (-\infty, b) &= \{x \in \mathbb{R} : x < b\}, \\ (-\infty, b] &= \{x \in \mathbb{R} : x \leq b\}, \\ (a, \infty) &= \{x \in \mathbb{R} : x > a\}, \\ [a, \infty) &= \{x \in \mathbb{R} : x \geq a\}, \\ (-\infty, \infty) &= \{x : x \in \mathbb{R}\} = \text{ the real line.} \end{aligned}$$

More general subsets of $\mathbb{R}$ that we often use may be obtained by taking a finite or infinite union of intervals or a finite or infinite intersection of intervals. Finally, for $\delta > 0$ and $a \in \mathbb{R}$, we call

$$(a-\delta, a+\delta) = \{x \in \mathbb{R} : |x-a| < \delta\}$$

a *δ-neighborhood* of a; it consists of all points x that are within distance δ of a.

Example 1.20. If $A = [0,1)$, then we see that $\sup A = 1$. Indeed, 1 is an upper bound for A. To show that 1 is the least upper bound, it suffices to prove that each $M' < 1$ is not an upper bound of A. In order to do this, we must find an element $x \in [0,1)$ with $x > M'$. But we know that for every $M' < 1$, there exists an x, say $x = (M'+1)/2$, with

$$M' < x < 1.$$

This inequality clearly implies that M' cannot be an upper bound for $[0,1)$, i.e., $M = 1$ is the least upper bound. ●

Using the same procedure as in the above example, we have the following:

(1) If $A = \{1,2,3\}$, then $\inf A = 1$ and $\sup A = 3$.
(2) If $A = \{x : -1 \leq x < 3\}$, then $\inf A = -1$ and $\sup A = 3$.
(3) If $A = \{x : x > 3\}$, then A has no upper bound, so that $\sup A = \infty$. Also, $\inf A = 3$.
(4) If $A = \{x : x < 1\}$, then A has no lower bound, so that $\inf A = -\infty$. Also, $\sup A = 1$.
(5) The sets $\mathbb{Z}$ and $\mathbb{Q}$ are neither bounded above nor bounded below. On the other hand, the set $\mathbb{N}$ is bounded below but not bounded above. In fact, 1 is a lower bound for $\mathbb{N}$ and so is any number less than 1. Moreover, $\inf \mathbb{N} = 1$.

Definition 1.21. *Let $A \subset \mathbb{R}$. If the least upper bound M of A belongs to A, then we say that A has a largest element. The smallest element of A may be defined similarly.*

If a set A has a largest element M, then we call M the *maximum element* of the set A, and we write $M = \max A$. Similarly, if A has a *smallest element* m, we call m the minimum element of A and write $m = \min A$. In this case, we have $\inf A = \min A$ and $\sup A = \max A$.

If $a, b \in \mathbb{Q}$, then so is its average $(a+b)/2$, which lies between a and b. Thus, between any two rational numbers there are infinitely many rational numbers. This shows that given a rational number, we cannot talk about the "next largest rational number." This observation and the above discussion imply that the rational number system has certain gaps. The real number system fills those gaps. Moreover, a convenient way of representing rational numbers is geometrically, as points on a number line (see Figure 1.3).

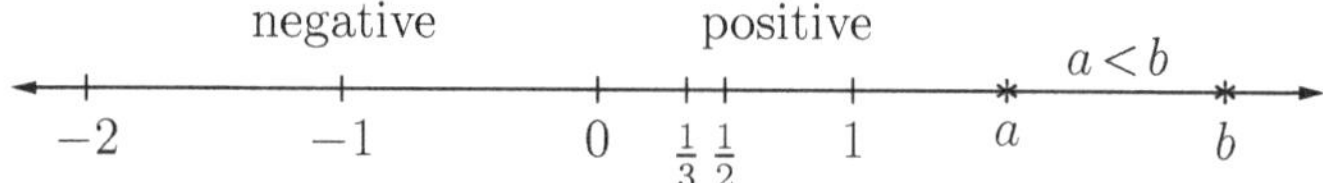

Fig. 1.3. The number line.

This geometric representation of rational numbers shows that the set of rational numbers has a natural order on the number line. If a lies to the left of b, then we write

$$a < b \quad \text{or} \quad b > a,$$

and say that "a is less than b" or "b is greater than a." For example,

$$\frac{1}{2} < \frac{2}{3} < 1.$$

The completeness axioms help us to conclude that there are both rational and irrational numbers between any two distinct real numbers (in fact, there are infinitely many of each).

1.2.2 Functions

Let X and Y be two nonempty subsets of a universal set, for example, $\mathbb{R}$. A *function* or *mapping*[2] f from X to Y is a rule, or formula, or assignment, or relation of association that assigns to each $x \in X$ a unique element $y \in Y$. We write

$$f : X \to Y \tag{1.1}$$

to denote the mapping f from X to Y. To be more precise about the rule of association, we say that a function from X to Y is a set f of ordered pairs in $X \times Y$ such that for each $x \in X$ there exists a unique element $y \in Y$ such that $(x, y) \in f$; i.e., if $(x, y) \in f$ and $(x, y') \in f$, then $y = y'$.

The set X on which the function f is defined is called the *domain* of f, and we write dom (f) for X. We call the set Y the *codomain* of f. When we define a map by describing its effect on the individual elements, we use the symbol $\mapsto$; thus "the mapping $x \mapsto y$ of X into Y" means that f is a mapping of X into Y taking each element x of X into a unique element y of Y. In practice, we denote the unique y by $f(x)$ and say that $f(x)$ is the image of x under f, or the value of f at x. Thus, when we use the notation $(x, y) \in f$, we write $y = f(x)$. For instance, if $X = \{a, b, c, d\}$ and $Y = \{1, 2, 3\}$, then the rule

$$a \mapsto 1, \quad b \mapsto 1, \quad c \mapsto 2, \quad d \mapsto 1,$$

defines a function $f \colon X \to Y$, because it assigns a unique element of Y to each element of X.

---

[2] The terms mapping, function, and transformation are frequently used synonymously.

In this book we will be concerned mainly with functions for which $X \subset \mathbb{R}$ and $Y = \mathbb{R}$, i.e., f is a real-valued function of a real variable x. However, when we discuss Fourier series, we will be dealing with complex-valued functions defined on a real variable t, although we shall not pay much attention to this. If a function is defined without its domain being indicated, then it is understood to be the largest subset on which the function is well defined. For example, if functions f, g, h are defined by

$$f(x) = \frac{1}{x}, \quad g(x) = \frac{1}{1+x^2}, \quad \text{and} \quad h(x) = \sqrt{1-x^2},$$

then $\operatorname{dom}(f) = \mathbb{R}\backslash\{0\}$, $\operatorname{dom}(g) = \mathbb{R}$, and $\operatorname{dom}(h) = [-1, 1]$. If f is defined on X and $S \subset X$, we can have $f\colon S \to Y$, and we call this new function the restriction of f in (1.1) to S and denote it by $f|_S$. Moreover,

$$f(S) := \{f(x) : x \in S\} = \{y \in Y\colon \text{there exists an } x \in S \text{ with } f(x) = y\}$$

is called the *image* of the set S under f. Clearly, $f(S)$ is a subset of the codomain Y, and $f(S)$ may be a proper subset of Y even if $X = S$. The subset $f(X)$ is called the *range* of f. For instance, if $X = \{a, b, c\}$, $Y = \{1, 2, 3, 4\}$, and $f\colon X \to Y$ is a function defined by the rule

$$a \mapsto 1, \quad b \mapsto 4, \quad c \mapsto 1,$$

then $\operatorname{dom}(f) = X$ and $f(X) = \{1, 4\}$ is the range of f, but $f(X) \neq Y$. Also, we remark that the notation $f(p)$ has two possible meanings, depending on whether p is an element of X or a subset of X. However, the standard practice of using lowercase letters for members of X and uppercase letters for sets makes the situation clear.

If $Y_1 \subset Y$, then the inverse image of Y_1 under f, denoted by $f^{-1}(Y_1)$, is the subset of X defined by

$$f^{-1}(Y_1) = \{x \in X : f(x) \in Y_1\}.$$

Also,

$$f(x) \in Y_1 \iff x \in f^{-1}(Y_1).$$

Thus, $f^{-1}(Y_1) \subset X$ for $Y_1 \subset Y$. If $Y_1 = \{y\} \subset Y$, then we write $f^{-1}(y)$ instead of $f^{-1}(\{y\})$. The notation f^{-1} is used for two different purposes but is universally accepted. However, the context of the usage of f^{-1} will always be made clear, so there should be no confusion about it.

For two mappings $f\colon X \to Y$ and $g\colon Y \to Z$ for which $f(X) \subset Y$, we can define the *composite* mapping $g \circ f\colon X \to Z$ by

$$(g \circ f)(x) = g(f(x)).$$

Composition is an associative operator, i.e.,

$$(g \circ f) \circ h = g \circ (f \circ h).$$

In general, composition is not commutative. For example, consider $f, g\colon \mathbb{R} \to \mathbb{R}$ by

$$f(x) = x^2 - 2x \quad \text{and} \quad g(x) = x^3 - 2.$$

Then

$$f(g(x)) = (x^3 - 2)^2 - 2(x^3 - 2) \quad \text{and} \quad g(f(x)) = (x^2 - 2x)^3 - 2.$$

Clearly, $g \circ f \neq f \circ g$, and so composition is not commutative.

The mapping $f\colon X \to Y$ is said to map X *onto* Y if the codomain and the range are equal, i.e., $f(X) = Y$. Therefore, in order to prove that f is onto, one must start with an arbitrary $y \in Y$ and then show that there is at least one $x \in X$ such that $f(x) = y$.

The mapping $f\colon X \to Y$ is said to be 1-to-1 (*one-to-one*) if it maps distinct elements into distinct elements, i.e., $f(x_1) \neq f(x_2)$ for all $x_1, x_2 \in X$ with $x_1 \neq x_2$. More formally, f is one-to-one if for $x_1, x_2 \in X$,

$$f(x_1) = f(x_2) \Longrightarrow x_1 = x_2.$$

A mapping that is both one-to-one and onto is called *bijective*.[3]

Example 1.22. Consider $X = \{-2, -1, 0, 1, 2, 3, 4\}$, $Y = \{0, 1, 4, 9, 16\}$, and the function $f\colon X \to Y$ given by $f(x) = x^2$. Then f is onto, but is not one-to-one because, for example, $f(-1) = f(1)$. ●

More generally, we have the following.

Example 1.23. Consider the mapping $f : A \to B$, $x \mapsto x^2$, where A and B are subsets of $\mathbb{R}$. Then

$$f(x_1) = f(x_2) \Longrightarrow (x_1 + x_2)(x_1 - x_2) = 0 \Longrightarrow x_1 = x_2 \quad \text{if } x_1 + x_2 \neq 0.$$

Therefore, we have the following results.

(a) Let $A = \mathbb{R}$ and $B = \mathbb{R}_0^+$, the set of all nonnegative real numbers. Since there exist $x_1, x_2 \in A$ such that $x_1 + x_2 = 0$, f is not one-to-one in this case. Similarly, if $A = B = \mathbb{Z}$, then f is not one-to-one, which can be shown by similar reasoning.

(b) Let $A = B = \mathbb{R}^+$, the set of all positive real numbers. Then for each $x_1, x_2 \in A$, we have $x_1 + x_2 \neq 0$, and therefore f is one-to-one in this case. Similarly, we see that if $A = B = \mathbb{N}$, the set of natural numbers, then f is one-to-one.

(c) If $A = B = \mathbb{R}$, then f is not onto, because the set of all real numbers is not the image of $\mathbb{R}$ under our mapping. Also, if $A = B = \mathbb{N}$, then f is not onto. However, if $A = \mathbb{R}$ and $B = \mathbb{R}_0^+$, then f is onto. In fact, when $A = B = \mathbb{R}^+$, f is bijective. ●

[3] The terms "one-to-one," "onto,", and "one-to-one correspondence" are sometimes referred as "injective," "surjective," and "bijective" mappings, respectively.

If $f\colon X \to Y$ is bijective, then we may define a function $g = f^{-1}$ by the rule

$$f^{-1}(x) = y \Longleftrightarrow f(y) = x.$$

We call f^{-1} the *inverse* of the function f. Also, we have

$$f \circ f^{-1}(y) = y \quad \text{for all } y \in Y \quad \text{and} \quad f^{-1} \circ f(x) = x \quad \text{for all } x \in X.$$

Example 1.24. Consider $f(x) = x^2 - 2x$ and $g(x) = x^3 - 2$. The function f is not one-to-one, because $f(0) = f(2) = 0$. On the other hand, g is bijective, because for each $y \in \mathbb{R}$, $y = x^3 - 2$ has the unique solution $x = (y + 2)^{1/3}$. ●

If $f\colon X \to Y$ is a function, then the inverse relation f^{-1} defined above is not, in general, a function. More about onto functions and related inverses will be addressed in Chapter 4.

Definition 1.25. *A function $f : I \to \mathbb{R}$ is said to be*

- bounded above *if there exists an $M \in \mathbb{R}$ such that $f(x) \le M$ for all $x \in I$,*
- bounded below *if there exists an $m \in \mathbb{R}$ such that $f(x) \ge m$ for all $x \in I$,*
- bounded *if it is bounded both below and above, that is, if there exists an $R > 0$ such that $|f(x)| \le R$ for all $x \in I$,*

where I is some interval in $\mathbb{R}$ or some subset of $\mathbb{R}$.

1.2.3 Equivalent and Countable Sets

Suppose that A and B are two sets. We say that A is equivalent to B, written $A \sim B$, if there is a bijective (i.e., one-to-one and onto) mapping from A to B. If $A \sim B$, then we say that A and B have a one-to-one correspondence between them. The following theorem is easy to prove.

Theorem 1.26. *Given three sets A, B, and C, we have*

(a) $A \sim A$,
(b) $A \sim B \Longrightarrow B \sim A$, *and*
(c) $A \sim B,\ B \sim C \Longrightarrow A \sim C$.

In view of this theorem, we can now reformulate equivalents in the following form.

Definition 1.27. *Two sets A and B are said to be equivalent, written $A \sim B$, if there is a bijection from A to B. The sets A and B are then said to have the same cardinality.*

Example 1.28. Define $f : [0, 1] \to [a, b]$ $(a < b)$ by $f(x) = (1 - x)a + xb$. We see that f is bijective, and therefore $[0, 1]$ and $[a, b]$ are equivalent. ●

Definition 1.29. *A set S is finite if either $S = \emptyset$ or S is equivalent to $\{1, 2, \ldots, n\}$ for some $n \in \mathbb{N}$. A set that is not finite is called infinite. A set S is said to be countable or denumerable if either it is finite or $S \sim \mathbb{N}$, i.e., if there exists a one-to-one correspondence between $\mathbb{N}$ and the set S. A set is said to be uncountable or nondenumerable if it is not countable.*

Example 1.30. The set $\mathbb{N}$ is countable, since the bijection $f(x) = x$ does the job. In order to prove that the set $\mathbb{Z}$ is countable, we just need to notice that elements of $\mathbb{Z}$ can be written as a list of

$$0, -1, 1, -2, 2, -3, 3, \ldots.$$

This amounts to defining a bijection $f : \mathbb{N} \to \mathbb{Z}$ by

$$f(n) = \begin{cases} \dfrac{n-1}{2} & \text{if } n = 1, 3, 5, \ldots, \\ -\dfrac{n}{2} & \text{if } n = 2, 4, 6, \ldots. \end{cases}$$

Therefore, $\mathbb{Z}$ is countable. Similarly, the set of all even positive integers is countable. ●

We may now state without proof the following.

Theorem 1.31. (a) *Every subset of a countable set is countable.*
(b) *The set $\mathbb{Q}$ of rationals is countable.*
(c) *A countable union of countable sets is countable.*
(d) *The Cartesian product $A \times B$ of countable sets A and B is countable.*
(e) *The set $\mathbb{R}$ is uncountable.*

Since $\mathbb{R}$ is uncountable, the set $\mathbb{R} \setminus \mathbb{Q}$ of all irrational numbers is uncountable. In particular, any interval that contains more than one point is uncountable. Indeed, the fact that there are uncountably many real numbers in $(0, 1)$ follows from constructing, for example, the set of all infinite sequences of 0's and 1's, which can be shown to be uncountable. Therefore, we have a natural question to look at: Are there *other familiar sets* that are uncountable? We note that according to the definition, the counting convention is via bijections, and the set of real numbers actually has in some sense many more numbers than the set of rational numbers. In general, given a set X, does there exist a method of constructing another set from X that will contain more elements than X? If X is countable (finite or infinite), then the answer is trivial, because if X is finite, then one can obtain a new set simply by adding one more element that does not belong to X. However, if X is countably infinite, then a new set obtained by adding a finite number of elements or even a countably infinite number of elements to X will again be countable. Hence, we have to think of some other method. Indeed, a method of getting bigger and bigger sets follows from the definition of *power set*. Thus, the notion of cardinality of a set X will play an important role.

If a set S is finite, then the number of elements of X is defined to be the cardinality of S, denoted by $|S|$ or $\operatorname{card} S$. Thus, two finite sets A and B have the same size, i.e., $\operatorname{card} A = \operatorname{card} B$, if they contain the same number of elements. An important question is how to carry the notion of equal size over to infinite sets such as $\mathbb{N}$ and $\mathbb{Z}$? We have the following definition. Given two arbitrary sets A and B (finite or infinite), then we say that $\operatorname{card} A = \operatorname{card} B$ if there exists a bijection between them. In particular, the notion of equal size is an equivalence relation, and we then associate a number called the *cardinal number* to every class of equal-sized sets. At this point, it is important to note that it is often difficult to find the cardinal number of a set, since the definition requires a function that is both one-to-one and onto. We note that it is usually easier to find one-to-one functions than onto functions. Therefore, we may make use of the following theorem, due to Cantor and Bernstein, which we state without proof.

Theorem 1.32 (Cantor–Bernstein). *Let A and B be two sets. If there exists a one-to-one function $f : A \to B$ and another one-to-one function $g : B \to A$, then* $\operatorname{card} A = \operatorname{card} B$.

This theorem can be used to show, for example, that

$$\operatorname{card}(\mathbb{R} \times \mathbb{R}) = \operatorname{card} \mathbb{R}.$$

Moreover, the fact that $\mathbb{Q}$ is countable can also be obtained by showing that $\mathbb{Q}$ and $\mathbb{Z} \times \mathbb{Z}$ have the same cardinality.

1.2.4 Questions and Exercises

Questions 1.33.

1. Should a nonempty bounded set in $\mathbb{R}$ have a maximum? minimum?
2. Suppose that A is a nonempty set in $\mathbb{R}$ and $-A = \{-x : x \in A\}$. What are the relations among $\inf A$, $\sup A$, $\inf(-A)$, and $\sup(-A)$?
3. What will happen if we divide an inequality by a negative real number?
4. Let A and B be two nonempty subsets of $\mathbb{R}$ such that $A \cap B$ is nonempty. How are $\inf(A \cup B)$, $\min\{\inf A, \inf B\}$, $\sup(A \cup B)$, $\max\{\sup A, \sup B\}$, $\inf(A \cap B)$, and $\sup(A \cap B)$ related?
5. Let A and B be two nonempty bounded sets of positive real numbers and $C = \{xy : x \in A \text{ and } y \in B\}$. Must $\sup C = (\sup A)(\sup B)$? If so, what if either A or B contains negative real numbers?
6. Does the completeness axiom hold for $\mathbb{Q}$?
7. Does there exist a bijection from the interval $(0, 1)$ to $\mathbb{R}$?
8. Does there exist a bijection from the interval $(0, 1)$ to $[0, 1)$?
9. Is the composition of one-to-one (respectively onto, bijective) mappings one-to-one (onto, bijective)?

10. Suppose that f and g are functions such that $g \circ f$ is onto. Must g be onto? Should f be onto?
11. Suppose that f and g are functions such that $g \circ f$ is one-to-one. Must f be one-to-one? Should g be one-to-one?
12. If $f : A \to B$ and $g : B \to C$ are such that $f(A) = B$ and $g(B) = C$, should $(g \circ f)(A) = C$?
13. Which one of the following is not true?
    (a) $f(A \cup B) = f(A) \cup f(B)$. (b) $f^{-1}(A \cup B) = f^{-1}(A) \cup f^{-1}(B)$.
    (c) $f(A \cap B) = f(A) \cap f(B)$. (d) $f^{-1}(A \cap B) = f^{-1}(A) \cap f^{-1}(B)$.
14. Can a finite set be equivalent to a proper subset of itself?
15. Must a set be infinite if it is equivalent to a proper subset of itself?

Exercises 1.34.

1. Let A consist of all positive rational numbers x whose square is less than 2, and let B consist of all positive rational numbers y such that $y^2 > 2$. Show that A contains no largest number and B contains no smallest number.
2. For $x, y \in \mathbb{R}$, show that
$$\max\{x, y\} = \frac{x + y + |x - y|}{2} \quad \text{and} \quad \min\{x, y\} = \frac{x + y - |x - y|}{2}.$$
3. Let A and B be two nonempty subsets of $\mathbb{R}$ such that $B \subset A$ is nonempty. Prove that
$$\inf A \le \inf B \le \sup B \le \sup A.$$
4. Let A and B be two nonempty bounded sets of positive real numbers. Set $S_1 = A \cup B$, $S_2 = A \cap B$, $S_3 = \{x + y : x \in A \text{ and } y \in B\}$, $S_4 = \{x + a : x \in A\}$ for some $a > 0$, and $S_5 = \{xa : x \in A\}$ for some $a > 0$. Determine a relationship among
    (a) $\sup A$, $\sup B$, and $\sup S_1$. (b) $\sup A$, $\sup B$, and $\sup S_2$.
    (c) $\sup A$, $\sup B$, and $\sup S_3$. (d) $\inf A$, $\inf B$, and $\inf S_3$.
    (e) $\sup A$ and $\sup S_4$. (f) $\sup A$ and $\sup S_5$.
5. Using the completeness properties (see Definitions 1.18 and 1.19), prove the following version of the archimedean property of $\mathbb{R}$: *If a and b are positive real numbers, then there exists a positive integer n such that $na > b$.*
6. Suppose that $x, y \in \mathbb{R}$ are such that $y > x$. Use the previous exercise to prove that there exist a rational number and an irrational number strictly between x and y.
7. Determine the domain of each of the following functions:
    (a) $f(x) = \sqrt{x(x^2 - 1)}$. (b) $f(x) = \dfrac{x}{[x]}$. (c) $f(x) = \sqrt{\dfrac{x-1}{x-4}}$.
8. Prove that $(-\pi/2, \pi/2)$ and $\mathbb{R}$ are equivalent.

9. Explain why the mapping $f : \mathbb{N} \to \mathbb{N}$, $n \mapsto 2n - 1$, is not a one-to-one correspondence.
10. Consider the function

$$f(x) = \frac{ax + b}{cx + d}, \quad x \in \mathbb{R}\backslash\{-d/c\}.$$

Determine conditions on a, b, c, d such that f is its own inverse.

2

Sequences: Convergence and Divergence

In Section 2.1, we consider (infinite) sequences, limits of sequences, and bounded and monotonic sequences of real numbers. In addition to certain basic properties of convergent sequences, we also study divergent sequences and in particular, sequences that tend to positive or negative infinity. We present a number of methods to discuss convergent sequences together with techniques for calculating their limits. Also, we prove the *bounded monotone convergence theorem* (BMCT), which asserts that every bounded monotone sequence is convergent. In Section 2.2, we define the limit superior and the limit inferior. We continue the discussion with Cauchy sequences and give examples of sequences of rational numbers converging to irrational numbers. As applications, a number of examples and exercises are presented.

2.1 Sequences and Their Limits

An infinite *(real) sequence* (more briefly, a sequence) is a nonterminating collection of (real) numbers consisting of a first number, a second number, a third number, and so on:

$$a_1, a_2, a_3, \ldots .$$

Specifically, if n is a positive integer, then a_n is called the nth term of the sequence, and the sequence is denoted by

$$\{a_1, a_2, \ldots, a_n, \ldots\} \quad \text{or, more simply,} \quad \{a_n\}.$$

For example, the expression $\{2n\}$ denotes the sequence $2, 4, 6, \ldots$. Thus, a sequence of real numbers is a special kind of function, one whose domain is the set of all positive integers or possibly a set of the form $\{n : n \geq k\}$ for some fixed $k \in \mathbb{Z}$, and the range is a subset of $\mathbb{R}$. Let us now make this point precise.

Definition 2.1. *A* real sequence $\{a_n\}$ *is a real-valued function f defined on a set $\{k, k+1, k+2, \ldots\}$. The functional values*

S. Ponnusamy, *Foundations of Mathematical Analysis*,
DOI 10.1007/978-0-8176-8292-7_2,

$$f(k), f(k+1), f(k+2), \ldots$$

are called the terms *of the sequence. It is customary to write* $f(n) = a_n$ *for* $n \geq k$*, so that we can denote the sequence by listing its terms in order; thus we write a sequence as*

$$\{a_n\}_{n \geq k} \quad or \quad \{a_{n+k-1}\}_{n=1}^{\infty} \quad or \quad \{a_n\}_{n=k}^{\infty} \quad or \quad \{a_k, a_{k+1}, \ldots\}.$$

The number a_n is called the *general term* of the sequence $\{a_n\}$ (nth term, especially for $k = 1$). The set $\{a_n : n \geq k\}$ is called the *range of the sequence* $\{a_n\}_{n \geq k}$. Sequences most often begin with $n = 0$ or $n = 1$, in which case the sequence is a function whose domain is the set of nonnegative integers (respectively positive integers). Simple examples of sequences are the sequences of positive integers, i.e., the sequence $\{a_n\}$ for which $a_n = n$ for $n \geq 1$, $\{1/n\}$, $\{(-1)^n\}$, $\{(-1)^n + 1/n\}$, and the constant sequences for which $a_n = c$ for all n. The *Fibonacci sequence* is given by

$$a_0, \ a_1 = 1, \ a_2 = 2, \ a_n = a_{n-1} + a_{n-2} \quad \text{for } n \geq 3.$$

The terms of this Fibonacci sequence are called *Fibonacci numbers*, and the first few terms are

$$1, \ 1, \ 2, \ 3, \ 5, \ 8, \ 13, \ 21.$$

2.1.1 Limits of Sequences of Real Numbers

A fundamental question about a sequence $\{a_n\}$ concerns the behavior of its nth term a_n as n gets larger and larger. For example, consider the sequence whose general term is

$$a_n = \frac{n+1}{n} = 1 + \frac{1}{n}.$$

It appears that the terms of this sequence are getting closer and closer to the number 1. In general, if the terms of a sequence can be made as close as we please to a number a for n sufficiently large, then we say that the sequence *converges to* a. Here is a precise definition that describes the behavior of a sequence.

Definition 2.2 (Limit of a sequence). *Let $\{a_n\}$ be a sequence of real numbers. We say that the sequence $\{a_n\}$ converges to the real number a, or tends to a, and we write*

$$a = \lim_{n \to \infty} a_n \quad \text{or simply} \quad a = \lim a_n,$$

if for every $\epsilon > 0$, there is an integer N such that

$$|a_n - a| < \epsilon \quad \text{whenever } n \geq N.$$

In this case, we call the number a a limit of the sequence $\{a_n\}$. We say that the sequence $\{a_n\}$ converges (or is convergent or has limit) if it converges to some number a. A sequence diverges (or is divergent) if it does not converge to any number.

For instance, in our example above we would expect

$$\lim_{n\to\infty} \frac{n+1}{n} = 1.$$

The notions of convergence and limit of a sequence play a fundamental role in analysis.

If $a \in \mathbb{R}$, other notations for the convergence of $\{a_n\}$ to a are

$$\lim_{n\to\infty} (a_n - a) = 0 \quad \text{and} \quad a_n \to a \quad \text{as } n \to \infty.$$

The notation $a = \lim a_n$ means that *eventually* the terms of the sequence $\{a_n\}$ can be made as close to a as may be desired by taking n sufficiently large. Note also that

$$|a_n - a| < \epsilon \quad \text{for } n \geq N \iff a_n \in (a - \epsilon, a + \epsilon) \quad \text{for } n \geq N.$$

That is, a sequence $\{a_n\}$ converges to a if and only if every neighborhood of a contains all but a finite number of terms of the sequence. Since N depends on ϵ, sometimes it is important to emphasize this and write $N(\epsilon)$ instead of N. Note also that the definition requires some N, but not necessarily the smallest N that works. In fact, if convergence works for some N then any $N_1 > N$ also works.

To motivate the definition, we again consider $a_n = (n+1)/n$. Given $\epsilon > 0$, we notice that

$$\left| \frac{n+1}{n} - 1 \right| = \frac{1}{n} < \epsilon \quad \text{whenever } n > \frac{1}{\epsilon}.$$

Thus, N should be some natural number larger than $1/\epsilon$. For example, if $\epsilon = 1/99$, then we may choose N to be any positive integer bigger than 99, and we conclude that

$$\left| \frac{n+1}{n} - 1 \right| < \epsilon = \frac{1}{99} \quad \text{whenever } n \geq N = 100.$$

Similarly, if $\epsilon = 2/999$, then $1/\epsilon = 499.5$, so that

$$\left| \frac{n+1}{n} - 1 \right| < \epsilon = \frac{2}{999} \quad \text{whenever } n \geq N = 500.$$

Thus, N clearly depends on ϵ.

The definition of limit makes it clear that changing a finite number of terms of a given sequence affects neither the convergence nor the divergence of the sequence. Also, we remark that the number ϵ provides a quantitative measure of "closeness," and the number N a quantitative measure of "largeness."

We now continue our discussion with a fundamental question: *Is it possible for a sequence to converge to more than one limit?*

Theorem 2.3 (Uniqueness of limits). *The limit of a convergent sequence is unique.*

Proof. Suppose that $a = \lim a_n$ and $a' = \lim a_n$. Let $\epsilon > 0$. Then there exist two numbers N_1 and N_2 such that

$$|a_n - a| < \epsilon \quad \text{for } n \geq N_1 \quad \text{and} \quad |a_n - a'| < \epsilon \quad \text{for } n \geq N_2.$$

In particular, these two inequalities must hold for $n \geq N = \max\{N_1, N_2\}$. We conclude that

$$|a - a'| = |a - a_n - (a' - a_n)| \leq |a_n - a| + |a_n - a'| < 2\epsilon \quad \text{for } n \geq N.$$

Since this inequality holds for every $\epsilon > 0$, and $|a - a'|$ is independent of ϵ, we must have $|a - a'| = 0$, i.e., $a = a'$. ∎

Also, as a direct consequence of the definition we obtain the following: If $a_n \to a$, then $a_{n+k} \to a$ for any fixed integer k. Indeed, if $a_n \to a$ as $n \to \infty$, then for a given $\epsilon > 0$ there exists an $N \in \mathbb{N}$ such that $|a_n - a| < \epsilon$ for all $n \geq N$. That is,

$$|a_{n+k} - a| < \epsilon \quad \text{for all } n + k \geq N + k = N_1 \quad \text{or} \quad |a_m - a| < \epsilon \quad \text{for } m \geq N_1,$$

which is same as saying that $a_m \to a$ as $m \to \infty$.

Definition 2.4. *A sequence $\{a_n\}$ that converges to zero is called a* null sequence.

Examples 2.5. (i) The sequence $\{n\}$ diverges because no matter what a and ϵ we choose, the inequality

$$a - \epsilon < n < a + \epsilon, \quad \text{i.e., } |n - a| < \epsilon,$$

can hold only for finitely many n. Similarly, the sequence $\{2^n\}$ diverges.

(ii) The sequence defined by $\{(-1)^n\}$ is $\{-1, 1, -1, 1, \ldots\}$, and this sequence diverges by oscillation because the nth term is always either 1 or -1. Thus a_n cannot approach any one specific number a as n grows large. Also, we note that if a is any real number, we can always choose a positive number ϵ such that at least one of the inequalities

$$a - \epsilon < -1 < a + \epsilon \quad \text{or} \quad a - \epsilon < 1 < a + \epsilon$$

is false. For example, the choice $\epsilon = |1 - a|/2$ if $a \neq 1$, and $\epsilon = |1 + a|/2$ if $a \neq -1$, will do. If $a = 1$ or -1, choose ϵ to be any positive real number less than 1. Thus the inequality $|(-1)^n - a| < \epsilon$ will be false for infinitely many n. Hence $\{(-1)^n\}$ diverges.

(iii) The sequence $\{\sin(n\pi/2)\}_{n \geq 1}$ diverges because the sequence is

$$\{1, 0, -1, 0, 1, 0, \ldots\},$$

and hence it does not converge to any number, by the same reasoning as above.

(iv) The sequence $\{(-1)^n/n\}$ converges to zero, and so it is a null sequence. •

Definition 2.6. *A sequence* $\{a_n\}$ *is* bounded *if there exists an* $R > 0$ *such that* $|a_n| \leq R$ *for all* n. *A sequence is* unbounded *if it is not bounded.*

Since a convergent sequence eventually clusters about its limit, it is fairly evident that a sequence that is not bounded cannot converge, and hence the next theorem is not too surprising; it will be used in the proof of Theorem 2.8.

Theorem 2.7. *Every convergent sequence is bounded. The converse is not true.*

Proof. Let $\{a_n\}_{n\geq 1}$ converge to a. Then there exists an $N \in \mathbb{N}$ such that $|a_n - a| < 1 = \epsilon$ for $n \geq N$. It follows that $|a_n| < 1 + |a|$ for $n \geq N$. Define $M = \max\{1 + |a|, |a_1|, |a_2|, \dots |a_{N-1}|\}$. Then $|a_n| < M$ for every $n \in \mathbb{N}$.

To see that the converse is not true, it suffices to consider the sequence $\{(-1)^n\}_{n\geq 1}$, which is bounded but not convergent, although the odd terms and even terms both form convergent sequences with different limits. ■

2.1.2 Operations on Convergent Sequences

The sum of sequences $\{a_n\}$ and $\{b_n\}$ is defined to be the sequence $\{a_n + b_n\}$. We have the following useful consequences of the definition of convergence that show how limits team up with the basic algebraic operations.

Theorem 2.8 (Algebra of limits for convergent sequences). *Suppose that* $\lim_{n\to\infty} a_n = a$ *and* $\lim_{n\to\infty} b_n = b$, *where* $a, b \in \mathbb{R}$. *Then*

- $\lim_{n\to\infty}(ra_n + sb_n) = ra + sb$, $r, s \in \mathbb{R}$. **[Linearity rule for sequences]**
- $\lim_{n\to\infty}(a_n b_n) = ab$. **[Product rule for sequences]**
- $\lim_{n\to\infty} a_n/b_n = a/b$, *provided* $b \neq 0$. **[Quotient rule for sequences]**
- $\lim_{n\to\infty} \sqrt[m]{a_n} = \sqrt[m]{a}$, *provided* $\sqrt[m]{a}_n$ *is defined for all* n *and* $\sqrt[m]{a}$ *exists.*

Proof. The linearity rule for sequences is easy to prove. The quotient rule for sequences is easy if we prove the product rule for sequences (see also Questions 2.44(33) and 2.44(34)). We provide a direct proof.

We write

$$a_n b_n - ab = (a_n - a)b_n + (b_n - b)a.$$

Since every convergent sequence must be bounded, there exists an $M > 0$ such that $|b_n| \leq M$ (say), for all n. Let $\epsilon > 0$ be given. Again, since $b_n \to b$ as $n \to \infty$, there exists an N_2 such that

$$|b_n - b| < \frac{\epsilon}{2(|a| + 1)} \quad \text{for } n \geq N_2.$$

(We remark that we could not use $\epsilon/2|a|$ instead of $\epsilon/[2(|a| + 1)]$ because a could be zero.)

Also by the hypothesis that $a_n \to a$ as $n \to \infty$, there exists an N_3 such that

$$|a_n - a| < \frac{\epsilon}{2M} \quad \text{for } n \geq N_3.$$

Finally, for $n \geq \max\{N_2, N_3\} = N$, we have

$$\begin{aligned}|a_n b_n - ab| &\leq |a_n - a|\,|b_n| + |b_n - b|\,|a| \\ &< \frac{\epsilon}{2M} M + \frac{\epsilon}{2(|a|+1)}|a| < \frac{\epsilon}{2} + \frac{\epsilon}{2} = \epsilon.\end{aligned}$$

The product rule clearly follows.

The proof of third part follows from Lemma 2.9. The proof of the final part is left as a simple exercise (see Questions 2.44(16)). ∎

Lemma 2.9 (Reciprocal rule). *If* $\lim_{n\to\infty} b_n = b$ *and* $b \neq 0$*, then the reciprocal rule holds:*

$$\lim_{n\to\infty} \frac{1}{b_n} = \frac{1}{b}.$$

Proof. The proof is easy, and so we leave it as a simple exercise. ∎

Note that if $a_n = (-1)^n$ and $b_n = (-1)^{n-1}$, then $\{a_n^2\}$ and $\{a_n + b_n\}$ both converge, although individual sequences $\{a_n\}$ and $\{b_n\}$ diverge.

Example 2.10. Find the limit of each of these convergent sequences:

(a) $\left\{\frac{1}{n^p}\right\}$ $(p > 0)$. **(b)** $\left\{\frac{n^2 - 2n + 3}{5n^3}\right\}$. **(c)** $\left\{\frac{n^6 + 3n^4 - 2}{n^6 + 2n + 3}\right\}$.

Solution. **(a)** As n grows arbitrarily large, $1/n$ (and hence $1/n^p$) gets smaller and smaller for $p > 0$. Thus, $\lim_{n\to\infty} 1/n^p = 0$. Also, we note that if $\epsilon > 0$, then $|(1/n^p) - 0| < \epsilon$ or $n > 1/(\epsilon^{1/p})$. Thus, if N is any integer greater than $1/(\epsilon^{1/p})$, then

$$|(1/n^p) - 0| < \epsilon \quad \text{for all } n \geq N.$$

Thus, for each $p > 0$, $n^{-p} \to 0$ as $n \to \infty$. That is, $\{1/n^p\}$ is a null sequence for each $p > 0$.

(b) We cannot use the quotient rule of Theorem 2.8 because neither the limit for the numerator nor that for the denominator exists. On the other hand, we can divide the numerator and denominator by n^3 and then use the linearity rule and the product rule. We then have

$$\frac{n^2 - 2n + 3}{5n^3} = \frac{1}{5}\left(\frac{1}{n} - \frac{2}{n^2} + \frac{3}{n^3}\right) \to 0 \quad \text{as } n \to \infty.$$

(c) Divide the numerator and denominator by n^6, the highest power of n that occurs in the expression, to obtain

$$\lim_{n\to\infty} \frac{n^6 + 3n^4 - 2}{n^6 + 2n + 3} = \lim_{n\to\infty} \frac{1 + \frac{3}{n^2} - \frac{2}{n^6}}{1 + \frac{2}{n^5} + \frac{3}{n^6}} = 1.$$

In fact, if we set

$$a_n = 1 + \frac{3}{n^2} - \frac{2}{n^6} \quad \text{and} \quad b_n = 1 + \frac{2}{n^5} + \frac{3}{n^6},$$

then the linearity rule gives that $a_n \to 1$ and $b_n \to 1$ as $n \to \infty$. Finally, the quotient rule gives the desired limit, namely,

$$\lim_{n\to\infty} \frac{a_n}{b_n} = 1. \qquad \bullet$$

Suppose that $\{a_n\}$ is a sequence of real numbers such that $a_n > 0$ for all but a finite number of n. Then there exists an N such that $a_n > 0$ for all $n \geq N$. If the new sequence $\{1/a_{n+N}\}_{n\geq 0}$ converges to zero, then we say that $\{a_n\}$ diverges to ∞ and write $\lim a_n = \infty$. Equivalently, if $\lim a_n$ does not exist because the numbers $a_n > 0$ become arbitrarily large as $n \to \infty$, we write $\lim_{n\to\infty} a_n = \infty$. We summarize the discussion as follows:

Definition 2.11 (Divergent sequence). *For given sequences $\{a_n\}$ and $\{b_n\}$, we have*

(a) $\lim_{n\to\infty} a_n = \infty$ *if and only if for each $R > 0$ there exists an $N \in \mathbb{N}$ such that $a_n > R$ for all $n \geq N$.*
(b) $\lim_{n\to\infty} b_n = -\infty$ *if and only if for each $R < 0$ there exists an $N \in \mathbb{N}$ such that $b_n < R$ for all $n \geq N$.*

We do not regard $\{a_n\}$ as a convergent sequence unless $\lim a_n$ exists as a finite number, as required by the definition. For instance,

$$\lim_{n\to\infty} n^3 = \infty, \quad \lim_{n\to\infty} (-n) = -\infty, \quad \lim_{n\to\infty} 3^n = \infty, \quad \lim_{n\to\infty} (\sqrt{n} + 5) = \infty.$$

We do not say that the sequence $\{n^2\}$ "converges to ∞" but rather that it "diverges to ∞" or "tends to ∞." To emphasize the distinction, we say that $\{a_n\}$ diverges to ∞ (respectively $-\infty$) if $\lim a_n = \infty$ (respectively $-\infty$). We note that $\lim(-1)^n n$ is unbounded but it diverges neither to ∞ nor to $-\infty$.

Definition 2.12 (Oscillatory sequence). *A sequence that neither converges to a finite number nor diverges to either ∞ or $-\infty$ is said to* oscillate *or* diverge by oscillation. *An oscillating sequence with finite amplitude is called a* finitely oscillating sequence. *An oscillating sequence with infinite amplitude is called an* infinitely oscillating sequence.

For instance,

$$\{(-1)^n\}, \quad \{1 + (-1)^n\}, \quad \{(-1)^n(1 + 1/n)\}$$

oscillate finitely. We remark that an unbounded sequence that does not diverge to ∞ or $-\infty$ oscillates infinitely. For example, the sequences

$$\{(-1)^n n\}, \quad \{(-1)^n n^2\}, \quad \{(-n)^n\}$$

are all unbounded and oscillate infinitely.

Example 2.13. Consider $a_n = (n^2+2)/(n+1)$. Then

$$a_n = n\left(\frac{1+\frac{2}{n^2}}{1+\frac{1}{n}}\right).$$

From the algebra of limits we observe that

$$\lim_{n\to\infty}\frac{1+\frac{2}{n^2}}{1+\frac{1}{n}} = 1.$$

On other hand, $\lim_{n\to\infty} a_n$ does not exist. Indeed, we can show that $a_n \to \infty$ as $n\to\infty$. According to the definition, we must show that for a given $R>0$, there exists an N such that $a_n > R$ for all $n \geq N$. Now we observe that

$$a_n > R \Longleftrightarrow n+1+\frac{3}{n+1} > R+2,$$

which helps to show that $a_n > R$ if $n \geq R+2$. So we can choose any positive integer N such that $N \geq R+2$. We then conclude that $a_n \to \infty$ as $n\to\infty$. Similarly, we easily have the following:

(1) As in Example 2.10(**c**), we write

$$\lim_{n\to\infty}\frac{n^7+2n^3-1}{n^6+n^2+3n+1} = \lim_{n\to\infty}\frac{1+\frac{2}{n^4}-\frac{1}{n^7}}{\frac{1}{n}+\frac{1}{n^5}+\frac{3}{n^6}+\frac{1}{n^7}}.$$

The numerator tends to 1 as $n\to\infty$, whereas the denominator approaches 0. Hence the quotient increases without bound, and the sequence must diverge. We may rewrite in the present notation,

$$\lim_{n\to\infty}\frac{n^7+2n^3-1}{n^6+n^2+3n+1} = \infty.$$

(2) $\{n/3+1/n\}$, $\{n^3-n\}$, $\{(n^2+1)/(n+1)\}$, and $\{(n^3+1)/(n+1)\}$ all diverge to ∞.
(3) $\{(-1)^n n^2\}$ diverges but neither to $-\infty$ nor to ∞.
(4) $a_n \to \infty \Longrightarrow a_n^2 \to \infty$.
(5) If $a_n > 0$ for all large values of n, then $a_n \to 0 \Longrightarrow 1/a_n \to \infty$. Is the converse true? ●

Finally, we let $a_n = \sqrt{n^2+5n} - n$ and consider the problem of finding $\lim a_n$. It would not be correct to apply the linearity property for sequences (because neither $\lim\sqrt{n^2+5n}$ nor $\lim n$ exists as a real number). At this place it important to remember that the linearity rule in Theorem 2.8 cannot be applied to $\{a_n\}$, since $\lim\sqrt{n^2+5n}=\infty$ and $\lim n = \infty$. It is also not correct to use this as a reason to say that the limit does not exist. The supporting argument is as follows. Rewriting a_n algebraically as

$$a_n = \left(\sqrt{n^2+5n} - n\right)\frac{\sqrt{n^2+5n}+n}{\sqrt{n^2+5n}+n} = \frac{5n}{\sqrt{n^2+5n}+n} = \frac{5}{\sqrt{1+\frac{5}{n}}+1},$$

we obtain $\lim_{n\to\infty}\left(\sqrt{n^2+5n} - n\right) = 5/2$.

Remark 2.14. We emphasize once again that Theorem 2.8 cannot be applied to sequences that diverge to ∞ or $-\infty$. For instance, if $a_n = n+1$, $b_n = n$, and $c_n = n^2$ for $n \geq 1$, then it is clear that the sequences $\{a_n\}$, $\{b_n\}$, and $\{c_n\}$ diverge to ∞, showing that the limits do not exist as real numbers. Also, it is tempting to say that

$$a_n - b_n \to \infty - \infty = 0 \quad \text{and} \quad c_n - b_n \to \infty - \infty = 0 \quad \text{as } n \to \infty.$$

Note that ∞ is not a real number, and so it cannot be treated like a usual real number. In our example, we actually have $a_n - b_n = 1$ for all $n \geq 1$, and

$$c_n - b_n = n(n-1) \to \infty \quad \text{as } n \to \infty.$$

•

2.1.3 The Squeeze/Sandwich Rule

In the following squeeze rule, the sequence $\{b_n\}$ is "sandwiched" between the two sequences $\{a_n\}$ and $\{c_n\}$.

Theorem 2.15 (Squeeze/Sandwich rule for sequences). *Let $\{a_n\}$, $\{b_n\}$, and $\{c_n\}$ be three sequences such that $a_n \leq b_n \leq c_n$ for all $n \geq N$ and for some $N \in \mathbb{N}$. If*

$$\lim_{n\to\infty} a_n = \lim_{n\to\infty} c_n = L,$$

then $\lim_{n\to\infty} b_n = L$. If $b_n \to \infty$, then $c_n \to \infty$. Also, if $c_n \to -\infty$, then $b_n \to -\infty$.

Proof. Let $\epsilon > 0$ be given. By the definition of convergence, there exist two numbers N_1 and N_2 such that

$$|a_n - L| < \epsilon \quad \text{for } n \geq N_1 \quad \text{and} \quad |c_n - L| < \epsilon \quad \text{for } n \geq N_2.$$

In particular, since $a_n \leq b_n \leq c_n$ for all $n \geq N$, we have

$$L - \epsilon < a_n \leq b_n \leq c_n < L + \epsilon \quad \text{for } n \geq N_3 = \max\{N, N_1, N_2\},$$

showing that $|b_n - L| < \epsilon$ for $n \geq N_3$, as required.

We leave the rest as a simple exercise. ∎

Corollary 2.16. *If $\{c_n\}$ is a null sequence of nonnegative real numbers, and $|b_n| \leq c_n$ for all $n \geq N$, then $\{b_n\}$ is a null sequence.*

For instance, since $\{1/\sqrt{n}\}$ is null and $1/(1+\sqrt{n}) < 1/\sqrt{n}$ for all $n \geq 1$, $\{1/(1+\sqrt{n})\}$ is also a null sequence. Similarly, comparing $1/3^n$ with $1/n$, it follows easily that $\{1/3^n\}$ is a null sequence.

Corollary 2.17. *If* $\lim_{n\to\infty} a_n = 0$ *and* $|b_n - L| \leq a_n$ *for all* $n \geq N$, *then* $\lim_{n\to\infty} b_n = L$.

Proof. By the last corollary, it follows that $\{b_n - L\}$ is a null sequence, and so the desired conclusion follows. Alternatively, it suffices to observe that

$$|b_n - L| \leq a_n \Longleftrightarrow L - a_n \leq b_n \leq L + a_n$$

and apply the squeeze rule. ∎

For instance, using the squeeze rule, we easily have the following:

(a) $\lim_{n\to\infty} \cos n^2/n = 0$, because $-(1/n) \leq \cos n^2/n \leq 1/n$. With the same reasoning, one has

$$\lim_{n\to\infty} \frac{\sin(n\pi/2)}{n} = 0.$$

(b) $\lim_{n\to\infty} \{\sqrt{n+1} - \sqrt{n}\} = 0$ and $\lim_{n\to\infty} \sqrt{n}(\sqrt{n+1} - \sqrt{n}) = 1/2$. Moreover,

$$0 < \sqrt{n+1} - \sqrt{n} = \frac{1}{\sqrt{n+1} + \sqrt{n}} < \frac{1}{2\sqrt{n}}.$$

Note that the above inequality is useful in estimating $\sqrt{n}$. For $n = 1$, this gives $\sqrt{2} < 1.5$, and for $n = 2, 4$, we have $\sqrt{3} < 1.875$ and $\sqrt{5} < 2.25$. Indeed, for $n = 2$, we have

$$\sqrt{3} < \sqrt{2} + \frac{\sqrt{2}}{4} = \frac{5\sqrt{2}}{4} < \frac{5 \times 1.5}{4} = \frac{7.5}{4} = 1.875.$$

(c) $\lim_{n\to\infty} n/2^n = 0$. Indeed, using induction we easily see that $2^n \geq n^2$ for $n \geq 4$, so that

$$0 < \frac{n}{2^n} \leq \frac{1}{n}.$$

(d) $\lim_{n\to\infty} b_n = 1$ if $b_n = 1/(\sqrt{n^2+1}) + 1/(\sqrt{n^2+2}) + \cdots + 1/(\sqrt{n^2+n})$. We note that

$$\frac{n}{\sqrt{n^2+n}} < b_n < \frac{n}{\sqrt{n^2+1}}, \quad \text{i.e.,} \quad \frac{1}{\sqrt{1+1/n}} < b_n < \frac{1}{\sqrt{1+1/n^2}}.$$

(e) $\lim_{n\to\infty} c_n = \infty$ if $c_n = 1/(\sqrt{n+1}) + 1/(\sqrt{n+2}) + \cdots + 1/(\sqrt{n+n})$. We note that

$$c_n > \frac{n}{\sqrt{n+n}} = \frac{\sqrt{n}}{\sqrt{2}} = b_n,$$

where $b_n \to \infty$ as $n \to \infty$.

Using the squeeze rule, Theorem 2.8, and a few standard examples allows one to calculate limits of important sequences.

Example 2.18. Show that

(a) $\lim_{n\to\infty} a^{1/n} = 1$ for $a > 0$. **(b)** $\lim_{n\to\infty} n^{1/n} = 1$. **(c)** $\lim_{n\to\infty} \frac{n!}{n^n} = 0$.

Solution. **(a)** We consider the cases $a > 1$ and $a < 1$, since there is nothing to prove if $a = 1$. Suppose first that $a > 1$. Then $a^{1/n} \geq 1$, and so

$$a^{1/n} = 1 + x_n$$

for some sequence $\{x_n\}$ of positive real numbers. Then by the binomial theorem,

$$a = (1 + x_n)^n \geq 1 + nx_n \quad \text{for all } n \geq 1,$$

which is equivalent to

$$0 < a^{1/n} - 1 \leq \frac{a-1}{n} \quad \text{for all } n \in \mathbb{N}.$$

Thus, $a^{1/n} \to 1$ as $n \to \infty$ if $a > 1$. For $0 < a < 1$, we have $(1/a)^{1/n} \to 1$ as $n \to \infty$, and therefore, by the reciprocal rule,

$$a^{1/n} = \frac{1}{(1/a)^{1/n}} \to \frac{1}{1} = 1 \quad \text{as } n \to \infty.$$

The sequence $\{a^{1/n}\}$ is referred to as the nth root sequence.

(b) Clearly $(1 + 1)^n \geq 1 + n > n$, so that $n^{1/n} - 1 < 1$ for $n \geq 1$. Also, for $n \geq 1$, we observe that $n^{1/n} \geq 1$, so that $n^{1/n} - 1 = x_n$ with $x_n \geq 0$. In particular, using the binomial theorem, we deduce that

$$n = (1 + x_n)^n \geq 1 + nx_n + \frac{n(n-1)}{2}x_n^2 \geq 1 + \frac{n(n-1)}{2}x_n^2,$$

which implies that

$$0 \leq x_n = n^{1/n} - 1 \leq \sqrt{\frac{2}{n}} \quad \text{for } n \geq 1.$$

By the squeeze rule, $x_n \to 0$ as $n \to 0$, since $1/\sqrt{n} \to 0$. We conclude that $n^{1/n} \to 1$ as $n \to \infty$, as desired.

(c) It follows that

$$0 < \frac{n!}{n^n} \leq \frac{1}{n}.$$

The second inequality is true because

$$n! = n(n-1)\cdots 2 \cdot 1 < n \cdot n \cdots n \cdot 1 = n^{n-1}.$$

The squeeze rule (with $a_n = 0, c_n = 1/n$) gives the desired conclusion. ●

Remark 2.19. We observe that case **(a)** of Example 2.18 may be obtained as a special case of case **(b)**. For instance, if $a \geq 1$, then for n large enough we have $1 \leq a < n$. Taking roots on both sides, we obtain

$$1 \leq a^{1/n} < n^{1/n} \quad \text{for large } n.$$

Again, by the squeeze rule, we see that $\lim_{n\to\infty} a^{1/n} = 1$.

As a consequence of **(a)** and **(b)** of Example 2.18 and the product rule for sequences, we can easily obtain that

$$\lim_{n\to\infty} (2n)^{1/n} = 1 \quad \text{and} \quad \lim_{n\to\infty} (3\sqrt{n})^{1/2n} = 1. \qquad \bullet$$

2.1.4 Bounded Monotone Sequences

Now we introduce some important terminology associated with sequences. A sequence $\{a_n\}$ is said to be

- *bounded above* if there exists an $M \in \mathbb{R}$ such that $a_n \leq M$ for all n,
- *bounded below* if there exists an $m \in \mathbb{R}$ such that $a_n \geq m$ for all n,
- *bounded* if it is bounded both below and above,
- *monotonically increasing* (or *simply increasing*) if $a_n \leq a_{n+1}$ for all n (see Figure 2.1),
- *monotonically decreasing* (or *simply decreasing*) if $a_n \geq a_{n+1}$ for all n (see Figure 2.2),

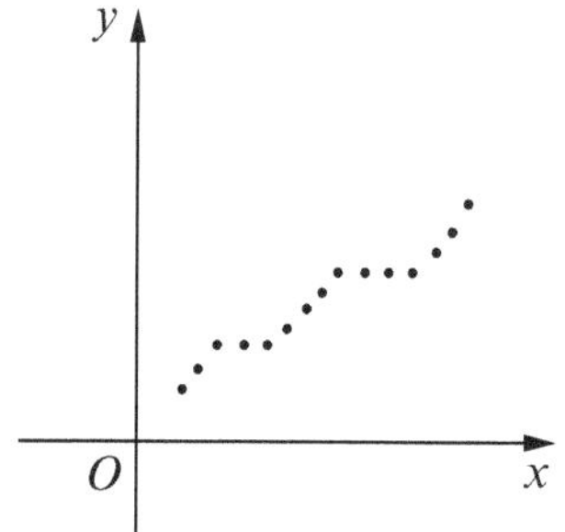

Fig. 2.1. An increasing sequence.

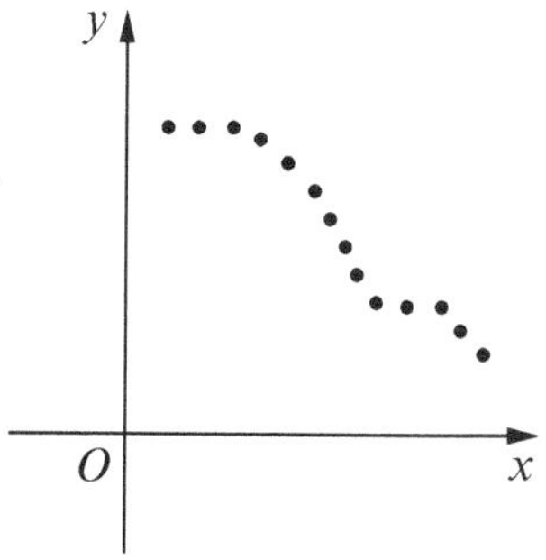

Fig. 2.2. A decreasing sequence.

- *strictly increasing* if $a_n < a_{n+1}$ for all n,
- *strictly decreasing* if $a_n > a_{n+1}$ for all n,
- *monotonic* if it is either increasing or decreasing,
- *strictly monotonic* if it is either strictly increasing or strictly decreasing,
- *alternating* if a_n changes sign alternately. In other words, a_n is of the form $a_n = (-1)^{n-1}b_n$ or $a_n = (-1)^n b_n (b_n \geq 0)$ for all n. That is, $a_n a_{n+1} < 0$ for all n.

Constant sequences are treated as both increasing and decreasing! We now demonstrate these definitions by giving several simple examples.

(1) $\{1/n\}_{n\geq1}$ is strictly decreasing and bounded.
(2) $\{n\}_{n\geq1}$ is strictly increasing and unbounded; however, it is bounded below by 1.
(3) $\{(-1)^{n-1}n\}_{n\geq1}$ is neither increasing nor decreasing. Also, it is unbounded.
(4) $\{(-1)^n\}_{n\geq1}$ is neither increasing nor decreasing nor convergent but is bounded.
(5) $\{(-1)^n/n\}_{n\geq1}$ is convergent but is neither increasing nor decreasing.
(6) If $a_n = 2$ for $1 \leq n \leq 5$ and $a_n = n$ for $n \geq 6$, then $\{a_n\}_{n\geq1}$ is increasing but not strictly.
(7) $\{n^{1/n}\}_{n\geq1}$ is not monotone, as can be seen by examining the first four terms of the sequence.
(8) $\{n!/n^n\}$ is decreasing and bounded.
(9) $\{a_n\}$, $a_n = 8^n/n!$, is neither increasing nor decreasing, because

$$\frac{a_{n+1}}{a_n} = \frac{8}{n+1} \begin{cases} \geq 1 & \text{if } n \leq 7 \\ \leq 1 & \text{if } n \geq 7. \end{cases}$$

On the other hand, if we ignore the first six terms, it follows that $\{a_n\}_{n\geq7}$ is decreasing. In such cases, we say that $\{a_n\}$ is eventually decreasing. Similarly, one can define eventually increasing sequences. Finally, we remark that (3)–(5) are examples of sequences that are alternating.

2.1.5 Subsequences

We now present two simple criteria that involve the notion of a subsequence for establishing that a sequence diverges. Let $\{a_n\}_{n\geq1}$ be a sequence and $\{n_k\}_{k\geq1}$ any strictly increasing sequence of positive integers; that is,

$$0 < n_1 < n_2 < n_3 < \cdots .$$

Then the sequence $\{a_{n_k}\}_{k\geq1}$, i.e., $\{b_k\}_{k\geq1}$, where $b_k = a_{n_k}$, is called a *subsequence* of $\{a_n\}_{n\geq1}$. That is, a subsequence is obtained by choosing terms from the original sequence, without altering the order of the terms, through the map $k \mapsto n_k$, which determines the indices used to pick out the subsequence. For instance, $\{a_{7k+1}\}$ corresponds to the sequence of positive integers $n_k = 7k+1$, $k = 1, 2, \ldots$. Observe that every increasing sequence $\{n_k\}$ of positive integers must tend to infinity, because

$$n_k \geq k \quad \text{for } k = 1, 2, \ldots.$$

The sequences

$$\left\{\frac{1}{k^2}\right\}_{k\geq1}, \quad \left\{\frac{1}{2k}\right\}_{k\geq1}, \quad \left\{\frac{1}{2k+1}\right\}_{k\geq1}, \quad \left\{\frac{1}{5k+3}\right\}_{k\geq1}, \quad \left\{\frac{1}{2^k}\right\}_{k\geq1}$$

are some subsequences of the sequence $\{1/k\}_{k\geq 1}$, formed by setting $n_k = k^2,\ 2k,\ 2k+1,\ 5k+3,\ 2^k$, respectively. Note that all the above subsequences converge to the same limit, 0, which is also the limit of the original sequence $\{1/k\}_{k\geq 1}$. *Can we conjecture that every subsequence of a convergent sequence must converge and converge to the same limit?* We have the following:

1. Every sequence is a subsequence of itself.
2. Let $a_k = 1+(-1)^k$, $k \geq 1$. Then $a_{2k} = 2$ and $a_{2k-1} = 0$, showing that the even sequence $\{a_{2k}\}$ and the odd sequence $\{a_{2k-1}\}$ are two convergent (constant) subsequences of $\{a_k\}$. Thus, a sequence may not converge yet have convergent subsequences with different limits.
3. Let $a_k = \sin(k\pi/2)$. Then $a_{2k-1} = (-1)^{k-1}$ and $a_{2k} = 0$ are two subsequences of a_k. Does the sequence $\{b_k^2\}$, where $b_k = (1+(-1)^{k-1})/2$, converge? Is $\{b_k\}$ a subsequence of $\{a_k\}$?

Definition 2.20 (Subsequential limits). *Let $\{a_k\}$ be a sequence. A subsequential limit is any real number or symbol ∞ or $-\infty$ that is the limit of some subsequence $\{a_{n_k}\}_{k\geq 1}$ of $\{a_k\}_{k\geq 1}$.*

For example, we have the following:

(1) 0 and 2 are subsequential limits of $\{1+(-1)^k\}$.
(2) $-\infty$ and ∞ are the only subsequential limits of $\{k(-1)^k\}$.
(3) $\{-\sqrt{3}/2, 0, \sqrt{3}/2\}$ is the set of subsequential limits of $\{a_k\}$, $a_k = \sin(k\pi/3)$. Here $\{a_{3k}\}$, $\{a_{3k+1}\}$, and $\{a_{3k+2}\}$ are convergent subsequences with limits 0, $-\sqrt{3}/2$, and $\sqrt{3}/2$, respectively.
(4) Every real number is a subsequential limit of some subsequence of the sequence of all rational numbers. Indeed, $\mathbb{R}\cup\{-\infty,\infty\}$ is the set of subsequential limits of the sequence of all rational numbers.

The following result, which shows that certain properties of sequences are inherited by their subsequences, is almost obvious.

Theorem 2.21 (Invariance property of subsequences). *If $\{a_n\}$ converges, then every subsequence $\{a_{n_k}\}$ of it converges to the same limit. Also, if $a_n \to \infty$, then $\{a_{n_k}\} \to \infty$ as well.*

Proof. Suppose that $\{a_{n_k}\}$ is a subsequence of $\{a_n\}$. Note that $n_k \geq k$. Let $L = \lim a_n$ and $\epsilon > 0$ be given. Then there exists an N such that

$$|a_k - L| < \epsilon \quad \text{for } k \geq N.$$

Now $k \geq N$ implies $n_k \geq N$, which in turn implies that

$$|a_{n_k} - L| < \epsilon \quad \text{for } n_k \geq N.$$

Thus, a_{n_k} converges to L as $k \to \infty$. The proof of the second part follows similarly. ∎

Here is an immediate consequence of Theorem 2.21.

Corollary 2.22. *The sequence $\{a_n\}$ is divergent if it has two convergent subsequences with different limits. Also, $\{a_n\}$ is divergent if it has a subsequence that tends to ∞ or a subsequence that tends to $-\infty$.*

In order to apply this corollary, it is necessary to identify convergent subsequences with different limits or subsequences that tend to ∞ or $-\infty$. Now the question is whether the converse of Theorem 2.21 also holds.

We can prove the divergence of a sequence if we are able to somehow prove that it is unbounded. For instance (see also Questions 2.44(8)), consider $a_n = \sum_{k=1}^{n} 1/k$. There are several ways one can see that the sequence diverges. Clearly, $a_n > 0$ for all $n \in \mathbb{N}$, $\{a_n\}$ is increasing, and

$$\begin{aligned} a_{2^n} &= 1 + \frac{1}{2} + \left(\frac{1}{3} + \frac{1}{4}\right) + \left(\frac{1}{5} + \frac{1}{6} + \frac{1}{7} + \frac{1}{8}\right) + \cdots + \left(\frac{1}{2^{n-1}+1} + \cdots + \frac{1}{2^n}\right) \\ &> 1 + \frac{n}{2}, \end{aligned}$$

so that $\{a_n\}_{n\geq 1}$ is increasing and not bounded above. Therefore, it cannot be convergent, and so it must diverge (see also the bounded monotone convergence theorem (BMCT), which is discussed later in this section). We remark that we may group the terms in a number of ways and obtain that $\{a_n\}_{n\geq 1}$ is unbounded, for example,

$$\begin{aligned} a_{10^n-1} &= \left(1 + \frac{1}{2} + \cdots + \frac{1}{9}\right) + \left(\frac{1}{10} + \cdots + \frac{1}{99}\right) \\ &\quad + \cdots + \left(\frac{1}{10^{n-1}} + \cdots + \frac{1}{10^n - 1}\right) \\ &> 9\left(\frac{1}{10}\right) + \frac{90}{100} + \cdots + \frac{9 \times 10^{n-1}}{10^n} = \left(\frac{9}{10}\right) n. \end{aligned}$$

We end this subsection with the following result, which is easy to prove.

Theorem 2.23. *A sequence is convergent if and only if there exists a real number L such that every subsequence of the sequence has a further subsequence that converges to L.*

Corollary 2.24. *If both odd and even subsequences of $\{a_n\}$ converge to the same limit l, then so does the original sequence.*

Note that $\{(-1)^n\}$ diverges, because it has two subsequences $\{(-1)^{2n}\}$ and $\{(-1)^{2n-1}\}$ converging to two different limits, namely 1 and -1.

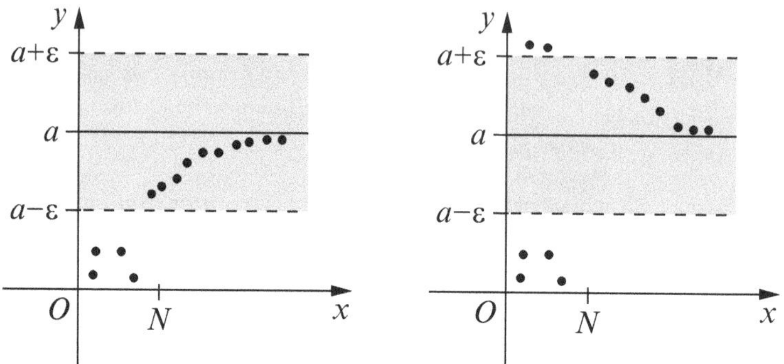

Fig. 2.3. Description for the bounded monotone convergence theorem.

2.1.6 Bounded Monotone Convergence Theorem

Until now, we have considered some basic techniques for finding the limit of a convergent sequence. In general, it is difficult to tell whether a given sequence converges. It is sometimes easy to show that a sequence is convergent even if we do not know its limit. For example, the following theorem is a starting point for our rigorous treatment of sequences and series, especially if we know that the given sequence is monotonic. However, we shall soon show that every bounded sequence has a convergent subsequence (see Theorem 2.42).

Theorem 2.25 (Monotone convergence theorem). *Every increasing sequence that is bounded above converges. Also, every decreasing sequence that is bounded below converges.*

Proof. Let $\{a_n\}_{n\geq 1}$ be an increasing sequence that is bounded above. According to the least upper bound property (Definition 1.18), since the range $A = \{a_n : n \in \mathbb{N}\}$ is bounded above, A has a least upper bound; call it a. We now prove that $a_n \to a$ as $n \to \infty$.

Clearly $a_n \leq a$ for all $n \in \mathbb{N}$, and by the definition of lub, given some $\epsilon > 0$ there exists an integer N such that $a_N > a - \epsilon$. Since $\{a_n\}$ is monotonically increasing,

$$a - \epsilon < a_N \leq a_n \leq a < a + \epsilon \quad \text{for } n \geq N.$$

That is, $|a_n - a| < \epsilon$ for $n \geq N$, and we conclude that $\{a_n\}$ converges to its least upper bound. That is, $\lim_{n\to\infty} a_n = a = \sup a_n$.

The proof for the case of decreasing sequences is identical, using the greatest lower bound instead of the least upper bound (see Figure 2.3).

Alternatively, it suffices to note that $\{b_n\}_{n\geq 1}$ is a decreasing sequence that is bounded below if and only if the sequence $\{-b_n\}_{n\geq 1}$ is increasing and bounded above. ∎

Remark 2.26. The monotonicity condition on the sequence $\{a_n\}$ in the above results need not be satisfied for all n. If this is true for all $n \geq N$, where N

is some suitably selected positive integer, then the conclusion of the above result is still true (see Figure 2.5). However, the tests in Theorem 2.25 tell us nothing about the limit, but they are often useful when we suspect that a sequence is convergent. •

For instance, we easily obtain the following simple examples:

(1) If $a_n = 1 + 1/n$, then $\{a_n\}$ is clearly decreasing and bounded below (by 1, for example), and so it is convergent by Theorem 2.25. In this case, of course, we know already that it converges to 1.
(2) If $a_n = 1/\sqrt{n}$, then $\{a_n\}$ is clearly decreasing for $n \geq 1$ and bounded by 1. Consequently, the sequence $\{1/\sqrt{n}\}$ must converge.
(3) If $a_n = (2n-7)/(3n+2)$, then

$$a_n = \frac{1}{3n+2}\left(\frac{2}{3}(3n+2) - 7 - \frac{4}{3}\right) = \frac{2}{3} - \frac{25}{3(3n+2)},$$

so that $a_n \leq 2/3$ and $\{a_n\}$ is increasing. By Theorem 2.25, the sequence $\{a_n\}_{n\geq 1}$ must converge. Indeed, $a_n \to 2/3$ as $n \to \infty$.
(4) Consider

$$a_n = \frac{1}{n+1} + \frac{1}{n+2} + \cdots + \frac{1}{2n}.$$

Then $0 < a_n \leq n/(n+1)$ for all $n \geq 1$, since each term (except the first) in the sum is strictly less than $1/(n+1)$, and so $\{a_n\}$ is a bounded sequence. Also, for $n \geq 1$,

$$\begin{aligned} a_{n+1} - a_n &= \frac{1}{2n+1} + \frac{1}{2(n+1)} - \frac{1}{n+1} \\ &= \frac{1}{2n+1} - \frac{1}{2(n+1)} \\ &= \frac{1}{2(2n+1)(n+1)} > 0. \end{aligned}$$

Thus, $\{a_n\}$ is a bounded monotone sequence, and so it converges by Theorem 2.25. What is the limit of the sequence $\{a_n\}$?

The following equivalent form of Theorem 2.25 is the key to many important results in analysis. We shall soon see its usefulness in our subsequent discussion.

Theorem 2.27 (BMCT: Bounded monotone convergence theorem). *Every bounded monotonic sequence of real numbers converges. Equivalently, a monotonic sequence converges if and only if it is bounded.*

Consider the sequence $\{a_n\}_{n\geq 1}$, where $a_n = \sum_{k=1}^{n} 1/k$. This is clearly an increasing sequence. Does there exist an upper bound for this sequence? In fact, we have already proved that $\{a_n\}_{n\geq 1}$ is unbounded (see also Questions 2.44(8)). We also remark that a bounded sequence can converge without being monotone. For example, consider $\{(-1/3)^n\}_{n\geq 1}$.

Example 2.28. Show that $\lim_{n\to\infty} r^n = 0$ if $|r| < 1$ (see also Theorem 2.34 and Example 2.43). Here $\{r^n\}$ is called a *power sequence.*

Solution. Observe that $-|r|^n \leq r^n \leq |r|^n$, and so it suffices to deal with $0 < r < 1$. In any case, define $a_n = |r|^n$ for $n \geq 1$. If $|r| < 1$, then we have

$$a_{n+1} = |r|a_n, \quad \text{i.e., } 0 \leq a_{n+1} < a_n,$$

showing that $\{a_n\}$ is decreasing and bounded below by 0. Therefore, $\{a_n\}$ converges, say to a. Allowing $n \to \infty$ in the last equality, we see that

$$a = |r|a, \quad \text{i.e., } (1 - |r|)a = 0,$$

which gives $a = 0$, since $|r| < 1$.

Alternatively, we first notice that there is nothing to prove if $r = 0$. Thus for $0 < |r| = c < 1$, we can write $|r|$ in the form $c = 1/(1 + a)$ for some $a > 0$, so that by the binomial theorem,

$$0 < c^n = \frac{1}{(1+a)^n} \leq \frac{1}{1+na} < \frac{1}{na},$$

and the result follows if we use the squeeze rule. •

Because every monotone sequence converges, diverges to ∞, or diverges to $-\infty$, we have the following analogue of Theorem 2.25 for unbounded monotone sequences.

Theorem 2.29. *Every increasing sequence that is not bounded above must diverge to ∞. Also, every decreasing sequence that is not bounded below must diverge to $-\infty$.*

Proof. Let $\{a_n\}_{n\geq 1}$ be an increasing sequence that is unbounded. Since the set $\{a_n : n \in \mathbb{N}\}$ is unbounded and it is bounded below by a_1, it must be unbounded above. Thus, given $R > 0$ there exists an integer N such that $a_N > R$. Since $\{a_n\}$ is monotonically increasing,

$$a_n \geq a_N > R \quad \text{for } n \geq N.$$

Since $R > 0$ is arbitrary, it follows that $\lim_{n\to\infty} a_n = \infty$.

The proof for decreasing sequences is identical and is left as an exercise. ∎

We may combine Theorems 2.27 and 2.29 in an equivalent form as follows.

Theorem 2.30. *Every monotone sequence converges, diverges to ∞, or diverges to $-\infty$. In other words, we say that $\lim_{n\to\infty} a_n$ is always meaningful for monotone sequences.*

Example 2.31. Set $a_n = (1 \cdot 3 \cdot 5 \cdots (2n-1))/(2 \cdot 4 \cdot 6 \cdots (2n))$. Then $\{a_n\}$ converges.

Solution. Note that $a_n > 0$ for all $n \geq 1$ and

$$a_{n+1} = a_n \left(\frac{2n+1}{2n+2} \right) < a_n.$$

Thus, $\{a_n\}$ is decreasing and bounded below by 0. Applying Theorem 2.25, we see that $\{a_n\}$ converges. Note also that $a_n < 1$ for $n \geq 1$. ●

Often sequences are defined by formulas. There is still another way of specifying a sequence, by defining its terms "inductively" or "recursively." In such cases, we normally specify the first term (or first several terms) of the sequence and then give a formula that specifies how to obtain all successive terms. We begin with a simple example and later present a number of additional examples (see Examples 2.39 and 2.58 and Exercises 2.45).

Example 2.32. Starting with $a_1 = 1$, consider the sequence $\{a_n\}$ with $a_{n+1} = \sqrt{2a_n}$ for $n \geq 1$. We observe that

$$a_1 = 1, \quad a_2 = \sqrt{2}, \quad a_3 = \sqrt{2\sqrt{2}}, \quad a_4 = \sqrt{2\sqrt{2\sqrt{2}}}, \ldots,$$

which seems to suggest that the given sequence is positive and increasing. Hence, the sequence must converge if it is bounded and increasing. It is not clear how to find an upper bound. However, the following observation might be useful. "If an increasing sequence converges, then the limit must be the least upper bound of the sequence" (see the proof of Theorem 2.25). As a consequence, if the given sequence converges to a, then the limit a must satisfy

$$a = \sqrt{2a}, \quad \text{i.e., } a(a-2) = 0,$$

so that $a = 2$, for $a = 0$ is not possible. By the method of induction, it is easy to prove that $0 < a_n \leq 2$ for all $n \geq 1$. Consequently,

$$a_{n+1} = \sqrt{2a_n} = a_n(\sqrt{2/a_n}) \geq a_n \quad \text{for all } n \geq 1,$$

showing that the sequence $\{a_n\}$ is bounded and increasing. Thus, $\{a_n\}$ converges and in fact converges to 2. ●

The BMCT is an extremely valuable theoretical tool, as we shall see by a number of examples below.

Example 2.33 (The number e). Let $a_n = (1+1/n)^n$, $n \geq 1$. The sequence $\{a_n\}$ is called *Euler's sequence*. Note that $(1+x)^n \geq 1 + nx$ for $x \geq 0$ and $n \geq 1$, so that for $x = 1/n$, this gives

$$\left(1 + \frac{1}{n}\right)^n \geq 2 \quad \text{for } n \geq 1.$$

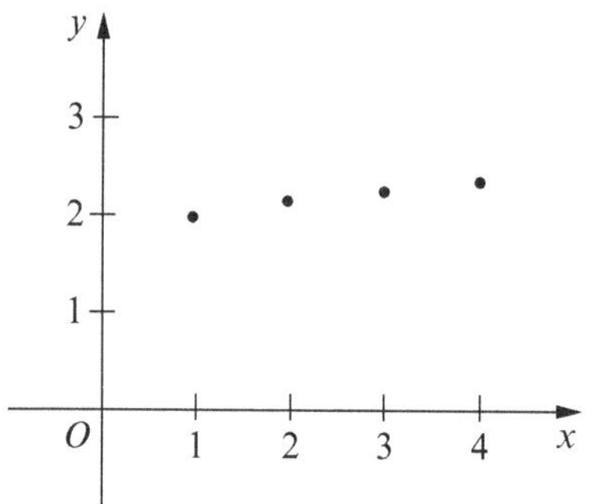

Fig. 2.4. Diagram for $a_n = (1 + 1/n)^n$.

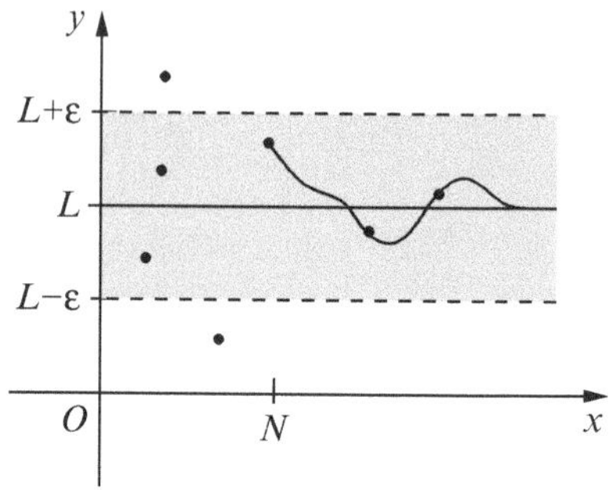

Fig. 2.5. a_n is eventually inside the strip.

If we plot the first few terms of this sequence on a sequence diagram, then it seems that the sequence $\{a_n\}$ increases and converges to a limit, which is less than 3 (see Figure 2.4).

First we show that the sequence is increasing (see Figure 2.4). This is an immediate consequence of the well-known arithmetic–geometric mean inequality

$$\left(\prod_{i=1}^{k} x_i\right)^{1/k} \leq \frac{1}{k}\sum_{i=1}^{k} x_i$$

if we choose $k = n+1$, $x_1 = 1$, and $x_i = 1 + 1/n$ for $i = 2, \ldots, n+1$. As an alternative proof, we may use the binomial theorem and obtain

$$\begin{aligned}
a_n &= 1 + \sum_{k=1}^{n} \binom{n}{k} \left(\frac{1}{n}\right)^k \\
&= 1 + \sum_{k=1}^{n} \frac{n(n-1)\cdots(n-k+2)(n-k+1)}{n^k} \frac{1}{k!} \\
&= 1 + 1 + \sum_{k=2}^{n} \left[1 \cdot \left(1 - \frac{1}{n}\right)\left(1 - \frac{2}{n}\right) \cdots \left(1 - \frac{k-2}{n}\right)\left(1 - \frac{k-1}{n}\right)\right] \frac{1}{k!} \\
&< 2 + \sum_{k=2}^{n} \left[\left(1 - \frac{1}{n+1}\right)\left(1 - \frac{2}{n+1}\right) \cdots \left(1 - \frac{k-2}{n+1}\right)\left(1 - \frac{k-1}{n+1}\right)\right] \frac{1}{k!} \\
&< 2 + \sum_{k=2}^{n+1} \left[\left(1 - \frac{1}{n+1}\right)\left(1 - \frac{2}{n+1}\right) \cdots \left(1 - \frac{k-2}{n+1}\right)\left(1 - \frac{k-1}{n+1}\right)\right] \frac{1}{k!} \\
&= a_{n+1},
\end{aligned}$$

and so $\{a_n\}$ is increasing. Next, we show that the sequence is bounded. Since $k! = 1 \cdot 2 \cdot 3 \cdots k \geq 1 \cdot 2 \cdot 2 \cdots 2 = 2^{k-1}$ for $k \geq 2$, we have

$$2 < a_n < 1 + \sum_{k=1}^{n} \frac{1}{k!} < 1 + \sum_{k=1}^{n} \frac{1}{2^{k-1}} = 1 + \frac{1-(1/2)^n}{1-(1/2)} < 1 + \frac{1}{1-1/2} = 3.$$

Thus, $\{a_n\}$ is an increasing bounded sequence. By BMCT, it follows that the sequence $\{a_n\}$ converges to a real number that is at most 3. It is customary to denote this limit by e, the base of the natural logarithm, a number that plays a significant role in mathematics. The above discussion shows that $2 < \mathrm{e} \leq 3$. The foregoing discussion allows us to make the following definition:

$$\mathrm{e} = \lim_{n\to\infty} \left(1 + \frac{1}{n}\right)^n.$$

Moreover, by considering the binomial expansion of $(1+x/n)^n$, the above discussion may be continued to make the following definition of e^x for $x > 0$:

$$\mathrm{e}^x = \lim_{n\to\infty} \left(1 + \frac{x}{n}\right)^n, \quad x > 0.$$

Later, we shall show that this limit actually exists also for $x < 0$ (see Theorem 5.7). Thus, we easily have

$$\lim_{n\to\infty} \left(1 - \frac{1}{3n}\right)^{n+2} = \lim_{n\to\infty} \left[\left(1 - \frac{1}{3n}\right)^{3n}\right]^{1/3} \left(1 - \frac{1}{3n}\right)^2 = \mathrm{e}^{-1/3} \cdot 1$$

and

$$\lim_{n\to\infty} \left(1 + \frac{5}{n}\right)^n = \lim_{n\to\infty} \left(1 + \frac{5}{5n}\right)^{5n} = \lim_{n\to\infty} \left[\left(1 + \frac{1}{n}\right)^n\right]^5 = \mathrm{e}^5.$$

Can we replace 5 in each step of the last of these equalities by a positive integer?

Moreover, by the product and the quotient rules for sequences, we have

$$\lim_{n\to\infty} \left(1 + \frac{1}{n+k}\right)^n = \lim_{n\to\infty} \frac{\left(1 + \frac{1}{n+k}\right)^{n+k}}{\left(1 + \frac{1}{n+k}\right)^k} = \frac{\lim_{n\to\infty} \left(1 + \frac{1}{n+k}\right)^{n+k}}{\lim_{n\to\infty} \left(1 + \frac{1}{n+k}\right)^k} = \mathrm{e},$$

where k is a fixed positive integer. Could k be any fixed integer? Could k be any positive real number? •

Theorem 2.34 (Convergence of a geometric sequence). *If r is a fixed number such that $|r| < 1$, then $\lim_{n\to\infty} r^n = 0$. Further, $\{r^n\}$ diverges if $|r| > 1$. At $r = 1$, the sequence converges, whereas it diverges for $r = -1$.*

Proof. We have already proved the first part in Example 2.28 (see also Example 2.43). If $r = 1$, the sequence reduces to a constant sequence and so converges to 1. If $r > 1$, then $r^n \to \infty$ as $n \to \infty$, so the sequence diverges. Indeed, if $r > 1$, then $1/r < 1$, and so

$$\frac{1}{r^n} = \left(\frac{1}{r}\right)^n \to 0 \quad \text{as } n \to \infty,$$

which implies that $r^n \to \infty$ as $n \to \infty$.

For $r = -1$, the sequence $\{(-1)^n\}$ diverges, and if $r < -1$, then $\{r^n\}$ diverges, since $|r|^n \to \infty$ as $n \to \infty$. ∎

Example 2.35. For $p > 0$, we easily have

$$\lim_{n\to\infty} \frac{r^n}{n^p} = \begin{cases} 0 & \text{if } |r| \leq 1, \\ \infty & \text{if } r > 1, \\ \text{does not exist} & \text{if } r < -1. \end{cases}$$

Indeed, for $|r| < 1$, let $a_n = r^n$ and $b_n = 1/n^p$. Then $\{a_n\}$ and $\{b_n\}$ are null sequences, and so is their product. For $r = 1, -1$, there is nothing to prove.

For $r > 1$, we write $r = 1 + x$ with $x > 0$. Let k be a positive integer such that $k > p$. Then for $n > 2k$,

$$(1+x)^n > \binom{n}{k} x^k = \frac{n(n-1)\cdots(n-k+1)}{k!} x^k > \left(\frac{n}{2}\right)^k \frac{x^k}{k!},$$

since $n - k + 1 > n/2$ for each k. Hence, since $k - p > 0$, it follows that

$$\frac{(1+x)^n}{n^p} > \frac{x^k}{2^k k!} n^{k-p} \to \infty \quad \text{as } n \to \infty.$$ ●

Example 2.36. Find $\lim_{n\to\infty} r^n/(1 + r^{2n})$ for various values of r.

Solution. Set $a_n = r^n/(1+r^{2n})$. We need to find $\lim_{n\to\infty} a_n$ for various values of r. For $r = 1$, we have $a_n = 1/2$, showing that $\lim_{n\to\infty} a_n = 1/2$. For $r = -1$, we have $a_n = (-1)^n/2$, so that $\{a_n\}$ diverges. On the other hand, for $|r| < 1$, let $c_n = 1 + r^{2n}$. By Theorem 2.34, $\lim_{n\to\infty} c_n = 1$ and $\lim_{n\to\infty} r^n = 0$. Therefore, by the quotient rule,

$$\lim_{n\to\infty} \frac{r^n}{1 + r^{2n}} = \frac{\lim_{n\to\infty} r^n}{\lim_{n\to\infty}(1 + r^{2n})} = \frac{0}{1} = 0.$$

Similarly for $|r| > 1$, we have $1/|r| < 1$, and so using the above argument, we see that

$$\lim_{n\to\infty} \frac{r^n}{1 + r^{2n}} = \lim_{n\to\infty} \frac{1/r^n}{1 + 1/r^{2n}} = \frac{0}{1} = 0.$$

We conclude that $\{a_n\}_{n\geq 1}$ converges for all $r \neq -1$. ●

Theorem 2.37. *Let $\{a_n\}$ and $\{b_n\}$ be two convergent sequences such that $a_n \to L$ and $b_n \to M$ as $n \to \infty$. We have*

(a) $|a_n| \to |L|$ *as* $n \to \infty$*;*
(b) *if* $a_n \leq b_n$ *for all* $n \geq N_0$*, then* $L \leq M$*.*

Here **(b)** *is often referred to as the limit inequality rule.*

Proof. We prove case **(b)** by contradiction. Suppose that $a_n \to L$, $b_n \to M$, and $L > M$. Then with $\epsilon = (L - M)/2$, there exists an N such that

$$L - \epsilon < a_n < L + \epsilon \quad \text{and} \quad M - \epsilon < b_n < M + \epsilon \quad \text{for all } n \geq N.$$

In particular,

$$b_n < M + \epsilon = \frac{L+M}{2} = L - \epsilon < a_n \quad \text{for all } n \geq N,$$

which is a contradiction to the hypothesis that $a_n \leq b_n$ for all $n \geq N_0$. Therefore, our assumption is wrong, and hence we must have $L \leq M$.

The proof of case **(a)** follows from the fact that $\big||a_n| - |L|\big| \leq |a_n - L|$. ∎

Corollary 2.38. *Let $\{b_n\}$ be a convergent sequence such that $b_n \to M$ as $n \to \infty$, and $b_n \geq 0$ for all sufficiently large n. Then $M \geq 0$.*

Proof. Set $a_n = 0$ for all n in Theorem 2.37. ∎

Example 2.39. Consider the following sequences $\{a_n\}_{n \geq 1}$:

(a) $a_n = 1/n^2 + 1/(n+1)^2 + \cdots + 1/(2n)^2$;
(b) $a_1 = 1,\ a_{n+1} = \sqrt{2 + a_n}$ for $n \geq 1$;
(c) $a_1 = 2,\ a_{n+1} = (1/2)(a_n + 2/a_n)$ for $n \geq 2$;
(d) $a_1 = \alpha$ and $a_{n+1} = (a_n + \beta/a_n)/2$ for $n \geq 1$, where $\alpha > 0$ is arbitrary and β is a fixed positive number.

In each case, determine whether the sequence converges.

Solution. **(a)** Clearly $0 < a_n < (n+1)/n^2$ for all $n \geq 1$, since each term (except the first) in the sum is strictly less than $1/n^2$, and so $\{a_n\}$ is a bounded sequence. Also, for $n \geq 1$,

$$a_{n+1} - a_n = \frac{1}{(2n+1)^2} + \frac{1}{(2n+2)^2} - \frac{1}{n^2} < \frac{1}{4n^2} + \frac{1}{4n^2} - \frac{1}{n^2} = -\frac{1}{2n^2} < 0,$$

that is, $a_{n+1} < a_n$ for all $n \geq 1$. Thus, $\{a_n\}$ is a bounded monotone sequence and so converges by Theorem 2.27.

Alternatively, we observe that for all $n \geq 1$,

$$\frac{n+1}{(2n)^2} \leq a_n \leq \frac{n+1}{n^2},$$

and so by the squeeze rule, we see that $\lim_{n \to \infty} a_n = 0$.

(b) Clearly $a_n > 0$ for all $n \geq 1$. Since $a_1 < 2$, by induction we obtain that $a_{n+1} = \sqrt{2 + a_n} < \sqrt{2+2} = 2$ for all $n \geq 1$. Since

$$a_{n+1} - a_n = \sqrt{2 + a_n} - a_n \geq 0 \iff (2 - a_n)(1 + a_n) \geq 0,$$

and since $a_n \leq 2$, it follows that the sequence $\{a_n\}$ is monotonically increasing and bounded; hence it is convergent. We see that

$$a = \lim_{n\to\infty} a_{n+1} = \lim_{n\to\infty} \sqrt{2 + a_n} = \sqrt{2 + a},$$

which gives $(a-2)(a+1) = 0$, or $a = 2$.

(c) First we observe that if the given sequence were convergent, then we would obtain its limit by allowing $n \to \infty$ in the given recurrence relation:

$$a = \frac{1}{2}\left(a + \frac{2}{a}\right), \quad \text{i.e., } a^2 = 2 \text{ or } a = \sqrt{2}.$$

Now we show that the given sequence indeed converges to $\sqrt{2}$. We have $a_1 = 2 > \sqrt{2}$, $a_n > 0$, and for $n \geq 1$,

$$a_{n+1} - \sqrt{2} = \frac{(a_n - \sqrt{2})^2}{2a_n} \geq 0.$$

(We remind the reader that it does not matter what positive value is assigned to a_1.) Thus, $a_n \geq \sqrt{2}$ for all $n \geq 2$, and therefore,

$$\frac{a_{n+1}}{a_n} = \frac{1}{2}\left(1 + \frac{2}{a_n^2}\right) \leq \frac{1}{2}(1+1) = 1, \quad \text{i.e., } a_{n+1} \leq a_n \text{ for } n \geq 2,$$

showing that $\{a_n\}$ is monotonically decreasing and bounded below by 0; hence it is convergent.

(d) Since α and β are positive and $a_1 > 0$ (arbitrary), the principle of induction shows that $a_n > 0$ for all $n \geq 2$. Next for $n \geq 1$, we have

$$a_{n+1}^2 - \beta = \frac{1}{4}\left(a_n + \frac{\beta}{a_n}\right)^2 - \beta = \frac{(a_n^2 - \beta)^2}{4a_n^2} \geq 0,$$

so that $a_{n+1}^2 \geq \beta$ for all $n \geq 1$. Also, for $n \geq 2$,

$$a_n - a_{n+1} = a_n - \frac{1}{2}\left(a_n + \frac{\beta}{a_n}\right) = \frac{a_n^2 - \beta}{2a_n} \geq 0,$$

showing that $\{a_n\}_{n\geq 2}$ is decreasing and bounded below (since all terms are positive). By Theorem 2.25, we are assured that the sequence converges; call the limit L. Since $a_{n+1}^2 \geq \beta$ and $a_n > 0$, we must have $a_{n+1} \geq \sqrt{\beta}$ for $n \geq 1$ and hence $L \geq \sqrt{\beta}$ (see Theorem 2.37). Since $a_n \to L$ as $n \to \infty$, $a_{n+1} \to L$ as $n \to \infty$. Thus, by the linearity rule,

$$L = \lim_{n\to\infty} a_{n+1} = \lim_{n\to\infty} \frac{1}{2}\left(a_n + \frac{\beta}{a_n}\right) = \frac{1}{2}\left(L + \frac{\beta}{L}\right), \quad \text{i.e., } L = \sqrt{\beta}. \quad \bullet$$

Remark 2.40. Example 2.39(**c**) (also 2.39(**d**) with $\beta = 2$ and Exercise 2.68(10)) provides a proof that there is a sequence of rational numbers that converges to the irrational number $\sqrt{2}$. Moreover, using the a_n from Example 2.39(**c**), we note that

$$a_1 = 2, \quad a_2 = \frac{3}{2}, \quad a_3 = \frac{1}{2}\left(\frac{3}{2} + \frac{4}{5}\right) = \frac{17}{12} \text{ and } a_4 = \frac{1}{2}\left(\frac{17}{12} + \frac{24}{17}\right) = \frac{577}{408},$$

so that a_4^2 is approximately 2.0006. Thus, the sequence $\{a_n\}$ defined in Example 2.39(**c**) provides a practical way of computing a rational approximation to $\sqrt{2}$. •

2.1.7 The Bolzano–Weierstrass Theorem

It is useful to have necessary and sufficient conditions for the convergence of sequences. For monotone sequences, BMCT (see Theorem 2.27) shows that boundedness is such a condition. On the other hand, for general sequences, boundedness is necessary but not sufficient for convergence. Indeed, we have seen examples of bounded sequences that do not converge yet have convergent subsequences. To show that this is true in general, we need to prove a lemma. It is convenient first to introduce a definition. We say that $n \in \mathbb{N}$ is a *peak point* of $\{a_n\}$ if

$$a_n \geq a_k \quad \text{for all } k \geq n.$$

Lemma 2.41. *Every sequence of real numbers contains a monotonic subsequence.*

Proof. Let $\{a_n\}_{n\geq 1}$ be a sequence of real numbers. We need to construct a monotone subsequence. Then either the sequence $\{a_n\}$ has infinitely many peak points or it has only finitely many peak points.

Assume that there are infinitely many peak points n. Let n_1 be the first such n with this property (i.e., the smallest peak point) and n_2 the second (i.e., the smallest peak point with $n_2 > n_1$), etc. Thus,

(i) $a_{n_1} \geq a_k$ for all $k \in \mathbb{N}$ with $k \geq n_1$;
(ii) $a_{n_2} \geq a_k$ for all $k \in \mathbb{N}$ with $k \geq n_2 \, (> n_1)$.

From (i) and (ii), it follows that

$$a_{n_1} \geq a_{n_2}.$$

We now introduce n_{k+1} inductively as the smallest peak point such that $n_{k+1} > n_k$. Consequently,

$$a_{n_k} \geq a_{n_{k+1}},$$

and so $\{a_{n_k}\}_{k\geq 1}$ is a monotonically decreasing subsequence of $\{a_n\}$.

On the other hand, if there are only finitely many n such that

$$a_n \geq a_k \quad \text{for all } k \in \mathbb{N} \text{ with } k \geq n,$$

then we can choose an integer m_1 greater than all peak points, so that no terms of the sequence

$$\{a_{m_1}, a_{m_1+1}, a_{m_1+2}, \dots\}$$

have this property. Because m_1 itself is not a peak point, there exists an m_2 with $m_2 > m_1$ for which

$$a_{m_1} < a_{m_2}.$$

Again, m_2 is not a peak point bigger than all peak points, and so there exists an m_3 with $m_3 > m_2$ and

$$a_{m_3} > a_{m_2}.$$

Continuing the process, we obtain a sequence $\{a_{m_k}\}_{k\geq 1}$ that is a monotonically increasing subsequence of $\{a_n\}$. This completes the proof. ■

We see that if a sequence is bounded, then even though it may diverge, it cannot behave "too badly." This fact follows from Lemma 2.41 together with BMCT.

Theorem 2.42 (Bolzano–Weierstrass). *Every bounded sequence of real numbers has a convergent subsequence (a subsequence with a limit in $\mathbb{R}$). That is, if $\{a_n\}$ is a sequence such that $|a_n| \leq M$ for all $n \geq N$, then there exist a number l in the interval $[-M, M]$ and a subsequence $\{a_{n_k}\}$ such that $\{a_{n_k}\}$ converges to l.*

Proof. Let $\{a_n\}$ be a bounded sequence of real numbers. By Lemma 2.41, it has a monotonic subsequence, say $\{a_{n_k}\}$. Because $\{a_n\}$ is bounded, so is every subsequence of $\{a_n\}$. Hence by BMCT, $\{a_{n_k}\}$ converges. ■

Next we remark that $\{\sin n\}$ is a bounded sequence. What is the behavior of $\sin n$ as $n \to \infty$? According to Theorem 2.42, there must exist at least one number l in $[-1, 1]$ such that some subsequences $\{\sin n_k\}$ will converge to l. A discussion of this surprising fact is beyond the scope of this book. However, we can prove that every number l in $[-1, 1]$ has this property.

We note that the Bolzano–Weierstrass theorem says nothing about uniqueness, for if $a_n = (-1)^n$, then $a_{2n} \to 1$ and $a_{2n-1} \to -1$ as $n \to \infty$.

Example 2.43. Fix r such that $0 < r < 1$, and consider the sequence $\{a_n\}_{n\geq 1}$, where $a_n = r^n$. Then $a_n > 0$ for all $n \geq 1$, and the sequence is decreasing, because

$$a_n - a_{n+1} = (1 - r)r^n > 0.$$

Thus, $\{a_n\}$, being a decreasing sequence that is bounded below by zero, converges; call the limit a. Also, since

$$a_{2n} = (r^n)(r^n),$$

$\{a_{2n}\}$ converges to a^2. On the other hand, $\{a_{2n}\}$ is a subsequence of $\{a_n\}$, and hence by the uniqueness of the limit, we have $a^2 = a$, i.e., $a = 0$ or 1. Clearly $a \neq 1$, since $\{r^n\}$ is decreasing and $r < 1$. Hence $\{r^n\}$ converges to 0 whenever $0 < r < 1$ (see also Theorem 2.34).

By the squeeze rule, the inequalities

$$-|r|^n \leq r^n \leq |r|^n$$

show that $\lim_{n\to\infty} r^n = 0$ for $-1 < r < 0$ also.

The same idea may be used to show that $\lim_{n\to\infty} a^{1/n} = 1$ for $0 < a < 1$ (see also Example 2.18(**a**)). ●

2.1.8 Questions and Exercises

Questions 2.44.

1. If $a_n \to a$ as $n \to \infty$, must the set $\{n : a_n \notin (a - \epsilon, a + \epsilon)\}$, where $\epsilon > 0$, be finite?
2. Is it true that a sequence $\{a_n\}$ is null iff $\{|a_n|\}$ is null?
3. Is every convergent sequence null? How about the converse?
4. Is the sum of two null sequences always null?
5. Does an alternating sequence always converge? Does it always diverge?
6. Is every convergent sequence monotone? Is every monotone sequence convergent?
7. Can a bounded sequence be convergent without being monotone?
8. Does every divergent increasing sequence diverge to ∞? How about a divergent decreasing sequence?
9. Can we say that $\{a_5, a_4, a_1, a_2, a_3, a_6, a_7, \ldots\}$ is a subsequence of $\{a_n\}_{n\geq 1}$?
10. Does every sequence have at most a countable number of subsequences? Does there exist a sequence with an uncountable number of subsequences?
11. Suppose that $\{a_n\}$ and $\{b_n\}$ are two sequences such that one converges to 0 while the other is bounded. Does $\{a_n b_n\}$ converge? If so, to what limit?
12. Suppose that $\{a_n\}$ is bounded and $\alpha \in (0,1)$ is fixed. Does $\{\alpha^n a_n\}$ converge? If so, does it converge to 0?
13. Suppose that $\{a_n\}$ is a bounded convergent sequence such that $|a_n| \leq M$ and the sequence has limit a. Must $|a| \leq M$?
14. Suppose that $\{a_n\}$ is increasing and bounded above by M. Must we have $a_n \to L$ for some L? Must $L \leq M$?
15. Suppose that $\{a_n\}$ is decreasing and bounded below by m. Must we have $a_n \to l$ for some l? Must $l \geq m$?

16. Let $\{a_n\}$ be a sequence of nonnegative real numbers, $p \in \mathbb{N}$, and $a \in [0, \infty)$. Is it true that $\{a_n\}$ converges to a if and only if $\{a_n^{1/p}\}$ converges to $a^{1/p}$?
17. Let $\{a_n\}$ be a null sequence of nonnegative real numbers, and $p \in \mathbb{R}$. Must $\{a_n^p\}$ be a null sequence? Is $\{1/n^p\}$ a null sequence?
18. Let $\{a_n\}$ be a sequence of positive real numbers. Is it true that $\{a_n\}$ diverges to ∞ if and only if $\{1/a_n\}$ converges to 0?
19. If $\{a_n\}$ is a sequence of real numbers such that $\{a_n/n\}$ converges to l for some $l \neq 0$, must $\{a_n\}$ be unbounded?
20. If $\{a_n\}$ converges to 0, must $\{(-1)^n a_n\}$ converge to 0?
21. If $\{a_n\}$ converges to a nonzero real number a, must $\{(-1)^n a_n\}$ oscillate?
22. If $\{a_n\}$ diverges to ∞, must $\{(-1)^n a_n\}$ oscillate?
23. If $\{|a_n|\}$ converges to $|a|$, must $\{a_n\}$ be convergent either to a or to $-a$? How about when $a = 0$? Does the sequence $\{(-1)^n\}$ address your concern for this question?
24. If $\{a_n\}$ converges and $\{b_n\}$ diverges, must $\{a_n b_n\}$ be divergent? Must $\{a_n + b_n\}$ be divergent?
25. If $\{a_n\}$ and $\{b_n\}$ are divergent, must $\{a_n b_n\}$ be divergent? Must $\{a_n + b_n\}$ be divergent?
26. Suppose that $\{a_n\}$ is an unbounded sequence of nonzero real numbers. Does $\{a_n\}$ diverge to ∞ or $-\infty$? Must $\{|a_n|\}$ be divergent to ∞? Must $\{1/a_n\}$ be bounded?
27. Suppose that $\{a_n\}$ is bounded. Must $\{1/a_n\}$ be bounded? Must $\{a_n/n\}$ be convergent?
28. If $\{a_n\}$ and $\{a_n b_n\}$ are both bounded, must $\{b_n\}$ be bounded?
29. If $a_1 = 1$ and $a_{n+1} = a_n + (1/a_n)$ for $n \geq 1$, must $\{a_n\}$ be bounded?
30. If $\{a_n\}$ and $\{b_n\}$ are both increasing, must $\{a_n b_n\}$ be increasing?
31. Suppose that $\{a_n\}$ and $\{b_n\}$ are two sequences of real numbers such that $|a_n - b_n| < 1/n$ for large n, and $a_n \to a$ as $n \to \infty$. Does $b_n \to a$ as $n \to \infty$?
32. If $\{a_n\}$ is a sequence such that $\{(a_n - 1)/(a_n + 1)\}$ converges to zero, does $\{a_n\}$ converge?
33. If $\{a_n\}$ converges to a, must $\{a_n^2\}$ converge to a^2? Does $\{a_n^p\}$ converge to a^p if $p \in \mathbb{N}$?
34. Suppose that $b_n \to b$ as $n \to \infty$ and $b \neq 0$. Must there exist an $R > 0$ and a positive integer N such that $|b_n| \geq R$ for all $n \geq N$?
35. If $\{a_n^2\}$ converges, must $\{a_n\}$ be convergent?
36. Suppose that $\{a_n^2\}$ converges and $a_n > 0$. Can $\{a_n\}$ be convergent? Can $\{a_n\}$ be convergent?
37. If $\{a_n^2\}$ converges to a, must $\{|a_n|\}$ converge to $\sqrt{a}$?
38. If $\{a_n^3\}$ converges to a^3, must $\{a_n\}$ converge to a?
39. Can there exist a divergent sequence that is monotone?
40. Can there exist a divergent sequence $\{s_n\}$ such that $s_{n+1} - s_n \to 0$ as $n \to \infty$?

41. If $\{a_n\}$ is an increasing sequence of real numbers that is bounded above and $L = \lim_{n\to\infty} a_n$, must we have $a_n \leq L$ for all n?
42. If $\{a_n\}$ is a decreasing sequence of real numbers that is bounded below and $L = \lim_{n\to\infty} a_n$, must we have $a_n \geq L$ for all n?
43. If $0 < a < 1$, does it follow that $\lim_{n\to\infty} a^{1/2^n} = 1$? Does it follow that $\lim_{n\to\infty} a^{1/3^n} = 1$?
44. Let $a_n = (1+1/n)^n$ and $b_n = (1+1/n)^{n+k}$, where k is a fixed integer. Do we have $\lim_{n\to\infty} a_n = \lim_{n\to\infty} b_n = e$?

Exercises 2.45.

1. Show that

$$\lim_{n\to\infty} \frac{n}{2n+3} = \frac{1}{2}, \quad \lim_{n\to\infty} \frac{3n+1}{2n+1} = \frac{3}{2}, \quad \text{and} \quad \lim_{n\to\infty} \frac{n^3-3}{n^4} = 0.$$

If $\epsilon = 0.001$ is chosen, find N in each case such that for $n \geq N$ we have

$$\left|\frac{n}{2n+3} - \frac{1}{2}\right| < 0.001, \quad \left|\frac{3n+1}{2n+1} - \frac{3}{2}\right| < 0.001, \quad \text{and} \quad \left|\frac{n^3-3}{n^4}\right| < 0.001.$$

2. Construct three sequences such that $a_n \leq b_n \leq c_n$ for all $n \geq N$, $\lim_{n\to\infty} a_n = L$ and $\lim_{n\to\infty} c_n = M$ for some real numbers L, M, but $\lim_{n\to\infty} b_n$ does not exist.
3. Suppose that $\{a_n\}_{n\geq 1}$ and $\{b_n\}_{n\geq 1}$ are two sequences of real numbers such that $\lim_{n\to\infty} a_n = \infty$ and $\lim_{n\to\infty} b_n = L$, where $0 < L \leq \infty$. Show that $\lim_{n\to\infty} a_n b_n = \infty$. Using this, show that

$$\lim_{n\to\infty} \frac{n^3-3}{n+2} = \infty \quad \text{and} \quad \lim_{n\to\infty} \frac{3^n}{n^2+(-1)^n} = \infty.$$

4. Which of the following sequences are monotone? bounded? convergent?

$$\left\{\frac{(-1)^n(n+2)}{n}\right\}, \ \left\{2^{(-1)^n}\right\}, \ \left\{\frac{n}{2^n}\right\}, \ \{\log(n+1) - \log n\}, \ \left\{\frac{3n-5}{2^n}\right\}.$$

5. For $p > 0$ and $|c| < 1$, prove that $\{c^n\}$, $\{n^p c^n\}$, and $\{n^p/n!\}$ are all null sequences.
6. Using BMCT, show that $a^{1/n} \to 0$ as $n \to \infty$, where $0 < a < 1$. Is it possible to use BMCT to show that $n^{1/n} \to 1$ as $n \to \infty$?
7. Which is larger in each of the following:
   (i) 1000^{1000} or 1001^{999}? **(ii)** $\left(1 + \frac{1}{100000}\right)^{100000}$ or 2?
8. Define a_n recursively by $a_1 = \sqrt{2}$ and $a_{n+1} = \sqrt{2+\sqrt{a_n}}$ for all $n \geq 1$. Show that the sequence $\{a_n\}_{n\geq 1}$ is convergent. Find its limit.
9. Define a_n recursively by $a_1 = \sqrt{2}$ and $a_{n+1} = \sqrt{2+a_n}$ for all $n \geq 1$. Show that the sequence $\{a_n\}_{n\geq 1}$ converges to 2.

10. For each of the following sequences, show that there is a number L such that $a_n \to L$. Find also the value of L.
    (a) $\{a_n\}$, where $a_1 = 1$ and $a_{n+1} = 1 + \sqrt{a_n}$ for $n \geq 1$.
    (b) $\{a_n\}$, where $a_1 = 3$ and $a_{n+1} = 3 + \sqrt{a_n}$ for $n \geq 1$.
    (c) $\{a_n\}$, where $a_1 = L$ $(L > 1)$ and $a_{n+1} = \sqrt{a_n}$ for $n \geq 1$.
    (d) $\{a_n\}$, where $a_1 > 0$, $a_2 > 0$, and $a_{n+2} = \sqrt{a_n} + \sqrt{a_{n+1}}$ for $n \geq 1$.
    (e) $\{a_n\}$, where $a_1 = 1$ and $a_{n+1} = \frac{1}{4}(2a_n + 3)$ for $n \geq 1$.
    (f) $\{a_n\}$, where $a_1 = 1$ and $a_{n+1} = a_n/(1 + a_n)$ for $n \geq 1$.
    (g) $\{a_n\}$, where $a_1 = \alpha > 0$ and $a_{n+1} = \sqrt{(\alpha\beta^2 + a_n^2)/(\alpha + 1)}$ $(\beta > \alpha)$.
11. Suppose that a sequence $\{a_n\}$ of real numbers satisfies $7a_{n+1} = a_n^3 + 6$ for $n \geq 1$. If $a_1 = \frac{1}{2}$, prove that the sequence increases and find its limit. What happens if $a_1 = \frac{3}{2}$ or $a_1 = \frac{5}{2}$?
12. Test each of the sequences given below for convergence. Find its limit if it converges.
    (a) $a_1 = 1$ and $a_{n+1} = \sqrt{5a_n}$. **(b)** $a_1 = 1$ and $a_{n+1} = \sqrt{5}\, a_n$.
    (c) $a_1 = 1$ and $a_{n+1} = \sqrt{5 + a_n}$.
13. Show that if $a_1 > b_1 > 0$, $a_{n+1} = \sqrt{a_n b_n}$, and $b_{n+1} = (a_n + b_n)/2$, then $\{a_n\}$ and $\{b_n\}$ both converge to a common limit.
14. Let $\{a_n\}$ be a sequence of positive real numbers such that $a_{n+1} \leq r a_n$ for some $r \in (0, 1)$ and for all n. Prove that $\{a_n\}$ converges to 0.
15. In the following problems, state whether the given sequence $\{a_n\}$ is convergent or divergent. If it is convergent, then determine its limit. Here a_n equals

(a) $2 + (-1)^n$. **(b)** $n(2 + (-1)^n)$ **(c)** $n \cos\left(\dfrac{n\pi}{2}\right)$.

(d) $2^{2008/n}$. **(e)** $\dfrac{3n^2 - \log n}{n^2 + 3n^{3/2}}$. **(f)** $\sqrt{n + 3\sqrt{n}} - \sqrt{n}$.

(g) $n^{2008/n}$. **(h)** $n^{1/(n+2008)}$. **(i)** $(n+1)^{1/(\log(1+n))}$.

(j) $\dfrac{5^n + 6^n}{1 + 7^n}$. **(k)** $(\log n)^{1/n}$. **(l)** $\sqrt{n(n+1)} - n$.

(m) $\dfrac{(n!)^{1/n}}{n}$. **(n)** $\log n - \log(n+1)$. **(o)** $\dfrac{1}{n} \sin\left(\dfrac{n\pi}{6}\right) + \dfrac{5n+1}{7n+6}$.

(p) $(an + 7)^{1/n}$. **(q)** $\dfrac{a^n - a^{-n}}{a^n + a^{-n}}$. **(r)** $(n + 2008)^{1/n}$.

(s) $\dfrac{a^n + n}{a^n - n}$. **(t)** $\dfrac{a^n}{n!}$ $(a \in \mathbb{R})$. **(u)** $n(a^{1/n} - 1)$.

2.2 Limit Inferior, Limit Superior, and Cauchy Sequences

Consider a sequence of real numbers $\{a_n\}_{n\geq 1}$. Then for each fixed $k \in \mathbb{N}$, let

$$M_k = \sup\{a_k, a_{k+1}, \ldots\} := \sup\{a_n : n \geq k\}$$

if the sequence is bounded above, and $M_k = \infty$ if it is not bounded above. Clearly, $M_k \geq M_{k+1}$ for every k. Similarly, let

$$m_k = \inf\{a_k, a_{k+1}, \ldots\} := \inf\{a_n : n \geq k\}$$

if the sequence is bounded below, and $m_k = -\infty$ if it is not bounded below. Clearly, $m_k \leq m_{k+1}$ for every k. Consequently,

$$m_1 \leq m_2 \leq \cdots \leq m_k \leq m_{k+1} \leq \cdots \leq M_{k+1} \leq M_k \leq \cdots \leq M_2 \leq M_1.$$

Since every monotone sequence *has a limit* (see Theorem 2.30 if we also allow $\pm\infty$), the limits

$$M = \lim_{k\to\infty} M_k \quad \text{and} \quad m = \lim_{k\to\infty} m_k$$

both exist. So $m \leq M$. We call M and m the limit superior and the limit inferior, respectively, of $\{a_n\}$. We denote these limits by

$$M = \limsup_{n\to\infty} a_n \text{ and } \overline{\lim_{n\to\infty}}\, a_n, \quad \text{and} \quad m = \liminf_{n\to\infty} a_n \text{ or } \underline{\lim}_{n\to\infty} a_n,$$

respectively. Thus,

$$\limsup_{n\to\infty} a_n = \lim_{k\to\infty} \sup_{n\geq k} a_n \quad \text{and} \quad \liminf_{n\to\infty} a_n = \lim_{k\to\infty} \inf_{n\geq k} a_n.$$

The right-hand sides of these are always meaningful, provided it is understood that the values of ∞ and $-\infty$ are allowed. Note that

$$\begin{cases} M = \infty & \text{if } \{a_n\} \text{ is not bounded above,} \\ m = -\infty & \text{if } \{a_n\} \text{ is not bounded below,} \\ M = -\infty & \text{if } \lim\limits_{n\to\infty} a_n = -\infty, \\ m = \infty & \text{if } \lim\limits_{n\to\infty} a_n = \infty. \end{cases}$$

For instance:

(a) For the sequence $\{a_n\}_{n\geq 1}$, where $a_n = 1/n$, we have

$$m_1 = \inf\{1, 1/2, 1/3, \ldots\} = 0, \quad m_2 = \inf\{1/2, 1/3, 1/4, \ldots\} = 0,$$

and $m_k = 0$ for each $k \geq 1$. Therefore, it is clear that

$$m = \lim m_k = 0, \quad \text{i.e., } \liminf a_n = 0.$$

Similarly, we see that

$$M_1 = \sup\{1, 1/2, 1/3, \ldots\} = 1, \quad M_2 = \sup\{1/2, 1/3, 1/4, \ldots\} = \frac{1}{2},$$

and $M_k = 1/k$ for each $k \geq 1$. Therefore,

$$M = \lim M_k = 0, \quad \text{i.e., } \limsup a_n = 0.$$

(b) $\limsup_{n\to\infty}(-1)^n = 1$ and $\liminf_{n\to\infty}(-1)^n = -1$.
(c) $\lim_{n\to\infty} n^2 = \infty$, and so $\limsup_{n\to\infty} n^2 = \liminf_{n\to\infty} n^2 = \infty$.
(d) $\limsup_{n\to\infty}(-n) = -\infty$ and $\limsup_{n\to\infty} n = \infty$.
(e)

$$\limsup_{n\to\infty} r^n = \begin{cases} \infty & \text{if } |r| > 1, \\ 1 & \text{if } |r| = 1, \\ 0 & \text{if } |r| < 1, \end{cases} \quad \text{and} \quad \liminf_{n\to\infty} r^n = \begin{cases} \infty & \text{if } r > 1, \\ 1 & \text{if } r = 1, \\ 0 & \text{if } |r| < 1, \\ -1 & \text{if } r = -1, \\ -\infty & \text{if } r < -1. \end{cases}$$

(f) If $a_n = (-1)^n(1 + 1/n)$, then $\limsup_{n\to\infty} a_n = 1$ and $\liminf_{n\to\infty} a_n = -1$. Also, we note that $a_{2n} \to 1$, $a_{2n-1} \to -1$ as $n \to \infty$, and the sequence $\{a_n\}$ has no subsequences that can converge to a limit other than 1 or -1. Note also that

$$\sup\{a_n : n \geq 1\} = \frac{3}{2} \quad \text{and} \quad \inf\{a_n : n \geq 1\} = -2.$$

The reader is warned not to confuse the supremum of a set with the limit superior of a sequence, and similarly the infimum of a set with the limit inferior of a sequence.
(g) $\limsup_{n\to\infty} (-1)^n/n = 0 = \liminf_{n\to\infty} (-1)^n/n$, because for $k \geq 1$,

$$M_k = \sup\left\{\frac{(-1)^k}{k}, \frac{-(-1)^k}{k+1}, \frac{(-1)^k}{k+2}, \ldots\right\} = \begin{cases} \dfrac{1}{k+1} & \text{if } k \text{ is odd}, \\ \dfrac{1}{k} & \text{if } k \text{ is even}, \end{cases}$$

and

$$m_k = \begin{cases} -\dfrac{1}{k} & \text{if } k \text{ is odd}, \\ -\dfrac{1}{k+1} & \text{if } k \text{ is even}, \end{cases}$$

so that $M_k \to 0$ and $m_k \to 0$ as $k \to \infty$.
(h) For the sequence $\{(-1)^n n\}_{n\geq 1} = \{\ldots, -5, -3, -1, 2, 4, 6, \ldots\}$, we have

$$\inf\{(-1)^n n : n \in \mathbb{N}\} = -\infty \quad \text{and} \quad \liminf(-1)^n n = -\infty$$

and

$$\sup\{(-1)^n n : n \in \mathbb{N}\} = \infty \quad \text{and} \quad \limsup(-1)^n n = \infty.$$

Lemma 2.46. *Suppose that* $\{a_n\}$ *is a sequence of real numbers with*

$$L = \limsup_{n\to\infty} a_n \quad \text{and} \quad \ell = \liminf_{n\to\infty} a_n.$$

Then for every $\epsilon > 0$ *there exist integers* N_1 *and* N_2 *such that*

$$\begin{cases} a_n - L < \epsilon & \text{for all } n \geq N_1, \\ a_n - L > -\epsilon & \text{for infinitely many } n \geq N_1, \end{cases}$$

and

$$\begin{cases} a_n - \ell > -\epsilon & \text{for all } n \geq N_2, \\ a_n - \ell < \epsilon & \text{for infinitely many } n \geq N_2, \end{cases}$$

respectively.

Proof. By the definition of the limit superior, since $L = \lim_{k\to\infty} M_k$, there exists an integer N_1 such that

$$|\sup\{a_k, a_{k+1}, \ldots\} - L| = |M_k - L| < \epsilon \quad \text{for all } k \geq N_1,$$

so that

$$a_k \leq \sup\{a_k, a_{k+1}, \ldots\} < L + \epsilon \quad \text{for all } k \geq N_1.$$

That is,

$$a_k < L + \epsilon \quad \text{for all } k \geq N_1.$$

Again, since $M_k \geq M_{k+1}$ for every $k \geq 1$, we have

$$L \leq \sup_{k\geq 1} M_k. \tag{2.1}$$

In particular, this gives

$$L \leq M_1 = \sup\{a_1, a_2, a_3, \ldots\}.$$

Thus, by the definition of supremum, there exists an n_1 such that $a_{n_1} > M_1 - \epsilon$, so that

$$a_{n_1} > L - \epsilon.$$

Now taking $k = n_1$ in (2.1), we obtain that

$$L \leq M_{n_1} = \sup\{a_{n_1}, a_{n_1+1}, \ldots\},$$

and so there exists an n_2 such that

$$a_{n_2} > M_{n_1} - \epsilon > L - \epsilon.$$

Proceeding indefinitely, we obtain integers $n_1 < n_2 < \cdots < n_k < \cdots$ such that

$$a_{n_k} > L - \epsilon \quad \text{for all } k \in \mathbb{N},$$

which proves the second inequality for the case of limit superior.

Similarly, since $\ell = \lim_{k\to\infty} m_k$, there exists an integer N_2 such that

$$a_k \geq \inf\{a_k, a_{k+1}, \ldots\} > L - \epsilon \quad \text{for all } k \geq N_2. \qquad \blacksquare$$

Theorem 2.47. *For any sequence of real numbers* $\{a_n\}$*, we have*

$$\lim_{n\to\infty} a_n = L \quad \textit{if any only if} \quad \limsup_{n\to\infty} a_n = \liminf_{n\to\infty} a_n = L.$$

Proof. If $L = \pm\infty$, then the equivalence is a consequence of the definitions of limit superior and limit inferior. Therefore, we assume that $\lim a_n = L$, where L is finite.

$\Rightarrow$: Given $\epsilon > 0$, there exists an $N \in \mathbb{N}$ such that

$$|a_n - L| < \epsilon, \quad \text{i.e., } L - \epsilon < a_n < L + \epsilon \text{ for all } n \geq N,$$

and so

$$L - \epsilon < M_N = \sup\{a_N, a_{N+1}, \ldots\} \leq L + \epsilon.$$

Thus, $\{M_k\}_{k\geq N}$ is a bounded monotone sequence and hence converges. That is,

$$L - \epsilon \leq \lim_{N\to\infty} M_N = \limsup_{n\to\infty} a_n \leq L + \epsilon.$$

Since ϵ is arbitrary, $\limsup_{n\to\infty} a_n = L$. A similar argument gives $\liminf_{n\to\infty} a_n = L$.

$\Leftarrow$: Conversely, suppose that $L = \limsup_{n\to\infty} a_n = \liminf_{n\to\infty} a_n = \ell$. Since $\ell = L$, by Lemma 2.46 we conclude that there exists $N = \max\{N_1, N_2\}$ such that

$$L - \epsilon < a_k < L + \epsilon \quad \text{for all } k \geq N.$$

This proves that $\lim_{k\to\infty} a_k = L$, as desired. ∎

For any bounded sequence $\{a_n\}$, we see that $\{M_k - m_k\}$ is increasing and converges to $M - m$. Thus, using Theorem 2.47, we may formulate the definition of convergence of a sequence as follows.

Theorem 2.48. *A sequence* $\{a_n\}$ *of real numbers is convergent if and only if it is bounded and* $\{M_k - m_k\}$ *converges to zero, where* $M_k = \sup\{a_n \colon n \geq k\}$ *and* $m_k = \inf\{a_n \colon n \geq k\}$.

Alternatively, Theorem 2.42 can be seen (without using Lemma 2.41) as an immediate consequence of the following result, which in particular, shows that there are subsequences converging to m and M. Moreover, m and M are, respectively, the smallest and the largest possible limits for convergent subsequences.

Theorem 2.49. *Let* $\{a_n\}$ *be a bounded sequence of real numbers and let*

$$S = \{x \in \mathbb{R} : a_{n_k} \to x \textit{ for some subsequence } a_{n_k}\}.$$

If $m = \liminf a_n$ *and* $M = \limsup a_n$*, then* $\{m, M\} \subset S \subset [m, M]$.

Proof. First we prove that $M \in S$. For this, we need to show that there exists a subsequence $\{a_{n_k}\}_{k\geq 1}$ such that for each given $\epsilon > 0$, there exists an integer N such that

$$|a_{n_k} - M| < \epsilon \quad \text{for all } k \geq N.$$

By Lemma 2.46, there exists an integer N_1 such that

$$a_k < M + \epsilon \ \text{ for all } k \geq N_1 \tag{2.2}$$

and $n_1 < n_2 < \cdots < n_k < \cdots$ such that

$$a_{n_k} > M - \epsilon \quad \text{for all } k \in \mathbb{N}. \tag{2.3}$$

Combining (2.2) and (2.3), we infer that

$$M - \epsilon < a_{n_k} < M + \epsilon, \quad \text{i.e., } |a_{n_k} - M| < \epsilon \text{ for all } n_k \geq N,$$

and so M is the limit of a subsequence of $\{a_n\}$. The assertion about m has a similar proof. Thus, $\{m, M\} \subset S$.

Next we prove that $S \subset [m, M]$. We assume that $a_{n_k} \to x$ as $k \to \infty$. We shall show that $x \in [m, M]$. Equation (2.2) shows that

$$a_n < M + \epsilon \quad \text{for sufficiently large } n,$$

and so

$$a_{n_k} < M + \epsilon \quad \text{for sufficiently large } k.$$

The limit inequality rule gives that

$$x \leq M + \epsilon,$$

and since $\epsilon > 0$ is arbitrary, it follows that $x \leq M$. The proof for $m \leq x$ is similar. ∎

Corollary 2.50. *A sequence $\{a_n\}$ of real numbers converges if and only if S is a singleton set. That is,* $\lim a_n$ *exists.*

In view of Theorem 2.49, we have the following equivalent definition: *If $\{a_n\}$ is a bounded sequence of real numbers, then M and m, the limit superior and the limit inferior of $\{a_n\}$, are respectively the greatest and the least subsequential limits of $\{a_n\}$.*

Theorem 2.51. *Suppose that $\{a_n\}_{n\geq 1}$ and $\{b_n\}_{n\geq 1}$ are two bounded sequences of real numbers. Then we have the following:*

(a) $\limsup_{n\to\infty}(a_n + b_n) \leq \limsup_{n\to\infty} a_n + \limsup_{n\to\infty} b_n$.
(b) $\liminf_{n\to\infty}(a_n + b_n) \geq \liminf_{n\to\infty} a_n + \liminf_{n\to\infty} b_n$.
(c) $\limsup_{n\to\infty} a_n \leq \limsup_{n\to\infty} b_n$ *and* $\liminf_{n\to\infty} a_n \leq \limsup_{n\to\infty} b_n$ *if* $a_n \leq b_n$ *for all* $n \geq 1$.

(d) $\limsup_{n\to\infty}(a_nb_n) \leq \left(\limsup_{n\to\infty} a_n\right)\left(\limsup_{n\to\infty} b_n\right)$ *if* $a_n > 0$, $b_n > 0$.

(e) $\liminf_{n\to\infty}(a_nb_n) \geq \left(\liminf_{n\to\infty} a_n\right)\left(\liminf_{n\to\infty} b_n\right)$ *if* $a_n > 0$, $b_n > 0$.

Proof. **(a)** and **(b)**:

Method 1: As usual, for each fixed $k \in \mathbb{N}$, let

$$M_k = \sup\{a_k, a_{k+1}, \ldots\} \quad \text{and} \quad P_k = \sup\{b_k, b_{k+1}, \ldots\}.$$

Then

$$a_n \leq M_k \quad \text{and} \quad b_n \leq P_k \quad \text{for all } n \geq k,$$

and therefore

$$a_n + b_n \leq M_k + P_k \quad \text{for all } n \geq k,$$

which shows that $M_k + P_k$ is an upper bound for

$$\{a_k + b_k, a_{k+1} + b_{k+1}, \ldots\}.$$

Consequently,

$$\sup\{a_k + b_k, a_{k+1} + b_{k+1}, \ldots\} \leq M_k + P_k,$$

and thus

$$\lim_{k\to\infty}\sup\{a_k + b_k, a_{k+1} + b_{k+1}, \ldots\} \leq \lim_{k\to\infty}(M_k + P_k) = \lim_{k\to\infty} M_k + \lim_{k\to\infty} P_k,$$

which, by the definition, is equivalent to **(a)**. The proof of **(b)** is similar and so will be omitted.

Method 2: Since $\{a_n+b_n\}_{n\geq 1}$ is a bounded sequence (by hypothesis), Lemma 2.46 shows that there exist integers N_1, N_2, N_3, and N_4 such that

$$a_k < L_a + \epsilon/2 \quad \text{for all } k \geq N_1 \quad \text{and} \quad a_k > \ell_a - \epsilon/2 \quad \text{for all } k \geq N_2$$

and

$$b_k < L_b + \epsilon/2 \quad \text{for all } k \geq N_3 \quad \text{and} \quad b_k > \ell_b - \epsilon/2 \quad \text{for all } k \geq N_4,$$

respectively. Here

$$L_a = \limsup a_n, \quad \ell_a = \liminf a_n, \quad L_b = \limsup b_n, \quad \text{and} \quad \ell_b = \liminf b_n.$$

Thus,

$$a_k + b_k < L_a + L_b + \epsilon \quad \text{for all } k \geq \max\{N_1, N_3\}$$

and

$$a_k + b_k > \ell_a + \ell_b - \epsilon \quad \text{for all } k \geq \max\{N_2, N_4\}.$$

Since $\epsilon > 0$ is arbitrary, **(a)** and **(b)** follow.

(c) Since $a_n \le b_n$ for all $n \ge 1$, it follows that

$$M_k \le P_k \quad \text{and} \quad m_k \le p_k,$$

where $m_k = \inf\{a_k, a_{k+1}, \ldots\}$ and $p_k = \inf\{b_k, b_{k+1}, \ldots\}$. Taking the limit as $k \to \infty$ yields the desired conclusion. ■

Observe that if $a_n = (-1)^n$ and $b_n = (-1)^{n+1}$, then we have

$$a_n + b_n = 0 \quad \text{for all } n \ge 0, \quad \limsup a_n = 1 = \limsup b_n.$$

We may also consider

$$a_n = \begin{cases} 0 & \text{if } n = 2k, \\ (-1)^{k+1} & \text{if } n = 2k-1, \end{cases} \quad \text{and} \quad b_n = \begin{cases} (-1)^k & \text{if } n = 2k, \\ 0 & \text{if } n = 2k-1, \end{cases}$$

so that

$$a_n + b_n = \begin{cases} (-1)^k & \text{if } n = 2k, \\ (-1)^{k+1} & \text{if } n = 2k-1. \end{cases}$$

In either case, the equalities in **(a)** and **(b)** of Theorem 2.51 do not always hold.

If

$$a_n = \begin{cases} 1 & \text{if } n \text{ is odd}, \\ 2 & \text{if } n \text{ is even}, \end{cases} \quad \text{and} \quad b_n = \begin{cases} 2 & \text{if } n \text{ is odd}, \\ 1 & \text{if } n \text{ is even}, \end{cases}$$

we see that equality in each of **(d)** and **(e)** of Theorem 2.51 does not hold.

2.2.1 Cauchy Sequences

If a sequence $\{a_n\}$ of real numbers converges to a number a, then the terms a_n of the sequence are close to a for large n, and hence the terms of the sequence themselves are close to each other "near a." This intuition led to the concept of Cauchy[1] sequence, which helps us in deducing the convergence of a sequence without necessarily knowing its limit. Moreover, unlike theorems (such as BMCT) that deal only with monotone sequences, we have theorems on Cauchy sequences that deal with sequences that are not necessarily monotone.

Definition 2.52 (Cauchy sequence). *A sequence $\{a_n\} \subset \mathbb{R}$ is called a Cauchy sequence if for each $\epsilon > 0$ there is a positive integer N such that $m, n \ge N$ implies $|a_n - a_m| < \epsilon$. Equivalently, we say that a sequence $\{a_n\}$ is Cauchy if for each $\epsilon > 0$ there is a positive integer N such that*

$$|a_{n+p} - a_n| < \epsilon \quad \textit{for all } n \ge N \textit{ and for all } p \in \mathbb{N}.$$

[1] Augustin-Louis Cauchy (1789–1857) is one of the important mathematicians who placed analysis on a rigorous footing.

For example, if $a_n = (-1)^{n-1}/n$, then $\{a_n\}$ is Cauchy; for

$$|a_n - a_m| = \left| \frac{(-1)^{n-1}}{n} - \frac{(-1)^{m-1}}{m} \right| \leq \frac{1}{n} + \frac{1}{m} < \frac{2}{n} \quad \text{if } m > n.$$

Our first result is algebraic.

Theorem 2.53. *Every convergent sequence is a Cauchy sequence.*

Proof. Suppose that $a_n \to a$ as $n \to \infty$, and let $\epsilon > 0$ be given. Then there exists an N such that

$$|a_n - a| < \frac{\epsilon}{2} \quad \text{for all } n \geq N.$$

Therefore, for $m, n \geq N$, we must have

$$|a_n - a_m| = |(a_n - a) - (a_m - a)| \leq |a_n - a| + |a_m - a| < \frac{\epsilon}{2} + \frac{\epsilon}{2} = \epsilon,$$

and hence $\{a_n\}$ is a Cauchy sequence. ∎

Theorem 2.53 gives a necessary condition for convergence. Equivalently, if a sequence is not Cauchy, then it cannot be convergent. Thus, Theorem 2.53 can be used to show the divergence of several nontrivial sequences. For example, we have the following:

(a) Neither $\{n\}_{n\geq 1}$ nor $\{1 + (-1)^n\}_{n\geq 1}$ is Cauchy.

(b) If $s_n = \sum_{k=1}^{n} 1/k$, then $\{s_n\}_{n\geq 1}$ is not Cauchy, because for any $n \in \mathbb{N}$ (with $m = 2n$),

$$s_{2n} - s_n = \sum_{k=1}^{2n} \frac{1}{k} - \sum_{k=1}^{n} \frac{1}{k} = \frac{1}{n+1} + \frac{1}{n+2} + \cdots + \frac{1}{2n} > n\left(\frac{1}{2n}\right) = \frac{1}{2}.$$

Thus, the sequence $\{s_n\}$ is not convergent.

(c) Similarly, if $s_n = \sum_{k=1}^{n} 1/(2k-1)$, then $\{s_n\}_{n\geq 1}$ is not Cauchy (and hence is not convergent), because for any $n \in \mathbb{N}$,

$$\begin{aligned} s_{2n} - s_n &= \sum_{k=1}^{2n} \frac{1}{2k-1} - \sum_{k=1}^{n} \frac{1}{2k-1} \\ &= \frac{1}{2n+1} + \frac{1}{2n+3} + \cdots + \frac{1}{2n+2n-1} \\ &> n\left(\frac{1}{4n-1}\right) > n\left(\frac{1}{4n}\right) = \frac{1}{4}. \end{aligned}$$

(d) Finally, consider the sequence $\{x_n\}$ given by

$$x_0 = 0 \quad \text{and} \quad x_{n+1} = \frac{10x_n + 6}{5} \quad \text{for } n \geq 0.$$

Then $\{x_n\}$ does not converge, because it is not Cauchy. Indeed,

$$x_n > 0 \quad \text{for all } n \geq 1 \quad \text{and} \quad x_{n+1} - x_n = x_n + \frac{6}{5} > \frac{6}{5},$$

showing that $\{x_n\}$ is not Cauchy.

We also remark that a sequence $\{s_n\}$ that satisfies the condition

$$s_{n+1} - s_n \to 0 \quad \text{as } n \to \infty$$

is not necessarily a Cauchy sequence (e.g., s_n as above or $s_n = \log n$).

Theorem 2.54. *Cauchy sequences are bounded.*

Proof. The proof is similar to that of the corresponding result for convergent sequences (see Theorem 2.7). For the sake of completeness we include a proof here. Consider a Cauchy sequence $\{a_n\}_{n\geq 1}$. Then by definition, there exists a positive integer $N \in \mathbb{N}$ such that

$$|a_m - a_n| < \epsilon = 1 \quad \text{for all } n > m \geq N.$$

That is, with $m = N$, we have $|a_n| < 1 + |a_N|$ for all $n > N$. We conclude that $\{a_n\}_{n\geq 1}$ is bounded. ∎

An interesting fact which that Cauchy sequences important is that the converse of Theorem 2.53 is also true. Our next task is to prove this result, which is also called the *general principle of convergence.*

Theorem 2.55 (Completeness criterion for sequences). *A sequence is convergent if and only if it is a Cauchy sequence.*

Proof. The first half of the theorem has already been proved. Thus, we have to show that every Cauchy sequence of real numbers converges. To do this, we begin with a Cauchy sequence $\{a_n\}$. Then $\{a_n\}$ is bounded by Theorem 2.54. Let $\epsilon > 0$. Then there exists an $N = N(\epsilon)$ such that

$$|a_n - a_m| < \frac{\epsilon}{2} \quad \text{whenever } n > m \geq N. \tag{2.4}$$

Method 1: In particular, taking $m = N$ in (2.4), it follows that

$$|a_n - a_N| < \frac{\epsilon}{2}, \quad \text{i.e., } -\frac{\epsilon}{2} + a_N < a_n < \frac{\epsilon}{2} + a_N \text{ for all } n > N.$$

This shows that $a_N - (\epsilon/2)$ and $a_N + (\epsilon/2)$ are, respectively, lower and upper bounds for the set

$$X_n = \{a_n, a_{n+1}, \ldots\} \quad \text{if } n > N.$$

Note that $X_n \supseteq X_{n+1} \supseteq \cdots$ and if $M_n = \sup X_n$, then $M_n \geq M_{n+1} \geq \cdots$. Thus, for $n > N$,

$$\underbrace{a_N - \frac{\epsilon}{2} \leq \inf\{a_n, a_{n+1}, \ldots\}} \leq \sup\{a_n, a_{n+1}, \ldots\} \leq a_N + \frac{\epsilon}{2},$$

which gives

$$\sup\{a_n, a_{n+1}, \ldots\} \leq \underbrace{a_N + \frac{\epsilon}{2} \leq \inf\{a_n, a_{n+1}, \ldots\} + \frac{\epsilon}{2} + \frac{\epsilon}{2}},$$

so that for $n > N$,

$$\sup\{a_n, a_{n+1}, \ldots\} \leq \inf\{a_n, a_{n+1}, \ldots\} + \epsilon.$$

Thus, by definition,

$$\limsup a_n \leq \sup\{a_n, a_{n+1}, \ldots\} \leq \inf\{a_n, a_{n+1}, \ldots\} + \epsilon \leq \liminf a_n + \epsilon.$$

Since this holds for every $\epsilon > 0$, we have

$$\limsup_{n\to\infty} a_n \leq \liminf_{n\to\infty} a_n.$$

The reverse inequality always holds, so that

$$\limsup_{n\to\infty} a_n = \liminf_{n\to\infty} a_n.$$

Hence $\{a_n\}$ converges by Theorem 2.47.

Method 2: Assume that $\{a_n\}$ is a Cauchy sequence. Then by the Bolzano–Weierstrass theorem (Theorem 2.42), $\{a_n\}$ has a convergent subsequence, say $\{a_{n_k}\}$. Let $a = \lim_{k\to\infty} a_{n_k}$. Then there exists an N_1 such that

$$|a_{n_k} - a| < \frac{\epsilon}{2} \quad \text{whenever } k > N_1.$$

We need to show that $a = \lim_{n\to\infty} a_n$. Choose k large enough that $n_k > N$ and $k > N_1$. Then because $\{a_n\}$ is Cauchy, (2.4) is also satisfied with $m = n_k$. Thus, $\{a_n\}$ converges, because

$$|a_n - a| \leq |a_n - a_{n_k}| + |a_{n_k} - a| < \frac{\epsilon}{2} + \frac{\epsilon}{2} = \epsilon \quad \text{whenever } n > N. \qquad \blacksquare$$

Definition 2.56 (Contractive sequence). *A sequence* $\{a_n\}_{n\geq 1}$ *is said to be* contractive *if there exists a constant* $\lambda \in (0,1)$ *such that* $|a_{n+1} - a_n| \leq \lambda |a_n - a_{n-1}|$ *for all* $n \geq 2$.

Theorem 2.57. *Every contractive sequence is Cauchy (and hence convergent by Theorem 2.55). What happens if one allows* $\lambda = 1$*?*

Proof. Assume that $\{a_n\}_{n\geq 1}$ is a contractive sequence. We find that $a_1 \neq a_2$; otherwise, $\{a_n\}$ reduces to a zero sequence, which converges trivially. We see that

$$|a_{n+1} - a_n| \leq \lambda^{n-1}|a_2 - a_1|,$$

and so for $m > n \geq N$, we have

$$\begin{aligned}
|a_m - a_n| &= |(a_m - a_{m-1}) + (a_{m-1} - a_{m-2}) + \cdots + (a_{n+1} - a_n)| \\
&= [\lambda^{m-2} + \lambda^{m-3} + \cdots + \lambda^{n-1}]|a_2 - a_1| \\
&= \frac{\lambda^{n-1}(1-\lambda^{m-n})}{1-\lambda}|a_2 - a_1| \\
&< \frac{\lambda^{n-1}}{1-\lambda}|a_2 - a_1| \leq \frac{\lambda^{N-1}}{1-\lambda}|a_2 - a_1|.
\end{aligned}$$

Since $\lambda \in (0,1)$, given $\epsilon > 0$, we can choose $N = N(\epsilon)$ such that

$$\frac{\lambda^{N-1}}{1-\lambda}|a_2 - a_1| < \epsilon,$$

showing that $|a_m - a_n| < \epsilon$ for all $m > n \geq N$. Thus $\{a_n\}$ is a Cauchy sequence and hence converges.

Note that if $a_n = \sqrt{n}$, then

$$a_{n+1} - a_n = \sqrt{n+1} - \sqrt{n} = \frac{1}{\sqrt{n+1}+\sqrt{n}} < \frac{1}{\sqrt{n}+\sqrt{n-1}} = a_n - a_{n-1},$$

but $\{\sqrt{n}\}$ is not a Cauchy sequence. ∎

Example 2.58. Define a_n inductively by

$$a_{n+1} = \tfrac{1}{2}(a_n + a_{n-1}) \quad \text{for } n \geq 2,$$

where a_1 and a_2 are fixed real numbers. Does the sequence $\{a_n\}$ converge? If it converges, what is its limit?

Solution. For definiteness, we may assume that $a_1 < a_2$. For $n \geq 2$, we have

$$a_{n+1} - a_n = -\frac{1}{2}(a_n - a_{n-1}) = \cdots = \left(-\frac{1}{2}\right)^{n-1}(a_2 - a_1). \tag{2.5}$$

Method 1: If n is even, then the factor on the right, namely $(-1/2)^{n-1}(a_2 - a_1)$, is negative, and so $a_{n+1} - a_n < 0$, and if n is odd, this factor is positive, and so the reverse inequality holds. Thus $\{a_{2n}\}$ is decreasing, whereas $\{a_{2n+1}\}$ is increasing. Observe that $\{a_n\}$ is not a monotone sequence but is bounded. By BMCT, both $\{a_{2n+1}\}$ and $\{a_{2n}\}$ converge. In order to show that $\{a_n\}$

converges, it suffices to prove that these odd and even sequences converge to the same limit. We now begin by observing that (2.5) gives

$$a_{2n+1} = a_{2n} + \left(-\frac{1}{2}\right)^{2n-1}(a_2 - a_1),$$

showing that $\lim_{n\to\infty} a_{2n+1} = \lim_{n\to\infty} a_{2n}$. Therefore, $\{a_n\}$ converges to a limit l, say. To obtain the limit, it suffices to note from the definition that

$$a_{n+1} + \frac{a_n}{2} = a_n + \frac{a_{n-1}}{2} = \cdots = a_2 + \frac{a_1}{2}.$$

Now allow $n \to \infty$ and get that

$$l + \frac{l}{2} = a_2 + \frac{a_1}{2}, \quad \text{i.e.,} \quad l = \frac{2a_2 + a_1}{3}.$$

Method 2: One could directly prove the convergence of $\{a_n\}$ by showing that it is Cauchy. Indeed, using (2.5), it follows that for $m > n \geq 2$,

$$\begin{aligned} |a_m - a_n| &\leq |a_m - a_{m-1}| + \cdots + |a_{n+1} - a_n| \\ &= (a_2 - a_1)\left[\frac{1}{2^{m-2}} + \frac{1}{2^{m-1}} + \cdots + \frac{1}{2^{n-1}}\right] \\ &= \frac{a_2 - a_1}{2^{n-1}}\left[1 + \frac{1}{2} + \frac{1}{2^{m-n-1}}\right] \\ &= \frac{a_2 - a_1}{2^{n-1}}\left[\frac{1 - (1/2)^{m-n}}{1 - (1/2)}\right] < \frac{a_2 - a_1}{2^{n-2}}. \end{aligned}$$

Now let $\epsilon > 0$ be given. Choose N large enough that

$$\frac{a_2 - a_1}{2^{N-2}} < \epsilon.$$

Thus for all $m > n \geq N$, we have

$$|a_m - a_n| < \epsilon,$$

showing that $\{a_n\}$ is a Cauchy sequence and therefore converges. To get the limit value, by (2.5), we may write a_{n+1} as

$$\begin{aligned} a_{n+1} &= a_1 + (a_2 - a_1) + (a_3 - a_2) + \cdots + (a_{n+1} - a_n) \\ &= a_1 + (a_2 - a_1)\left[1 - \frac{1}{2} + \cdots + \left(-\frac{1}{2}\right)^{n-1}\right] \\ &\to a_1 + (a_2 - a_1)\left(\frac{1}{1 + 1/2}\right) = \frac{2a_2 + a_1}{3} \quad \text{as } n \to \infty, \end{aligned}$$

so that $\{a_n\}$ converges to $(2a_2 + a_1)/3$, as desired. ●

Lemma 2.59. *Let $\{a_n\}$ be a sequence of positive numbers. Then we have*

$$\liminf_{n\to\infty} \frac{a_{n+1}}{a_n} \le \liminf_{n\to\infty} a_n^{1/n} \le \alpha := \limsup_{n\to\infty} a_n^{1/n} \le L := \limsup_{n\to\infty} \frac{a_{n+1}}{a_n}.$$

Proof. We need to prove that $\alpha \le L$. This is obvious if $L = \infty$, and so we assume that $0 \le L < \infty$. To prove $\alpha \le L$, it suffices to show that

$$\alpha \le \lambda \quad \text{for any } \lambda \text{ with } L < \lambda. \tag{2.6}$$

So we let $L < \lambda$. Then since

$$L = \limsup \frac{a_{n+1}}{a_n} = \lim_{k\to\infty} \left[\sup\left\{ \frac{a_{n+1}}{a_n} : n \ge k \right\} \right] < \lambda,$$

there exists a natural number N such that

$$\sup\left\{ \frac{a_{n+1}}{a_n} : n \ge N \right\} < \lambda,$$

which gives

$$\frac{a_{n+1}}{a_n} < \lambda \quad \text{for all } n \ge N,$$

so that for $n \ge N$,

$$a_n = a_N \left(\frac{a_{N+1}}{a_N} \right) \left(\frac{a_{N+2}}{a_{N+1}} \right) \cdots \left(\frac{a_n}{a_{n-1}} \right) < \lambda^{n-N} a_N.$$

Therefore,

$$a_n^{1/n} < \lambda^{1-N/n} a_N^{1/n} \quad \text{for } n \ge N,$$

where λ and a_N are fixed. Since $\lim_{n\to\infty} a^{1/n} = 1$ for $a > 0$ (see Example 2.18(**a**)), it follows that

$$\alpha = \limsup a_n^{1/n} \le \lambda.$$

Consequently, (2.6) holds. The proof for the first inequality in the statement is similar, whereas the middle inequality in Lemma 2.59 is trivial. ■

Corollary 2.60. *Let $\{a_n\}$ be a sequence of positive numbers. If $L = \lim_{n\to\infty} \frac{a_{n+1}}{a_n}$, then $\lim_{n\to\infty} a_n^{1/n} = L$.*

Example 2.61. Consider a_n defined by

$$a_n = \frac{n^n}{(n+1)(n+2)\cdots(n+n)}.$$

Suppose we wish to compute $\lim a_n^{1/n}$ (see also Example 7.16(**a**)). It is easier to apply Corollary 2.60. Now we have (by Example 2.33)

$$\frac{a_{n+1}}{a_n} = \frac{(n+1)^n(n+1)^2}{n^n(2n+1)(2n+2)} = \frac{(1+1/n)^n(1+1/n)^2}{(2+1/n)(2+2/n)} \to \frac{\mathrm{e}}{4} \quad \text{as } n \to \infty,$$

and so $\lim a_n^{1/n} = \mathrm{e}/4$. Similarly, it is easy to see that

$$\lim_{n\to\infty} \frac{(n!)^{1/n}}{n} = \frac{1}{\mathrm{e}}.$$

•

We shall provide a direct proof of Corollary 2.60 later, in Section 8.1. However, it is natural to ask the following: if $a_n > 0$ for all n and $\lim_{n\to\infty} a_n^{1/n}$ exists, does $\lim_{n\to\infty} a_{n+1}/a_n$ exist? Clearly not. For example, set

$$a_n = 3^{-n+(-1)^n}.$$

Then $a_n > 0$ and $a_n^{1/n} = 3^{c_n/n} = \mathrm{e}^{(c_n/n)\log 3}$, where

$$\frac{c_n}{n} = \frac{-n+(-1)^n}{n} = -1 + \frac{(-1)^n}{n} \to -1 \quad \text{as } n \to \infty,$$

which shows that $a_n^{1/n} \to e^{-\log 3} = 1/3$. On the other hand,

$$\frac{a_{n+1}}{a_n} = \frac{3^{c_{n+1}}}{3^{c_n}} = 3^{c_{n+1}-c_n} = 3^{-1-2(-1)^n} = \begin{cases} 3 & \text{if } n \text{ is odd,} \\ 3^{-3} & \text{if } n \text{ is even.} \end{cases}$$

This shows that

$$\frac{1}{27} = \liminf_{n\to\infty} \frac{a_{n+1}}{a_n} < 1 < \limsup_{n\to\infty} \frac{a_{n+1}}{a_n} = 3,$$

and $\lim_{n\to\infty} a_{n+1}/a_n$ does not exist. The above construction helps to generate many more examples. For instance, consider $a_n = 2^{-n+(-1)^n}$.

2.2.2 Summability of Sequences

Our aim here is to attach "in some sense" a limit to divergent sequences, while realizing at the same time that any "new limit" we define must agree with the limit in the ordinary sense when it is applied to a convergent sequence. More precisely, if $\{s_n\}$ possibly diverges, we introduce "another method of summation" by replacing $\lim_{n\to\infty} s_n$ by

$$\lim_{n\to\infty} \sigma_n, \quad \text{where } \sigma_n = \frac{1}{n}\sum_{k=1}^{n} s_k.$$

Here the $\{\sigma_n\}$ are called *Cesàro means*[2] (of order 1). Note that $\{\sigma_n\}$ is precisely the average of the first n terms of the sequence $\{s_n\}$, and hence $\{\sigma_n\}$ is also called a *sequence of averages*.

[2] Ernesto Cesàro (1859–1906) was an Italian mathematician who worked on this problem in early stage of his career.

Definition 2.62. *If $\{s_n\}_{n\geq 1}$ is a sequence of real numbers, then we say that $\{s_n\}_{n\geq 1}$ is $(C,1)$ summable to L if the new sequence $\{\sigma_n\}_{n\geq 1}$ converges to L, where*

$$\sigma_n = \frac{1}{n}\sum_{k=1}^{n} s_k.$$

In this case, we write

$$s_n \to L \quad (C,1) \quad \text{or} \quad s_n \to L \quad (\text{Cesàro}) \quad \text{or} \quad \lim_{n\to\infty} s_n = L \quad (C,1).$$

Next, consider a sequence $\{s_n\}$ of real numbers such that $\sigma_n \to 0$ as $n \to \infty$ but $\{s_n\}$ is not convergent.

Example 2.63. Suppose that $s_n = (-1)^{n-1}$ for $n \geq 1$. Then

$$\sigma_n = \begin{cases} 0 & \text{if } n \text{ is even} \\ \dfrac{1}{n} & \text{if } n \text{ is odd} \end{cases}, \qquad n \in \mathbb{N},$$

and so $\sigma_n \to 0$ as $n \to \infty$. Thus, $\{(-1)^{n-1}\}_{n\geq 1}$ is $(C,1)$ summable to 0, and we write

$$\lim_{n\to\infty} (-1)^{n-1} = 0 \quad (C,1).$$

•

All convergent sequences are $(C,1)$ summable to their limits. More precisely, we have the following result.

Theorem 2.64. *If $s_n \to x$, then $s_n \to x$ $(C,1)$.*

Proof. Suppose that $s_n \to x$ as $n \to \infty$. We need to prove that

$$\sigma_n = \frac{1}{n}\sum_{k=1}^{n} s_k \to x \quad \text{as } n \to \infty.$$

Clearly, it suffices to prove the theorem for the case $x = 0$. So we assume that $s_n \to 0$. Then given $\epsilon > 0$, there exists an $N \in \mathbb{N}$ such that $|s_n| < \epsilon/2$ for all $n > N$. Now for $n > N$,

$$\begin{aligned} |\sigma_n| = \left|\frac{1}{n}\sum_{k=1}^{n} s_k\right| &\leq \frac{1}{n}\left[\sum_{k=1}^{N}|s_k| + \sum_{k=N+1}^{n}|s_k|\right] \\ &= \frac{1}{n}\left(\sum_{k=1}^{N}|s_k|\right) + \frac{1}{n}(n-N)\frac{\epsilon}{2} < \frac{M}{n} + \frac{\epsilon}{2}, \quad M = \sum_{k=1}^{N}|s_k|. \end{aligned}$$

Note that M is independent of n and $1/n \to 0$ as $n \to \infty$. Consequently, given $\epsilon > 0$, there exists an N_1 such that

$$\left|\frac{1}{n}\sum_{k=1}^{n} s_k\right| < \frac{\epsilon}{2} + \frac{\epsilon}{2} = \epsilon \quad \text{for all } n \geq N_1,$$

and so $\sigma_n \to 0$ whenever $s_n \to 0$. ∎

As a consequence of Theorem 2.64, we easily have

(a) $\lim_{n\to\infty}(1/n)\sum_{k=1}^{n} k^{1/k} = 1$;
(b) $\lim_{n\to\infty}(1/n)\sum_{k=1}^{n} n/(\sqrt{n^2+k}) = 1$;
(c) $\lim_{n\to\infty}(1/n)\sum_{k=1}^{n} 1/(2k-1) = 0$.

Theorem 2.64 can also be obtained as a consequence of the following result.

Theorem 2.65. *Let $\{s_n\}$ be a sequence of real numbers and $\{\sigma_n\}$ its Cesàro means of order* 1. *Then we have*

$$\liminf_{n\to\infty} s_n \le \liminf_{n\to\infty} \sigma_n \le \alpha := \limsup_{n\to\infty} \sigma_n \le L := \limsup_{n\to\infty} s_n. \tag{2.7}$$

In particular, Theorem 2.64 *holds.*

Proof. We need to prove that $\alpha \le L$. This is obvious if $L = \infty$, and so we assume that $L < \infty$. In order to prove $\alpha \le L$, it suffices to show that

$$\alpha \le \lambda \quad \text{for any } \lambda \text{ with } L < \lambda.$$

So we let $L < \lambda$. By the definition of L, it follows that there exists an N such that $s_n < \lambda$ for all $n > N$. Now for $n \ge N$,

$$\sigma_n = \frac{1}{n}\left[\sum_{k=1}^{N} s_k + \sum_{k=N+1}^{n} s_k\right] < \frac{M}{n} + \frac{1}{n}(n-N)\lambda, \quad M = \sum_{k=1}^{N} s_k.$$

Fix N, and allow $n \to \infty$, and take limit superior on each side to obtain

$$\alpha \le \lambda \quad \text{for any } \lambda \text{ with } L < \lambda.$$

It follows that $\alpha \le L$. The proof for the first inequality in (2.7) is similar, whereas the middle inequality in (2.7) is trivial.

In particular, if $\lim_{n\to\infty} s_n$ exists, then so does $\lim_{n\to\infty} \sigma_n$, and they are equal, proving the second assertion. ∎

Now we ask whether a sequence $\{s_n\}$ that diverges to ∞ can be $(C,1)$ summable.

Example 2.66 (Not all divergent sequences are $(C,1)$ summable). For instance, consider $a_n = 1$ for all $n \ge 1$. Then

$$s_n = \sum_{k=1}^{n} a_k = n \quad \text{and} \quad \sigma_n = \frac{1}{n}\sum_{k=1}^{n} s_k = \frac{1}{n}\sum_{k=1}^{n} k = \frac{n+1}{2}.$$

Note that $\{s_n\}$ is a divergent sequence. Since $\{\sigma_n\}$ is not convergent, it follows that $\{s_n\}$ is not $(C,1)$ summable. •

We have seen examples of divergent series that are not $(C,1)$ summable, but repeating the process of following arithmetic means may lead to a convergent sequence. This idea leads to $(C,2)$ summable sequences, and further extension leads to (C,k) summable sequences. We shall discuss this briefly in Chapter 9.

2.2.3 Questions and Exercises

Questions 2.67.

1. Is every convergent sequence bounded? Is every bounded sequence convergent?
2. Do sequences always have a convergent subsequence?
3. Must a scalar multiple of a Cauchy sequence be Cauchy? Must a sum of two Cauchy sequences always be Cauchy?
4. If $\{a_{3n-2}\}$, $\{a_{3n-1}\}$, and $\{a_{3n}\}$ converge to the same limit a, must $\{a_n\}$ converge to a?
5. Can an unbounded sequence have a convergent subsequence? Can it have many convergent subsequences?
6. Let $\{a_n\}$ be a Cauchy sequence that has a subsequence $\{a_{n_k}\}$ converging to a. Must we have $a_n \to a$?
7. Suppose that we are given a sequence of rational numbers that converges to an irrational number r. Is it possible to obtain many such sequences each converging to the same limit r?
8. Suppose that $\beta > 0$ is given. Is it possible to construct a sequence of rational numbers converging to $\sqrt{\beta}$?
9. Does there exist an example of a bounded sequence having four subsequences converging to different limits?
10. Let $a_n = (-1)^n$. For each fixed N, do we have $|a_n - a_N| = 0$ for infinitely many values of n? Does $\{a_n\}$ satisfy the Cauchy criterion for convergence?
11. Let $a_n = \sqrt{n}$ and $p \in \mathbb{N}$ be fixed. Then
$$a_{n+p} - a_n = \sqrt{n+p} - \sqrt{n} = \frac{p}{\sqrt{n+p} + \sqrt{n}} \to 0 \quad \text{as } n \to \infty.$$
Does $\{a_n\}$ satisfy the Cauchy criterion for convergence?
12. Is every bounded monotone sequence Cauchy? Is every Cauchy sequence monotone?
13. Is the sequence $\{a_n\}$, $a_n = 1 + \frac{1}{2^2} + \frac{1}{3^2} + \cdots + \frac{1}{n^2}$, Cauchy?
14. If $a_{n+1} - a_n \to 0$ as $n \to \infty$, must $\{a_n\}$ be convergent?
15. Does $\lim_{n\to\infty}(1/n)\sum_{k=1}^{n}(1/k)$ exist? If so, what is this limit? If not, must it be ∞?
16. Does $\lim_{n\to\infty}(1/\sqrt{n})\sum_{k=1}^{n}(1/\sqrt{k})$ exist? If so, what is this limit?
17. Must a constant sequence be $(C, 1)$ summable?

Exercises 2.68.

1. Suppose that p is an integer. Show that if $|r| < 1$, then the sequence $\{n^p r^n\}_{n\geq 1}$ converges to zero. In particular, $r^n \to 0$ as $n \to \infty$ if $|r| < 1$.
2. Construct three divergent sequences each having a convergent subsequence.
3. If the subsequences $\{a_{2n}\}$ and $\{a_{2n+1}\}$ converge to a, prove that $\{a_n\}$ also converges to a.

4. Suppose that $\{a_n\}$ is a sequence of real numbers and $\lim_{n\to\infty} a_n = a$, $a \neq 0$. For any sequence $\{b_n\}$, show that
   (a) $\limsup_{n\to\infty}(a_n + b_n) = \lim_{n\to\infty} a_n + \limsup_{n\to\infty} b_n$.
   (b) $\liminf_{n\to\infty}(a_n + b_n) = \lim_{n\to\infty} a_n + \liminf_{n\to\infty} b_n$.
   (c) $\limsup_{n\to\infty} a_n b_n = \lim_{n\to\infty} a_n \limsup_{n\to\infty} b_n$.
   (d) $\liminf_{n\to\infty} a_n b_n = \lim_{n\to\infty} a_n \liminf_{n\to\infty} b_n$.
5. If $\{a_{2n}\}$ and $\{a_{2n+1}\}$ are both Cauchy, then show that $\{a_n\}$ need not be Cauchy. How about if $\{a_{2n}\}$ and $\{a_{2n+1}\}$ both converge to the same limit?
6. Show that the following sequences are Cauchy:
   (a) $a_n = \displaystyle\sum_{k=0}^{n-1} \frac{(-1)^k}{k!}$. **(b)** $a_n = \displaystyle\sum_{k=0}^{n} \frac{1}{k!}$. **(c)** $a_n = \displaystyle\sum_{k=1}^{n} \frac{(-1)^{k-1}}{2k-1}$.
7. Define $a_n = \sin(n\pi/2)$. Extract subsequences of $\{a_n\}$ each having the stated property below:
   (a) converging to 1. **(b)** converging to -1.
   (c) converging to 0. **(d)** divergent.
8. Suppose that $\{a_n\}$ is a sequence such that
$$|a_{n+2} - a_{n+1}| \leq \frac{3}{n}|a_{n+1} - a_n| \quad \text{for } n \geq 1.$$
   Show that $\{a_n\}$ is Cauchy.
9. If $|a_n| < 1/2$ and $|a_{n+1} - a_{n+2}| \leq (1/8)|a_{n+1}^2 - a_n^2|$ for all $n \in \mathbb{N}$, prove that the sequence $\{a_n\}$ converges.
10. Let $a_1 = 1$ and $a_{n+1} = 1 + 1/(1 + a_n)$ for all $n \geq 1$. Is $\{a_n\}$ a Cauchy sequence? If so, find its limit.
11. Define $a_1 = 1$ and $a_{n+1} = 1/(3 + a_n)$ for $n \geq 1$. Show that $\{a_n\}$ converges. Also, find the limit of the sequence.
12. If $\{x_n\}$ is a sequence of real numbers such that $x_{n+1} - x_n \to x$, show that $x_n/n \to x$.
13. Show that
   (a) $\displaystyle\lim_{n\to\infty} \frac{1}{n}\prod_{k=1}^{n}(2n+k)^{1/n} = \frac{27}{4\mathrm{e}}$. **(b)** $\displaystyle\lim_{n\to\infty} \frac{1}{n}\prod_{k=1}^{n}(a+k)^{1/n} = \frac{1}{\mathrm{e}}$.
14. Show that if $\{s_n\}$ and $\{t_n\}$ are $(C,1)$ summable to S and T, respectively, then $\{s_n \pm t_n\}$ is $(C,1)$ summable to $S \pm T$.

3

Limits, Continuity, and Differentiability

The key underlying ideas of this chapter are the notion of continuity and the principles of differentiability. These are two important concepts in analysis. In Section 3.1, we include an explicit similarity between the definition of limit of a sequence and limit of a function (see Theorem 3.4). Section 3.1 gives some basic results relating to limits. In Section 3.2, we study properties of continuous and uniformly continuous functions. We define continuity in terms of sequences and then show that our definition is equivalent to the classical ϵ-δ definition. In addition to continuous functions, differentiable functions are important in calculus. We discuss differentiability in Section 3.3 and deal with the derivative of functions, and establish some of the algebraic rules of calculus, such as how to differentiate sums, products, quotients, and compositions of functions. The limit concept enables us to study derivatives, and hence maxima and minima, asymptotes, improper integrals, and many other mathematical concepts. Many applications of differentiable functions are presented in the following chapter.

3.1 Limit of a Function

This section introduces the concepts of neighborhood and limit; neighborhoods are important in understanding limits, and limits are in turn needed to understand continuity and differentiability.

3.1.1 Limit Point of a Set

We begin by formulating the concept of the neighborhood of a point in $\mathbb{R}$.

Definition 3.1 (Neighborhood). *A neighborhood of a point $x_0 \in \mathbb{R}$ is an open interval containing x_0. For $\epsilon > 0$, an ϵ-neighborhood of a point x_0 is the interval $(x_0 - \epsilon, x_0 + \epsilon)$; it is denoted by $B(x_0; \epsilon)$. Then the set $B(x_0; \epsilon) \backslash \{x_0\}$ defined by*

S. Ponnusamy, *Foundations of Mathematical Analysis*,
DOI 10.1007/978-0-8176-8292-7_3,

$$B(x_0;\epsilon)\backslash\{x_0\} = \{x : |x - x_0| < \epsilon\}\backslash\{x_0\} := \{x : 0 < |x - x_0| < \epsilon\}$$

is called a deleted ϵ-neighborhood of x_0.

Let $A \subset \mathbb{R}$. A point $x_0 \in A$ is called an *interior point* of A if there exists a δ-neighborhood $B(x_0;\delta)$ contained in A. A point $x_0 \in \mathbb{R}$ is a *limit point* of A if every ϵ-neighborhood $B(x_0;\epsilon)$ of x_0 contains a point of A other than x_0. The point x_0 itself may or may not belong to the set A. For example,

(a) $x_0 = 0$ is a limit point of
$$A = \left\{1, \frac{1}{2}, \frac{1}{3}, \dots, \frac{1}{n}, \dots\right\},$$
but $0 = x_0 \notin A$.

(b) Each point x such that $|x| \leq 1$ is a limit point of $A = \{x : |x| < 1\}$, but the boundary points -1 and 1 do not belong to the set A.

(c) The limit points of a closed interval $I = [a, b]$ are precisely the points of I.

(d) Every real number is a limit point of $\mathbb{Q}$.

Note also that a finite set has no limit points.

Theorem 3.2. *Let $A \subset \mathbb{R}$. A point $x_0 \in \mathbb{R}$ is a limit point of A if and only if there exists a sequence $\{x_n\}$ in A with $x_n \neq x_0$ for all $n \in \mathbb{N}$ such that $x_n \to x_0$ as $n \to \infty$.*

Proof. Choose $x_n \in \{B(x_0; 1/n)\backslash\overline{B(x_0; 1/(n+1))}\} \cap A$. ∎

The concept of a limit of a function as $x \to x_0$ also involves the idea of closeness. In terms of the limit of sequences, we can now state a precise definition of the limit of a function at a point.

3.1.2 Sequential Characterization of Limits

Definition 3.3. *Let $A \subset \mathbb{R}$ and let $x_0 \in \mathbb{R}$ be a limit point of A. Suppose that f is a function defined on A except possibly at x_0. Then f is said to have limit ℓ as $x \to x_0$, and we write*

$$\lim_{x \to x_0} f(x) = \ell \quad \text{or} \quad f(x) \to \ell \quad \text{as } x \to x_0$$

if $f(x_n) \to \ell$ for each sequence $\{x_n\}_{n\geq 1}$ in A with $x_n \neq x_0$ for all $n \in \mathbb{N}$ and $x_n \to x_0$ as $n \to \infty$.

It is straightforward to state

$$\lim_{x \to x_0} f(x) = \ell \iff \lim_{x \to x_0} |f(x) - \ell| = 0.$$

Less precisely stated, this means that if x gets close to x_0 but $x \neq x_0$, then $f(x)$ gets close to ℓ. More precisely, this geometric intuition can be stated in the following form, which is often used as a definition of the limit of a function.

Theorem 3.4. *Let $A \subset \mathbb{R}$ and let $x_0 \in \mathbb{R}$ be a limit point of A, and $f\colon A \to \mathbb{R}$. Then the following are equivalent:*

(a) $\lim_{x\to x_0} f(x) = \ell$.
(b) *For every $\epsilon > 0$, there exists a $\delta = \delta(\epsilon, x_0) > 0$ such that*

$$|f(x) - \ell| < \epsilon \quad \textit{whenever} \quad x \in A \textit{ and } 0 < |x - x_0| < \delta.$$

Proof. **(a)** $\Longrightarrow$ **(b)** : Assume that $\lim_{x\to x_0} f(x) = \ell$. We will now use a proof by contradiction. We suppose that **(b)** is not true. Then there must exist some $\epsilon > 0$ such that for every $\delta > 0$ there corresponds a point x such that

$$0 < |x - x_0| < \delta \quad \text{and} \quad |f(x) - f(x_0)| \geq \epsilon.$$

Fix such an ϵ. Then for each $n \in \mathbb{N}$ there exists an $x \in A \cap B(x_0; 1/n)$, denoted by x_n, such that

$$0 < |x_n - x_0| < \frac{1}{n} \quad \text{and} \quad |f(x_n) - f(x_0)| \geq \epsilon.$$

So $x_n \to x_0$ but $f(x_n) \not\to f(x_0)$ as $n \to \infty$. Because of the contradiction stemming from the assumption, the assumption must be false, and so the desired conclusion holds.

(a) $\Longleftarrow$ **(b)** : Assume that **(b)** holds. Consider a sequence $\{x_n\}$ in A such that $x_n \neq x_0$ and $x_n \to x_0$ as $n \to \infty$. Let $\epsilon > 0$ be given. Then by **(b)**, there exists a $\delta > 0$ such that

$$|f(x) - \ell| < \epsilon \quad \text{whenever } x \in A \text{ and } 0 < |x - x_0| < \delta.$$

Now choose N such that

$$0 < |x_n - x_0| < \delta \quad \text{for all } n > N.$$

It follows from **(b)** that

$$|f(x_n) - f(x_0)| < \epsilon \quad \text{for all } n > N,$$

from which it follows that $f(x_n) \to f(x_0)$, as desired. ∎

First, it should be noted that the function need not be defined at x_0 in order to have a limit at x_0, and so $\lim_{x\to x_0} f(x) = \ell$ does not depend on $f(x_0)$ even if f is defined at x_0. Second, it is only a deleted neighborhood $B(x_0; \delta) \setminus \{x_0\}$ of x_0 that is involved. So x_0 need not be in A. Third, even if the condition $x_0 \in A$ holds, we may have $f(x_0) \neq \ell$. Also, we note that a point $x \to x_0$ can approach x_0 in the following ways:

(a) x approaches x_0 with $x < x_0$ (from the left).
(b) x approaches x_0 with $x > x_0$ (from the right).
(c) x can approach x_0 in an oscillating manner (from both left and right).

If $f(x)$ has a limit ℓ as $x \to x_0$, then we also say that $f(x)$ approaches ℓ as x approaches x_0.

As x approaches x_0, the values of $f(x)$ may not get close to any particular number. In that case, we say that $\lim_{x\to x_0} f(x)$ does not exist. For instance, we shall see that $f : \mathbb{R}\backslash\{0\} \to \mathbb{R}$ defined by $f(x) = |x|/x$ illustrates a case in which a limit does not exist as $x \to 0$ (see Example 3.12). Moreover, if the limit exists, then it must be unique. Suppose that

$$\lim_{x\to x_0} f(x) = \ell_1 \quad \text{and} \quad \lim_{x\to x_0} f(x) = \ell_2.$$

Then for a given $\epsilon > 0$, there exist $\delta_1, \delta_2 > 0$ such that

$$|f(x) - \ell_j| < \epsilon/2 \quad \text{whenever } x \in A \text{ and } 0 < |x - x_0| < \delta_j \quad \text{for } j = 1, 2.$$

Therefore, whenever $x \in A$ and $0 < |x - x_0| < \delta = \min\{\delta_1, \delta_2\}$,

$$|\ell_1-\ell_2| = |(f(x)-\ell_2)-(f(x)-\ell_1)| \le |f(x)-\ell_2|+|f(x)-\ell_1| < (\epsilon/2)+(\epsilon/2) = \epsilon.$$

Both the left and right sides of the above inequality are independent of δ. Since ϵ is arbitrary, the inequality holds if and only if $\ell_1 = \ell_2$.

Theorem 3.5 (Divergence criteria). *Let $A \subset \mathbb{R}$, let $x_0 \in \mathbb{R}$ be a limit point of A, and suppose $f\colon A \to \mathbb{R}$. Let $\ell \in \mathbb{R}$ be given. Then $f(x) \not\to \ell$ as $x \to x_0$ iff there exists a sequence $\{x_n\}$ in A with $x_n \neq x_0$ for all $n \in \mathbb{N}$ such that $x_n \to x_0$ as $n \to \infty$, but $f(x_n) \not\to \ell$ as $n \to \infty$.*

We may now formulate

Definition 3.6 (ϵ-δ definition of limit). *Let f be defined in some neighborhood of $x_0 \in \mathbb{R}$, except possibly at x_0. We say that $\lim_{x\to x_0} f(x)$ exists if there exists a real number ℓ satisfying the following condition: for every given $\epsilon > 0$ there exists $\delta = \delta(\epsilon, x_0) > 0$ such that*

$$|f(x) - \ell| < \epsilon \quad \textit{whenever } 0 < |x - x_0| < \delta.$$

According to the ϵ-δ definition of limit, in order to show that $f(x) \not\to \ell$ as $x \to x_0$, it suffices to find an $\epsilon > 0$ for which there is no $\delta > 0$ such that

$$|f(x) - \ell| < \epsilon \quad \text{whenever } 0 < |x - x_0| < \delta.$$

In other words, $f(x) \not\to \ell$ as $x \to x_0$ if there exists an $\epsilon > 0$ such that for every $\delta > 0$, there exists an $x_\delta \in A$ such that $0 < |x_\delta - x_0| < \delta$ but $|f(x_\delta) - \ell| \ge \epsilon$.

Finally, we remark that Definition 3.3 and the statement of Theorem 3.5 are equivalent because of Theorem 3.4.

Examples 3.7. Using the ϵ-δ definition, show that:

(a) $\lim_{x\to 0} \sin(1/x)$ does not exist. **(b)** $\lim_{x\to 3} x^2 = 9$. **(c)** $\lim_{x\to a} x^2 = a^2$.

Solution. **(a)** Suppose to the contrary that $\lim_{x\to 0}\sin(1/x)$ exists; call the limit ℓ. Then for each $\epsilon > 0$, there exists a $\delta > 0$ such that

$$|\sin(1/x) - \ell| < \epsilon \quad \text{for all } x \text{ satisfying } 0 < |x - 0| < \delta. \tag{3.1}$$

Now we choose

$$x_n = \frac{1}{n\pi} \quad \text{and} \quad y_n = \frac{1}{\left(n + \frac{1}{2}\right)\pi}, \qquad n \in \mathbb{N}.$$

Then for large n, we have $0 < |x_n| < \delta$ and $0 < |y_n| < \delta$. Moreover, $f(x_n) = 0$ and $f(y_n) = (-1)^n$ for all n. Consequently, for each $\epsilon > 0$,

$$|f(x_n) - \ell| < \epsilon \quad \text{and} \quad |f(y_n) - \ell| < \epsilon \quad \text{for large } n,$$

so that

$$|f(x_n) - f(y_n)| \le |f(x_n) - \ell| + |f(y_n) - \ell| < 2\epsilon.$$

That is, $1 < 2\epsilon$, which is obviously a contradiction, because this inequality cannot be true for every $\epsilon > 0$. For example, $\epsilon = 1/4$ does not satisfy the inequality. It follows that no such ℓ and δ exist satisfying the inequality (3.1), and the desired conclusion follows (see Figure 3.1).

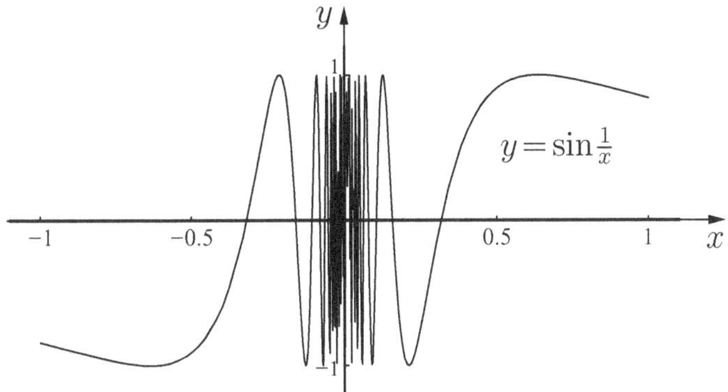

Fig. 3.1. Graph of $f(x) = \sin(1/x)$ on $[-1, 1]$.

(b) Let $\epsilon > 0$ be given. We must prove that there exists a $\delta > 0$ such that

$$|x^2 - 9| = |x - 3|\,|x + 3| < \epsilon \quad \text{for all } x \text{ with } |x - 3| < \delta.$$

With $\delta \le 1$, $|x - 3| < \delta \le 1$ gives $x \in (2, 4)$, and so $x + 3 \in (5, 7)$. This gives

$$|x^2 - 9| < 7|x - 3|.$$

Thus, if we choose $\delta = \min\{1, \epsilon/7\}$, then for any given $\epsilon > 0$,

$$|x - 3| < \delta \quad \text{implies that } |x^3 - 9| < \epsilon.$$

Note that **(b)** is a particular case of **(c)**. Perhaps we can imitate the method used in **(b)** to prove **(c)**.

(c) Again with $\delta \le 1$, $|x-a| = |x+a-2a| < \delta \le 1$ gives $|x+a| < 1+2|a|$, and so

$$|x^2 - a^2| = |x-a|\,|x+a| < (1+2|a|)|x-a|.$$

Thus, for a given $\epsilon > 0$ there exists $\delta = \min\{1, \epsilon/(1+2|a|)\}$ such that

$$|x^2 - a^2| < \epsilon \quad \text{whenever } |x-a| < \delta.$$

Note that δ depends on ϵ and a. •

3.1.3 Properties of Limits of Functions

In order to carry out computations with limits, it will be helpful to have some elementary rules and properties for limits. Suppose that f and g are two functions defined on $I \subset \mathbb{R}$. Then $f+g$ is the function defined on I by

$$(f+g)(x) = f(x) + g(x), \quad x \in I.$$

Similarly, we define $(fg)(x) = f(x)g(x)$ for $x \in I$, and when $g(x) \neq 0$ in I, we have

$$\left(\frac{f}{g}\right)(x) = \frac{f(x)}{g(x)}, \quad x \in I.$$

Theorem 3.8 (Combination rule). *Let $A \subset \mathbb{R}$ and let $x_0 \in \mathbb{R}$ be a limit point of A. Suppose that f and g are defined on A with*

$$\lim_{x \to x_0} f(x) = \ell \quad \text{and} \quad \lim_{x \to x_0} g(x) = \ell'.$$

Then we have the following:

(a) $\lim_{x \to x_0} [f(x) + g(x)] = \ell + \ell'$;
(b) $\lim_{x \to x_0} [f(x)g(x)] = \ell\ell'$;
(c) $\lim_{x \to x_0} [f(x)/g(x)] = \ell/\ell'$ *if* $\ell' \neq 0$.

In particular, for a, b real constants, $\lim_{x \to x_0}(ax+b) = ax_0 + b$.

Proof. The result follows at once from corresponding experience with the algebra of limits for convergent sequences and some standard arguments (see Theorem 2.8).

Note that in **(c)**, it follows by definition that $\lim_{x \to x_0} g(x) \neq 0$ ensures that $g(x) \neq 0$ in some deleted neighborhood of x_0 (see Corollary 3.9). ∎

Corollary 3.9 (Sign-preserving property). *Let $\lim_{x \to x_0} f(x) = \ell \neq 0$. Then there exists a deleted neighborhood $B(x_0;\delta) \setminus \{x_0\}$ on which $f(x) \neq 0$. Moreover, $f(x)$ has the same sign as ℓ on $B(x_0;\delta) \setminus \{x_0\}$.*

Proof. For $\epsilon = |\ell|/2$, there exists a $\delta > 0$ such that

$$|f(x) - \ell| < \epsilon = \frac{|\ell|}{2} \quad \text{whenever } 0 < |x - x_0| < \delta,$$

or equivalently,

$$\ell - \frac{|\ell|}{2} < f(x) < \ell + \frac{|\ell|}{2} \quad \text{for all } x \in (x_0 - \delta, x_0 + \delta) \backslash \{x_0\}. \tag{3.2}$$

Suppose first that $\ell > 0$. Then the left-hand inequality in (3.2) shows that $f(x) > 0$ for $0 < |x - x_0| < \delta$.

Suppose next that $\ell < 0$. Then the right-hand inequality in (3.2) implies that $f(x) < 0$ for $0 < |x - x_0| < \delta$. In either case, $f(x) \neq 0$. ∎

To prove that a specific function has a limit, we can avail ourselves of the above results. For instance, it follows from parts **(a)** and **(b)** of Theorem 3.8 that if $p(x)$ is a polynomial, then $\lim_{x \to x_0} p(x) = p(x_0)$. The quotient of two polynomials is called a rational function, which we may write thus:

$$r(x) = \frac{a_0 + a_1 x + \cdots + a_n x^n}{b_0 + b_1 x + \cdots + b_m x^m} \quad (b_m \neq 0).$$

The domain of r is the set of points where the denominator is nonzero. Thus, if $r(x)$ is a rational function (i.e., the quotient p/q of two polynomials $p(x)$ and $q(x)$), then

$$\lim_{x \to x_0} r(x) = r(x_0),$$

for all x_0 at which the rational function is defined.

Theorem 3.10 (Squeeze/Sandwich rule for functions). *Let f, g, and h be defined in a deleted neighborhood of x_0 such that*

(a) *$g(x) \leq f(x) \leq h(x)$ for all x in a neighborhood of x_0, $x \neq x_0$;*
(b) *$\lim_{x \to x_0} g(x) = \lim_{x \to x_0} h(x) = \ell$.*

Then $\lim_{x \to x_0} f(x) = \ell$.

Proof. The theorem follows as an application of the sandwich theorem for sequences (see Theorem 2.15). Indeed, the condition **(a)** implies that

$$g(x) - \ell \leq f(x) - \ell \leq h(x) - \ell, \tag{3.3}$$

for all x in a neighborhood of x_0, $x \neq x_0$. The condition **(b)** shows that given $\epsilon > 0$, there exists a $\delta > 0$ such that

$$0 < |x - x_0| < \delta \quad \text{implies} \quad |g(x) - \ell| < \epsilon \quad \text{and} \quad |h(x) - \ell| < \epsilon.$$

So if we allow $x \to x_0$ in (3.3) and use the last relation, we get that given $\epsilon > 0$, there exists a $\delta > 0$ such that

$$0 < |x - x_0| < \delta \quad \text{implies} \quad -\epsilon < f(x) - \ell < \epsilon,$$

as required. ∎

An equivalent formulation of this theorem follows.

Theorem 3.11. *Suppose that f, g, and h are functions defined on $I \subset \mathbb{R}$ with*

$$g(x) \le f(x) \le h(x) \quad \text{for all } x \in I.$$

If x_0 is a limit point of I and $\lim_{x\to x_0} g(x) = \ell = \lim_{x\to x_0} h(x)$, then $\lim_{x\to x_0} f(x) = \ell$.

3.1.4 One-Sided Limits

Let f be defined on $A = (x_0, x_0 + \delta)$ for some $\delta > 0$. We say that $f(x)$ approaches the limit ℓ as x approaches x_0 from the right, and write

$$\lim_{x\to x_0+} f(x) = \ell,$$

if for each sequence $\{x_n\}_{n\ge 1}$ in A such that $x_n \to x_0$ one has $f(x_n) \to \ell$. We also write $f(x) \to \ell$ as $x \to x_0+$, and the right-hand limit ℓ is often denoted by $f(x_0+)$.

Similarly, if f is defined on $A = (x_0 - \delta, x_0)$ for some $\delta > 0$, then we say that $f(x)$ approaches the limit ℓ as x approaches x_0 from the left, and write

$$\lim_{x\to x_0-} f(x) = \ell,$$

if for each sequence $\{x_n\}_{n\ge 1}$ in A such that $x_n \to x_0$, one has $f(x_n) \to \ell$. We also write

$$f(x) \to \ell \quad \text{as } x \to x_0-,$$

and the left-hand limit ℓ is often denoted by $f(x_0-)$.

Example 3.12. Consider $f(x) = |x|/x$ for $x \ne 0$. Suppose that $\{x_n\}$ is a sequence in $(0, 1/2)$ such that $x_n \to 0$ as $n \to \infty$ and $\{y_n\}$ is a sequence $(-1/2, 0)$ such that $y_n \to 0$ as $n \to \infty$. Then $f(x_n) = 1$, $f(y_n) = -1$, so that

$$f(x_n) \to 1 \quad \text{and} \quad f(y_n) \to -1 \quad \text{as } n \to \infty. \qquad \bullet$$

The following result is easy to prove.

Theorem 3.13. *Let f be defined in a deleted neighborhood of x_0. Then*

$$\lim_{x\to x_0} f(x) = \ell \Longleftrightarrow \lim_{x\to x_0+} f(x) \text{ and } \lim_{x\to x_0-} f(x) \text{ both exist and are equal to } \ell.$$

A function $f : [a, b] \to \mathbb{R}$ is said to have a "simple jump discontinuity" at a point $x_0 \in (a, b)$ if both

$$f(x_0+) := \lim_{\substack{h\to 0\\ h>0}} f(x_0 + h) \quad \text{and} \quad f(x_0-) := \lim_{\substack{h\to 0\\ h<0}} f(x_0 + h) = \lim_{\substack{k\to 0\\ k>0}} f(x_0 - k)$$

exist but are unequal. Recall that the limit values $f(x_0+)$ and $f(x_0-)$ are called the *right-hand limit* of f at x_0 and the *left-hand limit* of f at x_0, respectively.

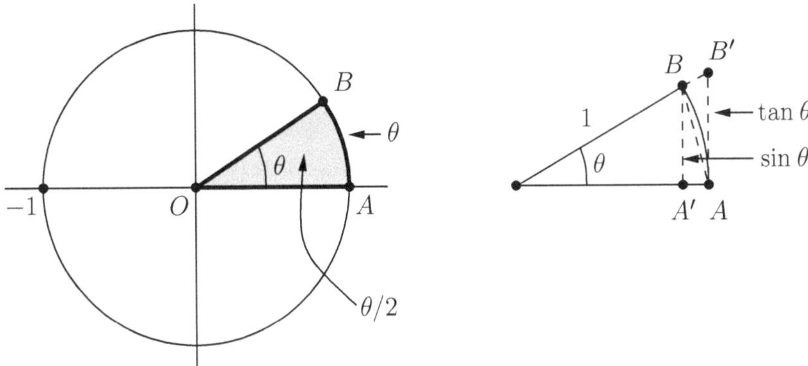

Fig. 3.2. The limit of the quotient $(\sin\theta)/\theta$ when θ is small and positive.

Example 3.14. Show that $\lim_{n\to\infty} n\sin(1/n) = 1$ by finding an estimate for $\sin\theta$ when θ is small and positive. Conclude that

$$\lim_{\theta\to 0} f(\theta) = 1, \quad f(\theta) = \frac{\sin\theta}{\theta}.$$

Solution. Since $f(\theta) = f(-\theta)$, it suffices to show that $\lim_{\theta\to 0+} f(\theta) = 1$. Let θ be an arbitrarily small angle such that $0 < \theta < \pi/2$. Consider a sector of the unit circle with angle θ as in Figure 3.2 and B a point on the circle.

Extend the segment OB from the origin to the point B, and form a right triangle $\Delta OAB'$. Clearly, the sectorial area is larger than the area of $\Delta OA'B$ and is smaller than the area of $\Delta OAB'$. Since the length of the line segment $A'B$ is $\sin\theta$ and that of AB' is $\tan\theta$, we have the inequalities (see also Example 4.24)

$$\frac{\sin\theta}{2} < \frac{\theta}{2} < \frac{\tan\theta}{2}, \quad \text{i.e., } \cos\theta < \frac{\sin\theta}{\theta} < 1.$$

Since $\cos\theta = OA'/OB$, which approaches 1 as $\theta \to 0+$, we have

$$\lim_{\theta\to 0^+} f(\theta) = 1.$$

Since $f(\theta) = f(-\theta)$ for all θ, we can conclude that

$$\lim_{\theta\to 0^-} f(\theta) = \lim_{\theta\to 0^+} f(-\theta) = \lim_{\theta\to 0^+} f(\theta) = 1,$$

showing that $\lim_{\theta\to 0} \sin\theta/\theta = 1$. ●

3.1.5 Infinite Limits

Let $A \subset \mathbb{R}$ and let $x_0 \in \mathbb{R}$ be a limit point of A, and suppose $f : A \to \mathbb{R}$. Then it may happen that f is not bounded on $B(x_0;\delta) \cap A$ for some $\delta > 0$, and so it is possible that $\lim_{x\to x_0} f(x)$ does not exist. On the other hand, we write $\lim_{x\to x_0} f(x) = \infty$ if for every positive $R > 0$ there exists a $\delta > 0$ such that

$$f(x) > R \quad \text{for all } x \in (B(x_0;\delta)\backslash\{x_0\}) \cap A.$$

In this case, we say that $f(x)$ tends to ∞ as $x \to x_0$. Note that if $f(x) \to \infty$ as $x \to x_0$, it follows that

$$\frac{1}{f(x)} \to 0 \quad \text{as } x \to x_0.$$

Similarly, we say that $f(x)$ tends to $-\infty$ as $x \to x_0$ if for every $R > 0$ there exists a δ such that

$$f(x) < -R \quad \text{for all } x \in (B(x_0;\delta)\backslash\{x_0\}) \cap A.$$

In this case, we write $\lim_{x\to x_0} f(x) = -\infty$. An equivalent sequential version of this definition may now be formulated.

Definition 3.15. *Let $A \subset \mathbb{R}$, and let $x_0 \in \mathbb{R}$ be a limit point of A, and suppose $f : A \to \mathbb{R}$. Then $\lim_{x\to x_0} f(x) = \infty$ if and only if for every sequence $\{x_n\}$ in $A\backslash\{x_0\}$ with $x_n \to x_0$, the sequence $\{f(x_n)\}$ diverges to ∞.*

Similarly, one can define

$$\lim_{x\to x_0} f(x) = -\infty, \qquad \lim_{x\to x_0+} f(x) = \infty, \qquad \lim_{x\to x_0+} f(x) = -\infty, \ldots.$$

As with sequences, we have the following theorem.

Theorem 3.16 (Reciprocal rule). *Suppose that $f(x) > 0$ in some deleted neighborhood of x_0 such that $f(x) \to 0$ as $x \to x_0$. Then*

$$\frac{1}{f(x)} \to \infty \quad \textit{as } x \to x_0.$$

Examples 3.17. Now we can easily see the following:

(a) $\lim_{x\to 0}(1/x^2) = \infty$.
(b) $\lim_{x\to 0+}(1/x) = \infty$ and $\lim_{x\to 0-}(1/x) = -\infty$.
(c) $\lim_{x\to(\pi/2)+} \tan x = -\infty$ and $\lim_{x\to(\pi/2)-} \tan x = \infty$. •

3.1.6 Limits at Infinity

Let $f : (a,\infty) \to \mathbb{R}$ for some $a \in \mathbb{R}$. We say that f approaches ℓ as $x \to \infty$ if for a given $\epsilon > 0$ there exists an $R > a$ such that

$$|f(x) - \ell| < \epsilon \quad \text{whenever } x > R.$$

In this case we write $\lim_{x\to\infty} f(x) = \ell$.

Similarly, suppose that $f : (-\infty, a) \to \mathbb{R}$ for some $a \in \mathbb{R}$. We say that f approaches ℓ as $x \to -\infty$ if for a given $\epsilon > 0$ there exists an $M < a$ such that

$$|f(x) - \ell| < \epsilon \quad \text{whenever } x < M.$$

In this case, we write $\lim_{x\to-\infty} f(x) = \ell$.

Example 3.18. For instance, $\lim_{x\to\infty} \sin x$ does not exist. This also follows from the fact that $\lim_{x\to0+} \sin(1/x)$ does not exist (see Example 3.7). Similarly, $\lim_{x\to\infty} \cos x$ does not exist. •

As before, sequential definitions for these two cases may be formulated.

Examples 3.19. (a) $\lim_{x\to\infty} x \sin x$ does not exist, as the sequences $x_n = n\pi$ and $y_n = ((4n \pm 1)/2)\pi$ suggest.

(b) $\lim_{x\to\infty} 1/x = 0$. •

The proof of the following corollary (see Theorem 3.11) is easy, and so we invite the reader to prove it.

Corollary 3.20. *Let f, g, and h be defined on (a, ∞) for some $a \in \mathbb{R}$ such that*

(a) *$g(x) \leq f(x) \leq h(x)$ for all x with $x > R$ $(> a)$;*

(b) $\lim_{x\to\infty} g(x) = \lim_{x\to\infty} h(x) = \ell$.

Then $\lim_{x\to\infty} f(x) = \ell$.

Theorem 3.21. *Suppose that $f(x) \geq 0$ for x near x_0 (respectively for x near ∞) and $\lim_{x\to x_0} f(x) = \ell$ (respectively $\lim_{x\to\infty} f(x) = \ell$). Then $\ell \geq 0$.*

Proof. Suppose that $\ell < 0$. Then for all x in a deleted neighborhood of x_0 (respectively for all x with $x > R$ for some $R > 0$),

$$|f(x) - \ell| \geq |\ell| > 0.$$

But then $f(x) \not\to \ell$ as $x \to x_0$ (respectively $x \to \infty$), which is a contradiction. Consequently, $\ell \geq 0$. ∎

Corollary 3.22. *Let h and g be defined in a deleted neighborhood of x_0 such that*

(i) *$h(x) \leq g(x)$ for all x near x_0, $x \neq x_0$;*

(ii) *$\lim_{x\to x_0} h(x) = \ell$ and $\lim_{x\to x_0} g(x) = \ell'$.*

Then $\ell \leq \ell'$. (Similar statements can be made when f and g are defined on $(-\infty, a)$ or (b, ∞) for some $a, b \in \mathbb{R}$.)

Proof. Given $f(x) = g(x) - h(x) \geq 0$ for x near x_0 $(x \neq x_0)$ such that $\lim_{x\to x_0} f(x) = \ell' - \ell$, by Theorem 3.21, $\ell' - \ell \geq 0$, as required. ∎

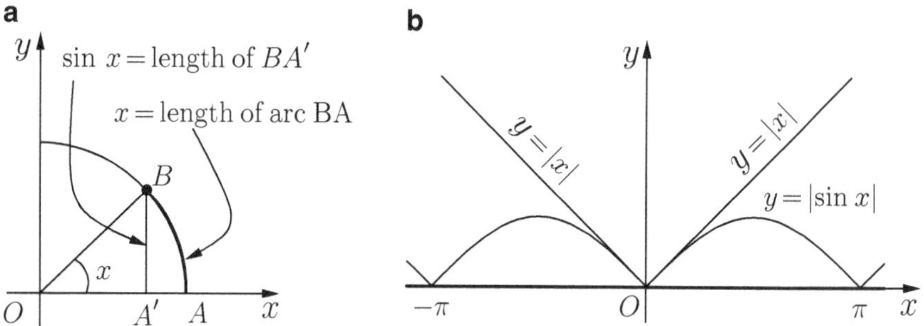

Fig. 3.3. Depiction of $|\sin x| \leq |x|$ for $x \in \mathbb{R}$.

Example 3.23. We have the sine inequality $|\sin x| \leq |x|$ for $x \in \mathbb{R}$. In particular,

$$\lim_{x \to a} \sin x = \sin a.$$

Solution. There is nothing to prove if $x = 0$. For $0 < x \leq \pi/2$, consider the part of the unit circle in the first quadrant. From Figure 3.3, it is clear that (see also Examples 3.14 and 4.24)

$$0 < \sin x \leq x \quad \text{for } 0 < x \leq \pi/2.$$

For $-\pi/2 \leq x < 0$, the desired inequality follows from the fact that $\sin(-x) = -\sin x$. For $|x| > \pi/2$, the inequality is trivial because

$$|\sin x| \leq 1 < \pi/2 < |x|.$$

The desired first inequality follows. Finally,

$$|\sin x - \sin a| = 2\left|\sin\left(\frac{x-a}{2}\right)\right|\left|\cos\left(\frac{x+a}{2}\right)\right| \leq 2\left|\sin\left(\frac{x-a}{2}\right)\right| \leq |x-a|,$$

and the second part follows. ●

As in the last example, we see that $\lim_{x\to 0} \cos x = 1$, because

$$|\cos x - 1| = |2\sin^2(x/2)| \leq 2|x/2|^2 = x^2/2.$$

3.1.7 Questions and Exercises

Questions 3.24.

1. Is the intersection of a finite number of neighborhoods of a point c a neighborhood of c?
2. Is the intersection of an infinite number of neighborhoods of a point c a neighborhood of c?

3. Can a finite set have a limit point? Does $\mathbb{N}$ have a limit point?
4. If $A \subseteq \mathbb{R}$, is every interior point of A a limit point of A?
5. Suppose that f is a continuous function such that

$$S_1 = \{x \in \mathbb{R} : f(x) > 0\}, \quad S_2 = \{x \in \mathbb{R} : f(x) < 0\},$$
$$S_3 = \{x \in \mathbb{R} : f(x) = 0\}, \quad S_4 = \{x \in \mathbb{R} : f(x) \neq 0\}.$$

   Is either of these sets open in $\mathbb{R}$?
6. Is it true that if $\lim_{x\to x_0} f(x)$ does not exist and $\lim_{x\to x_0} g(x)$ does not exist, then $\lim_{x\to x_0}(f+g)(x)$ does not exist? How about the existence of $\lim_{x\to x_0} f(x)g(x)$?
7. When we deal with a quotient f/g of functions, what is the domain of f/g?
8. Is it true that $f(x) \to 0$ as $x \to x_0$ if and only if $|f(x)| \to 0$ as $x \to x_0$?
9. Suppose that $\lim_{x\to x_0} f(x)$ exists but $\lim_{x\to x_0} g(x)$ does not exist. Can $\lim_{x\to x_0}(f+g)(x)$ exist?
10. If $f(x) < g(x)$ in a deleted neighborhood of x_0, do we have $\lim_{x\to x_0} f(x) < \lim_{x\to x_0} g(x)$?
11. Suppose that $f(x)$ is a function defined on $\mathbb{R}$ such that $\lim_{x\to x_0} |f(x)|$ exists for each $x_0 \in \mathbb{R}$. Must $\lim_{x\to x_0} f(x)$ exist?
12. Does $\lim_{x\to 0} \sqrt{x}$ exist? Does $\lim_{x\to 0+} \sqrt{x}$ exist?
13. If $\lim_{x\to a} f(x) = \ell > 0$, do we have $\lim_{x\to a} \sqrt{f(x)} = \sqrt{\ell}$?
14. Let f be defined on (a, ∞) for some $a \in \mathbb{R}$ and $\ell \in \mathbb{R}$. Must

$$\lim_{x\to\infty} f(x) = \ell \iff \lim_{x\to 0+} f\left(\frac{1}{x}\right) = \ell?$$

15. Let f be defined on $(-\infty, a)$ for some $a \in \mathbb{R}$ and $\ell \in \mathbb{R}$. Must

$$\lim_{x\to -\infty} f(x) = \ell \iff \lim_{x\to 0-} f\left(\frac{1}{x}\right) = \ell?$$

16. Suppose that f and g are defined on (c, ∞) for some $c \in \mathbb{R}$ such that $\lim_{x\to\infty} f(x) = \ell$ for some real ℓ and $\lim_{x\to\infty} g(x) = \infty$. Does the limit $\lim_{x\to\infty}(f \circ g)(x)$ exist? If so, what is the limit?
17. Suppose that f and g are defined in a deleted neighborhood of a such that $\lim_{x\to a} f(x) = \ell$ for some nonnegative real number $\ell \geq 0$ and $\lim_{x\to a} g(x)$. Must $\lim_{x\to a} f(x)g(x)) = \infty$ if $\ell > 0$? What can you say about the limit $\lim_{x\to a} f(x)g(x)$ if $\ell = 0$?
18. Suppose that $\lim_{x\to 0} f(x) = 0$ and $\lim_{x\to 0} g(x) = \infty$. Does either $\lim_{x\to 0} f(x)g(x) = 0$ or $\lim_{x\to 0} f(x)g(x) = \infty$ hold?
19. Suppose that $\lim_{x\to 0} f(x) = \infty = \lim_{x\to 0} g(x)$. Must $\lim_{x\to 0}(f(x) - g(x)) = \infty$?
20. For what values of α does $\lim_{x\to\infty}(\sin x/|x|^\alpha)$ exist? When does it not?
21. Does the sequence $\{\sin n\}_{n\geq 1}$ converge? Does it have a convergent subsequence?

22. For what values of $t \in \mathbb{R}$ does $\{\sin(nt)\}$ converge?
23. For what values of $t \in \mathbb{R}$ does $\{\cos(nt)\}$ converge?
24. Does $\{\cos(n\pi)\}$ converge?
25. Does $\{\sin(n\pi/2)\}$ converge? Does $\{(1/n)\sin(n\pi/2)\}$ converge? How about the sequences $\{(1/n)\sin(n\pi/4)\}$ and $\{(1/n)\sin(n\pi/5)\}$?
26. What can be said about the convergence of the sequences $\{(1/n)\sin n\}$ and $\{(1/n)\cos n\}$?
27. What is meant by a limit point of a sequence? How does a limit point differ from a limit of sequence?

Exercises 3.25.

1. For each of the following sets determine the set of all limit points: $\mathbb{N}$, $\mathbb{Z}$, $\mathbb{Q}$, $\mathbb{R}$, $\emptyset$, $\mathbb{R}\backslash\mathbb{Q}$.
2. Show that $\lim_{x\to 0} 3^{1/x}$ does not exist.
3. Define $f(x) = (1/x)\sin(1/x)$ for $x \neq 0$. Determine $\lim_{x\to 0} f(x)$ if it exists. If not, explain why the limit does not exist.
4. Suppose that f is bounded and monotone on (a, b) and $c \in (a, b)$. Show that $\lim_{x\to c+} f(x)$ and $\lim_{x\to c-} f(x)$ both exist.
5. Draw the graph of
$$f(x) = \begin{cases} |x| + \dfrac{x}{|x|} & \text{for } x \neq 0, \\ 0 & \text{for } x = 0, \end{cases}$$
for $x \in \mathbb{R}$. Determine $\lim_{x\to 0} f(x)$ if it exists. If not, explain why it does not exist.
6. State and prove the squeeze rule for functions f, g, and h defined on (a, ∞) (respectively $(-\infty, a)$).
7. Compute the following limits if they exist:
(a) $\displaystyle\lim_{x\to\infty} \frac{x + \sin x}{x}$. (b) $\displaystyle\lim_{x\to 2} \frac{1}{(1-x)^2}$. (c) $\displaystyle\lim_{x\to 0} \frac{\sin 3x - 3x}{x^3}$.
(d) $\displaystyle\lim_{x\to 0} \frac{\cos(|x|) - 1}{x}$. (e) $\displaystyle\lim_{x\to 0} \frac{1}{|x|}$. (f) $\displaystyle\lim_{x\to 0} \frac{1}{x}$.
(g) $\displaystyle\lim_{x\to 0} \frac{x\mathrm{e}^{1/x}}{1 + \mathrm{e}^{1/x}}$. (h) $\displaystyle\lim_{x\to 0} \frac{\sin x}{|x|}$. (i) $\displaystyle\lim_{x\to 0} \frac{\mathrm{e}^{1/x} - \mathrm{e}^{-1/x}}{\mathrm{e}^{1/x} + \mathrm{e}^{-1/x}}$.
In the cases in which the limit does not exist, determine the left- and right-hand limits if they exist.
Note: Those who are not familiar with the exponential function can wait until we introduce $\exp x$.
8. Draw the graph of
$$f(x) = \frac{1}{1 + \mathrm{e}^{1/x}}$$
and determine the following limits:
(a) $\displaystyle\lim_{x\to 0-} f(x)$. (b) $\displaystyle\lim_{x\to 0+} f(x)$.

9. Let f be defined in a deleted neighborhood B' of x_0. Prove or disprove the following: $\lim_{x\to x_0} f(x)$ exists if and only if given $\epsilon > 0$, there exists a $\delta > 0$ such that $|f(x) - f(y)| < \epsilon$ for every pair of points x, y in B' such that $|x - y| < \delta$.

3.2 Continuity

Definition 3.26. *Let I be an open interval containing x_0, and let $f : I \to \mathbb{R}$. Then f is said to be continuous at $x_0 \in I$ if $\lim_{x\to x_0} f(x) = f(x_0)$.*

In the case of the boundary points a and b of $I = (a, b)$, we can talk only about right and left continuity, respectively. If f is continuous at each point of I, then f is said to be continuous on I. The function f is said to be continuous if it is continuous on the domain of f.

Intuitively, our definition implies that the values $f(x_n)$ are close to $f(x_0)$ when the values of x_n are close to x_0. Then the following theorem shows that the sequential definition is equivalent to the ϵ-δ definition of continuity of f given in many calculus texts.

Theorem 3.27. *Let f be a real-valued function defined on an interval I that contains x_0 as an interior point of I. Then the following are equivalent:*

(a) *f is continuous at x_0.*
(b) *For a given $\epsilon > 0$, there exists a $\delta = \delta(x_0, \epsilon) > 0$ such that*

$$|f(x) - f(x_0)| < \epsilon \quad \text{whenever } x \in I \text{ and } |x - x_0| < \delta. \tag{3.4}$$

(c) *The following three conditions hold:*

$$f(x_0) \text{ is defined}, \quad \lim_{x\to x_0} f(x) \text{ exists}, \quad \text{and} \quad \lim_{x\to x_0} f(x) = f(x_0).$$

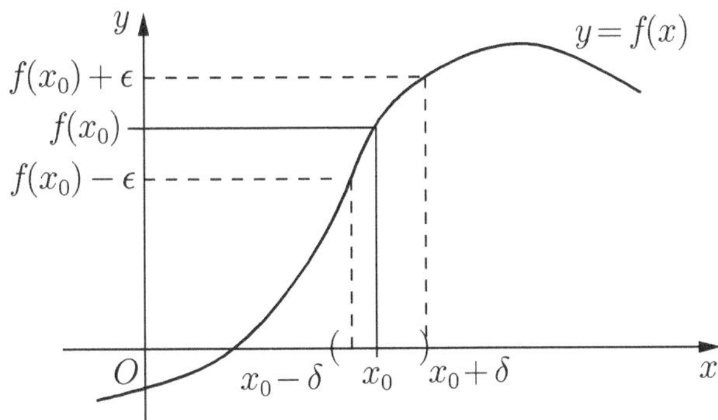

Fig. 3.4. Depiction of the continuity of f at x_0.

Proof. The proof is clear and is simply a reformulation of Theorem 3.4 with $\ell = f(x_0)$ (see Figure 3.4). ∎

Equivalently, if f is continuous, then the graph of f is an unbroken curve; that is, the graph of f could be traced by a particle in motion or by a moving pencil point without being lifted from the paper.

Examples 3.28. **(a)** If $f(x) = c$ for all $x \in \mathbb{R}$, then f is continuous on $\mathbb{R}$, because for each $x_0 \in \mathbb{R}$,

$$|f(x) - f(x_0)| = 0 < \epsilon \quad \text{for } |x - x_0| < \delta$$

holds for any given $\epsilon > 0$.

(b) If $f(x) = x$ for $x \in \mathbb{R}$, then $\lim_{x \to x_0} f(x) = x_0$ for each $x_0 \in \mathbb{R}$. In particular, f is continuous on $\mathbb{R}$.

(c) The function $f(x) = x^k$ is continuous for every natural number k. This follows from the fact that $x_n \to x_0$ implies that $x_n^k \to x_0^k$ as $n \to \infty$ and since $f(x) = x$ is continuous at each $x_0 \in \mathbb{R}$. For a direct approach, we refer to Example 3.7.

(d) The functions $\sin x$ and $\cos x$ are continuous on $\mathbb{R}$; see Example 3.23. ●

Corollary 3.9 implies the following:

Corollary 3.29 (Sign-preserving property for continuous functions). *Let f be continuous at x_0 and $f(x_0) \neq 0$. Then there exists a neighborhood $B(x_0; \delta)$ of x_0 on which $|f(x)| > |f(x_0)|/2$, and $f(x)$ has the same sign as $f(x_0)$.*

This corollary says that if a continuous function f does not vanish at a point x_0, then there is an interval containing x_0 in which f does not vanish.

3.2.1 Basic Properties of Continuous Functions

As a result of translating the properties of limits (see Theorem 3.8) into terms of continuity, we obtain the following result, which is often useful in proving that certain specific functions are continuous.

Theorem 3.30 (Algebra of continuous functions). *Suppose that f and g are defined on an interval I. If f, g are continuous at $x_0 \in I$, then their sum $f + g$, product fg, quotient f/g where $g(x_0) \neq 0$, and $|f|$ are also continuous at x_0. In particular, every polynomial $a_0 + a_1 x + \cdots + a_n x^n$ is continuous on $\mathbb{R}$.*

Composition is another basic operation that can be performed on functions. If f maps I to J and g maps J to $\mathbb{R}$, then the composition of f with g, denoted by $g \circ f$, maps I to $\mathbb{R}$ by sending $x \mapsto g(f(x))$; see Figure 3.5. Thus $g \circ f$ is defined by

$$(g \circ f)(x) = g(f(x)), \quad x \in I.$$

It follows that composition is not commutative.

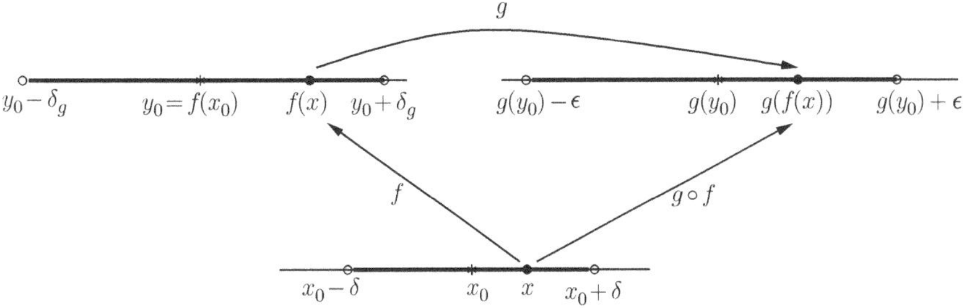

Fig. 3.5. Composition of continuous functions at x_0.

Theorem 3.31 (Composition rule). *If* $\lim_{x\to x_0} f(x) = y_0$ *and* g *is a function that is continuous at the point* y_0*, then* $\lim_{x\to x_0}(g\circ f)(x) = g(y_0)$.

Proof. Let $\epsilon > 0$ be given. Then the continuity of g at y_0 implies that there exists a $\delta_g > 0$ such that

$$|g(y) - g(y_0)| < \epsilon \quad \text{whenever } |y - y_0| < \delta_g. \tag{3.5}$$

Further, since $\lim_{x\to x_0} f(x) = y_0$, for this $\delta_g > 0$, there exists a $\delta > 0$ such that

$$|f(x) - y_0| < \delta_g \quad \text{whenever } x \in B(x_0;\delta)\setminus\{x_0\}.$$

Now if we let $y = f(x)$ in (3.5), we see that for all $x \in B(x_0;\delta)\setminus\{x_0\}$,

$$|(g\circ f)(x) - g(y_0)| = |g(f(x)) - g(y_0)| < \epsilon,$$

from which we obtain the required conclusion. ∎

Corollary 3.32. *If* $f : I \to J$ *is continuous at* $x_0 \in I$ *and if* $g : J \to \mathbb{R}$ *is continuous at* $y_0 = f(x_0)$*, then* $g\circ f$ *defined by* $(g\circ f)(x) = g(f(x))$ *is continuous at* x_0*. That is, the composition of two continuous functions is continuous.*

Proof. The proof is a consequence of Theorem 3.31; see Figure 3.5. ∎

Example 3.33. Suppose that $f : \mathbb{R} \to \mathbb{R}$ is continuous on $\mathbb{R}$ and satisfies $f(x + y) = f(x) + f(y)$ for all $x, y \in \mathbb{R}$. Show that $f(x) = f(1)x$.

Solution. Setting $x = y = 0$ gives $f(0) = 0$. Setting $y = -x$ gives

$$f(x) = -f(-x). \tag{3.6}$$

If $x = n \in \mathbb{N}$, then

$$f(x) = f(1 + \cdots + 1) = nf(1) = xf(1).$$

Similarly, if $-x = n \in \mathbb{N}$, then by (3.6),

$$f(x) = f(-1 - \cdots - 1) = nf(-1) = -nf(1) = xf(1).$$

Thus, $f(x) = xf(1)$ when $x \in \mathbb{Z}$.

Next, let x be a rational number p/q (in lowest terms), $p \in \mathbb{Z}$, $q \in \mathbb{N}$. Then $p = qx$, and thus

$$f(p) = f(qx) = f(x + \cdots + x) = qf(x),$$

and from the discussion, we have

$$f(p) = pf(1).$$

Combining the last two equalities, we obtain

$$qf(x) = pf(1), \quad \text{i.e., } f(p/q) = \frac{p}{q}f(1), \text{ i.e., } f(x) = xf(1) \text{ when } \; x \in \mathbb{Q}.$$

Finally, let $x = \alpha$ be an irrational number, and $\{x_n\}$ a sequence of rational numbers converging to α. Since f is continuous at α, $f(x_n) \to f(\alpha)$. But

$$\lim_{n\to\infty} f(x_n) = \lim_{n\to\infty} x_n f(1) = f(1)\alpha.$$

Consequently, $f(\alpha) = f(1)\alpha$. In conclusion, $f(x) = xf(1)$ for all $x \in \mathbb{R}$. •

3.2.2 Squeeze Rule and Examples of Continuous Functions

Corollary 3.34 (Squeeze rule for continuous functions). *Let f, g, and h be defined in a neighborhood of x_0 such that*

(i) *$g(x) \leq f(x) \leq h(x)$ for all x in a neighborhood of x_0;*
(ii) *g and h are continuous at a and $g(x_0) = f(x_0) = h(x_0)$.*

Then f is continuous at x_0.

Proof. The proof of the corollary is a consequence of Theorem 3.10 (see Figure 3.6). ∎

A function $f : I \to \mathbb{R}$ is discontinuous (or has a discontinuity) at a point x_0 if f is not continuous at x_0.

Example 3.35 (Dirichlet's function on $\mathbb{R}$). Define $f : \mathbb{R} \to \mathbb{R}$ by

$$f(x) = \begin{cases} 1 & \text{for } x \in \mathbb{Q}, \\ 0 & \text{for } x \in \mathbb{R}\backslash\mathbb{Q}. \end{cases}$$

Then f is discontinuous at each point of $\mathbb{R}$. In order to prove this, suppose that $a \in \mathbb{R}$ and $\lim_{x\to a} f(x) = \ell$. Then for a given $\epsilon > 0$ there exist a $\delta > 0$ and points $x_n \in \mathbb{Q}$ and $y_n \in \mathbb{R}\backslash\mathbb{Q}$ such that $x_n \to a$ and $y_n \to a$,

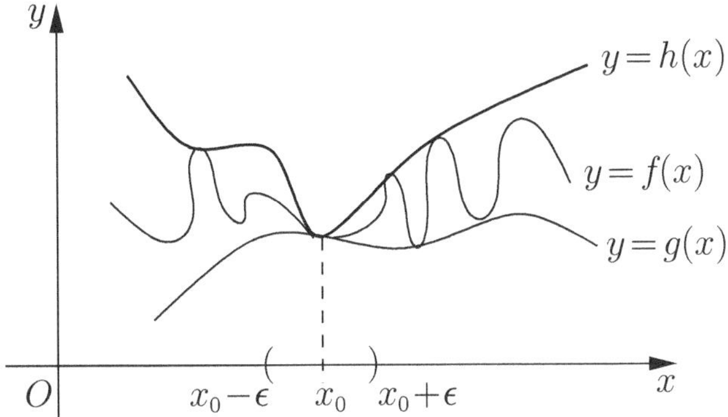

Fig. 3.6. Squeeze rule for continuous functions.

$$|f(x_n) - \ell| < \frac{\epsilon}{2} \quad \text{when } 0 < |x_n - a| < \delta,$$

and

$$|f(y_n) - \ell| < \frac{\epsilon}{2} \quad \text{when } 0 < |y_n - a| < \delta.$$

Hence, there exists $N \in \mathbb{N}$ such that for $n > N$,

$$1 = |f(x_n) - f(y_n)| \leq |f(x_n - \ell| + |f(y_n) - \ell| < \frac{\epsilon}{2} + \frac{\epsilon}{2} = \epsilon,$$

because $f(x_n) = 1$ and $f(y_n) = 0$. This is a contradiction to our assumption (especially when $\epsilon < 1$). Thus, $f(x)$ does not approach a limit as $x \to a$ whether a is rational or irrational. It follows that f is discontinuous at each point of $\mathbb{R}$. ●

Example 3.36 (Riemann function on $\mathbb{R}$). Define $f : \mathbb{R} \to \mathbb{R}$ by

$$f(x) = \begin{cases} \dfrac{1}{q} \text{ for } x = \dfrac{p}{q} \in \mathbb{Q}\backslash\{0\} \text{ in its lowest terms, with } q > 0, \\ 0 \text{ for } x \in \mathbb{R}\backslash\mathbb{Q},\ x = 0. \end{cases}$$

Show that f is continuous at each irrational point of $\mathbb{R}$ and is discontinuous at each rational point of $\mathbb{R}$.

Solution. For a rough sketch (because of we cannot accurately graph this function) of this function on $[0, 1]$, see Figure 3.7. It is not clear from the rough sketch of the graph of f whether f is continuous at any point of $\mathbb{R}$. Let $a = p/q$ be an arbitrary nonzero rational number, $q > 0$, such that

$$f(a) = \frac{p}{q}.$$

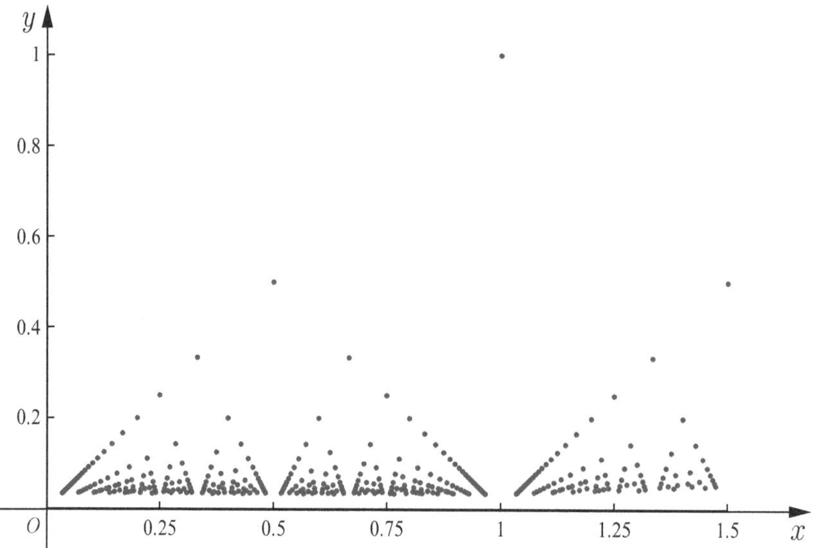

Fig. 3.7. The Riemann function.

Accordingly, every neighborhood of p/q (e.g., $0 < |x - a| < 1/n$) contains irrational numbers y_n such that

$$0 = f(y_n) \neq f(a) = \frac{1}{q}.$$

Thus, $y_n \to a$ but $f(y_n) \not\to f(a)$. Alternatively, since

$$|f(y_n) - f(a)| = |0 - 1/q| = 1/q,$$

for $\epsilon < 1/q$ there cannot exist a neighborhood $B(a;\delta)$ of p/q such that for every $y \in B(a;\delta) \cap (\mathbb{R} \setminus \mathbb{Q})$,

$$|f(y) - f(a)| = \frac{1}{q} < \epsilon.$$

By either argument, f is discontinuous at $a = p/q$.

Next, let a be any irrational number or 0. Then $f(a) = 0$, and we must show that, given $\epsilon > 0$, there exists a $\delta > 0$ such that

$$f(x) = |f(x) - f(a)| < \epsilon \quad \text{for all } x \text{ satisfying } 0 < |x - a| < \delta.$$

Note that $f(x) \geq 0$ on $\mathbb{R}$, by the definition of f.

Now given $\epsilon > 0$, let N be a positive number such that $N > 1/\epsilon$. Since a is irrational, there exists a $\delta_1 > 0$ such that the interval $(a - \delta_1, a + \delta_1)$ contains no integers. Likewise, there exists a $\delta_2 > 0$ such that the interval

$(a-\delta_2, a+\delta_2)$ contains no rational number of the form $p/2$ in lowest terms. Continue the process, and introduce

$$\delta = \delta(\epsilon) = \min\{\delta_1, \dots, \delta_N\}.$$

Then $\delta > 0$. Then the interval $(a-\delta, a+\delta)$ contains no rational numbers of the form p/q in lowest terms, $0 < q \leq N$. It follows that if $|x-a| < \delta$, then either x is irrational, so that $f(x) = 0$ $(< \epsilon)$, or else x is rational, so that $x = p/q$ with $q > N$, and

$$f(x) = \frac{1}{q} < \frac{1}{N} < \epsilon.$$

In either case, $f(x) < \epsilon$ and f is continuous at a. •

3.2.3 Uniform Continuity

We say that f is *uniformly continuous* on I if for a given $\epsilon > 0$, there exists a $\delta = \delta(\epsilon)$ depending only on ϵ such that

$$|f(x) - f(y)| < \epsilon \quad \text{whenever } x, y \in I \text{ and } |x-y| < \delta.$$

Examples of functions that are uniformly continuous:

- Every linear function $f(x) = ax + b$ is uniformly continuous on $\mathbb{R}$.
- $f(x) = \sin x$ is uniformly continuous on $\mathbb{R}$ because for every $x, y \in \mathbb{R}$, $|\sin x - \sin y| \leq |x-y|$ (see Example 3.23).
- $f(x) = x^2$ is uniformly continuous on $[0, b]$, because $|x^2 - y^2| \leq 2b|x-y|$.
- $f(x) = 1/x$ is uniformly continuous on $[b, \infty)$ $(b > 0)$, because

$$\left|\frac{1}{x} - \frac{1}{y}\right| = \frac{|x-y|}{|xy|} \leq \frac{|x-y|}{b^2}.$$

Uniform continuity is a property involving a function f and a set I on which it is defined. It makes no sense to speak of f being uniformly continuous at a point of I (except perhaps when I consists of a single point!). Clearly, every uniformly continuous function on I is continuous on I, but the converse is not true in general.

Examples 3.37 (Continuous functions that are not uniformly continuous). Define $f(x) = \cos(1/x)$ for $x > 0$. Clearly, f is continuous on $(0, \infty)$. We show that f fails to be uniformly continuous on any interval $(0, b)$, where $b > 0$ is fixed. In order to show this, we proceed as follows. We shall show that for $\epsilon = 1$, there exists no $\delta > 0$ such that x and y in $(0, b)$ such that

$$|x-y| < \delta \quad \text{implies that } |f(x) - f(y)| < 1.$$

Let $\delta > 0$ be arbitrary. We now wish to find two points x and y in $(0, b)$ such that $|x-y| < \delta$ and $|f(x) - f(y)| = 2$. To do this, we choose x_n and

y_n such that $x_n \to 0$ and $y_n \to 0$ as $n \to \infty$ but $|f(x_n) - f(y_n)| = 2$ for all $n \in \mathbb{N}$. For example, if x_n and y_n are such that

$$\frac{1}{x_n} = 2n\pi \quad \text{and} \quad \frac{1}{y_n} = (2n+1)\pi,$$

then for all $n \in \mathbb{N}$,

$$|f(x_n) - f(y_n)| = |\cos 2n\pi - \cos(2n+1)\pi| = 2$$

and

$$x_n - y_n = \frac{1}{2n\pi} - \frac{1}{(2n+1)\pi} = \frac{1}{2n(2n+1)\pi}.$$

Now we can choose a natural number N to ensure that

$$\frac{1}{2n(2n+1)\pi} < \delta \quad \text{for all } n > N.$$

This observation shows that for $\epsilon = 1 > 0$, there exists no positive number δ (independent of x, y) such that

$$|f(x) - f(y)| < \epsilon$$

whenever $|x - y| < \delta$. So f is not uniformly continuous on $(0, b)$. Note that f is bounded and oscillates, and the trouble occurs near the origin. However, a function can be uniformly continuous on an unbounded set. For example, $f(x) = \cos x$ is uniformly continuous on $\mathbb{R}$.

Next we present another example. Define $f(x) = x^2$ for $x \in \mathbb{R}$. Clearly, f, being a power function, is continuous on $\mathbb{R}$. In order to show that f is not uniformly continuous on $\mathbb{R}$, for any $\delta > 0$, we write

$$|f(x) - f(y)| = |x^2 - y^2| = (|x+y|)\,(|x-y|) = 2 = 6\left(\frac{1}{\delta}\right)\left(\frac{\delta}{3}\right)$$

by choosing x and y such that

$$x + y = \frac{1}{\delta} \quad \text{and} \quad x - y = \frac{\delta}{3}.$$

Solving the last two equations, we get

$$x = \frac{1}{2}\left(\frac{1}{\delta} + \frac{\delta}{3}\right) \quad \text{and} \quad y = \frac{1}{2}\left(\frac{1}{\delta} - \frac{\delta}{3}\right).$$

Thus, for any positive real number δ, we have found two numbers x and y such that

$$|x - y| < \delta/3 < \delta \quad \text{and} \quad |f(x) - f(y)| = 2 > 1.$$

In other words, no δ "works" for every pair of real numbers x and y such that

$$|f(x) - f(y)| < \epsilon = 1 \quad \text{whenever } |x - y| < \delta.$$

So f is not uniformly continuous on $\mathbb{R}$. Observe that f grows too quickly.

Similarly, it can be easily seen that $f : (0, 1) \to \mathbb{R}$ defined by $f(x) = 1/x$ is not uniformly continuous. In fact, given any $\delta > 0$, choose x such that $0 < x < \min\{1, \delta\}$ and $y = x/2$. Then $|x - y| = x/2 < \delta$ but

$$\left|\frac{1}{y} - \frac{1}{x}\right| = \frac{2}{x} - \frac{1}{x} = \frac{1}{x} > 1,$$

which clearly shows that f is not uniformly continuous on $(0, 1)$. ●

The following result is one of the most important results in analysis.

Theorem 3.38. *Every continuous function f on a bounded closed interval $[a, b]$ is uniformly continuous therein.*

Proof. Assume that f is continuous on $[a, b]$. Suppose on the contrary that f is not uniformly continuous on $[a, b]$. Then there exists an $\epsilon > 0$ such that for each $\delta > 0$ there are two points $x, y \in [a, b]$ such that

$$|f(x) - f(y)| \geq \epsilon \quad \text{and} \quad |x - y| < \delta.$$

In particular, for each $n \in \mathbb{N}$, we can define two sequences $\{x_n\}$ and $\{y_n\}$ in $[a, b]$ such that for every $n \geq 1$,

$$|x_n - y_n| < \frac{1}{n} \quad \text{and} \quad |f(x_n) - f(y_n)| \geq \epsilon.$$

Since $\{x_n\}$ is bounded, by the Bolzano–Weierstrass theorem, it contains a convergent subsequence x_{n_k} that converges to some point $c \in [a, b]$ as $k \to \infty$. Now,

$$|y_{n_k} - c| \leq |y_{n_k} - x_{n_k}| + |x_{n_k} - c| \to 0 \quad \text{as } k \to \infty,$$

so that the sequence $\{y_{n_k}\}$ converges to c as $k \to \infty$. Since f is continuous at c, we must have

$$f(x_{n_k}) - f(y_{n_k}) \to f(c) - f(c) = 0 \quad \text{as } k \to \infty.$$

This contradicts the fact that $|f(x_{n_k}) - f(y_{n_k})| \geq \epsilon$ for all $n \in \mathbb{N}$. This contradiction shows that f must be uniformly continuous on $[a, b]$. ■

3.2.4 Piecewise Continuous Functions

A function $f : [a, b] \to \mathbb{R}$ is said to be *piecewise continuous* if it has at most a finite number of discontinuities on $[a, b]$ and the one-sided limits exist at each point of discontinuity. Here we allow the possibility that the function may not be defined at the points of discontinuity. Every point at which the one-sided limits are not equal is called a *jump point* or *removable discontinuity* of f. At points x_0 where $f(x)$ is continuous, each of the one-sided limits is of course equal to $f(x_0)$. Thus, we have the following definition.

Definition 3.39 (Piecewise continuous function). *A function* $f : [a, b] \to \mathbb{R}$ *is said to be* piecewise continuous *if there exists a partition* $P = \{x_0, x_1, \ldots, x_n\}$ *of* $[a, b]$ *such that*

- f *is continuous on each subinterval* (x_{k-1}, x_k), $1 \le k \le n$;
- $f(x_k+)$ *for* $0 \le k \le n-1$ *and* $f(x_k-)$ *for* $1 \le k \le n$ *exist.*

In particular, every continuous function on $[a, b]$ is piecewise continuous. Graphs of some piecewise continuous functions are given in Figures 3.8 and 3.9.

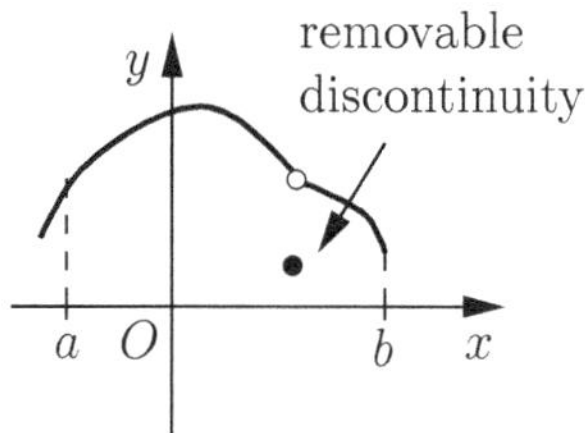

Fig. 3.8. Examples of piecewise continuous functions.

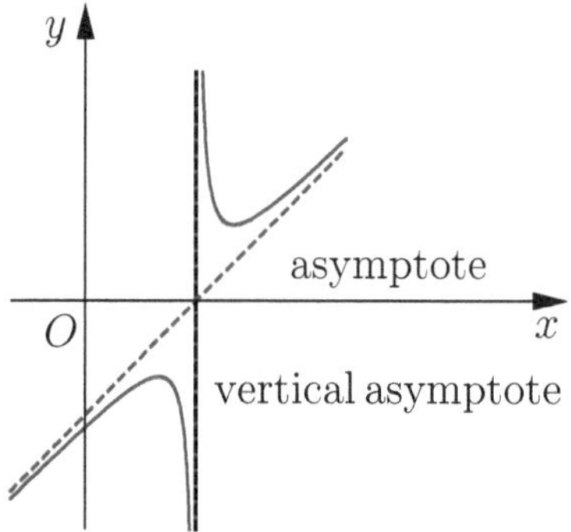

Fig. 3.9. Piecewise continuous functions may not have vertical asymptotes.

3.2.5 Questions and Exercises

Questions 3.40.

1. Can every continuous function be accurately graphed?
2. Suppose that f is continuous at a point a. Does this imply that

$$\lim_{h \to 0} (f(a+h) - f(a-h)) = 0?$$

   How about the converse?

3. Suppose that f is a continuous function in a neighborhood $B(0;\delta)$ of the origin and $f(x) = f(x^2)$ on $B(0;\delta)$. What can be said about f?
4. If f is continuous on $\mathbb{R}$, must $f(x+a)$ be continuous?
5. Suppose that f is a nonnegative continuous function in $\mathrm{dom}\,(f)$, the domain of f. Must $\sqrt{f(x)}$ be continuous on $\mathrm{dom}\,(f)$?
6. Must every continuous function on a bounded interval $[a,b]$ be bounded?
7. Let f be a real-valued function that is continuous at $x_0 \in \mathbb{R}$. Must $|f|$ defined by $|f|(x) = |f(x)|$ be continuous at x_0?
8. Suppose f and g are continuous at $x_0 \in \mathbb{R}$. Must $\phi = \max\{f, g\}$, where $\phi(x) = \max\{f(x), g(x)\}$, be continuous at x_0? How about the continuity of $\min\{f, g\}$?
9. Suppose f is continuous on $[a,\infty)$ for some $a \in \mathbb{R}$, and $\lim\limits_{x\to\infty} f(x) = \ell$ for some $\ell \in \mathbb{R}$. Must f be bounded on $[a,\infty)$?
10. Suppose that f is continuous on $(-\infty, a]$ and $\lim_{x\to-\infty} f(x)$ is finite. Must f be bounded on $(-\infty, a]$?
11. Suppose that f is continuous on $\mathbb{R}$ such that $f(x) = 1$ for $x \in \mathbb{Q}$. What can be said about f on $\mathbb{R}$?
12. Is $[x]$ piecewise continuous on $[0,3]$? Is $[x]$ one-to-one on $\mathbb{R}$?
13. If $f(x) = [x] - [x/3]$ on $[-1,4]$, is f continuous at $x = 3$?
14. Suppose $f : I \to \mathbb{R}$ is a function such that $|f|$ is continuous on I. Must f be continuous on I?
15. Suppose that f and g are discontinuous on $\mathbb{R}$. Can the composition $g \circ f$ be continuous on $\mathbb{R}$? Can the sum $f+g$ and product fg be continuous on $\mathbb{R}$?
16. Can there exist a function that is discontinuous only at $1/n$, $n \in \mathbb{N}$, and nowhere else?
17. Suppose that $f : [0,1] \to \mathbb{Q}$ is continuous such that $f(1/3) = 1/3$. Must $f(x) = 1/3$ on $[0,1]$?
18. Does the sequence $\{\sin(\pi/n)\}_{n\geq 1}$ converge to zero?
19. Are the sum and product of two uniformly continuous functions uniformly continuous?
20. Suppose that $f : [a,b] \to \mathbb{R}$ is uniformly continuous. Must f be bounded? How about if $[a,b]$ is replaced by a bounded subset of $\mathbb{R}$?
21. Can a product of two functions that are not uniformly continuous be uniformly continuous?
22. Can there exist an unbounded function that is uniformly continuous?

Exercises 3.41.

1. Let $f : \mathbb{R} \to \mathbb{R}$ by

$$f(x) = \begin{cases} \dfrac{1}{x} \sin\left(\dfrac{1}{x^2}\right) & \text{for } x \neq 0, \\ 0 & \text{for } x = 0. \end{cases}$$

Prove that f is not continuous at the origin.

2. Using the ϵ-δ definition, show that $f(x) = \sqrt{x+2}$ is continuous at $x = 2$.
3. Give an example of a function that is continuous on $[0, 2)$ and $(2, 3]$, but not on any open interval containing 2.
4. Prove that the following functions are continuous at the indicated points by finding δ for a given $\epsilon > 0$:

   (a) $f(x) = \sqrt{x}$ at $x = 4$. (b) $f(x) = \sqrt{x^2 - 9}$ at $x = 3$.

   (c) $f(x) = \dfrac{1}{x}$ at $x = a \neq 0$. (d) $f(x) = \dfrac{1}{\sqrt{x}}$ at $x = a > 0$.
5. For $x > 0$ or $-1 < x < 0$, prove the Bernoulli inequality
$$(1 + x)^n > 1 + nx \quad \text{for } n \geq 2.$$
   Using this, prove that $\{r^n\}$ converges if and only if $-1 < r \leq 1$ (see also Theorem 2.34).
6. For what values of a does $\lim_{x \to a}[x]$ exist? Determine the domain where $[x]$ is continuous.
7. If $f(x) = \sqrt{x - [x]}$ on $(0, 2)$, determine $\lim_{x \to 0+} f(x)$ and $\lim_{x \to 0-} f(x)$. Determine whether f is continuous at $x = 1$.
8. Define
$$f(x) = \frac{(1 + \sin \pi x)^n - 1}{(1 + \sin \pi x)^n + 1}, \qquad x \in \mathbb{R}.$$
   Determine points where f is discontinuous.
9. Determine the constants a and b such that f defined by
$$f(x) = \begin{cases} ax + 3 & \text{for } x > 4, \\ 7 & \text{for } x = 4, \\ x^2 + bx + 3 & \text{for } x < 4, \end{cases}$$
   is continuous on $\mathbb{R}$.
10. Suppose that f is uniformly continuous on a set E and $\{x_n\}$ is a Cauchy sequence in E. Show that $\{f(x_n)\}$ is a Cauchy sequence. Using this, show that $f(x) = 1/x^2$ is not uniformly continuous on $(0, 1)$.
11. Consider $f : \mathbb{R} \to \mathbb{R}$ by $f(x) = 3|x|$. Let $\epsilon > 0$ be given. Find $\delta(\epsilon) > 0$ such that $|x - y| < \delta(\epsilon)$ implies $|f(x) - f(y)| < \epsilon$ for all $x, y \in \mathbb{R}$.
12. Suppose that f is continuous on $[0, \infty)$ such that $\lim_{x \to \infty} f(x) = \ell$ for some $\ell \in \mathbb{R}$. Prove that f is uniformly continuous.
13. Suppose that f is continuous on $\mathbb{R}$ and
$$\lim_{x \to \infty} f(x) = 0 = \lim_{x \to -\infty} f(x).$$
   Prove that f is uniformly continuous on $\mathbb{R}$.
14. Define
$$f(x) = \begin{cases} x \sin(1/x) & \text{for } x \neq 0, \\ 0 & \text{for } x = 0, \end{cases} \quad \text{and} \quad g(x) = \begin{cases} \sin(1/x) & \text{for } x \neq 0, \\ 0 & \text{for } x = 0. \end{cases}$$
   Prove or disprove the following: f is uniformly continuous on $\mathbb{R}$, but g is not.

3.3 Differentiability

Though the concepts and results in Sections 3.1 and 3.2 help us to understand functions somewhat, the concept of the derivative of a function is needed in order to understand its rate of change, maxima, minima, and other important features.

Let I be an open interval in $\mathbb{R}$ and $a \in I$. Since I is open, for all h such that $|h|$ is small ($h \neq 0$), the point $a+h$ also lies in I. Suppose that $f : I \to \mathbb{R}$. We define the slope of the graph of f at $(a, f(a))$ to be the limit as $x \to a$ of the slope of the chord through the points $(a, f(a))$ and $(x, f(x))$. The slope of the chord is "the difference quotient for f at a" and is given by

$$\frac{f(x) - f(a)}{x - a},$$

and so the slope of the graph of f at $(a, f(a))$ is

$$\lim_{x \to a} \frac{f(x) - f(a)}{x - a} = \lim_{h \to 0} \frac{f(a+h) - f(a)}{h}, \tag{3.7}$$

provided the limit exists and is finite. This limit is denoted by $f'(a)$ and is called the derivative of f at a. If f has a derivative at a, then the function f is said to be *differentiable* at a. This is same as saying that the graph $y = f(x)$ has a tangent at the point $(a, f(a))$ with slope $f'(a)$ (see Figure 3.10). If f is

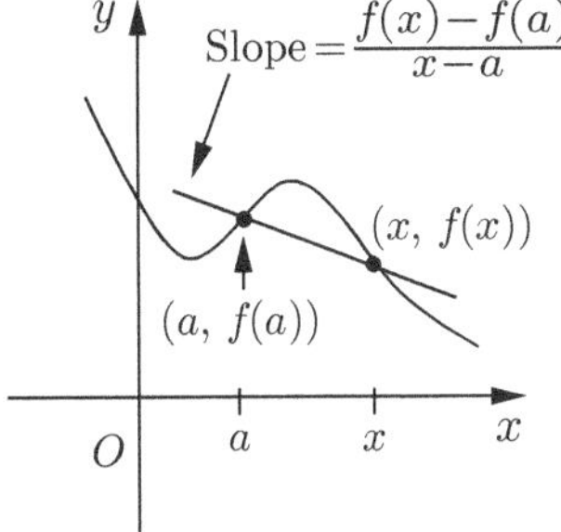

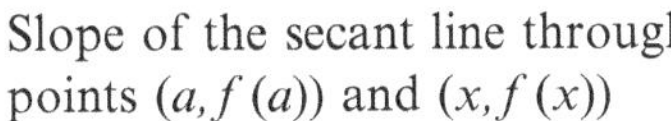

Slope of the secant line through points $(a, f(a))$ and $(x, f(x))$

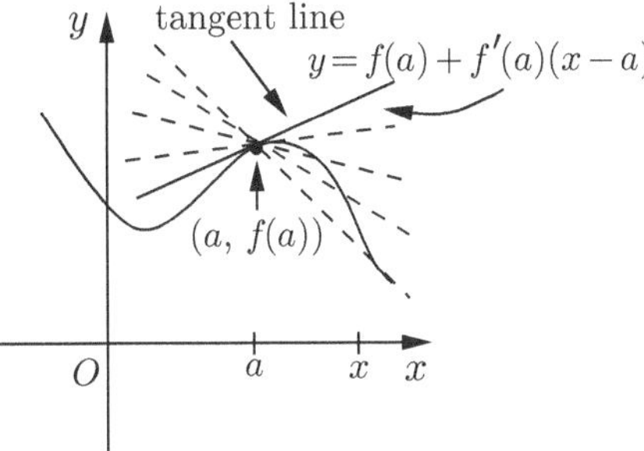

Secant lines tend to tangent line when $x \to a$

Fig. 3.10. Depiction of differentiability of f at a.

differentiable at every point of I (i.e., $f'(x)$ exists on I), then f is said to be differentiable on I, and the function $f' : x \mapsto f'(x)$ is called the *differentiable function*. The operation of obtaining $f'(x)$ from $f(x)$ is called *differentiation*. We remark that any one of the following notations may also be used instead of $f'(x)$ for the derivative of $y = f(x)$:

$$Df(x), \quad \frac{\mathrm{d}y}{\mathrm{d}x}, \quad y', \quad f^{(1)}(x).$$

It is important to remark that the symbol $\mathrm{d}y/\mathrm{d}x$ is purely a notation and does not mean some quantity $\mathrm{d}y$ "divided by" another quantity $\mathrm{d}x$. Clearly, by (3.7), many results obtained for limits and continuity can be used to prove analogous results for derivatives.

Suppose that f is defined on $[a, b)$. Then as in the case of one-sided continuity, the right derivative of f at a is defined to be the limit

$$\lim_{x\to a+} \frac{f(x) - f(a)}{x - a} = \lim_{h\to 0+} \frac{f(a+h) - f(a)}{h} \quad (h = x - a > 0),$$

provided the limit exists. It is customary to denote this limit by $f'_+(a)$.

If f is defined on $(a, b]$, then the left derivative of f at b, denoted by $f'_-(b)$, is defined to be

$$f'_-(b) := \lim_{x\to b-} \frac{f(x) - f(b)}{x - b} = \lim_{h\to 0-} \frac{f(b+h) - f(b)}{h} \quad (h = x - b < 0),$$

provided the limit exists. The left and right derivatives are called *one-sided derivatives* (see Figure 3.11).

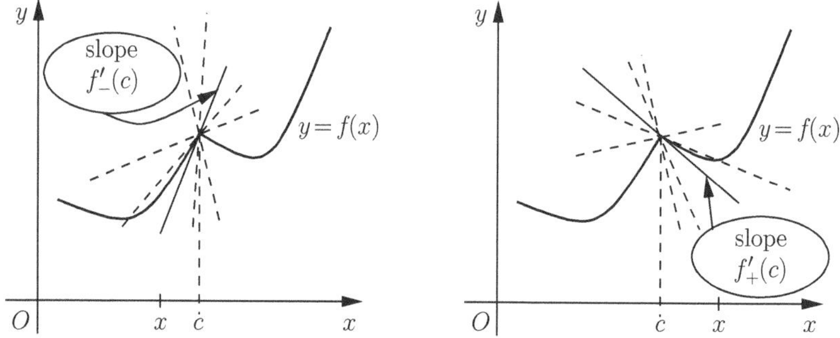

Fig. 3.11. Left and right derivatives.

The following result, which provides the connection between the definitions of derivative and one-sided derivatives, is rather obvious.

Theorem 3.42. *Suppose that f is defined on (a, b) and $c \in (a, b)$. Then f is differentiable at c if and only if both $f'_+(c)$ and $f'_-(c)$ exist and $f'_+(c) = f'_-(c)$.*

As in the case of continuous functions, we have the following definition.

Definition 3.43. *We say that $f : [a, b] \to \mathbb{R}$ is differentiable (on $[a, b]$) if $f'(x)$ exists on (a, b) and both $f'_+(a)$ and $f'_-(b)$ exist.*

3.3.1 Basic Properties of Differentiable Functions

The first of the following basic theorems relates differentiability and continuity.

Theorem 3.44. *Let f be defined on an open interval (a,b), and $x \in (a,b)$. If f is differentiable at x, then f is continuous at x.*

Proof. For $h \neq 0$, we consider the identity

$$f(x+h) - f(x) = \left(\frac{f(x+h)-f(x)}{h}\right) h \quad (h \neq 0).$$

By the sum and product rules for limits,

$$\lim_{h\to 0}(f(x+h)-f(x)) = \lim_{h\to 0}\left(\frac{f(x+h)-f(z)}{h}\right) h = f'(x)\cdot 0 = 0,$$

so that $\lim_{h\to 0} f(x+h) = f(x)$. Thus, f is continuous at x. ∎

The continuity of f does not necessarily imply differentiability of f, so differentiability is stronger than continuity. Here is a simple example to demonstrates this.

Example 3.45. Consider $f(x) = |x|$ on $(-1,1)$. If $h \neq 0$, then

$$\frac{f(h)-f(0)}{h-0} = \frac{|h|}{h} = \begin{cases} 1 & \text{for } h > 0, \\ -1 & \text{for } h < 0. \end{cases}$$

Thus, $f(x) = |x|$ is not differentiable at $x = 0$ but is continuous at 0. Is f differentiable at other points in $\mathbb{R}$? Note that the graph of f on $(-1,1)$ does not have a tangent at the origin; no line through $(0,0)$ is a tangent to the graph of $y = f(x) = |x|$ at $x = 0$.

One could also use the sequential version of the definition. Clearly, $\{(-1)^n/n\}$ converges to zero. Set

$$x_{2n} = \frac{1}{2n} \quad \text{and} \quad x_{2n-1} = -\frac{1}{2n-1}.$$

Then $x_n \to 0$ as $n \to \infty$, and

$$T(x_n) = \frac{f(x_n)-f(0)}{x_n - 0} = \begin{cases} 1 & \text{for } n \text{ even}, \\ -1 & \text{for } n \text{ odd}, \end{cases}$$

showing that $\{T(x_n\}$ does not converge. It follows that $|x|$ is not differentiable at the origin. ●

Theorem 3.44 may be used as a test for nondifferentiability. For example, if $f(x) = [x]$ denotes the integer part of x, then the function is not continuous at integer points, and hence it is not differentiable at integer points.

The function $f(x) = x^2$ is everywhere differentiable, because

$$\lim_{h\to 0} \frac{f(x+h) - f(x)}{h} = \lim_{h\to 0} \frac{(x+h)^2 - x^2}{h} = \lim_{h\to 0} \frac{2xh + h^2}{h} = 2x.$$

More generally, if $f(x) = x^n$, $n \in \mathbb{N}$, then $f'(x) = nx^{n-1}$. In fact, for each fixed $x_0 \in \mathbb{R}$ and $n \geq 2$, we have

$$f(x) - f(x_0) = x^n - x_0^n = (x - x_0)(x^{n-1} + x^{n-2}x_0 + \cdots + x_0^{n-1}),$$

so that

$$\lim_{x\to x_0} \frac{f(x) - f(x_0)}{x - x_0} = \lim_{x\to x_0} [x^{n-1} + x^{n-2}x_0 + \cdots + x_0^{n-1}] = nx_0^{n-1}.$$

Since x_0 is an arbitrary point of $\mathbb{R}$, we can write this in the form

$$f'(x) = \frac{\mathrm{d}}{\mathrm{d}x}(x^n) = nx^{n-1}, \quad x \in \mathbb{R},$$

for $n = 1, 2, \ldots$.

Example 3.46. Consider

$$f(x) = \begin{cases} 1 + x^2 & \text{for } -1 \leq x \leq 0, \\ \cos x & \text{for } 0 \leq x \leq 2\pi. \end{cases}$$

Determine whether f is differentiable at the origin and whether f has right and left derivatives at the endpoints -1 and 2π, respectively.

Solution. At $c = -1$, for $0 < h < 1$, we easily have

$$\frac{f(-1+h) - f(-1)}{h} = h - 2 \to -2 \quad \text{as } h \to 0+,$$

showing that $f'_+(-1)$ exists and equals -2. Also for $-1 < h < 0$, we have

$$\frac{f(h) - f(0)}{h} = h \to 0 \quad \text{as } h \to 0-,$$

showing that $f'_-(0) = 0$. Similarly, for $0 < h < \pi/2$,

$$\frac{f(h) - f(0)}{h} = \frac{\cos h - 1}{h} = -\frac{2\sin^2(h/2)}{(h/2)^2}\left(\frac{h}{2}\right),$$

which approaches zero as $h \to 0+$. That is, $f'_+(0) = 0$. Since the left and right derivatives at 0 are equal, it follows that f is differentiable at 0 and $f'(0) = 0$. Finally, for $-(\pi/2) < h < 0$, we have (see Example 3.23)

$$\frac{f(2\pi + h) - f(2\pi)}{h} = \frac{\cos h - 1}{h},$$

which approaches zero as $h \to 0-$. Consequently, f is differentiable from the left at 2π and $f'_-(2\pi) = 0$. ●

Example 3.47. Consider $f(x) = \sin x$ and arbitrary $x_0 \in \mathbb{R}$. Then, using the addition formula $\sin(x_0 + h) = \sin x_0 \cos h + \cos x_0 \sin h$, we have

$$\frac{f(x_0 + h) - f(x_0)}{h} = \sin x_0 \left(\frac{\cos h - 1}{h} \right) + \cos x_0 \left(\frac{\sin h}{h} \right),$$

so that (see Examples 3.23 and 4.24)

$$\begin{aligned}\lim_{h \to 0} \frac{f(x_0 + h) - f(x_0)}{h} &= \sin x_0 \lim_{h \to 0} \frac{\cos h - 1}{h} + \cos x_0 \lim_{h \to 0} \frac{\sin h}{h} \\ &= \sin x_0 \cdot 0 + \cos x_0 \cdot 1.\end{aligned}$$

It follows that f is differentiable at x_0, and $f'(x_0) = \cos x_0$. Since x_0 is arbitrary, $f'(x) = \cos x$ on $\mathbb{R}$. ●

By the definition of differentiability and theorems on limit and continuity, we have the following analogous results for derivatives.

Theorem 3.48. *If f and g are differentiable at x_0, then their sum $f + g$, difference $f - g$, product fg, quotient f/g (where $g(x_0) \neq 0$), and the scalar multiplication cf are also differentiable at x_0, and*

$$(f + g)' = f' + g', \quad (fg)' = f'g + fg', \quad \left(\frac{f}{g} \right)' = \frac{f'g - fg'}{g^2}, \quad (cf)' = cf',$$

where c is a real constant.

More generally, finite linear combinations (of the form $\alpha_1 f_1 + \alpha_2 f_2 + \cdots + \alpha_n f_n$, $\alpha_j \in \mathbb{R}$, $j = 1, 2, \ldots, n$) and finite products of functions differentiable at x_0 are also differentiable at x_0.

Proof. For $x \neq x_0$, consider the following:

$$\frac{(f + g)(x) - (f + g)(x_0)}{x - x_0} = \left\{ \frac{f(x) - f(x_0)}{x - x_0} \right\} + \left\{ \frac{g(x) - g(x_0)}{x - x_0} \right\},$$

$$\frac{(fg)(x) - (fg)(x_0)}{x - x_0} = f(x) \left\{ \frac{g(x) - g(x_0)}{x - x_0} \right\} + g(x_0) \left\{ \frac{f(x) - f(x_0)}{x - x_0} \right\},$$

and for $g(x_0) \neq 0$,

$$\frac{\left(\frac{f}{g} \right)(x) - \left(\frac{f}{g} \right)(x_0)}{x - x_0} = \frac{g(x_0) \left(\frac{f(x) - f(z_0)}{x - x_0} \right) - f(x_0) \left(\frac{g(x) - g(x_0)}{x - x_0} \right)}{g(x)g(x_0)}.$$

The assertions then follow from the above equalities and the properties of the limit by letting $x \to x_0$ and noting that if f, g are differentiable at x_0, they are continuous at x_0, so that $f(x) \to f(x_0)$, $g(x) \to g(x_0)$ as $x \to x_0$. A similar argument takes care of the general result. ■

Corollary 3.49. *Every polynomial $p(x)$ defined by*

$$p(x) = a_0 + a_1 x + \cdots + a_n x^n,$$

where $a_0, \dots, a_n \in \mathbb{R}$, is differentiable on $\mathbb{R}$, and its derivative is given by

$$p'(x) = a_1 + 2a_2 x + \cdots + n a_n x^{n-1}, \quad x \in \mathbb{R}.$$

As a consequence of Theorem 3.48, we see that every rational function $r(x)$ is differentiable on the domain of $r(x)$.

Suppose that f is differentiable at x_0. Then in terms of ϵ-δ notation, given any $\epsilon > 0$, there exists a $\delta = \delta(\epsilon, x_0) > 0$ such that

$$\left| \frac{f(x) - f(x_0)}{x - x_0} - f'(x_0) \right| < \epsilon \quad \text{whenever} \quad 0 < |x - x_0| < \delta.$$

In view of the limit $\lim_{x \to x_0} f(x) = f(x_0)$, we may let

$$\eta(x) = \begin{cases} \dfrac{f(x) - f(x_0)}{x - x_0} - f'(x_0) & \text{for } 0 < |x - x_0| < \delta, \\ 0 & \text{for } x = x_0, \end{cases}$$

and observe that $\lim_{x \to x_0} \eta(x) = 0 = \eta(x_0)$. Therefore η is continuous at x_0, and we get an explicit expression for $f(x)$ in the form

$$f(x) = f(x_0) + f'(x_0)(x - x_0) + (x - x_0)\eta(x) \tag{3.8}$$

for $|x - x_0| < \delta$. In conclusion, we have the following result.

Proposition 3.50. *Suppose that f is differentiable at x_0. Then there exists a function η that is continuous at x_0 and satisfies (3.8) for all x in some neighborhood $B(x_0; \delta)$. Equivalently, f is differentiable at x_0 if and only if*

$$f(x) = f(x_0) + f'(x_0)(x - x_0) + E(x), \tag{3.9}$$

where E is a function defined in a neighborhood of x_0 such that

$$\lim_{x \to x_0} \frac{E(x)}{x - x_0} = 0.$$

By (3.9), we obtain a linear function

$$L(x) = f(x_0) + f'(x_0)(x - x_0)$$

that approximates $f(x)$ up to an "error" term $E(x)$, which is small in absolute value in comparison with $|x - x_0|$ for x close to x_0 (see Figure 3.12).

Here is the rule for differentiating a composite function. The proof requires substantial understanding of the abstract ϵ-δ formulation of the limit.

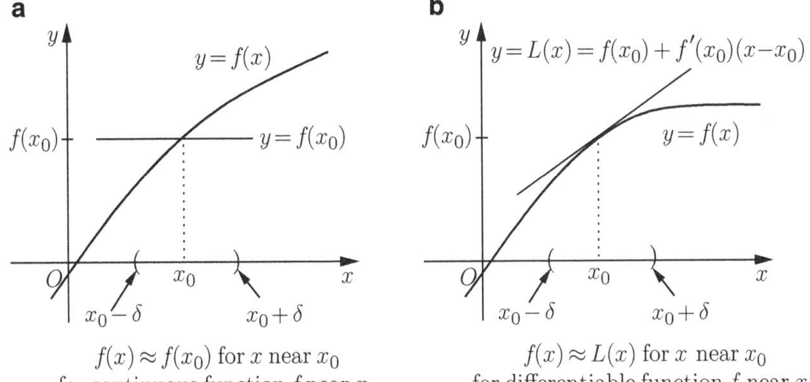

Fig. 3.12. Tangent approximation to $f(x)$ at x_0.

Theorem 3.51 (Chain rule). *Let $f : I \to \mathbb{R}$ and $g : J \to \mathbb{R}$ be such that $f(I) \subseteq J$, where I and J are some open intervals containing x_0 and $y_0 = f(x_0)$, respectively. If f is differentiable at x_0 and g is differentiable at $f(x_0)$, then the composition $(g \circ f)(x) = g(f(x))$ is differentiable at x_0 and*

$$(g \circ f)'(x_0) = g'(y_0)f'(x_0) = (g' \circ f)(x_0)f'(x_0). \tag{3.10}$$

Further, if f is differentiable on I and g is differentiable on J, then $g \circ f$ is differentiable on I and (3.10) *holds for each $x_0 \in I$.*

Proof. We remind the reader that the hypothesis that f is differentiable at x_0 implies that x_0 is a limit point of I. Similarly, y_0 is a limit point of J.

Let $y = f(x)$, $x \in I$. Since f is differentiable at x_0, by (3.8), there exists a $\delta_1 > 0$ such that

$$f(x) - f(x_0) = (x - x_0)\eta_f(x) \quad \text{for } x \in B(x_0; \delta_1) \subset I,$$

where η_f ($\equiv f'(x_0) + \eta(x)$ in (3.8)) is continuous in $B(x_0; \delta_1)$ with

$$\lim_{x \to x_0} \eta_f(x) = f'(x_0).$$

Further, since g is differentiable at y_0, there exists a $\delta_2 > 0$ such that

$$g(y) - g(y_0) = (y - y_0)\eta_g(y) \quad \text{for } y \in B(y_0; \delta_2) \subset J,$$

where η_g is continuous in $B(y_0; \delta_2)$ with $\lim_{y \to y_0} \eta_g(y) = g'(y_0)$. Now choose $\delta > 0$ such that $\delta < \delta_1$ and

$$|x - x_0| < \delta_1 \quad \text{implies} \quad |f(x) - f(x_0)| < \delta_2.$$

Then for $x \in B(x_0;\delta)$, we have by substitution

$$g(f(x)) - g(f(x_0)) = (f(x) - f(x_0))\eta_g(f(x)) = (x - x_0)\eta_f(x) \cdot \eta_g(f(x)),$$

so that

$$\frac{g(f(x)) - g(f(x_0))}{x - x_0} = \eta_f(x) \cdot \eta_g(f(x)),$$

where $\eta_f(x) \cdot \eta_g(f(x))$ is continuous at x_0 and approaches $f'(x_0) \cdot g'(y_0)$ as $x \to x_0$. The assertion now follows. ■

3.3.2 Smooth and Piecewise Smooth Functions

Suppose that f is differentiable on an interval I. Then we obtain a new function f' (whose domain may be a subset of I). Even if $f'(x)$ exists on I, the derived function f' need not be continuous on I (see Example 3.54).

Definition 3.52. *We say that f is of class C^1 on I, denoted by $f \in C^1(I)$, if f is differentiable on I and f' is continuous on I. If $f \in C^1(I)$, then we often say that f is continuously differentiable on I.*

In particular, we say that f is continuously differentiable on $[a,b]$ if f is differentiable on $[a,b]$, f' is continuous on (a,b), and

$$f'_+(a) = \lim_{x\to a+} f'(x) =: f'(a+) \quad \text{and} \quad f'_-(b) = \lim_{x\to b-} f'(x) =: f'(b-).$$

Definition 3.53 (Piecewise smooth function). *A function $f : [a,b] \to \mathbb{R}$ is said to be* piecewise smooth *if there exists a partition $P = \{x_0, x_1, \dots, x_n\}$ of $[a,b]$ such that f is continuously differentiable on each subinterval (x_{k-1}, x_k), $1 \le k \le n$, where $x_0 = a$ and $x_n = b$.*

In particular, every $f \in C^1([a,b])$ (i.e., every continuously differentiable function) is piecewise smooth. We remark that functions in $C^1([a,b])$ are called *smooth* functions..

Suppose that f is differentiable on an open interval I and $a \in I$. Then it is natural ask whether f' differentiable at a. If so, we denote the derivative of f' at a by $f''(a)$. This is called the *second derivative* of f at a, and is also denoted by $f^{(2)}(a)$. The higher-order derivatives $f^{(n)}(a)$ may be defined similarly.

Example 3.54. Define $f, g, \phi : \mathbb{R} \to \mathbb{R}$ by

$$f(x) = \begin{cases} \sin(1/x) & \text{for } x \neq 0, \\ 0 & \text{for } x = 0, \end{cases}$$

$g(x) = xf(x)$, and $\phi(x) = x^2 f(x)$. Then we have the following:

(a) f is not continuous on $\mathbb{R}$.

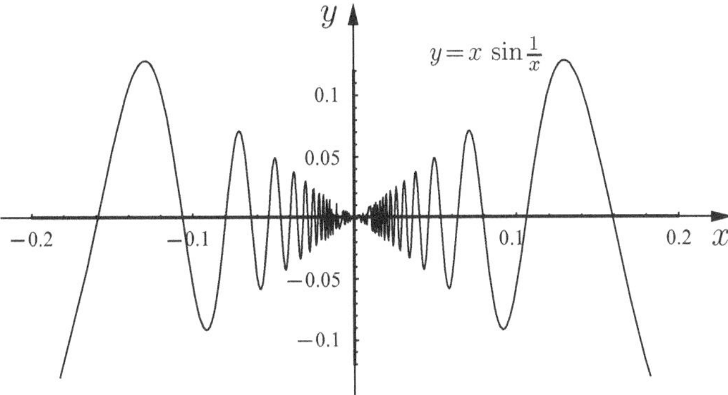

Fig. 3.13. Graph of $f(x) = x \sin(1/x)$ on $[-0.2, 0.2]$.

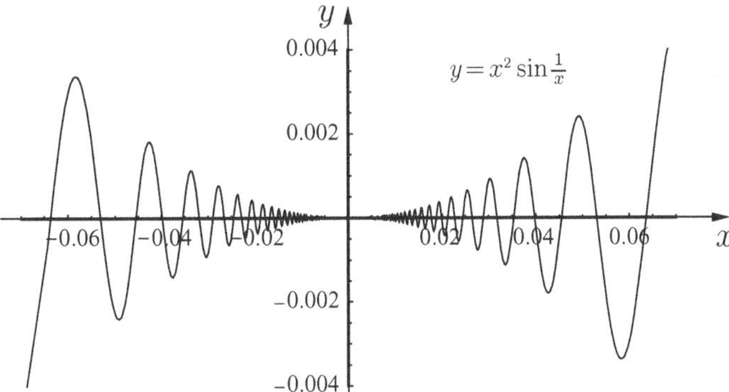

Fig. 3.14. Graph of $f(x) = x^2 \sin(1/x)$ on $[-0.07, 0.07]$.

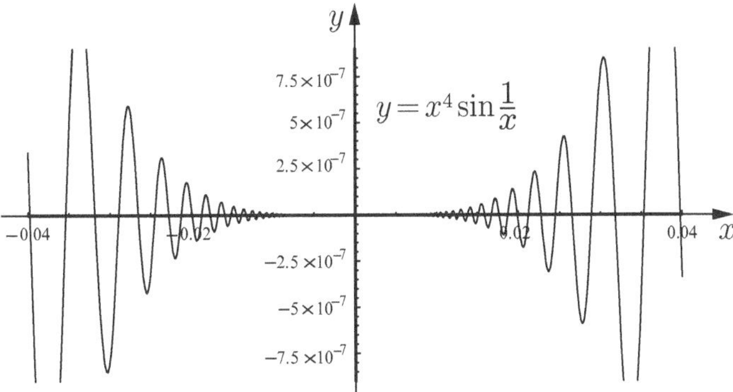

Fig. 3.15. Graph of $f(x) = x^4 \sin(1/x)$ on $[-0.04, 0.04]$.

(b) g is continuous on $\mathbb{R}$ but not differentiable at 0.
(c) ϕ is differentiable on $\mathbb{R}$ but not continuously differentiable on $\mathbb{R}$.

The graphs of f, g, ϕ, and $x^4 f(x)$ are pictured in Figures 3.1 and 3.13–3.15.

First we notice that it is a routine matter to show that each of f, g, and ϕ is differentiable on $\mathbb{R}\setminus\{0\}$. At the end, we shall see that ϕ is also differentiable at the origin. In order to prove that f is not continuous at $x = 0$, it suffices to observe that $\lim_{x\to 0} \sin(1/x)$ does not exist (see Example 3.7). Alternatively, consider

$$|f(x) - f(0)| = |\sin(1/x)|, \quad x \neq 0.$$

In any interval $(-\delta, \delta)$, no matter how small $\delta > 0$ is, there are points at which

$$|\sin(1/x)| = 1,$$

showing that f is not continuous at 0. In fact, we observe that every interval $(-\delta, \delta)$ contains points x of the form $x = 1/n\pi$ for large n. Indeed, if

$$x_n = \frac{1}{(2n + 1/2)\pi} \quad \text{and} \quad y_n = \frac{1}{(2n + 3/2)\pi},$$

then $f(x_n) = 1$ and $f(y_n) = -1$, showing that f does not tend to a limit as $x \to 0$. To prove the continuity of g at 0, let $\epsilon > 0$. Then

$$|g(x) - g(0)| = |g(x)| \leq |x| \quad \text{for all } x.$$

Since we want this to be less than ϵ, we let $\delta = \epsilon$. Then

$$|g(x) - g(0)| < \epsilon \quad \text{for all } |x - 0| < \delta,$$

and so according to the ϵ-δ property, g is continuous at 0. A similar argument with $\delta = \sqrt{\epsilon}$ shows that ϕ is continuous at $x = 0$.

To check the differentiability of g at 0, we compute

$$\lim_{h\to 0} \frac{g(h) - g(0)}{h} = \lim_{h\to 0} \sin(1/h),$$

showing that g is not differentiable at 0.

Further, since $|h \sin(1/h)| \leq |h|$ for all $h \neq 0$, it follows that

$$\lim_{h\to 0} \frac{\phi(h) - \phi(0)}{h} = \lim_{h\to 0} h \sin(1/h) = 0,$$

and therefore we have

$$\phi'(x) = \begin{cases} 2x \sin(1/x) - \cos(1/x) & \text{for } x \neq 0, \\ 0 & \text{for } x = 0. \end{cases}$$

Thus, ϕ is differentiable for all x. For $x \neq 0$, the first term in the last expression has limit 0 as $x \to 0$, whereas the second term takes values between -1 and $+1$ in every neighborhood of $x = 0$. Thus $\lim_{h\to 0} \phi'(h)$ does not exist, even though $\phi'(0) = 0$. Hence, $\phi'(x)$ is not continuous at $x = 0$. •

3.3.3 L'Hôpital's Rule

In Theorem 3.8, we proved that

$$\lim_{x\to a}\frac{f(x)}{g(x)}=\frac{\lim\limits_{x\to a}f(x)}{\lim\limits_{x\to a}g(x)},$$

provided both $\lim_{x\to a}f(x)$ and $\lim_{x\to a}g(x)$ exist, and $\lim_{x\to a}g(x)\neq 0$. However, in curve sketching and other applications, one encounters evaluation of a limit of the form

$$\lim_{x\to a}\frac{f(x)}{g(x)},$$

where $\lim_{x\to a}f(x)$ and $\lim_{x\to a}g(x)$ are either both zero or both infinite. Such limits are called $0/0$ or ∞/∞ indeterminate forms, respectively. The rule to evaluate such limits if they exist is known as l'Hôpital's rule. The numerator and denominator may have limits, but the quotient need not. We also meet other indeterminate forms such as $\infty-\infty$, 1^∞, ∞^0, 0^0, and $\infty\cdot 0$. In this subsection we state another important consequence of differentiability for computing limits of the indeterminate form $0/0$, and the corresponding rule for the remaining cases may be reformulated from the result for the form $0/0$. Again, we emphasize that there are several versions of the rule. However, one can make appropriate modifications to state a more general result of the following form for differentiable functions as in Exercise 3.59(13).

Theorem 3.55 (L'Hôpital's rule). *Let $f(x)$ and $g(x)$ be differentiable at x_0, with $f(x_0)=g(x_0)=0$. If $g'(x_0)\neq 0$, then*

$$\lim_{x\to x_0}\frac{f(x)}{g(x)}=\frac{f'(x_0)}{g'(x_0)}.$$

Proof. Since $g'(x_0)\neq 0$, by Corollary 3.29, there is an open interval containing x_0 such that $g'(x)\neq 0$ for all x in this interval. Consequently, $g(x)-g(x_0)\neq 0$ for all x in an open interval containing x_0. Finally, the result is a consequence of the definition of derivative and the algebra of limits, for

$$\frac{f'(x_0)}{g'(x_0)}=\frac{\lim_{x\to x_0}\dfrac{f(x)-f(x_0)}{x-x_0}}{\lim_{x\to x_0}\dfrac{g(x)-g(x_0)}{x-x_0}}=\lim_{x\to x_0}\frac{\dfrac{f(x)-f(x_0)}{x-x_0}}{\dfrac{g(x)-g(x_0)}{x-x_0}}=\lim_{x\to x_0}\frac{f(x)}{g(x)}.$$ ∎

Examples 3.56. (a) Let $f(x)=x^3-x^2$ and $g(x)=x$. Then

$$\lim_{x\to 0}\frac{f(x)}{g(x)}=\frac{f'(0)}{g'(0)}=\frac{0}{1}=0.$$

Note that for $x\neq 0$, $f(x)/g(x)$ equals x^2-x.

(b) Let $f(x) = x^8 - 1$ and $g(x) = x^2 - 1$. Then

$$\lim_{x\to 1} \frac{f(x)}{g(x)} = \lim_{x\to 1} \frac{8x^7}{2x} = 4.$$

(c) For $f(x) = \sin x$, we have

$$\lim_{x\to 0} \frac{f(ax)}{f(x)} = \lim_{x\to 0} \frac{\sin ax}{\sin x} = a\frac{\cos 0}{\cos 0} = a,$$

where a is any real number.

(d) Let $f(x) = 1 - \cos x$ and $g(x) = \sin^2 x$. Then $g(n\pi) = 0$, $f(2n\pi) = 0$,

$$f'(x) = \sin x \quad \text{and} \quad g'(x) = 2\sin x \cos x,$$

so that $f'(n\pi) = 0$ and $g'(n\pi) = 0$. So Theorem 3.55 is not applicable in the present form to compute $\lim_{x\to 2n\pi} f(x)/g(x)$. Since $g'(2n\pi) = 0$ for each $n \in \mathbb{Z}$, Theorem 3.55 is not applicable. But using the fact that

$$\sin^2 x = 1 - \cos^2 x = (1 - \cos x)(1 + \cos x),$$

we have for each $n \in \mathbb{Z}$,

$$\lim_{x\to 2n\pi} \frac{1 - \cos x}{\sin^2 x} = \lim_{x\to 2n\pi} \frac{1 - \cos x}{1 - \cos^2 x} = \lim_{x\to 2n\pi} \frac{1}{1 + \cos x} = \frac{1}{2}.$$ •

3.3.4 Limit of a Sequence from a Continuous Function

The graph of a sequence consists of a succession of isolated points. For example, the graph of $y = \sqrt{x^2 + 5x} - x$, $x \geq 1$, is clearly a continuous curve. The only difference between $\lim_{n\to\infty} a_n = a$ and $\lim_{x\to a} f(x) = a$ is that n is required to be an integer with $f(n) = a_n$. We already know quite a bit about the limit of a sequence and the limit of a function. We are now confronted with a situation that is very similar. So our knowledge of functions and the theory of limits developed earlier carry over immediately to some sequences and their limits.

Theorem 3.57 (Limit of a Sequence from the limit of a continuous function). *Given the sequence $\{a_n\}$, let f be a continuous function such that $a_n = f(n)$ for large n. If $\lim_{x\to\infty} f(x) = L$, where L is in the extended limits, then $\lim_{n\to\infty} a_n = L$.*

Proof. Let $\epsilon > 0$ and $L \in (-\infty, \infty)$. Because $\lim_{x\to\infty} f(x) = L$, there exists a number $N > 0$ such that

$$|f(x) - L| < \epsilon \quad \text{whenever } x \geq N.$$

In particular, we may choose N large enough that if $n \geq N$, then

$$|f(n) - L| = |a_n - L| < \epsilon,$$

and so $\{a_n\}$ converges.

We leave the proof for the cases $L = \pm\infty$ as exercises. ∎

We note that Theorem 3.57 does not say that if $\lim_{n\to\infty} a_n = L$, then one has $\lim_{x\to\infty} f(x) = L$. We present some simple examples.

(1) Consider $a_n = n^3/(1+\mathrm{e}^n)$. If we let $f(x) = x^3/(1+\mathrm{e}^x)$, then f is continuous for all x, and $f(n) = a_n$ for $n = 1, 2, \ldots$. Theorem 3.57 tells us that $\lim_{n\to\infty} a_n$ is the same as $\lim_{x\to\infty} f(x)$, provided the latter limit exists. According to l'Hôpital's rule,

$$\lim_{x\to\infty} f(x) = \lim_{x\to\infty} \frac{x^3}{1+\mathrm{e}^x} = \lim_{x\to\infty} \frac{3x^2}{\mathrm{e}^x} = \lim_{x\to\infty} \frac{6}{\mathrm{e}^x} = 0.$$

Thus by Theorem 3.57, $\lim_{n\to\infty} a_n = \lim_{x\to\infty} f(x) = 0$.

(2) Consider $a_n = n\sin(1/n)$. If we set $f(x) = (1/x)\sin x$, then $a_n = f(1/n)$, and we observe that (for instance by l'Hôpital's rule)

$$f(x) \to 1 \quad \text{as } x \to 0+.$$

Theorem 3.57 then shows that $a_n \to 1$.

3.3.5 Questions and Exercises

Questions 3.58.

1. Are there nonconstant functions f for which $f'(x) = 0$ for all $x \in \operatorname{dom}(f)$?
2. Suppose that f is differentiable on (a, c) and $[c, b)$. Must f be differentiable at c?
3. Suppose that $|f|$ is differentiable at a point c. Must f be differentiable at c?
4. Suppose $f'_+(a)$ and $f'_-(a)$ both exist. Must f be continuous at a?
5. If f is continuous at the origin, must $g(x) = xf(x)$ be differentiable at the origin?
6. Is $f(x) = |x|\sin x$ differentiable at 0?
7. Suppose that $f(x) = x^{1/3} - x^{4/3}$ on $[0, 1]$. Is f continuous on $[0, 1]$? Is f differentiable on $(0, 1)$? Does there exist a point $c \in (0, 1)$ such that $f'(c) = 0$? If so, at which c does this hold?
8. Must the derivative of an odd function be even?
9. Suppose that f and g are differentiable on (a, b) such that $f(c) = g(c)$ for some $c \in (a, b)$ and $f(x) \geq g(x)$ on (a, b). Must $f'(c) = g'(c)$?
10. Suppose that $f : \mathbb{R} \to \mathbb{R}$ such that $|f(x) - f(y)| \leq (x - y)^2$ for all $x, y \in \mathbb{R}$. What can be said about the function f? For example, can it be differentiable on $\mathbb{R}$? If so, what is its derivative?
11. Suppose that $f : [a, b] \to [a, b]$ is such that $|f(x) - f(y)| < |x - y|$ for all $x, y \in [a, b]$, $x \neq y$. Must there exist a number $M < 1$ such that

$$|f(x) - f(y)| < M|x - y| \quad \text{for all } x, y \in [a, b]?$$

12. If $f:(a,b)\to\mathbb{R}$ is differentiable, must the derivative $f'(c)$ at each $c\in(a,b)$ be given by
$$\lim_{h\to 0}\frac{f(c+h)-f(c-h)}{2h}\ ?$$
Does the limit exist if $f(x)=|x|$?
13. Suppose that f is differentiable at c, and α, β are two nonzero real numbers. Must the limit
$$\lim_{h\to 0}\frac{f(c+\alpha h)-f(c+\beta h)}{h}$$
exist?
14. Suppose that f is differentiable at a such that $f(a)=0$. Let $g(x)=|f(x)|$. Is it true that g is differentiable at a if and only if $f'(a)=0$?
15. If f is differentiable at $x=c$, what can be said about $xf(c)-cf(x)$?
16. Suppose that f is differentiable on $\mathbb{R}$ such that $f(x+y)=f(x)f(y)$ for all x and y in $\mathbb{R}$. If $f(a)$ and $f'(1-a)$ are given for some $a\in\mathbb{R}$, what is the value of $f'(0)$? If $f(1)=3$ and $f'(2)=1$, do we have $f'(3)=3$?

Exercises 3.59.

1. Show that $f(x)=\cos x$ is differentiable on $\mathbb{R}$.
2. Show that $f(x)=x^{1/3}$ is not differentiable at the origin.
3. If $f:(a,b)\to\mathbb{R}$ is twice differentiable at $c\in(a,b)$, then show that
$$f''(c)=\lim_{h\to 0}\frac{f(c+h)+f(c-h)-2f(c)}{h^2}.$$
If $f(x)=x|x|$ for $x\in\mathbb{R}$, then show that the limit on the right exists for $c=0$, but $f''(0)$ does not exist.
4. If $f:(a,b)\to\mathbb{R}$ is thrice differentiable at $c\in(a,b)$, then show that
$$\frac{f'''(c)}{3}=\lim_{h\to 0}\left[\frac{f(c+h)-f(c-h)-2hf'(c)}{h^3}\right].$$
Find an example of a function f for which the limit on the right exists, but $f'''(c)$ does not.
5. Define $f:\mathbb{R}\to[-1,1]$ by $f(x)=\sqrt{1+\cos x}$. Determine $f'_+(x)$ and $f'_-(x)$. Also, determine points where f fails to be differentiable.
6. Suppose that f is differentiable on (a,b) and continuous on $[a,b]$. Do $f'_+(a)$ and $f'_-(b)$ exist?
7. If $f(x)=|x(x-1)|$, determine the left and right derivatives of f at $x=0,1$.
8. Show that $f(x)=|x|x$ is differentiable on $\mathbb{R}$, but its derived function is not differentiable at 0. How about $f(x)=|x|x^n$? (see Figures 3.16 and 3.17).

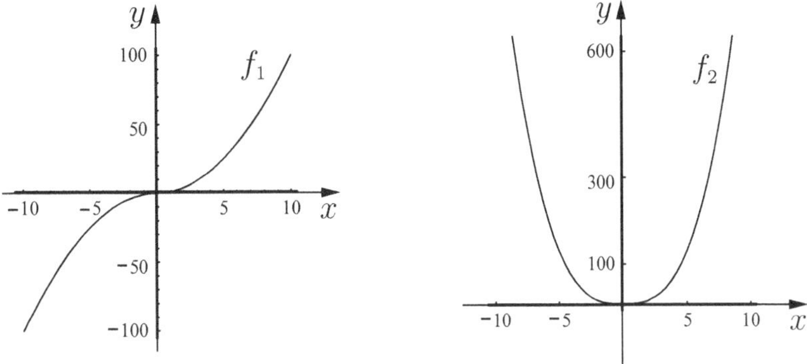

Fig. 3.16. Graph of $f_n(x) = |x|x^n$ when $x \neq 0$ and 0 when $x = 0$, for $n = 1, 2$.

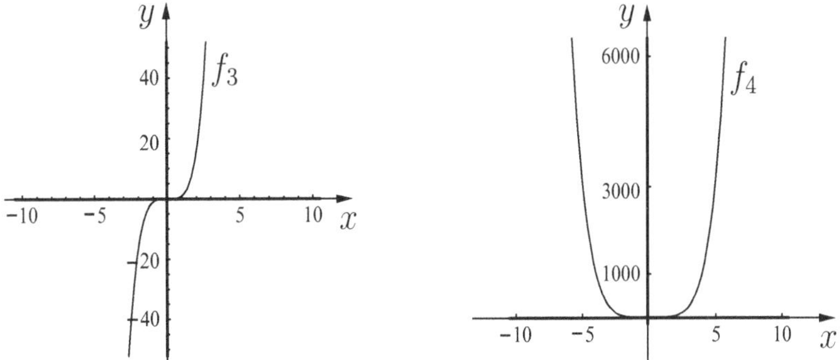

Fig. 3.17. Graph of $f_n(x) = |x|x^n$ when $x \neq 0$ and 0 when $x = 0$, for $n = 3, 4$.

9. Define $f(x) = x - [x]$ on $[0, 4]$ and $g(x) = |2f(x) - 1|$ on $[0, 4]$. Draw the graphs of f and g. Prove the following:
   (a) f is piecewise continuous on $[0, 4]$.
   (b) g is piecewise differentiable on $[0, 4]$.
10. Define
$$f(x) = \begin{cases} x^2 \sin(1/x) & \text{for } x < 0, \\ x^3 & \text{for } x \geq 0. \end{cases}$$
Verify whether f is differentiable at the origin by computing $f'_+(0)$ and $f'_-(0)$ if they exist. Also determine $f'(0+)$ if it exists.
11. Define $f : \mathbb{R} \to \mathbb{R}$ by
$$f(x) = \begin{cases} \sin x & \text{for } x \in \mathbb{Q}, \\ x & \text{for } x \in \mathbb{R} \backslash \mathbb{Q}. \end{cases}$$
Show that f is differentiable at the origin.
12. Suppose that f is differentiable at a such that $f(a) \neq 0$. Show that g defined by $g(x) = |f(x)|$ is differentiable at a. Also, determine $g'(a)$.

13. If $f(x_0) = g(x_0) = 0$, $f'(x_0)$, and $g'(x_0)$ exist with $g'(x_0) \neq 0$, do we have

$$\lim_{x \to x_0} \frac{f(x)}{g(x)} = \lim_{x \to x_0} \frac{f'(x)}{g'(x)}?$$

More generally, if f, g, and their $(n-1)$st derivatives are zero at x_0 and $g^{(n)}(x_0) \neq 0$, does it follow that

$$\lim_{x \to x_0} \frac{f(x)}{g(x)} = \frac{f^{(n)}(x_0)}{g^{(n)}(x_0)}?$$

Note: Yes. This is called a general l'Hôpital's rule.

14. Using the chain rule, determine the derivative of
    (a) $f(x) = \sqrt{x + \sqrt{x + \sqrt{x}}}$ on $[0, \infty)$.
    (b) $h(x) = \sin^2(x) \sin(1/\sin x)$ on $(0, \pi)$.
15. If f and g have derivatives of all orders at a point $a \in \mathbb{R}$, determine the derivative formula for $h^{(n)}(a)$, where $h = fg$.
16. Suppose that

$$f(x) = \begin{cases} x & \text{for } x \in \mathbb{Q}, \\ \sin x & \text{for } x \in \mathbb{R} \setminus \mathbb{Q}. \end{cases}$$

Show that f is differentiable at the origin. Also, determine $f'(0)$.

17. Draw the graph of

$$f(x) = \begin{cases} x^3 & \text{for } x > 0, \\ x^2 & \text{for } x \leq 0, \end{cases} \quad \text{and} \quad g(x) = \begin{cases} x^3 & \text{for } x < 1, \\ 3x - 2 & \text{for } x \geq 1. \end{cases}$$

Does $f'(x)$ exist on $\mathbb{R}$? If so, is f' differentiable at the origin? Does $g'(x)$ exist on $\mathbb{R}$? If so, is g' differentiable at the origin?

18. Find the set of points where the following functions are not differentiable:
    (a) $\sin(|x|)$. (b) $|x| + |x-1|$. (c) $|x| + |x-1| + |x-2|$.
    (d) $|x^2 - 9|$. (e) $|x^3 - 27|$.
19. Consider

$$f(x) = \begin{cases} x^4 & \text{for } x < 1, \\ ax + b & \text{for } x \geq 1. \end{cases}$$

For what values of a and b is f continuous at the point $x = 1$? Is it differentiable at the point $x = 1$? If yes, what is $f'(1)$? If not, determine the left and right derivatives at $x = 1$.

20. Find the following limits if they exist:
    (a) $\displaystyle\lim_{x \to 0} \frac{e^{3x} - \cos 2x}{x}$. (b) $\displaystyle\lim_{x \to 0} \frac{\sqrt{2+x} - \sqrt{2-x}}{x}$.
    (c) $\displaystyle\lim_{x \to 0} \frac{x^3}{\sin x - x}$. (d) $\displaystyle\lim_{x \to 0} \frac{x^4}{1 - \cos 3x - 9x^2}$.
    (e) $\displaystyle\lim_{x \to 0} \frac{x(e^x - 1)}{x - e^x + 1}$. (f) $\displaystyle\lim_{x \to -2} \frac{x+2}{x^2 + x - 2}$.

21. Draw the graph of f defined by

$$f(x) = \begin{cases} 1 & \text{for } x < 0, \\ \cos x & \text{for } 0 \leq x \leq \pi, \\ -1 & \text{for } x > \pi. \end{cases}$$

Is f continuous on $\mathbb{R}$? Is f differentiable on $\mathbb{R}$? If so, determine whether $f'(x)$ is continuous on $\mathbb{R}$. If not, explain at what points f is not differentiable.

4

Applications of Differentiability

In Section 4.1, we introduce standard properties associated with functions and define inverse functions. In this section, we will learn the importance of the inversion process. In Section 4.2, we begin the discussion with local and global extrema and then continue to derive sufficient conditions for the existence of local extrema. In this section, we also prove two important theorems in calculus, namely Rolle's theorem and the mean value theorem.

4.1 Basic Concepts of Injectivity and Inverses

To begin with, let us introduce some important terminology associated with functions. Let I be an interval and $f\colon I \to \mathbb{R}$ a given function. We say that f is

- *monotonically increasing* (or *increasing*) on I if $f(x) \leq f(y)$ for all $x, y \in I$ with $x < y$;
- *monotonically decreasing* (or *decreasing*) on I if $f(x) \geq f(y)$ for all $x, y \in I$ with $x < y$;
- *strictly increasing* if $f(x) < f(y)$ for all $x, y \in I$ with $x < y$;
- *strictly decreasing* on I if $f(x) > f(y)$ for all $x, y \in I$ with $x < y$;
- *monotone* if it is either increasing or decreasing on I;
- *strictly monotone* if it is either strictly increasing or strictly decreasing on I.

We have already encountered functions $f : \mathbb{R} \to \mathbb{R}$ for which there is no inverse function $g : \mathbb{R} \to \mathbb{R}$ because f is not one-to-one, that is, that some $y_0 \in \mathbb{R}$ is the image of two different numbers $x_1, x_2 \in \mathbb{R}$, namely $f(x_1) = f(x_2) = y_0$, with $x_1 \neq x_2$. Even if f is one-to-one, if it is not also onto, that is, if there is at least one number y_0 that is the image of no real number x, then f will have no inverse function whose domain is all of $\mathbb{R}$.

In such cases we may require to restrict the domain and the codomain so that the modified function has one or more inverses. This leads to a discussion on "local inverses".

S. Ponnusamy, *Foundations of Mathematical Analysis*,
DOI 10.1007/978-0-8176-8292-7_4,

For the definitions of one-to-one and onto functions, we refer to Section 1.2.2.

Example 4.1. Suppose that I and I' are subsets of $\mathbb{R}$. Then the graph of $f : I \to I'$ helps to determine whether f is one-to-one: *a function f is one-to-one iff every line $y = c$ intersects the graph in at most one point*; see Figures 4.1 and 4.3.

For instance, consider the graph of $y = f(x) = x^2$ for all x in some neighborhood I of the origin, say $I = (-2, 2)$. Then we observe that for $y = 1$ we have $x^2 = 1$, which has the two solutions $x = \pm 1$. That is, we cannot define $f^{-1}(1)$ uniquely. Thus, $f(x) = x^2$ on $(-2, 2)$ fails to have an inverse function, because it is not one-to-one. For instance, for $y = -1$, we have $x^2 = -1$, which has no real solution. In particular, $f : \mathbb{R} \to \mathbb{R}$ defined by $f(x) = x^2$ has no inverse. However, it is a simple exercise to see that $g : [0, \infty) \to [0, \infty)$ defined by $g(x) = x^n$, $n \in \mathbb{N}$, is one-to-one, and hence it does have an inverse function $g^{-1} : [0, \infty) \to [0, \infty)$ defined by

$$g^{-1}(y) = y^{1/n} \quad \text{for every } y \in [0, \infty).$$

In other words, if $x \geq 0$ and $y \geq 0$, then for each fixed $n \in \mathbb{N}$,

$$y = x^n \iff x = y^{1/n}.$$

How about $f_n : \mathbb{R} \to \mathbb{R}$ defined by $f_n(x) = x^{2n+1}$? (See Figures 4.3 and 4.4.) ●

Example 4.2. The graph of $y = \sin x$ shows that it is one-to-one on the interval

$$\left(-\frac{\pi}{2} + k\pi, \frac{\pi}{2} + k\pi\right)$$

for each fixed $k \in \mathbb{Z}$ (Figures 4.1 and 4.2). Similarly, by drawing the graph of $g(x) = \cos x$, it can be easily seen that g is one-to-one on each interval

$$(k\pi, (k+1)\pi), \quad k \in \mathbb{Z},$$

and hence it has an inverse function $\cos^{-1}(y)$ defined from $(-1, 1)$ onto $(k\pi, (k+1)\pi)$; see Examples 4.7. ●

Unfortunately, it is in general not possible to solve the equation $y = f(x)$ for x in terms of y. However, in order to prove that f has an inverse, the following simple result is useful. We invite the reader to prove this result.

Proposition 4.3. *Let $f : (a, b) \to \mathbb{R}$, $a < b$, be continuous. Then the following statements are equivalent:*

(a) *f is one-to-one on (a, b).*
(b) *$f(I)$ is an open interval whenever I is an open interval in (a, b).*
(c) *f is strictly monotone on (a, b).*

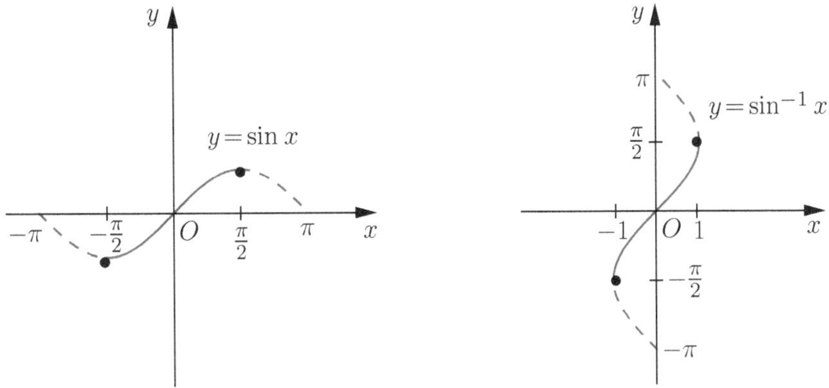

Fig. 4.1. Graphs of $y = \sin x$ on $[-\pi/2, \pi/2]$ and $y = \sin^{-1} x$ on $[-1, 1]$.

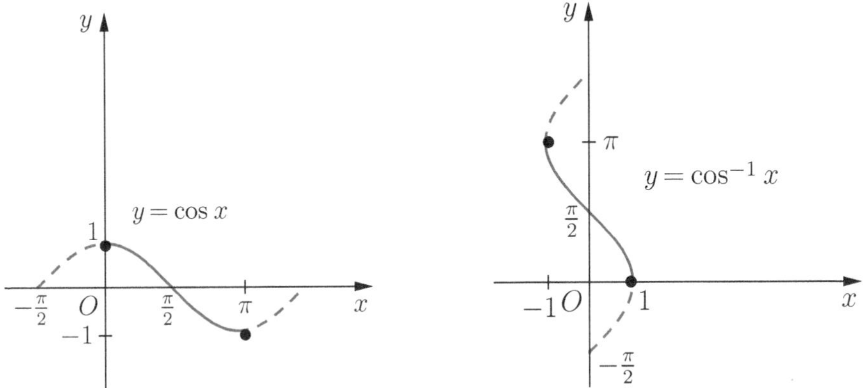

Fig. 4.2. Graphs of $y = \cos x$ on $[0, \pi]$ and $y = \cos^{-1} x$ on $[-1, 1]$.

For instance if $f(x) = x^7 + 2x + 1$ $(x \in \mathbb{R})$, then for $x_1 < x_2$, we have $x_1^7 < x_2^7$ and

$$x_1^7 + 2x_1 + 1 < x_2^7 + 2x_2 + 1,$$

so that f is strictly increasing. Therefore, f is one-to-one on $\mathbb{R}$.

At this point, we emphasize that by Proposition 4.3, every strictly monotone continuous function $f : (a, b) \to \mathbb{R}$ has an inverse function f^{-1} with domain $J = f((a, b))$. Moreover, f^{-1} is also continuous and strictly monotone. For example, $f(x) = \tan x$ for $x \in (-\pi/2, \pi/2)$ has a strictly increasing continuous inverse function with domain $f(-\pi/2, \pi/2) = \mathbb{R}$, denoted by $\tan^{-1}$ (see Figure 4.5).

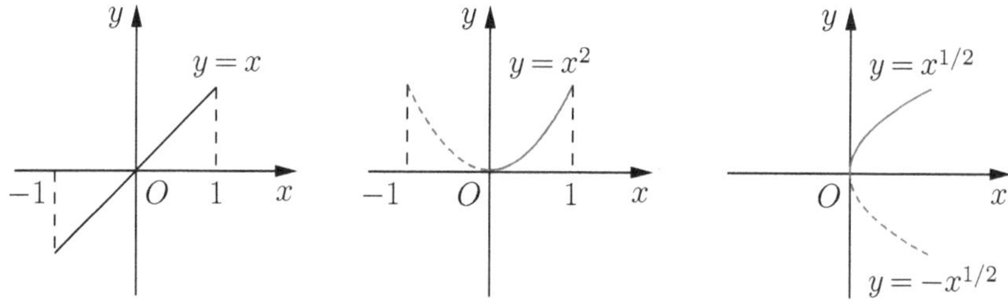

Fig. 4.3. Graphs of $y = x$, $y = x^2$ on $[-1, 1]$, and $y = x^{1/2}$ on $[0, 1]$.

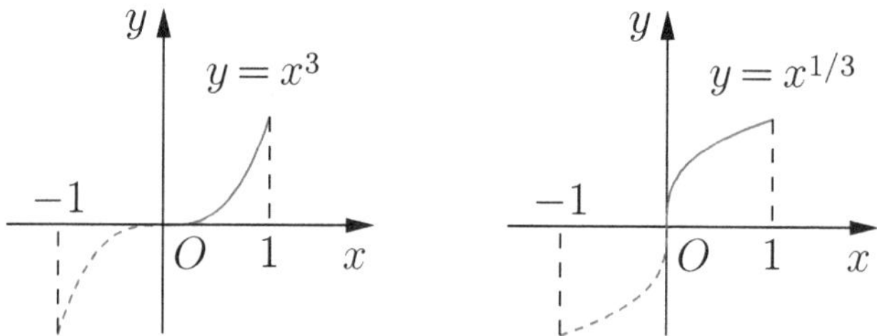

Fig. 4.4. Graphs of $y = x^3$ and $y = x^{1/3}$ on $[-1, 1]$.

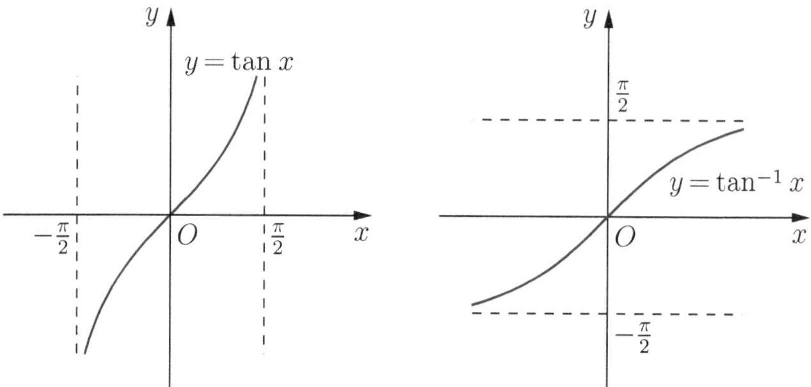

Fig. 4.5. Graphs of $y = \tan x$ on $(-\pi/2, \pi/2)$ and its inverse.

4.1.1 Basic Issues about Inverses on $\mathbb{R}$

Suppose that Ω and Ω' are two subsets of $\mathbb{R}$, and $g : \Omega' \to \Omega$ is an inverse of $f : \Omega \to \Omega'$. Then we can express this by writing

$$g \circ f = I_\Omega, \quad \text{i.e., } (g \circ f)(x) = g(f(x)) = x \text{ for all } x \in \Omega,$$

and

$$f \circ g = I_{\Omega'}, \quad \text{i.e., } (f \circ g)(y) = f(g(y)) = y \text{ for all } y \in \Omega'.$$

Here $I_\Omega : \Omega \to \Omega$ and $I_{\Omega'} : \Omega' \to \Omega'$ are the identity mappings on Ω and Ω', respectively; that is, for all $x \in \Omega$ and $y \in \Omega'$, one has

$$I_\Omega(x) = x \quad \text{and} \quad I_{\Omega'}(y) = y.$$

Remark 4.4. **(a)** The inverse of f is often denoted by f^{-1} (which should not be confused with $1/f$).

(b) Further, the notion of inverse f^{-1} of $f : \Omega \to \Omega'$ should not be confused with the inverse image set $f^{-1}(B)$ for $B \subset \Omega'$. The latter is a set that exists for every function f, while the former is a function that exists only when f is bijective; see Proposition 4.5. ●

To summarize, given a function $f : \Omega \to \Omega'$, if there exists a function $g : \Omega' \to \Omega$ such that $y = f(x) \iff x = g(y)$, then the function g is called the *inverse function* of f.

Clearly, g is the inverse of f iff f is the inverse of g. It is in fact an easy exercise to prove the following result.

Proposition 4.5. *Let Ω and Ω' be two subsets of $\mathbb{R}$. Then the function $f : \Omega \to \Omega'$ has an inverse iff f is bijective.*

Proof. ($\Rightarrow$): Suppose that f has an inverse, say $g : \Omega' \to \Omega$. Let $x_1, x_2 \in \Omega$ be such that $f(x_1) = f(x_2)$. Then by the definition of inverse,

$$x_1 = g(f(x_1)) = g(f(x_2)) = x_2, \quad \text{i.e., } f \text{ is one-to-one.}$$

To show that f is onto, choose an arbitrary point y in Ω'. Then $g(y) \in \Omega$ because $f(g(y)) = y$.

($\Leftarrow$): Conversely, let f be bijective. Because f is onto, for each $y \in \Omega'$ there is an $x \in \Omega$ such that $y = f(x)$. We denote the element x by setting $x = g(y)$. Moreover, because f is one-to-one, there will be exactly one element of Ω that f maps to a given element $y \in \Omega'$. Now,

$$f(g(y)) = f(x) = y.$$

Also, for each $x \in \Omega$, $g(f(x))$ is an element of Ω such that

$$f(g(f(x))) = f(x),$$

and because f is one-to-one, the last equation gives

$$g(f(x)) = x.$$

Thus, $g : \Omega' \to \Omega$ can be defined by $g(y) = x$ whenever $f(x) = y$. ■

4.1.2 Further Understanding of Inverse Mappings

Suppose that I is an open interval and $f : I \subset \mathbb{R} \to \mathbb{R}$ belongs to $C^1(I)$ such that $f'(a) \neq 0$ for some point $a \in I$. If $f'(a) > 0$, then from the continuity of f' it follows that there is an open interval I_1 containing a such that $f'(x) > 0$

on I_1. A similar statement holds when $f'(a) < 0$. Thus, f is strictly monotone on I_1, and by Proposition 4.3, f is one-to-one on I_1 with an inverse function $g = f^{-1}$ defined on some open interval J_1 containing $f(a)$. Moreover,

$$g(y) = x \quad \text{with} \quad y = f(x) \quad \text{for } x \in I_1,\ y \in J_1. \tag{4.1}$$

Example 4.6. We illustrate Proposition 4.3 with a simple function. Define $f : \mathbb{R} \to \mathbb{R}^+$ by $f(x) = (x-3)^2 + 1$. Then

$$f(\mathbb{R}) = [1, \infty), \quad f((1,4)) = [1,5) \quad \text{and} \quad f((3,4)) = (1,2),$$

from which one can obtain that f is not one-to-one on the whole of $\mathbb{R}$. On the other hand,

$$f'(x) = 2(x-3) \begin{cases} > 0 & \text{for } x > 3, \\ < 0 & \text{for } x < 3, \end{cases}$$

showing that f is one-to-one on each of $(-\infty, 3)$ and $(3, \infty)$. Note that $f'(3) = 0$, and f is strictly decreasing on the interval $(-\infty, 3)$ and strictly increasing on $(3, \infty)$. Therefore, f cannot be one-to-one on any open interval that contains the point 3. ●

Examples 4.7. Some familiar examples of functions and their inverses are as follows:

(a) Consider (see also Figure 4.3)

$$\begin{cases} f_1(x) = x^2 & \text{for } x \in \Omega = \{x : x \geq 0\}, \\ g_1(y) = \sqrt{y} & \text{for } y \in \Omega. \end{cases}$$

Note that

$$(g_1 \circ f_1)(x) = g_1(f_1(x)) = \sqrt{x^2} = x$$

and

$$(f_1 \circ g_1)(y) = f_1(g_1(y)) = f_1(\sqrt{y}) = y.$$

The inverse function $g_1 (= f_1^{-1})$ in this case is called the *positive square root function* of f_1.

(b) If $f_2(x) = x^2$ for $x \in \Omega = \{x : x \leq 0\}$, then g_2 defined by

$$g_2(y) = -\sqrt{y}, \quad y \in \Omega' = \{x : x \geq 0\},$$

is the inverse of f_2 and is called the *negative square root function of* f_2. However, $h : \mathbb{R} \to \mathbb{R}^+$ defined by $h(x) = x^2$ has no inverse, since $h(-x) = h(x)$. That is, h is not one-to-one.

(c) Our next example of a function f and its inverse function g is given by

$$\begin{cases} f(x) = \mathrm{e}^x & \text{for } x \in \Omega = \mathbb{R}, \\ g(y) = \log y & \text{for } y \in \Omega' = \{y : y > 0\}. \end{cases}$$

Note that f is increasing and one-to-one on $\mathbb{R}$. Also, $f(\mathbb{R}) = (0, \infty)$ and $f((0,a)) = (1, \mathrm{e}^a)$ for $a > 0$.

(d) Finally, we consider (see also Figure 4.1)

$$f(x) = \sin x, \quad x \in \Omega = [-\pi/2, \pi/2].$$

Then $f'(x) = \cos x > 0$ for $x \in (-\pi/2, \pi/2)$. So the inverse $g = f^{-1}$ exists with domain $[-1, 1]$. Further,

$$g'(y) = \frac{1}{f'(x)} = \frac{1}{\cos x} = \frac{1}{\sqrt{1 - \sin^2 x}} = \frac{1}{\sqrt{1 - f^2(x)}} = \frac{1}{\sqrt{1 - y^2}}.$$

Here g is called the *inverse sine function*. In other words, $f(x) = \sin x$ on $[-\pi/2, \pi/2]$ has a strictly increasing continuous inverse function, denoted by

$$f^{-1} = \sin^{-1} := \text{Arcsin},$$

with domain $\Omega' = [-1, 1]$ and

$$\frac{d}{dy}(\text{Arcsin}\, y) = \frac{1}{\sqrt{1 - y^2}}.$$

Note that the function $\sin x$ is of course not one-to-one on $\mathbb{R}$, since $\sin n\pi = 0$ for all $n \in \mathbb{Z}$ (compare with Proposition 4.3). Other trigonometric functions can be handled in a similar fashion.

(e) If $f(x) = \cos x$ for $x \in [0, \pi]$, then $f'(x) = -\sin x$ for $x \in [0, \pi]$, so that f is one-to-one on $[0, \pi]$, and hence it has an inverse (see also Figure 4.2). Thus, $f(x) = \cos x$ on $[0, \pi]$ has a strictly decreasing continuous inverse function, denoted by

$$f^{-1} = \cos^{-1} := \text{Arccos},$$

with domain $\Omega' = [-1, 1]$ and

$$\frac{\mathrm{d}}{\mathrm{d}y}(\text{Arccos}\, y) = -\frac{1}{\sqrt{1 - y^2}}.$$

(f) Similarly, the function $f(x) = \tan x$ is one-to-one on $(-\pi/2, \pi/2)$, and the inverse tangent function, denoted by $\text{Arctan}\, y$ or $\tan^{-1} y$, is given by

$$x = f^{-1}(y) = \text{Arctan}\, y,$$

with domain $\mathbb{R}$. Again f^{-1} is a strictly increasing continuous function on $\mathbb{R}$. ●

Having seen a number of examples of one-to-one and inverse functions, we now ask the following question, which has not been considered so far.

Problem 4.8. *If a differentiable function f has an inverse g, is the inverse function necessarily differentiable? If so, is it possible to obtain $g'(y)$ from $f'(x)$, where $y = f(x)$?*

Suppose that

$$g(y) = x \quad \text{with} \quad y = f(x) \quad \text{for } x \in I, \ y \in J,$$

where I and J are some open intervals. If g is differentiable, then by differentiating it with respect to x, we see that

$$g'(y)f'(x) = 1,$$

so that

$$\left(f^{-1}\right)'(y) = \frac{1}{f'\left(f^{-1}(y)\right)} \quad \text{for } y \in J.$$

Equivalently,

$$(f^{-1})'(f(x))f'(x) = 1 \quad \text{for } x \in I \quad \text{or} \quad g'(y) = [f'(x)]^{-1},$$

so that the derivative of an inverse function $g = f^{-1}$ is the reciprocal of the derivative of the original function f. Finally, we invite the reader to show that g is actually differentiable on J (see Exercise 4.10(11)).

4.1.3 Questions and Exercises

Questions 4.9.

1. Is the composition (respectively product, sum) of two decreasing functions monotone?
2. Is the product of two strictly increasing functions necessarily increasing?
3. Must every continuous function on $[a, b]$ be monotone on $[a, b]$?
4. Must every monotone function on $[a, b]$ be continuous on $[a, b]$?
5. Suppose that f and g are two positive increasing functions defined on an interval $[a, b]$. Must the product fg be increasing on $[a, b]$?
6. How are one-to-one and monotone properties of a function related?
7. Suppose that $f\colon \mathbb{R} \to \mathbb{R}$ is continuous and one-to-one. Must f be monotone?
8. Suppose that $f : [0, 2] \to [0, 1]$ is monotone and bijective. Must f and f^{-1} both be continuous?
9. Suppose that f is monotone on (a, b) and we have $c \in (a, b)$. Must $\lim_{x\to c+} f(x)$ and $\lim_{x\to c-} f(x)$ both exist?
10. Suppose that f and g are monotone on (a, b), $h = f - g$, and $c \in (a, b)$. Must $\lim_{x\to c+} h(x)$ and $\lim_{x\to c-} h(x)$ both exist?
11. Let f be increasing and bounded above on (a, ∞) for some $a \in \mathbb{R}$. Must $\lim_{x\to\infty} f(x)$ exist?
12. Suppose that $f : (a, b) \to \mathbb{R}$ is continuous. Must $f(a, b)$ be an open interval in $\mathbb{R}$? If not, when is this possible?

Exercises 4.10.

1. Let $f : (-\infty, a) \to (-\infty, 0)$ be defined by $f(x) = 1/(x - a)$. Prove that f has an inverse function f^{-1} given by $f^{-1}(y) = a + (1/y)$.
2. Show that $f\colon \mathbb{R} \to \mathbb{R}$ defined by $f(x) = x/(1 + |x|)$ is one-to-one but not onto.

3. Give an example of a discontinuous function that is one-to-one.
4. Give an example of a continuous function that is neither one-to-one nor onto.
5. Determine whether the following functions are one-to-one on $(0, \infty)$:
   (a) $f(x) = x^4 - (1/x^2)$. **(b)** $x^4 + 3x + 1$. **(c)** x^n $(n \in \mathbb{Z})$.
6. Set $f(x) = (1 + 1/x)^x$ for $x > 0$. Show that f is increasing on $(0, \infty)$. In particular, $f(n+1) > f(n)$ for all $n \geq 1$ (see Example 2.33).
7. Let $f : I \to J$ be one-to-one and have an inverse $g : J \to I$. If g is differentiable at $y_0 = f(x_0)$, $x_0 \in I$, where f is continuous at x_0 and $g'(y_0) \neq 0$, then show that $f'(x_0)$ exists, and
$$f'(x_0) = \frac{1}{g'(y_0)}.$$
8. Show that $x/\sin x$ is increasing on $(0, \pi/2)$.
9. Consider $f : (0, \infty) \to \mathbb{R}$ by $f(x) = x(3 + \sin(\log x^2))$. Determine whether f is a monotone function.
10. Suppose that f is monotone on the open interval (a, b).
    (a) If f is bounded above on (a, b), then show that $\lim_{x\to b-} f(x)$ exists.
    (b) If f is bounded below on (a, b), then show that $\lim_{x\to a+} f(x)$ exists.
    Note: This is an analogue of the bounded monotone convergence theorem for sequences.
11. Suppose that f is one-to-one and continuous on (a, b) and $c \in (a, b)$. If f is differentiable at c and $f'(c) \neq 0$, then show that f^{-1} is differentiable at $f(c)$ and $(f^{-1})'(f(c)) = 1/f'(c)$. This is referred to as the *inverse function theorem.*

4.2 Differentiability from the Geometric View Point

In this section, we discuss a few fundamental results from the theory of functions of a single variable that are motivated by the geometric consideration of differentiability in terms of tangents.

4.2.1 Local Extremum Theorem

Definition 4.11. *Let f be defined on an interval I.*

- *A point $x_0 \in I$ is called a point of local minimum of f, or equivalently, we say that f has a local minimum $f(x_0)$ at x_0, if there exists a neighborhood B of x_0 such that*
$$f(x) \geq f(x_0) \quad \textit{for all } x \in B \cap I.$$

- *A point $x_0 \in I$ is a local maximum of f, or equivalently, f has a local maximum $f(x_0)$ at x_0, if there exists a neighborhood B of x_0 such that*

$$f(x) \le f(x_0) \quad \textit{for all } x \in B \cap I.$$

  We say that f has a local extremum at x_0 if x_0 is a point of local minimum or a point of local maximum.
- *We say that f has a minimum (or global minimum) at x_0 if $f(x) \ge f(x_0)$ for all $x \in I$. The notion of global maximum is defined similarly.*
- *A point x_0 is called a critical point of f if either f is not differentiable at x_0 or if it is, $f'(x_0) = 0$.*
- *A critical point that is not a local extremum is called a saddle point.*

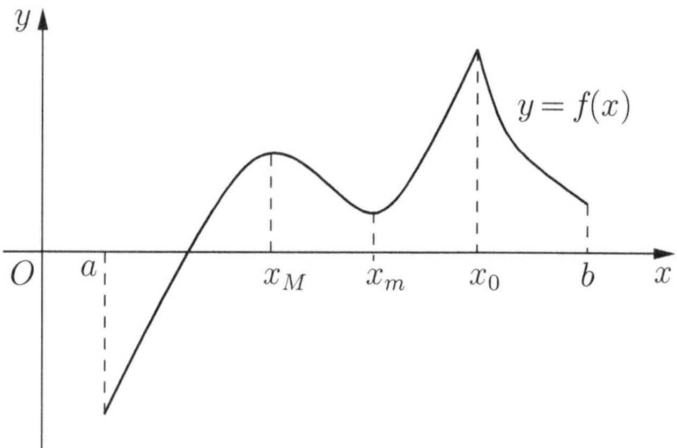

Fig. 4.6. Nondifferentiable function with local extrema.

In Figure 4.6, for instance, $y = f(x)$ has an absolute minimum at $x = a$ and a local maximum at x_M. Also, it has a local minimum at x_m and an absolute maximum x_0. In Figure 4.7, we present graphs of functions that have no extrema.

We now establish a necessary condition for the existence of local extrema for differentiable functions (see Figure 4.8).

Theorem 4.12 (Local extremum theorem). *Let $f\colon I \to \mathbb{R}$, where I is a neighborhood of c, e.g., $I = [a, b]$ with $c \in (a, b)$. If f has a local extremum at c and $f'(c)$ exists, then $f'(c) = 0$.*

Proof. Suppose that f has a local maximum (the proof for a local minimum is similar). Then there exists a $\delta > 0$ such that

$$f(x) \le f(c) \quad \text{for all } x \in \{x : |x - c| < \delta\} \subset I.$$

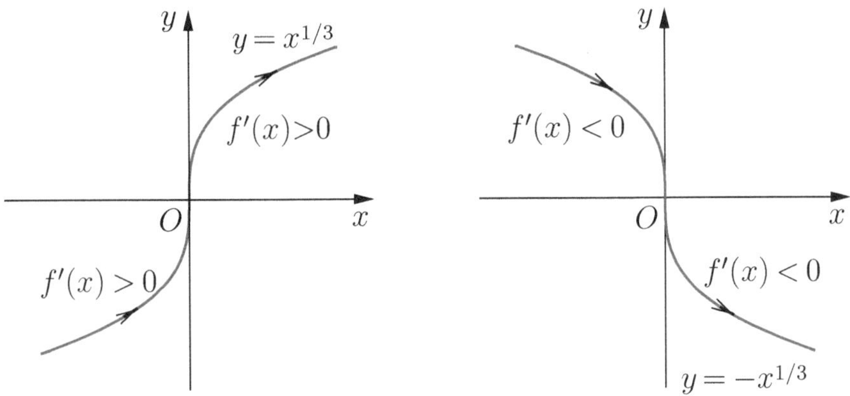

Fig. 4.7. No extrema.

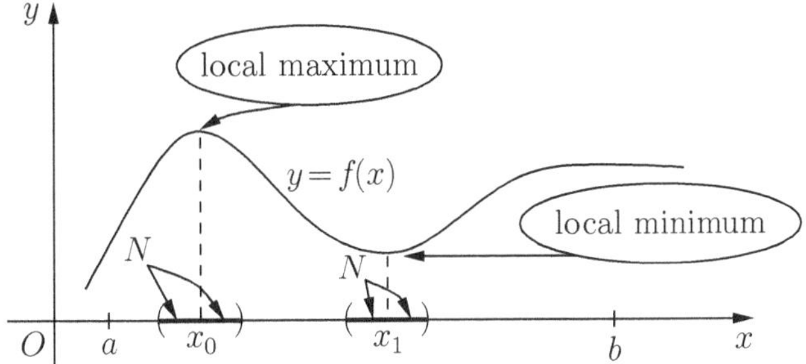

Fig. 4.8. Existence of local extrema.

First, considering the points to the left of c, we have (see Figure 4.9)

$$\frac{f(x)-f(c)}{x-c} \geq 0 \quad \text{for } c-\delta < x < c,$$

so that

$$f'_{-}(c) = \lim_{x \to c-} \frac{f(x)-f(c)}{x-c} \geq 0.$$

Next, considering points to the right of c, we have

$$\frac{f(x)-f(c)}{x-c} \leq 0 \quad \text{for } c < x < c+\delta,$$

so that (see Figure 4.9)

$$f'_{+}(c) = \lim_{x \to c+} \frac{f(x)-f(c)}{x-c} \leq 0.$$

Since $f'(c)$ exists, the left and the right derivatives at c exist and are equal. Consequently, $f'(c) = 0$. ∎

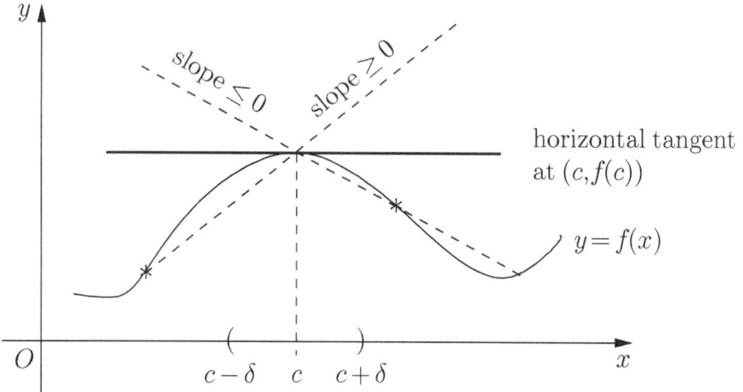

Fig. 4.9. Left and right hand derivatives.

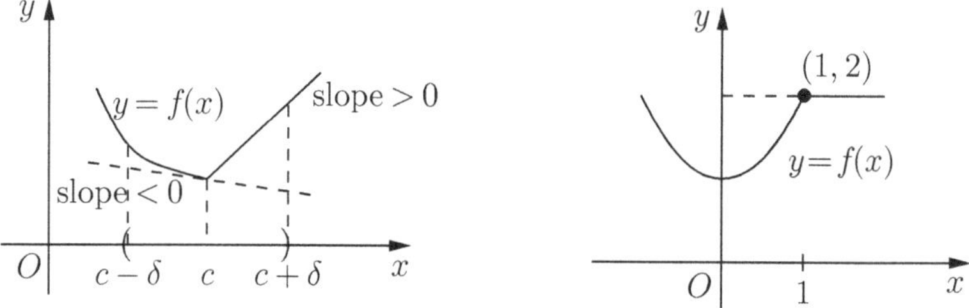

Fig. 4.10. f is not differentiable at $x = c$.

Fig. 4.11. Graph of f given by (4.2).

For instance, the function $f(x) = x + 1/x$ $(x \neq 0)$ has a local minimum at 1, since

$$x + \frac{1}{x} \geq 2 \quad \text{for all } x \text{ near } 1.$$

For $x < 0$, this inequality is no longer valid. Clearly, the function $f(x) = x + 1/x$ has no global minimum and no global maximum.

Remark 4.13. (a) The function $f(x) = |x|$ has a local minimum at 0 although f is not differentiable at the origin. This demonstrates that a function may have a local extremum at a point without the function being differentiable at that point.

(b) The function $f(x) = x^3$, $x \in [-1, 1]$, does not have a local extremum at the origin although $f'(0) = 0$. We see that Theorem 4.12 does not assert that a point c where $f'(c) = 0$ is necessarily a local extremum. That is, the converse of Theorem 4.12 is false. ●

Figure 4.10 shows that f is not differentiable at c, since $f'_-(c)$ and $f'_+(c)$ are not equal.

Now, we consider the function (see Figure 4.11)

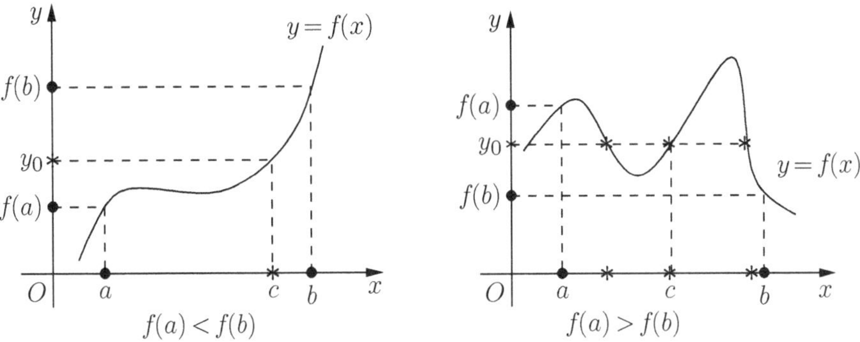

Fig. 4.12. Sketch for the intermediate value theorem.

$$f(x) = \begin{cases} x^2 + 1 & \text{for } x < 1, \\ 2 & \text{for } x \geq 1. \end{cases} \tag{4.2}$$

Then f is continuous on $\mathbb{R}$ and differentiable everywhere except at $x = 1$. We see that

$$f'(x) = \begin{cases} 2x & \text{for } x < 1, \\ 0 & \text{for } x > 1, \end{cases}$$

whereas

$$f'_{-}(1) = \lim_{h \to 0-} \frac{f(1+h) - f(1)}{h} = \lim_{h \to 0-} \frac{(1+h)^2 + 1 - 2}{h} = 2$$

and

$$f'_{+}(1) = \lim_{h \to 0+} \frac{f(1+h) - f(1)}{h} = \lim_{h \to 0+} \frac{0}{h} = 0.$$

Does it have a local extremum at 1?

4.2.2 Rolle's Theorem and the Mean Value Theorem

First we recall the intermediate value theorem (see Figure 4.12), which is one of the most important theoretical tools used to prove a number of results in calculus. We omit its proof, since it depends on the completeness property of the real numbers. However, a geometric proof suggests the truth of it.

Theorem 4.14 (Darboux/Intermediate value property). *Let $a < b$ and and let $f : [a, b] \to \mathbb{R}$ be continuous. Suppose that y_0 is a point that lies in the open interval with endpoints $f(a)$ and $f(b)$. Then there exists at least one point $c \in (a, b)$ such that $f(c) = y_0$.*

For example, by Theorem 4.14, for each $\alpha > 1$ the equation $\mathrm{e}^x = \alpha$ has a solution (it suffices to observe that $\mathrm{e}^0 = 1 < \alpha = y_0 < \mathrm{e}^\alpha$ and $f(x) = \mathrm{e}^x$).

Theorem 4.14 explains why the graphs of differentiable functions possess certain geometric properties.

The mean value theorem plays an important role in the differential and integral calculus of a single variable. We state both Rolle's theorem and the mean value theorem.

Theorem 4.15 (Rolle's theorem). *Suppose $f : [a,b] \to \mathbb{R}$ is continuous on the (closed, bounded) interval $[a,b]$ and differentiable on (a,b). If $f(a) = f(b)$, then there exists at least one point $c \in (a,b)$ with $f'(c) = 0$.*

Proof. The proof of this theorem is simple. Consider (see Figure 4.15)

$$F(x) = f(x) - f(a).$$

Then F is continuous on $[a,b]$, differentiable on (a,b), and $F(a) = F(b) = 0$. If $F(x) = 0$ on (a,b), then $F'(x) = f'(x) = 0$ for all $x \in (a,b)$, so that the result is trivial, and in this case, we may choose c to be any point in (a,b).

Suppose $F(x) \neq 0$ for some $x \in (a,b)$. Then F, being continuous on a closed and bounded interval $[a,b]$, assumes its maximum and minimum, say at x_1 and x_2, respectively. Since $F(x)$ is not identically zero on $[a,b]$, and $F(a) = F(b) = 0$, at least one of x_1 and x_2 must belong to (a,b), say $x_1 \in (a,b)$. The local extremum theorem applied to the point x_1 shows that $F'(x_1) = f'(x_1) = 0$, as desired. ∎

Remark 4.16. We observe the following:

- Rolle's theorem guarantees that the point c exists somewhere, although it gives no indication of how to find such a point. Figure 4.15 makes this point geometrically: if in the graph of f, the line segment connecting $(a, f(a))$ and $(b, f(b))$ is parallel to the horizontal line, then so is the tangent to the graph of f at some point on the interval (a,b).
- If $f(x)$ is a polynomial such that $f(a) = f(b) = 0$, then $f'(x) = 0$ has a root in (a,b).
- In the statement of Rolle's theorem, the differentiability of f on (a,b) is essential (see Figure 4.13). For instance, if

$$f(x) = 1 - |x| \quad \text{for } x \in [-1,1],$$

  then f is continuous on $[-1,1]$ and differentiable everywhere on $(-1,1)$ except at the interior point $x = 0$. However, there exists no point x such that $f'(x) = 0$.
- In the statement of Rolle's theorem, the continuity of f on the closed interval $[a,b]$ is essential. In particular, continuity at the endpoints is necessary (see Figure 4.14). •

For example, if we apply Rolle's theorem to $f(x) = (x-1)e^x - x$ on $[0,1]$, it follows that the equation $xe^x - 1 = 0$ has exactly one root in the interval $(0,1)$.

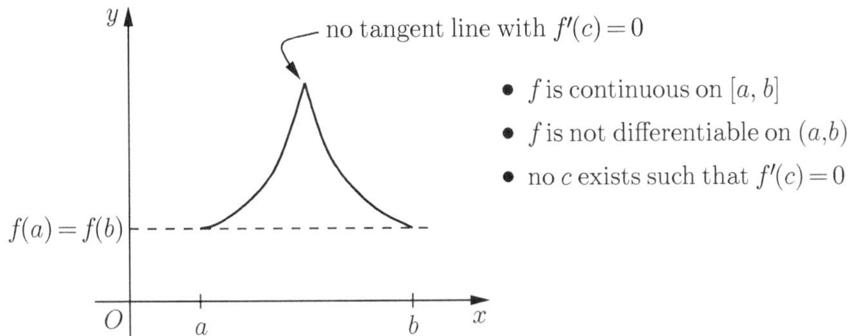

Fig. 4.13. Differentiability on (a, b) is necessary in Rolle's theorem.

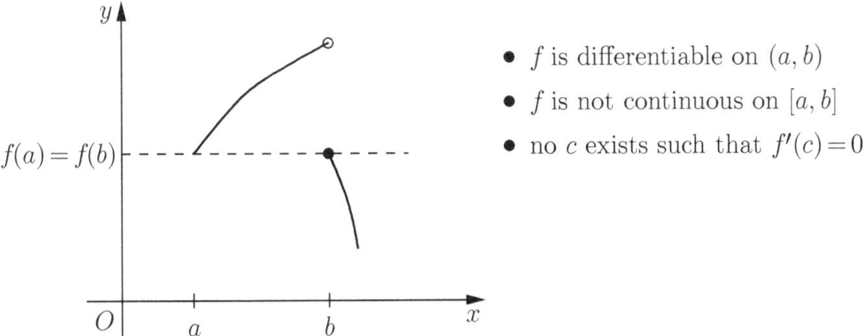

Fig. 4.14. Continuity on $[a, b]$ is necessary in Rolle's theorem.

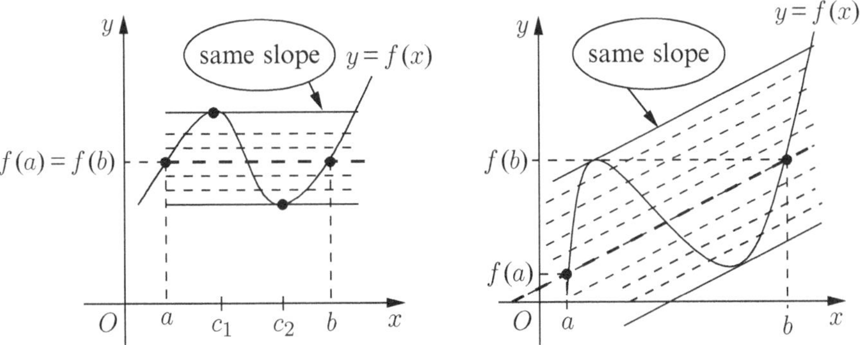

Fig. 4.15. Sketch for Rolle's theorem and the mean value theorem.

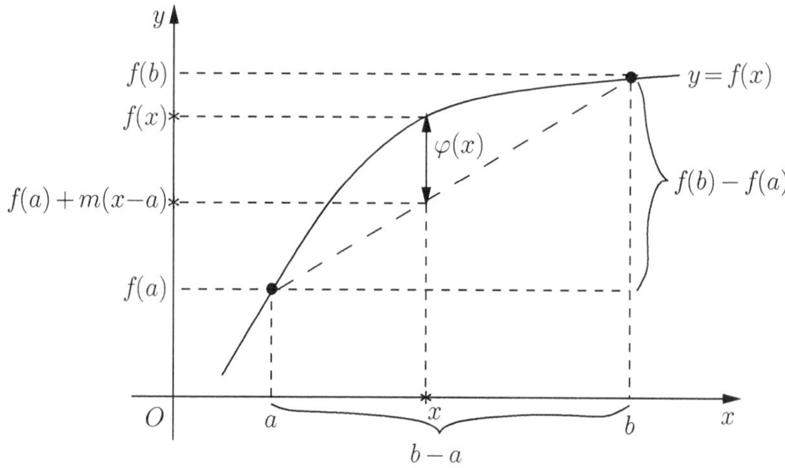

Fig. 4.16. Sketch for the proof of the mean value theorem.

Theorem 4.17 (Mean value theorem). *If $f : [a, b] \to \mathbb{R}$ $(a < b)$ is differentiable on (a, b) and continuous on the closed interval $[a, b]$, then there exists at least one point $c \in (a, b)$ such that*

$$\frac{f(b) - f(a)}{b - a} = f'(c). \tag{4.3}$$

Proof. To apply Rolle's theorem, we just need a linear function that maps the line through the points $(a, f(a))$ and $(b, f(b))$ to the points $(a, 0)$ and $(b, 0)$, respectively. The required function is given by

$$y = f(a) + m(x - a), \quad m = \frac{f(b) - f(a)}{b - a},$$

which is indeed the equation of the chord joining the points $(a, f(a))$ and $(b, f(b))$. The number m is the slope of this chord. Now define (see Figures 4.15 and 4.16)

$$\phi(x) = f(x) - [f(a) + m(x - a)].$$

Then ϕ is continuous on $[a, b]$ and differentiable on (a, b) with $\phi(a) = \phi(b) = 0$. By Rolle's theorem, $\phi'(c) = 0$ for some $c \in (a, b)$. This condition gives the desired result. ∎

Example 4.18. Consider $f(x) = |x| + |x - 1|$ for $x \in [-2, 2]$. Then (see Figure 4.17) we may simplify it as

$$f(x) = \begin{cases} 1 - 2x & \text{for } -2 \le x \le 0, \\ 1 & \text{for } 0 \le x \le 1, \\ 2x - 1 & \text{for } 1 \le x \le 2. \end{cases}$$

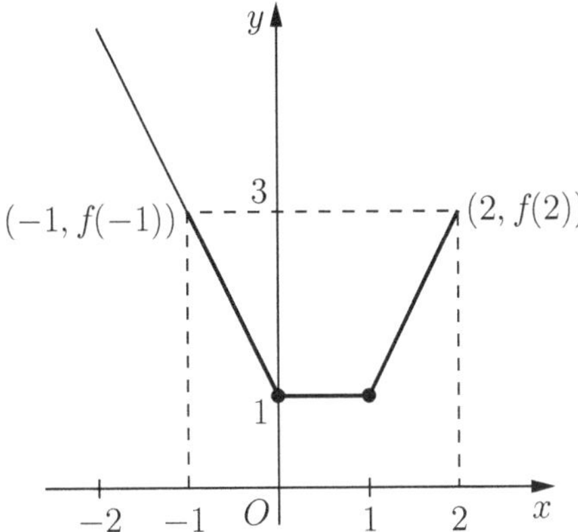

Fig. 4.17. The graph of $f(x) = |x| + |x-1|$ for $x \in [-2, 2]$.

Clearly, f is continuous on $[-2, 2]$ but is differentiable on $(-2, 2)$ except at 0 and 1. Note that

$$\frac{f(2)-f(-2)}{2-(-2)} = -\frac{1}{2}, \quad f'(x) = \begin{cases} -2 & \text{for } -2 < x < 0, \\ 0 & \text{for } 0 < x < 1, \\ 2 & \text{for } 1 < x < 2. \end{cases}$$

There exists no $c \in (-2, 2)$ satisfying (4.3). On the other hand,

$$f(2) = f(-1) = 3 \quad \text{and} \quad f'(x) = 0 \quad \text{on } (0, 1),$$

showing that $f'(x) = 0$ for points on $(-1, 2)$ without satisfying all the required conditions of Rolle's theorem for f on $[-1, 2]$. ●

Theorem 4.17 involves a single function, namely f, and is due to Lagrange (1736–1813). There is another result (see Theorem 4.26) that involves two functions and is due to Cauchy (1789–1857). Consider the expression (4.3). The term on the left-hand side of (4.3), namely $f'(c)$, is the slope of the tangent line at $(c, f(c))$, where c is some point in (a, b), whereas the expression on the right-hand side, namely the number $(f(b) - f(a))/(b - a)$, is the slope of the secant line that passes through the endpoints $(a, f(a))$ and $(b, f(b))$. This equation says that the slopes are the same. That is, the secant line (chord) between the endpoints $(a, f(a))$ and $(b, f(b))$ and the tangent line at $(c, f(c))$ are parallel; see Figure 4.15. Note that we are given no indication how to find the point c.

Remark 4.19. Consider an equation of motion

$$s = f(t), \quad t \in [a, b],$$

so that $f(t)$ represents the position of a moving point at time t. Then $\Delta s = f(b) - f(a)$ represents the change in s corresponding to $\Delta t = b - a$, the change in time from $t = a$ to $t = b$, so that

$$\frac{\Delta s}{\Delta t} = \frac{f(b) - f(a)}{b - a},$$

which is the average velocity over the time interval $[a, b]$. The mean value theorem then asserts that there is an instant $t = c$ between a and b at which the instantaneous velocity $f'(c)$ at c equals the average velocity. For example, by the mean value theorem, we see that if a motorist makes a trip with average velocity 30 kilometers per hour, then at least once during the trip, his speedometer must have registered precisely 30 kilometers per hour. ●

Also, we observe the following important points:

- There may be more than one value of c satisfying the conclusion of the mean value theorem (as in Figure 4.18).

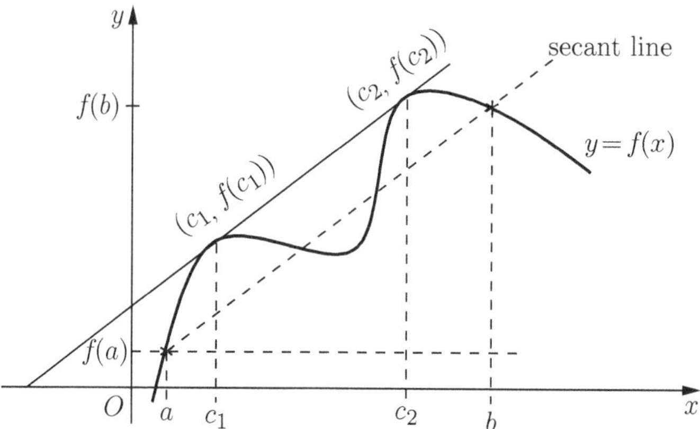

Fig. 4.18. The existence of more than one c in the mean value theorem.

- The conclusion may be wrong if not all the hypotheses are met. For example, the graph of f described in Figure 4.19 is not continuous at a, and no tangent line can be parallel to the secant line. In Figure 4.20, the graph of f is not continuous at b, and there exists no tangent line parallel to the secant line. In Figure 4.21, the secant line is parallel to the x-axis, whereas there exists no horizontal tangent. So the mean value theorem fails if f is not continuous at all points of $[a, b]$. In Figure 4.22, f is not differentiable at 0, whereas the secant line is horizontal, but at no point does there exist a horizontal tangent.
- If $c \in (a, b)$, then $c - a \in (0, b - a)$, and therefore

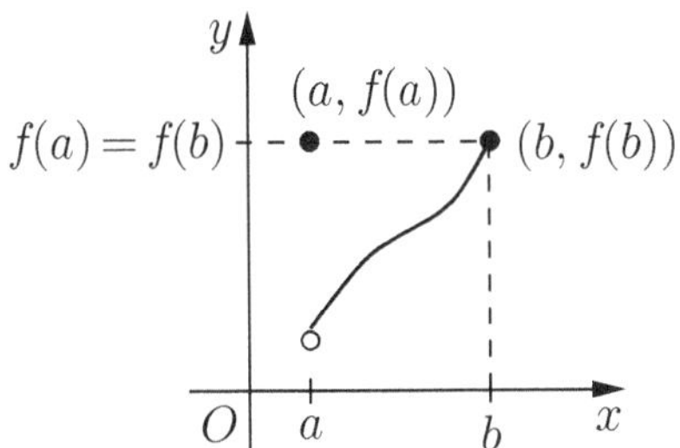

- slope of the secant line is $\frac{f(b)-f(a)}{b-a} = 0$, i.e. secant line is parallel to x-axis
- there exists no c in (a, b) such that $f'(c) = 0$

Fig. 4.19. Necessity of the continuity of f at the endpoints.

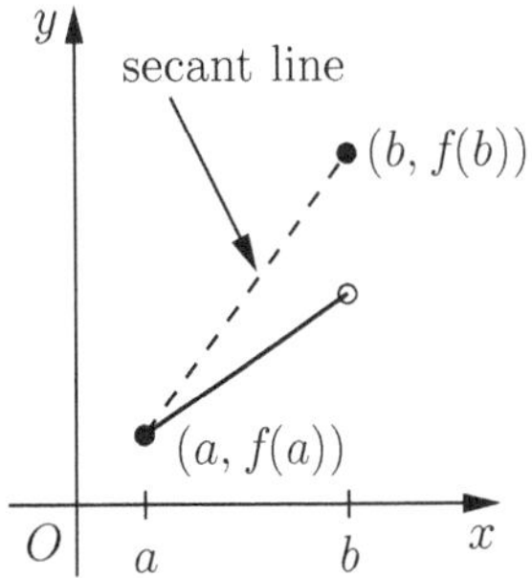

- $f'(x)$ is constant on (a, b)
- $\frac{f(b)-f(a)}{b-a}$ depends on the choice of $f(b)$ which may be at our disposal

Fig. 4.20. Necessity of the continuity of f at the endpoints.

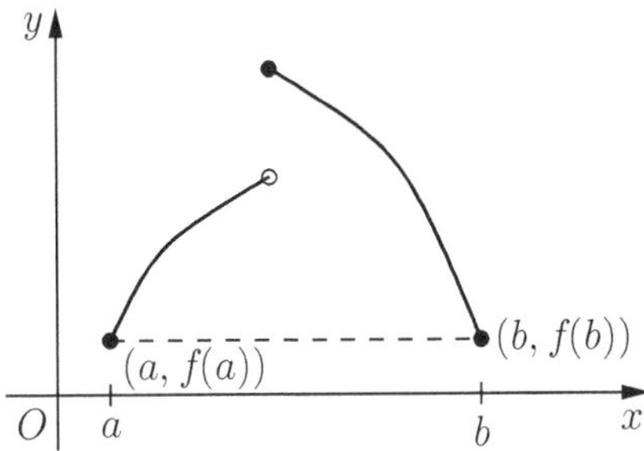

Fig. 4.21. Continuity of f inside $[a, b]$ is essential.

$$c - a = \theta(b - a) \quad \text{or} \quad c = a + \theta(b - a),$$

where θ is some real number in the interval $(0, 1)$. But then the conclusion of the mean value theorem may be written as

$$\frac{f(b) - f(a)}{b - a} = f'(a + \theta(b - a)) \quad \text{for some } \theta \in (0, 1).$$

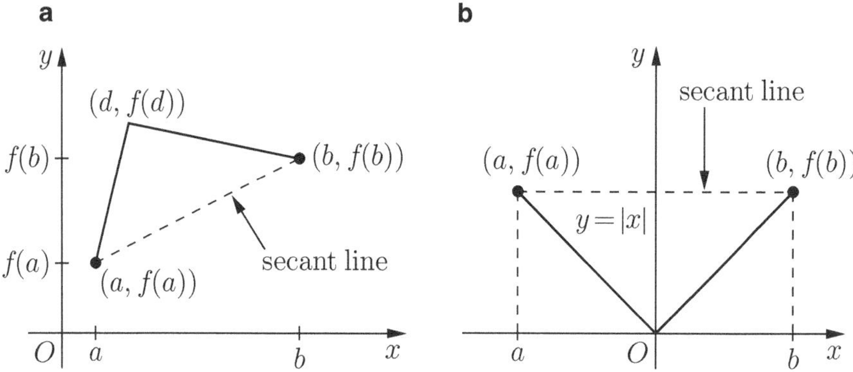

Fig. 4.22. Differentiability of f on (a, b) is essential.

Example 4.20. Consider $f(x) = 1/x$ on $[-1, 1]\backslash\{0\}$. Then

$$\frac{f(b) - f(a)}{b - a} = \frac{f(1) - f(-1)}{1 - (-1)} = 1 \quad \text{and} \quad f'(c) = -\frac{1}{c^2}.$$

On the other hand, there exists no c in $(-1, 1)\backslash\{0\}$ such that $-1/c^2 = 1$. Thus (4.3) has no solution. Does this contradict the mean value theorem by chance? If not, what does this convey? ●

Corollary 4.21. *If f is differentiable on (a, b) such that $f'(x) = 0$ on (a, b), then f is constant on (a, b).*

Proof. Let x_1 and x_2 be any two points in (a, b) such that

$$a < x_1 < x_2 < b.$$

Then f is differentiable on (x_1, x_2) and continuous on $[x_1, x_2]$, and so by the mean value theorem,

$$f(x_2) - f(x_1) = (x_2 - x_1)f'(c) \tag{4.4}$$

for some $c \in (x_1, x_2)$. This relationship, because $f'(c) = 0$, gives

$$f(x_2) = f(x_1),$$

which is true for arbitrary points x_1 and x_2 in (a, b). Thus, f is constant on (a, b). ■

Example 4.22. Suppose that $f : \mathbb{R} \to \mathbb{R}$ is differentiable on $\mathbb{R}$ and satisfies $f(x + y) = f(x) + f(y)$ for all $x, y \in \mathbb{R}$. Show that $f(x) = f'(0) \cdot x$ on $\mathbb{R}$.

Solution. Consider $f(x+y) = f(x) + f(y)$. If we take $x = 0 = y$, we have $f(0) = 2f(0)$, i.e., $f(0) = 0$. Next,

$$\lim_{h\to 0} \frac{f(x+h) - f(x)}{h} = \lim_{h\to 0} \frac{f(x) + f(h) - f(x)}{h} = \lim_{h\to 0} \frac{f(h) - f(0)}{h} = f'(0),$$

so that $f'(x) = f'(0)$. This gives $f(x) = f'(0)x$. ●

From Corollary 4.21, we can conclude that if $f'(x)$ exists for each $x \in [a,b]$ and if $f'(x) \neq 0$ for all $x \in [a,b]$, then f is one-to-one on $[a,b]$. There are other consequences of the mean value theorem.

Corollary 4.23 (First derivative test). *Suppose that f is differentiable on (a,b). Then we have the following:*

(a) *If $f'(x) > 0$ on (a,b), then f is strictly increasing on (a,b).*
(b) *If $f'(x) < 0$ on (a,b), then f is strictly decreasing on (a,b).*
(c) *If $f'(x) \geq 0$ on (a,b), then f is increasing on (a,b).*
(d) *If $f'(x) \leq 0$ on (a,b), then f is decreasing on (a,b).*

Proof. Apply (4.4). ■

Using this corollary, we see that $\sin x \leq x$ for all $x \geq 0$. We now include a few more examples to appreciate Rolle's theorem and the mean value theorem geometrically.

Example 4.24. For $x \in (0, \pi/2)$, we have (see also Example 3.14)

$$\cos x < \frac{\sin x}{x} < 1.$$

In particular, by the squeeze rule,

$$\lim_{x\to 0+} \frac{\sin x}{x} = 1 \quad \text{and} \quad \lim_{x\to 0-} \frac{\sin x}{x} = 1.$$

Solution. Consider $f(x) = \sin x - x$ on $[0, \pi/2]$. Then f is continuous on $[0, \pi/2]$ and differentiable on $(0, \pi/2)$ with

$$f'(x) = \cos x - 1 < 0 \quad \text{on } (0, \pi/2),$$

so that f is decreasing on $(0, \pi/2)$. Thus,

$$f(x) < f(0), \quad \text{i.e.,} \quad \frac{\sin x}{x} < 1 \text{ on } (0, \pi/2).$$

Similarly, if $g(x) = \tan x - x$ for $x \in [0, \pi/2)$, then $g'(x) = \sec^2 x - 1 > 0$ on $(0, \pi/2)$, so that

$$g(x) > g(0) = 0, \quad \text{i.e.,} \ \tan x > x \text{ or } \frac{\sin x}{x} > \cos x \text{ on } (0, \pi/2).$$

The inequality also holds on $(-\pi/2, 0)$ because $\cos x$ and $(\sin x)/x$ are even functions. ●

Example 4.25. Consider $f(x) = x^3$ on $[-1, 1]$. Then

$$m = \frac{f(b) - f(a)}{b - a} = \frac{f(1) - f(-1)}{1 - (-1)} = 1 \quad \text{and} \quad f'(c) = 3c^2.$$

The mean value theorem shows the existence of $c \in (-1, 1)$ satisfying the condition (4.3), i.e., $f'(c) = m$. This gives $3c^2 = 1$, i.e., $c = \pm 1/\sqrt{3}$. It follows that there are two values of c, namely $c_1 = 1/\sqrt{3}$ and $c_2 = -1/\sqrt{3}$, satisfying the condition

$$\frac{f(1) - f(-1)}{1 - (-1)} = f'(c).$$

Note that the tangent to the graph of f at $(c_1, f(c_1))$ and $(c_2, f(c_2))$ is parallel to the chord joining $(-1, f(-1))$ and $(1, f(1))$; see Figure 4.23. ●

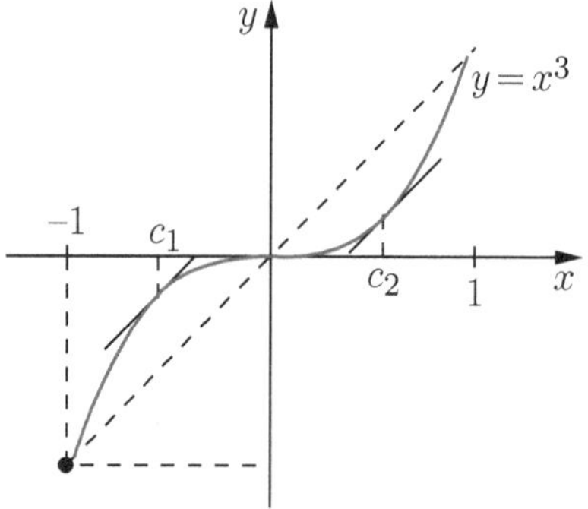

Fig. 4.23. The graph of $f(x) = x^3$ on $[-1, 1]$.

Rolle's theorem and the Langrange mean value theorem together with the following generalization of the mean value theorem are all logically equivalent statements:

Theorem 4.26 (Generalized mean value theorem). *If $f, g : [a, b] \to \mathbb{R}$ $(a < b)$ are differentiable on (a, b) and continuous at a and b, then there exists a point $c \in (a, b)$ such that*

$$f'(c) \cdot (g(b) - g(a)) = g'(c) \cdot (f(b) - f(a)).$$

Proof. The proof follows from an application of Rolle's theorem to

$$h(x) = f(x)(g(b) - g(a)) - g(x)(f(b) - f(a)).$$

■

Theorem 4.26 is also called *Cauchy mean value theorem*. It is worth noting that the Lagrange mean value theorem is a special case of the Cauchy mean value theorem for $g(x) = x$. In particular, if $g'(x) \neq 0$ for $x \in (a,b)$, then the mean value theorem shows that $g(b) - g(a) \neq 0$, and therefore the Cauchy mean value theorem can be written in the form

$$\frac{f(b) - f(a)}{g(b) - g(a)} = \frac{f'(c)}{g'(c)},$$

since the quotients make sense in this case. Among the important and immediate consequences of these theorems are the following. The reader should have no trouble in providing a detailed proof.

- Choosing $g(x) = x$ in Theorem 4.26 recovers the classical mean value theorem.
- If $f'(x) = g'(x)$ on an open interval, then f and g must differ by a constant value on that interval.

At this point it would be appropriate to pose the following problems.

Problem 4.27. *Are there any other applications of the mean value theorem?*

Theorem 4.28. *Suppose that f is differentiable on $\mathbb{R}$ such that $|f'(x)| \leq \lambda < 1$ for all $x \in \mathbb{R}$. Then $f(x) = x$ has a unique solution.*

Proof. Suppose that there exist x and x' such that $f(x) = x$ and $f(x') = x'$, so that by the mean value theorem,

$$x - x' = f'(c)(x - x')$$

for some c between x and x'. Since $|f'(c)| < 1$, we must have $x = x'$. Thus, the solution is unique.

To prove the existence of the solution, we begin with an arbitrary x_0 and consider $x_n = f(x_{n-1})$ for $n \geq 1$. The mean value theorem then gives

$$x_{n+1} - x_n = f(x_n) - f(x_{n-1}) = f'(c_n)(x_n - x_{n-1}),$$

so that

$$|x_{n+1} - x_n| \leq \lambda |x_n - x_{n-1}|.$$

By Theorem 2.57, $\{x_n\}$ converges to x, say. The continuity of f implies that $x = f(x)$. ∎

4.2.3 L'Hôpital's Rule: Another Form

Theorem 4.29 (L'Hôpital's rule of $0/0$ form). *Let f and g be differentiable on $(a, \delta]$ such that $g'(x) \neq 0$ on $(a, \delta]$, and*

$$\lim_{x \to a+} f(x) = 0 = \lim_{x \to a+} g(x). \tag{4.5}$$

Then

$$\lim_{x\to a+}\frac{f(x)}{g(x)}=\lim_{x\to a+}\frac{f'(x)}{g'(x)},$$

provided the limit on the right exists.

Proof. We assume that

$$\lim_{x\to a+}\frac{f'(x)}{g'(x)}=\ell \quad \text{for some } \ell\in\mathbb{R}.$$

By (4.5), both f and g will be continuous at a if we define $f(a)=0=g(a)$. It follows from the generalized mean value theorem that there exists a point $c\in(a,x)$ such that

$$\frac{f'(c)}{g'(c)}=\frac{f(x)-f(a)}{g(x)-g(a)}=\frac{f(x)}{g(x)},$$

where $x\in(a,\delta)$, and c of course depends on x.

Now let $x\to a+$. Since $a<c<x$, it follows that $c\to a+$ too. Consequently,

$$\lim_{x\to a+}\frac{f(x)}{g(x)}$$

exists and has the value ℓ. The theorem follows. ■

It is easy to formulate another variant of l'Hôpital's rule (see Theorem 3.55).

Theorem 4.30 (L'Hôpital's rule of 0/0 form). *Let f and g be differentiable on (a,b) such that $g'(x)\neq 0$ on (a,b), and*

$$\lim_{x\to b-}f(x)=0=\lim_{x\to b-}g(x).$$

Then

$$\lim_{x\to b-}\frac{f(x)}{g(x)}=\lim_{x\to b-}\frac{f'(x)}{g'(x)},$$

provided the limit on the right exists.

Remark 4.31. **(a)** In the hypothesis of Theorem 4.29, a may be $-\infty$, and in the hypothesis of Theorem 4.30, b may be ∞.

(b) In the conclusion of Theorem 4.29, the limit $\ell=\lim_{x\to a+}(f'(x)/g'(x))$ need not be finite. A similar observation holds for $\lim_{x\to b-}(f'(x)/g'(x))$ in Theorem 4.30.

(c) There are many other forms of l'Hôpital's rule. However, the forms stated here are sufficient for most applications. ●

Also, it is easy to derive the following form.

Theorem 4.32 (L'Hôpital's rule of ∞/∞ form). *Let f and g be differentiable on (a,b) such that $g'(x) \neq 0$ on (a,b), and*

$$\lim_{x\to a+} f(x) = \infty = \lim_{x\to a+} g(x).$$

Then

$$\lim_{x\to a+} \frac{f(x)}{g(x)} = \lim_{x\to a+} \frac{f'(x)}{g'(x)},$$

provided the limit on the right exists.

Proof. We leave the proof as a simple exercise. ∎

Example 4.33. For $x > 0$, let $f(x) = x^{1/x} = \exp((1/x)\log x)$. Then f is continuous on $(0,\infty)$. By l'Hôpital's rule, we obtain

$$\lim_{x\to\infty} \frac{\log x}{x} = \lim_{x\to\infty} \frac{1/x}{1} = 0,$$

and since the exponential function is continuous, we conclude that

$$\lim_{x\to\infty} f(x) = \mathrm{e}^0 = 1.$$

Thus,

$$\lim_{n\to\infty} f(n) = \lim_{n\to\infty} n^{1/n} = 1.$$ •

4.2.4 Second-Derivative Test and Concavity

The second-derivative test introduced below can be used to determine whether a value of c such that $f'(c) = 0$ provides a local minimum or a local maximum for $f(x)$.

Theorem 4.34 (Second-derivative test for relative extrema). *Let f be a function defined in an open interval containing c such that $f'(c) = 0$. Then we have the following:*

- *If $f''(c) > 0$, then $f(c)$ is a local minimum for f,*
- *If $f''(c) < 0$, then $f(c)$ is a local maximum for f,*
- *If $f''(c) = 0$, then the test is inconclusive (a maximum or a minimum or neither may occur).*

Proof. Suppose that f is defined in a neighborhood of c such that

$$f'(c) = 0 \quad \text{and} \quad f''(c) > 0.$$

Because $(f')'(c) > 0$, there exists a punctured neighborhood N of c (say $N = (c-\delta, c) \cup (c, c+\delta)$) such that

$$\frac{f'(x)}{x-c} = \frac{f'(x)-f'(c)}{x-c} > 0 \quad \text{for all } x \in N.$$

That is

$$\begin{cases} f'(x) < 0 & \text{for } c-\delta < x < c, \\ f'(x) = 0 & \text{for } x = c, \\ f'(x) > 0 & \text{for } c < x < c+\delta. \end{cases}$$

It follows that f has a local minimum at c (see Figure 4.24).

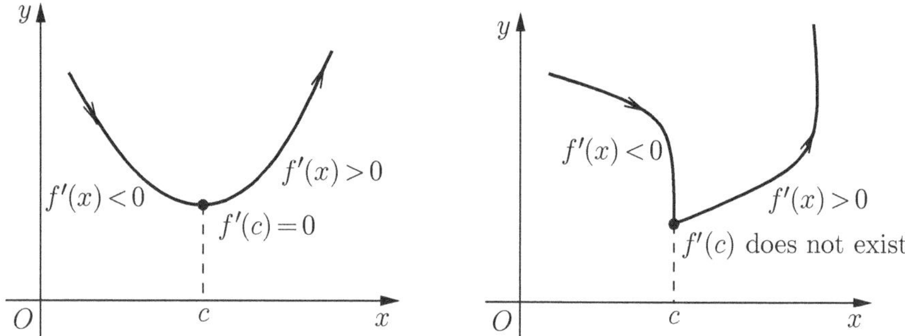

Fig. 4.24. A relative minimum.

A similar argument (or apply the above arguments for $-f$) shows that if $f'(c) = 0$ and $f''(c) < 0$, then f has a local maximum at c (see Figure 4.25). ∎

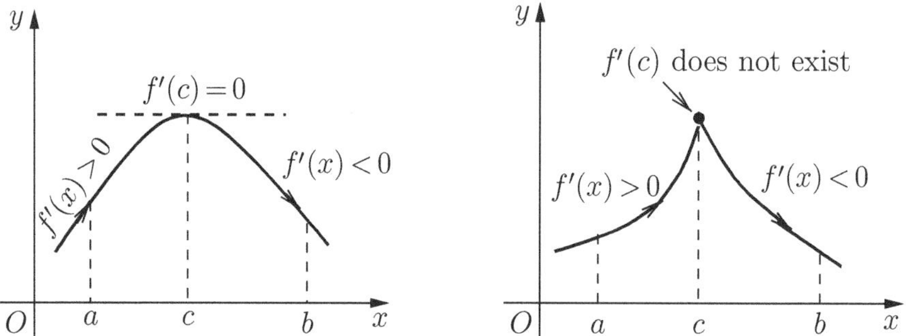

Fig. 4.25. A relative maximum.

Definition 4.35 (Concavity and inflection point). *We consider a curve $y = f(x)$. If the curve $y = f(x)$ faces up on (a, b) (i.e., if all points of the curve lie above any tangent to it on the interval), then we say that the curve* *is concave up (or equivalently convex downward). Similarly, we say that the curve $y = f(x)$ is concave down (or equivalently convex upward) on (a, b) if all points of the curve lie below any tangent to it on the interval (see Figure 4.26).*

A point on the graph of $y = f(x)$ at which the concavity changes is a point of inflection.

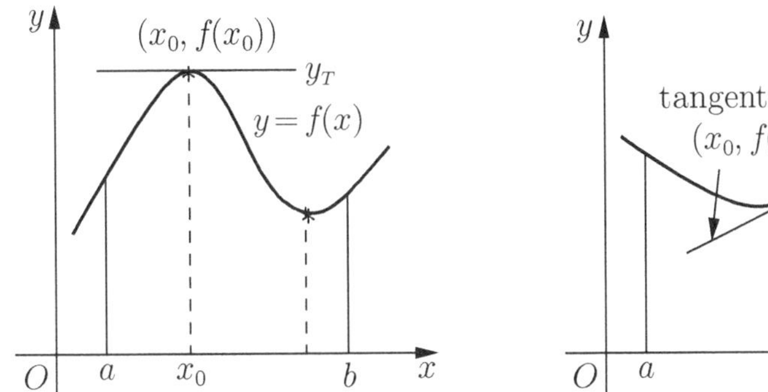

Fig. 4.26. Point of inflection at x_0.

Example 4.36. Consider $f(x) = 2x^3 + 3x^2 + 1$. Then

$$f'(x) = 6x(x+1), \quad f''(x) = 6(2x+1),$$

so that $x = 0$ and $x = -1$ are the critical values of f. Since $f(0) = 1$ and $f(-1) = 2$, $(0, 1)$ and $(-1, 2)$ are the critical points of f. Observe that

$$f''(0) = 6 > 0, \quad f''(-1) = -6 < 0,$$

and so by the second-derivative test, $(0, 1)$ is a point of local minimum and $(-1, 2)$ is a point of local maximum. Moreover,

$$\begin{cases} f'(x) > 0 & \text{for } x \in (-\infty, -1) \cup (0, \infty), \\ f'(x) < 0 & \text{for } x \in (-1, 0). \end{cases}$$

Finally, $f''(x) > 0$ for $x > -1/2$ and $f''(x) < 0$ for $x < -1/2$, and $f''(1/2) = 0$. Note that the inflection points occur where the sign of $f''(x)$ changes. Thus, there is a point of inflection on the curve at $x = -1/2$, with coordinates $(-1/2, f(-1/2))$ (see Figure 4.27). •

For instance, if $f(x) = x^2$, then $f''(x) = 2 > 0$ on $(-\infty, \infty)$, and so the curve $y = f(x)$ is concave up on $(-\infty, \infty)$.

Theorem 4.37 (Second-derivative test for concavity). *The graph of a twice differentiable function $y = f(x)$ is*

(a) *concave up on an interval I if $f''(x) > 0$ on I;*

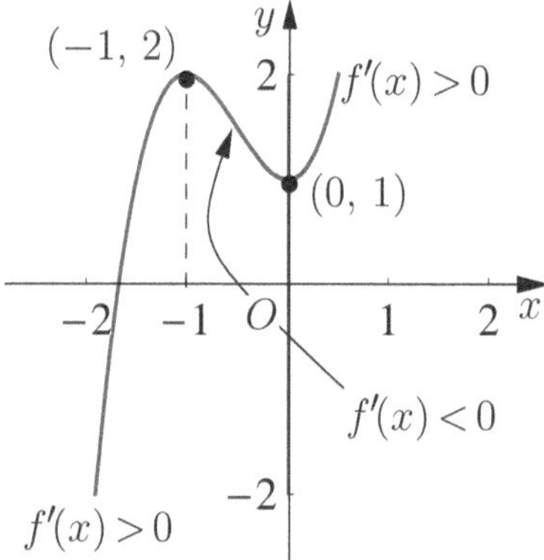

Fig. 4.27. Local extremum behavior of $f(x) = 2x^3 + 3x^2 + 1$ at $(-1, 2)$ and $(0, 1)$.

(b) *convex down on an interval I if $f''(x) < 0$ on I.*

Proof. Assume that $f''(x) > 0$ on $I = (a, b)$ and $x_0 \in (a, b)$ is an arbitrary point. The equation of the tangent to the curve $y = f(x)$ at $(x_0, f(x_0))$ is given by

$$y_T = f(x_0) + f'(x_0)(x - x_0),$$

so that

$$y - y_T = f(x) - f(x_0) - f'(x_0)(x - x_0).$$

By the mean value theorem, the last equation takes the form

$$y - y_T = (f'(c) - f'(x_0))(x - x_0),$$

where c is a point between x_0 and x. Applying the mean value theorem for f' on the interval with endpoints c and x_0 yields that

$$y - y_T = f''(c_1)\underbrace{(c - x_0)(x - x_0)},$$

where c_1 is a point lying between c and x_0. In either case (as shown in Figure 4.28), we see that the factor $(c - x_0)(x - x_0)$ is positive. Consequently (see Figure 4.26),

$$\begin{cases} y - y_T > 0 \iff f''(c_1) > 0, \\ y - y_T < 0 \iff f''(c_1) < 0, \end{cases}$$

and the result follows. ■

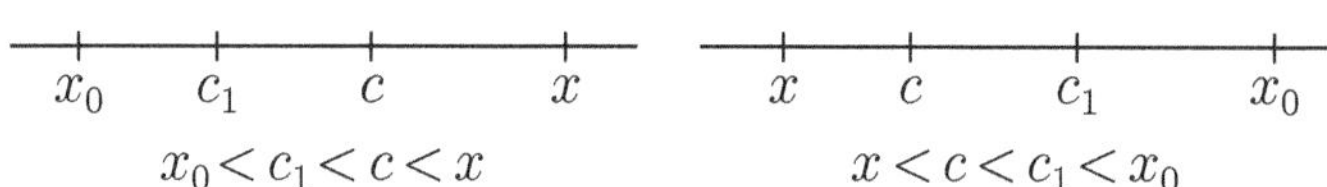

Fig. 4.28. Sign of $(c - x_0)(x - x_0)$.

Theorem 4.38. *Suppose that the graph of a continuous function $y = f(x)$ is such that $f''(a) = 0$ or $f''(a)$ does not exist, and the derivative $f''(x)$ changes sign when passing through $x = a$. Then the point $(a, f(a))$ is a point of inflection.*

Proof. Obvious from the hypothesis. ∎

For instance, we have the following:

(a) If $f(x) = x^{1/3}$, then for $x \neq 0$,

$$f'(x) = \frac{1}{3}x^{-2/3} \quad \text{and} \quad f''(x) = -\frac{2}{9}x^{-5/3} = \begin{cases} > 0 & \text{for } x < 0, \\ < 0 & \text{for } x > 0, \end{cases}$$

so that $f''(x)$ does not exist at $x = 0$ and therefore $y = x^{1/3}$ has a point of inflection at $x = 0$.

(b) For the function $f(x) = x^4$, $x = 0$ is not an inflection point although $f''(0) = 0$. We observe that $f''(x) = 12x^2$ does not change sign.

4.2.5 Questions and Exercises

Questions 4.39.

1. Does every continuous function on a closed and bounded interval $[a, b]$ attain global extrema?
2. Assume that $f, g : [a, b] \to \mathbb{R}$ are continuous functions such that $f(a) < g(a)$ and $f(b) > g(b)$. Does there exist a point $c \in (a, b)$ such that $f(c) = g(c)$?
3. Suppose that $x = a$ is a point of discontinuity of a function $f(x)$. Can both $f(a+)$ and $f(a-)$ exist? If both limits exist, can $f(a+) = f(a-)$ be true?
4. Suppose that f' does not change sign. Can a maximum or a minimum occur?
5. If f' changes its sign from positive to negative, can there be a maximum?
6. If f' changes its sign from negative to positive, can f have a minimum?
7. Suppose that $f : (a, b) \to \mathbb{R}$ is continuous and has no local extrema on (a, b). Must f be strictly monotone on (a, b)?
8. Suppose that $f : [a, b] \to [a, b]$ is continuous. Can there exist a point c in $[a, b]$ such that $f(c) = c$?
9. Must there exist real polynomials p and q such that

$$q(1) \neq q(0) \quad \text{and} \quad \frac{p(1) - p(0)}{q(1) - q(0)} = \frac{p'(c)}{q'(c)}$$

for every $c \in (0, 1)$?

10. Suppose that f and g are C^1 functions on $(0,1)$ such that $\lim_{x\to 0} f(x) = \lim_{x\to 0} g(x) = 0$. Also, assume that g and g' are nonvanishing on $(0,1)$ such that $\lim_{x\to 0}(f(x)/g(x))$ exists. What can be said about the existence of the limit $\lim_{x\to 0}(f'(x)/g'(x))$?
11. Suppose that $f\colon \mathbb{R} \to \mathbb{R}$ is continuous and periodic (i.e., $f(x+\omega) = f(x)$ for some $\omega \in \mathbb{R} \setminus \{0\}$). Does f attain both a maximum and a minimum?
12. In the intermediate value property (see Theorem 4.14), is the continuity of f at each point of $[a,b]$ essential?
13. Is there a number c between 0 and x such that
$$(1+x)^n = 1 + nx(1+c)^{n-1} \quad (n \in \mathbb{N})?$$
If so, does it imply
$$\lim_{x\to 0} \frac{(1+x)^n - 1}{x} = 1?$$
14. Suppose that $f(x) = x^{n+1}$ and $g(x) = x^n$ $(n \in \mathbb{N})$ on $[a,b]$ in the generalized mean value theorem (see Theorem 4.26). What can be said about the quotient
$$\frac{n(b^{n+1} - a^{n+1})}{(n+1)(b^n - a^n)}?$$
15. Suppose $p(x) = a_0 + a_1 x + \cdots + a_n x^n$ $(a_n \neq 0)$ such that
$$a_0 + \frac{a_1}{2} + \cdots + \frac{a_n}{n+1} = 0.$$
Does there exist a point $c \in (0,1)$ such that $p(c) = 0$?
16. Let $f : \mathbb{R} \to \mathbb{R}$ be infinitely differentiable such that $f(1) = 0$, and $f^{(k)}(0) = 0$ for $k = 0, 1, \ldots, n$. Must $f^{(n+1)}(c) = 0$ for some c in $(0,1)$?
17. Suppose that $f''(x) \geq 0$ on $[a,b]$. Must
$$f\left(\frac{x+y}{2}\right) \leq \frac{f(x)+f(y)}{2} \quad \text{for all } x, y \in [a,b]?$$
18. Suppose that f is differentiable on $\mathbb{R}$ such that $f'(x) = cf(x)$ for all $x \in \mathbb{R}$ and for some real constant c. What can be said about f?
19. Suppose that f is differentiable for each $x > 0$ and $\lim_{x\to\infty} f'(x) = 0$. Must $\lim_{x\to\infty}(f(x+1) - f(x)) = 0$?
20. Suppose that f is twice differentiable on $[a,b]$ such that there exists a point $c \in (a,b)$ with $f(a) = f(b) = f(c)$. Can there exist a point $\zeta \in (a,b)$ such that $f''(\zeta) = 0$?
21. Suppose that $f : [0,1] \to \mathbb{R}$ is a function such that $f^{(n)}(x)$ exists on $[0,1]$ for $n = 1, 2, 3$ and $f(0) = f'(1) = f(1) = f'(1) = 0$. Can $f''(x)$ have a zero in (a,b)?

Exercises 4.40.

1. Let $f(x) = x^5 - 5x$, $x \in \mathbb{R}$. Determine the points of local maximum and local minimum. Does the function have a global maximum or minimum?
2. Draw the graphs of $f(x) = |x+1|$ and $g(x) = |x-1|$ on $[-3, 3]$. Also, determine the critical values of f and g.
3. Prove that among all rectangles of a given perimeter, the square has the greatest area.
4. Suppose that
$$f(x) = \begin{cases} \dfrac{1}{x} & \text{for } 0 < |x| \le 1, \\ 0 & \text{for } x = 0. \end{cases}$$
Then $f(-1) = 1$ and $f(1) = 1$. Is there a number c in $(-1, 1)$ such that $f'(c) = 1/3$? If so, prove it. If not, explain why it does not contradict the intermediate value property.
5. Consider the following functions:
   (a) $f(x) = x^3 - 3x + 1$ for $x \in [-1, 1]$.
   (b) $f(x) = 1 - 1/x$ for $1 \le x \le 9/4$.
   Verify in each case whether the mean value theorem is applicable. If yes, determine c in $(-1, 1)$ such that the tangent to the graph of f is parallel to the chord joining $(-1, f(-1))$ and $(1, f(1))$.
6. Suppose that $g(x) = f(x) + f(1-x)$ and $f''(x) > 0$ on $[0, 1]$. Show that g is decreasing on $(0, 1/2)$ and increasing on $(1/2, 1)$.
7. Suppose that $f(x) = (x-1)(x-3)$ on $[1, 3]$. Find a suitable point c for Rolle's theorem, i.e., such that $f'(c) = 0$.
8. Suppose that f and g are differentiable on (α, β) such that $[a, b] \subset (\alpha, \beta)$, $a < b$, and $f(a) = f(b) = 0$. Show that there exists a point $c \in (a, b)$ such that $f'(c) + f(c)g'(c) = 0$.
9. Consider $f(x) = \sqrt{x}$ on $[100, 102]$. Using the mean value theorem, show that
$$\frac{111}{11} < \sqrt{102} < \frac{101}{10}.$$
Apply the same principle to the interval $[100, 105]$ and compute an estimate for $\sqrt{105}$.
10. Suppose that f is differentiable on $(-1, 3)$ and $f(0) = 0, f(1) = 2 = f(2)$. Show that there exists a point $c \in (0, 2)$ such that $f'(c) = 1$, and $c' \in (0, 1)$ such that $f'(c') = 2$.
11. Use the mean value theorem to prove that
    (a) $|\cos x - \cos y| \le |x - y|$ for all $x, y \in \mathbb{R}$.
    (b) $|\sin x - \sin y| \le |x - y|$ for all $x, y \in \mathbb{R}$.
12. Suppose that f is differentiable on $\mathbb{R}$ such that $f'(x) \in [1, 2]$ for all $x \in \mathbb{R}$ and $f(0) = 0$. Must $f(x) \in [x, 2x]$ for all $x \ge 0$?
13. Show that $\sin x = x^3 - x$ has at least one solution in the interval $(\pi/4, \pi/2)$.
14. Using Rolle's theorem with $f(x) = (x-1)\sin x$ on $[0, 1]$, show that $\tan x + x - 1 = 0$ has at least one solution $x \in (0, 1)$.

15. Prove or disprove the following:
   (a) For any $\lambda \in \mathbb{R}$, the equation $x^3 - 3x^2 + \lambda = 0$ has two distinct zeros on $[0, 1]$.
   (b) For any $\lambda \in \mathbb{R}$, the equation $x^3 + 2x + \lambda = 0$ has exactly one real root in $\mathbb{R}$.
   (c) The equation $3x^3 + \sin x - 1 = 0$ has a root in the interval $[-1, 1]$.
   (d) The equation $2\sin^2 x - 2x - 1 = 0$ has a root in $\mathbb{R}$.
   (e) The equation $\cot x = x$ has a solution in the interval $(0, \pi/2)$.
16. Use the mean value theorem to prove l'Hôpital's rule (see Theorem 3.55).
17. Examine the critical values, inflection points, and local extrema of $f(x) = 3x^5 - 5x^3 + 1$.
18. Suppose that f is differentiable on $[a, b]$ such that $f'(a) \neq f'(b)$ and λ is a real number between $f'(a)$ and $f'(b)$. Prove that there exists a point $c \in (a, b)$ such that $f'(c) = \lambda$?

5

Series: Convergence and Divergence

The main goal of this chapter is to examine the theory and applications of infinite sums, which are known as *infinite series*. In Section 5.1, we introduce the concept of convergent infinite series, and discuss geometric series, which are among the simplest infinite series. We also discuss general properties of convergent infinite series and applications of geometric series. In Section 5.2, we examine various tests for convergence so that we can determine whether a given series converges or diverges without evaluating the limit of its partial sums. Our particular emphasis will be on divergence tests, and series of nonnegative numbers, and harmonic p-series. In Section 5.3, we deal with series that contain both positive and negative terms and discuss the problem of determining when such a series is convergent. In addition, we look at what can happen if we rearrange the terms of such a convergent series. We ask, *Does the new series obtained by rearrangement still converge*? A remarkable result of Riemann on conditionally convergent series answers this question in a more general form. Finally, we also deal with Dirichlet's test and a number of consequences of it.

5.1 Infinite Series of Real Numbers

We know how to add finitely many numbers. Now we are concerned with examining the existence and meaning of the value of the sum of the terms of an infinite sequence of real numbers, $\{a_n\}_{n\geq 1}$. The formal expression

$$a_1 + a_2 + a_3 + a_4 + \cdots ,$$

denoted by

$$\sum_{k=1}^{\infty} a_k,$$

is called an *infinite series* (or simply a series), with a_k the kth term of the series. We shall be able to "add" infinitely many numbers, not by usual addition, but rather by a method of finding a limit. Thus, we need to give a

S. Ponnusamy, *Foundations of Mathematical Analysis*,
DOI 10.1007/978-0-8176-8292-7_5,

precise meaning to the notion of infinite sum. For instance, if $a_k = 1/3^k$ for $k \geq 0$, then it is not hard to see that the sum of the first n terms is given by

$$S_n = 1 + \frac{1}{3} + \frac{1}{3^2} + \cdots + \frac{1}{3^{n-1}} = \frac{1-(1/3)^n}{1-1/3} = \frac{3}{2}\left(1 - \frac{1}{3^n}\right).$$

Since $1/3^n \to 0$ as $n \to \infty$, we have $S_n \to 3/2$ as $n \to \infty$, and it seems reasonable and sensible to write

$$1 + \frac{1}{3} + \frac{1}{3^2} + \cdots + \frac{1}{3^{n-1}} + \cdots = \frac{3}{2}.$$

We use this approach to define a convergent infinite series. Thus, to study the properties of an infinite series, it is natural to examine the convergence of the sequence of "partial sums" $\{S_n\}_{n\geq 1}$ defined by

$$S_n = \sum_{k=1}^{n} a_k, \quad n = 1, 2, 3, \ldots.$$

Here the finite sum S_n is called the nth partial sum of the series $\sum_{k=1}^{\infty} a_k$. Motivated by the above example, we ask, *Is it possible to associate a numerical value to the infinite sum* $\sum_{k=1}^{\infty} a_k$*?* If so, we would expect the sequence of the partial sums $\{S_n\}_{n\geq 1}$ to approach that value. It is customary to make this idea precise in the following form, which defines the behavior of an infinite series in terms of its sequence of partial sums.

Definition 5.1 (Infinite series). *The series $\sum_{k=1}^{\infty} a_k$ is said to be convergent, or to converge to S, if the sequence of partial sums $\{S_n\}_{n\geq 1}$, $S_n = \sum_{k=1}^{n} a_k$, converges to S. In this case, we say that the series converges to S, and we write*

$$\sum_{k=1}^{\infty} a_k = \lim_{n\to\infty} S_n = S.$$

Here S is referred to as the sum of the series $\sum_{k=1}^{\infty} a_k$. If the sequence $\{S_n\}$ does not converge, then we say that the series $\sum_{k=1}^{\infty} a_k$ diverges and has no sum. In particular, if $\lim_{n\to\infty} S_n = \infty$ or $-\infty$, we say that the series diverges to ∞ or $-\infty$, and write

$$\sum_{k=1}^{\infty} a_k = \infty \text{ or } \sum_{k=1}^{\infty} a_k = -\infty.$$

A divergent series that does not diverge to $\pm\infty$ is said to oscillate or be oscillatory.

As an example of the latter case, the series $\sum_{k=1}^{\infty} k$ diverges, because

$$S_n = 1 + 2 + 3 + \cdots + n = \frac{n(n+1)}{2} \to \infty \quad \text{as } n \to \infty.$$

Remark 5.2. 1. Suppose that we are given a series $\sum_{k=1}^{\infty} a_k$ with $a_k = 0$ for $k \geq N+1$. Then $S_k = S_N$ for all $k \geq N$, and so $\{S_n\}_{n=1}^{\infty}$ converges to $S_N = \sum_{k=1}^{N} a_k$. In this case, the infinite series is actually a finite sum. Thus, an infinite series can be viewed as a generalization of a finite sum.

2. We remark that the statement "$\sum_{k=1}^{\infty} a_k$ converges" refers to the behavior of the sequence of its partial sums $\{S_n\}_{n\geq 1}$ and does not directly say anything about the sequence $\{a_n\}_{n\geq 1}$.
3. We will use the symbol $\sum_{k=1}^{\infty} a_k$ regardless of whether this series converges or diverges. However, if the sequence of partial sums $\{S_n\}$ converges, then the symbol $\sum_{k=1}^{\infty} a_k$ plays a dual role: it represents *both* the series and its sum. If the series neither converges to a finite value nor diverges to ∞ or $-\infty$, then the symbol $\sum_{k=1}^{\infty} a_k$ continues to represent the infinite series, but it does not represent an (extended) real number.
4. The convergence or divergence of a series is independent of whether the summation index begins with $k = 1$ or with $k = m$, for some integer m. Thus, altering a finite number of terms of a series in any fashion whatsoever has no effect on the convergence of the original series, though it will generally affect the sum if the series converges.
5. If a series is given in the form $\sum_{k=m}^{\infty} a_k$, then the sequence $\{S_n\}_{n=m}^{\infty}$ of partial sums is defined by

$$S_n = \sum_{k=1}^{n} a_{m+k-1} \quad \text{or sometimes even by} \quad S_n = \sum_{k=m}^{n} a_k.$$

   We note that in the latter form, we have not used the first $m-1$ terms of the series to denote S_n. In either case, the limits of these two sequences will be the same if the series converges.
6. We see that to each series there corresponds a sequence, whose limit, if it exists, is the sum of the series. However, for a given sequence $\{S_n\}_{n\geq 1}$ with $S_n \to S$ we can associate a series $\sum_{k=1}^{\infty} a_k$ with sum exactly S. For a proof, we simply set $a_1 = S_1$ and $a_n = S_n - S_{n-1}$, so that $S_n = \sum_{k=1}^{n} a_k$. Thus, the sequences $\{a_n\}$ and $\{S_n\}$ determine each other uniquely. •

In Chapter 2, we considered some sequences that were indeed sequences of partial sums of important series. However, a general problem in the study of series is to determine whether a given series is convergent and in some cases to evaluate the sum of the series. We begin our discussion with geometric series.

5.1.1 Geometric Series

A *geometric series* is an infinite series in which the ratio of successive terms in the series is constant. If this constant ratio is r, then the series has the form

$$\sum_{k=0}^{\infty} ar^k = a + ar + ar^2 + ar^3 + \cdots + ar^n + \cdots.$$

We note that the convergence is obvious if $a = 0$. So throughout the discussion we assume that $a \neq 0$. For instance, ordinary division leads to

$$\frac{1}{3} = 0.3333\ldots = \frac{3}{10} + \frac{3}{10^2} + \frac{3}{10^3} + \cdots,$$

which is an example of a geometric series with $a = 3/10$ and $r = 1/10$. Geometric series occur in many applications, most interestingly in relation to series of functions (e.g., power series, Fourier series, orthogonal series). Geometric series also arise when one wishes to compute the total distance vertically traveled by a ball that is dropped from a height of a feet, assuming that each time the ball strikes the ground after falling a distance a it rebounds a distance ar. The total up-and-down distance the ball travels is

$$a + 2ar + 2ar^2 + 2ar^3 + \cdots$$

(see Figure 5.1).

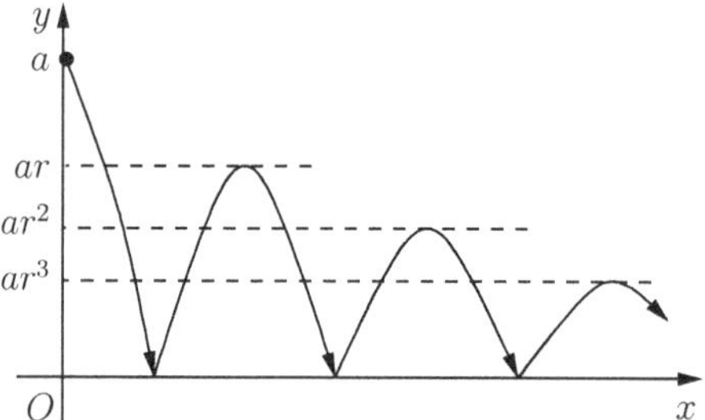

Fig. 5.1. Example of a geometric series.

In the following theorem we discuss this problem for geometric series, which are perhaps the most important type of convergent series.

Theorem 5.3 (Geometric series). *The geometric series* $\sum_{k=0}^{\infty} ar^k$ *with* $a \neq 0$ *converges if* $|r| < 1$*, with sum*

$$\sum_{k=0}^{\infty} ar^k = \frac{a}{1-r}, \tag{5.1}$$

and diverges if $|r| \geq 1$.

Proof. Suppose $r \neq 1$. Then the nth partial sum S_n is given by

$$S_n = a[1 + r + r^2 + \cdots + r^{n-1}],$$

so that

$$rS_n = a[r + r^2 + r^3 + \cdots + r^n].$$

By subtracting rS_n from S_n, we find that

$$(1-r)S_n = a(1-r^n), \text{ i.e., } S_n = \frac{a(1-r^n)}{1-r} \quad (r \neq 1).$$

Now we take the limit as $n \to \infty$. If $|r| < 1$, Theorem 2.34 tells us that $r^n \to 0$ as $n \to \infty$, and so we have

$$\lim_{n\to\infty} S_n = \frac{a}{1-r},$$

which establishes the formula (5.1).

The divergence part can be deduced immediately from a divergence test (see Corollary 5.19) that appears in Section 5.2. However, it is appropriate to present an independent direct proof here.

If $r = 1$, the divergence of the geometric series is clear, because for $r = 1$, we have $S_n = an$. If $r = -1$, we note that the nth partial sum is

$$S_n = \begin{cases} a & \text{if } n \text{ is odd,} \\ 0 & \text{if } n \text{ is even.} \end{cases}$$

Because the sequence $\{S_n\}$ has no limit, the series $\sum a(-1)^k$ must diverge. In this case, $\{S_n\}$ oscillates finitely between a and 0.

Finally, if $|r| > 1$, then $|r|^n$, and so $\{S_n\}$ has no limit. It follows that the series diverges if $|r| > 1$. Moreover,

$$\sum_{k=0}^{\infty} ar^k = \begin{cases} \infty & \text{if } a > 0 \text{ and } r > 1, \\ -\infty & \text{if } a < 0 \text{ and } r > 1. \end{cases}$$

For $r < -1$, $\sum ar^k$ oscillates infinitely and $\{S_n\}$ alternates in sign. ∎

Some applications of Theorem 5.3 follow:

1. $\displaystyle\sum_{k=0}^{\infty}(-1)^k 3^k r^k = \frac{1}{1+3r}$ for all r with $-1/3 < r < 1/3$, because the series is exactly (5.1) but with $-3r$ in place of r, and with $a = 1$. Similarly, we see that

$$\sum_{k=0}^{\infty} 5^k (r-1)^k = \frac{1}{1-5(r-1)} = \frac{1}{6-5r} \quad (|r-1| < 1/5).$$

2. The series $\sum_{k=2}^{\infty}\left(-\frac{1}{5}\right)^k$ converges with sum

$$S = \frac{a}{1-r} = \frac{1/25}{1-(-1/5)} = \frac{1}{30}$$

(since $a = 1/25$ and $r = -1/5$).

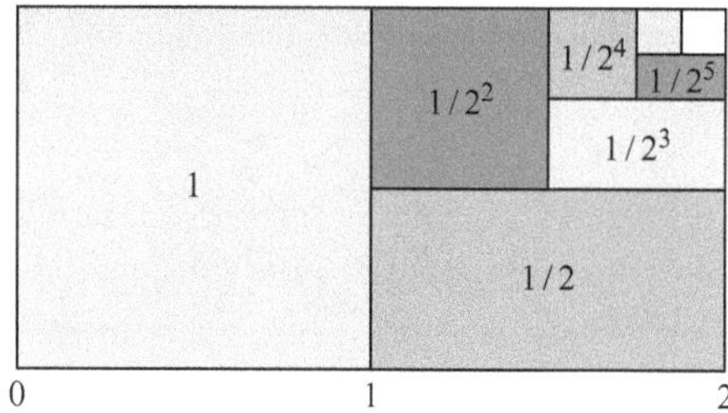

Fig. 5.2. Geometric proof for $\sum_{k=0}^{\infty}\frac{1}{2^k}=2$ via length.

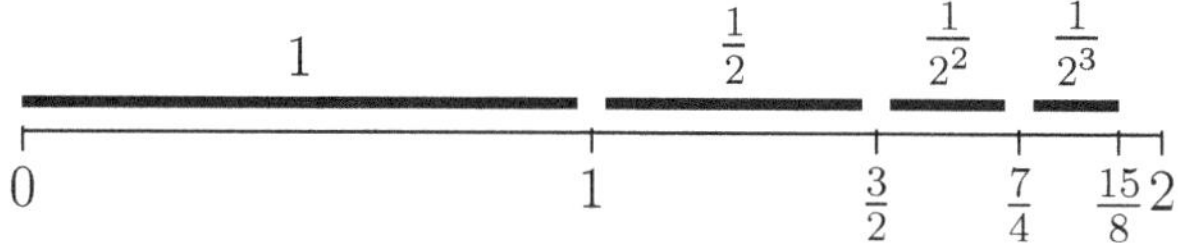

Fig. 5.3. Geometric proof for $\sum_{k=0}^{\infty}\frac{1}{2^k}=2$ via area.

3. Set $a=1$ and $r=1/2$, so that $S_n=2(1-1/2^n)\to 2=S$ as $n\to\infty$. We conclude that the series $\sum_{k=0}^{\infty}\frac{1}{2^k}$ converges and $\sum_{k=0}^{\infty}\frac{1}{2^k}=2$. This fact can be seen geometrically; see Figure 5.2.
4. The series $\sum_{k=0}^{\infty}(3/2)^k$ clearly diverges.

Geometric proof of Theorem 5.3. It suffices to prove the theorem for $a=1$. We begin with $a=1$, and $r=1/2$ to show geometrically that

$$1+\frac{1}{2}+\frac{1}{2^2}+\frac{1}{2^3}+\cdots=2.$$

See Figures 5.2 and 5.3, which is self-explanatory. Note that in Figure 5.3, the distance from 0 to 2 can be split into infinite sequences of lengths $1, 1/2, 1/2^2, \ldots$, and so it is reasonable to write as above at the first instance itself. For $0<x<1$, we refer to Figure 5.4, and the proof is clear. We remark that for this geometric proof one is not really required to know the sum in advance.

5.1.2 Decimal Representation of Real Numbers

By a (positive) infinite decimal, we mean

$$0.a_1a_2a_3\ldots=\lim_{n\to\infty}\sum_{k=1}^{n}\frac{a_k}{10^k},$$

provided the limit exists, where each a_k $(k\geq 1)$ is an integer such that $0\leq a_k\leq 9$. Actually, we shall see in a moment that the limit always exists. Geometric series provide another interpretation of the decimal representation of rational numbers. For example, consider

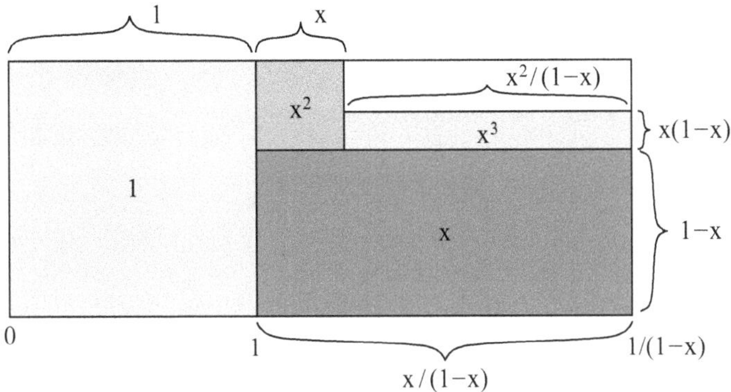

Fig. 5.4. Geometric proof for $\sum_{k=0}^{\infty} x^k = 1/(1-x)$.

$$\frac{1}{3} = 0.3333\cdots := 0.\bar{3}.$$

Note that another way of interpreting the symbol 0.333... is by the infinite series

$$\frac{3}{10} + \frac{3}{10^2} + \frac{3}{10^3} + \cdots,$$

which is a geometric series with the first term $a = 3/10$ and common ratio $r = 1/10$. From our earlier discussion, we see that this series converges to

$$\frac{3/10}{1 - 1/10} = \frac{1}{3}.$$

We shall now show that every real number has at least one decimal expansion, and conversely, every decimal represents a real number. For our discussion, we consider the series

$$a_0 + \sum_{k=1}^{\infty} \frac{a_k}{10^k}, \tag{5.2}$$

where $a_0 \in \mathbb{Z}$ and each a_k $(k \geq 1)$ is an integer with $0 \leq a_k \leq 9$. We need first to show that the series converges. To do this, we introduce

$$S_n = a_0 + \sum_{k=1}^{n} \frac{a_k}{10^k}.$$

Then $\{S_n\}$ is clearly an increasing sequence of real numbers. Further, since

$$\sum_{k=1}^{n} \frac{a_k}{10^k} \leq 9 \sum_{k=1}^{n} \frac{1}{10^k} < 9 \sum_{k=1}^{\infty} \frac{1}{10^k} = 9\left(\frac{1/10}{1 - 1/10}\right) = 1,$$

it follows that $\{S_n\}$ is bounded above by $a_0 + 1$. Consequently, $\{S_n\}$, and hence the series (5.2) converges to a real number x with $a_0 \leq x \leq 1 + a_0$. Thus, we have the following theorem.

Theorem 5.4. *If $\{a_n\}_{n\geq 1}$ is a sequence of integers with $0 \leq a_n \leq 9$ for all $n \geq 1$, then $\sum_{k=0}^{\infty} a_k 10^{-k}$ converges to a real number x with $a_0 \leq x \leq a_0 + 1$.*

The series (5.2) may be denoted by

$$a = a_0.a_1a_2a_3\ldots,$$

and we call the series as a *decimal expansion* of the number a. Here a_n is often called the nth digit of the decimal expansion. Note that some numbers have more than one decimal expansion. For example $0.999\ldots$ represents

$$\lim_{n\to\infty} S_n, \quad \text{where } S_n = \frac{9}{10} + \frac{9}{10^2} + \cdots + \frac{9}{10^n} \to 1.$$

Consequently, $0.9999\ldots$ (all 9's) and $1.000\ldots$ (all 0's) are two different decimal expansions that represent the same real number 1. Similarly, $0.5000\ldots$ and $0.49999\ldots$ are two different decimal expansions of $1/2$.

The converse of Theorem 5.4 also holds. Recall that for every $x \in \mathbb{R}$ there exists an integer a such that $a \leq x \leq a+1$. Consequently, it suffices to prove the following.

Theorem 5.5. *For each x with $0 \leq x \leq 1$, there is a decimal expansion converging to x. That is, there is a sequence $\{a_n\}_{n\geq 1}$ of integers such that $0 \leq a_n \leq 9$ for all n and $\sum_{k=1}^{\infty} a_k 10^{-k}$ converges to x.*

Proof. Suppose that $x \in [0,1]$. Divide $[0,1]$ into ten equal subintervals. Then x lies in a subinterval

$$\left[\frac{a_1}{10}, \frac{a_1+1}{10}\right]$$

for some integer a_1 in $\{0,1,2,\ldots,9\}$, and so $0 \leq (x - a_1/10)10 = y \leq 1$.

Again if we subdivide $[0,1]$ into ten subintervals, we see that

$$\frac{a_2}{10} \leq y = \left(x - \frac{a_1}{10}\right)10 \leq \frac{a_2+1}{10}, \quad \text{i.e.,} \quad \frac{a_1}{10} + \frac{a_2}{10^2} \leq x \leq \frac{a_1}{10} + \frac{a_2+1}{10^2},$$

for some a_2 in $\{0,1,2,\ldots,9\}$. Continuing the above process, we see that

$$\frac{a_1}{10} + \frac{a_2}{10^2} + \cdots + \frac{a_n}{10^n} \leq x \leq \frac{a_1}{10} + \frac{a_2}{10^2} + \cdots + \frac{a_n}{10^n} + \frac{1}{10^n}$$

for some integers $a_1, a_2, \ldots, a_n$ in $\{0,1,2,\ldots,9\}$. The last inequality is equivalent to

$$x - \frac{1}{10^n} \leq S_n \leq x, \quad S_n = \sum_{k=1}^{n} \frac{a_k}{10^k}.$$

Thus, by the squeeze/sandwich rule, it follows that $\lim_{n\to\infty} S_n = x$. ∎

Remark 5.6. 1. If $x \geq 1$ and $a \in \mathbb{N}$ is such that $a \leq x < a + 1$, then $x - a$ belongs to $[0, 1)$, and so

$$x - a = 0.a_1a_2 \ldots a_n, \quad \text{or} \quad x = a.a_1a_2a_3 \ldots,$$

where the decimal expansion of $x - a$ is as in Theorem 5.5. We may treat negative numbers similarly.

2. If $x = \sum_{k=1}^{\infty} \frac{a_k}{10^k}$, where the a_k are given by Theorem 5.5, then it is a simple exercise to see that $a_n = 0$ for all n whenever $x = 0$, and $a_n = 9$ for all n whenever $x = 1$. ●

5.1.3 The Irrationality of e

We defined e in Example 2.33 as the limit of $\left(1 + \frac{1}{n}\right)^n$. Now we have the following theorem.

Theorem 5.7. $\mathrm{e} = \displaystyle\sum_{k=0}^{\infty} \frac{1}{k!}$.

Proof. Set $S_n = \sum_{k=0}^{n} \frac{1}{k!}$. We have shown in Example 2.33 that

$$2 < a_n := \left(1 + \frac{1}{n}\right)^n < S_n < 3,$$

and $\{S_n\}$, being an increasing bounded sequence, converges. Thus, we have

$$\lim_{n\to\infty} a_n \leq \lim_{n\to\infty} S_n, \quad \text{i.e., } \mathrm{e} \leq \lim_{n\to\infty} S_n. \tag{5.3}$$

Moreover, for $n \geq m$,

$$\begin{aligned} a_n &\geq 1 + \sum_{k=1}^{m} \binom{n}{k} \frac{1}{n^k} \\ &= 1 + 1 + \frac{1}{2!}\left(1 - \frac{1}{n}\right) + \cdots + \frac{1}{m!}\left(1 - \frac{1}{n}\right) \cdots \left(1 - \frac{m-1}{n}\right). \end{aligned}$$

Now letting $n \to \infty$, keeping m fixed, we obtain $e \geq S_m$. Now again, allowing $m \to \infty$ in this inequality, we finally get

$$\mathrm{e} \geq \lim_{m\to\infty} S_m. \tag{5.4}$$

The desired conclusion follows from (5.3) and (5.4). ∎

Corollary 5.8. e *is irrational.*

Proof. Suppose, for a contradiction, that e is rational. Then $\mathrm{e} = p/q$, where p and q are positive integers. Choose n such that $n > \max\{q, 3\}$. With S_n as above, we see that $n!e$ and $n!S_n$ are clearly integers, and so $n!(\mathrm{e} - S_n)$ is again a positive integer. Now

$$\mathrm{e} - S_n = \frac{1}{(n+1)!} + \frac{1}{(n+2)!} + \cdots,$$

so that

$$\begin{aligned} n!(\mathrm{e} - S_n) &= \frac{1}{n+1}\left[1 + \frac{1}{n+2} + \frac{1}{(n+2)(n+3)} + \cdots\right] \\ &< \frac{1}{n+1}\left[1 + \frac{1}{n+1} + \frac{1}{(n+1)^2} + \cdots\right] \\ &= \frac{1}{n+1}\left[\frac{1}{1 - (1/(n+1))}\right] = \frac{1}{n}, \end{aligned}$$

and thus by the last inequality, it follows that

$$0 < n!(\mathrm{e} - S_n) < 1/n,$$

implying the existence of an integer in the interval $(0, 1)$, a contradiction. ∎

5.1.4 Telescoping Series

Geometric series (as in the above examples) are easy to deal with because we can find a closed-form expression for the nth partial sum S_n of the given series and hence are able to find its sum. Often it is difficult or even impossible to find a simple formula for S_n. In normal circumstances, on the other hand, it is not hard to determine whether a given series converges. Consequently, there is a need to develop efficient techniques for determining whether a given series is convergent. Before we proceed to develop them, it might be good motivation to mention another series for which we can find a formula for S_n explicitly.

A series is called a *telescoping series* if there is internal cancellation in the partial sums. For instance, for a sequence of real numbers $\{a_n\}$, the series

$$\sum_{k=1}^{\infty} (a_{k+1} - a_k)$$

is a telescoping series. We now illustrate this concept by the following simple examples:

1. Consider the series $\sum_{k=1}^{\infty} \log(1 + 1/k)$. The nth partial sum of the given series can be represented as

$$
\begin{aligned}
S_n &= \sum_{k=1}^{n} (\log(k+1) - \log k) \\
&= (\log 2 - \log 1) + (\log 3 - \log 2) + \cdots + (\log(n+1) - \log n) \\
&= \log(n+1).
\end{aligned}
$$

Thus, the sequence of partial sums $\{\log(n+1)\}_{n\geq 1}$ is unbounded, and hence the given series is not convergent. Note that the kth term $a_k = \log((k+1)/k)$ converges to $\log 1 = 0$ as $k \to \infty$.

2. Consider the series

$$\sum_{k=1}^{\infty} \frac{1}{k^2+k}.$$

We can easily see that the series converges with sum 1. Indeed, we find that

$$\frac{1}{k^2+k} = \frac{1}{k} - \frac{1}{k+1} \quad \text{and} \quad S_n = 1 - \frac{1}{n+1}.$$

Thus, $\{S_n\}$ converges to 1 as $n \to \infty$, and hence the given series converges with sum $S = 1$. Again, we note that the kth term of the series converges to 0 as $k \to \infty$. Also, we have

$$\frac{1}{k} - \frac{1}{k+1} = \frac{1}{k(k+1)} < \frac{1}{k^2} < \frac{2}{k(k+1)} = 2\Big(\frac{1}{k} - \frac{1}{k+1}\Big),$$

and so

$$1 - \frac{1}{n+1} \leq \sum_{k=1}^{n} \frac{1}{k^2} \leq 2\Big(1 - \frac{1}{n+1}\Big),$$

showing that $\sum_{k=1}^{\infty}(1/k^2)$ converges and

$$1 \leq \sum_{k=1}^{\infty} \frac{1}{k^2} \leq 2.$$

The method of proof discussed here may be used to show that

$$2 < \sum_{k=0}^{\infty} \frac{1}{k!} < 3. \tag{5.5}$$

Indeed, for $k \geq 1$, we have

$$\frac{k}{(k+1)!} = \frac{k+1-1}{(k+1)!} = \frac{1}{k!} - \frac{1}{(k+1)!},$$

so that

$$T_n = \sum_{k=1}^{n} \frac{k}{(k+1)!} = 1 - \frac{1}{(n+1)!} \to 1 \text{ as } n \to \infty, \quad \text{i.e.,} \sum_{k=1}^{\infty} \frac{k}{(k+1)!} = 1.$$

We also note that for $k \geq 1$,

$$\frac{k+1}{(k+1)!} = \frac{1}{k!} > \frac{k}{(k+1)!}; \quad \text{i.e.,} \quad \sum_{k=1}^{\infty} \frac{1}{k!} > \sum_{k=1}^{\infty} \frac{k}{(k+1)!} = 1.$$

Also,

$$\sum_{k=1}^{\infty} \frac{1}{k!} = 1 + \sum_{k=2}^{\infty} \frac{1}{k!} < 1 + \sum_{k=2}^{\infty} \frac{k-1}{k!} = 1 + \sum_{k=1}^{\infty} \frac{k}{(k+1)!} = 2,$$

and therefore, by combining the last two inequalities, we obtain (5.5). Consequently, the value of the irrational number e lies between 2 and 3.

3. Finally, consider the series

$$\sum_{k=1}^{\infty} \frac{3}{(k+a)(k+a+3)}, \qquad \frac{3}{(k+a)(k+a+3)} = \frac{1}{k+a} - \frac{1}{k+a+3},$$

where $a > -1$ is a fixed real number. It is then a simple exercise to see that

$$\begin{aligned} S_n &= \sum_{k=1}^{n} \frac{3}{(k+a)(k+a+3)} \\ &= \frac{1}{1+a} + \frac{1}{2+a} + \frac{1}{3+a} - \frac{1}{n+a+1} - \frac{1}{n+a+2} - \frac{1}{n+a+3}, \end{aligned}$$

which clearly shows that $\{S_n\}$ converges with sum

$$\sum_{k=1}^{\infty} \frac{3}{(k+a)(k+a+3)} = \frac{1}{1+a} + \frac{1}{2+a} + \frac{1}{3+a}.$$

In particular, for $a = 0$, this gives

$$\sum_{k=1}^{\infty} \frac{1}{k^2+3k} = \frac{1}{3}\left(1 + \frac{1}{2} + \frac{1}{3}\right) = \frac{11}{18}.$$

The last three examples above fall under the following general category (see also Exercise 5.17(2)).

Theorem 5.9 (Telescoping series). *Suppose that* $\{a_k\}_{k\geq 1}$ *is a convergent sequence with limit* A. *Then we have*

(a) $\sum_{k=1}^{\infty}(a_k - a_{k+1})$ *converges with sum* $a_1 - A$.
(b) $\sum_{k=1}^{\infty}(a_k - a_{k+2}) = a_1 + a_2 - 2A$.

Proof. The proof follows from observing that

$$\sum_{k=1}^{n}(a_k - a_{k+1}) = a_1 - a_{n+1} \quad \text{and} \quad \sum_{k=1}^{n}(a_k - a_{k+2}) = a_1 + a_2 - a_{n+1} - a_{n+2}$$

and then using the definitions of convergence of series and sequence, respectively. ■

5.1.5 Operations and Convergence Criteria in Series

Often we will be concerned with general properties of a series rather than its sum. Therefore, *whenever the starting term of a series is not important, we may simply write* $\sum a_k$ *rather than* $\sum_{k=m}^{\infty} a_k$. Most basic statements about series can be reinterpreted as statements about sequences, by considering series in terms of the corresponding sequences of partial sums. For example, by the uniqueness property and the linearity rule for sequences in Theorem 2.8, we easily obtain the basic properties of series that convergent series can be added, subtracted, and multiplied by constants.

Theorem 5.10 (Uniqueness of sum and linearity of infinite series). *The sum of a convergent series is unique. Moreover, if* $\sum a_k$ *and* $\sum b_k$ *are two convergent series with sums* A *and* B*, respectively, then for any pair of constants* α *and* β*, the series* $\sum(\alpha a_k + \beta b_k)$ *also converges with sum* $\alpha A + \beta B$*; that is,*

$$\sum(\alpha a_k + \beta b_k) = \alpha \sum a_k + \beta \sum b_k = \alpha A + \beta B.$$

For example, by Theorem 5.10, we now conclude that

$$\sum_{k=1}^{\infty} \left[\frac{3}{k^2 + 3k} - \frac{5}{3^{k+1}} \right] = 3 \sum_{k=1}^{\infty} \frac{1}{k^2 + 3k} - \frac{5}{3} \sum_{k=1}^{\infty} \frac{1}{3^k} = 3\left(\frac{11}{18}\right) - \frac{5}{3}\left(\frac{1}{2}\right) = 1.$$

Remark 5.11. Suppose that $a_k = k$ and $b_k = -k$. Then $\sum a_k$ and $\sum b_k$ both diverge. But $\sum(a_k + b_k)$ converges to 0. •

On the other hand, Theorem 5.10 also provides a useful result about a series of the form $\sum(\alpha a_k + \beta b_k)$.

Theorem 5.12. *If either* $\sum a_k$ *or* $\sum b_k$ *diverges and the other converges, then the series* $\sum(a_k + b_k)$ *must diverge.*

Proof. Assume that $\sum a_k$ diverges and $\sum b_k$ converges. Suppose to the contrary that the series $\sum(a_k + b_k)$ converges. Then by the linearity property, the series

$$\sum[(a_k + b_k) - b_k] = \sum a_k$$

must converge, contradicting the hypothesis that $\sum a_k$ diverges. It follows that the series $\sum(a_k + b_k)$ diverges. ∎

For example, by Theorem 5.12, each of the series

$$\sum_{k=1}^{\infty} \left[\frac{1}{k^2 + 3k} + (-1)^k \right], \quad \sum_{k=1}^{\infty} \left[\frac{1}{k^2 + k} - (-1)^k \right] \text{ and } \sum_{k=1}^{\infty} \left[\frac{1}{k^2 + k} - \frac{1}{k} \right]$$

diverges, because $\sum_{k=1}^{\infty}(-1)^k$ and $\sum_{k=1}^{\infty} \frac{1}{k}$ diverge.

We recall that a sequence is a succession of terms, whereas a series is a sum of such terms, and so these two concepts have very different properties. For example, a sequence of terms may converge, but the series of the same terms may diverge:

- Both $\{1 - 1/n\}$ and $\{1 + 1/3^n\}$ converge to 1.
- Both $\sum_{k=1}^{\infty}(1 - 1/k)$ and $\sum_{k=1}^{\infty}\left(1 + 1/3^k\right)$ diverge.

How about the converse? That is, if a series converges, must the sequence of terms of the series converge?

The following test for convergence is an immediate consequence of Cauchy's convergence criterion for sequences (see Theorem 2.55).

Theorem 5.13 (Cauchy's convergence criterion for series). *A series $\sum_{k=1}^{\infty} a_k$ is convergent if and only if the sequence of partial sums is a Cauchy sequence, i.e., given $\epsilon > 0$ there exists an integer N such that*

$$|S_m - S_n| = \left|\sum_{k=1}^{m} a_k - \sum_{k=1}^{n} a_k\right| = |a_{n+1} + \cdots + a_m| < \epsilon \quad \textit{if } m > n \geq N,$$

or equivalently,

$$|S_{n+p} - S_n| = \left|a_{n+1} + a_{n+2} + \cdots + a_{n+p}\right| < \epsilon \quad \textit{if } n \geq N \textit{ and } p > 0.$$

In particular, from Theorem 5.13, we conclude the following: *if a series $\sum_{k=1}^{\infty} a_k$ is convergent, then*

$$S_{2n} - S_n = a_{n+1} + a_{n+2} + \cdots + a_{2n} \to 0 \quad \textit{as } n \to \infty.$$

In order to apply the last situation, we consider the harmonic series

$$\sum_{k=1}^{\infty} \frac{1}{k} = 1 + \frac{1}{2} + \frac{1}{3} + \frac{1}{4} + \cdots,$$

which arises in connection with overtones produced by a vibrating string. The fact that the harmonic series is divergent can be seen in a number of ways (see also Remark 5.25), for example, by proving that the sequence of partial sums $\{S_n\}$ is either unbounded or is not Cauchy. Indeed, with $S_n = \sum_{k=1}^{n}(1/k)$, we have

$$S_{2n} - S_n = \sum_{k=1}^{n} \frac{1}{n+k} > n\left(\frac{1}{n+n}\right) = \frac{1}{2},$$

showing that $\{S_n\}$ is not Cauchy (a fact that was verified in Section 2.1) and hence is not convergent.

5.1.6 Absolutely and Conditionally Convergent Series

Given a series $\sum a_k$, we may form a new series $\sum |a_k|$. If this new series is convergent, then we say that the original series $\sum a_k$ is *absolutely convergent.* If a series is convergent but not absolutely convergent, then it is said to be *conditionally convergent.* Now, because

$$|a_{n+1} + a_{n+2} + \cdots + a_{n+p}| \leq |a_{n+1}| + |a_{n+2}| + \cdots + |a_{n+p}| \text{ if } n \geq N \text{ and } p > 0,$$

the Cauchy convergence criterion (Theorem 5.13) gives the following theorem, which explains the importance of absolutely convergent series.

Theorem 5.14 (The absolute convergence test). *An absolutely convergent series is convergent.*

For an alternative proof of Theorem 5.14, we refer to Exercise 5.43(1). Thus, one way to determine whether a general series converges is to determine whether the corresponding series of absolute values is convergent. But then the problem of whether a series converges absolutely is simply the problem of whether a series of nonnegative terms converges. Convergence of such series will be discussed in detail in Section 5.2.

Example 5.15. The converse of Theorem 5.14 is not true, as we see below from the *alternating harmonic series*

$$\sum_{k=1}^{\infty} \frac{(-1)^{k-1}}{k}.$$

This is an example of a conditionally convergent series. First, we note that the given series of absolute values is the harmonic series, which is divergent. Next, we let

$$S_n = \sum_{k=1}^{n} \frac{(-1)^{k-1}}{k},$$

the nth partial sum of the alternating harmonic series.

Method 1: We calculate the first few terms of S_n and plot them on a sequence diagram:

$$S_1 = 1, \; S_3 = \frac{1}{2} + \frac{1}{3} = \frac{5}{6}, \; S_5 = \frac{7}{12} + \frac{1}{5} = \frac{47}{60}, \; \ldots$$

and

$$S_2 = \frac{1}{2}, \; S_4 = \frac{5}{6} - \frac{1}{4} = \frac{7}{12}, \; S_6 = \frac{47}{60} - \frac{1}{6} = \frac{37}{60}, \; \ldots .$$

Note that S_{2n-1} appears to be decreasing with all terms positive, and also S_{2n} appears to be increasing with all terms positive; see Figure 5.5. Thus, it appears that S_n converges. To prove this fact, we write the even partial sums $\{S_{2n}\}$ in two different ways:

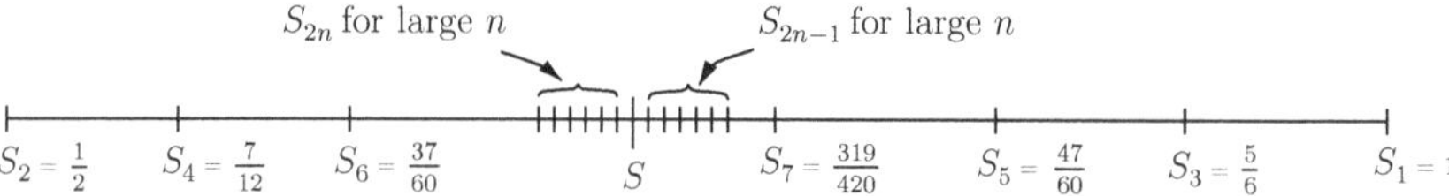

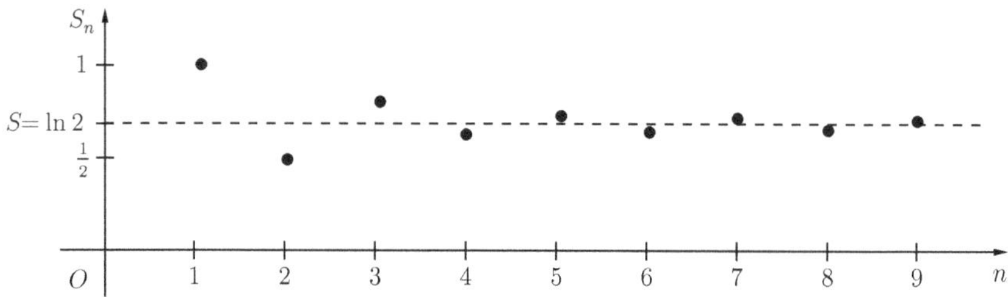

Fig. 5.5. Convergence of the alternating harmonic series.

$$\begin{aligned}S_{2n} &= \left(1-\frac{1}{2}\right)+\left(\frac{1}{3}-\frac{1}{4}\right)+\cdots+\left(\frac{1}{2n-1}-\frac{1}{2n}\right)\\ &= 1-\left(\frac{1}{2}-\frac{1}{3}\right)-\left(\frac{1}{4}-\frac{1}{5}\right)-\cdots-\left(\frac{1}{2n-2}-\frac{1}{2n-1}\right)-\frac{1}{2n}.\end{aligned}$$

This shows that $\{S_{2n}\}$ is increasing and bounded above by 1. Hence $\{S_{2n}\}$ is convergent to a limit, by BMCT. Further, because

$$S_{2n} = S_{2n-1} - \frac{1}{2n},$$

it follows that $\{S_{2n-1}\}$ is also convergent to the same limit. Consequently, $\{S_n\}$ is convergent. Later we shall use the same method to prove the alternating series test (Theorem 5.44).

Method 2: We shall now use Cauchy's convergence criterion. To apply this, we compute

$$S_{n+p} - S_n = (-1)^n\left[\frac{1}{n+1}-\frac{1}{n+2}+\cdots+\frac{(-1)^{p-1}}{n+p}\right] =: (-1)^n T_{n,p}.$$

We shall first show that

$$0 < T_{n,p} < \frac{1}{n+1}. \tag{5.6}$$

Indeed, if p is even, then $T_{n,p}$ may be written in two different ways:

$$\begin{aligned}T_{n,p} &= \left(\frac{1}{n+1}-\frac{1}{n+2}\right)+\left(\frac{1}{n+3}-\frac{1}{n+4}\right)+\cdots+\left(\frac{1}{n+p-1}-\frac{1}{n+p}\right)\\ &= \frac{1}{n+1}-\left(\frac{1}{n+2}-\frac{1}{n+3}\right)-\cdots-\left(\frac{1}{n+p-2}-\frac{1}{n+p-1}\right)-\frac{1}{n+p}.\end{aligned}$$

Similarly, if p is odd, then we may write

$$T_{n,p} = \left(\frac{1}{n+1} - \frac{1}{n+2}\right) + \cdots + \left(\frac{1}{n+p-2} - \frac{1}{n+p-1}\right) + \frac{1}{n+p}$$
$$= \frac{1}{n+1} - \left(\frac{1}{n+2} - \frac{1}{n+3}\right) - \cdots - \left(\frac{1}{n+p-1} - \frac{1}{n+p}\right).$$

Since each term in the expressions in parentheses is positive, it follows that (5.6) holds for all arbitrary positive integers n and p. Therefore,

$$|S_{n+p} - S_n| < \frac{1}{n+1} \quad \text{for all } n \geq 1 \text{ and } p \geq 1.$$

Since $1/(n+1) \to 0$ as $n \to \infty$, the alternating harmonic series satisfies Cauchy's convergence criterion and hence converges.

Method 3: Using elementary arguments (as earlier) it is possible to obtain that the sum of the alternating harmonic series is $\log 2$. Recall that

$$\frac{1}{1+r} = 1 - r + r^2 - r^3 + \cdots + (-1)^{n-1}r^{n-1} + (-1)^n \frac{r^n}{1+r},$$

which, by integration from 0 to x, gives,

$$\log(1+x) = x - \frac{x^2}{2} + \frac{x^3}{3} - \frac{x^4}{4} + \cdots + (-1)^{n-1}\frac{x^n}{n} + R_n,$$

where

$$|R_n| = \left|(-1)^n \int_0^x \frac{r^n}{1+r}\,dr\right| = \int_0^x \frac{r^n}{1+r}\,dr < \int_0^x r^n\,dr = \frac{x^{n+1}}{n+1},$$

which approaches zero as $n \to \infty$ if $-1 < x \leq 1$. Here we assume that the reader is familiar with integration theory, although we discuss it in detail only in a later chapter. Hence

$$\log(1+x) = \sum_{k=1}^{\infty} \frac{(-1)^{k-1}}{k} x^k \quad \text{for } -1 < x \leq 1.$$

In particular,

$$\log 2 = \sum_{k=1}^{\infty} \frac{(-1)^{k-1}}{k}.$$

The previous equation can also be written in the equivalent form

$$-\log(1-x) = \sum_{k=1}^{\infty} \frac{x^k}{k} \quad \text{for } -1 \leq x < 1.$$

If we let $x = 1 - 1/m$ $(m > 1)$ on the right-hand side, then

$$\sum_{k=1}^{\infty} \frac{1}{k} \geq \sum_{k=1}^{\infty} \frac{1}{k}\left(1 - \frac{1}{m}\right)^k = -\log(1/m) = \log m \quad \text{for each } m > 1.$$

Hence, since the sequence of partial sums of the harmonic series is increasing and unbounded above, the last inequality yields another proof for

$$\sum_{k=1}^{\infty} \frac{1}{k} = \infty.$$

Method 4: It turns out that the alternating harmonic series converges by the alternating series test (see Section 5.3). ●

5.1.7 Questions and Exercises

Questions 5.16.

1. What are the main differences between sequences and series?
2. What is a geometric series? Where does the name come from?
3. Must the sequence of partial sums of a convergent series be bounded?
4. Suppose that $\sum a_k$ and $\sum b_k$ are both divergent series. Must $\sum(a_k + b_k)$ be divergent? Must $\sum(a_k - b_k)$ be divergent?
5. Suppose that $\sum(a_k+b_k)$ converges. What can be said about the individual series $\sum a_k$ and $\sum b_k$?
6. Suppose that $\sum(a_k + b_k)$ diverges. Must one of the series be convergent? Must one of the series be divergent? Must both series be divergent?
7. If the series $\sum_{k=1}^{\infty}(a_{2k-1} + a_{2k})$ converges, what can be said about the series $\sum_{k=1}^{\infty} a_k$?
8. If $\sum_{k=1}^{\infty} a_k$ converges, what can be said about $\sum_{k=1}^{\infty}(a_{2k-1} + a_{2k})$?
9. Suppose that the series $\sum a_k$ is absolutely convergent. Must every subseries $\sum a_{n_k}$ also be absolutely convergent?
10. Suppose that $a_k \neq 0$ for all $k \geq 1$ such that $\{a_{k+1}/a_k\}_{k\geq 1}$ is a constant sequence. Must $\sum_{k=1}^{\infty} a_k$ be a geometric series?
11. Suppose that $a_k \geq 0$ for all $k \geq 1$ and $\sum_{k=1}^{\infty} a_k$ converges. If $p > 1$, must $\sum_{k=1}^{\infty} a_k^p$ be convergent? Must $\sum_{k=1}^{\infty} a_k a_{k+1}$ be convergent?
12. Suppose that $m > 1$ is a fixed positive integer. Is $\sum_{k=1}^{\infty} a_k$ convergent (respectively divergent) if and only if $\sum_{k=m}^{\infty} a_k$ is convergent (respectively divergent)?
13. Can there exist a divergent series $\sum a_k$ such that $\sum |a_k|$ is convergent?
14. Suppose that $\sum |a_k|$ diverges. Must $\sum a_k$ be divergent?
15. Suppose that $\sum a_k$ and $\sum b_k$ both converge. Must $\sum a_k b_k$ be convergent?
16. Is it possible that $\sum_{k=0}^{\infty} a_k^2 = \sum_{k=0}^{\infty} b_k^2 = 4$ and $\sum_{k=0}^{\infty} a_k b_k = 5$?
17. Suppose that $\sum_{k=1}^{\infty} a_k$ converges with sum A. Must $\sum_{k=1}^{\infty}(a_k + a_{k+1})$ be convergent? If so, what is its sum?
18. Suppose that $\{a_n\}_{n\geq 1}$ is defined by

$$a_n = \frac{1}{n^2} + \frac{1}{(n+1)^2} + \cdots + \frac{1}{(2n)^2}.$$

Does it converge to 0?

19. Is the absolute convergence of $\sum a_k$ equivalent to saying that $\sum |a_k| < \infty$?
20. If $0 < a < 1$, must $\sum_{k=1}^{\infty} a^{k!}$ be convergent?
21. Can different decimal expansions represent the same real number?

Exercises 5.17.

1. Explain the fallacy in the following arguments:
   (a) Set $S = 1+2+4+8+16+32+\cdots$. If we multiply through by 2, we obtain
   $$2S = 2+4+8+16+32+\cdots = S-1,$$
   which we can rewrite in the form $2S = S-1$. This gives $S=-1$.
   (b) Set
   $$S_1 = 1+\frac{1}{3}+\frac{1}{5}+\frac{1}{7}+\cdots \quad \text{and} \quad S_2 = \frac{1}{2}+\frac{1}{4}+\frac{1}{6}+\frac{1}{8}+\cdots.$$
   Then
   $$2S_2 = 1+\frac{1}{2}+\frac{1}{3}+\frac{1}{4}+\cdots,$$
   which can be rewritten as $S_1+S_2 = 2S_2$, so that $S_2 = S_1$. However, each term of S_1 is greater than the corresponding term of S_2, so $S_1 > S_2$.
2. Suppose that $\lim_{k\to\infty} a_k = A$, where A is finite. Is it possible to find the sum $\sum_{k=1}^{\infty}(a_k - a_{k+3})$ in terms of a_1, a_2, a_3, and A? If so, find the sum; if not, explain why it is impossible.
3. Using Theorem 5.9 or otherwise, evaluate the series
   (a) $\displaystyle\sum_{k=0}^{\infty} \frac{1}{(a+k)(a+k+1)}$ $(-a \notin \mathbb{N}_0)$. (b) $\displaystyle\sum_{k=2}^{\infty} \frac{1}{k^2-1}$.
   (c) $\displaystyle\sum_{k=0}^{\infty} \frac{1}{(a+k)(a+k+2)}$ $(-a \notin \mathbb{N}_0)$. (d) $\displaystyle\sum_{k=1}^{\infty} \left[k^{1/k} - (k+2)^{1/(k+2)}\right]$.
4. Determine whether each of the following is a geometric series. If so, determine whether the series converges or diverges. Find also the sum of each convergent series.
   (a) $\displaystyle\sum_{k=1}^{\infty} \frac{2^k}{3^{k+5}}$. (b) $\displaystyle\sum_{k=2}^{\infty} \frac{(-5)^{k-3}}{7^{k+2}}$. (c) $\displaystyle\sum_{k=5}^{\infty} e^{-0.5k}$.
5. For each of the following, determine whether the given series is convergent.
   (a) $\displaystyle\frac{1}{5}-\frac{1}{5^2}+\frac{1}{5^3}-\frac{1}{5^4}+\cdots$.
   (b) $1+e+e^2+e^3+\cdots$.
   (c) $\displaystyle\frac{1}{7}+\left(\frac{1}{7}\right)^4+\left(\frac{1}{7}\right)^7+\left(\frac{1}{7}\right)^{10}+\cdots$.
   (d) $\displaystyle\frac{3}{4}-\left(\frac{3}{4}\right)^3+\left(\frac{3}{4}\right)^5-\left(\frac{3}{4}\right)^7+\cdots$.
   (e) $\displaystyle 2+\sqrt{2}+1+\frac{1}{\sqrt{2}}+\frac{1}{2}+\cdots$.

(f) $3 - \sqrt{3} + 1 - \dfrac{1}{\sqrt{3}} + \dfrac{1}{3} - \cdots$.

(g) $(\sqrt{2}+1) + 1 + (\sqrt{2}-1) + (3-2\sqrt{2}) + \cdots$.

6. In each of the given telescoping series, determine whether the series converges or diverges by examining the limit of the nth partial sums. Find also the sum of the series if the series is convergent.

(a) $\displaystyle\sum_{k=1}^{\infty} \frac{k-1}{2^{k+1}}$. (b) $\displaystyle\sum_{k=1}^{\infty} \frac{\sqrt{k+1}-\sqrt{k}}{\sqrt{k^2+k}}$. (c) $\displaystyle\sum_{k=1}^{\infty} \frac{\log\left(k^{k+1}/(k+1)^k\right)}{k(k+1)}$.

(d) $\displaystyle\sum_{k=1}^{\infty} \frac{4k-1}{5^{k+1}}$. (e) $\displaystyle\sum_{k=1}^{\infty} \frac{3k+1}{k^2(2k+1)^2}$. (f) $\displaystyle\sum_{k=1}^{\infty} \left[\frac{1}{k^{\alpha}} - \frac{1}{(k+1)^{\alpha}}\right]$.

7. Find $\displaystyle\sum_{k=0}^{\infty}(3a_k + 3^{-k})$ given that $\displaystyle\sum_{k=0}^{\infty} a_k = 1$.

8. Evaluate

(a) $\displaystyle\sum_{k=1}^{\infty} \sin\frac{1}{3^n}\cos\frac{2}{3^n}$. (b) $\displaystyle\sum_{k=0}^{\infty}\left(\frac{1}{3^k}+\frac{1}{5^k}\right)^2$.

9. Suppose that $\{a_n\}$ is a sequence such that $na_n \to 0$ as $n \to \infty$. Then show that $\sum_{k=1}^{\infty} a_k$ is convergent if and only if $\sum_{k=1}^{\infty} k(a_k - a_{k+1})$ is convergent. Will the two series have the same sum?

10. Let T be an equilateral triangle with sides of length 1 unit. Remove the middle triangle formed by joining the midpoints of the sides of T (see Figure 5.6(a)). This leaves three equal triangular regions (Figure 5.6(b)), and the next step is to remove the middle triangle from each of these (Figure 5.6(c)). If one continues the process indefinitely, what remains is called the Sierpiński triangle (Figure 5.6(d)). Show that the Sierpiński triangle has area 0.

11. Consider a square $ABCD$ having sides of unit length. Form a square $EFGH$ by connecting the midpoints of the sides of the first square, as shown in the first diagram in Figure 5.7.

Fig. 5.6. Sierpinski gasket.

Assume that the pattern of shaded regions in the square is continued indefinitely. Find the total area of the shaded regions at the nth stage and in the infinite limit.

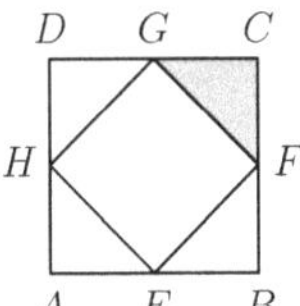

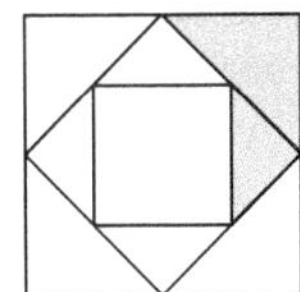
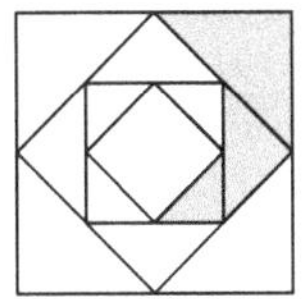
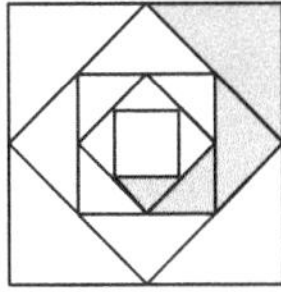

Fig. 5.7. Shaded regions at the nth stage, $n = 1, 2, 3, 4$.

12. Evaluate $\displaystyle\sum_{k=1}^{\infty} \frac{20^k}{(5^k - 4^k)(5^{k+1} - 4^{k+1})}$.
13. Interpret $0.\overline{12}$ as an infinite series, and find the value of $0.\overline{12}$ as a fraction.

5.2 Convergence and Divergence Tests for Series

In investigating the series $\sum a_k$, the most important problem is to find tests for convergence. In general, it is not possible to obtain a convenient formula for the nth partial sum of a given infinite series. On the other hand, in a series, the sequence $\{a_k\}$ of the terms of the series is generally more accessible than the sequence of partial sums $\{S_n\}$, and so it would be convenient if the question of convergence of $\{S_n\}$ could be settled by investigating the limiting behavior of the general term a_k. This is illustrated in the following test.

5.2.1 Basic Divergence Tests

Theorem 5.18. *A necessary, but not a sufficient, condition for a series* $\sum a_k$ *to converge is that* $a_k \to 0$ *as* $k \to \infty$.

Proof. The proof follows if we let $m = n + 1$ (or $p = 1$) in Theorem 5.13.

For a direct proof, let $\sum a_k$ be convergent. Suppose that the sequence of partial sums $\{S_n\}$ converges to the limit L. Then

$$S_n - S_{n-1} = a_n \ (n \geq 2).$$

It follows that

$$\lim a_n = \lim (S_n - S_{n-1}) = \lim S_n - \lim S_{n-1} = L - L = 0.$$

Thus, for the convergent series $\sum a_k$, we must have $\lim a_k = 0$, as desired. ∎

For instance, for $p \leq 0$, the series $\sum_{k=1}^{\infty} \frac{(-1)^{k-1}}{k^p}$ is divergent, because the general term does not approach zero.

An equivalent formulation of Theorem 5.18 gives a particular test that enables us to test the divergence of a variety of series.

Corollary 5.19 (The divergence test). *If* $\lim a_k$ *does not exist, or exists but is not equal to zero, then the series* $\sum a_k$ *must diverge.*

We remind the reader that this test *cannot* be used to show convergence of a series. That is, the converse of Theorem 5.18 is false. There are sequences for which $a_k \to 0$ as $k \to \infty$ but $\sum a_k$ diverges. For instance, the harmonic series $\sum_{k=1}^{\infty}(1/k)$ is known to be divergent even though the general term $1/k$ tends to 0 as $k \to \infty$.

There are a number of ways to show that the sequence $\{a_k\}$ does not tend to zero. For example, if

- either $\{a_k\}$ has a convergent subsequence with nonzero limit
- or $\{a_k\}$ has a subsequence that tends to ∞ or a subsequence that tends to $-\infty$,

then $\sum a_k$ is divergent, by the divergence test.

In Example 7.27, we shall use the integral test to show that the harmonic series diverges. Also, $\sum_{k=1}^{\infty} \frac{1}{2k-1}$ diverges even though the general term of the series approaches zero as $k \to \infty$. However, using the above divergence test, we present below some simple examples:

1. the series $\displaystyle\sum_{k=1}^{\infty} \frac{k}{2k+1}$ diverges because $\displaystyle\lim_{k\to\infty} \frac{k}{2k+1} = \frac{1}{2}$.
2. the series
$$\sqrt{\frac{1}{8}} + \sqrt{\frac{2}{10}} + \sqrt{\frac{3}{12}} + \cdots + \sqrt{\frac{k}{2(k+3)}} + \cdots$$
diverges because the general term approaches $1/\sqrt{2}$ as $k \to \infty$.
3. the series $\sum_{k=1}^{\infty} \cos(1/k)$ is divergent because $\cos(1/k) \to \cos 0 = 1$ as $k \to \infty$.
4. the geometric series $\sum_{k=0}^{\infty} ar^k$ with $a \neq 0$ and $|r| \geq 1$ diverges because the general term $a_k = ar^k$ does not converge to zero as $k \to \infty$. Equivalently, we say that the geometric sequence $\{ar^k\}_{k\geq 1}$ for $|r| \geq 1$ is not summable unless $a = 0$.
5. the series $\sum_{k=1}^{\infty} k^{1/k}$ is not convergent because $k^{1/k} \to 1 \neq 0$ as $k \to \infty$.
6. The series $\sum_{k=1}^{\infty} \frac{k^k}{k!}$ diverges, because
$$\frac{k^k}{k!} = \left(\frac{k}{1}\right)\left(\frac{k}{2}\right)\left(\frac{k}{3}\right)\cdots\left(\frac{k}{k}\right) \geq 1 \text{ for all } k \geq 1,$$
so that the general term does not approach zero.

5.2.2 Tests for Series of Nonnegative Terms

Series whose terms are all nonnegative numbers play a crucial role in the general discussion of series and in applications. In a series $\sum_{k\geq 1} a_k$ with $a_k \geq 0$ for all $k \geq 1$, we have

$$S_n = S_{n-1} + a_n,$$

and so the sequence of partial sums $\{S_n\}$ is increasing, and therefore the bounded monotone convergence theorem (see Theorem 2.25) immediately allows us to formulate the following convergence criterion for series with nonnegative terms.

Theorem 5.20. *Suppose that $a_k \geq 0$ for all $k \geq 1$. Then the series $\sum a_k$ either converges or diverges to ∞. In particular, if the sequence of partial sums $\{S_n\}$ is bounded above, then $\sum a_k$ converges, and in this case $\sum a_k = \sup\{S_n : n \geq 1\}$.*

An immediate consequence of Theorem 5.20 is the following.

Corollary 5.21. *Suppose that $a_k \geq 0$ for all k. Then the series $\sum a_k$ converges if and only if the sequence of partial sums is bounded.*

If $a_k \geq 0$ for all k, then we write $\sum a_k < \infty$ if the series converges and $\sum a_k = \infty$ if it diverges.

Note that if $a_k \leq 0$ for all k, we can still apply the above results after making a sign change. On the other hand, it is harder to determine the behavior of a series with both positive and negative terms, because the sequence $\{S_n\}$ of partial sums is not monotone. However, the above results can be used if $\{a_k\}$ contains only finitely many negative terms.

Next, we note that Theorem 5.20 may fail if the a_k are not nonnegative for all k. For example, if $a_k = (-1)^{k-1}$, then the sequence of partial sums $\{S_n\}$ of $\sum a_k$ is either 0 or 1. Here, even though $\{S_n\}$ is bounded above by 1, $\{S_n\}$ is not convergent.

Note also that if $a_k = 1+(-1)^{k-1}$, then $a_k \geq 0$ for all k, and the sequence of partial sums $\{S_n\}$ of $\sum_{k=1}^{\infty} a_k$ diverges to ∞.

Example 5.22. Prove that the series $\sum_{k=1}^{\infty}(k/(k+3))x^k$ converges for each x in $[0,1)$.

Solution. Fix $x \in [0,1)$ and set $a_k(x) = kx^k/(k+3)$. Then $a_k(x) \geq 0$ for all fixed $x \geq 0$, and for each fixed x, consider $\sum_{k=1}^{\infty} a_k(x)$ as a numerical series. Further, since $k/(k+3) < 1$ for all $k \geq 1$,

$$a_k(x) \leq x^k \quad \text{for } x \geq 0.$$

Thus, for $x \in [0,1)$,

$$0 \leq S_n = \sum_{k=1}^{n} a_k(x) \leq \sum_{k=1}^{n} x^k = \frac{x(1-x^n)}{1-x} \leq \frac{x}{1-x},$$

so the sequence of partial sums is increasing and bounded by $x/(1-x)$ for each $x \in [0,1)$. By Theorem 5.20, the series converges. We shall soon see that the series also converges for $-1 < x < 0$, showing that the given series is actually convergent for $x \in (-1,1)$. ●

5.2.3 Abel–Pringsheim Divergence Test

We can strengthen Theorem 5.18 in the following way.

Theorem 5.23 (Abel–Pringsheim test). *Suppose that $\{a_k\}$ is a decreasing sequence of positive real numbers. Then a necessary condition for the series $\sum a_k$ to converge is that $ka_k \to 0$ as $k \to \infty$.*

Proof. Let $\sum_{k=1}^{\infty} a_k$ be convergent, where $a_k \geq 0$ and $a_k \geq a_{k+1}$ for all k. Then the sequence of partial sums $\{S_n\}$ is convergent, and hence it is Cauchy. Because it is Cauchy, given $\epsilon > 0$ there exists an N such that

$$\sum_{k=n+1}^{m} a_k = |S_m - S_n| < \epsilon \quad (m > n \geq N).$$

Since $\{a_k\}$ is a decreasing sequence of positive real numbers, it follows that for $n \geq N$ (choose $m = 2n$),

$$na_{2n} \leq a_{n+1} + a_{n+2} + \cdots + a_{2n} < \epsilon \quad (n \geq N),$$

which implies that

$$\lim_{n\to\infty} 2na_{2n} = 0.$$

Again, because $\{a_k\}$ is decreasing,

$$(2n+1)a_{2n+1} \leq (2n+1)a_{2n} \leq (2n+n)a_{2n} = 3na_{2n},$$

which implies that

$$\lim_{n\to\infty} (2n+1)a_{2n+1} = 0.$$

Thus, we must have $\lim ka_k = 0$, as desired. ∎

In Theorem 5.23, the condition $ka_k \to 0$ as $k \to \infty$ is not sufficient for the convergence of the series $\sum a_k$. For example, consider $a_k = 1/(k \log k)$. Then $\lim_{k\to\infty} ka_k = 0$, but $\sum_{k=2}^{\infty} a_k$ diverges (see Example 7.29(c) with $p = 1$).

The assumption that $\{a_k\}$ is decreasing is essential (see Question 5.42(3)). That is, there are convergent positive series $\sum a_k$ such that $ka_k \not\to 0$ as $k \to \infty$. Theorem 5.23 gives another useful result for testing the divergence of a variety of series.

Corollary 5.24 (The divergence test). *If $\{a_k\}$ is a decreasing sequence of positive real numbers and $\lim ka_k$ does not exist or exists but is not equal to zero, then the series $\sum a_k$ must diverge.*

Remark 5.25. Corollary 5.24 shows immediately that the harmonic series $\sum a_k$ $(a_k = 1/k)$ diverges, since $\{a_k\}$ is a decreasing positive sequence and $ka_k = 1$ does not approach zero. ●

Example 5.26. Set $a_1 = 1$ and $a_{n+1} = \frac{2n-1}{2n}a_n$ for all $n \geq 1$. Show that $\sum_{k=1}^{\infty} a_k$ diverges.

Solution. Clearly, $\{a_n\}$ is decreasing. Also, we see that (as $a_1 = 1$)

$$\begin{aligned} a_{n+1} &= \frac{2n-1}{2n} \cdot \frac{2n-3}{2(n-1)} \cdots \frac{3}{4} \cdot \frac{1}{2} \\ &= \frac{1}{2n}\left(\frac{2n-1}{2n-2}\right)\left(\frac{2n-3}{2n-4}\right) \cdots \frac{5}{4} \cdot \frac{3}{2} \cdot 1 \\ &\geq \frac{1}{2n} \quad \text{for each } n \geq 2. \end{aligned}$$

Therefore, $(n+1)a_{n+1}$ does not converge to zero. By Corollary 5.24, the series $\sum a_k$ diverges. Also, the comparison test below gives the desired conclusion.

Alternatively, we may note that

$$a_{n+1} = \left(1 - \frac{1}{2n}\right) a_n < a_n,$$

showing that $\{a_n\}$ is decreasing. Moreover, by induction it is easy to show that

$$a_{n+1} \geq \frac{1}{n+1} \quad \text{for all } n \geq 1.$$

Indeed, if $a_1 = 1$, $a_2 = 1/2$ and if $a_n \geq 1/n$, then

$$a_{n+1} \geq \left(\frac{2n-1}{2n}\right)\frac{1}{n},$$

which is obviously greater than $1/(n+1)$, and the claim is true. Therefore, $(n+1)a_{n+1} \geq 1$, so that $(n+1)a_{n+1}$ does not approach 0 as $n \to \infty$. By Corollary 5.24, $\sum_{k=1}^{\infty} a_k$ diverges. ●

5.2.4 Direct Comparison Test

There are many tests for series of nonnegative terms. We discuss here two fundamental convergence criteria called the *direct comparison test* and the *limit comparison test*. They are in fact an important consequence of Theorem 5.20. A few other tests will be given in the subsequent sections.

Theorem 5.27 (Direct comparison test). *Suppose there exist numbers $N \in \mathbb{N}$ and $M > 0$ such that $0 \leq a_k \leq Mc_k$ for all $k \geq N$. Then we have*

- *if $\sum c_k$ converges, then $\sum a_k$ also converges and $\sum_{k \geq N} a_k \leq M \sum_{k \geq N} c_k$;*
- *if $\sum a_k$ diverges, then $\sum c_k$ also diverges.*

Proof. Assume that the series $\sum_{k=1}^{\infty} c_k$ is convergent. Then $\sum_{k=N}^{\infty} c_k$ also converges; let the sum be S. Then $S_n = \sum_{k=1}^{N-1} a_k + T_n$, where

$$T_n := \sum_{k=N}^{n} a_k \leq M \sum_{k=N}^{n} c_k \leq M \sum_{k=N}^{\infty} c_k = \text{MS} \quad \text{(say)},$$

and so the sequence of partial sums $\{T_n\}$ of the series $\sum_{k=N}^{\infty} a_k$ is increasing (because $a_k \geq 0$ for all $k \geq N$) and bounded above by MS. Thus, by BMCT (Theorem 2.25), the series $\sum_{k=N}^{\infty} a_k$ converges, and hence $\sum_{k=1}^{\infty} a_k$ converges.

To prove the second part, assume that the series $\sum a_k$ diverges. Then, since $a_k \geq 0$, it follows that

$$\sum_{k=N}^{\infty} a_k = \infty.$$

Thus,

$$\sum_{k=N}^{\infty} c_k \geq \frac{1}{M} \sum_{k=N}^{\infty} a_k,$$

and so $\sum c_k$ must diverge to ∞. ∎

By Theorem 5.27, we conclude that $\sum(k/(k^2+4))$ diverges, because

$$\frac{k}{k^2+4} \geq \frac{1}{2k} \quad \text{for } k \geq 2$$

and the harmonic series diverges.

Example 5.28. We have already shown by a direct method that the harmonic p-series $\sum_{k=1}^{\infty} \frac{1}{k^p}$ converges for $p = 2$ and diverges for $p = 1$. Using these two facts, one can easily provide an independent proof (e.g., without using the integral test) that the p-harmonic series converges if $p > 1$ and diverges if $p \leq 1$. In order to achieve this, we observe that

$$\frac{1}{k^p} \begin{cases} \geq \dfrac{1}{k} & \text{for } p \leq 1, \\[2mm] \leq \dfrac{1}{k^2} & \text{for } p \geq 2. \end{cases}$$

The direct comparison test gives the conclusion except when $1 < p < 2$. The following argument works for $p > 1$, although in view of this observation, it suffices to deal with the case $1 < p < 2$. Now,

$$S_{2^n-1} = 1 + \left(\frac{1}{2^p} + \frac{1}{3^p}\right) + \left(\frac{1}{4^p} + \frac{1}{5^p} + \frac{1}{6^p} + \frac{1}{7^p}\right) + \cdots + \left(\frac{1}{(2^{n-1})^p} + \cdots + \frac{1}{(2^n-1)^p}\right).$$

There are $2^n - 2^{n-1} = 2^{n-1}$ terms in the last bracketed term of the above expression on the right, each less than $1/2^{(n-1)p}$. Thus, we have

$$\begin{aligned} S_{2^n-1} &< 1 + \frac{2}{2^p} + \frac{4}{4^p} + \cdots + \frac{2^{n-1}}{2^{(n-1)p}} \\ &= 1 + \frac{1}{2^{p-1}} + \left(\frac{1}{2^{p-1}}\right)^2 + \cdots + \left(\frac{1}{2^{p-1}}\right)^{n-1}. \end{aligned}$$

Since $p > 1$, we have $1/2^{p-1} < 1$, and so

$$S_{2^n-1} < \frac{1}{1 - 1/2^{p-1}} = M, \quad \text{say.}$$

Moreover, for each fixed $N \geq 1$, there exists an n such that $2^n > N$. Therefore,

$$S_N \leq S_{2^n-1} < M,$$

so that $\{S_N\}$ is an increasing sequence of partial sums bounded above by M. Hence, by BMCT (see Theorem 2.27), $\{S_N\}_{N\geq 1}$ is convergent. ●

Assuming only the divergence of the harmonic series and the convergence part of the geometric series, we now discuss some simple examples.

Example 5.29. Test each of the following series for convergence:

(a) $\displaystyle\sum_{k=1}^{\infty} \frac{1}{5^k + 1}$. (b) $\displaystyle\sum_{k=2}^{\infty} \frac{2}{2\sqrt{k} - 3}$. (c) $\displaystyle\sum_{k=1}^{\infty} \frac{1}{k!}$. (d) $\displaystyle\sum_{k=1}^{\infty} \frac{1}{\sqrt{k^3 + 1}}$.

Solution. We apply the direct comparison test with $M = 1$. First we see that

$$\frac{1}{5^k + 1} < \frac{1}{5^k}, \quad \frac{2}{2\sqrt{k} - 3} > \frac{1}{\sqrt{k}}, \quad \frac{1}{k!} \leq \frac{1}{2^{k-1}}, \quad \frac{1}{\sqrt{k^3 + 1}} < \frac{1}{\sqrt{k^3}}$$

for all $k > 3$, and note that $k! \geq 2^{k-1}$.

(a) The first series is convergent because it is *dominated by* the convergent geometric series $\displaystyle\sum \frac{1}{5^k}$.

(b) We know that $\sum_{k=1}^{\infty} \frac{1}{k^{1/2}}$ diverges. The direct comparison test tells us that the given series must diverge.

(c) The given series is *dominated by* the convergent geometric series $\sum \frac{1}{2^{k-1}}$. Therefore, the given series must also converge.

(d) Because of the last inequality in the above list of inequalities, the given series converges. ●

The direct comparison test (see Theorem 5.27) suggests the following.

Problem 5.30. *Can we replace the existence of the constant $M > 0$ in Theorem 5.27 by a bounded sequence $\{b_k\}$ and conclude something about the behavior of the product sequence $\sum a_k b_k$?*

Theorem 5.31. *Suppose that the series* $\sum a_k$ *converges absolutely and the sequence* $\{b_k\}$ *is bounded. Then* $\sum a_k b_k$ *converges absolutely.*

Proof. Suppose that M is bounded for the sequence $\{b_k\}$. Then since $|a_k b_k| \leq M|a_k|$, the conclusion follows from the comparison test. ∎

The conclusion of Theorem 5.31 fails in general if $\{b_n\}$ is not bounded. For instance, if $b_k = k$ and $a_k = 1/k^{p+1}$ $(0 < p \leq 1)$, then $\sum a_k$ converges absolutely, whereas $\sum a_k b_k = \sum(1/k^p)$ diverges for $0 < p \leq 1$.

Also, Theorem 5.31 fails in general if we replace the assumption that $\sum a_k$ converges absolutely by the assumption that $\sum a_k$ converges. This may be seen by choosing

$$b_k = (-1)^{k-1} \text{ and } a_k = \frac{(-1)^{k-1}}{k^p} \quad (0 < p \leq 1).$$

Corollary 5.32. *If* $\sum a_k$ *is a convergent series of nonnegative terms and* $\{b_k\}$ *is a sequence of nonnegative real numbers with an upper bound, then the series* $\sum a_k b_k$ *is convergent.*

Applying Theorem 5.31 with $a_k = 1/k^p$ $(p > 1)$ and

$$b_k = (-1)^{k-1}, \quad k^{1/k}, \quad (k + 1/k)^k, \quad (1 + 1/k)^{2k}, \quad \sin kx \ (x \in \mathbb{R}),$$

it follows that each of the following series converges:

$$\sum_{k=1}^{\infty} \frac{(-1)^{k-1}}{k^p}, \quad \sum_{k=1}^{\infty} \frac{k^{1/k}}{k^p}, \quad \sum_{k=1}^{\infty} \frac{(1 + 1/k)^k}{k^p}, \quad \sum_{k=1}^{\infty} \frac{(1 + 1/k)^{2k}}{k^p}, \quad \sum_{k=1}^{\infty} \frac{\sin kx}{k^p}.$$

5.2.5 Limit Comparison Test

It is not always easy or even possible to make a suitable direct comparison between two similar series. This is clearly a disadvantage of the comparison test, since it demands a "known" series in order to apply the comparison test. However, it might still be possible to guess the series' behavior by examining the "order of magnitude" of the general term as given in the limit comparison test below. For example, it is natural to expect that the series $\sum 1/(3^k - 5)$ converges by comparing the kth term of the convergent geometric series $\sum 1/3^k$. We first note that $k \geq 2$,

$$0 \leq \frac{1}{3^k} < \frac{1}{3^k - 5}.$$

Although $\sum 1/3^k$ is convergent, the comparison test with $M = 1$ *cannot* be used to determine the convergence of $\sum 1/(3^k - 5)$ by directly comparing it with $\sum 1/3^k$. On the other hand, if $k \geq 3$, we have

$$0 < \frac{1}{3^k - 5} \leq \frac{2}{3^k},$$

and so by a comparison with the convergent series $2\sum 1/3^k$, the given series $\sum 1/(3^k - 5)$ converges. So we need to pin down the underlying idea from the direct comparison test. Here is a simple reformulation of the comparison test that is often useful in practice.

Theorem 5.33 (Limit comparison test). *Suppose $a_k > 0$ and $b_k > 0$ for all $k \geq N_0$ such that*

$$\lim_{k\to\infty} \frac{a_k}{b_k} = L.$$

Then we have the following:

(a) *If $0 < L < \infty$, then the two series $\sum a_k$ and $\sum b_k$ either both converge or both diverge.*
(b) *If $L = 0$, then the series $\sum a_k$ converges whenever $\sum b_k$ converges.*
(c) *If $L = \infty$, then the series $\sum a_k$ diverges whenever $\sum b_k$ diverges.*

Proof. Since a_k and b_k are positive for $k \geq N_0$, we observe that $L \geq 0$ or $L = \infty$.

(a) Let $0 < L < \infty$. Then since $\lim(a_k/b_k) = L$, given $\epsilon = L/2 > 0$ there exists a positive integer N such that for all $k \geq N$ $(\geq N_0)$ we have

$$\left.\begin{array}{l} L - \dfrac{a_k}{b_k} \\ \dfrac{a_k}{b_k} - L \end{array}\right\} \leq \left|\frac{a_k}{b_k} - L\right| < \frac{L}{2}, \quad \text{i.e.,} \quad \frac{L}{2} \leq \frac{a_k}{b_k} < \frac{3L}{2}.$$

This amounts to

$$0 < \frac{L}{2} b_k \leq a_k < \frac{3L}{2} b_k \quad \text{for } k \geq N.$$

Now the comparison test yields the desired conclusion.

(b) Let $L = 0$. Then there exists an N such that

$$\frac{a_k}{b_k} < 1, \quad \text{i.e.,} \quad 0 < a_k < b_k, \quad \text{for all } k \geq N \ (\geq N_0).$$

The direct comparison test gives the conclusion.

(c) Let $L = \infty$. Then there exists an N such that

$$\frac{a_k}{b_k} > 1, \quad \text{i.e.,} \quad a_k > b_k, \quad \text{for all } k \geq N \ (\geq N_0).$$

Again the conclusion follows from the direct comparison test. ∎

Corollary 5.34. *Suppose that $\{a_k\}$ and $\{b_k\}$ are two sequences of real numbers such that*

$$\lim_{k\to\infty}\left|\frac{a_k}{b_k}\right| = L$$

exists and $0 < L < \infty$. Then the series $\sum a_k$ converges absolutely if and only if $\sum b_k$ converges absolutely.

Remark 5.35. When $L = 0$ and $L = \infty$ in Theorem 5.33, then nothing more can be concluded. We illustrate this fact by some simple examples.

1. Set $a_k = 1/k^p$ $(p > 1)$ and $b_k = 1/k$ for $k \geq 1$. Then

$$\lim_{k\to\infty}\frac{a_k}{b_k} = \lim_{k\to\infty}\frac{1}{k^{p-1}} = 0 \text{ and } \lim_{k\to\infty}\frac{b_k}{a_k} = \lim_{k\to\infty}k^{p-1} = \infty,$$

where $\sum a_k$ converges but $\sum b_k$ diverges.

2. Set $a_k = 1/k^p$ and $b_k = 1/k^q$, where $0 < p < q < 1$. Then

$$\lim_{k\to\infty}\frac{a_k}{b_k} = \lim_{k\to\infty}k^{q-p} = \infty \text{ and } \lim_{k\to\infty}\frac{b_k}{a_k} = \lim_{k\to\infty}\frac{1}{k^{q-p}} = 0,$$

where both $\sum a_k$ and $\sum b_k$ diverge.

3. Set $a_k = 1/k^p$ and $b_k = 1/k^q$, where $p > q > 1$. Then

$$\lim_{k\to\infty}\frac{a_k}{b_k} = \lim_{k\to\infty}\frac{1}{k^{p-q}} = 0 \text{ and } \lim_{k\to\infty}\frac{b_k}{a_k} = \lim_{k\to\infty}k^{p-q} = \infty,$$

where both $\sum a_k$ and $\sum b_k$ converge. •

According to the limit comparison test, it is possible to try to determine whether a_k is comparable with the kth term of some familiar series whose convergence (or divergence) is known.

Example 5.36. Investigate the convergence of the series $\sum_{k=1}^{\infty} a_k$, where a_k equals

(a) $\sqrt{\dfrac{1+5^k}{1+7^k}}$. **(b)** $\dfrac{7k^2+4}{\sqrt{k}(k^2-15)}$. **(c)** $\dfrac{2k+700}{e^{k/7}-90}$. **(d)** $\sin(1/k)$.

Solution. **(a)** Set $b_k = \sqrt{5^k/7^k}$. We compute the limit

$$L = \lim_{k\to\infty}\frac{a_k}{b_k} = \lim_{k\to\infty}\sqrt{\frac{1+5^k}{1+7^k}\times\frac{7^k}{5^k}} = \lim_{k\to\infty}\sqrt{\frac{(1/5^k)+1}{(1/7^k)+1}} = 1.$$

So the limit comparison test tells us that the given series converges because $\sum_{k=1}^{\infty} b_k$ converges.

(b) Set

$$a_k = \frac{7k^2+4}{\sqrt{k}(k^2-15)} \quad \text{and} \quad b_k = \frac{k^2}{\sqrt{k}(k^2)} = \frac{1}{\sqrt{k}}$$

and compute the limit

$$\lim_{k\to\infty} \frac{a_k}{b_k} = \lim_{k\to\infty} \frac{7k^2+4}{k^2-15} = 7.$$

Since $\sum b_k$ diverges, by the limit comparison test, we conclude that the given series diverges.

(c) If we let

$$a_k = \frac{2k+700}{e^{k/7}-90} \quad \text{and} \quad b_k = \frac{k}{e^{k/7}},$$

then we find that $\lim_{k\to\infty} a_k/b_k = 2$. In Example 7.29, we shall show that the series $\sum k e^{-k/7}$ converges. We see that by the limit comparison test, the given series also converges.

(d) Set $a_k = \sin(1/k)$ and $b_k = 1/k$. Then $a_k > 0$ and

$$L = \lim_{k\to\infty} \frac{a_k}{b_k} = \lim_{k\to\infty} \frac{\sin(1/k)}{1/k} = \lim_{x\to 0} \frac{\sin x}{x} = 1,$$

showing that $\sum \sin(1/k)$ is divergent, because $\sum b_k$ is divergent. •

Example 5.37. We show that the p-log *series* defined by $\sum_{k=2}^{\infty} (\log k)/k^p$ converges if $p > 1$, and diverges if $p \leq 1$.

First we suppose that $p > 1$. Then there exists a number q such that $1 < q < p$. Indeed, $q = (1+p)/2$ satisfies the condition. Next, set $a_k = (\log k)/k^p$ and $b_k = 1/k^q$. Then, using l'Hôpital's rule for ∞/∞ form, we compute

$$\lim_{k\to\infty} \frac{a_k}{b_k} = \lim_{k\to\infty} \frac{\log k}{k^{p-q}} = \lim_{k\to\infty} \frac{1/k}{(p-q)k^{p-q-1}} = \lim_{k\to\infty} \frac{1}{(p-q)k^{p-q}} = 0$$

as $p - q > 0$. Because the harmonic q-series $\sum 1/k^q$ converges for $q > 1$, the given series converges by the limit comparison test for all $p > 1$.

Next we suppose that $p \leq 1$. Set $B_k = 1/k^p$. Then $a_k \geq B_k$ for all $k \geq 3$ and

$$\lim_{k\to\infty} \frac{a_k}{B_k} = \lim_{k\to\infty} \log k = \infty.$$

Again, because $\sum B_k = \sum \frac{1}{k^p}$ diverges for $p \leq 1$, it follows from the comparison test that the given series diverges when $p \leq 1$. •

Example 5.38. Discuss the convergence of $\sum_{k=1}^{\infty} a_k$ when a_k equals

(a) $\log\left(1+\frac{1}{k}\right) - \frac{1}{k}$. **(b)** $k^{1/k} - 1$. **(c)** $a^{1/k} - 1$ for $a > 0$.

Solution. **(a)** Set $a_k = \frac{1}{k} - \log(1+1/k)$ and $b_k = 1/k^2$. Then

$$\lim_{k\to\infty} \frac{a_k}{b_k} = \lim_{x\to 0} \frac{x - \log(1+x)}{x^2} = \lim_{x\to 0} \frac{1 - 1/(1+x)}{2x} = \lim_{x\to 0} \frac{1}{2(1+x)} = \frac{1}{2},$$

and so $\sum_{k=1}^{\infty} a_k$ converges, since $\sum_{k=1}^{\infty} b_k$ converges.

(b) Set $a_k = k^{1/k} - 1$ and $b_k = 1/k$. Then $a_k \to 0$ as $k \to \infty$. Since $e^x \geq 1+x$ for $x \geq 0$, it follows that $x \geq \log(1+x)$ for $x \geq 0$. Consequently, with $x = a_k$, the last inequality gives

$$\frac{\log k}{k} = \log(k^{1/k}) \leq k^{1/k} - 1, \quad \text{i.e.,} \quad a_k \geq \frac{\log k}{k} > \frac{1}{k} \quad \text{for} \quad k > 3,$$

which implies that $\sum_{k=1}^{\infty} a_k$ diverges.

(c) Set $a_k = a^{1/k} - 1$ and $b_k = 1/k$. Then for $a > 1$, we have

$$\lim_{k\to\infty} \frac{a_k}{b_k} = \lim_{x\to 0} \frac{a^x - 1}{x} = \lim_{x\to 0} \frac{e^{x\log a} - 1}{x} = \lim_{x\to 0} \frac{(\log a)e^{x\log a}}{1} = \log a.$$

If $0 < a < 1$, then set $b = 1/a > 1$ and observe that

$$\lim_{k\to\infty} \frac{a_k}{-b_k} = \lim_{k\to\infty} \frac{b^{-1/k} - 1}{-1/k} = \lim_{y\to 0^-} \frac{b^y - 1}{y} = \log b = \log(1/a).$$

Consequently, $\sum_{k=1}^{\infty} a_k$ diverges to ∞ if $a > 1$ and diverges to $-\infty$ if $0 < a < 1$. ●

5.2.6 Cauchy's Condensation Test

Our next result is useful for testing the convergence of certain series painlessly.

Theorem 5.39 (Cauchy's condensation test). *Suppose $\{a_n\}_{n\geq 1}$ is a decreasing sequence of nonnegative terms. Then the series $\sum_{k=1}^{\infty} a_k$ and $\sum_{k=1}^{\infty} 2^k a_{2^k}$ are either both convergent or both divergent.*

Proof. Denote by S_n and T_n the nth partial sums of the series $\sum_{k=1}^{\infty} a_k$ and $\sum_{k=1}^{\infty} 2^k a_{2^k}$, respectively. That is, $S_n = \sum_{k=1}^{n} a_k$ and $T_n = \sum_{k=1}^{n} 2^k a_{2^k}$. The proof depends on the following two inequalities: for $n \geq 4$, we have

$$\begin{aligned}\frac{1}{2}\sum_{k=1}^{n} 2^k a_{2^k} &= a_2 + 2a_4 + 4a_8 + \cdots + 2^{n-1}a_{2^n} \\ &< a_1 + a_2 + (a_3 + a_4) + (a_5 + a_6 + a_7 + a_8) + \cdots \\ &\quad + (a_{2^{n-1}+1} + a_{2^{n-1}+2} + \cdots + a_{2^n}).\end{aligned}$$

Also, for each fixed N, choose n such that $2^n > N$. Then

$$\sum_{k=1}^{2^{n+1}-1} a_k = a_1 + (a_2 + a_3) + (a_4 + a_5 + a_6 + a_7) + \cdots$$
$$+(a_{2^n} + a_{2^n+1} + \cdots + a_{2^{n+1}-1})$$
$$\leq a_1 + 2a_2 + 2^2 a_4 + \cdots + 2^n a_{2^n} = a_1 + T_n.$$

We have thus obtained

$$T_n < 2S_{2^n} \quad \text{and} \quad S_{2^{n+1}-1} \leq a_1 + T_n,$$

and both $\{S_n\}$ and $\{T_n\}$ are increasing sequences, by hypothesis. Moreover, $\{S_n\}$ is bounded if and only if $\{T_n\}$ is bounded. By BMCT, $\sum_{k=1}^{\infty} a_k$ converges if and only if $\sum_{k=0}^{\infty} 2^k a_{2^k}$ converges. ∎

Consider $\{a_n\}$ as follows:

$$a_n = \begin{cases} 0 & \text{if } n \neq 2^m, \\ 1/2^n & \text{if } n = 2^m, \end{cases}$$

for some nonnegative integer m. Then $a_1 = 1$, $a_2 = 1/2$, $a_3 = 0$, $a_4 = 1/2^2$, $a_5 = 0$, showing that the sequence $\{a_n\}$ is not decreasing. However,

$$\sum_{k=1}^{\infty} a_k = 1 + \frac{1}{2} + \frac{1}{2^2} + \cdots,$$

which is a convergent geometric series, whereas

$$\sum_{k=1}^{\infty} 2^k a_{2^k} = \sum_{k=1}^{\infty} 2^k \left(\frac{1}{2^k}\right) = \sum_{k=1}^{\infty} 1,$$

which is divergent. This observation shows that the hypothesis that $\{a_n\}$ is decreasing cannot be dropped from Theorem 5.39.

We remark that Cauchy's condensation test is useful because it allows us to investigate the convergence of series just by considering small subsets of the terms of the given series.

Example 5.40. 1. Using the Cauchy condensation test it is easy to see that the harmonic p-series $\sum_{k=1}^{\infty} k^{-p}$ converges if and only if $p > 1$ (see also Example 5.28). Indeed, for $p > 0$, $k^p < (k+1)^p$, and therefore $\{1/k^p\}$ is a decreasing sequence of positive terms. If $a_k = 1/k^p$, then

$$\sum_{k=0}^{\infty} 2^k a_{2^k} = \sum_{k=0}^{\infty} 2^k \left(\frac{1}{2^{kp}}\right) = \sum_{k=0}^{\infty} (2^{1-p})^k,$$

which is a convergent geometric series if and only if $2^{1-p} < 1$, i.e., $p > 1$. Recall that $k^{-p} \to \infty$ for $p < 0$, and $k^{-p} = 1 \neq 0$ if $p = 0$. Consequently, $\sum_{k=1}^{\infty} k^{-p}$ converges if and only if $p > 1$.

2. If $a_k = 1/[k(\log k)^p]$ for $k \geq 2$, then

$$2^k a_{2^k} = 2^k \frac{1}{2^k (\log 2^k)^p} = \frac{1}{(\log 2)^p}\left(\frac{1}{k^p}\right),$$

and so $\sum_{k=2}^{\infty} a_k$ converges if and only if $p > 1$. ●

The following general result holds.

Theorem 5.41. *Suppose that $\{a_n\}_{n\geq 1}$ is a decreasing sequence of nonnegative terms and $r \in \mathbb{N} \setminus \{1\}$. Then the series $\sum_{k=1}^{\infty} a_k$ and $\sum_{k=1}^{\infty} r^k a_{r^k}$ are either both convergent or both divergent.*

Proof. We observe that

$$\frac{r-1}{r}\sum_{k=1}^{n} r^k a_{r^k} = (r-1)a_r + (r-1)ra_{r^2} + (r-1)r^2 a_{r^3} + \cdots + (r-1)r^{n-1}a_{r^n},$$

and we leave the rest of the argument as a simple exercise. ∎

5.2.7 Questions and Exercises

Questions 5.42.

1. For a series $\sum a_n$ with positive terms, is it true that $\sum a_n < \infty$ if and only if the corresponding sequence of partial sums $\{S_n\}$ is bounded?
2. If $\sum_{k=1}^{\infty} a_k$ converges, must $\lim_{n\to\infty}(a_n + a_{n+1} + \cdots + a_{n+p}) = 0$ for every $p > 0$? How about the converse?
3. Suppose that $\sum_{k=1}^{\infty} a_k$ is a convergent series of positive terms. Is it necessary that $na_n \to 0$ as $n \to \infty$?
4. Suppose $a_k > 0$ for all $k \geq 1$ and that $\sum_{k=1}^{\infty} a_k$ converges. Must $\sum_{k=1}^{\infty}(1/a_k)$ be divergent?
5. Suppose that $0 < a_k < 1$ for all k, and that $\sum a_k$ converges. Must $\sum a_k^2$ be convergent?
6. Suppose that $0 < a_k < 1$ for all $k \geq 1$, and that $\sum_{k=1}^{\infty} a_k$ diverges. Can $\sum_{k=1}^{\infty} a_k^2$ be convergent?
7. If $\sum_{k=1}^{\infty} a_k^2$ and $\sum_{k=1}^{\infty} b_k^2$ both are convergent, must $\sum_{k=1}^{\infty} a_k b_k$ be convergent? If $p > 1/2$, what can be said about the convergence of $\sum_{k=1}^{\infty}(a_k/k^p)$?
8. What is a harmonic p-series? What is an alternating harmonic p-series?
9. Suppose that $a_k > 0$ and $b_k > 0$ for all $k \geq 1$ such that $\sum_{k=1}^{\infty} a_k$ is convergent and $\{b_k\}_{k\geq 1}$ is bounded. Must $\sum_{k=1}^{\infty} a_k b_k$ be divergent?
10. Suppose that $\sum_{k=1}^{\infty} a_k$ is a convergent series of positive terms and $\sum_{k=1}^{\infty} b_k$ diverges such that $b_k \to 0$ as $k \to \infty$. Must $\sum_{k=1}^{\infty} a_k b_k$ be divergent? How about the series $\sum_{k=1}^{\infty} k^{1/k} a_k$?

11. We know that $\lim_{n\to\infty} n^{1/n} = 1$, $\sum(1/n)$ diverges, and $\lim_{n\to\infty} \frac{n^{1/n}}{n} = 0$. Must $\sum_{n=1}^{\infty} n^{-1-1/n}$ be divergent? Must $\sum_{n=1}^{\infty} n^{-2-1/n}$ be convergent?
12. Suppose that $\sum a_k$ is a convergent series of positive terms and $b_k \to b \neq 0$ as $k \to \infty$. Must $\sum b_k$ and $\sum a_k b_k$ converge or diverge together?
13. Suppose that $\sum_{k=1}^{\infty} a_k$ converges and $\sum_{k=1}^{\infty} b_k$ diverges such that $b_k \to 0$ as $k \to \infty$. What can be said about the convergence of $\sum_{k=1}^{\infty} a_k b_k$?
14. Suppose that $\lim a_k = 0 = \lim b_k$ and both $\sum a_k$ and $\sum b_k$ diverge. Must $\sum a_k b_k$ be divergent? How about the series $\sum \frac{1}{\sqrt{k}\,2^k}$ and $\sum \frac{1}{\sqrt{k}\,3^k}$? How about the series $\sum a_k b_k$ with $a_k = b_k = 1/\sqrt{k}$?
15. Suppose that $0 \leq a_k \leq A$ for some number A and $b_k \geq k^2$. Must $\sum(a_k/b_k)$ be convergent?
16. Suppose that $a_k \to 0$ as $k \to \infty$. Can there exist an $N \in \mathbb{N}$ such that $ka_k < 1$ for all $k \geq N$?
17. Suppose that $\{a_n\}$ is a sequence of positive terms such that the limit $\lim_{k\to\infty} k^p a_k$ exists. Must $\sum a_k$ be convergent if $p > 1$? How about $p \leq 1$?
18. Must a positive-term series (i.e., a series all of whose terms are positive) either converge or else diverge to ∞?
19. Must a negative-term series (i.e., a series all of whose terms are negative) either converge or else diverge to $-\infty$?
20. Suppose that $\{a_n\}$ is a sequence of positive terms such that $\sum_{k=1}^{\infty} a_k$ converges. Does $\sum_{k=1}^{\infty} \sqrt{a_k a_{k+1}}$ converge? Does there exist a sequence of positive terms $\{c_n\}$ such that $c_n \to \infty$ as $n \to \infty$, and $\sum_{k=1}^{\infty} a_k c_k$ converges?

Exercises 5.43.

1. Using the direct convergence test, prove Theorem 5.14.
2. For what values of the real number a does the series
$$\sum_{k=1}^{\infty} \frac{\sqrt{k+1} - \sqrt{k}}{k^a}$$
converge?
3. Let $a_k > 0$ for all $k \geq 1$, and suppose $\sum_{k=1}^{\infty} a_k$ converges. If $r_n = \sum_{k=n}^{\infty} a_k$, show that
(a) $\displaystyle\sum_{k=1}^{\infty} \frac{a_k}{r_k}$ diverges. (b) $\displaystyle\sum_{k=1}^{\infty} \frac{a_k}{\sqrt{r_k}}$ converges.
4. Show that if $a_k > 0$ for all $k \geq 1$ and $b_k = a_k/(1 + a_k)$, then $\sum_{k=1}^{\infty} a_k$ converges if and only if $\sum_{k=1}^{\infty} b_k$ converges.
5. Suppose that $a_k \geq 0$ for all $k \geq 1$ and α is a positive real number. Prove or disprove that $\sum_{k=1}^{\infty} a_k$ converges if and only if $\sum_{k=1}^{\infty}(a_k/(1 + \alpha a_k))$ converges.
6. Suppose that $a_k \geq 0$ for all $k \geq 1$. Show that if $\sum_{k=1}^{\infty} a_k$ converges, then $\sum_{k=1}^{\infty}((k+1)/k)a_k^2$ converges.

7. Suppose that $\{a_n\}$ and $\{b_n\}$ are two sequences of positive numbers such that $\lim_{n\to\infty} b_n = B \neq 0$. Show that $\sum a_k$ converges if and only if $\sum a_k b_k$ converges.
8. Show that the series $\displaystyle\sum_{k=1}^{\infty} \frac{1}{(\log k)^q k^p}$ converges if and only if either $p > 1$ with any real q, or $p = 1$ with $q > 1$.
9. Examine the convergence of the following series:

   (a) $\dfrac{1}{1\cdot 2\cdot 3} + \dfrac{3}{2\cdot 3\cdot 4} + \dfrac{5}{3\cdot 4\cdot 5} + \dfrac{7}{4\cdot 5\cdot 6} + \cdots$.

   (b) $1 - \dfrac{1}{3\cdot 2^2} + \dfrac{1}{5\cdot 3^2} - \dfrac{1}{7\cdot 4^2} + \cdots$.

   (c) $\dfrac{1}{\sqrt{1\cdot 2}} + \dfrac{1}{\sqrt{2\cdot 3}} + \dfrac{1}{\sqrt{3\cdot 4}} + \cdots$.

   (d) $1 + \dfrac{1}{4^{2/3}} + \dfrac{1}{9^{2/3}} + \dfrac{1}{16^{2/3}} + \cdots$.

   (e) $\dfrac{1}{\sqrt{1}+\sqrt{2}} + \dfrac{1}{\sqrt{2}+\sqrt{3}} + \cdots$.

   (f) $\dfrac{\sqrt{2}-1}{3^3-1} + \dfrac{\sqrt{3}-1}{4^3-1} + \dfrac{\sqrt{4}-1}{5^3-1} + \cdots$.

10. Suppose that $\{a_n\}_{n\geq 1}$ and $\{b_n\}_{n\geq 1}$ are two sequences of positive numbers such that $\frac{a_{n+1}}{a_n} \leq \frac{b_{n+1}}{b_n}$ for all $n \geq 1$. Show that if $\sum_{k=1}^{\infty} b_k$ converges, then $\sum_{k=1}^{\infty} a_k$ converges. Also, conclude that $\sum_{k=1}^{\infty} b_k$ diverges whenever $\sum_{k=1}^{\infty} a_k$ diverges.
11. Suppose $a_n > 0$ and $b_n > 0$ for all $n \geq N_0$ such that
$$\limsup_{n\to\infty} \frac{a_n}{b_n} = L \quad \text{and} \quad \liminf_{n\to\infty} \frac{a_n}{b_n} = l.$$
Then prove or disprove the following (see Theorem 5.33):
   (a) If $0 \leq L < \infty$, then the series $\sum a_k$ converges whenever $\sum b_k$ converges.
   (b) If $0 < l \leq \infty$, then the series $\sum a_k$ diverges whenever $\sum b_k$ diverges.
12. In the following problems either use the divergence test to show that the given series diverges or show that the divergence test does not apply.

   (a) $\displaystyle\sum_{k=1}^{\infty} \frac{k}{k+1}$. (b) $\displaystyle\sum_{k=1}^{\infty} \left(1+\frac{1}{k}\right)^{-k}$. (c) $\displaystyle\sum_{k=2}^{\infty} \frac{k}{\sqrt{k^2-1}}$.

   (d) $\displaystyle\sum_{k=1}^{\infty} \frac{1}{k^{1/2}2^k}$. (e) $\displaystyle\sum_{k=0}^{\infty} \frac{1}{e^k+e^{-k}}$. (f) $\displaystyle\sum_{k=0}^{\infty} \sin^k(\pi/6)$.
13. Examine the convergence of the series $\sum a_k$, where a_k equals

(a) $\dfrac{1}{k^p + 2\cos(k\pi)}$ $(p > 0)$.

(b) $\dfrac{\sqrt{k^2+2k+3} - \sqrt{k^2-2k+3}}{k}$.

(c) $k^{(2-3k)/k}$.

(d) $\dfrac{1}{\sqrt{k} + \sqrt{k+1} + \sqrt{k+2}}$.

(e) $\dfrac{k+3}{(2k^2+1)^p(\log k)^{20}}$.

(f) $\sqrt[p]{k^p+1} - k$ $(p > 0)$.

(g) $\dfrac{1}{k^{p+1/k}}$.

(h) $\dfrac{(k+1)^q}{k^p}$ $(p, q \in \mathbb{R})$.

(i) $\dfrac{1}{\sqrt{k} + 5\sqrt[4]{k} - 1}$.

(j) $\dfrac{k^2-3k+5}{5k^5-2k}$.

(k) $\dfrac{\cos^2(2k)}{k^3}$.

(l) $\dfrac{1}{k^p + \sqrt{k}}$ $(p > 1)$.

(m) $\dfrac{1}{(1+k^2+k^4+k^6)^{1/4}}$.

(n) $\dfrac{2^k + k}{3^k - 2k}$.

5.3 Alternating Series and Conditional Convergence

We next consider a special type of series called *alternating series*. These are series whose successive terms alternate in sign. Such series will be of the form

$$\sum (-1)^{k-1} a_k \quad \text{with } a_k \geq 0 \text{ for all } k.$$

Consider the alternating harmonic series defined by

$$\sum_{k=1}^{\infty} \frac{(-1)^{k-1}}{k} = 1 - \frac{1}{2} + \frac{1}{3} - \cdots . \tag{5.7}$$

We have already shown that the series is convergent but is not absolutely convergent. For an alternative proof, we observe the following:

- The sequence of partial sums is no longer monotone, in contrast to the harmonic series.
- Although the sequence of partial sums can be shown to be bounded, it is not possible to use BMCT (the bounded monotone convergence theorem) to obtain convergence for a nonmonotone sequence.
- If we let

  $$a_k = \frac{1}{k} \text{ and } S_n = \sum_{k=1}^{n} \frac{(-1)^{k-1}}{k},$$

  then the sequence $\{a_k\}$ has the following properties:

  $$\lim_{k\to\infty} a_k = 0, \quad \{a_k\} \text{ is decreasing, and } a_k > 0 \text{ for all } k. \tag{5.8}$$

We have already verified that S_{2n} is increasing, while S_{2n-1} is decreasing. In fact,

$$\frac{1}{2} = S_2 < S_4 < S_6 < \quad \cdots \quad < S_5 < S_3 < S_1 = 1.$$

Now we prove that the series (5.7) converges, and later we use exactly the same idea to show that $\sum_{k=1}^{\infty}(-1)^{k-1}a_k$ converges under the condition (5.8) for a general a_k, rather than $a_k = 1/k$. To present our geometric proof of the convergence of the alternating harmonic series (5.7), we group the partial sums as

$$\begin{aligned} S_{2n} &= 1 - \frac{1}{2} + \frac{1}{3} - \frac{1}{4} + \cdots + \frac{1}{2n-1} - \frac{1}{2n} \\ &= \left(1 + \frac{1}{2} + \frac{1}{3} + \frac{1}{4} + \cdots + \frac{1}{2n-1} + \frac{1}{2n}\right) - 2\left(\frac{1}{2} + \frac{1}{4} + \cdots + \frac{1}{2n}\right) \\ &= \left(1 + \frac{1}{2} + \frac{1}{3} + \frac{1}{4} + \cdots + \frac{1}{2n}\right) - \left(1 + \frac{1}{2} + \cdots + \frac{1}{n}\right), \end{aligned}$$

so that

$$S_{2n} = \frac{1}{n+1} + \frac{1}{n+2} + \cdots + \frac{1}{n+n}.$$

Also, we have

$$S_{2n-1} = S_{2n} + \frac{1}{2n}. \tag{5.9}$$

The linearity rule (for sequences) applied to (5.9) shows that

$$\lim_{n\to\infty} S_{2n-1} = \lim_{n\to\infty} S_{2n} + \lim_{n\to\infty} \frac{1}{2n} = \log 2.$$

We conclude that $\lim S_n$ exists and equals $\log 2$.

Alternatively, we may rewrite S_{2n} as

$$S_{2n} = \frac{1}{n}\sum_{k=1}^{n} \frac{1}{1+k/n},$$

which by item 5 in Remark 7.26, gives $\lim S_{2n} = \log 2$. As will be seen in Chapter 6, the sequence S_{2n} can be recognized as a lower Riemann sum associated with the continuous function

$$f(x) = \frac{1}{1+x} \text{ on } [0,1] \quad \left(\text{or } g(x) = \frac{1}{x} \text{ on } [1,2]\right),$$

and therefore $\lim S_{2n}$ exists and

$$\lim_{n\to\infty} S_{2n} = \int_0^1 \frac{dx}{1+x} = \log 2.$$

By (5.9), $\{S_{2n-1}\}$ also converges to $\log 2$. Thus $\{S_n\}$ converges to $\log 2$.

5.3.1 Alternating Series Test

In general, although the condition $\lim a_k = 0$ is necessary for the convergence of the series $\sum a_k$, this condition tells us very little about the convergence of the series $\sum a_k$. However, as in the above example, it turns out that an alternating series must converge if the absolute value of its terms decreases monotonically toward zero. Now we extend this idea and prove the following result, established by Leibniz in the seventeenth century. This test is sometimes called the *Leibniz test.*

Theorem 5.44 (Alternating series test). *An alternating series*

$$\sum_{k=1}^{\infty}(-1)^{k-1}a_k,$$

where $a_k \geq 0$ for all $k \geq 1$, converges if the following two conditions are satisfied:

(1) $\lim a_n = 0$;
(2) $\{a_n\}_{n\geq 1}$ *is a decreasing sequence; that is, $a_{n+1} \leq a_n$ for all $n \geq 1$.*

Proof. Let $S_n = a_1 - a_2 + a_3 - a_4 + \cdots + (-1)^{n-1}a_n$. We need to prove that $\{S_n\}$ converges. Our strategy is to follow the same method of proof used to prove that the alternating harmonic series converges. First, we note that for $n > 1$,

$$S_{2n} = (a_1 - a_2) + (a_3 - a_4) + \cdots + (a_{2n-1} - a_{2n}) \geq S_{2(n-1)},$$

because each pair of quantities in parentheses is nonnegative (since $\{a_k\}$ is a decreasing). It follows that $\{S_{2n}\}$ is *increasing*. Moreover,

$$S_{2n} = a_1 - (a_2 - a_3) - \cdots - (a_{2n-2} - a_{2n-1}) - a_{2n} \leq a_1,$$

showing that $\{S_{2n}\}$ is bounded above by a_1. Consequently, by BMCT (see Theorem 2.27), $\{S_{2n}\}$ converges to some number, say to S. Further, since

$$S_{2n-1} = S_{2n} + a_{2n} \quad \text{and} \quad \lim_{n\to\infty} a_n = 0,$$

we see that

$$\lim_{n\to\infty} S_{2n-1} = \lim_{n\to\infty} S_{2n} + \lim_{n\to\infty} a_{2n} = S - 0 = S.$$

Thus, the odd and even subsequences of $\{S_n\}$ both converge to the same limit S, and so $\{S_n\}$ itself converges to S. Hence, $\sum_{k=1}^{\infty}(-1)^{k-1}a_k$ is convergent, with sum S. Note also that

$$\begin{aligned} S_{2n+1} &= S_{2n} + a_{2n+1} \\ &= a_1 - (a_2 - a_3) - \cdots - (a_{2n-2} - a_{2n-1}) - (a_{2n} - a_{2n+1}) \\ &\leq S_{2n-1}, \end{aligned}$$

showing that $\{S_{2n-1}\}$ is *decreasing*; see Figure 5.8. ∎

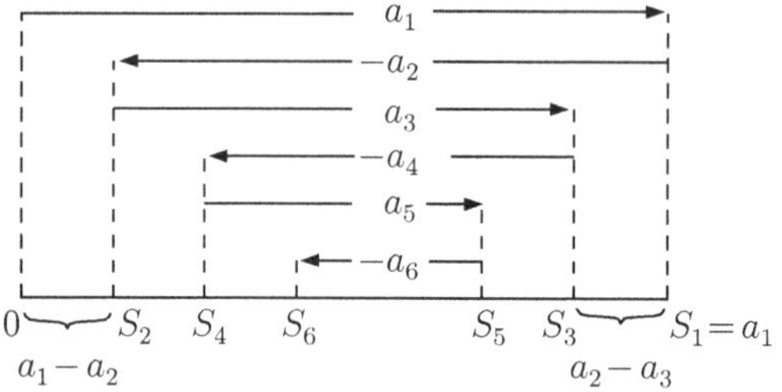

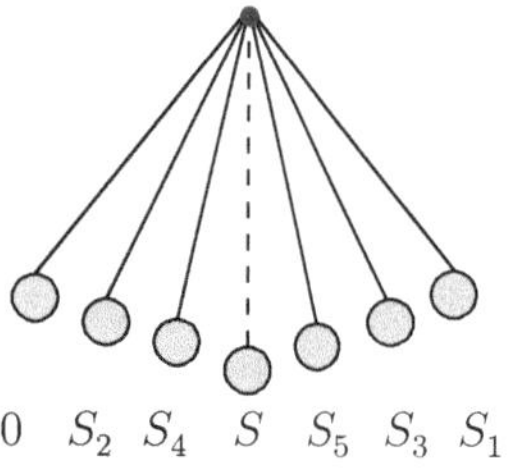

Fig. 5.8. Description for the partial sums in $\sum_{k=1}^{\infty}(-1)^{k-1}a_k$.

Remark 5.45. 1. In the case $a_k > 0$, $k \geq N$, it is often easier to check that $\{a_k\}_{k\geq N}$ is decreasing by verifying that $\{\frac{1}{a_k}\}_{k\geq N}$ is increasing.

2. Observe that in the alternating series $\sum_{k=1}^{\infty}(-1)^{k-1}a_k$, where $a_k > 0$ for all $k \geq 1$, the partial sums $\{S_n\}$ are reminiscent of a swinging (simple) pendulum that is slowly coming to rest at a fixed position that is equivalent to the sum of the series; see Figure 5.8 for the behavior of the partial sums of an alternating convergent series.
3. Note that for each $n \geq 1$,

$$S_2 \leq S_4 \leq \cdots \leq S_{2n} = S_{2n-1} - a_{2n} \leq S_{2n-1} \leq \cdots \leq S_5 \leq S_3 \leq S_1 = a_1.$$

Since $\{S_{2n}\}$ is increasing to S and $\{S_{2n-1}\}$ is decreasing to S, it follows that

$$S_{2n} \leq S \leq S_{2n-1} \quad \text{for } n = 1, 2, 3, \ldots,$$

so that the sum S lies between two consecutive partial sums. ●

We easily have the following:

1. $\sum_{k=1}^{\infty} \frac{(-1)^{k-1}\sin\sqrt{k}}{k^p}$ converges absolutely for all $p > 1$, since

$$\left|(-1)^{k-1}\frac{\sin\sqrt{k}}{k^p}\right| \leq \frac{1}{k^p}.$$

2. $\sum_{k=1}^{\infty}(-1)^{k-1}\frac{k}{k^2+p}$ $(p > 0)$ is conditionally convergent. Indeed, it is not absolutely convergent, because

$$\left|(-1)^{k-1}\frac{k}{k^2+p}\right| = \frac{k}{k^2+p} > \frac{p}{k(p+1)}$$

for all $k > p^2$ and $\sum \frac{1}{k}$ diverges. Moreover, it is convergent because $a_k = k/(k^2+p)$ is positive, $a_k \to 0$ as $k \to \infty$, and $\{a_k\}$ is decreasing for all $k \geq \sqrt{p}$, and so the alternating series test is applicable.
3. $\sum_{k=1}^{\infty}(-1)^{k-1}\frac{1}{k+p^2}$ $(p \in \mathbb{R})$ is conditionally convergent.

Corollary 5.46 (The error estimate for an alternating series). *Suppose an alternating series $\sum_{k=1}^{\infty}(-1)^{k-1}a_k$ satisfies conditions* (1) *and* (2) *of Theorem* 5.44. *If the series has the sum S, then*

$$|S - S_n| \leq a_{n+1},$$

where S_n is the nth partial sum of the series.

Proof. Let $S_n = \sum_{k=1}^{n}(-1)^{k-1}a_k$. Then, as in the discussion of Example 5.15, we easily get

$$S_{n+p} - S_n = (-1)^n T_{n,p} \text{ with } 0 \leq T_{n,p} \leq a_{n+1},$$

so that

$$|S_{n+p} - S_n| \leq a_{n+1}.$$

Thus $\{S_n\}$ is Cauchy and hence converges to S (say). Finally, fixing n and allowing $p \to \infty$ gives $|S - S_n| \leq a_{n+1}$, and the proof is complete. ∎

From the last corollary, we conclude that if an alternating series satisfies the conditions of the alternating series test, we can approximate the sum of the series with the nth partial sum, and the error will have absolute value no greater than the first term omitted (namely, a_{n+1}).

5.3.2 Rearrangement of Terms in a Series

The algebraic manipulation of terms in a series has to be done with more care than in finite sums. For instance, it is meaningful to write

$$1 + 2 + (3 + 5) + 6 = (1 + 2 + 3) + (5 + 6) = (1 + 2) + (3 + 5 + 6) = \cdots,$$

so that the terms of a finite sum can be grouped by inserting parentheses arbitrarily. On the other hand, there are obvious problems in trying to do the same with series of real numbers. To begin with, we consider the series $\sum_{k=1}^{\infty}(-1)^{k-1}$. It is tempting to pair the terms of this series as

$$(1 - 1) + (1 - 1) + \cdots = 0 + 0 + \cdots = 0,$$

and equally tempting to pair the terms of the same series as

$$1 - (1 - 1) - (1 - 1) - \cdots = 1 - 0 - 0 - \cdots = 1.$$

We see that inserting parentheses in two different ways results in two convergent series with different limits. But we know that the original series is divergent. What is wrong? A natural question is whether a series can be grouped by inserting parentheses arbitrarily with

(a) divergent series that are divergent to $\pm\infty$;
(b) divergent series that are not divergent to $\pm\infty$;

(c) absolutely convergent series;
(d) conditionally convergent series.

Definition 5.47 (Rearrangement of series). *Suppose that* $\sum_{k=1}^{\infty} a_k$ *is a given series. Let* $\{n_k\}$ *be a sequence of positive integers such that each positive integer occurs exactly once in the sequence. That is, there exists a bijective map* $f : \mathbb{N} \to \mathbb{N}$ *with* $f(k) = n_k$, $k \in \mathbb{N}$, *so that each term in the series* $\sum_{k=1}^{\infty} b_k$ $(b_k = a_{n_k})$ *is also a term in* $\sum_{k=1}^{\infty} a_k$, *but occurs in different order. The series* $\sum_{k=1}^{\infty} b_k$ *is called a rearrangement of* $\sum_{k=1}^{\infty} a_k$.

Suppose that $\sum_{k=1}^{\infty} b_k$ is a rearrangement of $\sum_{k=1}^{\infty} a_k$. Then by definition, since $b_k = a_{n_k} = a_{f(k)}$, we have $a_k = b_{f^{-1}(k)}$, and thus $\sum_{k=1}^{\infty} a_k$ is also a rearrangement of $\sum_{k=1}^{\infty} b_k$.

Example 5.48. Consider

$$a_n = \begin{cases} 1 & \text{if } n \text{ is odd,} \\ \dfrac{-(2^{n-1}-1)}{2^{n-1}} & \text{if } n \text{ is even,} \end{cases}$$

for $n \in \mathbb{N}$. Then $a_{2n-1} \to 1$ and $a_{2n} \to -1$ as $n \to \infty$, so that $\lim a_n$ does not exist, showing that $\sum_{k=1}^{\infty} a_k$ diverges, and indeed is an oscillatory series. However, if we introduce parentheses,

$$\begin{aligned}
\sum_{k=1}^{\infty} a_k &= (a_1 + a_2) + (a_3 + a_4) + \cdots \\
&= \left(1 - \frac{1}{2}\right) + \left(1 - \frac{3}{4}\right) + \left(1 - \frac{7}{8}\right) + \cdots \\
&= \frac{1}{2} + \frac{1}{2^2} + \frac{1}{2^3} + \cdots,
\end{aligned}$$

the result is a convergent series. ●

Next we consider the alternating harmonic series (conditionally convergent) (5.7). We have already shown that the series converges to $S = \log 2$. Suppose we rearrange this series as

$$\begin{aligned}
&\left(1 - \frac{1}{2}\right) - \frac{1}{4} + \left(\frac{1}{3} - \frac{1}{6}\right) - \frac{1}{8} + \left(\frac{1}{5} - \frac{1}{10}\right) - \frac{1}{12} + \cdots . \\
&\qquad = \frac{1}{2} - \frac{1}{4} + \frac{1}{6} - \frac{1}{8} + \frac{1}{10} - \frac{1}{12} + \cdots \\
&\qquad = \frac{1}{2}\left(1 - \frac{1}{2} + \frac{1}{3} - \frac{1}{4} + \frac{1}{5} - \frac{1}{6} + \cdots\right) \\
&\qquad = \frac{1}{2} \log 2
\end{aligned}$$

We now have a rearranged series of the alternating harmonic series converging to $(1/2)\log 2$.

In general, it can be shown that if $\sum a_k$ converges conditionally, there is a rearrangement of the terms of $\sum a_k$ such that the sum of the series is equal to any specified real number.

Let us now rearrange the terms of the series (5.7) so that the sum is $(3/2)\log 2$. For this, because (5.7) is convergent, we have

$$\frac{1}{2}\log 2 = \sum_{k=1}^{\infty} \frac{(-1)^{k-1}}{2k} = \frac{1}{2} - \frac{1}{4} + \frac{1}{6} - \frac{1}{8} + \frac{1}{10} - \frac{1}{12} + \cdots . \tag{5.10}$$

Next, we introduce

$$a_k = \begin{cases} 0 & \text{if } k \text{ is odd,} \\ \dfrac{(-1)^{(k-2)/2}}{k} & \text{if } k \text{ is even,} \end{cases}$$

and consider the series

$$\sum_{k=1}^{\infty} a_k = 0 + \frac{1}{2} + 0 - \frac{1}{4} + 0 + \frac{1}{6} + 0 - \frac{1}{8} + \cdots , \tag{5.11}$$

which is obtained by inserting 0 into (5.10) between pairs of terms. Also, we observe that if S_n and T_n are the nth partial sums of (5.10) and (5.11), respectively, then we have

$$T_1 = 0, \quad T_{2n} = S_n = T_{2n+1} \text{ for } n \geq 1,$$

showing that $\lim T_n = (1/2)\log 2 = \lim S_n$. Thus, adding (5.7) and (5.11) gives a new convergent series with sum $(3/2)\log 2$. That is,

$$\frac{3}{2}\log 2 = \sum_{k=1}^{\infty} \left(\frac{(-1)^{k-1}}{k} + a_k \right) = 1 + 0 + \frac{1}{3} - \frac{1}{2} + \frac{1}{5} + 0 + \frac{1}{7} - \frac{1}{4} + \cdots .$$

If we delete 0 from the series on the right-hand side, the resulting series is actually a rearrangement of (5.7) that is convergent to $(3/2)\log 2$. More generally, we have the following.

Example 5.49. Consider the alternating harmonic series $\sum_{k=1}^{\infty} \frac{(-1)^{k-1}}{k}$. Show that if the first p *positive terms* are followed by the first q *negative terms*, the next p positive terms followed by the next q negative terms, and so on, then the resulting rearranged series will converge to $\log(2\sqrt{p/q})$.

Solution. Recall from Remark 7.26(4) that if $\gamma_n = \sum_{k=1}^{n} \frac{1}{k} - \log n$, then we have $0 < \gamma_n < 1$ and $\gamma_n \to \gamma$, where γ is the Euler constant. These facts about γ_n can be verified directly, without the using integration theory. We may write

$$\gamma_n = \gamma + R_n, \quad \text{or} \quad H_n := \sum_{k=1}^{n} \frac{1}{k} = \log n + \gamma + R_n, \tag{5.12}$$

where $R_n \to 0$ as $n \to \infty$ and H_n is the nth partial sum of the harmonic series $\sum_{k=1}^{\infty} \frac{1}{k}$. If we consider the first p *positive terms* in the given alternating series and use (5.12), then we find that

$$1 + \frac{1}{3} + \frac{1}{5} + \cdots + \frac{1}{2p-1} = \sum_{k=1}^{2p} \frac{1}{k} - \sum_{k=1}^{p} \frac{1}{2k}$$
$$= \log 2p - \frac{1}{2}\log p + \frac{\gamma}{2} + R_{2p} - \frac{R_p}{2}. \tag{5.13}$$

Similarly, if we sum the first q *negative terms* of the given series, we have

$$-\left[\frac{1}{2} + \frac{1}{4} + \frac{1}{6} + \cdots + \frac{1}{2q}\right] = -\frac{1}{2}H_q = -\frac{1}{2}(\log q + \gamma + R_q). \tag{5.14}$$

Adding (5.13) and (5.14) yields

$$\sum_{k=1}^{p} \frac{1}{2k-1} - \sum_{k=1}^{q} \frac{1}{2k} = \log\left(\frac{2p}{\sqrt{pq}}\right) + R_{2p} - \frac{R_p}{2} - \frac{R_q}{2}$$
$$= \log\left(2\sqrt{\frac{p}{q}}\right) + R(p,q),$$

where $R(p,q) \to 0$ as $p, q \to \infty$. The result follows. ●

We have the following from the last example.

Example 5.50. (1) For $p = q = 1$, we get

$$\sum_{k=1}^{\infty} \frac{(-1)^{k-1}}{k} = 1 - \frac{1}{2} + \frac{1}{3} - \cdots = \log 2.$$

(2) For $p = 1,\ q = 2$, we get the series

$$1 - \frac{1}{2} - \frac{1}{4} + \frac{1}{3} - \frac{1}{6} - \frac{1}{8} + \cdots = \frac{1}{2}\log 2.$$

(3) For $p = 1$ and $q = 4$, we get the series

$$1 - \frac{1}{2} - \frac{1}{4} - \frac{1}{6} - \frac{1}{8} + \frac{1}{3} - \frac{1}{10} - \frac{1}{12} - \frac{1}{14} - \frac{1}{16} + \frac{1}{5} - \cdots,$$

whose sum is $\log\left(2\sqrt{\frac{1}{4}}\right) = 0$.

(4) For $p = 2$ and $q = 1$, we get

$$1 + \frac{1}{3} - \frac{1}{2} + \frac{1}{5} + \frac{1}{7} - \frac{1}{4} + \cdots = \log(2\sqrt{2}) = \frac{3}{2}\log 2.$$

(5) For $p = 3,\ q = 2$, we get a series with sum $\log\left(2\sqrt{\frac{3}{2}}\right) = \frac{1}{2}\log 6$.
(6) For $p = 3 = q$, we see that

$$1 + \frac{1}{3} + \frac{1}{5} - \frac{1}{2} - \frac{1}{4} - \frac{1}{6} + \ + \ + \ \cdots = \log 2.$$

For a direct proof of case (4), we may proceed as follows. Consider

$$1 + \frac{1}{3} - \frac{1}{2} + \frac{1}{5} + \frac{1}{7} - \frac{1}{4} + \cdots$$

and let its partial sum be denoted by S_n. Then

$$\begin{aligned}
S_{3n} &= 1 + \frac{1}{3} - \frac{1}{2} + \cdots + \frac{1}{4n-3} + \frac{1}{4n-1} - \frac{1}{2n} \\
&= \left(1 + \frac{1}{3} + \frac{1}{5} + \cdots + \frac{1}{4n-3} + \frac{1}{4n-1}\right) - \left(\frac{1}{2} + \frac{1}{4} + \cdots + \frac{1}{2n}\right) \\
&= \left(1 + \frac{1}{2} + \frac{1}{3} + \frac{1}{4} + \cdots + \frac{1}{4n-1} + \frac{1}{4n}\right) - \frac{1}{2}\left(1 + \frac{1}{2} + \cdots + \frac{1}{2n}\right) - \frac{1}{2}H_n \\
&= H_{4n} - \frac{1}{2}H_{2n} - \frac{1}{2}H_n.
\end{aligned}$$

As noticed earlier,

$$T_{2n} = \sum_{k=1}^{2n}(-1)^{k-1}\frac{1}{k} = \sum_{k=1}^{2n}\frac{1}{k} - 2\sum_{k=1}^{n}\frac{1}{2k} = H_{2n} - H_n.$$

Consequently,

$$S_{3n} = (H_{4n} - H_{2n}) + \frac{1}{2}(H_{2n} - H_n),$$

so that

$$S_{3n} = T_{4n} + \frac{1}{2}T_{2n}.$$

Also,

$$S_{3n-1} = S_{3n} + \frac{1}{2n} \quad \text{and} \quad S_{3n-2} = S_{3n} + \frac{1}{2n} - \frac{1}{4n-1}.$$

We know that $T_n \to \log 2$ as $n \to \infty$, and so $T_{2n} \to \log 2$ and $T_{4n} \to \log 2$ as $n \to \infty$. The last three relations for the subsequences $\{S_{3n}\}$, $\{S_{3n-1}\}$, and $\{S_{3n-2}\}$ show that $\{S_n\}$ converges to $\frac{3}{2}\log 2$. ●

The following results show that the sum of an absolutely convergent series is independent of the order of the terms, but in the case of conditionally convergent series, the situation can be entirely different, and the order of the terms is crucial.

Theorem 5.51 (Rearrangement of absolutely convergent series). *If $\sum_{k=1}^{\infty} a_k$ converges absolutely with sum S, then every series $\sum_{k=1}^{\infty} b_k$ obtained by rearranging its terms also converges absolutely to the same sum S.*

Proof. Let S_n and T_n denote the nth partial sums of $\sum a_k$ and $\sum b_k$, respectively. Suppose that $\sum |a_k| = A$. Then, since each term of $\sum b_k$ is a term of $\sum a_k$, we have

$$B_n := \sum_{k=1}^{n} |b_k| \leq \sum_{k=1}^{\infty} |a_k| = A \quad \text{for each } n \geq 1,$$

so that A is an upper bound for the increasing sequence of partial sums $\{B_n\}$. Therefore, by BMCT, $\sum b_k$ is absolutely convergent.

Next we must show that $\sum_{k=1}^{\infty} b_k = S$. Since $\sum a_k$ is convergent to the sum S and is also absolutely convergent, it follows that for a given $\epsilon > 0$ there exists an N such that

(i) $|S - S_n| < \epsilon/2$ for $n \geq N$.
(ii) $A - \sum_{k=1}^{n} |a_k| < \epsilon/2$ for $n \geq N$. In particular, for $n = N$ this gives

$$\sum_{k=N+1}^{\infty} |a_k| < \frac{\epsilon}{2}.$$

Further, since each term of $\sum a_k$ is a term of $\sum b_k$, there exists an integer N_1 such that terms in $\{a_1, a_2, \ldots, a_N\}$ are included among the terms in $\{b_1, b_2, \ldots, b_{N_1}\}$. If $n \geq N_1$, since S_n and T_n both include the terms $a_1, a_2, \ldots, a_N$ in their finite sum, we write

$$\sum_{k=1}^{n} b_k = \sum_{k=1}^{N} a_k + E,$$

where E is a finite sum of terms of $\sum a_k$ each of which occurs after a_N. Now we observe that from **(ii)**, $|E| < \epsilon/2$, and from **(i)**, we have

$$\left| S - \sum_{k=1}^{n} b_k \right| \leq \left| S - \sum_{k=1}^{N} a_k \right| + |E| < \frac{\epsilon}{2} + \frac{\epsilon}{2} = \epsilon \quad \text{for } n \geq N_1.$$

Thus $\sum b_k$ converges to the sum S. ∎

5.3.3 Riemann's Theorem on Conditionally Convergent Series

We present an important theorem due to Georg Friedrich Bernhard Riemann (1826–1866) that shows that every conditionally convergent series can be rearranged so that the resulting series has as its sum any preassigned real number. This remarkable result is a consequence of our next result. Also, every conditionally convergent series can be made to be divergent to ∞ or $-\infty$, or oscillatory (finite or infinite).

Theorem 5.52 (Riemann's rearrangement theorem). *Assume that the series $\sum_{k=1}^{\infty} a_k$ is conditionally convergent, with nth partial sum s_n. Let α and β be given such that $-\infty \le \alpha \le \beta \le \infty$. Then there exists a rearrangement $\sum_{k=1}^{\infty} d_k$ of $\sum_{k=1}^{\infty} a_k$ such that*

$$\liminf_{n\to\infty} t_n = \alpha \quad \text{and} \quad \limsup_{n\to\infty} t_n = \beta, \tag{5.15}$$

where $t_n = \sum_{k=1}^{n} d_k$. In particular, we have the following:

(a) *There is a rearranged series diverging to ∞.*
(b) *There is a rearranged series diverging to $-\infty$.*
(c) *For any given real number S, there is a rearranged series converging to S.*

Moreover, there is a rearranged series such that its partial sums oscillate finitely or infinitely.

Proof. Without loss of generality we assume that $a_k \neq 0$ for all $k \ge 1$, since the terms of the series that are zero do not affect its convergence or divergence. Let b_n and $-c_n$ denote the nth positive term and the nth negative term of $\sum a_k$, respectively. Then both $\sum b_k$ and $\sum c_k$ become positive-term series, and because $\sum a_k$ is conditionally convergent, we have

$$\sum |a_k| = \sum b_k + \sum c_k = \infty,$$

which implies that either $\sum b_k$ or $\sum c_k$ diverges.

Claim: Both $\sum b_k$ and $\sum c_k$ are divergent.

Suppose this is not the case. Then one of these two series is convergent. Without loss of generality, we assume that $\sum c_k$ converges to C ($C > 0$) and $\sum b_k = \infty$. Since $\sum b_k$ is a divergent series of positive terms, given $R > 0$, there exists an N_1 such that

$$\sum_{k=1}^{N_1} b_k > R + C.$$

Choose N_2 large enough that the first N_2 terms of $\{a_n\}$ contain the first N_1 terms of $\{b_n\}$. Then for all $n \ge N_2$,

$$s_n = \sum_{k=1}^{n} a_k \ge \sum_{k=1}^{N_1} b_k - \sum_{k=1}^{\infty} c_k > R + C - C = R,$$

showing that the sequence $\{s_n\}$ of partial sums is unbounded, and hence $\sum a_k$ diverges, a contradiction. Thus both $\sum b_k$ and $\sum c_k$ must be divergent. The claim is proved, and it follows that

$$\sum b_k = \infty \quad \text{and} \quad \sum c_k = \infty.$$

Further, because $\sum a_k$ is convergent, $a_n \to 0$, and so

$$\lim_{n\to\infty} b_n = 0 = \lim_{n\to\infty} c_n.$$

We now introduce

$$B_n = \sum_{k=1}^{n} b_k \quad \text{and} \quad C_n = \sum_{k=1}^{n} c_k$$

and describe a method that tells us when to take positive terms and when to take negative terms.

(a) Since $\sum b_k$ diverges to ∞, the sequence of its partial sums is therefore unbounded. Thus, there exists an m_1 such that

$$B_{m_1} > 1 + c_1.$$

Again, choose m_2 large enough that

$$B_{m_2} - (1 + c_1) > 1 + c_2, \quad \text{i.e.,} \quad B_{m_2} - (c_1 + c_2) > 2.$$

If we continue the process in this way, we obtain a sequence $\{m_p\}$ of positive integers such that

$$B_{m_p} - C_p = \sum_{k=1}^{m_p} b_k - \sum_{k=1}^{p} c_k > p,$$

and hence we obtain a rearrangement of $\sum a_k$ diverging to ∞.

(b) A very similar argument with the roles of b_n and c_n reversed allows us to build a rearrangement for which there is a sequence of partial sums diverging to $-\infty$.

(c) Now we fix α and β with $\alpha \leq \beta$. Note that we do not exclude the following possibilities:

(i) $\alpha = \beta = \infty$; **(ii)** $\alpha = \beta = -\infty$; **(iii)** $\alpha = -\infty$, $\beta = \infty$.
We have already taken care of the cases in which $\alpha = \beta = S$, with $S = \pm\infty$. Thus, the rearranged series diverges to ∞ or to $-\infty$ depending on whether $\alpha = \beta = \infty$ or $\alpha = \beta = -\infty$, respectively. Clearly, **(i)** and **(ii)** follow from the proof of **(iii)**, but just for the sake of clarity and for getting used to the idea of specifying a rearrangement by a qualitative process, we have included the proof for **(i)**. Finally, if $\alpha = \beta = S$, a finite real number, then our procedure shows that the rearranged series converges to S.

Now we present our argument for the general case (5.15), and it is very much the same. Let $\{x_n\}$ and $\{y_n\}$ be two sequences of real numbers such that

$$\lim_{n\to\infty} x_n = \alpha, \quad \lim_{n\to\infty} y_n = \beta, \quad \text{with } x_n < y_n \text{ and } y_1 > 0.$$

Choose just enough (say m_1) positive terms so that

$$T_1 = B_{m_1} > y_1.$$

Again, choose just enough (say n_1) negative terms so that

$$S_1 = B_{m_1} - C_{n_1} < x_1.$$

Next, we continue the process and choose just enough further positive terms and negative terms so that

$$T_2 = B_{m_2} - C_{n_1} > y_2 \quad \text{and} \quad S_2 = B_{m_2} - C_{n_2} < x_2.$$

If we continue the process, we obtain the following rearrangements of $\sum a_k$:

$$\underbrace{\underbrace{b_1 + \cdots + b_{m_1}} - c_1 - \cdots - c_{n_1} + b_{m_1+1} + \cdots + b_{m_2}} - c_{n_1+1} - \cdots - c_{n_2} + \cdots$$

and

$$\underbrace{\underbrace{b_1 + \cdots + b_{m_1} - c_1 - \cdots - c_{n_1}} + b_{m_1+1} + \cdots + b_{m_2} - c_{n_1+1} - \cdots - c_{n_2}} + \cdots.$$

Also, we observe that

$$T_k = B_{m_k} - C_{n_{k-1}} > y_k, \quad \text{and} \quad S_k = B_{m_k} - C_{n_k} < x_k,$$

where m_k and n_k are increasing sequences that are the least positive integers greater than m_{k-1} and n_{k-1}, respectively. Here $C_{n_0} = 0$. The last two inequalities may be combined as

$$S_k < x_k < y_k < T_k.$$

(See Figure 5.9.)

Also note that

$$\begin{aligned} T_k &= B_{m_k} - C_{n_{k-1}} \\ &= S_k + (c_{n_{k-1}+1} + \cdots + c_{n_k}) \\ &= S_{k-1} + (B_{m_k} - B_{m_{k-1}}) \\ &= S_{k-1} + (b_{m_{k-1}+1} + \cdots + b_{m_k}). \end{aligned}$$

Fig. 5.9. Positions of S_k, x_k, y_k and T_k's.

Thus, T_k and S_k are respectively the kth partial sums of the two rearranged series of $\sum a_k$ with last terms b_{m_k} and $-c_{n_k}$. We leave the rest of the argument as a simple exercise. ■

5.3.4 Dirichlet Test

Definition 5.53 (Bounded variation for a sequence). *A sequence* $\{a_n\}_{n\geq 0}$ *of real numbers is said to be of* bounded variation *if the series* $\sum_{k=1}^{\infty} |a_k - a_{k-1}|$ *converges.*

It is easy to see that the following statements are true:

- Every sequence of bounded variation is convergent.
- Not every convergent sequence is of bounded variation.
- Every bounded monotone sequence is of bounded variation, for

$$\sum_{k=1}^{n} |a_k - a_{k-1}| = |a_0 - a_n| \to |a_0 - \lim_{n\to\infty} a_n|.$$

- A linear combination of two sequences of bounded variation is of bounded variation.

Theorem 5.54 (Generalized Dirichlet test). *Let* $\{a_n\}_{n\geq 1}$ *and* $\{b_n\}_{n\geq 1}$ *be sequences of real numbers such that*

(i) $|s_n| \leq M$ *for* $n \geq 1$, $s_n = \sum_{k=1}^{n} a_k$; *i.e.,* $\{a_n\}$ *has bounded partial sums;*
(ii) $b_n \to 0$ *and* $\sum_{k=1}^{\infty} |b_k - b_{k+1}| < \infty$; *i.e.,* $\{b_n\}$ *is of bounded variation converging to* 0.

Then $\sum_{k=1}^{\infty} a_k b_k$ *converges.*

Proof. We have $a_1 = s_1$, and $a_k = s_k - s_{k-1}$ for $k \geq 2$. Also we see that

$$\begin{aligned}
\sum_{k=1}^{n} a_k b_k &= a_1 b_1 + \sum_{k=2}^{n} (s_k - s_{k-1}) b_k \\
&= s_1 b_1 + \sum_{k=2}^{n} s_k b_k - \sum_{k=1}^{n} s_k b_{k+1} + s_n b_{n+1} \\
&= \sum_{k=1}^{n} s_k (b_k - b_{k+1}) + s_n b_{n+1},
\end{aligned}$$

which is often called a *formula for summation by parts.*

Assume the hypotheses that $\{s_n\}$ is bounded and $\{b_n\}$ is of bounded variation such that $b_n \to 0$ as $n \to \infty$. Consequently:

- $s_n b_{n+1} \to 0$ as $n \to \infty$.

- Because $\sum_{k=1}^{\infty}(b_k - b_{k+1})$ is absolutely convergent and $\{s_n\}$ is bounded, it follows that

$$\sum_{k=1}^{n} |s_k(b_k - b_{k+1})| \le M \sum_{k=1}^{n} |b_k - b_{k+1}| \le M \sum_{k=1}^{\infty} |b_k - b_{k+1}|,$$

and so $\sum_{k=1}^{\infty} s_k(b_k - b_{k+1})$ is absolutely convergent. In particular, the series $\sum_{k=1}^{\infty} s_k(b_k - b_{k+1})$ converges.

Thus the formula for summation by parts gives the desired conclusion. ∎

Corollary 5.55 (Dirichlet's test). *Suppose that $\{a_n\}_{n\ge 1}$ is a sequence of real numbers with bounded partial sums, and $\{b_n\}_{n\ge 1}$ is a decreasing sequence of nonnegative real numbers with limit zero. Then $\sum_{k=1}^{\infty} a_k b_k$ converges.*

Proof. Since every bounded monotone sequence is of bounded variation, the hypothesis implies that $\{b_n\}$ is of bounded variation. The result follows from Theorem 5.54. ∎

For instance, we have the following:

- The choice $a_k = (-1)^{k-1}$ gives the Leibniz alternating series test (see Theorem 5.44) as a special case. Thus, Dirichlet's test generalizes the Leibniz test.
- The choice $a_k = (-1)^{k-1}$ gives that s_n is either 1 or 0, and so $b_k = \frac{1}{k^p}$ $(p > 0)$ in Dirichlet's test shows that the alternating harmonic p-series $\sum_{k=1}^{\infty} \frac{(-1)^{k-1}}{k^p}$ converges for $p > 0$.

Example 5.56. Decide whether the series $\sum_{k\ge 2} \frac{\sin k}{\log k}$ converges but not absolutely.

Solution. Let $a_n = \sin n$ and $b_n = 1/\log n$ for $n \ge 2$. Then $\{b_n\}$ is a decreasing sequence converging to zero and

$$\begin{aligned} S_n = \sum_{k=1}^{n} a_k &= \frac{1}{2\sin(\frac{1}{2})} \sum_{k=1}^{n} \left[\cos\left(k - \frac{1}{2}\right) - \cos\left(k + \frac{1}{2}\right)\right] \\ &= \frac{1}{2\sin(\frac{1}{2})} \left[\cos\left(\frac{1}{2}\right) - \cos\left(n + \frac{1}{2}\right)\right], \end{aligned}$$

so that $|S_n| \le 1/\sin(1/2)$. Thus, Dirichlet's test is applicable. To prove that the series is conditionally convergent, it suffices to show that $\sum |a_k b_k|$ behaves like $\sum \frac{1}{\log k}$. Since $|\sin x| + |\sin(x+1)| > 0$ for $x \in \mathbb{N}$, there exists a number $m > 0$ such that $m = \inf_{x\in\mathbb{N}}(|\sin x| + |\sin(x+1)|)$. Consequently,

$$\begin{aligned}\sum_{k=2}^{\infty}\frac{|\sin k|}{\log k}&=\sum_{k=2}^{\infty}\left(\frac{|\sin(2k-1)|}{\log(2k-1)}+\frac{|\sin(2k-2)|}{\log(2k-2)}\right)\\&>\sum_{k=2}^{\infty}\frac{1}{\log(2k-1)}\left(|\sin(2k-1)|+|\sin(2k-2)|\right)\\&=m\sum_{k=2}^{\infty}\frac{1}{\log(2k-1)},\end{aligned}$$

and so the given series does not converge absolutely. ●

Corollary 5.57. *If $\sum_{k=1}^{\infty} a_k$ is convergent and if $\{b_n\}$ is of bounded variation, then $\sum_{k=1}^{\infty} a_k b_k$ is convergent.*

Proof. Set $s_n = \sum_{k=1}^{n} a_k$. Then by hypothesis, $\{s_n\}$ is convergent (and therefore bounded). Since $\{b_n\}$ is of bounded variation, $\sum_{k=1}^{\infty} |b_k - b_{k-1}|$ is convergent, so that the series $\sum_{k=1}^{\infty}(b_k - b_{k+1})$ is convergent. This gives that the sequence of its partial sums is $\{b_1 - b_{n+1}\}$, which is clearly convergent. Thus, the series $\sum_{k=1}^{\infty} s_k(b_k - b_{k+1})$ and the sequence $\{b_n\}$ are convergent. The result follows from Abel's formula for summation by parts. ∎

Since every monotone bounded sequence is of bounded variation, we also have the following corollary, which is referred to as Abel's test (compare with Theorem 5.31 and Corollary 5.32).

Corollary 5.58 (Abel's test). *If $\sum a_k$ is convergent and $\{b_n\}$ is monotone and bounded, then $\sum a_k b_k$ is convergent.*

Neither of the two hypotheses in Corollary 5.57 can be dropped. For instance, consider the following pairs of choices:

(a) $a_n = (-1)^{n-1}$, $b_n = (-1)^n/n$;
(b) $a_n = (-1)^n/\sqrt{n}$, $b_n = (-1)^n/\sqrt{n}$;
(c) $a_n = (-1)^n$, $b_n = 1 + (1/n)$;
(d) $a_n = 1$, $b_n = 1/n$.

For each pair of choices, $\sum_{k=1}^{\infty} a_k b_k$ diverges.

Corollary 5.59. *Let $\{b_n\}_{n\geq 1}$ be a monotone sequence such that $b_n \to 0$ as $n \to \infty$. Then:*

(i) *$\sum_{k=1}^{\infty} b_k \sin kx$ converges for each $x \in \mathbb{R}$.*
(ii) *$\sum_{k=1}^{\infty} b_k \cos kx$ converges for each $x \in \mathbb{R}$ with $x \neq 2m\pi$, $m \in \mathbb{Z}$.*

Proof. **(i)** Fix x and let $a_n = \sin nx$ for $n \geq 1$. Then (as in Example 5.56)

$$S_n = \sum_{k=1}^{n} a_k = \frac{1}{2\sin(\frac{x}{2})}\left[\cos\left(\frac{x}{2}\right) - \cos\left(\frac{2k+1}{2}\right)x\right].$$

Note that if $x = 2m\pi$ for some $m \in \mathbb{Z}$, then $a_n = 0$ for each $n \geq 1$, and so $S_n = 0$ for each $n \geq 1$. Consequently, for $\sin(x/2) \neq 0$, it follows that

$$|S_n| \leq \frac{1}{|\sin(x/2)|} \quad \text{for each } n \geq 1.$$

Thus $\{S_n\}$ is bounded. The result follows from Dirichlet's test.

(ii) Let $a_n = \cos nx$ for x with $x \neq 2m\pi$ for any $m \in \mathbb{Z}$. We see that

$$\begin{aligned} S_n(x) &= \frac{1}{2\sin(x/2)} \sum_{k=1}^{n} 2\sin(x/2)\cos(kx) \\ &= \frac{1}{2\sin(x/2)} \sum_{k=1}^{n} [\sin(k+1/2)x - \sin(k-1/2)x] \\ &= \frac{1}{2\sin(x/2)} [\sin(n+1/2)x - \sin(x/2)]. \end{aligned}$$

Since $\sin(x/2) \neq 0$, the result follows from **(i)**. ∎

Again we remark that the Leibniz test is a special case Corollary 5.59 (choose $x = \pi$).

Example 5.60. If $x = \pi$ and $b_k = \frac{1}{\sqrt{k}}, \frac{1}{\log(k+1)}$, Corollary 5.59 gives that both

$$\sum_{k=1}^{\infty} \frac{(-1)^{k-1}}{\sqrt{k}} \quad \text{and} \quad \sum_{k=1}^{\infty} \frac{(-1)^{k-1}}{\log(k+1)}$$

are convergent. •

We next state the following theorem, the proof of which follows directly from the definition and the hypothesis.

Theorem 5.61. *If the sequence of partial sums of $\sum_{k=1}^{\infty} a_k$ is bounded and $\{b_n\}$ is a decreasing null sequence of positive numbers, then $\sum_{k\geq 1} a_k b_k$ converges.*

Proof. By hypothesis, there exists an $M > 0$ such that $|s_n| \leq M$ for $n \geq 1$, where $s_n = \sum_{k=1}^{n} a_k$. Since $\{b_n\}$ is a null sequence of positive numbers, given any $\varepsilon > 0$, there exists an N such that

$$b_n < \frac{\varepsilon}{2M} \text{ for all } n \geq N.$$

The formula for summation by parts (see the proof of Theorem 5.54) gives

$$\sum_{k=n+1}^{m} a_k b_k = \sum_{k=n+1}^{m} s_k(b_k - b_{k+1}) + s_m b_{m+1} - s_n b_{n+1},$$

so that

$$\left|\sum_{k=n+1}^{m} a_k b_k\right| \leq M \left[\sum_{k=n+1}^{m} (b_k - b_{k+1}) + (b_{n+1} + b_{m+1})\right]$$
$$\leq 2Mb_{n+1} < \varepsilon \text{ for all } n \geq N.$$

Thus $\{\sum_{k=1}^{n} a_k b_k\}$ is a Cauchy sequence, and the result follows by Cauchy's criterion. ■

5.3.5 Cauchy Product

If we formally consider the product of two power series $\sum_{k\geq 0} a_k x^k$ and $\sum_{k\geq 0} b_k x^k$ and organize the resulting series in powers of x, then we end up with

$$\Big(\sum_{k\geq 0} a_k x^k\Big)\Big(\sum_{k\geq 0} b_k x^k\Big) = \sum_{n\geq 0} c_n x^n, \quad c_n = \sum_{k=0}^{n} a_k b_{n-k},\ n = 0, 1, 2, \ldots.$$

A motivation for the introduction of the Cauchy product may be seen by substituting $x = 1$ in the above identity and making the following definition: *The Cauchy product of two convergent infinite series $\sum_{k\geq 0} a_k$, $\sum_{k\geq 0} b_k$ is the series*

$$\sum_{n\geq 0} c_n.$$

The result one might hope for is that $\sum_{n\geq 0} c_n$ converges. It turns out that this is false. We then ask, under what conditions does the series $\sum_{n\geq 0} c_n$ converge?

Theorem 5.62 (Mertens test). *If $A = \sum_{k\geq 0} a_k$ and $B = \sum_{k\geq 0} b_k$ are two convergent series, then $AB = \sum_{n\geq 0} c_n$ (meaning that the series converges), provided that at least one of the series is absolutely convergent.*

Proof. Let $\sum a_k$ be absolutely convergent. Define

$$A_n = \sum_{k=0}^{n} a_k,\ B_n = \sum_{k=0}^{n} b_k,\ C_n = \sum_{k=0}^{n} c_k \text{ and } a = \sum_{k\geq 0} |a_k|.$$

Then we note that

$$\begin{aligned} C_n &= c_0 + c_1 + \cdots + c_n \\ &= a_0 b_0 + (a_0 b_1 + a_1 b_0) + (a_0 b_2 + a_1 b_1 + a_2 b_0) + \cdots + (a_0 b_n + \cdots + a_n b_0) \\ &= a_0 B_n + a_1 B_{n-1} + \cdots + a_n B_0 \\ &= [a_0 B + a_1 B + \cdots + a_n B] - [a_0(B - B_n) + \cdots + a_n(B - B_0)] \\ &= A_n B - D_n, \end{aligned}$$

where

$$D_n = a_0(B - B_n) + a_1(B - B_{n-1}) + \cdots + a_n(B - B_0).$$

Note that

$$A_n B \longrightarrow AB \text{ as } n \to \infty. \tag{5.16}$$

Since $\sum_{k\geq 0} |a_k|$ is convergent, given $\varepsilon > 0$, there exists an N_1 such that

$$\sum_{k\geq N_1} |a_k| < \varepsilon \text{ for } n \geq N_1.$$

As $B_n - B \to 0$, for a given $\varepsilon > 0$, there exists an N_2 such that

$$|B_n - B| < \varepsilon \text{ for } n \geq N_2.$$

Therefore, for $n > \max\{N_1, N_2\} = N$ and $d_n = B - B_n$,

$$\begin{aligned} |D_n| &\leq (|a_0| + \cdots + |a_{n-N}|)\varepsilon + (|a_{n-N+1}| + \cdots + |a_n|) \sup_{n\in\mathbb{N}\cup\{0\}} \{|d_n|\} \\ &\leq \left(a + \sup_{n\in\mathbb{N}\cup\{0\}} \{|d_n|\}\right) \varepsilon \text{ with } a = \sum_{n\geq 0} |a_n|, \end{aligned}$$

which approaches 0 as $n \to \infty$, since $d_n \to 0$, and so $\sup_{n\in\mathbb{N}\cup\{0\}}\{|d_n|\} < \infty$. Hence by (5.16), $C_n \to AB$. This completes the proof. ∎

Remark 5.63. In Theorem 5.62, it is essential that one of the two series be absolutely convergent. For instance, if we consider

$$a_k = b_k = \frac{(-1)^{k+1}}{\sqrt{k+1}}, \quad k \geq 0,$$

then each of $\sum_{k\geq 0} a_k$ and $\sum_{k\geq 0} b_k$ converges (by the alternating series test) but not absolutely. On the other hand, we see that

$$\begin{aligned} |c_n| &= \left| (-1)^n \sum_{k=0}^{n} \frac{1}{\sqrt{k+1}\sqrt{n-k+1}} \right| \\ &= \frac{1}{\sqrt{1}\sqrt{n+1}} + \frac{1}{\sqrt{2}\sqrt{n}} + \cdots + \frac{1}{\sqrt{n+1}\sqrt{1}} \\ &\geq \sum_{k=0}^{n} \frac{1}{\sqrt{n+1}\sqrt{n+1}} = \frac{n+1}{n+1} = 1. \end{aligned}$$

Since the general term in the Cauchy product $\sum_{n\geq 0} c_n$ does not approach 0, the Cauchy product of the two chosen series does not converge. This example demonstrates that the Cauchy product of two conditionally convergent series may fail to be convergent.

The same conclusion can also be drawn if we choose

$$a_k = b_k = \frac{(-1)^k}{\log(k+2)}, \quad k \geq 0.$$

In this choice, we see that

$$|c_n| \geq \frac{n}{(\log(n+2))^2} \to \infty \text{ as } n \to \infty.$$

•

In view of Remark 5.63, it is natural to ask whether the series $\sum c_n$, if convergent, must have the sum AB. The answer is affirmative, see Theorem 9.54.

5.3.6 $(C, 1)$ Summability of Series

There are instances in which divergent series may be viewed as convergent series. The theory of summability methods for series primarily concerns itself with the question whether in some sense, a "sum" may be assigned to a series $\sum a_k$ even when it is divergent. At the same time, any "new sum" we define must agree with the sum in the ordinary sense, namely

$$\lim_{n\to\infty} s_n = \sum_{k=1}^{\infty} a_k$$

when the series is convergent.

Definition 5.64 ($(C,1)$ summable series). *If $\{s_n\}_{n\geq 1}$ is the sequence of partial sums of the series $\sum_{k=1}^{\infty} a_k$, then we say that the series $\sum_{k=1}^{\infty} a_k$ is $(C,1)$ summable to L, or $(C,1)$ summable with sum L, if $s_n \to L \quad (C,1)$. In this case, we write (see Definition* 2.62*)*

$$\sum_{k=1}^{\infty} a_k = L \quad (C,1) \quad \text{or} \quad \sum_{k=1}^{\infty} a_k = L \quad (\text{Cesàro}).$$

Example 5.65. Set $a_n = (-1)^{n-1}$ for $n \geq 1$. Then

$$s_n = \begin{cases} 0 & \text{if } n \text{ is even,} \\ 1 & \text{if } n \text{ is odd,} \end{cases}$$

which gives

$$\sigma_{2n} = \frac{n}{2n} = \frac{1}{2} \quad \text{and} \quad \sigma_{2n-1} = \frac{n}{2n-1}, \quad \text{where} \quad \sigma_n = \frac{1}{n}\sum_{k=1}^{n} s_k,$$

and so $\sigma_n \to 1/2$ as $n \to \infty$. Thus,

$$\sum_{n=1}^{\infty}(-1)^{n-1} = \frac{1}{2} \ (C,1).$$

•

Now we ask whether every divergent series can be $(C,1)$ summable.

Example 5.66 (Not all divergent series are $(C,1)$ summable). Show that the harmonic series $\sum_{k=1}^{\infty}(1/k)$ is not $(C,1)$ summable.

Solution. We have $a_k = 1/k$ and

$$\begin{aligned}
\sigma_n &= \frac{1}{n}\Big(s_1 + s_2 + \cdots + s_n\Big), \quad s_k = 1 + \frac{1}{2} + \cdots + \frac{1}{k} \\
&= \frac{1}{n}\left[n\cdot 1 + (n-1)\cdot\frac{1}{2} + (n-2)\cdot\frac{1}{3} + \cdots + 2\cdot\frac{1}{n-1} + \frac{1}{n}\right] \\
&= 1 + \left(1 - \frac{1}{n}\right)\frac{1}{2} + \left(1 - \frac{2}{n}\right)\frac{1}{3} + \cdots + \left(1 - \frac{n-1}{n}\right)\frac{1}{n} \\
&= s_n - \frac{1}{n}\left(\frac{1}{2} + \frac{2}{3} + \frac{3}{4} + \cdots + \frac{n-1}{n}\right) \\
&=: s_n - b_n, \text{ say,}
\end{aligned}$$

where

$$0 < b_n < \frac{1}{n}\left[(n-1)\left(\frac{n-1}{n}\right)\right] < 1 \text{ for all } n > 1.$$

Since $\{s_n\}$ is divergent, it follows that $\{\sigma_n\}$ is a divergent sequence of positive numbers. Consequently, the harmonic series is not $(C,1)$ summable. ●

If $a_n = c$ for some nonzero constant and for all $n \geq 1$, then $s_n = cn$, so that $\sigma_n = c(n+1)/2$, and so the series whose terms are some nonzero constant is not $(C,1)$ summable.

In our next theorem, we present a simple condition that makes $(C,1)$ summable series become convergent. An analogue of this theorem for Abel summable series is given in Theorem 9.57.

Theorem 5.67. *Suppose that $\sum_{k=0}^{\infty} a_k$ is $(C,1)$ summable to A, and $na_n \to 0$ as $n \to \infty$. Then $\sum_{k=0}^{\infty} a_k$ is convergent with sum A.*

Proof. Set $s_n = \sum_{k=1}^{n} a_n$, so that $n\sigma_n = \sum_{k=1}^{n} s_n$ and $\sigma_n \to A$ as $n \to \infty$. Further, since $na_n \to 0$ as $n \to \infty$, Theorem 2.64 shows that

$$T_n = \frac{1}{n}\sum_{k=1}^{n} ka_k \to 0 \quad \text{as } n \to \infty.$$

We need to show that $\{s_n\}$ converges to A. For this, we write

$$\begin{aligned} nT_n &= \sum_{k=1}^{n} k(s_k - s_{k-1}) \quad (s_0 = 0) \\ &= \sum_{k=1}^{n} (ks_k - (k-1)s_{k-1}) - \sum_{k=1}^{n} s_{k-1} \\ &= ns_n - \left(\sum_{k=1}^{n} s_k - s_n \right), \end{aligned}$$

and therefore

$$nT_n = (n+1)s_n - n\sigma_n \quad \text{or} \quad s_n = \frac{n}{n+1}\sigma_n + \frac{n}{n+1}T_n.$$

Since $\sigma_n \to A$ and $T_n \to 0$, it follows that $s_n \to A$ as $n \to \infty$, and the proof is complete. ∎

5.3.7 Questions and Exercises

Questions 5.68.

1. Does a sum of a convergent and a divergent series converge? Does $\sum_{k=1}^{\infty} \left(\frac{1}{\sqrt{k}} + \frac{(-1)^{k-1}}{k} \right)$ diverge?
2. Must an alternating series whose general term approaches zero be convergent? Does $\sum_{k=1}^{\infty} (-1)^{k-1} \left(1 - 3^{1/k} \right)$ converge?
3. Does a sum of two alternating convergent series converge?
4. Does the series $\sum_{k=1}^{\infty} (-1)^{k-1} \left(\frac{1}{\sqrt{k}} + \frac{(-1)^{k-1}}{k} \right)$ converge?
5. What (if anything) is wrong with the following computation?

$$\begin{aligned} 1 &- \frac{1}{2} + \frac{1}{3} - \frac{1}{4} + \frac{1}{5} - \frac{1}{6} + \cdots \\ &= 1 + \left(\frac{1}{2} - 1 \right) + \frac{1}{3} + \left(\frac{1}{4} - \frac{1}{2} \right) + \frac{1}{5} + \left(\frac{1}{6} - \frac{1}{3} \right) + \cdots \\ &= \left(1 + \frac{1}{2} + \frac{1}{3} + \frac{1}{4} + \cdots \right) - 1 - \frac{1}{2} - \frac{1}{3} - \frac{1}{4} - \cdots \\ &= \left(1 + \frac{1}{2} + \frac{1}{3} + \cdots \right) - \left(1 + \frac{1}{2} + \frac{1}{3} + \cdots \right) = 0. \end{aligned}$$

6. Suppose that $\sum a_k^2$ is convergent. Must $\sum a_k$ be divergent? Must $\sum a_k$ be convergent?
7. Suppose that $\sum a_k$ is convergent. Must $\sum a_k^2$ be divergent? Can $\sum a_k^2$ be convergent?

8. Does there exist a divergent alternating series $\sum(-1)^{k-1}a_k$ such that $\lim_{k\to\infty} a_k = 0$, but $\{a_k\}$ is not decreasing?
9. Suppose that $\sum_{k=1}^{\infty} a_k$ is absolutely convergent. Can
$$\sum_{k=1}^{\infty} a_k = \sum_{k=1}^{\infty} a_{2k-1} + \sum_{k=1}^{\infty} a_{2k}?$$
10. Suppose that $\sum_{k=1}^{\infty} a_k$ is conditionally convergent. Must $\sum_{k=1}^{\infty} a_k = \sum_{k=1}^{\infty} a_{2k-1} + \sum_{k=1}^{\infty} a_{2k}$?
11. Must a product of two sequences of bounded variation be of bounded variation?
12. Can we drop the condition $b_n \to 0$ in Theorem 5.54(2)?
13. If $\sum_{k=1}^{\infty} a_k$ is absolutely convergent, must $\{a_n\}_{n\geq 1}$ be of bounded variation? How about if "absolutely convergent" is replaced by "convergent"?
14. If $\{b_n\}$ is a decreasing sequence of bounded variation, i.e., $\sum_{k=1}^{\infty} |b_k - b_{k+1}| < \infty$, must it be convergent?
15. Must every monotone bounded sequence be of bounded variation?
16. Is the series $\sum_{k=1}^{\infty} k(-1)^{k-1}$ Cesàro summable?

Exercises 5.69.

1. Let $\{a_n\}$ be a decreasing sequence of nonnegative real numbers with $a_n \to 0$. Set
$$b_n = \frac{1}{n}\sum_{k=1}^{n} a_k, \quad c_n = \frac{1}{n}\sum_{k=1}^{n} a_{2k-1} \quad \text{and} \quad d_n = \frac{1}{2n-1}\sum_{k=1}^{n} a_{2k-1}.$$
Show that if A_n is either b_n or c_n or d_n, then $\sum_{k=1}^{\infty}(-1)^{k-1}A_k$ is convergent.
2. Test the series $\sum_{k=1}^{\infty} a_k$ for convergence, where a_k equals
$$\sqrt{2}\frac{\sin(k\pi/2 - \pi/4)}{k}.$$
3. Using the alternating series test or otherwise, test $\sum_{k=1}^{\infty}(-1)^{k-1}a_k$ for convergence, where a_k is given by

(a) $\dfrac{1}{k^{1/3}+k^{1/2}}$. (b) $\dfrac{1}{(k+1)\log(k+1)}$. (c) $\dfrac{k+2^k}{3^k+5}$.

(d) $\dfrac{k}{k+2}$. (e) $\dfrac{1\cdot 3\cdot \cdots \cdot (2k-1)}{2\cdot 4\cdot \cdots \cdot 2k}$. (f) $\dfrac{1}{k+a^2}$.

4. At least how many terms are to be considered from the series $\sum_{k=1}^{\infty} \frac{(-1)^{k-1}}{2k-1}$ so that the error does not exceed 0.0001?

5. Show that the series

$$1+\frac{1}{2}+\frac{1}{3}-\frac{1}{4}-\frac{1}{5}-\frac{1}{6}+\frac{1}{7}+\frac{1}{8}+\frac{1}{9}-\quad-\quad-\quad+\quad+\quad+\quad\cdots$$

converges and find its sum.

6. Does the series

$$1-1+\frac{1}{2}-\frac{1}{2}+\frac{1}{3}-\frac{1}{3}+\frac{1}{4}-\frac{1}{4}+\cdots$$

converge? If so, does it have a divergent rearrangement? How about the rearrangement of the series

$$1-1+\frac{1}{2}-\frac{1}{2}+\frac{1}{3}+\frac{1}{4}-\frac{1}{3}+\frac{1}{5}+\frac{1}{6}+\frac{1}{7}+\frac{1}{8}-\frac{1}{4}+\cdots?$$

7. Using the fact that

$$\frac{1}{3n-2}+\frac{1}{3n-1}-\frac{1}{3n}>\frac{1}{3n}\quad\text{for } n=1,2,\ldots,$$

deduce that the series

$$1+\frac{1}{2}-\frac{1}{3}+\frac{1}{4}+\frac{1}{5}-\frac{1}{6}+\cdots$$

is divergent.

8. Show that the series

$$\left(1-\frac{1}{2}\right)+\left(1-\frac{3}{4}\right)+\left(1-\frac{7}{8}\right)+\cdots$$

is convergent. Show also that when the parentheses are removed, it oscillates.

9. Show that

(i) $\displaystyle\frac{\pi}{8}=\frac{1}{1\cdot 3}+\frac{1}{5\cdot 7}+\frac{1}{9\cdot 11}+\cdots.$

(ii) $\displaystyle\log 2=\frac{1}{1\cdot 2}+\frac{1}{3\cdot 4}+\frac{1}{5\cdot 6}+\cdots.$

10. Suppose that $\sum a_k$ is convergent (e.g., $a_k=(-1)^{k-1}/k$). Using Abel's test (Corollary 5.58) or otherwise, test the convergence of $\sum a_k b_k$ when b_k equals

(a) $k^{1/k}$. **(b)** $\displaystyle\frac{1}{\log k}$. **(c)** $\displaystyle\left(1+\frac{1}{k}\right)^k$. **(d)** k^{-p} $(p>0)$.

11. Test the convergence of the series $\sum_{k=1}^{\infty} a_k(x)$ whose kth terms are given below:

(a) $\displaystyle\frac{\cos kx}{\log(k+1)}$. **(b)** $\displaystyle\frac{\sin kx}{(\log k)^{\alpha}}$. **(c)** $\displaystyle\frac{\cos kx}{k^{\alpha}}$.

12. Consider the following series:

    (a) $\sum_{k=1}^{\infty} \sin(\pi/k)$. (b) $\sum_{k=1}^{\infty} (-1)^{k-1} \cos(\pi/k)$.

    Verify whether each of these converges or diverges.
13. If $a_n = \frac{(-1)^n}{n+1}$, $n \geq 0$, determine the Cauchy product of $\sum_{k=0}^{\infty} a_k$ with itself. Verify whether the Cauchy product series converges.
14. Give an example such that
    (a) $\{b_n\}$ is bounded but is not of bounded variation.
    (b) $\{b_n\}$ is convergent but is not of bounded variation.
    (c) $\{b_n\}$ is of bounded variation but is not monotone.
15. Show that $\sum_{k=1}^{\infty} \sin(k\pi/2)$ is $(C, 1)$ summable to $1/2$.
16. Show that a divergent series of positive numbers cannot be Cesàro summable.
17. Suppose that the series $\sum_{k=1}^{\infty} a_k$ is $(C, 1)$ summable. Show that
    (a) the sequence $\{a_n\}_{n\geq 1}$ is $(C, 1)$ summable.
    (b) the sequence $\{s_n/n\}_{n\geq 1}$ converges to 0.

6

Definite and Indefinite Integrals

In Section 6.1, we define the definite integral, called the *Riemann integral*, using Riemann sums. Then we study the properties of the Riemann integral. The fundamental result that we prove in this section states that every bounded function on an interval $[a, b]$ is integrable if it is either monotone on $[a, b]$ (see Theorem 6.20) or continuous on $[a, b]$ (see Theorem 6.21). In Section 6.1, we show that the definite integral $\int_a^b f(x)\,dx$ exists if $f(x)$ is a piecewise continuous function defined on a bounded interval $[a, b]$.

In Section 6.2, we meet the fundamental theorem of calculus, which connects the integral of a function and its antiderivative. Later, we will move on to the Riemann–Stieltjes integral. Basic linearity properties of definite and indefinite integrals are presented. In Section 6.2, we present an important theorem in integral calculus, namely the mean value theorem for integrals, and as a consequence, we define the average value of a function. In addition, using the fundamental theorem of integral calculus, we introduce the logarithmic and exponential functions and develop their principal properties.

In a later chapter (see Section 9.2), we examine a close relationship between uniform convergence and integration, and then later between uniform convergence and differentiation.

6.1 Definition and Basic Properties of Riemann Integrals

We begin our discussion with an arbitrary bounded function $f(x)$ defined on a closed interval $[a, b]$. A *partition* $P = \{x_0, x_1, \dots, x_n\}$ of an interval $[a, b]$ is a finite set of points arranged in such a way that

$$a = x_0 < x_1 < x_2 < \cdots < x_{n-1} < x_n = b.$$

The partition P defines n closed subintervals

$$[x_0, x_1],\ [x_1, x_2],\ \dots,\ [x_{k-1}, x_k],\ \dots,\ [x_{n-1}, x_n]$$

S. Ponnusamy, *Foundations of Mathematical Analysis*,
DOI 10.1007/978-0-8176-8292-7_6,

of $[a,b]$. The typical closed subinterval $[x_{k-1},x_k]$ is called the *kth subinterval representative* of the partition P. The length of the kth subinterval is

$$\Delta x_k = x_k - x_{k-1}, \quad k=1,2,\ldots,n.$$

The largest of the lengths of these subintervals is called the *norm* (sometimes called the *mesh* or *width*) of the partition P and is denoted by $\|P\|$; that is,

$$\|P\| = \max_{k=1,2,\ldots,n} \Delta x_k := \max_{k=1,2,\ldots,n} (x_k - x_{k-1}).$$

A *standard partition* or *equally spaced partition* is a partition all of whose subintervals are of equal length. For an arbitrary interval $[a,b]$, the standard partition is given by

$$P = \{x_0, x_1, \ldots, x_n\}, \quad \text{where } x_k = a + \frac{k(b-a)}{n} \quad \text{for } k=0,1,2,\ldots,n.$$

The family of all partitions of $[a,b]$ will be denoted by $\mathcal{P}[a,b]$ or simply by $\mathcal{P}$ when the interval under discussion is clear. For each $k=1,2,\ldots,n$, choose an arbitrary point $x_k^* \in [x_{k-1},x_k]$ (Figure 6.1). On each subinterval, we form the product

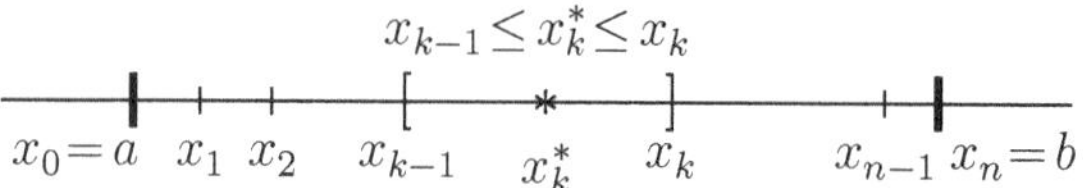

Fig. 6.1. Partition of $[a,b]$.

$$A_k = f(x_k^*)\Delta x_k$$

and the sum

$$S_n = \sum_{k=1}^{n} A_k.$$

This sum, which depends on the partition P and the choice of the points $x_1^*, x_2^*, \ldots, x_n^*$, is called the *integral sum* (also called *Riemann sum*) of f over the interval $[a,b]$ with respect to P and points $x_k^* \in [x_{k-1},x_k]$, $k=1,2,\ldots,n$. Also, if we let (see Figure 6.2)

$$m_k = \inf_{x\in[x_{k-1},x_k]} f(x), \quad M_k = \sup_{x\in[x_{k-1},x_k]} f(x),$$

then each partition determines two sums that correspond to overestimates and underestimates of the possible area:

$$\overline{S_n} := \sum_{k=1}^{n} M_k \Delta x_k \quad \text{and} \quad \underline{s_n} := \sum_{k=1}^{n} m_k \Delta x_k. \tag{6.1}$$

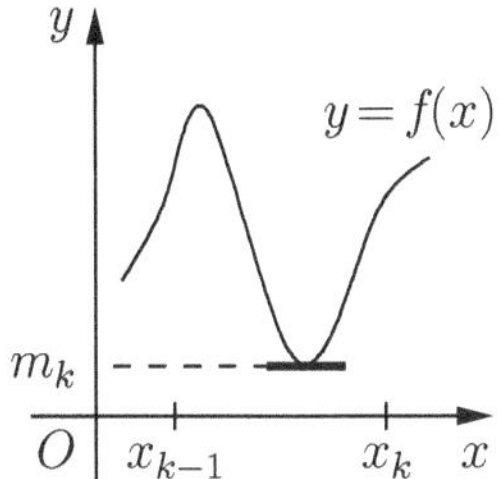

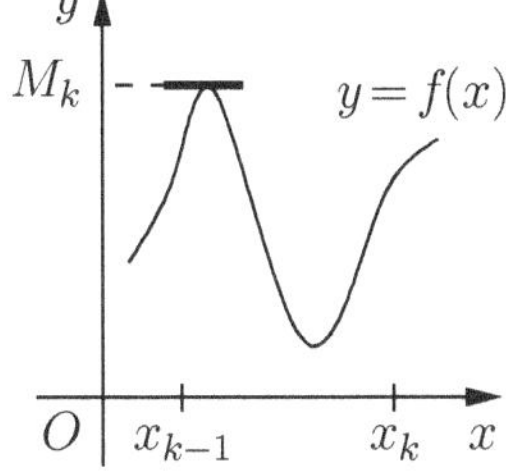

Fig. 6.2. Description for m_k and M_k.

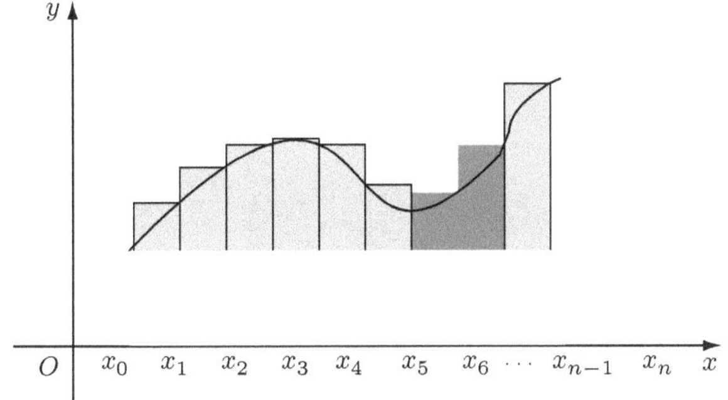

Fig. 6.3. Upper Riemann sum.

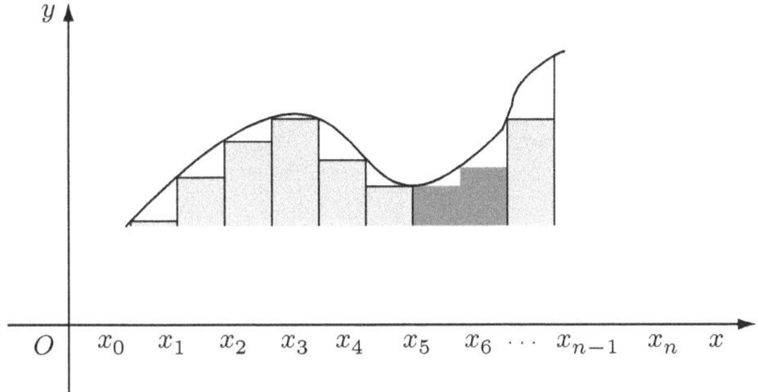

Fig. 6.4. Lower Riemann sum.

Here $\overline{S_n}$ and $\underline{s_n}$ are referred to as an *upper sum* and a *lower sum* of f on $[a, b]$, respectively (Figures 6.3 and 6.4). We observe that as n becomes larger, the subdivision of $[a, b]$ corresponding to the partition P becomes finer.

For the purpose of visualizing the notion of area of the region R bounded by the x-axis and the curve

$$y = f(x), \quad \text{and the lines } x = a,\ x = b,$$

it is worthwhile for the moment to consider the special case in which f is continuous on $[a,b]$ and $f(x) \geq 0$ on $[a,b]$. Then A_k defined above is the area of the kth rectangle, and the area in question must lie between $\overline{S_n}$ and $\underline{s_n}$, showing that the upper sum overestimates the area, while the lower sum underestimates it. If the number of partition points is increased such that $\|P\|$ is sufficiently small, then the upper sum decreases, while the lower sum increases. In other words, as $\|P\| \to 0$, both $\overline{S_n}$ and $\underline{s_n}$ should converge to the area under the curve. Thus, by the squeeze/sandwich rule, we express the area under the curve $y = f(x)$ ($f(x) \geq 0$, $x = a$, $x = b$, and the x-axis) as

$$A = \lim_{\|P\| \to 0} S_n,$$

provided the limit exists.

6.1.1 Darboux Integral

Definition 6.1. *Suppose that $f\colon [a,b] \to \mathbb{R}$ is bounded and $P = \{x_0, x_1, \ldots, x_n\}$ is a partition and $x_k^* \in [x_{k-1}, x_k]$ $(k = 1, 2, \ldots, n)$ is arbitrary. Then the quantities $\overline{S_n}$, $\underline{s_n}$, and S_n are called the upper sum (or upper Darboux sum or upper integral sum), the lower sum (or lower Darboux sum or lower integral sum), and the Riemann sum, respectively, of the function f associated with the partition P. These are usually denoted by*

$$\overline{S_n} := U(P,f), \quad \underline{s_n} := L(P,f), \quad \text{and} \quad S_n := \sigma(P,f,x^*).$$

For a bounded function f on $[a,b]$, we define

$$m = \inf_{x \in [a,b]} f(x) \ \text{ and } \ M = \sup_{x \in [a,b]} f(x),$$

and so

$$m \leq m_k \leq f(x_k^*) \leq M_k \leq M.$$

If we multiply the inequalities above by $x_k - x_{k-1}$ and sum over $k = 1, 2, \ldots, n$, then we obtain

$$m(b-a) \leq \underbrace{\underline{s_n} \leq S_n \leq \overline{S_n}} \leq M(b-a);$$

that is,

$$m(b-a) \leq \underbrace{L(P,f) \leq \sigma(P,f,x^*) \leq U(P,f)} \leq M(b-a)$$

holds for every partition P and for every choice of the points $x_k^* \in [x_{k-1}, x_k]$, $k = 1, 2, \ldots, n$ (Figure 6.5). In other words,

$$\{U(P,f) : P \in \mathcal{P}[a,b]\} \quad \text{and} \quad \{L(P,f) : P \in \mathcal{P}[a,b]\}$$

form bounded sets. The above discussion leads to the following definition.

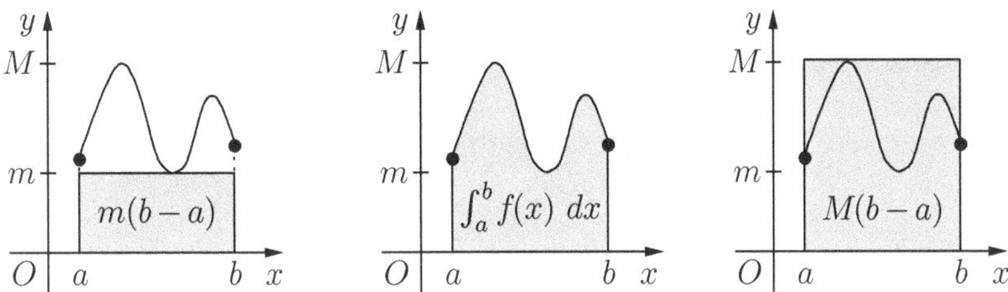

Fig. 6.5. Inequalities for integrals.

Definition 6.2 (Darboux integral). *The upper (Darboux) integral of f on $[a,b]$ is defined by*

$$U(f) := \overline{\int_a^b} f(x)\,dx = \inf\{U(P,f) : P \in \mathcal{P}[a,b]\},$$

and the lower (Darboux) integral of f on $[a,b]$ is defined by

$$L(f) := \underline{\int_a^b} f(x)\,dx = \sup\{L(P,f) : P \in \mathcal{P}[a,b]\}$$

(as a finite number). A bounded function f defined on $[a,b]$ is said to be (Darboux) integrable on $[a,b]$ if the upper and the lower (Darboux) integrals are equal, i.e., if $U(f) = L(f)$. The common value is called the integral of f on $[a,b]$ or the definite integral of f from a to b and is denoted by

$$\int_a^b f(x)\,dx.$$

In this case, we say that f is Darboux integrable. If $U(f) > L(f)$, then we say that f is not Darboux integrable.

Note that if $\int_a^b f(x)\,dx$ exists, then

$$L(P,f) \leq \int_a^b f(x)\,dx \leq U(Q,f) \quad \text{for every } P, Q \in \mathcal{P}[a,b].$$

Recall that in defining the quantities $U(f)$ and $L(f)$, we have assumed the completeness property of the real numbers (see Definitions 1.18 and 1.19). We follow the convention that whenever an interval $[a,b]$ is employed, we assume that $a < b$ and therefore $\int_a^b f(x)\,dx$ for $a < b$ only. If $a = b$, we set $\int_a^b f(x)\,dx = 0$.

Nonspecialists actually call the lower Darboux integral and upper Darboux integral the lower Riemann integral and the upper Riemann integral, respectively. Also, $\int_a^b f(x)\,dx$ is often referred to as the Riemann integral of f on

$[a, b]$. This is because Riemann's definition of integrability is slightly different (see Definition 6.14). However, this will be clear once we prove that these two definitions are actually equivalent.

The function f that is being integrated is called the *integrand*, the interval $[a, b]$ is the *interval of integration*, and the endpoints a and b are called, respectively, the *lower and upper limits of integration*.

Example 6.3. A partition of $[0, 1]$ is

$$P = \left\{0, \frac{3}{7}, \frac{1}{2}, \frac{3}{4}, 1\right\},$$

so that

$$\Delta x_1 = \frac{3}{7}, \quad \Delta x_2 = \frac{1}{14}, \quad \Delta x_3 = \frac{1}{4}, \quad \Delta x_4 = \frac{1}{4},$$

and the norm of the partition P is

$$\|P\| = \max\left\{\frac{3}{7}, \frac{1}{14}, \frac{1}{4}, \frac{1}{4}\right\} = \frac{3}{7}.$$

Note that P is not a standard partition of $[a, b]$, because its subintervals are not all of equal length. ●

While forming the integral sum, f must be assumed to be bounded, since every integrable function is necessarily bounded; see, for example, Remark 6.5.

Example 6.4 (Not every bounded function is integrable). Consider the Dirichlet function f defined over the interval $[0, 1]$:

$$f(x) = \begin{cases} 0 & \text{if } x \text{ is irrational} \\ 1 & \text{if } x \text{ is rational} \end{cases}, \quad x \in [0, 1].$$

Let $P = \{x_0, x_1, \ldots, x_n\}$ be any partition of $[0, 1]$. Since every interval $[x_{k-1}, x_k]$ contains both rational and irrational points, we have (see Figure 6.6)

$$m_k = \inf_{x_{k-1} \leq x \leq x_k} f(x) = 0 \quad \text{and} \quad M_k = \sup_{x_{k-1} \leq x \leq x_k} f(x) = 1.$$

We compute the upper and lower integral sums

$$U(P, f) = \sum_{k=1}^{n} 1.(x_k - x_{k-1}) = x_n - x_0 = 1 \quad \text{and} \quad L(P, f) = 0.$$

Therefore, since the above is true for any $P \in \mathcal{P}[a, b]$, we have

$$\overline{\int_0^1} f(x)\, dx = 1 \quad \text{and} \quad \underline{\int_0^1} f(x)\, dx = 0,$$

and hence the Dirichlet function f is not integrable on $[0, 1]$. This example shows that not every bounded function is integrable. ●

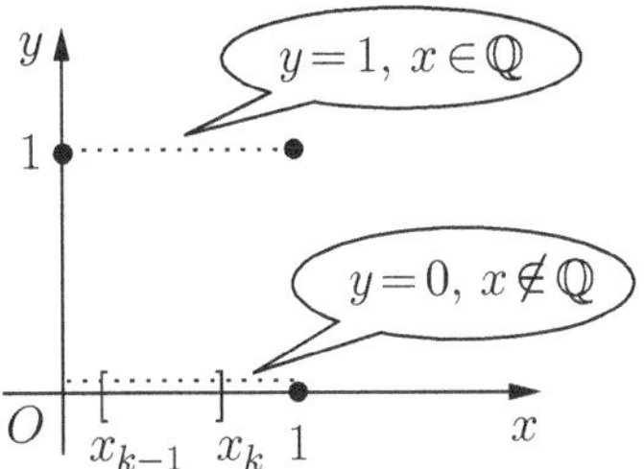

Fig. 6.6. Each $[x_{k-1}, x_k]$ contains both rational and irrational points.

Remark 6.5 (Unbounded functions). We encounter some difficulties if we try to apply the definition for unbounded functions. For example, consider

$$f(x) = \begin{cases} \dfrac{1}{x} & \text{for } x \in (0,1] \\ 0 & \text{for } x = 0 \end{cases} \quad \text{and} \quad g(x) = \begin{cases} \dfrac{1}{\sqrt{x}} & \text{for } x \in (0,1], \\ 0 & \text{for } x = 0. \end{cases}$$

Then both f and g are unbounded on $[0,1]$ with

$$\lim_{x \to 0+} f(x) = \infty = \lim_{x \to 0+} g(x).$$

Let $P = \{x_0, x_1, \dots, x_n\}$ be a partition. If in the sum $S_n = \sum_{k=1}^{n} f(x_k^*) \Delta x_k$, we have $x_1^* = 0$, then $f(x_1^*) = 0$, and so

$$S_n := \sigma(P, f, x^*) = \sum_{k=2}^{n} f(x_k^*) \Delta x_k.$$

But if $x_1^* > 0$, then because $\sup_{x \in [0, x_1^*]} f(x) = \infty$, no matter how small the first subinterval length $\Delta x_1 = x_1 - x_0 = x_1$, we can make $f(x_1^*)$ as large as possible by choosing x_1^* sufficiently close to zero. Thus, $U(P, f)$ does not exist, and this observation forces us to avoid unbounded functions on any closed interval when considering integrability. ●

We remark that the upper and lower sums depend on the choice of partitions, while the upper and lower integrals are independent of the partitions.

Example 6.6 (Integrability of a constant function). We begin our discussion with $f(x) = c$ (constant) for all $x \in [a, b]$. Then every Riemann sum of f over any partition P becomes

$$\sigma(P, f, x^*) = \sum_{k=1}^{n} f(x_k^*)\, \Delta x_k = c \sum_{k=1}^{n} (x_k - x_{k-1}) = c(x_n - x_0) = c(b - a),$$

which is a constant. In particular,

$$c(b - a) = L(P, f) \le L(f) \le U(f) \le U(P, f) = c(b - a),$$

so that $L(f) = U(f)$, showing the integrability of f, and $\int_a^b c\, dx = c(b - a)$. ●

6.1.2 Basic Properties of Upper and Lower Sums

We wish to establish some basic properties of the Riemann integral and obtain different types of classes of functions that are Riemann integrable. In particular, we need to address the following fundamental questions:

- Is every monotone function on $[a, b]$ Riemann integrable?
- Is every continuous function on $[a, b]$ Riemann integrable?
- Is every bounded function that has a finite number of discontinuities in $[a, b]$ Riemann integrable?
- Is every bounded function that has an infinite number of discontinuities in $[a, b]$ Riemann integrable?
- Is every monotone function that has an infinite number of discontinuities in $[a, b]$ Riemann integrable?

In order to answer these questions and to build our presentation in understanding Riemann's approach to integration, we need some preparations.

If P_1 and P_2 are two partitions of $[a, b]$ such that every division point of P_1 is also a division point of P_2, then we say that P_2 is a refinement of P_1 (or that P_2 refines P_1 or P_2 is finer than P_1), and write $P_1 \subseteq P_2$ or $P_2 \supseteq P_1$. Thus if

$$P_1 = \{x_0, x_1, x_2, \ldots, x_n\} \text{ and } P_2 = \{y_0, y_1, y_2, \ldots, y_m\},$$

where $a = x_0 = y_0$ and $b = x_n = y_m$, then $P_1 \subseteq P_2$ means that $m \geq n$ and

$$\{x_0, x_1, x_2, \ldots, x_n\} \subseteq \{y_0, y_1, y_2, \ldots, y_m\}.$$

Note that if $P_1 \subseteq P_2$, then $||P_1|| \geq ||P_2||$. In other words, refinement of a partition decreases its norm, but the converse does not necessarily hold. Further, a partition R of $[a, b]$ is called a *common refinement* of two partitions P_1 and P_2 if $P_1 \cup P_2 \subseteq R$. Here the partition $P_1 \cup P_2$ is obtained by taking into account of all the partition points of P_1 and P_2.

For instance, if

$$P_1 = \left\{0, \frac{1}{2}, \frac{3}{5}, \frac{3}{4}, \frac{4}{5}, 1\right\} \text{ and } P = \left\{0, \frac{1}{2}, \frac{3}{5}, \frac{4}{5}, 1\right\}$$

are two partitions of $[0, 1]$, then P_1 is a refinement of P, since $P_1 \supseteq P$. Similarly if

$$P_2 = \left\{0, \frac{1}{4}, \frac{1}{3}, \frac{1}{2}, \frac{3}{5}, \frac{3}{4}, \frac{4}{5}, 1\right\},$$

then P_2 is a refinement of both P_1 and P. However, the partition

$$Q = \left\{0, \frac{1}{5}, \frac{1}{2}, 1\right\}$$

of $[0,1]$ is not a refinement of P, because the partition points of Q do not include all partition points of P. Also, the common refinement R_1 of P and P_1 is

$$R_1 = \left\{0, \frac{1}{2}, \frac{3}{5}, \frac{3}{4}, \frac{4}{5}, 1\right\},$$

and the common refinement of P and $P_3 = \left\{0, \frac{1}{3}, \frac{5}{6}, \frac{6}{7}, 1\right\}$ is

$$R_2 = \left\{0, \frac{1}{3}, \frac{1}{2}, \frac{3}{5}, \frac{4}{5}, \frac{5}{6}, \frac{6}{7}, 1\right\}.$$

Note that P_3 is not a refinement of P.

Finally, we remark that for any bounded function f on an interval I, the definitions of supremum and infimum give that

$$\sup_{x\in I} f(x) \geq \sup_{x\in J} f(x) \quad \text{and} \quad \inf_{x\in I} f(x) \leq \inf_{x\in J} f(x) \quad \text{whenever } J \subseteq I.$$

The following is useful in understanding lower sums and upper sums.

Lemma 6.7. *Let f be a bounded function on $[a,b]$, and let P and Q be two partitions of $[a,b]$. Then we have the following:*

(a) $L(P,f) \leq L(Q,f) \leq U(Q,f) \leq U(P,f)$ *if* $P \subseteq Q$.
(b) $L(P,f) \leq U(Q,f)$ *for any P and Q.*
(c) $L(f) \leq U(f)$.

Proof. **(a)** The middle inequality is obvious and has been discussed earlier. The first inequality says that a refinement can only increase (or leave fixed) the lower sum, and the third inequality conveys that a refinement can only decrease (or leave fixed) the upper sum. We shall prove the first inequality. There is nothing to prove if $P = Q$.

First we observe that if Q has r $(r \geq 1)$ points not in P, we can start at P and arrive at Q after r steps by adjoining extra points, say $c_1, c_2, \ldots, c_r$. So we start with the partition $P = \{x_0, x_1, \ldots, x_n\}$ of $[a,b]$. Let P_1 be a new partition formed by adding one extra point, say $c = c_1 \in (x_{k-1}, x_k)$, so that

$$P_1 = \{x_0, x_1, \ldots, \underbrace{x_{k-1}, c, x_k}, x_{k+1}, \ldots, x_n\}.$$

Clearly, the only contribution to the lower and the upper sums that may differ for P and P_1 is from the interval $[x_{k-1}, x_k]$. Since f is bounded, we have $|f(x)| \leq K$, i.e., $\sup_{x\in[a,b]} |f(x)| < K$, for some $K > 0$. Also, we define

$$m_k' = \inf_{x\in[x_{k-1},c]} f(x) \quad \text{and} \quad m_k'' = \inf_{x\in[c,x_k]} f(x).$$

Clearly, $m_k' \geq m_k$, $m_k'' \geq m_k$, and

$$\begin{aligned} L(P_1, f) &= \sum_{j=1,\ j\neq k}^{n} m_j \Delta x_j + \underbrace{m_k'(c - x_{k-1}) + m_k''(x_k - c)} \\ &\geq \sum_{j=1,\ j\neq k}^{n} m_j \Delta x_j + \underbrace{m_k(c - x_{k-1}) + m_k(x_k - c)} \\ &= \sum_{j=1}^{n} m_j \Delta x_j = L(P, f). \end{aligned}$$

Similar arguments prove that

$$U(P_1, f) \leq U(P, f),$$

with the obvious notation $M_k' \leq M_k$ and $M_k'' \leq M_k$. By applying this fact a finite number of times (or by the method of induction), we obtain the desired inequalities **(a)**.

Also, we observe that

$$\begin{aligned} L(P_1, f) - L(P, f) &= \underbrace{m_k'(c - x_{k-1}) + m_k''(x_k - c)} - m_k(x_k - x_{k-1}) \\ &\leq \underbrace{K(c - x_{k-1}) + K(x_k - c)} + K(x_k - x_{k-1}) \\ &= 2K(x_k - x_{k-1}), \end{aligned}$$

so that

$$L(P_1, f) - L(P, f) \leq 2K\|P\|. \tag{6.2}$$

More generally, if P' is obtained by adjoining r points that are not in P, an induction argument clearly shows that

$$L(P', f) - L(P, f) \leq 2rK\|P\|.$$

Similarly, we can obtain $U(P', f) - U(P, f) \geq -2rK\|P\|$.

(b) Consider $P \cup Q$, which is a partition of $[a, b]$. Then $P \subseteq P \cup Q$ and $Q \subseteq P \cup Q$. Note that $P \cup Q$ is obtained by lumping together all of the points of P and Q. By (a),

$$L(P, f) \leq L(P \cup Q, f) \leq U(P \cup Q, f) \leq U(Q, f)$$

for any two arbitrary partitions P and Q.

(c) Finally, we show that for any partition $P \in \mathcal{P}[a, b]$, we have

$$L(P, f) \leq L(f) = \underline{\int_a^b} f(x)\, dx \leq U(f) = \overline{\int_a^b} f(x)\, dx \leq U(P, f).$$

Since

$$m(b-a) \le L(P,f) \le U(P,f) \le M(b-a),$$

the existence of $L(f)$ and $U(f)$ is obvious. Indeed, $L(P,f)$ is a lower bound for the set $\{U(P,f) : P \in \mathcal{P}[a,b]\}$, and therefore

$$L(P,f) \le \inf\{U(P,f) : P \in \mathcal{P}[a,b]\} = U(f),$$

showing that $U(f)$ serves as an upper bound for the set $\{L(P,f) : P \in \mathcal{P}[a,b]\}$. Hence $\sup_P L(P,f)$ exists and

$$U(f) = \overline{\int_a^b} f(x)\,\mathrm{d}x \ge L(f) = \underline{\int_a^b} f(x)\,\mathrm{d}x.$$

The assertion follows. ■

6.1.3 Criteria for Integrability

Our first major criterion for integrability says nothing about the value of the integral but uses only the difference between the upper and lower sums.

Theorem 6.8 (Riemann's criterion for integrability). *If f is a bounded function on $[a,b]$, then f is integrable on $[a,b]$ if and only if for each $\epsilon > 0$ there is a partition P of $[a,b]$ such that (see Figure 6.7)*

$$U(P,f) - L(P,f) < \epsilon. \tag{6.3}$$

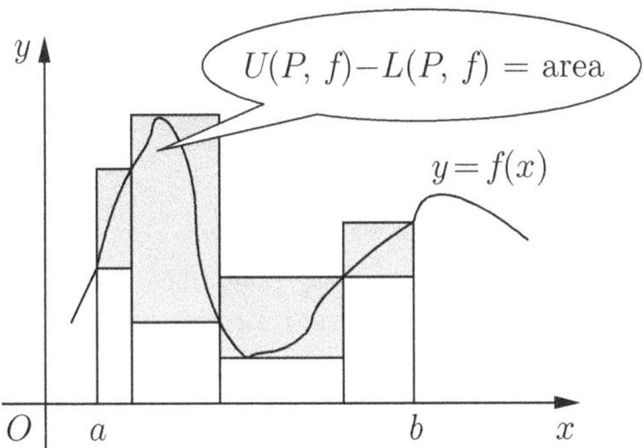

Fig. 6.7. $U(P,f) - L(P,f) = \text{area} \ge 0$.

Proof. (Necessity $\Longrightarrow$): Assume first that f is integrable on $[a,b]$, i.e., $U(f) = L(f)$. Denote the common value of these two quantities by α. Let $\epsilon > 0$ be given. Since $U(f)$ is the infimum of all upper sums, and $L(f)$ is the supremum of all lower sums, there exist two partitions P_1 and P_2 of $[a,b]$ for which

$$U(P_1,f) < U(f) + \frac{\epsilon}{2} \text{ and } L(P_2,f) > L(f) - \frac{\epsilon}{2}.$$

Set $P = P_1 \cup P_2$. By Lemma 6.7**(a)**, we obtain

$$L(f) - \frac{\epsilon}{2} < L(P_2,f) \leq L(P,f) \leq U(P,f) \leq U(P_1,f) < U(f) + \frac{\epsilon}{2},$$

so that

$$\alpha - \frac{\epsilon}{2} < L(P,f) \leq U(P,f) < \alpha + \frac{\epsilon}{2},$$

and the assertion (6.3) holds (see Figure 6.8).

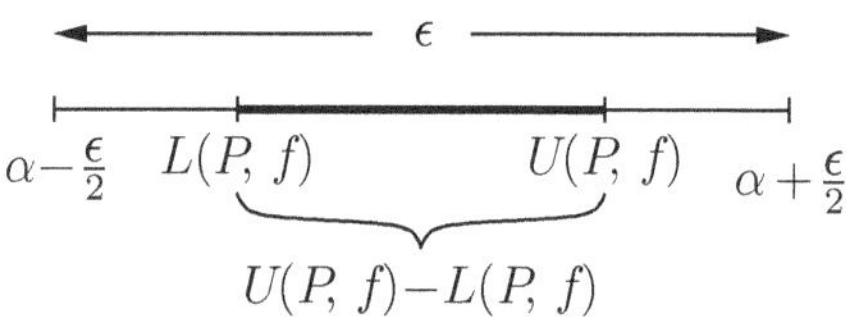

Fig. 6.8. Sketch for the quantity $U(P,f) - L(P,f)$.

(Sufficiency $\Longleftarrow$): Suppose that for each $\epsilon > 0$, the inequality (6.3) holds for some partitions P of $[a,b]$. Note that by the definitions of $L(f)$ and $U(f)$,

$$U(f) - L(f) \leq U(P,f) - L(P,f) < \epsilon,$$

and so

$$0 \leq U(f) - L(f) < \epsilon \quad \text{for every } \epsilon > 0.$$

Since $\epsilon > 0$ is arbitrary, we must have $U(f) = L(f)$, and so f is integrable on $[a,b]$, as required. ■

The next theorem gives another criterion for integrability.

Theorem 6.9. *If f is a bounded function on $[a,b]$, then f is integrable if and only if for each $\epsilon > 0$ there exists a $\delta > 0$ such that*

$$U(P,f) - L(P,f) < \epsilon \tag{6.4}$$

for all partitions P of $[a,b]$ for which $\|P\| < \delta$.

Proof. (Sufficiency $\Longleftarrow$): The sufficiency follows from Theorem 6.8.

(Necessity $\Longrightarrow$): Let f be integrable on $[a,b]$ and let $\epsilon > 0$. By Theorem 6.8, there exists a partition $Q = \{y_0, y_1, \dots, y_r\}$ of $[a,b]$ such that

$$U(Q,f) - L(Q,f) < \epsilon/2,$$

where r is the number of subintervals in Q. Since f is bounded, there exists a constant $K > 0$ such that $|f(x)| \le K$ for all $x \in [a,b]$. Set $\delta = \epsilon/(8rK)$. To verify (6.4), we consider a partition $P = \{x_0, x_1, \ldots, x_n\}$ with $\|P\| < \delta$. Define $P' = P \cup Q$ to be a common refinement of P and Q, and note that P' has at most r points that are not in P. As in the proof of Lemma 6.7 (see (6.2)),

$$L(P', f) - L(P, f) \le 2rK\|P\| < 2rK\delta = \frac{\epsilon}{4}.$$

Further, since $Q \subset P'$, we also have (by Lemma 6.7(a))

$$L(Q, f) \le L(P', f) < L(P, f) + \frac{\epsilon}{4},$$

which implies that

$$L(Q, f) - L(P, f) < \frac{\epsilon}{4}.$$

Similarly,

$$U(P, f) - U(Q, f) < \frac{\epsilon}{4}.$$

Adding the last two inequalities gives

$$U(P, f) - L(P, f) < U(Q, f) - L(Q, f) + \frac{\epsilon}{2} < \frac{\epsilon}{2} + \frac{\epsilon}{2} = \epsilon,$$

and (6.4) is valid. ∎

Example 6.10 (Integrable functions that are not continuous). Consider the functions

$$f(x) = \begin{cases} 1 & \text{if } x \in [0,1] \smallsetminus \{1/2\}, \\ 0 & \text{if } x = 1/2, \end{cases} \quad \text{and} \quad g(x) = \begin{cases} 0 & \text{if } 0 \le x \le 1/2, \\ 1 & \text{if } 1/2 < x \le 1. \end{cases}$$

Note that both f and g are bounded with a single jump discontinuity on $[0,1]$, and in addition, g is piecewise constant on $[0,1]$. It is easy to show that f and g are integrable on $[0,1]$.

Note that f is discontinuous at $x = \frac{1}{2}$. If $P = \{x_0, x_1, \ldots, x_n\}$ is a partition of $[0,1]$, then $1/2$ must belong to $[x_{k-1}, x_k]$ for some k, $1 \le k \le n$, and therefore

$$\begin{cases} m_j = M_j = 1 & \text{for } j \ne k, \\ m_j = 0,\ M_j = 1 & \text{for } j = k, \end{cases}$$

$j = 1, 2, \ldots, n$. Now

$$U(P, f) - L(P, f) = \sum_{j=1}^{n} (M_j - m_j)\, \Delta x_j = \Delta x_k,$$

showing that f is integrable on $[0,1]$, by Theorem 6.9. To determine $\int_0^1 f(x)\,dx$, it is enough to consider either

$$U(P,f) = \sum_{j=1}^{n} M_j\,\Delta x_j = \sum_{j=1}^{n} 1\cdot \Delta x_j = x_n - x_0 = 1-0 = 1$$

or

$$L(P,f) = \sum_{j=1}^{n} m_j\,\Delta x_j = \sum_{j=1,\ j\neq k}^{n} 1\cdot \Delta x_j = x_n - x_0 - \Delta x_k = 1 - \Delta x_k.$$

In either case, it follows that $\int_a^b f(x)\,dx = 1$, and thus f is integrable. We leave the proof of the integrability of g as an exercise. ●

Our last criterion for integrability may be phrased in terms of a sequence of partitions.

Theorem 6.11 (Sequential version of integrability). *Suppose that f is a bounded function on $[a,b]$. We have the following:*

(i) *If f is integrable on $[a,b]$, then there exists a sequence $\{Q_n\}$ of partitions in $[a,b]$ such that*

$$\lim_{n\to\infty} U(Q_n,f) = \alpha = \lim_{n\to\infty} L(Q_n,f), \quad \alpha = \int_a^b f(x)\,dx.$$

(ii) *If there exists a sequence $\{Q_n\}$ of partitions on $[a,b]$ such that*

$$\lim_{n\to\infty} U(Q_n,f) = \lim_{n\to\infty} L(Q_n,f),$$

then f is integrable on $[a,b]$, and the common value of these two limits is α.

Proof. **(i)** Assume first that f is integrable on $[a,b]$. Then corresponding to each integer $n\geq 1$, there exist two partitions P_n and P_n' of $[a,b]$ such that

$$U(P_n,f) < U(f) + \frac{1}{n} \quad\text{and}\quad L(P_n',f) > L(f) - \frac{1}{n}.$$

Set $Q_n = P_n \cup P_n'$. As in the proof of Theorem 6.8, it follows that

$$\alpha - \frac{1}{n} < L(Q_n,f) \leq U(Q_n,f) < \alpha + \frac{1}{n}.$$

The squeeze rule for sequences yields the desired conclusion.

(ii) Assume the hypothesis. Then we have

$$\lim_{n\to\infty} \left(U(Q_n,f) - L(Q_n,f)\right) = 0$$

for some sequence $\{Q_n\}$ of partitions on $[a,b]$ mentioned in the hypothesis. Then, for $\epsilon > 0$, there exists an $N = N(\epsilon)$ such that

$$0 \leq U(Q_n, f) - L(Q_n, f) < \epsilon \quad \text{for all } n \geq N,$$

and the integrability of f follows from Theorem 6.8. Next, because f is integrable, by definition, we have

$$L(Q_n, f) \leq \int_a^b f(x)\,dx \leq U(Q_n, f) \quad \text{for all } n.$$

The desired result follows if we allow $n \to \infty$ and observe that the common value of the two limits is $\int_a^b f(x)\,dx$. ∎

Example 6.12. Define f on $[0,1]$ by

$$f(x) = \begin{cases} 0 & \text{if } x \text{ is irrational,} \\ 1/q & \text{if } x = p/q \text{ is rational expressed with no common factors.} \end{cases}$$

Show that f is integrable on $[0,1]$ and $\int_0^1 f(x)\,dx = 0$.

Solution. Let $\epsilon > 0$ be given. For any partition Q of $[0,1]$, we always have $m_k = 0$, so that $L(Q, f) = 0$. Therefore, to complete the proof, it suffices to find a partition P of $[0,1]$ such that $U(P, f) < \epsilon$. To do this, for each $n \geq 2$, we consider the partition $P_n = \{\frac{p}{q} \in [0,1] : p \leq q \leq n\}$. For instance, for $n = 4$, we have

$$P_4 = \left\{0, \frac{1}{4}, \frac{1}{3}, \frac{1}{2}, \frac{2}{3}, \frac{3}{4}, 1\right\}.$$

On the other hand, with respect to the partition P_n, we have

$$M_k = \sup_{x_{k-1} \leq x \leq x_k} f(x) < \frac{1}{n}$$

and $||P_n|| \to 0$ as $n \to \infty$. Also,

$$U(P_n, f) - L(P_n, f) < \sum_{k=1}^{n} \frac{1}{n}(x_k - x_{k-1}) - 0 = \frac{1}{n}.$$

Thus f is integrable on $[0,1]$. Note that this function is continuous at 0 and at every irrational point in $[0,1]$ and discontinuous at every nonzero rational point on $[0,1]$. ●

In Example 6.12 we observe that although f is nonnegative and takes positive values at every rational point in $[0,1]$, we have $\int_0^1 f(x)\,dx = 0$.

The following result is often useful in practice.

Theorem 6.13. *Suppose that f is integrable on $[a,b]$ and $\{P_n\}$ is a sequence of partitions of $[a,b]$ such that $\|P_n\| \to 0$ as $n \to \infty$. Then we have*

$$\lim_{n\to\infty} U(P_n, f) = \alpha = \lim_{n\to\infty} L(Q_n, f) = \int_a^b f(x)\,dx.$$

Proof. This is a consequence of Theorem 6.9. We need only to observe the following fact: since $\|P_n\| \to 0$ as $n \to \infty$, given $\delta > 0$, there exists an N such that $\|P_n\| < \delta$ for all $n \geq N$. This means that

$$U(f) \leq U(P_n, f) < U(f)+\epsilon \text{ and } L(f) \geq L(P_n, f) > L(f)-\epsilon \quad \text{for all } n \geq N,$$

and the desired conclusion follows from this if we allow $n \to \infty$. ∎

The definition of Riemann integrability may now be framed as follows:

Definition 6.14 (Riemann integrability). *A bounded function f defined on $[a,b]$ is said to be Riemann integrable on $[a,b]$ if there exists a number I with the following property: For every $\epsilon > 0$ there exists a $\delta > 0$ such that*

$$|\sigma(P, f, x^*) - I| < \epsilon$$

for every *Riemann sum $\sigma(P, f, x_k^*)$ of f associated with a partition P of $[a,b]$ for which $\|P\| < \delta$. In this case, we write*

$$\lim_{\|P\|\to 0} \sigma(P, f, x^*) = I.$$

Formally, the quantity I is the definite integral of f on $[a,b]$.

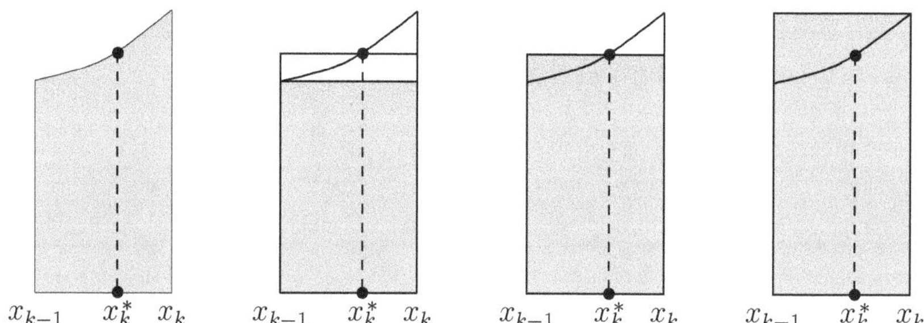

Fig. 6.9. Actual area, and areas corresponding to lower, arbitrary, and upper rectangles.

It is important to emphasize that I is independent of the particular way in which the partitions of $[a,b]$ and the subinterval representatives x_k^* are chosen (see Figure 6.9). In Theorem 6.15, we actually show that the number I is the *definite integral* of f on $[a,b]$, and so $I = \int_a^b f(x)\,dx$.

Our next result establishes the equivalence of Riemann's and Darboux's definitions of integrability.

Theorem 6.15 (Equivalence of the definitions of Riemann and Darboux). *If f is a bounded function on $[a,b]$, then f is Riemann integrable (in the sense of Definition 6.14) if and only if f is Darboux integrable (in the sense of Definition 6.2).*

Proof. (Sufficiency $\Longleftarrow$): Let f be (Darboux) integrable on $[a,b]$ in the sense of Definition 6.2, i.e., $\alpha := \int_a^b f(x)\,dx$ exists and $U(f) = L(f) = \alpha$. Let $\epsilon > 0$ be given. Then there exists a $\delta > 0$ such that

$$U(P,f) - L(P,f) < \epsilon \tag{6.5}$$

for every partition P with $\|P\| < \delta$. We need to show that

$$|\sigma(P,f,x^*) - \alpha| < \epsilon, \tag{6.6}$$

where

$$\sigma(P,f,x^*) = \sum_{k=1}^{n} f(x_k^*)(x_k - x_{k-1}),$$

for every partition P with $\|P\| < \delta$. As noted at the outset, we have

$$L(P,f) \leq \sigma(P,f,x^*) \leq U(P,f). \tag{6.7}$$

Because of (6.5) and the definition of $L(f)$, we also have

$$U(P,f) < L(P,f) + \epsilon \leq L(f) + \epsilon = \alpha + \epsilon,$$

so that (6.7) implies that

$$\sigma(P,f,x^*) \leq U(P,f) < \alpha + \epsilon.$$

Similarly, the other inequality of (6.7) gives, by (6.6),

$$\sigma(P,f,x^*) \geq L(P,f) > U(P,f) - \epsilon \geq U(f) - \epsilon = \alpha - \epsilon.$$

The last two inequalities imply that

$$\alpha - \epsilon < \sigma(P,f,x^*) < \alpha + \epsilon \quad \text{for} \quad \|P\| < \delta,$$

and hence (6.6) holds.

(Necessity $\Longrightarrow$): Conversely, suppose that f is Riemann integrable. Then

$$\lim_{\|P\| \to 0} \sigma(P,f,x^*) = I.$$

We need to show that $I = \alpha$. Fix $\epsilon > 0$. Then by the definition of Riemann integrability, there exists a $\delta > 0$ such that if $\|P\| < \delta$, then

$$I - \epsilon < \sigma(P,f,x^*) < I + \epsilon.$$

Now choose a partition P of $[a, b]$ with $\|P\| < \delta$ and $x_k^* \in [x_{k-1}, x_k]$, so that $f(x_k^*) < m_k + \epsilon$, i.e.,

$$\sum_{k=1}^{n} f(x_k^*)\Delta x_k < \sum_{k=1}^{n} m_k \Delta x_k + \epsilon \sum_{k=1}^{n} \Delta x_k = L(P, f) + \epsilon(b - a).$$

Then the Riemann sum, say $L_1(P, f)$, associated with this choice of x_k^* satisfies

$$L_1(P, f) \leq L(P, f) + \epsilon(b - a),$$

and by assumption,

$$|L_1(P, f) - I| < \epsilon, \quad \text{i.e.,} \quad I - \epsilon < L_1(P, f) < I + \epsilon.$$

It follows that

$$L(f) \geq L(P, f) \geq L_1(P, f) - \epsilon(b - a) > I - \epsilon - \epsilon(b - a).$$

Since $\epsilon > 0$ is arbitrary, $L(f) \geq I$. Similarly, we can show that $U(f) \leq I$. But $U(f) \geq L(f)$, and so

$$I \geq U(f) \geq L(f) \geq I,$$

showing that $U(f) = L(f) = I$, which means that f is Darboux integrable on $[a, b]$ with $I = \int_a^b f(x)\, dx$. ∎

6.1.4 Basic Examples of Integrable Functions

Example 6.16 (General approach). Consider now the function $f(x) = x$ on $[a, b]$. Then the Riemann sum takes the form

$$S_n := \sigma(P, f, x^*) = \sum_{k=1}^{n} x_k^*(x_k - x_{k-1}),$$

where $f(x_k^*) = x_k^*$ and $x_k^* \in [x_{k-1}, x_k]$ is arbitrary. Note that x_k^* is either the midpoint of the interval $[x_{k-1}, x_k]$ or to the left of it or to the right of it. Consequently, we write x_k^* conveniently as

$$x_k^* = \frac{x_{k-1} + x_k}{2} + \delta_k,$$

where

$$|\delta_k| \leq \frac{x_k - x_{k-1}}{2} \leq \frac{\|P\|}{2} \quad \text{for } k = 1, 2, \ldots, n.$$

In view of the representation of x_k^*, we write the Riemann sum in the form

$$\begin{aligned} S_n &= \sum_{k=1}^{n} \frac{x_{k-1}+x_k}{2}(x_k - x_{k-1}) + \sum_{k=1}^{n} \delta_k (x_k - x_{k-1}) \\ &= \frac{1}{2}\sum_{k=1}^{n}(x_k^2 - x_{k-1}^2) + E_n \\ &= \frac{b^2-a^2}{2} + E_n, \end{aligned}$$

where $E_n = \sum_{k=1}^{n} \delta_k(x_k - x_{k-1})$. We have

$$|E_n| \leq \sum_{k=1}^{n} |\delta_k|(x_k - x_{k-1}) \leq \frac{\|P\|}{2}\sum_{k=1}^{n}(x_k - x_{k-1}) \leq \frac{\|P\|}{2}(b-a) \to 0$$

as $\|P\| \to 0$. Hence, $\int_a^b x\, dx = (b^2-a^2)/2$.

Note also that

$$x_k^* = x_k \iff \delta_k = \frac{x_k - x_{k-1}}{2} \ \text{ and } \ x_k^* = x_{k-1} \iff \delta_k = -\frac{x_k - x_{k-1}}{2},$$

so that these choices correspond to the upper and the lower sums, respectively.

In particular, if $a > 0$, then the area in question is that of a trapezoid, agreeing with the results of elementary geometry. ●

Example 6.17 (Using Theorem 6.8 or Theorem 6.9). Consider the function $f(x) = x^2$ on $[0, b]$. Let $\epsilon > 0$ be given. Let $P = \{x_0, x_1, \ldots, x_n\}$ be any partition of $[0, b]$ such that $\|P\| < \epsilon/2b^2$. Then we have

$$x_k^2 - x_{k-1}^2 = (x_k + x_{k-1})(x_k - x_{k-1}) \leq 2b\|P\| < \frac{\epsilon}{b}.$$

Further, since $f(x)$ is increasing and continuous on $[0, b]$,

$$M_k = f(x_k) = x_k^2 \ \text{ and } \ m_k = f(x_{k-1}) = x_{k-1}^2,$$

so that

$$U(P, f) = \sum_{k=1}^{n} x_k^2 (x_k - x_{k-1}) \ \text{ and } \ L(P, f) = \sum_{k=1}^{n} x_{k-1}^2 (x_k - x_{k-1}).$$

Therefore (with $x_0 = 0$, $x_n = b$),

$$\begin{aligned} U(P, f) - L(P, f) &= \sum_{k=1}^{n} (x_k^2 - x_{k-1}^2)(x_k - x_{k-1}) \\ &< \frac{\epsilon}{b} \sum_{k=1}^{n} (x_k - x_{k-1}) \\ &= \frac{\epsilon}{b}(x_n - x_0) = \epsilon, \end{aligned}$$

showing that f is integrable, by Theorem 6.9, although we have no information yet as to the value of $\int_0^b x^2\,dx$.

Alternatively, we can consider a convenient partition P and apply Theorem 6.8 to obtain the integrability of f as well as the value of $\int_0^b x^2\,dx$. To do this, we choose $x_k = k(b/n)$. Then $\Delta x_k = b/n$, and so $\|P\| \to 0$ if $n \to \infty$. Further,

$$x_k^2(x_k - x_{k-1}) = \left(\frac{b}{n}\right)^3 k^2 \text{ and } x_{k-1}^2(x_k - x_{k-1}) = \left(\frac{b}{n}\right)^3 (k-1)^2,$$

so that

$$U(P,f) = \frac{b^3}{n^3}\sum_{k=1}^{n} k^2 = \frac{b^3}{n^3}\frac{n(n+1)(2n+1)}{6} = \frac{b^3}{6}\left(1+\frac{1}{n}\right)\left(2+\frac{1}{n}\right),$$

and similarly

$$L(P,f) = \frac{b^3}{n^3}\sum_{k=1}^{n}(k-1)^2 = \frac{b^3}{6}\left(1-\frac{1}{n}\right)\left(2-\frac{1}{n}\right).$$

This gives

$$U(P,f) - L(P,f) = \frac{b^3}{n},$$

and one can apply either Theorem 6.8 or Theorem 6.9 by choosing large n. It follows that f is integrable. Further, since

$$L(P,f) \le L(f) = \int_0^b x^2\,dx = U(f) \le U(P,f)$$

and

$$U(P,f) = \overline{S_n} \to \frac{b^3}{3} \text{ and } L(P,f) = \underline{s_n} \to \frac{b^3}{3} \quad \text{as } n \to \infty,$$

we conclude that

$$\int_0^b x^2\,dx = \frac{b^3}{3}. \qquad \bullet$$

Our next example gives us a method of evaluating the definite integral of an integrable function as the limit of a sequence. Because of its independent interest, we present a solution to this example that is key to solving a number of exercises, especially when the given function is continuous on $[a,b]$; see, for instance, Examples 6.24 and Exercises 6.32(10).

Example 6.18. Suppose that f is integrable on $[a,b]$. Then

$$\int_a^b f(x)\,dx = \lim_{n\to\infty} S_n, \quad S_n = \frac{b-a}{n}\sum_{k=1}^{n} f\left(a + \frac{k(b-a)}{n}\right).$$

In particular, if f is integrable on $[0,1]$, then we have

$$\int_0^1 f(x)\,dx = \lim_{n\to\infty} \frac{1}{n}\sum_{k=1}^{n} f\Big(\frac{k}{n}\Big),$$

or more generally, with $x_k^* \in [(k-1)/n, k/n]$, one has

$$\int_0^1 f(x)\,dx = \lim_{n\to\infty} \frac{1}{n}\sum_{k=1}^{n} f(x_k^*).$$

Solution. Suppose that f is integrable on $[a,b]$. Then we have $L(f) = U(f) = \int_a^b f(x)\,dx =: \alpha$. Choose the standard partition $P = \{x_0, x_1, \dots, x_n\}$ and let $h = (b-a)/n$. Consider $x_k = a + kh$, for $k = 0,1,2,\dots,n$, as points of division of $[a,b]$ into n equal parts of length $\Delta x_k = \Delta x = h$. Note that these points of division form an arithmetic progression. By the definitions of $L(P,f)$, $U(P,f)$, $L(f)$, $U(f)$, and $\sigma(P,f,x^*)$, it follows that

$$L(P,f) \le \sigma(P,f,x^*) \le U(P,f) \quad \text{and} \quad L(P,f) \le \alpha \le U(P,f),$$

which, in particular, implies that (see Figure 6.10)

$$|S_n - \alpha| \le U(P,f) - L(P,f).$$

Since f is integrable, by Theorem 6.8, for a given $\epsilon > 0$ there exists a $\delta > 0$

Fig. 6.10. Bounds for $S_n - \alpha$.

such that

$$U(Q,f) - L(Q,f) < \epsilon \ \text{ for all partitions } Q \text{ of } [a,b] \text{ for which } \|Q\| < \delta.$$

Note that for our partition, for any $\delta > 0$, there exists an N such that $\|P\| = h = (b-a)/n < \delta$ for all $n \ge N$. Consequently, given $\epsilon > 0$, there exists an N such that for all $n \ge N$,

$$|S_n - \alpha| \le U(P,f) - L(P,f) < \epsilon,$$

showing that $S_n \to \alpha$ as $n \to \infty$. ●

Theorems 6.8, 6.9, and 6.15 may be reformulated as follows.

Theorem 6.19. *If f is a bounded function on $[a,b]$, then the following are equivalent:*

(a) *f is integrable.*
(b) *For every $\epsilon > 0$ there is a partition P of $[a,b]$ such that*

$$U(P,f) - L(P,f) < \epsilon.$$

(c) *For every $\epsilon > 0$ there is a $\delta > 0$ such that every partition P of $[a,b]$ with $\|P\| < \delta$ satisfies $U(P,f) - L(P,f) < \epsilon$.*
(d) *For every $\epsilon > 0$ there exists a $\delta > 0$ such that*

$$\left| \sigma(P,f,x^*) - \int_a^b f(x)\,\mathrm{d}x \right| < \epsilon$$

for every Riemann sum $\sigma(P,f,x^*)$ of f associated with a partition P for which $\|P\| < \delta$.

6.1.5 Integrability of Monotone/Continuous Functions

Now we are in a position to attempt the remaining questions that we raised at the beginning of this subsection. Our next two results provide us with two different classes of integrable functions.

Theorem 6.20 (Integrability of monotone functions). *Every monotone function on $[a,b]$ is integrable on $[a,b]$. The converse is false.*

Proof. If the monotone function f is constant on $[a,b]$, it is certainly integrable on $[a,b]$. So we shall assume that f is nonconstant, and in particular, we have $f(a) \neq f(b)$. It suffices to consider the case in which f is monotonically increasing on $[a,b]$, so that $f(a) < f(b)$. A similar argument works if f is monotonically decreasing.

Since $f(a) \leq f(x) \leq f(b)$ for all $x \in [a,b]$, f is clearly bounded on $[a,b]$. Let $\epsilon > 0$ be given. In order to apply Theorem 6.19, we consider an arbitrary partition $P = \{x_0, x_1, \dots, x_n\}$ of $[a,b]$ with

$$\|P\| < \delta = \frac{\epsilon}{f(b) - f(a)}.$$

Since $x_{k-1} < x_k$ and f is increasing, we have for each $k \in \{1,2,\dots,n\}$,

$$M_k = \sup_{x \in [x_{k-1},x_k]} f(x) = f(x_k) \quad \text{and} \quad m_k = \inf_{x \in [x_{k-1},x_k]} f(x) = f(x_{k-1}).$$

Thus, with the usual notation for lower and upper sums,

$$\begin{aligned} U(P,f) - L(P,f) &= \sum_{k=1}^{n} (M_k - m_k)\Delta x_k \\ &= \sum_{k=1}^{n} \left(f(x_k) - f(x_{k-1})\right)(x_k - x_{k-1}) \\ &< \delta \sum_{k=1}^{n} \left(f(x_k) - f(x_{k-1})\right) \\ &= \delta\left(f(b) - f(a)\right) = \epsilon, \end{aligned}$$

which proves the existence of a partition P with $U(P,f) - L(P,f) < \epsilon$. Thus, by Theorem 6.21, it follows that f is integrable.

We see that $f(x) = \cos x$ is integrable on $[0, 2\pi]$, but is not monotone on $[0, 2\pi]$ (see Example 6.22(**d**)). ∎

For instance, define f on $[0,1]$ by

$$f(0) = 0 \text{ and } f(x) = \frac{1}{2^{k-1}} \text{ for } \frac{1}{2^k} < x \leq \frac{1}{2^{k-1}}, \; k = 1, 2, \ldots.$$

Then f is increasing and bounded on $[0,1]$. Therefore, f is integrable. Is f continuous on $[0,1]$? How about the function $g(x) = [x]$ over any interval $[a,b]$? Is g integrable on $[a,b]$?

We have constructed examples of functions that are discontinuous at a point on $[a,b]$ but may or may not be integrable on $[a,b]$. Next, to enlarge the class of integrable functions, we deal with continuous functions as well.

Theorem 6.21 (Integrability of continuous functions). *Every continuous function f on $[a,b]$ is integrable. The converse is false.*

Proof. Let $\epsilon > 0$ be given. Since f is continuous on the closed interval $[a,b]$, it is uniformly continuous on $[a,b]$. Therefore, for $\epsilon > 0$ there is a $\delta > 0$ such that

$$|f(x) - f(y)| < \frac{\epsilon}{b-a} \text{ whenever } |x - y| < \delta \text{ and } x, y \in [a,b].$$

Consider any partition $P = \{x_0, x_1, \ldots, x_n\}$ of $[a,b]$ with norm $\|P\| < \delta$. Now, if $x, y \in [x_{k-1}, x_k]$, then

$$|x - y| \leq x_k - x_{k-1} = \Delta x_k \leq \|P\| < \delta,$$

and so we have $|f(x) - f(y)| < \epsilon/(b-a)$. Since f assumes a maximum and minimum on each subinterval $[x_{k-1}, x_k]$, we have

$$0 \leq M_k - m_k = \max_{x \in [x_{k-1}, x_k]} f(x) - \min_{x \in [x_{k-1}, x_k]} f(x) < \frac{\epsilon}{b-a} \quad \text{for each } k,$$

and using this inequality, we get

$$U(P,f) - L(P,f) = \sum_{k=1}^{n}(M_k - m_k)\Delta x_k < \frac{\epsilon}{b-a}\sum_{k=1}^{n}(x_k - x_{k-1}) = \epsilon,$$

which shows that f is integrable, by Theorem 6.21.

For the converse, see Example 6.10. ∎

We remark that Theorem 6.21 does not say anything about the actual value of the definite integral. The hypothesis clearly implies that f is bounded on $[a,b]$, since every continuous function f on a compact set $[a,b]$ is bounded.

Examples 6.22 (The integral as a limit of Riemann sums). Using the summation formula, evaluate

(a) $\int_a^b x\,dx$; (b) $\int_a^b x^2\,dx$; (c) $\int_a^b e^x\,dx$; (d) $\int_a^b \cos x\,dx$.

Solution. Parts **(a)** and **(b)** have already been considered, but for the sake of looking at the problem from a different viewpoint, we present their solutions here.

By Theorem 6.21, the integral exists because the integrand in each case is continuous on $[a,b]$ (see Figure 6.11). Because the integral can be computed

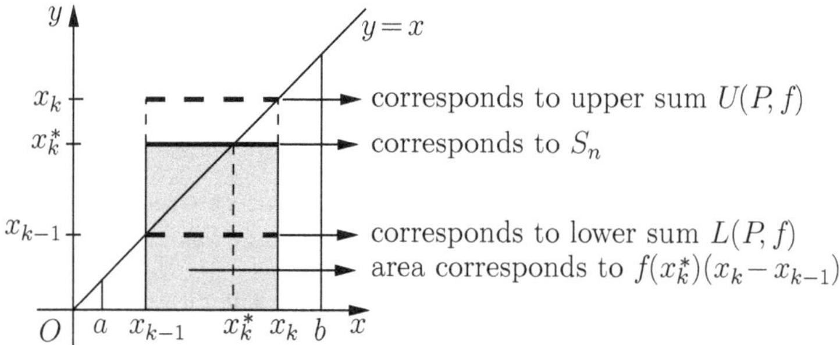

Fig. 6.11. Riemann sums associated with $f(x) = x$.

by any partition whose norm approaches 0, as in Example 6.18, we may use the standard partition. Using the standard partition, we set

$$x_k^* = x_k = a + k\Delta x = a + k\left(\frac{b-a}{n}\right), \quad k = 1, 2, \ldots, n,$$

and observe that $\|P\| = (b-a)/n \to 0$ as $n \to \infty$. We note that x_k^* here corresponds to the upper sum $U(P,f)$.

(a) A Riemann sum for $f(x) = x$ on $[a, b]$ is

$$\begin{aligned}\sum_{k=1}^{n} f(x_k^*)\Delta x &= \sum_{k=1}^{n} \left[a + \frac{k(b-a)}{n}\right]\frac{b-a}{n} \\ &= \frac{a(b-a)}{n}\sum_{k=1}^{n} 1 + \frac{(b-a)^2}{n^2}\sum_{k=1}^{n} k \\ &= a(b-a) + \frac{(b-a)^2}{2}\left(1 + \frac{1}{n}\right).\end{aligned}$$

Equivalently, we may set

$$x_k^* = x_{k-1} = a + (k-1)\left(\frac{b-a}{n}\right),$$

and with this choice, we obtain that

$$\sum_{k=1}^{n} f(x_k^*)\frac{b-a}{n} = a(b-a) + \frac{(b-a)^2}{2}\left(1 - \frac{1}{n}\right),$$

which is actually the lower sum $L(P, f)$. In either case, we have

$$\int_a^b x\,dx = \lim_{n\to\infty}\sum_{k=1}^{n} f(x_k^*)\Delta x = a(b-a) + \frac{(b-a)^2}{2} = \frac{b^2 - a^2}{2}.$$

(b) The integral in question is equal to the area under the parabola $y = x^2$ on $[a, b]$. A Riemann sum for the function $f(x) = x^2$ on $[a, b]$ is

$$S_n = \sum_{k=1}^{n} f(x_k^*)\Delta x,$$

which simplifies to

$$\begin{aligned}S_n &= \sum_{k=1}^{n}\left[a + \frac{k(b-a)}{n}\right]^2\frac{b-a}{n} \\ &= \frac{b-a}{n}\left[\sum_{k=1}^{n} a^2 + \frac{2a(b-a)}{n}\sum_{k=1}^{n} k + \frac{(b-a)^2}{n^2}\sum_{k=1}^{n} k^2\right] \\ &= \frac{b-a}{n}\left[a^2 n + \frac{2a(b-a)}{n}\frac{n(n+1)}{2} + \frac{(b-a)^2}{n^2}\frac{n(n+1)(2n+1)}{6}\right].\end{aligned}$$

Allowing $n \to \infty$, we conclude that

$$\int_a^b x^2\,dx = (b-a)\left[a^2 + a(b-a) + \frac{(b-a)^2}{3}\right] = \frac{b^3 - a^3}{3}.$$

(c) In the case of the continuous function $f(x) = \mathrm{e}^x$, the upper sum is

$$\begin{aligned} S_n &= \sum_{k=1}^{n} \mathrm{e}^{a+k\Delta x} \Delta x \\ &= \mathrm{e}^a \Delta x \sum_{k=1}^{n} \left(\mathrm{e}^{\Delta x}\right)^k = \mathrm{e}^a \Delta x \frac{\mathrm{e}^{\Delta x}(1 - \mathrm{e}^{n\Delta x})}{1 - \mathrm{e}^{\Delta x}} \\ &= \mathrm{e}^a (1 - \mathrm{e}^{b-a}) \left(\frac{\Delta x}{1 - \mathrm{e}^{\Delta x}}\right) \mathrm{e}^{\Delta x}. \end{aligned}$$

Note that $\|P\| = (b-a)/n = \Delta x \to 0$ iff $n \to \infty$. By l'Hôpital's rule, we see that

$$\lim_{n\to\infty} S_n = -\mathrm{e}^a (1 - \mathrm{e}^{b-a}) = \mathrm{e}^b - \mathrm{e}^a.$$

We leave **(d)** as a simple exercise. ●

Example 6.23. Consider $f(x) = x^p$ on $[a, b]$, where $p \neq -1$ and $0 < a < b$. Let $P = \{a = x_0, x_1, \ldots, x_n = b\}$ be a partition of $[a, b]$ with

$$x_k^* = x_k = ah^k \quad \text{with } h = (b/a)^{1/n}.$$

Then a Riemann sum S_n is given by

$$\begin{aligned} S_n &= \sum_{k=0}^{n-1} (ah^k)^p (ah^{k+1} - ah^k) \\ &= a^{p+1}(h-1) \sum_{k=0}^{n-1} h^{k(p+1)} \\ &= a^{p+1}(h-1) \left(\frac{h^{(p+1)n} - 1}{h^{p+1} - 1}\right) \\ &= a^{p+1} \left[\left(\frac{b}{a}\right)^{p+1} - 1\right] \left[\frac{h-1}{h^{p+1} - 1}\right]. \end{aligned}$$

Note that $(b/a)^{1/n} \to 1$ as $n \to \infty$ is equivalent to $h \to 1$. Consequently,

$$\begin{aligned} \lim_{n\to\infty} S_n &= (b^{p+1} - a^{p+1}) \lim_{h\to 1} \frac{h-1}{h^{p+1} - 1} \\ &= (b^{p+1} - a^{p+1}) \lim_{h\to 1} \frac{1}{(p+1)h^p} \\ &= \frac{b^{p+1} - a^{p+1}}{p+1}, \end{aligned}$$

which is equivalent to

$$\int_a^b x^p \, dx = \frac{b^{p+1} - a^{p+1}}{p+1}. \qquad \bullet$$

Examples 6.24. Using the definition of the definite integral as the limit of Riemann (integral) sums, evaluate the following:

(a) $\lim_{n\to\infty} \sum_{k=1}^{n} \frac{1}{n+ck}$ for $c > 0$.
(b) $\lim_{n\to\infty} \frac{1}{n^{3/2}} \sum_{k=1}^{n} (n+k)^{1/2}$.

(c) $\lim_{n\to\infty} \sum_{k=1}^{n} \frac{k^a}{n^{a+1}}$ for $a > -1$.
(d) $\lim_{n\to\infty} \sum_{k=1}^{n} \frac{n}{n^2+k^2}$.

(e) $\lim_{n\to\infty} \frac{1}{n} \sum_{k=1}^{n} \cos\left(\frac{2k-1}{2n}\right)$.
(f) $\lim_{n\to\infty} \sum_{k=1}^{n} \frac{\cos(\log(k+n) - \log(n))}{n+k}$.

Solution. In all these cases, we consider the standard partition

$$P = \left\{0, \frac{1}{n}, \frac{2}{n}, \dots, 1\right\}$$

of $[0,1]$ with $x_k^* = k/n$, and so $\Delta x_k = 1/n$. So we may skip the further details, since we just need to recognize each of the given quantities/sequences as a Riemann sum associated with a suitable function f. We have, as $n \to \infty$, the following:

(a) $S_n = \frac{1}{n} \sum_{k=1}^{n} \frac{1}{1+c(k/n)} \to S = \int_0^1 \frac{dx}{1+cx} = \frac{1}{c} \log(1+c)$.

(b) $S_n = \frac{1}{n} \sum_{k=1}^{n} \left(1+\frac{k}{n}\right)^{1/2} \to S = \int_0^1 (1+x)^{1/2} \, dx = (2/3)[2^{3/2} - 1]$.

(c) $S_n = \frac{1}{n} \sum_{k=1}^{n} \left(\frac{k}{n}\right)^a \to S = \int_0^1 x^a \, dx = \frac{1}{1+a}$.

(d) $S_n = \frac{1}{n} \sum_{k=1}^{n} \frac{1}{1+(k^2/n^2)} \to S = \int_0^1 \frac{dx}{1+x^2} = \tan^{-1} x \Big|_0^1 = \frac{\pi}{4}$.

(e) Using the cosine summation formula, the given Riemann sum may be written as

$$\frac{1}{n} \sum_{k=1}^{n} \cos\left(\frac{2k-1}{2n}\right) = \cos\left(\frac{1}{2n}\right) \frac{1}{n} \sum_{k=1}^{n} \cos\left(\frac{k}{n}\right) + \sin\left(\frac{1}{2n}\right) \frac{1}{n} \sum_{k=1}^{n} \sin\left(\frac{k}{n}\right),$$

which clearly converges to

$$S = \int_0^1 \cos x \, dx + 0. \int_0^1 \sin x \, dx = \sin 1.$$

(f) Again, rewriting the given Riemann sum, we get

$$\frac{1}{n}\sum_{k=1}^{n}\frac{\cos(\log(1+k/n))}{1+k/n}\to S=\int_0^1\frac{\cos(\log(1+x))}{1+x}\,dx,$$

from which we obtain $S=\sin(\log 2)$. ●

6.1.6 Basic Properties of Definite Integrals

Given a bounded function f on $[a,b]$, we can consider $|f|$ as the composition $h\circ f$ with $h(x)=|x|$ for all x, and f^2 as the composition $h\circ f$ with $h(x)=x^2$ for all x. In both cases, h will be uniformly continuous on the range of $f(x)$. It would be interesting to know whether $h\circ f$ is integrable on $[a,b]$ whenever f is. The following elementary properties of the integral are useful in the computation. Since the integral is the limit of Riemann sums, each of these properties can be obtained using the algebraic properties of sequences and their limits. Of course, there are also other ways of proving them.

Theorem 6.25 (General properties of the definite integrals). *Suppose that f and g are integrable on $[a,b]$. We have the following:*

(a) *c_1f+c_2g is integrable for constants c_1 and c_2 and*

$$\int_a^b[c_1f(x)+c_2g(x)]\,dx=c_1\int_a^b f(x)\,dx+c_2\int_a^b g(x)\,dx.$$

This is called the linearity rule for integrals.

(b) *If $f(x)\le g(x)$ on $[a,b]$, then*

$$\int_a^b f(x)\,dx\le\int_a^b g(x)\,dx.$$

This is called the dominance rule for integrals.

(c) *If $m\le f(x)\le M$ for $x\in[a,b]$ and h is continuous on $[m,M]$, then ϕ defined by $\phi(x)=h(f(x))$ is integrable on $[a,b]$. In particular,*

$$m(b-a)\le\int_a^b f(x)\,dx\le M(b-a).$$

Proof. Each of these properties is easy to establish using the linearity property of sums or limits with the definition of the definite integral.

(a) Using standard notation, we note that any Riemann sum of c_1f+c_2g can be expressed as

$$\sum_{k=1}^{n}[c_1f(x_k^*)+c_2g(x_k^*)]\Delta x_k=c_1\sum_{k=1}^{n}f(x_k^*)\Delta x_k+c_2\sum_{k=1}^{n}g(x_k^*)\Delta x_k,$$

and the desired linearity property follows by taking the limit on each side of this equation as the norm of the partition tends to 0.

Alternatively, we may just consider the case of $c_1 = c_2 = 1$ but with a different method of proof. For a partition $P = \{x_0, x_1, \dots, x_n\}$ of $[a, b]$, we have

$$\inf_{x \in [x_{k-1}, x_k]} (f+g)(x) \geq \inf_{x \in [x_{k-1}, x_k]} f(x) + \inf_{x \in [x_{k-1}, x_k]} g(x)$$

and

$$\sup_{x \in [x_{k-1}, x_k]} (f+g)(x) \leq \sup_{x \in [x_{k-1}, x_k]} f(x) + \sup_{x \in [x_{k-1}, x_k]} g(x).$$

Therefore, as usual, we easily see that

$$L(P, f) + L(P, g) \leq L(P, f+g) \leq U(P, f+g) \leq U(P, f) + U(P, g),$$

so that

$$U(P, f+g) - L(P, f+g) \leq (U(P, f) - L(P, f)) + (U(P, g) - L(P, g)),$$

which holds for any partition P of $[a, b]$.

Now we assume that f and g are integrable on $[a, b]$, and so the above inequalities help to prove that $f + g$ is integrable on $[a, b]$. By Theorem 6.8, for each $\epsilon > 0$, there are partitions P_1 and P_2 of $[a, b]$ such that

$$U(P_1, f) - L(P_1, f) < \frac{\epsilon}{2} \text{ and } U(P_2, g) - L(P_2, g) < \frac{\epsilon}{2},$$

respectively. Set $Q = P_1 \cup P_2$. Adding the last two inequalities gives

$$U(Q, f+g) - L(Q, f+g) < \epsilon,$$

showing that $f + g$ is integrable on $[a, b]$, by Riemann's criterion.

Moreover, the integrability of f and g shows that there exist two partitions P_1 and P_2 of $[a, b]$ such that

$$U(P_1, f) < \int_a^b f(x)\, dx + \frac{\epsilon}{2} \text{ and } U(P_2, g) < \int_a^b g(x)\, dx + \frac{\epsilon}{2}.$$

Since $Q = P_1 \cup P_2$ is a refinement of both P_1 and P_2, we have

$$U(Q, f) \leq U(P_1, f) \text{ and } U(Q, g) \leq U(P_2, g),$$

and therefore, since $U(Q, f+g) \leq U(Q, f) + U(Q, g)$, we have

$$U(Q, f+g) \leq U(P_1, f) + U(P_2, g) \leq \int_a^b f(x)\, dx + \int_a^b g(x)\, dx + \epsilon.$$

Similarly,

$$L(Q, f+g) \geq L(P_1, f) + L(P_2, g) \geq \int_a^b f(x)\, \mathrm{d}x + \int_a^b g(x)\, \mathrm{d}x - \epsilon.$$

Since $\epsilon > 0$ is arbitrary, the above inequalities yield the desired result.

(b) Since $g(x) - f(x) \geq 0$ on $[a, b]$, every lower sum of $g - f$ with respect to any partition of $[a, b]$ is nonnegative. Thus, $L(g - f) \geq 0$ and $g - f$ is integrable by **(a)**. Therefore, by the linearity property,

$$\int_a^b g(x)\, \mathrm{d}x - \int_a^b f(x)\, \mathrm{d}x = \int_a^b (g - f)(x)\, \mathrm{d}x = L(g - f) \geq 0,$$

which yields the desired dominance rule.

(c) We leave the proof of the last case as a simple exercise. ∎

A standard induction argument enables us to expand the linearity property to a finite linear combination of integrable functions. It is not obvious that the integrability of f on $[a, b]$ implies that f is integrable on $[a, c]$ and on $[c, b]$, if $c \in (a, b)$. The next result establishes this fact.

Theorem 6.26 (Subinterval property for the integral). *Assume that f is bounded on $[a, b]$ and $c \in (a, b)$. We have the following:*

(a) *If f is integrable on $[a, b]$, then f is integrable on $[a, c]$ and on $[c, b]$. Moreover,*

$$\int_a^b f(x)\, \mathrm{d}x = \int_a^c f(x)\, \mathrm{d}x + \int_c^b f(x)\, \mathrm{d}x. \tag{6.8}$$

(b) *Conversely, if f is integrable on $[a, c]$ and on $[c, b]$, then f is integrable on $[a, b]$ and* (6.8) holds.

Proof. **(a)** We use standard notation. Let P_n be a sequence of partitions of $[a, b]$ such that c is a partition point of each P_n with $\|P_n\| \to 0$ as $n \to \infty$.

Let Q_n consist of the partition points of P_n that lie in $[a, c]$ and let Q_n' consist of the partition points of P_n that lie in $[c, b]$ (see Figure 6.12). Thus,

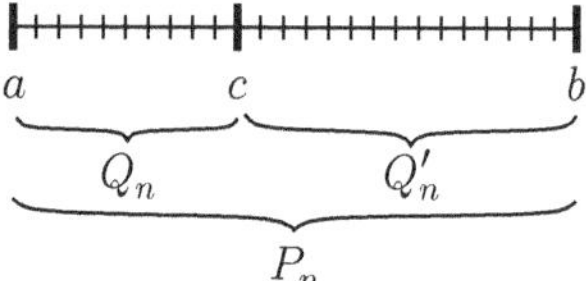

Fig. 6.12. Partition of $[a, b]$.

we have two partitions, Q_n of $[a,c]$ and Q'_n of $[c,b]$, such that $\|Q_n\| \to 0$ and $\|Q'_n\| \to 0$ as $n \to \infty$. By the definition of Riemann sums,

$$U(Q_n, f) - L(Q_n, f) \leq U(P_n, f) - L(P_n, f) \to 0 \quad \text{as } n \to \infty,$$

showing that f is integrable on $[a,c]$. Here we have used the fact that $U(Q_n, f) - L(Q_n, f)$ represents the sum of those terms in the sum

$$\sum_{k=1}^{n} (M_k - m_k)\, \Delta x_k$$

that correspond to the subintervals that lie in $[a,c]$. A similar argument shows that f is integrable on $[c,b]$. Using the sequential version of the integrability theorem (see Theorem 6.11), we obtain that

$$L(Q_n, f) \to \int_a^c f(x)\,dx \quad \text{and} \quad L(Q'_n, f) \to \int_c^b f(x)\,dx,$$

and so letting $n \to \infty$ in the obvious identity

$$L(P_n, f) = L(Q_n, f) + L(Q'_n, f),$$

it follows that

$$\int_a^b f(x)\,dx = \int_a^c f(x)\,dx + \int_c^b f(x)\,dx.$$

(b) For the converse part, assume that f is integrable on $[a,c]$ and on $[c,b]$. Then by Theorem 6.8, given $\epsilon > 0$, there exist two partitions, Q of $[a,c]$ and Q' of $[c,b]$, such that

$$U(Q, f) - L(Q, f) < \frac{\epsilon}{2} \quad \text{and} \quad U(Q', f) - L(Q', f) < \frac{\epsilon}{2}.$$

Then $P = Q \cup Q'$ is a partition of $[a,b]$, and we have the identity

$$L(P, f) = L(Q, f) + L(Q', f) \quad \text{and} \quad U(P, f) = U(Q, f) + U(Q', f).$$

It turns out that

$$\begin{aligned} U(P, f) - L(P, f) &= (U(Q, f) - L(Q, f)) + (U(Q', f) - L(Q', f)) \\ &< \frac{\epsilon}{2} + \frac{\epsilon}{2} = \epsilon, \end{aligned}$$

and so f is integrable on $[a,b]$ by Theorem 6.8. We now complete the proof by establishing (6.8). We have

$$\begin{aligned} \int_a^b f(x)\,dx &\leq U(P, f) = U(Q, f) + U(Q', f) \\ &< \left(L(Q, f) + \frac{\epsilon}{2}\right) + \left(L(Q', f) + \frac{\epsilon}{2}\right) \\ &\leq \int_a^c f(x)\,dx + \int_c^b f(x)\,dx + \epsilon. \end{aligned}$$

Similarly, we see that

$$\int_a^b f(x)\,\mathrm{d}x > \int_a^c f(x)\,\mathrm{d}x + \int_c^b f(x)\,\mathrm{d}x - \epsilon.$$

Since $\epsilon > 0$ is arbitrary, the last two inequalities yield (6.8). ∎

Again, a standard induction argument implies that Theorem 6.26 holds for a decomposition of the interval $[a, b]$ into a finite union of nonoverlapping intervals. The next result guarantees the existence of the integral for a very large class of functions.

Corollary 6.27 (Product and modulus properties for the integral). *Let f, g be integrable on $[a, b]$. Then the product fg and the modulus function $|f|$ are integrable on $[a, b]$. Also,*

$$\left| \int_a^b f(x)\,\mathrm{d}x \right| \leq \int_a^b |f(x)|\,\mathrm{d}x.$$

In particular, if in addition, $|f(x)| \leq K$ on $[a, b]$, we have

$$\left| \int_a^b f(x)\,\mathrm{d}x \right| \leq K(b-a).$$

(This also follows from Theorem 6.25**(c)***).*

Proof. Define $h(x) = x^2$. Then h is continuous on $\mathbb{R}$, and so

$$h(f+g) = (f+g)^2 \quad \text{and} \quad h(f-g) = (f-g)^2$$

are integrable by Theorem 6.25(c). Therefore,

$$fg = \left[(f+g)^2 - (f-g)^2\right]/4$$

is also integrable on $[a, b]$. Finally, we note that $h(x) = |x|$ is continuous on $\mathbb{R}$. Therefore, $h(f(x)) = |f(x)|$ is integrable on $[a, b]$. Also, since

$$\left.\begin{matrix} f(x) \\ -f(x) \end{matrix}\right\} \leq |f(x)| = |f|(x) \quad \text{for all } x \in [a, b],$$

it follows that

$$\left.\begin{matrix} \displaystyle\int_a^b f(x)\,\mathrm{d}x \\ \displaystyle -\int_a^b f(x)\,\mathrm{d}x \end{matrix}\right\} \leq \int_a^b |f(x)|\,\mathrm{d}x,$$

which gives the desired inequality. ∎

Theorem 6.28. *Let f and g be bounded on $[a,b]$ and suppose that $f(x) = g(x)$ except possibly for a finite number of points on $[a,b]$. Then*

$$L(f) = L(g) \quad \text{and} \quad U(f) = U(g).$$

In particular, f is integrable on $[a,b]$ if and only if g is integrable on $[a,b]$.

Proof. We will prove only $L(f) = L(g)$, since the equality for the upper integral follows analogously. By hypothesis, $|f(x)| \leq K$ and $|g(x)| \leq K$ for some $K > 0$.

Let r be the number of points in $[a,b]$ at which f and g differ. Consider a sequence of partitions P_n such that $\|P_n\| \to 0$ as $n \to \infty$. Then there are at most $2r$ subintervals on which

$$M_k = \sup_{x \in [x_{k-1}, x_k]} f(x) \neq M_k' = \sup_{x \in [x_{k-1}, x_k]} g(x).$$

On $[x_{k-1}, x_k]$, we have $|M_k - M_k'| \leq |M_k| + |M_k'| \leq 2K$, and so

$$|L(P_n, f) - L(P_n, g)| \leq 2rK\|P_n\|.$$

Now

$$\begin{aligned}
|L(f) - L(g)| &\leq |L(f) - L(P_n, f)| + |L(P_n, f) - L(P_n, g)| + |L(g) - L(P_n, g)| \\
&\leq |L(f) - L(P_n, f)| + 2rK\|P_n\| + |L(g) - L(P_n, g)|.
\end{aligned}$$

Allowing $n \to \infty$ yields that $L(f) = L(g)$, and we have completed the proof. ∎

The following result is a simple consequence of the previous result.

Corollary 6.29. *If f is a bounded function on $[a,b]$ and is continuous on $[a,b]$ except for a finite number of points on $[a,b]$, then f is integrable on $[a,b]$. In particular, every piecewise continuous function is integrable.*

We end the section with the following remark.

Remark 6.30. Usually, we find area by evaluating a definite integral, but if it is possible to recognize the integral as the area of some known geometric figure, then we can use the known formula instead of the definite integral. For instance, to evaluate

$$I = \int_{-a}^{a} \sqrt{a^2 - x^2}\, dx, \quad a > 0,$$

we may set $f(x) = \sqrt{a^2 - x^2}$ and observe that the curve $y = \sqrt{a^2 - x^2}$ is a semicircle centered at the origin of radius a. Thus, the given integral can be interpreted as the area under the semicircle on the interval $[-a,a]$. From geometry, we know that the area of the semicircle is $a^2\pi/2$, which is in fact the value of the given integral. ●

Throughout the exercises below, integrable means Riemann (Darboux) integrable.

6.1.7 Questions and Exercises

Questions 6.31.

1. In the definition of m_k and M_k, why do we use inf and sup instead of min and max, respectively?
2. When is a function Riemann integrable?
3. How is the Riemann sum S_n connected with the Darboux integral?
4. Why is the norm of a partition so important in the theory of Riemann integration?
5. Suppose that P and Q are two partitions of $[a,b]$. Must $P \cup Q$ always be a refinement of both P and Q?
6. Let P and Q be two partitions of $[a,b]$.
   (a) If Q is a refinement of P, how are the norms $\|P\|$ and $\|Q\|$ related? Is $\|Q\| \leq \|P\|$?
   (b) If $\|Q\| \leq \|P\|$, must Q be a refinement of P?
7. Suppose that f is bounded on $[a,b]$, P is a partition of $[a,b]$, and P' is a refinement of P. Must the addition of just one point to P increase the lower integral sum and decrease the upper integral sum? That is, must we have $L(P,f) \leq L(P',f)$? Must we have $U(P',f) \leq U(P,f)$?
8. What is the relationship between finding an area and evaluating a definite integral?
9. Suppose that f is bounded on $[a,b]$ and continuous on (a,b). Must f be integrable on $[a,b]$?
10. Suppose that f is bounded on $[a,b]$ and continuous on (a,b) except at $c \in (a,b)$. Must f be integrable on $[a,b]$?
11. Suppose that f is bounded on $[a,b]$ and has an infinite number of points on $[a,b]$ at which f is discontinuous. Can f still be integrable on $[a,b]$?
12. Must every integrable function on $[a,b]$ be bounded?
13. Are there integrable functions on $[a,b]$ that are neither continuous nor monotone on $[a,b]$?
14. Suppose that f is bounded on $[a,b]$ and $c \in \mathbb{R}$ is fixed. When do we have $L(cf) = c\,U(f)$ and $U(cf) = c\,L(f)$? When do we have $L(cf) = c\,L(f)$ and $U(cf) = c\,U(f)$?
15. Does $\int_a^b cf(x)\,dx$ always represent the area of a region?
16. Suppose that f is a nonnegative integrable function on $[a,b]$, and $\alpha > 0$. Must f^α be integrable on $[a,b]$? Is it true tha
$$\int_a^b f(x)\,dx = 0 \text{ if and only if } \int_a^b f^\alpha(x)\,dx = 0?$$
17. Must the composition of integrable functions be integrable?
18. Can a monotone function f have an infinite number of discontinuities on a bounded interval $[a,b]$? If yes, can it be integrable?

19. Suppose that $f(x) = x$ on $[a, b]$. Can there exist a partition $P = \{x_0, x_1, \ldots, x_n\}$ such that the corresponding Riemann sum has the constant value $S_n = (b^2 - a^2)/2$? Does this mean that $\int_a^b x\,dx = (b^2 - a^2)/2$?
20. Suppose that $f(x) = x^2$ on $[a, b]$ and $P = \{x_0, x_1, \ldots, x_n\}$ is a partition such that $x_k^* = \sqrt{(x_{k-1}^2 + x_{k-1}x_k + x_k^2)/3}$. Must the corresponding Riemann sum have the constant value $S_n = (b^3 - a^3)/3$? Does this show that $\int_a^b x^2\,dx = (b^3 - a^3)/3$?
21. Suppose that $f : [a, b] \to \mathbb{R}$ is defined by

$$f(x) = \begin{cases} 2 & \text{if } x \text{ is rational in } [a, b], \\ 3 & \text{if } x \text{ is irrational in } [a, b]. \end{cases}$$

Does there exist a partition P of $[a, b]$ such that $U(P, f) = L(P, f)$? If not, must $2U(P, f) = 3L(P, f)$ always be true?
22. Assume that f is continuous on $\mathbb{R}$. Is it true that f is even on $\mathbb{R}$ if and only if

$$\int_{-x}^{x} f(t)\,dt = 2\int_0^x f(t)\,dt \quad \text{on } \mathbb{R}?$$

Exercises 6.32.

1. Let $P = \{0, 1/2, 3/4, 1\}$ and $Q = \{0, 1/2, 3/4, 7/8, 1\}$ be two partitions of $[0, 1]$, and $f(x) = x^2$. Find the lower and upper sums corresponding to the partitions P and Q of $[0, 1]$. Verify that

$$L(P, f) \le L(Q, f) \le U(Q, f) \le U(P, f).$$

2. Consider

$$f(x) = \begin{cases} 2x & \text{if } x \in [0, 1) \setminus \{1/2\}, \\ 0 & \text{if } x = 1/2, \\ 1 & \text{if } x = 1, \end{cases} \qquad \text{and } g(x) = 2x - 1 \text{ on } [0, 1].$$

Sketch the graph of $f(x)$ on $[0, 1]$ and $g(x)$ on $[0, 1]$. Evaluate $L(P, f)$, $U(P, f)$, $L(Q, g)$, and $U(Q, g)$ corresponding to the partitions

(a) $P = \left\{0, \frac{1}{3}, \frac{3}{4}, 1\right\}$. **(b)** $Q = \left\{0, \frac{1}{4}, \frac{1}{3}, \frac{2}{3}, 1\right\}$.

3. Let $f, g : [a, b] \to \mathbb{R}$ be defined by

$$f(x) = \begin{cases} 3 & \text{if } x \text{ is rational}, \\ 1 & \text{if } x \text{ is irratioanl}, \end{cases} \quad \text{and } g(x) = \begin{cases} 0 & \text{if } x \in [a, b] \setminus \{\frac{a+b}{2}\}, \\ 1 & \text{if } x = \frac{a+b}{2}, \end{cases}$$

and let P be an arbitrary partition of $[a, b]$. Verify that

(a) $U(P, f) = 3L(P, f)$ for $a > 0$. **(b)** $U(P, g) = L(P, g) = 0$.

4. For $n \geq 3$, consider $f(x) = x^n$ on $0 < a < b$. Show directly that f is Riemann integrable on $[a, b]$. Also, show that

$$\int_a^b x^n \, dx = \frac{b^{n+1} - a^{n+1}}{n+1}.$$

5. Suppose that f is integrable on $[a, b]$, $0 < a < b$, and $h = (b/a)^{1/n}$. Then show that

$$\int_a^b f(x) \, dx = \lim_{n\to\infty} S_n, \quad S_n = \sum_{k=1}^{n} a f(ah^{k-1}) \left(h^k - h^{k-1}\right).$$

6. Let $f : [a, b] \to \mathbb{R}$ be integrable, and let P be a partition of $[a, b]$ given by

$$P = \{x_0, x_1, \ldots, x_n\}, \quad x_k = a + k\left(\frac{b-a}{n}\right) \text{ for } k = 1, 2, \ldots, n.$$

Define the trapezoidal rule by

$$T_n(P, f) = \frac{b-a}{n} \sum_{k=1}^{\infty} \left(\frac{f(x_{k-1}) + f(x_k)}{2}\right).$$

Show that $\lim_{n\to\infty} T_n(P, f) = \int_a^b f(x) dx$.

7. Find the exact area under the graph of $f(x) = 3x + 1$ between $x = 0$ and $x = 2$. Repeat the question with $f(x) = x^2 + 3$.
8. Corresponding to $\epsilon = 0.01$, find a partition P of $[0, 1]$ such that

$$U(P, f) - L(P, f) < \epsilon$$

when $f(x)$ equals
(1) x^2. (2) $\sin x$. (3) $2x + 3$.

9. Let $f(x) = \sin x$ on $[0, \pi/2]$, and let P be the standard partition of $[0, \pi/2]$ given by

$$P = \{x_0, x_1, \ldots, x_n\}, \quad x_k = k\left(\frac{\pi}{2n}\right) \text{ for } k = 1, 2, \ldots, n.$$

Compute $U(P, f)$ and prove that $\lim_{n\to\infty} U(P, f) = 1$.

10. Using the limit process described in Example 6.18, calculate $\lim_{n\to\infty} a_n$ in each of the following cases by expressing it as a definite integral of some continuous function and then using the fundamental theorem of calculus:
(a) $a_n = \dfrac{[(n+1)(n+2)\cdots(n+n)]^{1/n}}{n}$.
(b) $a_n = \dfrac{1}{n+1} + \dfrac{1}{n+2} + \cdots + \dfrac{1}{n+n}$.
(c) $a_n = \dfrac{1}{n} + \dfrac{n^2}{(n+1)^3} + \dfrac{n^2}{(n+2)^3} + \cdots + \dfrac{1}{8n}$.

(d) $a_n = \left[\prod_{k=1}^{n}\left(1+\frac{k^2}{n^2}\right)\right]^{1/n}$.

(e) $a_n = \frac{1}{n+1} + \frac{1}{n+2} + \cdots + \frac{1}{pn}$, where $p \in \mathbb{N} \setminus \{1\}$ is fixed.

(f) $a_n = \sum_{k=1}^{n} \frac{1}{\sqrt{2nk-k^2}}$.

(g) $a_n = \sum_{k=1}^{n} \frac{k^2}{n^3+k^3}$.

(h) $a_n = \prod_{k=1}^{n}\left(1+\frac{k}{n}\right)^{1/k}$.

(i) $a_n = \sum_{k=1}^{n-1} \frac{n^2}{(n^2+k^2)^{3/2}}$.

(j) $a_n = \sum_{k=1}^{n} \frac{n+k}{n^2+k^2}$.

(k) $a_n = \frac{3}{n}\left[1+\sqrt{\frac{n}{n+3}}+\sqrt{\frac{n}{n+6}}+\cdots+\sqrt{\frac{n}{n+3(n-1)}}\right]$.

(l) $a_n = \frac{\pi}{n}\sum_{k=1}^{n} \sin\left(\frac{k\pi}{n}\right)$.

(m) $a_n = \frac{\pi}{2n}\sum_{k=0}^{n-1} \cos\left(\frac{k\pi}{2n}\right)$.

(n) $a_n = \frac{1}{n}\sum_{k=1}^{n} e^{ak/n} \quad (a \in \mathbb{R})$.

(o) $a_n = \frac{\pi}{2n}\sum_{k=1}^{n} \sin^{2p}\left(\frac{k\pi}{2n}\right)$, $p \in \mathbb{N}$ is fixed.

(p) $a_n = \frac{1}{n^2}\sum_{k=1}^{n} k\cos^2\left(\frac{k^2}{n^2}\right)$.

(q) $a_n = \sum_{k=0}^{n-1} \frac{1}{\sqrt{n^2-k^2}}$.

11. Identify the definite integral for which the lower Riemann sum is given by

$$s_n = \sum_{r=0}^{n-1} \frac{1}{\sqrt{4n^2-r^2}}.$$

Find also the upper Riemann sum S_n and then determine $\lim_{n\to\infty} S_n$.

12. Define $f : [0,1] \to \mathbb{R}$ by

$$f(x) = \begin{cases} x & \text{if } x \neq 0 \text{ and } 1/x \text{ is an integer,} \\ 0 & \text{otherwise.} \end{cases}$$

Show that f is not continuous at $1/n$, $n \in \mathbb{N}$. Is f integrable on $[0,1]$?

13. In each part below, give an example of a function satisfying the following:
    (a) bounded monotone but not continuous.
    (b) discontinuities at a finite number of points.
    (c) infinitely many discontinuities.

14. Show that the function $f : [0,1] \to \mathbb{R}$ defined by

$$f(x) = \begin{cases} \sqrt{1-x^2} & \text{if } x \in \mathbb{Q} \cap [0,1], \\ 1-x & \text{if } x \in \mathbb{Q}^c \cap [0,1], \end{cases}$$

is not integrable. Also, show that each of the functions $f_j : [0,1] \to \mathbb{R}$ ($j = 1,2$) defined by

$$f_1(x) = \begin{cases} x^2 & \text{if } x \in \mathbb{Q} \cap [0,1], \\ 1+x^2 & \text{if } x \in \mathbb{Q}^c \cap [0,1], \end{cases} \quad \text{and} \quad f_2(x) = \begin{cases} x & \text{if } x \in \mathbb{Q} \cap [0,1], \\ x^2 & \text{if } x \in \mathbb{Q}^c \cap [0,1], \end{cases}$$

is not integrable.

15. Prove that

$$f(x) = \begin{cases} x & \text{if } x \in \mathbb{Q} \cap [-1,1], \\ -x & \text{if } x \in \mathbb{Q}^c \cap [-1,1], \end{cases} \quad \text{and} \quad g(x) = \begin{cases} x^2 & \text{if } x \in \mathbb{Q} \cap [-1,1], \\ 0 & \text{if } x \in \mathbb{Q}^c \cap [-1,1], \end{cases}$$

are not integrable over $[-1,1]$.

16. Show that $f : [0,\pi/4] \to \mathbb{R}$ defined by

$$f(x) = \begin{cases} \sin x & \text{if } x \in \mathbb{Q} \cap [0,\pi/4], \\ \cos x & \text{if } x \in \mathbb{Q}^c \cap [0,\pi/4], \end{cases}$$

is not integrable.

17. Determine for each of the following functions whether it is integrable. Justify your answers.

(a) $f(x) = \begin{cases} 0 \text{ if } x \in \mathbb{Q} \cap [0,1], \\ 1 \text{ if } x \in \mathbb{Q}^c \cap [0,1]. \end{cases}$ (b) $f(x) = \begin{cases} \dfrac{1}{3x-1} & \text{if } x \in [0,1] \backslash \{\frac{1}{3}\}, \\ 0 & \text{if } x = 1/3. \end{cases}$

(c) $f(x) = \dfrac{e^x - 1}{x^3 - x}$ on $(-1, 1)$. (d) $f(x) = \begin{cases} \sin \frac{1}{x} & \text{if } x \in \mathbb{Q}^c \cap [0,1], \\ 0 & \text{if } x \in \mathbb{Q}. \end{cases}$

(e) $f(x) = \begin{cases} x & \text{if } x \in [-1, \frac{1}{4}], \\ \dfrac{1}{x} & \text{if } x \in [\frac{1}{4}, 1]. \end{cases}$ (f) $f(x) = \begin{cases} x & \text{if } x \in \mathbb{Q} \cap [0,1], \\ 0 & \text{if } x \in \mathbb{Q}^c \cap [0,1]. \end{cases}$

(g) $f(x) = \cos(1/x)$ on $(0, 1]$. (h) $f(x) = \begin{cases} \dfrac{1}{1-x^2} & \text{if } x \in (-1,1), \\ 0 & \text{if } x = \pm 1. \end{cases}$

(i) $f(x) = \sin(1/x)$ on $(0, 1]$. (j) $f(x) = \begin{cases} 3x & \text{if } x \in [0, 1/3], \\ 1-x & \text{if } x \in (1/3, 1/2], \\ 1+x & \text{if } x \in (1/2, 1]. \end{cases}$

18. Is $f : [0,1] \to \mathbb{R}$ defined by

$$f(x) = \begin{cases} (-1)^{k-1} & \text{for } \dfrac{1}{k+1} < x \leq \dfrac{1}{k}, \ k \in \mathbb{N}, \\ 1 & \text{for } x = 0, \end{cases}$$

integrable on $[0, 1]$?

19. Suppose that f is a nonconstant bounded function on $[a, b]$ that satisfies one of the following:
(a) f is integrable on $[c, b]$ for every $c \in (a, b)$.
(b) f is integrable on $[a, d]$ for every $d \in (a, b)$.
(c) f is integrable on $[c, d]$ for every $[c, d] \subset (a, b)$.
Show that f is (Riemann) integrable on $[a, b]$.

20. Assume that f is integrable on $[0, b]$.
(a) Show that if f is an even function, then f is integrable on $[-b, b]$ and

$$\int_{-b}^{b} f(t)\, dt = 2 \int_{0}^{b} f(t)\, dt.$$

(b) Show that if f is an odd function, then f is integrable on $[-b, b]$ and

$$\int_{-b}^{b} f(t)\, dt = 0.$$

6.2 Fundamental Theorems

In this section, we discuss the first fundamental theorem of calculus (integral of a derivative) and, the second fundamental theorem of calculus (differentiation of an *indefinite integral*). These two theorems are known together as the fundamental theorem of calculus, although we make a distinction by calling them the first and the second fundamental theorems. This result provides a method of computing certain integrals and reveals the close relationship between integration and differentiation.

Definition 6.33. *Let f be a function defined on I. If there exists a differentiable function F such that $F'(x) = f(x)$ on I, then F is called an antiderivative or a primitive of f on I.*

If F is an antiderivative of f on I, then so is F plus a constant.

Example 6.34. Set

$$f(x) = \begin{cases} -\dfrac{2}{x}\cos(1/x^2) + 2x\sin(1/x^2) & \text{for } x \in [-1,1] \setminus \{0\}, \\ 0 & \text{for } x = 0. \end{cases}$$

Clearly f is continuous except at the origin. On the other hand, f admits a primitive F given by

$$F(x) = \begin{cases} x^2\sin(1/x^2) & \text{for } x \in [-1,1] \setminus \{0\}, \\ 0 & \text{for } x = 0. \end{cases}$$

Note that f is not bounded (why?) on $[-1,1]$ and hence is not integrable. This example illustrates the existence of a *nonintegrable function having primitives.* •

6.2.1 The Fundamental Theorems of Calculus

The following result is also known as *Newton–Leibniz formula*, not in the exact sense but because Newton and Leibniz were the first to establish a relationship between integration and differentiation.

Theorem 6.35 (The first fundamental theorem of calculus). *If f is integrable on $[a,b]$ and F is an antiderivative of f on $[a,b]$, then*

$$\int_a^b f(x)\,dx = F(b) - F(a).$$

Proof. Let $P = \{a = x_0, x_1, x_2, \ldots, x_n = b\}$ be a partition of the interval $[a,b]$. Note that F satisfies the hypotheses of the classical mean value theorem for differentiable functions on each subinterval $[x_{k-1}, x_k]$, $k = 1,2,\ldots,n$. Thus, the mean value theorem tells us that there is a point x_k^* in each open subinterval (x_{k-1}, x_k) for which

$$F(x_k) - F(x_{k-1}) = F'(x_k^*)(x_k - x_{k-1}) = f(x_k^*)\Delta x_k, \quad k = 1,2,\ldots,n.$$

This relation gives

$$\sum_{k=1}^{n} f(x_k^*)\Delta x_k = \sum_{k=1}^{n}[F(x_k) - F(x_{k-1})] = F(x_n) - F(x_0) = F(b) - F(a).$$

Finally, we take the limit of the left side as $\|P\| \to 0$ as $n \to \infty$. Because f is integrable, Theorem 6.19 implies that the left-hand side becomes $\int_a^b f(x)\,dx$ in the limit, whereas the right-hand side remains $F(b) - F(a)$, which is a constant. Hence we have

$$\lim_{n\to\infty} \sum_{k=1}^{n} f(x_k^*)\Delta x_k = \int_a^b f(x)\,dx = F(b) - F(a),$$

as required. ∎

In particular, the definite integral of any continuous function f on $[a,b]$ can be computed without calculating the Riemann sums and often without much effort, simply by finding an antiderivative F and evaluating it at the limits of integration a and b. We have then two fundamental questions. Under what conditions on f does it has an antiderivative? How do we find an antiderivative? The existence of an antiderivative is asserted by the second fundamental theorem of calculus, which will be discussed soon.

Theorem 6.35 provides a convenient practical method of evaluating definite integrals when the antiderivative of the integrand is easy to find. For instance, suppose we want to evaluate a definite integral

$$\int_2^3 x^p\,dx \quad (p > -1).$$

We note that $F(x) = x^{p+1}/(p+1)$ is an antiderivative of $f(x) = x^p$. Thus, by Theorem 6.35,

$$\int_2^3 x^p\,dx = F(3) - F(2) = \frac{3^{p+1} - 2^{p+1}}{p+1}.$$

Note that if we choose a *different* antiderivative, say $G(x) = F(x) + c$, then $G(3)-G(2) = F(3)-F(2)$ as well. In Example 6.23, this integral was evaluated using the definition of the Riemann sum.

Suppose f is a continuous function on $[a,b]$. Then f is integrable on $[a,b]$. In particular, for any c between a and b, f is integrable on $[a,c]$, and the value of the integral $\int_a^c f(t)\,dt$ varies as c varies. To emphasize that the upper limit of integration is a variable instead of a constant, we might use the letter x and consider

$$F(x) = \int_a^x f(t)\,dt,$$

which is clearly a function of the variable x, $x \in [a,b]$, because every value of x gives a single value. Thus, from the notation itself it is clear that integration can be regarded as a process of switching from one function to another. Observe that $F(a) = 0$. Note that it is less confusing to write $F(x)$ in the above form rather than $F(x) = \int_a^x f(x)\,dx$. Moreover, if f is a nonnegative continuous function and x lies to the right of a, then $f(x)$ is the area under the graph of $y = f(x)$ from a to x.

The conditions under which an antiderivative exists is provided in the next theorem.

Theorem 6.36 (The second fundamental theorem of calculus). *Let f be integrable on $[a, b]$ and define*

$$G(x) = \int_a^x f(t)\,\mathrm{d}t \quad \text{for } a \le x \le b. \tag{6.9}$$

Then we have the following:

(a) *G is continuous on $[a, b]$.*
(b) *If f is continuous at $c \in [a, b]$, then G is differentiable at c and $G'(c) = f(c)$. If f is continuous from the right at a, then $G'_+(a) = f(a)$. If f is continuous from the left at b, then $G'_-(b) = f(b)$.*

Proof. **(a)** We will first prove that G is continuous on $[a, b]$. Because f is bounded on $[a, b]$, there exists a $K > 0$ such that $|f(x)| \le K$ on $[a, b]$. If $x, y \in [a, b]$ and $x < y$, then it follows from the basic properties of the integral that

$$G(y) - G(x) = \int_a^y f(t)\,\mathrm{d}t - \int_a^x f(t)\,\mathrm{d}t = \int_a^y f(t)\,\mathrm{d}t + \int_x^a f(t)\,\mathrm{d}t = \int_x^y f(t)\,\mathrm{d}t,$$

so that $|G(y) - G(x)| \le K|y - x|$, and the continuity of G follows.

(b) We will next prove that G is differentiable at c and $G'(c) = f(c)$ in the case that $c \in (a, b)$. The cases $c = a$ and $c = b$ are similar, with the appropriate one-sided derivatives being used. To discuss the derivative of G at c, we choose $c \in (a, b)$ and fix it. We form the difference quotient (see Figure 6.13 when $f(x) \ge 0$ on $[a, b]$):

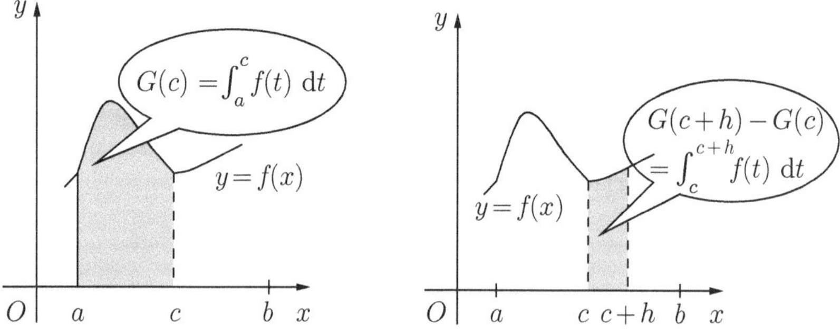

Fig. 6.13. Sketch for the second fundamental theorem of calculus.

$$\frac{G(c+h)-G(c)}{h} = \frac{1}{h}\int_c^{c+h} f(t)\,\mathrm{d}t,$$

where it is understood that h is always selected small enough that $c+h$ is also in (a,b) and h may be positive or negative. Since f is continuous at c, it is continuous on J, where J denotes the closed interval with endpoints c and $c+h$. Define

$$m = m(c) = \min_{t\in J} f(t) \quad \text{and} \quad M = M(c) = \max_{t\in J} f(t).$$

Clearly, we have that $m \le f(t) \le M$ for $t \in J$, so that by the dominance rule of the Riemann integral, we obtain

$$m = \frac{1}{h}\int_c^{c+h} m\,\mathrm{d}t \le \frac{1}{h}\int_c^{c+h} f(t)\,\mathrm{d}t \le \frac{1}{h}\int_c^{c+h} M\,\mathrm{d}t = M.$$

But since f is continuous at c, we know that $m = M = f(c)$ as $h \to 0$. Therefore,

$$G'(c) = \lim_{h\to 0} \frac{G(c+h)-G(c)}{h} = f(c),$$

which proves the assertion. ∎

Theorem 6.36(**b**) may be stated in the following form.

Corollary 6.37. *Suppose that f is continuous on $[a,b]$. Then G defined by (6.9) is an antiderivative of f on $[a,b]$.*

We remark that for f and G satisfying the last corollary, G necessarily has the form

$$G(x) = G(a) + \int_a^x f(t)\,\mathrm{d}t.$$

Thus the primitive G for which $G(a) = 0$ is referred to as the *indefinite integral* of f. Also, we have the following:

- If $f : [a,b] \to \mathbb{R}$ has a primitive in $[a,b]$, then $\int_a^b f(t)\,\mathrm{d}t$ is independent of the choice of the primitive.
- A function that has a primitive need not be continuous; see, for instance, Examples 6.34 and 6.38, and Exercise 6.57(1).

Example 6.38 (The indefinite integral of an integrable function is not necessarily differentiable). Define $f : [-2,2] \to \mathbb{R}$ by

$$f(x) = \begin{cases} 1 & \text{for } -2 \le x \le 0, \\ 0 & \text{for } 0 < x \le 2. \end{cases}$$

Clearly f is integrable on $[-2,2]$ (why?). If we define $F: [-2,2] \to \mathbb{R}$ by

$$F(x) = \int_{-2}^{x} f(t)\,\mathrm{d}t,$$

then we see that F is continuous on $[-2,2]$. We can now compute F explicitly. For $-2 \le x \le 0$,

$$F(x) = \int_{-2}^{x} f(t)\,\mathrm{d}t = \int_{-2}^{x} \mathrm{d}t = x+2,$$

and for $0 < x \le 2$, we have

$$F(x) = \int_{-2}^{x} f(t)\,\mathrm{d}t = \int_{-2}^{0} f(t)\,\mathrm{d}t + \int_{0}^{x} f(t)\,\mathrm{d}t = \int_{-2}^{0} \mathrm{d}t = 2.$$

Thus

$$F(x) = \begin{cases} x+2 & \text{for } -2 \le x \le 0, \\ 2 & \text{for } 0 < x \le 2, \end{cases}$$

which is clearly continuous on $[-2,2]$. Note that F is differentiable everywhere except at $x=0$ (where f is not continuous). •

Example 6.39 (sign(x) **is integrable).** The *signum function* is defined by

$$\operatorname{sign}(x) := \begin{cases} -1 & \text{if } x < 0, \\ 0 & \text{if } x = 0, \\ 1 & \text{if } x > 0. \end{cases}$$

The function $\operatorname{sign}(x)$ is monotone on $\mathbb{R}$ and hence integrable over any interval $[a,b]$. It is now easy to see that, for instance, for $a < 0$,

$$F(x) = \int_{a}^{x} \operatorname{sign}(t)\,\mathrm{d}t = \begin{cases} a - x & \text{if } x \le 0, \\ a + x & \text{if } x > 0, \end{cases}$$

and so $F(x) = a + |x|$, which is continuous on $\mathbb{R}$. •

The second fundamental theorem of calculus not only establishes the close relationship between the integration and differentiation, but also offers a useful method of evaluating integrals. Also, Theorem 6.36 implies that every continuous function f has an antiderivative. Moreover, an antiderivative is given by the integral $\int_a^x f(t)\,\mathrm{d}t$, which is called the *indefinite integral* of f on $[a,b]$. Some discontinuous functions have antiderivatives, and others do not. For example, consider

$$f(x) = \begin{cases} \sin(1/x) & \text{for } x \in [-1,1] \setminus \{0\}, \\ 0 & \text{for } x = 0, \end{cases}$$

and

$$g(x) = \begin{cases} 1 & \text{for } x \in [-1,1] \setminus \{0\}, \\ 0 & \text{for } x = 0. \end{cases}$$

It is easy to see that f has an antiderivative but g does not. Note that f is not monotonic on $[-1,1]$ but is continuous everywhere except at 0. It has been shown that f is integrable on $[-1,1]$.

Additional examples for Theorem 6.36 follow:

(a) $\dfrac{\mathrm{d}}{\mathrm{d}x}\left(\int_7^x (at+b)\,\mathrm{d}t\right) = ax+b.$

(b) To evaluate $\dfrac{\mathrm{d}}{\mathrm{d}x}\left(\int_7^{x^2} (t^2+t)\,\mathrm{d}t\right)$, we proceed with

$$\frac{\mathrm{d}}{\mathrm{d}x^2}\left(\int_7^{x^2} (t^2+t)\,\mathrm{d}t\right)\frac{\mathrm{d}(x^2)}{\mathrm{d}x} = \left((x^2)^2+x^2\right)(2x) = 2x^5+2x^3.$$

(c) Under the hypotheses of Theorem 6.36, we obtain

$$\frac{\mathrm{d}}{\mathrm{d}x}\left(\int_x^a f(t)\,\mathrm{d}t\right) = -f(x).$$

For instance, we have

$$\frac{\mathrm{d}}{\mathrm{d}x}\left(\int_x^5 \frac{\sin u}{u}\,\mathrm{d}u\right) = -\frac{\mathrm{d}}{\mathrm{d}x}\left(\int_5^x \frac{\sin u}{u}\,\mathrm{d}u\right) = -\frac{\sin x}{x}.$$

(d) If $\phi = \phi(x)$ is a differentiable function, then under the hypotheses of Theorem 6.36, one has

$$\frac{\mathrm{d}}{\mathrm{d}x}\left(\int_a^{\phi(x)} f(t)\,\mathrm{d}t\right) = \frac{\mathrm{d}}{\mathrm{d}\phi}\left(\int_a^{\phi} f(t)\,\mathrm{d}t\right)\frac{\mathrm{d}\phi(x)}{\mathrm{d}x} = f(\phi(x))\phi'(x).$$

For instance, in order to evaluate

(i) $\dfrac{\mathrm{d}}{\mathrm{d}x}\left(\int_{x^3}^{x^2} \dfrac{\mathrm{d}t}{\sqrt[3]{1+t^2}}\right)$, $x\in[0,1]$, **(ii)** $\dfrac{\mathrm{d}}{\mathrm{d}x}\left(\int_{x^2}^{x^3} \dfrac{\mathrm{d}t}{(1+t^2)^3}\right)$, $x\in[1,\infty)$,

we may just rewrite the two integrals as

$$\int_{x^3}^{x^2} \frac{\mathrm{d}t}{\sqrt[3]{1+t^2}} = -\int_0^{x^3} \frac{\mathrm{d}t}{\sqrt[3]{1+t^2}} + \int_0^{x^2} \frac{\mathrm{d}t}{\sqrt[3]{1+t^2}}$$

and

$$\int_{x^2}^{x^3} \frac{\mathrm{d}t}{(1+t^2)^3} = -\int_1^{x^2} \frac{\mathrm{d}t}{(1+t^2)^3} + \int_1^{x^3} \frac{\mathrm{d}t}{(1+t^2)^3},$$

and then proceed to use the general formula stated above.

Corollary 6.40. *The first fundamental theorem of calculus for continuous functions follows from Theorem 6.36.*

Proof. Assume that f is continuous on $[a, b]$. Then Theorem 6.36 shows that

$$G(x) = \int_a^x f(t)\,\mathrm{d}t$$

is a differentiable function with $G'(x) = f(x)$ on $[a, b]$. If F is any antiderivative of f, then $F'(x) = f(x)$, so that $(G - F)'(x) = 0$, and therefore $G(x) = F(x) + c$ for some constant c and for all x on the interval $[a, b]$. In particular, when $x = a$, we have

$$0 = G(a) = F(a) + c, \quad \text{i.e.,} \quad c = -F(a).$$

Consequently, $\int_a^x f(t)\,\mathrm{d}t = F(x) - F(a)$. Finally, by letting $x = b$, we obtain

$$\int_a^b f(t)\,\mathrm{d}t = F(b) - F(a),$$

as claimed by the first fundamental theorem of calculus. ∎

As a consequence of the first fundamental theorem of calculus, we have the following.

Corollary 6.41. *If $f \in C^1[a, b]$ (i.e., $f'(x)$ not only exists on $[a, b]$ but is also continuous on $[a, b]$), then*

$$\int_a^b f'(x)\,\mathrm{d}x = f(b) - f(a).$$

Using Theorems 6.35 and 6.36, we may now formulate the following useful version.

Theorem 6.42 (Fundamental theorem of calculus—combined form). *Let G and f be two continuous functions on $[a, b]$ such that $G(a) = 0$. Then we have*

$$G'(x) = f(x) \quad \text{on } [a, b] \quad \text{if and only if} \quad G(x) = \int_a^x f(t)\,\mathrm{d}t \quad \text{on } [a, b].$$

(At the endpoints, $G'(a)$ and $G'(b)$ refer to one-sided derivatives.)

We next use the fundamental theorem of calculus to prove formulas for integration by substitution and integration by parts.

Corollary 6.43 (Change of variables for integrals). *Suppose that $f : [a, b] \to \mathbb{R}$ is continuous and $g : [c, d] \to [a, b]$ is differentiable with continuous derivative. Then we have*

$$\int_{g(c)}^{g(d)} f(t)\,\mathrm{d}t = \int_c^d f(g(s))g'(s)\,\mathrm{d}s.$$

Proof. Set $F(x) = \int_a^x f(t)\,\mathrm{d}t$. Now

$$\begin{aligned}\int_{g(c)}^{g(d)} f(t)\,\mathrm{d}t &= F(g(d)) - F(g(c)) \\ &= \int_c^d (F \circ g)'(s)\,\mathrm{d}s \\ &= \int_c^d F'(g(s))g'(s)\,\mathrm{d}s \quad \text{(by the chain rule)} \\ &= \int_c^d f(g(s))g'(s)\,\mathrm{d}s,\end{aligned}$$

which completes the proof. ■

Corollary 6.44 (Integration by parts). *Suppose that $f : [a, b] \to \mathbb{R}$ has a continuous derivative and $g : [a, b] \to \mathbb{R}$ is continuous. Let $G : [a, b] \to \mathbb{R}$ be an indefinite integral of g, i.e., $G'(x) = g(x)$. Then we have*

$$\int_a^b f(x)g(x)\,\mathrm{d}x = f(x)G(x)\Big|_a^b - \int_a^b f'(x)G(x)\,\mathrm{d}x.$$

Proof. The proof of this result is left as an exercise. ■

As an application of integration by parts, one can obtain another version of Talyor's formula (see Theorem 8.43) with a remainder term in the form of an integral. We refer to Exercise 8.51(20).

6.2.2 The Mean Value Theorem for Integrals

The mean value theorem for derivatives implies that under certain conditions on f, there is at least one number c in the interval (a, b) such that

$$\frac{f(b) - f(a)}{b - a} = f'(c).$$

Our next result, which is in some sense analogous to the mean value theorem for derivatives, is especially useful in obtaining estimates for certain definite integrals rather than an exact value. It is one of the several forms of mean value theorems for integrals.

Theorem 6.45 (Mean value theorem for integrals). *If f is continuous on the interval $[a, b]$, then there is at least one number c on this interval such that*

$$\int_a^b f(x)\,\mathrm{d}x = f(c)(b - a). \tag{6.10}$$

Proof. Since f is continuous on the interval $[a, b]$, there exist M and m such that

$$m = \min_{t\in[a,b]} f(t) \le f(x) \le M = \max_{t\in[a,b]} f(t) \quad \text{when } a \le x \le b,$$

and so (by the dominance rule)

$$m(b-a) = \int_a^b m\,\mathrm{d}x \le \int_a^b f(x)\,\mathrm{d}x \le \int_a^b M\,\mathrm{d}x = M(b-a).$$

Dividing by the positive number $b - a$ does not change the inequalities, and so we see that there is a real number μ such that

$$m \le \mu \le M, \quad \mu = \frac{1}{b-a}\int_a^b f(x)\,\mathrm{d}x.$$

Now, f is continuous on the closed interval $[a, b]$, and the number μ lies between m and M. The intermediate value theorem for continuous functions implies that f must assume every value between m and M. Consequently, there exists a number c between a and b for which $f(c) = \mu$. Therefore, (6.10) holds. ∎

Example 6.46. Find a value of c guaranteed by the mean value theorem for integrals for the following functions:

(a) $f(x) = \sin x$ on $[0, \pi]$. **(b)** $f(x) = x^3$ on $[-1, 1]$.
(c) $f(x) = x^2$ on $[-1, 1]$.

Solution. **(a)** We know that (see Figure 6.14)

$$\int_0^\pi \sin x\,\mathrm{d}x = -\cos x\big|_0^\pi = 2.$$

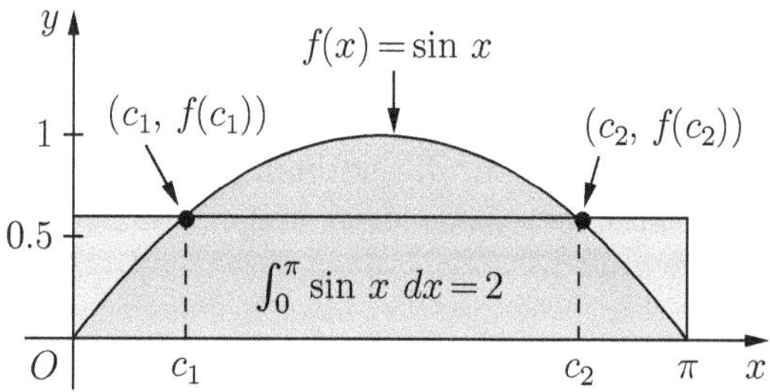

Fig. 6.14. Illustration for the mean value theorem for integrals with $f(x) = \sin x$ on $[0, \pi]$.

According to the mean value theorem, there exists a number c on $[0, \pi]$ such that

$$f(c)(b-a) = 2, \quad \text{i.e.,} \quad \sin c = 2/\pi.$$

We see that $c \approx c_1 = 0.690107$ or $c_2 = 2.451486$. Because both c_1 and c_2 lie between 0 and π, we have found these two (approximate) values of c.

(b) For $f(x) = x^3$, we have $\int_{-1}^{1} x^3\, dx = 0$. The mean value theorem for integrals asserts the existence of a point $c \in [-1, 1]$ such that

$$0 = f(c)[1-(-1)] = 2f(c), \quad \text{i.e.,} \quad f(c) = 0.$$

It is obvious in this case that $c = 0$.

(c) Finally, for $f(x) = x^2$, we have $\int_{-1}^{1} x^2\, dx = 2/3$, and so there exists a point c such that

$$2/3 = 2f(c), \quad \text{i.e.,} \quad c^2 = 1/3.$$

Clearly, $c = \pm 1/\sqrt{3}$ do the job. Each of these points lies in the interval $[-1, 1]$ and satisfies the condition of Theorem 6.45. •

Remark 6.47. 1. If $f'(x)$ exists and is continuous on $[a, b]$, then with $f'(x)$ in place of $f(x)$, we see that the mean value theorem for derivatives follows from the mean value theorem for integrals.

2. The mean value theorem for integrals has a geometric interpretation, especially when $f(x) \geq 0$. For example (see Figure 6.15), the theorem says that it is possible to find at least one number c on the interval (a, b) such that the area of the rectangle with height $f(c)$ and base $(b-a)$ has exactly the same area as the region under the curve $y = f(x)$ on $[a, b]$.
3. As in the case of the mean value theorem for derivatives, the mean value theorem for integrals gives no indication how to determine c. •

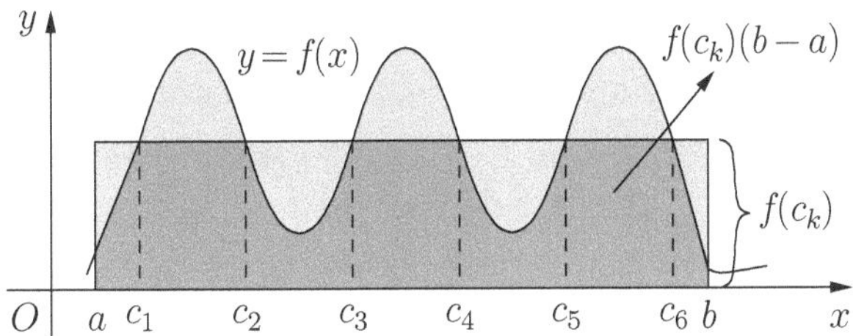

Fig. 6.15. Geometric interpretation of the mean value theorem for integrals.

Corollary 6.48 (Generalized mean value theorem for integrals). *Suppose that f and g are continuous on $[a, b]$ and $g(x) \geq 0$ on $[a, b]$. Then*

$$\int_a^b f(x)g(x)\, dx = f(c) \int_a^b g(x)\, dx \tag{6.11}$$

for some c in (a, b).

Proof. Using the notation of Theorem 6.45, we have $m \leq f(x) \leq M$ for $x \in [a, b]$, and since $g(x) \geq 0$ on $[a, b]$, we have

$$mg(x) \leq f(x)g(x) \leq Mg(x) \quad \text{for } x \in [a, b].$$

Equation (6.11) clearly holds for any c in (a, b) if $\int_a^b g(x)\,dx = 0$. Therefore, we assume that $\int_a^b g(x)\,dx \neq 0$, and so we have $\int_a^b g(x)\,dx > 0$, since $g(x)$ is a nonnegative function on $[a, b]$. The above inequalities then give

$$m \leq \mu = \frac{\int_a^b f(x)g(x)\,dx}{\int_a^b g(x)\,dx} \leq M,$$

and the desired conclusion follows from the intermediate mean value theorem. ∎

Example 6.49. Show that if f is continuous on $[0, 1]$, then

$$\lim_{n\to\infty} \int_0^1 \frac{nf(x)}{1+n^2x^2}\,dx = \frac{\pi}{2}f(0).$$

Solution. Using the generalized mean value theorem for integrals, we obtain that for each $n > 1$ there exists a $c_n \in (0, 1/\sqrt{n})$ such that

$$\begin{aligned} I_n := \int_0^{1/\sqrt{n}} \frac{nf(x)}{1+n^2x^2}\,dx &= f(c_n) \int_0^{1/\sqrt{n}} \frac{n\,dx}{1+(nx)^2} \\ &= f(c_n)\ \arctan(nx)\Big|_0^{1/\sqrt{n}} \\ &= f(c_n)\arctan(\sqrt{n}) \\ &\to f(0)\frac{\pi}{2} \quad \text{as } n \to \infty, \end{aligned}$$

and similarly, there exists a $d_n \in (1/\sqrt{n}, 1)$ such that

$$J_n := \int_{1/\sqrt{n}}^1 \frac{nf(x)}{1+n^2x^2}\,dx = f(d_n) \int_{1/\sqrt{n}}^1 \frac{n}{1+n^2x^2}\,dx,$$

so that

$$\begin{aligned} |J_n| &= |f(d_n)| \int_{1/\sqrt{n}}^1 \frac{n}{1+n^2x^2}\,dx \\ &= |f(d_n)|(\arctan(n) - \arctan(\sqrt{n})) \\ &\to 0 \ \text{ as } n \to \infty \text{ (because } f \text{ is bounded on } [0, 1]). \end{aligned}$$

Therefore,

$$\int_0^1 \frac{nf(x)}{1+n^2x^2}\,dx = I_n + J_n \to \frac{\pi}{2}f(0) \quad \text{as } n \to \infty.$$

•

What else can one learn from the mean value theorem for integrals? For instance, suppose that f is continuous on $[a, b]$ such that $\int_a^b f(x)\,dx = 0$. Then according to Theorem 6.45, there exists a point c in $[a, b]$ such that $f(c) = 0$. In other words, the graph of f must touch the x-axis at least once, i.e., $f(x) = 0$ has a solution in $[a, b]$. Secondly we ask, *Is continuity of f in the mean value theorem for integrals important?* Indeed it is. To see this, consider f defined on $[0, 2]$ such that

$$f(x) = \begin{cases} 0 & \text{for } x \in [0, 1], \\ 5 & \text{for } x \in (1, 2]. \end{cases}$$

Clearly, f is discontinuous at $x = 1$, and

$$\frac{1}{b-a} \int_a^b f(x)\,dx = \frac{1}{2-0} \int_0^2 f(x)\,dx = \frac{1}{2} \left[\int_0^1 0\,dx + \int_1^2 5\,dx \right] = \frac{5}{2}.$$

It follows that there exists no c such that $f(c) = 5/2$.

Corollary 6.50. *The second fundamental theorem of calculus follows from the mean value theorem for integrals.*

Proof. Let $f(t)$ be continuous on $[a, b]$ and define

$$G(x) = \int_a^x f(t)\,dt.$$

We first fix $x \in (a, b)$. For Δx such that $x + \Delta x \in [a, b]$, we write

$$G(x + \Delta x) - G(x) = \int_x^{x+\Delta x} f(t)\,dt.$$

Applying the mean value theorem for integrals to the integral on the right, we obtain that

$$G(x + \Delta x) - G(x) = (x + \Delta x - x) f(c), \quad \text{i.e.,} \quad \frac{G(x + \Delta x) - G(x)}{\Delta x} = f(c),$$

where c lies between x and $x + \Delta x$. But since $c \to x$ as $\Delta x \to 0$, we allow $\Delta x \to 0$ and obtain $G'(x) = f(x)$ on (a, b), since x is arbitrary. ■

6.2.3 Average Value of a Function

We are now interested in defining what is called an *average value* of a continuous function on an interval. Recall that the average value of n numbers $x_1, x_2, \ldots, x_n$ is defined by

$$\frac{1}{n} \sum_{k=1}^{n} x_k.$$

The question before us is this: *How do we define the notion of "average" if there are infinitely many numbers?* In particular, what is the average value of a continuous function $f(x)$ on $[a,b]$? The procedure to define this is as follows. Divide the interval $[a,b]$ into n equal subintervals

$$[x_{k-1}, x_k] = [a+(k-1)\Delta x, a+k\Delta x] \quad (k=1,2,\ldots,n),$$

each of width $\Delta x = (b-a)/n$. For $k=1,2,\ldots,n$, let x_k^* be a number chosen arbitrarily from the kth subinterval. Then the average value of an arbitrary continuous function f on $[a,b]$ is estimated by the average S_n of the n sampled values:

$$S_n = \frac{f(x_1^*)+f(x_2^*)+\cdots+f(x_n^*)}{n} = \frac{1}{n}\sum_{k=1}^{n} f(x_k^*) = \frac{1}{b-a}\sum_{k=1}^{n} f(x_k^*)\Delta x.$$

The sum on the right is indeed a Riemann sum for f on $[a,b]$ with norm $\|P\| = (b-a)/n$. Thus it is natural to define the average/mean value of a continuous function $f(x)$ defined on the interval $[a,b]$ by

$$\lim_{n\to\infty} \frac{1}{b-a}\sum_{k=1}^{n} f(x_k^*)\Delta x = \frac{1}{b-a}\int_a^b f(x)\,dx.$$

Also, we remark that the value of $f(c)$ in the mean value theorem for integrals is in some sense the average, or mean, height of $f(x)$ on $[a,b]$.

6.2.4 The Logarithmic and Exponential Functions

In earlier chapters we have used the logarithmic and exponential functions, which are undoubtedly familiar to the reader from precalculus mathematics. It may seem strange at first that we introduce the logarithmic function as a definite integral, but later we will see that our definition obeys the laws of the logarithmic and exponential functions considered in precalculus courses. We recall that the fundamental theorem of calculus concerns the function F defined by

$$F(x) = \int_a^x f(t)\,dt$$

under suitable conditions on f. If $f(t) = t^\alpha$, then

$$\int_a^x t^\alpha\,dt = \frac{1}{\alpha+1}\left(x^{\alpha+1} - a^{\alpha+1}\right) \quad \text{for } \alpha \in \mathbb{R}\setminus\{-1\}.$$

Clearly $\alpha = -1$ cannot be used. That is, we are unable to determine an antiderivative of $f(t) = 1/t$. Our next result will remedy this situation.

Note that since $f(t) = 1/t$ is continuous on $\mathbb{R}\setminus\{0\}$, f is integrable on any interval $[a,b]$ not containing 0. This observation helps to define a continuous function $L : (0,\infty) \to \mathbb{R}$ by

$$L(x) = \int_1^x \frac{dt}{t} \quad \text{for } x > 0. \tag{6.12}$$

The expression $L(x)$ is called the *natural logarithm* of x, denoted by $\log x$, which we shall soon clarify. Here the restriction $x > 0$ is necessary, because the integrand $1/t$ has an indefinite discontinuity when it is considered on the interval $(x, 1)$ with $x < 0$, and hence $\int_1^x (1/t)\, dt$ does not exist.

For $x = 1$, $L(1) = 0$. For $x > 1$, this integral represents the area of the region bounded by the curve $y = 1/t$ from $t = 1$ to $t = x$. Thus, $L(x) > 0$ for $x > 1$. If $0 < x < 1$, then

$$L(x) = -\int_x^1 \frac{dt}{t},$$

which is the negative of the area of the region under the curve $y = 1/t$ from $t = x$ to $t = 1$. Thus, $L(x) < 0$ for $0 < x < 1$. Consequently, (6.12) defines a computable function of x on $(0, \infty)$. Further, by the second fundamental theorem of calculus (see Theorem 6.36),

$$L'(x) = \frac{1}{x},$$

so that L is increasing on $(0, \infty)$, and thus L is one-to-one on $(0, \infty)$. Also, $L''(x) < 0$ on $(0, \infty)$, so that the graph of the curve $y = L(x)$ is concave downward on $(0, \infty)$; see Figure 6.16. We summarize the above discussion

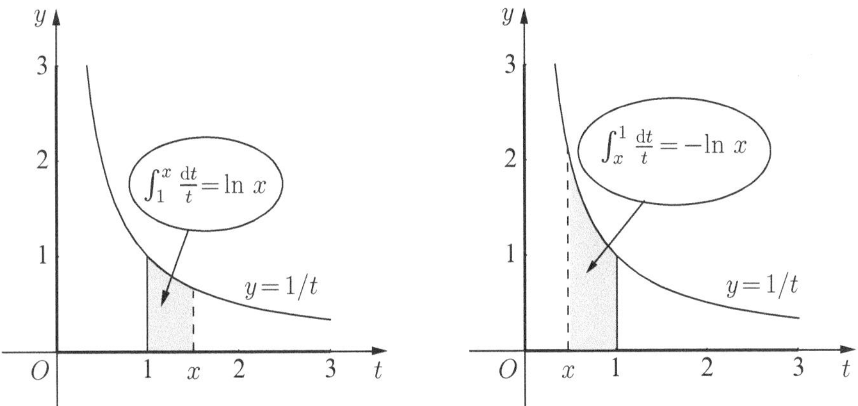

Fig. 6.16. The natural logarithm function.

together with a few additional properties.

Theorem 6.51 (Properties of the natural logarithm function). *The following statements hold:*

(a) $L(x) > 0$ *for* $x > 1$.

(b) $L(x) = 0$ *for* $x = 1$.
(c) $L(x) < 0$ *for* $0 < x < 1$.
(d) $\dfrac{x-1}{x} \le L(x) < x - 1$ *for* $x > 0$.
(e) $L(xy) = L(x) + L(y)$ *for* $x, y \in (0, \infty)$.
(f) $L(x/y) = L(x) - L(y)$ *for* $x, y \in (0, \infty)$.
(g) $L(1/x) = -L(x)$.
(h) $L(x^a) = aL(x)$ *for* $x > 0$ *and for every* $a \in \mathbb{R}$.
(i) L *is bijective on* $(0, \infty)$ *with the range of* L *is all of* $\mathbb{R}$.
(j) $L(x) \to \infty$ *as* $x \to \infty$.
(k) $L(x) \to -\infty$ *as* $x \to 0+$.

Proof. **(d)** To prove **(d)** above, it suffices to observe that

$$\begin{cases} 1 \le t \le x \iff \dfrac{1}{x} \le \dfrac{1}{t} \le 1 & \text{for } x > 1, \\ x \le t \le 1 \iff 1 \le \dfrac{1}{t} \le \dfrac{1}{x} & \text{for } 0 < x < 1. \end{cases}$$

(e) Fix $a > 0$ and consider $f(x) = L(ax)$. Then

$$f'(x) = aL'(ax) = \frac{1}{x} = L'(x), \quad \text{i.e.,} \quad (f - L)'(x) = 0,$$

so that $f - L$ is constant. In particular,

$$f(b) - L(b) = f(1) - L(1) = L(a),$$

and so $L(ab) = L(a) + L(b)$. Alternatively,

$$\begin{aligned} L(ab) &= \int_1^a \frac{dt}{t} + \int_a^{ab} \frac{dt}{t} \\ &= \int_1^a \frac{dt}{t} + \int_1^b \frac{du}{u} \quad \text{(by the change of variable } t = au\text{)} \\ &= L(a) + L(b). \end{aligned}$$

(f) We have

$$\begin{aligned} L(x/y) &= L(x) + L(1/y) \quad \text{(by (e))} \\ &= L(x) + \int_1^{1/y} \frac{dt}{t} \quad \text{(by definition)} \\ &= L(x) - \int_1^y \frac{dv}{v} \quad \text{(by the change of variable } t = 1/v\text{)} \\ &= L(x) - L(y). \end{aligned}$$

(g) Follows if we apply **(f)** with $x = 1$.

(h) Case 1: If $a = n \in \mathbb{N}$, use **(e)** and the method of induction.
Case 2: If $a = -n$, then by **(g)** and Case 1, one has

$$L(x^a) = L(x^{-n}) = L\Big(\frac{1}{x^n}\Big) = -L(x^n) = -nL(x) = aL(x).$$

Case 3: If $a > 0$ and $a \in \mathbb{Q}$, then a can be written as $a = m/n$ for some positive integers m and n, $n > 1$, and so by Case 1, we get

$$L(x^a) = L((x^{1/n})^m) = \frac{m}{n}[nL(x^{1/n})] = \frac{m}{n}L((x^{1/n})^n) = aL(x).$$

Case 4: If $a = -m/n$, then

$$L(x^a) = L\Big(\frac{1}{x^{-a}}\Big) = -L(x^{-a}) = -(-a)L(x) = aL(x).$$

Case 5: If a is an irrational number, then we consider a sequence $\{\alpha_n\}$ of rational numbers converging to a and obtain that

$$L(x^{\alpha_n}) = \alpha_n L(x).$$

Since L is continuous, we may allow $n \to \infty$, and statement **(h)** follows.

(i) The function L is known to be one-to-one because $L'(x) = 1/x > 0$ on $(0, \infty)$. By the intermediate value theorem, it follows that L maps $(0, \infty)$ onto $\mathbb{R}$.

(j) We observe that $L(3) > 0$ and that $L(3^n) = nL(3)$ (by **(h)**). Thus, $\lim_{n\to\infty} L(3^n) = \infty$, and since L is increasing, **(j)** follows.

(k) Since $L(1/3) = -L(3) < 0$ and $L(1/3^n) = nL(1/3) = -nL(3)$, **(k)** follows similarly. ●

Remark 6.52. By Property **(a)** above, it follows that there is a unique real number x such that $L(x) = 1$. Suppose we denote such an x by e. Then $L(\mathrm{e}) = 1$, and the reason for denoting it by e is justified as follows. Now, since $L'(x) = 1/x$, it follows that

$$L'(1) = 1, \quad \text{i.e.,} \quad \lim_{h\to 0} \frac{L(1+h) - L(1)}{h} = \lim_{h\to 0} \frac{L(1+h)}{h} = 1.$$

With $h = 1/n$, this reduces to

$$1 = \lim_{n\to\infty} nL\left(1 + \frac{1}{n}\right) = \lim_{n\to\infty} L\left(\left(1 + \frac{1}{n}\right)^n\right) = L\left(\lim_{n\to\infty}\left(1 + \frac{1}{n}\right)^n\right),$$

so that because L is one-to-one,

$$\lim_{n\to\infty}\left(1 + \frac{1}{n}\right)^n = \mathrm{e},$$

as defined earlier. ●

Since L is bijective with range $(-\infty, \infty)$, it follows that L has an inverse, which we denote by E. Thus,

$$E(L(x)) = x \quad \text{for } x \in (0, \infty) \quad \text{and} \quad L(E(y)) = y \quad \text{for } y \in \mathbb{R}.$$

The function E will be called the *exponential function* and will be denoted by exp; this definition is equivalent to the earlier definition of e^x.

Since $L(1) = 0$, we obtain $E(0) = 1$. Also, we see that the basic properties of L are also reflected in its inverse.

Theorem 6.53 (Properties of the exponential function). *The following statements hold:*

(a) $0 < E(x) < 1 \iff x < 0$.
(b) $E(x) = 1 \iff x = 0$.
(c) $E(x) > 1 \iff x > 0$.
(d) $E(x + y) = E(x)E(y)$ *for* $x, y \in \mathbb{R}$.
(e) $E(-x) = \dfrac{1}{E(x)}$ *for* $x \in \mathbb{R}$.
(f) $E(ax) = (E(x))^a$ *for* $x \in \mathbb{R}$ *and* $a \in \mathbb{Q}$.
(g) $\lim_{y \to 0} \dfrac{L(y + 1)}{y} = 1$.
(h) $\lim_{x \to 0} \dfrac{E(x) - 1}{x} = 1$.
(i) $E'(x) = E(x)$ *for* $x \in \mathbb{R}$.
(j) $E(x) \to \infty$ *as* $x \to \infty$.
(k) $E(x) \to 0$ *as* $x \to -\infty$.

Proof. **(d)** Since $L(E(x)) = x$ for all $x \in \mathbb{R}$,

$$\begin{aligned} E(x + y) &= E(L(E(x)) + L(E(y))) \\ &= E(L(E(x)E(y))), \quad \text{by Theorem 6.51(e)} \\ &= E(x)E(y), \quad \text{by the definition of inverse.} \end{aligned}$$

(e) This follows if we set $y = -x$ in **(d)**.
(f) Case (i): Let $a \in \mathbb{N}_0$. There is nothing to prove if $a = 0, 1$. If $a = 2$, then

$$E(2x) = E(x + x) = E(x)E(x) = (E(x))^2,$$

and so by the induction argument, one has

$$E(ax) = (E(x))^a \quad \text{for } a = n \in \mathbb{N}.$$

Case (ii) For $a = -n$, $n \in \mathbb{N}$, by **(e)**, we get

$$E(-nx) = \frac{1}{E(nx)} = \frac{1}{(E(x))^n} = (E(x))^{-n},$$

and so **(f)** holds for $a = -n$.

Case (iii) If $a = m/n$, where $m, n \in \mathbb{N}$, then by Case (i),

$$\left(E\left(\frac{m}{n}x\right)\right)^n = E\left(n\frac{m}{n}x\right) = E(mx) = (E(x))^m,$$

and because $E(x) > 0$ for all $x \in \mathbb{R}$, this gives

$$E\left(\frac{m}{n}x\right) = (E(x))^{m/n}.$$

Similarly, if $a = -m/n$, then we easily have

$$E\left(-\frac{m}{n}x\right) = (E(x))^{-m/n}.$$

This completes the proof for **(f)**.

(g) From Theorem 6.51**(d)**, we obtain that

$$\frac{1}{y+1} \leq \frac{L(y+1)}{y} < 1 \quad \text{for } y > -1.$$

Note that since $L(1) = 0$, the quotient $L(y+1)/y$ in question is in indeterminate form at $y = 0$ and lies between $1/(y+1)$ and 1. If we allow $y \to 0$, then the squeeze rule gives the result.

(h) Set $y = E(x) - 1$. Then $E(x) = y + 1$ or $x = L(y+1)$. Because $E(x)$ is continuous on $\mathbb{R}$ and $E(0) = 1$, we have $y \to 0$ whenever $x \to 0$. Consequently, by **(g)**,

$$\lim_{x\to 0} \frac{E(x) - 1}{x} = \lim_{y\to 0} \frac{y}{L(y+1)} = 1.$$

(i) Let $x \in \mathbb{R}$ be arbitrary. For $h \neq 0$, **(d)** gives

$$\frac{E(x+h) - E(x)}{h} = \frac{E(x)E(h) - E(x)}{h} = E(x)\left(\frac{E(h) - 1}{h}\right),$$

which by **(h)**, gives the result as $h \to 0$. Thus, $E'(x) = E(x)$ on $\mathbb{R}$. •

An expression such as a^x $(a > 0)$ for x rational can be defined by elementary means. For instance, if $x = m/n$, where $m, n \in \mathbb{N}$, then a^x may be written as

$$a^x = (a^m)^{1/n}.$$

On the other hand, expressions such as

$$2^{\sqrt{3}},\ (\sqrt{3})^\pi$$

cannot be given in such an elementary way. Thus we need a suitable definition to deal with such cases.

Definition 6.54. *If $a > 0$, then we define*

$$a^x = E(xL(a)) \quad \text{for } x \in \mathbb{R}. \tag{6.13}$$

Thus, we have (because $L = \mathrm{e}^{-1}$)

$$L(a^x) = xL(a) \quad \text{for all } a > 0 \text{ and } x \in \mathbb{R}.$$

If we differentiate (6.13), we obtain by Theorem 6.53**(i)**,

$$(a^x)' = E'(xL(a)).\, L(a) = E(xL(a))L(a) = a^x L(a).$$

Similarly, we may define

$$x^a = E(aL(x)) \quad \text{for } a \in \mathbb{R} \text{ and } x > 0,$$

and if we differentiate this, we obtain

$$(x^a)' = E'(aL(x))\frac{a}{x} = E(aL(x))\frac{a}{x} = x^a\left(\frac{a}{x}\right) = ax^{a-1}. \tag{6.14}$$

The well-known integration formula for x^a follows from (6.14) if $a \neq -1$.

Example 6.55. Prove the following:

(a) $\lim_{x\to\infty} x^{-a}L(x) = 0$ for all $a > 0$. **(b)** $\lim_{x\to\infty} \frac{x^a}{E(x)} = 0$ for all real a.

Solution. **(a)** To prove **(a)**, choose b such that $0 < b < a$ and $x > 1$. Then since $t^b > 1$ for $t > 1$, we have

$$x^{-a}L(x) = x^{-a}\int_1^x \frac{\mathrm{d}t}{t} < x^{-a}\int_1^x t^{b-1}\,\mathrm{d}t = x^{-a}\left(\frac{x^b - 1}{b}\right) < \frac{x^{b-a}}{b},$$

and the proof of **(a)** follows.
Since $L(x) \to \infty$ as $x \to \infty$, and $x^a \to \infty$ as $x \to \infty$ $(a > 0)$, from **(a)** we observe that $L(x)$ approaches ∞ "more slowly" than any positive power of x as $x \to \infty$.

(b) Set $y = L(x)$ for $x > 0$. Then $y \to \infty$ as $x \to \infty$, by Theorem 6.51**(j)**. Also, $x = E(y)$ for $y \in \mathbb{R}$, and so for $a > 0$, this observation together with **(a)** gives

$$0 = \lim_{x\to\infty} \frac{L(x)}{x^a} = \lim_{y\to\infty} \frac{y}{(E(y))^a} = \frac{1}{a}\lim_{y\to\infty} \frac{ay}{E(ay)},$$

and hence **(b)** follows if $a > 0$. If $a < 0$, set $a = -\alpha$, and obtain a proof for **(b)**. There is nothing to prove if $a = 0$, since $E(x) \to \infty$ as $x \to \infty$. ●

6.2.5 Questions and Exercises

Questions 6.56.

1. What is the difference between definite and indefinite integrals?
2. Suppose that f is integrable on $[a,b]$. Must f have a primitive on $[a,b]$? How about if f is continuous on $[a,b]$?
3. If f is continuous and nonnegative on $[a,b]$ such that $\int_a^b f(x)\,dx = 0$, must $f(x) = 0$ on $[a,b]$?
4. If f is continuous on $[a,b]$ such that $\int_a^b f(x)g(x)\,dx = 0$ for every continuous function g on $[a,b]$, must $f(x) = 0$ on $[a,b]$?
5. If f and g are two continuous functions on $[a,b]$ such that $\int_a^b f(x)\,dx = \int_a^b g(x)\,dx$, must $f(x) = g(x)$ have a solution in $[a,b]$?
6. Suppose that f is integrable on $[a,b]$ such that $f(x) \geq 0$ for all x except for a finite number of points $x \in [a,b]$, and $f(c) > 0$ for some $c \in (a,b)$ at which f is continuous. Must we have $\int_a^b f(x)\,dx > 0$?
7. If f is continuous on $[a,b]$ such that $f(x) = \int_a^x f(t)\,dt$, must we have $f(x) = 0$ on $[a,b]$?
8. Suppose that $f(x) = [x]$ on $[0,3]$. Can $\int_0^3 f(t)\,dt$ be evaluated using the fundamental theorem of calculus?
9. If one uses the first fundamental theorem of calculus, then one writes

$$\int_{-1}^{1} \frac{dx}{x^4} = \left[-\frac{3}{x^3}\right]\Bigg|_{-1}^{1} = -3 - 3 = -6$$

because $f(x) = 1/x^2$ has a primitive $F(x) = -3/x^3$ on $[-1,1]$. On the other hand, the function $y = 1/x^2$ is always positive. What is wrong with this "evaluation"?
10. Is the point c in the mean value theorem for integrals unique?
11. Does the generalized mean value theorem for integrals (i.e., Corollary 6.48) hold in the case $g(x) < 0$ on $[a,b]$?
12. Must we have $\sin x > x - x^3/6$ for all $x > 0$?
13. Suppose that f is continuous on $[0,1]$ and $n \in \mathbb{N}$. Is it true that $\int_0^1 x^n f(x)\,dx \to 0$ as $n \to \infty$?
14. Let f be continuous on $[-1,1]$. Is it true that $\int_0^{2\pi} f(\sin x)\,\cos x\,dx = 0$? Must we have $\int_0^{2\pi} f(\sin x)\,dx = \int_{\pi/2}^{\pi} f(\sin x)\,dx$?
15. Which is larger, π^3 or 3^π?

Exercises 6.57.

1. Define $f : \mathbb{R} \to \mathbb{R}$ by

$$f(x) = \begin{cases} 2x\sin(1/x) - \cos(1/x) & \text{for } x \neq 0, \\ 0 & \text{for } x = 0. \end{cases}$$

Determine the antiderivative of $f(x)$ and thereby evaluate $\int_0^1 f(x)\,dx$.

2. Let $f : \mathbb{R} \to \mathbb{R}$ be a continuous function and let $F : \mathbb{R} \to \mathbb{R}$ be defined by

$$F(x) = \int_0^{e^x} f(t)\, dt.$$

Show that F is differentiable on $\mathbb{R}$. Compute $F'(x)$. Repeat the exercise when e^x is replaced by $\sin x$.

3. If $f(x) = \int_0^{x^2} \sqrt{t - t^6}\, dt$ $(x > 0)$, find $f'(3)$.

4. Determine dG/dx, if $G(x)$ equals:

(a) $\int_2^x (2t+3)\, dt$. **(b)** $\int_{x^2}^4 \sec^2(t)\, dt$. **(c)** $\int_{-x}^x \cot t\, dt$.

(d) $\int_{\log x}^1 \frac{\cos t}{t}\, dt$. **(e)** $\int_{\sqrt{x}}^{\sqrt[3]{x}} t^2\, dt$. **(f)** $\int_x^{x^2} \cos(t^2)\, dt$.

(g) $\int_0^{x^3} e^{t^2}\, dt$. **(h)** $\int_x^{x^2} \sqrt{1+t^5}\, dt$. **(i)** $\int_{\sin x}^{\cos x} e^{\sqrt{t}}\, dt$.

5. For $c \neq 0$, define $f(x)$ by $f(0) = 0$ and

$$f(x) = 3x^2 \cos(c/x^2) + 2c \sin(c/x^2) \quad \text{for} \quad x \neq 0.$$

Show that f is integrable on $[-a, a]$ $(a > 0)$. Compute $\int_{-a}^a f(t)\, dt$.

6. Define $f : [0,3] \to \mathbb{R}$ by

$$f(x) = \begin{cases} 1 & \text{if } 0 \le x < 1, \\ 2 & \text{if } 1 \le x < 2, \\ 3 & \text{if } 2 \le x \le 3. \end{cases}$$

Show that f is integrable on $[0,3]$ and $\int_0^3 f(x)\, dx = 6$. Does f have a primitive on $[0,3]$?

7. Define $f : [0,2] \to \mathbb{R}$ by

$$f(x) = \begin{cases} 2x & \text{if } 0 \le x \le 1, \\ 2(x-1) & \text{if } 1 \le x \le 2. \end{cases}$$

Show that f is integrable on $[0,2]$.

(a) Determine the function $G(x) = \int_0^x f(t)\, dt$ on $[0,2]$.
(b) Sketch the graph of $y = G(x)$ on $[0,2]$.
(c) Where is $G(x)$ continuous?
(d) Where is $G(x)$ differentiable?
(e) Determine G' at the points of differentiability, in particular, $G'_+(0)$, $G'_-(2)$, $G'_+(1)$, and $G'_-(1)$. Determine whether $G'_+(1) = G'_-(1)$. (For the definitions of left and right derivatives, we refer to Section 3.3.2).

8. Consider

$$f(x) = \begin{cases} 1 & \text{for } x > 0, \\ 0 & \text{for } x = 0, \\ -1 & \text{for } x < 0, \end{cases}$$

and set $F(x) = |x|$ for $x \in \mathbb{R}$. Then $F'(x) = f(x)$ for $x \neq 0$, and F is not differentiable at $x = 0$. Show that

$$\int_{-1}^{0} f(x)\,\mathrm{d}x = F(0) - F(-1) \text{ and } \int_{0}^{1} f(x)\,\mathrm{d}x = F(1) - F(0).$$

Explain why neither form of the fundamental theorems can be applied directly to prove this.

9. State suitable conditions on $a(x)$, $b(x)$, and $f(x)$ for being able to determine

$$\frac{\mathrm{d}}{\mathrm{d}x}\left(\int_{a(x)}^{b(x)} f(t)\,\mathrm{d}t\right).$$

10. Prove that if $f(x)$ is monotonic on $[a, b]$, then there exists $c \in [a, b]$ such that

$$\int_{a}^{b} f(x)\,\mathrm{d}x = f(a)(c - a) + f(b)(b - c).$$

11. Show that there exists a real number c with $0 < c < 1$ such that

$$\lim_{n\to\infty}\left(\frac{1}{n}\sum_{k=1}^{n} \mathrm{e}^{\sqrt{k/n}}\right) = \mathrm{e}^{\sqrt{c}}.$$

12. Show that if $0 < a < b < \infty$, then

$$\left|\int_{a}^{b} \frac{\sin x}{x}\,\mathrm{d}x\right| \leq \frac{2}{a}.$$

13. Prove the following:

(a) $\displaystyle \frac{28}{81} < \int_{0}^{1/3} \mathrm{e}^{x^2}\,\mathrm{d}x < \frac{3}{8}$. (b) $\displaystyle \frac{1}{2} < \int_{0}^{1} \frac{\mathrm{d}x}{\sqrt{4 - x^2 + x^\alpha}} < \frac{\pi}{6} \quad (\alpha > 2)$.

(c) $\displaystyle \lim_{x\to 0} \frac{1}{x^2}\int_{0}^{x^2} \mathrm{e}^{\sqrt{1+t^2}}\,\mathrm{d}t = \mathrm{e}$. (d) $\displaystyle \frac{\pi^2}{9} < \int_{\pi/6}^{\pi/2} \frac{x}{\sin x}\,\mathrm{d}x < \frac{2\pi^2}{9}$.

14. Show that $\displaystyle \lim_{x\to 0} \frac{\int_{-2x}^{2x} f(t)\,\mathrm{d}t}{\int_{0}^{3x} f(t+2)\,\mathrm{d}t} = \frac{f(0)}{f(2)}$ if f is continuous on $\mathbb{R}$.

15. Compute the value of

$$\lim_{x\to 0} \frac{1}{x^4}\int_{0}^{x} \frac{t^3}{1 + t^2}\,\mathrm{d}t.$$

16. Suppose that f is continuous on $[0,1]$. Show that

$$\int_0^1 x^3 f(x)\, dx = \frac{1}{4} f(c)$$

for some $c \in [0,1]$.

17. Prove that if $f(x)$ is continuous and increasing on $[a,b]$ and $c = (a+b)/2$, then

$$f(a) + f(c) \le \frac{2}{b-a} \int_a^b f(x)\, dx \le f(c) + f(b).$$

18. Find c that satisfies the conclusion of the mean value theorem for integrals for the functions defined below. If you cannot find such a value, explain why the theorem does not apply. Find also the average/mean value of these functions on the prescribed interval:
   (a) $f(x) = \cos x$ on $\left[-\frac{\pi}{2}, \frac{\pi}{2}\right]$. **(b)** $f(x) = \tan x$ on $[0,2]$.

19. Using integration by parts, evaluate the following integrals:
   (a) $\displaystyle\int_1^3 x \log x\, dx$. **(b)** $\displaystyle\int_0^5 x^4 e^{x^3}\, dx$. **(c)** $\displaystyle\int_0^x e^t \cos t\, dt$.

20. Evaluate the following definite integrals:
   (a) $\displaystyle\int_1^2 \frac{x^3+1}{x^2}\, dx$. **(b)** $\displaystyle\int_1^2 \frac{x^2+x-1}{\sqrt{x}}\, dx$. **(c)** $\displaystyle\int_1^2 \frac{x^3-1}{x}\, dx$.
   (d) $\displaystyle\int_0^1 \frac{dx}{\sqrt{4-x^2}}$. **(e)** $\displaystyle\int_0^2 (2x - |x-1|)\, dx$. **(f)** $\displaystyle\int_{-1}^1 (x+|x|)\, dx$.

21. Using the integration by parts, evaluate the integral $\displaystyle\int_1^4 \log x\, dx$.

22. Using the definition of L and E, show that e lies between 2 and 3.

23. Using Definition 6.54 and Theorem 6.51**(d)**, can we conclude that

$$\frac{x-1}{x} \le L(x) \le x-1,$$

where $L(x) = \lim_{n\to\infty} \frac{x^{1/n}-1}{1/n}$?

7

Improper Integrals and Applications of Riemann Integrals

So far, our theory of integration has dealt with the definite integral

$$\int_a^b f(x)\,\mathrm{d}x,$$

where the integrand $f(x)$ is bounded on a closed and bounded interval $[a, b]$. Thus, definite integrals have finite limits of integration, and $f(x)$ has a finite range. However, in many interesting applications (e.g., in physics, economics, statistics, and other applied areas), we also encounter problems that fail to satisfy one or both of these conditions. So we need to discuss a way to integrate functions that are unbounded or are defined on an unbounded interval. Our investigation along such lines leads to what are called *improper integrals*. There are basically two types of improper integrals; others can be developed from them. In Section 7.1, we consider functions f defined on $[a, b)$ with $b = \infty$, or $(a, b]$ with $a = -\infty$ or $(-\infty, \infty)$ or functions f that are unbounded in a neighborhood of a finite number of points on the interval of integration $[a, b]$ with a and b finite. The main aim of this section is to discuss the convergence and divergence of the corresponding improper integrals $\int_a^b f(x)\,\mathrm{d}x$. At the end, we consider the most important and interesting examples of improper integrals, namely the *gamma function* and the *beta function*. These two functions play important roles in analysis. Finally, we discuss certain important integrals in connection with the convergence of series of nonnegative numbers. Our particular emphasize will be on the integral test, the convergence of harmonic p-series, and the Abel–Pringsheim divergence test. In Section 7.2, we deal with a number of applications of the Riemann integral, such as in finding areas of regions bounded by curves and the arc length of plane curves.

7.1 Improper Integrals

We consider integrals of the form $\int_a^b f(x)\,\mathrm{d}x$ having one of the following forms:

S. Ponnusamy, *Foundations of Mathematical Analysis*,
DOI 10.1007/978-0-8176-8292-7_7,

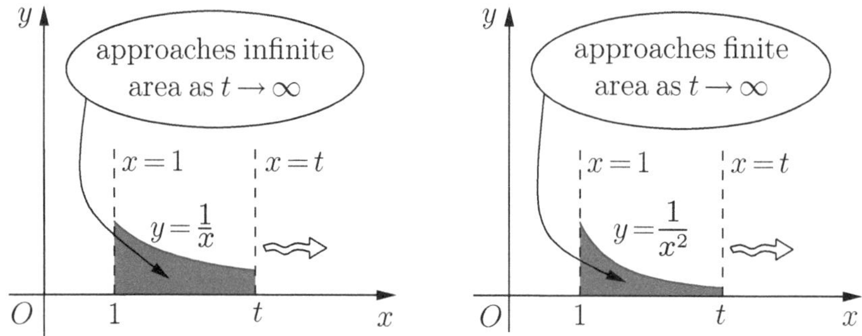

Fig. 7.1. On the convergence of $\int_1^\infty 1/x^p \, dx$ for $p = 1, 2$.

- a is finite but $b = \infty$, or $a = -\infty$ but b is finite, or $a = -\infty$ and $b = \infty$.
- f is unbounded near a finite number of points on the interval of integration $[a, b]$, where a and b are finite.

If one of the above cases occurs, then the integral in question is called an *improper integral.* Our main investigation in this section is to give meaning to such integrals and to evaluate them.

7.1.1 Improper Integrals over an Unbounded Interval

One of the central problems in analysis has to do with the definition of "infinity." A general question is how to deal with infinite quantities. To consider one such problem, we introduce the concept of an integral that is defined on $\mathbb{R}$, or on a half-line of the form $[a, \infty)$ or $(-\infty, a]$. If $f(x) \geq 0$, then $\int_a^\infty f(x)\, dx$ can be thought of as the area under the curve $y = f(x)$ on the unbounded interval $[a, \infty)$.

Before dealing with the general case, let us try to discuss the evaluation of the integral $\int_1^\infty (1/x^2)\, dx$. To do so, we need to find a reasonable strategy (see Figure 7.1).

Clearly, the region under the curve $y = 1/x^2$ for $x \geq 1$ is unbounded. Does this mean that the area is also infinite? A natural approach is to begin by computing the integral from 1 to t, where t is some "large" number and then see what happens as $t \to \infty$. We have

$$\int_1^t \frac{dx}{x^2} = -\frac{1}{x}\bigg|_1^t = -\frac{1}{t} + 1,$$

and we then take the limit as $t \to \infty$. This suggests that the region under $y = 1/x^2$ for $x \geq 1$ actually has finite area approaching 1 as t gets larger and larger. Analytically, it is reasonable to conclude that

$$\int_1^\infty \frac{dx}{x^2} = \lim_{t\to\infty} \int_1^t \frac{dx}{x^2} = \lim_{t\to\infty} \left[-\frac{1}{t} + 1\right] = 1.$$

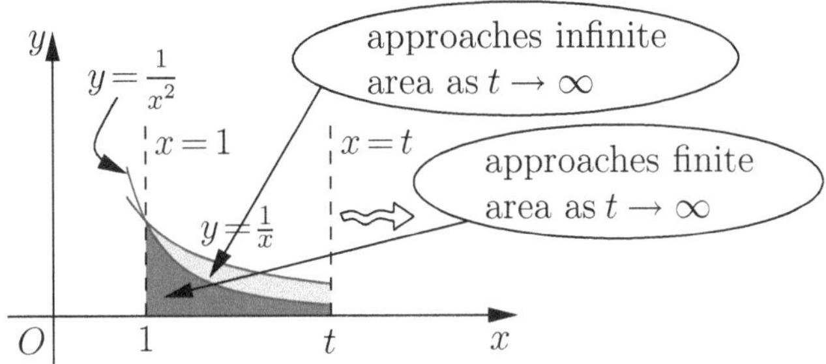

Fig. 7.2. A finite and an infinite area.

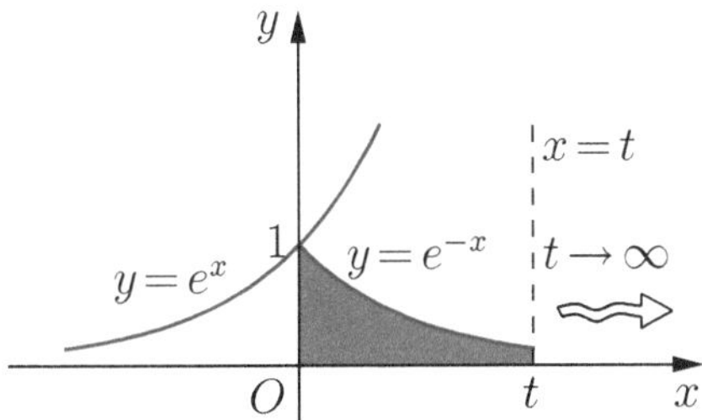

Fig. 7.3. Area under $y = \mathrm{e}^{-x}$ on $[0, t)$ for large t.

Similarly, if we use the same procedure, then we see that

$$\int_1^\infty \frac{\mathrm{d}x}{x} = \lim_{t\to\infty} \int_1^t \frac{\mathrm{d}x}{x} = \lim_{t\to\infty} \log x \,\Big|_1^t = \lim_{t\to\infty} [\log t - \log 1] = \infty$$

(see Figure 7.2). We also have (see Figure 7.3),

$$\int_0^\infty \mathrm{e}^{-x}\,\mathrm{d}x = \lim_{t\to\infty} \int_0^t \mathrm{e}^{-x}\,\mathrm{d}x = \lim_{t\to\infty} \left(-\mathrm{e}^{-x}\right)\Big|_0^t = \lim_{t\to\infty} \left(1 - \mathrm{e}^{-t}\right) = 1.$$

On the other hand, $\int_0^\infty \mathrm{e}^x\,\mathrm{d}x = \infty$. Now we give a definition using the idea of these examples.

Definition 7.1 (Improper integrals—first type). *Let a and b be fixed real numbers.*

(a) *Let f be a function having the property that $\int_a^N f(x)\,\mathrm{d}x$ exists as a Riemann integral for every N with $N \geq a$. Then if $\lim_{N\to\infty} \int_a^N f(x)\,\mathrm{d}x$ exists, we define the improper integral*

$$\int_a^\infty f(x)\,\mathrm{d}x := \lim_{N\to\infty} \int_a^N f(x)\,\mathrm{d}x. \tag{7.1}$$

As with infinite series, the improper integral is said to converge if this limit exists and is finite. If the limit in (7.1) *does not exist, or if it exists but is infinite, then we say that the improper integral is divergent. The improper integral* $\int_a^\infty f(x)\,\mathrm{d}x$ *is said to absolutely convergent if* $\int_a^\infty |f(x)|\,\mathrm{d}x$ *is convergent. If an improper integral converges but fails to converge absolutely, it is said to converge conditionally.*

(b) *If the interval of integration is* $(-\infty, b]$, *then* (7.1) *must be modified in an obvious way:*

$$\int_{-\infty}^{b} f(x)\,\mathrm{d}x := \lim_{M\to-\infty}\int_M^b f(x)\,\mathrm{d}x,$$

provided f *is a function having the property that* $\int_M^b f(x)\,\mathrm{d}x$ *exists for all* M *with* $M < b$. *The improper integral* $\int_{-\infty}^b f(x)\,\mathrm{d}x$ *is said to converge if the limit on the right exists (as a finite value).*

(c) *If the interval of integration is* $(-\infty, \infty)$, *then* (7.1) *is replaced by*

$$\int_{-\infty}^{\infty} f(x)\,\mathrm{d}x = \int_{-\infty}^{a} f(x)\,\mathrm{d}x + \int_a^\infty f(x)\,\mathrm{d}x, \tag{7.2}$$

provided both the integrals exist in the sense of the above two cases. Here we agree that $\infty + L = \infty$ *if* $L \neq -\infty$ *and* $-\infty + L = -\infty$ *if* $L \neq \infty$. *It is easy to see that every choice of* a *will give the same result.*

The terms convergent, divergent, absolutely convergent, conditionally convergent for the last two cases may be defined in a similar manner as in the first case.

Remark 7.2. It might happen that neither of the integrals on the right of (7.2) exists independently, but that the symmetric limit

$$\lim_{N\to\infty}\int_{-N}^{N} f(x)\,\mathrm{d}x \tag{7.3}$$

exists. This limit exists, for example, for any odd function whatsoever. The limit in (7.3), if it exists, is called the *Cauchy principal value* of $\int_{-\infty}^\infty f(x)\,\mathrm{d}x$, and is denoted by $PV\ \int_{-\infty}^\infty f(x)\,\mathrm{d}x$. Note that if $\int_{-\infty}^\infty f(x)\,\mathrm{d}x$ exists in the sense of (7.2), then it also exists in the sense of (7.3), i.e., Cauchy's principal value of $\int_{-\infty}^\infty f(x)\,\mathrm{d}x$ exists, and the two values are equal. However, the converse is not true in general. For instance, by our first definition, the improper integral $\int_{-\infty}^\infty x\,\mathrm{d}x$ diverges because $\int_0^\infty x\,\mathrm{d}x$ and $\int_{-\infty}^0 x\,\mathrm{d}x$ diverge. But since $\int_{-N}^N x\,\mathrm{d}x = 0$, the Cauchy principal value of $\int_{-\infty}^\infty x\,\mathrm{d}x$ is zero. This explains why we do not define

$$\int_{-\infty}^{\infty} f(x)\,\mathrm{d}x = \lim_{N\to\infty}\int_{-N}^{N} f(x)\,\mathrm{d}x$$

as the general definition of the improper integral. The problem here is that $\int_{-\infty}^{0} x\,dx = -\infty$ and $\int_{0}^{\infty} x\,dx = \infty$, and the expression $\infty - \infty$ is not defined. Similarly, since

$$\int_0^b \sin x\,dx = 1 - \cos b \quad \text{for all } b > 0$$

and the value of $1 - \cos b$ oscillates between 0 and 2 as $b \to \infty$, we see that neither $\int_0^\infty \sin x\,dx$ nor $\int_{-\infty}^0 \sin x\,dx$ exists. Consequently, $\int_{-\infty}^\infty \sin x\,dx$ does not converge. However, the limit

$$\lim_{N\to\infty} \int_{-N}^{N} \sin x\,dx$$

clearly exists and equals zero. Thus, 0 is the Cauchy principal value of $\int_{-\infty}^\infty \sin x\,dx$. Note that for any odd function f, the Cauchy principal value of $\int_{-\infty}^\infty f(x)\,dx$ is zero. ●

We have already shown that the improper integral $\int_1^\infty \frac{dx}{x^2}$ converges, whereas $\int_1^\infty \frac{dx}{x^2}$ diverges. Geometrically, this says that the area to the right of $x = 1$ under the curve $y = 1/x^2$ is finite, whereas the corresponding area under the curve $y = 1/x$ is infinite (see Figure 7.1).

Evaluation of improper integrals often uses L'Hôpital's rule, change of variable of integration, and integration by parts, as we shall see in a number of cases below.

Example 7.3. As illustrated above, to discuss the convergence of the integral $\int_a^\infty (1/x^p)\,dx$ where $a > 0$, we first evaluate it from a to N and then let N go to infinity. Indeed (see Figures 7.4 and 7.5),

$$\int_a^N \frac{dx}{x^p} = \begin{cases} \left.\dfrac{x^{-p+1}}{-p+1}\right|_a^N & \text{if } p \neq 1, \\ \log x \,\Big|_a^N & \text{if } p = 1, \end{cases} = \begin{cases} \dfrac{1}{1-p}\left[N^{1-p} - a^{1-p}\right] & \text{if } p \neq 1, \\ \log(N/a) & \text{if } p = 1. \end{cases}$$

If $p > 1$, then $N^{1-p} \to 0$ as $N \to \infty$. If $p < 1$, then $N^{1-p} \to \infty$ as $N \to \infty$. Consequently, for $a > 0$, we have

$$\int_a^\infty \frac{dx}{x^p} = \begin{cases} \dfrac{1}{(p-1)a^{p-1}} & \text{if } p > 1, \\ \infty & \text{if } p \leq 1. \end{cases}$$

Thus the improper integral diverges for $p \leq 1$, and converges for $p > 1$. ●

Example 7.4. Determine whether each of the following integrals converges or diverges

(a) $\displaystyle\int_0^\infty xe^{-4x}\,dx.$ (b) $\displaystyle\int_{-\infty}^\infty xe^{-x^2}\,dx.$ (c) $\displaystyle\int_1^\infty \frac{\log x}{x^2}\,dx.$

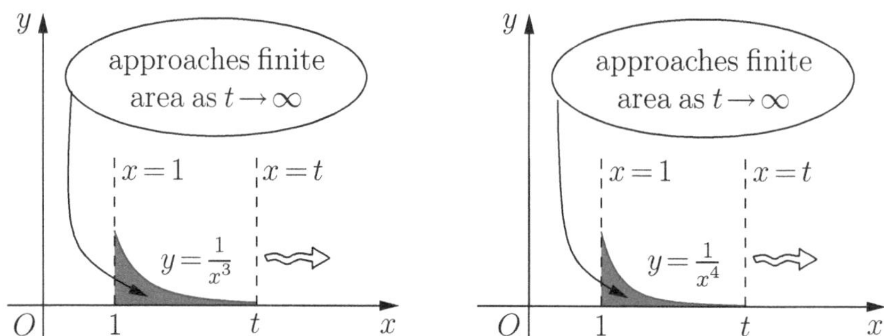

Fig. 7.4. Convergence of $\int_1^\infty 1/x^p \, dx$ for $p = 3, 4$.

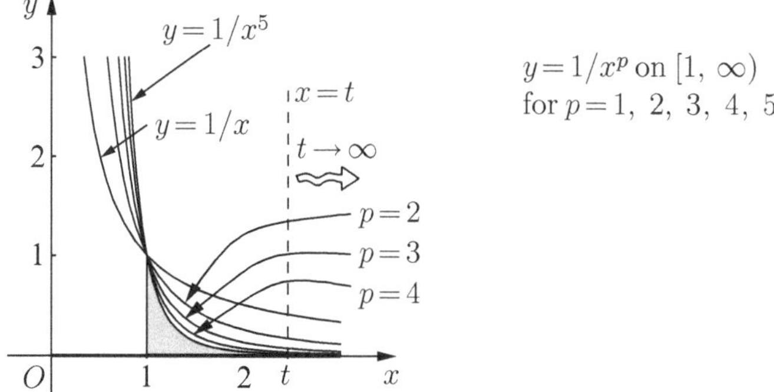

Fig. 7.5. Convergence of the integral $\int_1^\infty (1/x^p) \, dx$, for $p \in [1, \infty)$.

Solution. **(a)** We evaluate

$$\begin{aligned}
\int_0^\infty xe^{-4x} \, dx &= \lim_{N\to\infty} \int_0^N x d\left(\frac{e^{-4x}}{-4}\right) \\
&= \lim_{N\to\infty} \left[\left(-\frac{xe^{-4x}}{4}\right)\Big|_0^N + \int_0^N \frac{1}{4}e^{-4x} \, dx\right] \\
&= \lim_{N\to\infty} \left[-\frac{xe^{-4x}}{4} - \frac{e^{-4x}}{16}\right]\Big|_0^N \\
&= -\frac{1}{16} \lim_{N\to\infty} \left(\frac{4N+1}{e^{4N}}\right) + \frac{1}{16} \\
&= -\frac{1}{16} \lim_{N\to\infty} \left(\frac{4}{4e^{4N}}\right) + \frac{1}{16} = \frac{1}{16} \quad \text{(by l'Hôpital's rule.)}
\end{aligned}$$

(b) We see that $\int_{-\infty}^\infty xe^{-x^2} \, dx = 0$ because

$$\int_0^N xe^{-x^2} \, dx = \frac{1 - e^{-N^2}}{2} \quad \text{and} \quad \int_{-M}^0 xe^{-x^2} \, dx = \frac{e^{-M^2} - 1}{2}.$$

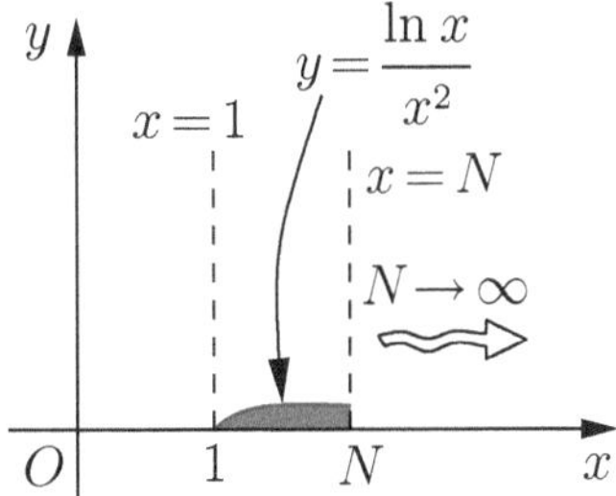

Fig. 7.6. Area under $y = (\ln x)/x^2$ on $[1, N)$ for large N.

(c) We see that (see Figure 7.6)

$$\int_1^N \frac{\log x}{x^2}\,\mathrm{d}x = \int_1^N \log x\,\mathrm{d}\left(-\frac{1}{x}\right) = \left[-\frac{\log x}{x} - \frac{1}{x}\right]\Bigg|_1^N = 1 - \frac{1}{N} - \frac{\log N}{N},$$

which approaches 1 as $n \to \infty$. Hence, the improper integral $\int_1^\infty (\log x/x^2)\,\mathrm{d}x$ converges and has the value 1. ●

In many cases it suffices to determine whether certain improper integrals of a nonnegative function converge or diverge, and to estimate their values. The following result is of course an analogue of the comparison tests for sequences and series (see Theorems 5.27 and 5.33).

Theorem 7.5 (Direct comparison test). *If f and g are two continuous functions on $[a, \infty)$ such that*

$$0 \le f(x) \le Mg(x) \quad \textit{for all } x \ge a,$$

for some constant $M > 0$, then we have the following:

- *$\int_a^\infty f(x)\,\mathrm{d}x$ converges if $\int_a^\infty g(x)\,\mathrm{d}x$ converges. In this case, we write $\int_a^\infty f(x)\,\mathrm{d}x \le M\int_a^\infty g(x)\,\mathrm{d}x$.*
- *$\int_a^\infty g(x)\,\mathrm{d}x$ diverges if $\int_a^\infty f(x)\,\mathrm{d}x$ diverges. In other words, $\int_a^\infty g(x)\,\mathrm{d}x = \infty$ if $\int_a^\infty f(x)\,\mathrm{d}x = \infty$.*

Proof. Note that for $b > a$,

$$\int_a^b f(x)\,\mathrm{d}x \le M\int_a^b g(x)\,\mathrm{d}x,$$

and therefore the conclusions follow by letting $b \to \infty$. ∎

Clearly, a similar statement holds for other types of improper integrals. We omit the corresponding formulations.

Theorem 7.6. *If the improper integral $\int_a^\infty |f(x)|\,\mathrm{d}x$ converges where f is bounded and integrable on $[a, N]$ for each $N > a$, then $\int_a^\infty f(x)\,\mathrm{d}x$ converges. The converse is not true.*

Proof. Since $0 \le |f(x)| - f(x) \le 2|f(x)|$ and

$$0 \le S_N = \int_a^N (|f(x)| - f(x))\,dx \le 2\int_a^N |f(x)|\,dx \le 2\int_a^\infty |f(x)|\,dx$$

for all $N \ge a$, by the direct comparison test, it is clear that absolute integrability of f implies that $|f(x)| - f(x)$ is integrable on $[a,N]$. Note that because $\{S_N\}$ is an increasing sequence bounded above, the sequence $\{S_N\}$ converges. Therefore, the improper integral $\int_a^\infty f(x)\,dx$ converges. For the converse, we refer to Example 7.8(**c**). ∎

Example 7.7. Examine the absolute convergence of $\int_1^\infty f(x)\,dx$ if

$$f(x) = \frac{\sin x}{\sqrt{(x-1)^3 + x - 1}}.$$

Solution. Since $|f(x)| \le \dfrac{1}{\sqrt{(x-1)^3 + x - 1}} = g(x)$ and $\int_1^\infty g(x)\,dx$ is convergent, $\int_1^\infty f(x)\,dx$ is convergent (absolutely). ●

Example 7.8. Prove that each of the following integrals converges:

(**a**) $\displaystyle\int_{-\infty}^\infty e^{-x^2}\,dx.$ (**b**) $\displaystyle\int_{-\infty}^\infty \frac{dx}{1+x^2}.$ (**c**) $\displaystyle\int_1^\infty \frac{\sin x}{x^p}\,dx$ for $p > 0$.

Show also that $\int_1^\infty \left|\frac{\sin x}{x^p}\right| dx$ converges for $p > 1$ and diverges for $0 < p \le 1$. This provides an example of a conditionally convergent improper integral.

Solution. (**a**) Since e^{-x^2} is continuous on $\mathbb{R}$, $\int_a^b e^{-x^2}\,dx$ is integrable for all finite a and b. Therefore, it suffices to discuss the convergence of $\int_1^\infty e^{-x^2}\,dx$ and $\int_{-\infty}^{-1} e^{-x^2}\,dx$. Again the change of variable $t = -x$ shows that

$$\int_{-N}^{-1} e^{-x^2}\,dx = \int_1^N e^{-t^2}\,dt,$$

and so we need to discuss only the integral $\int_1^\infty e^{-x^2}\,dx$. Note that if $x \ge 1$, then $x^2 \ge x$, and so $e^{-x^2} \le e^{-x}$. Since

$$\int_1^\infty e^{-x}\,dx = \lim_{N\to\infty} \left.\frac{e^{-x}}{-1}\right|_1^N = \lim_{N\to\infty} (e^{-1} - e^{-N}) = \frac{1}{e},$$

it follows by the direct comparison test that $\int_1^\infty e^{-x^2}\,dx$ converges. Hence $\int_{-\infty}^\infty e^{-x^2}\,dx$ converges.

(**b**) We see that

$$\int_0^N \frac{dx}{1+x^2} = \left.\tan^{-1} x\right|_0^N = \tan^{-1} N - \tan^{-1} 0 \to \frac{\pi}{2} \text{ as } N \to \infty$$

and similarly,

$$\int_{-M}^{0} \frac{dx}{1+x^2} = \tan^{-1} x\Big|_{-M}^{0} = 0 - \tan^{-1}(-M) \to \frac{\pi}{2} \text{ as } M \to \infty.$$

Consequently,

$$\int_{-\infty}^{\infty} \frac{dx}{1+x^2} = \lim_{M\to\infty} \int_{-M}^{0} \frac{dx}{1+x^2} + \lim_{N\to\infty} \int_{0}^{N} \frac{dx}{1+x^2} = \frac{\pi}{2} + \frac{\pi}{2} = \pi.$$

(c) For $x > 0$, we have

$$\left|\frac{\sin x}{x^p}\right| \le \frac{1}{x^p},$$

and the direct comparison test does not yield the desired result, since $\int_1^\infty \frac{dx}{x^p}$ diverges for $0 < p \le 1$. However, the comparison test implies that $\int_1^\infty |x^{-p} \sin x|\, dx$ is convergent for $p > 1$, which means that $\int_1^\infty x^{-p} \sin x\, dx$ is absolutely convergent (and hence convergent) for $p > 1$. On the other hand, to show that $\int_1^\infty \frac{\sin x}{x^p}\, dx$ converges for $0 < p \le 1$, we may use integration by parts and obtain

$$\begin{aligned}\int_1^N \frac{\sin x}{x^p}\, dx &= -\frac{\cos x}{x^p}\Big|_1^N - p\int_1^N \frac{\cos x}{x^{p+1}}\, dx \\ &= -\frac{\cos N}{N^p} + \cos 1 - p\int_1^N \frac{\cos x}{x^{p+1}}\, dx.\end{aligned}$$

Since $\int_1^\infty \frac{dx}{x^{p+1}}$ converges for $p > 0$ and

$$\left|\frac{\cos x}{x^{p+1}}\right| \le \frac{1}{x^{p+1}},$$

it follows that $\int_0^\infty \frac{\cos x}{x^{p+1}}\, dx$ converges (absolutely) for $p > 0$, and therefore, letting $N \to \infty$, we see that

$$\int_1^\infty \frac{\sin x}{x^p}\, dx = \cos 1 - p\int_1^\infty \frac{\cos x}{x^{p+1}}\, dx \quad \text{for } p > 0,$$

and the result follows. Note that the latter approach proves the convergence of $\int_1^\infty \frac{\sin x}{x^p}\, dx$ for all $p > 0$.

To show that $\int_1^\infty \frac{\sin x}{x^p}\, dx$ does not converge absolutely for $0 < p \le 1$, according to Theorem 7.5, it suffices to prove the result only for the case $p = 1$. Let $n \ge 3$. Then

$$\begin{aligned} I_n = \int_1^{n\pi} \frac{|\sin x|}{x}\, dx > \int_\pi^{n\pi} \frac{|\sin x|}{x}\, dx &= \sum_{k=1}^{n-1} \int_{k\pi}^{(k+1)\pi} \frac{|\sin x|}{x}\, dx \\ &> \sum_{k=1}^{n-1} \frac{1}{(k+1)\pi} \int_{k\pi}^{(k+1)\pi} |\sin x|\, dx \\ &= \frac{2}{\pi} \sum_{k=1}^{n-1} \frac{1}{k+1} \\ &\to \infty \quad \text{as } n \to \infty. \end{aligned}$$

Consequently, $\int_1^\infty \frac{|\sin x|}{x}\, dx$ diverges. Since $\frac{1}{x^p} \geq \frac{1}{x}$ for $p \leq 1$, by Theorem 7.5, we can now conclude that $\int_1^\infty \frac{|\sin x|}{x^p}\, dx$ diverges for $p \leq 1$. ●

Remark 7.9. We notice that $\lim_{x\to 0+} \frac{\sin x}{x^p}$ exists if $0 < p \leq 1$ (because $|\sin x| \leq |x|$ on $\mathbb{R}$), and therefore $\int_0^a \frac{\sin x}{x^p}\, dx$ converges for all $a > 0$ and for $0 < p \leq 1$. Thus,

$$\int_0^\infty \frac{\sin x}{x^p}\, dx = \int_0^1 \frac{\sin x}{x^p}\, dx + \int_1^\infty \frac{\sin x}{x^p}\, dx$$

converges for $0 < p \leq 1$, because the second integral on the right converges by Example **7.8(c)**. ●

7.1.2 Improper Integrals of Unbounded Functions

We often encounter integrals in which the integrand has a singularity somewhere in the domain of integration. So we move on to a discussion of the second type of improper integrals (with unbounded integrands). A function f is *unbounded* at c if $|f(x)|$ has arbitrarily large values near c. Geometrically, this occurs when the graph of f has a vertical asymptote at $x = c$. For example, if f is continuous on $(a, b]$ but $|f(x)| \to \infty$ as $x \to a+$, the graph of $y = f(x)$ approaches the vertical line $x = a$, as shown in Figure 7.7.

If f is unbounded at c and $a \leq c \leq b$, then the Riemann integral $\int_a^b f(x)\, dx$ is not even defined (because boundedness of f is essential for the integrability of f). However, it may still be possible to define $\int_a^b f(x)\, dx$ as an improper integral in certain cases. For example, consider

$$f(x) = 1/\sqrt{x} \quad \text{for} \quad 0 < x \leq 1.$$

Then f is unbounded at $x = 0$, and so $\int_0^1 f(x)\, dx$ is not defined. However, $f(x)$ is continuous on every interval $[t, 1]$ for $t > 0$, as shown in Figure 7.8. So our basic strategy is to evaluate $\int_t^1 \frac{dx}{\sqrt{x}}$ and see what happens as $t \to 0+$. Thus, if t is a small positive real number, then we have

$$I(t) = \int_t^1 x^{-1/2}\, dx = 2\sqrt{x}\Big|_t^1 = 2 - 2\sqrt{t}.$$

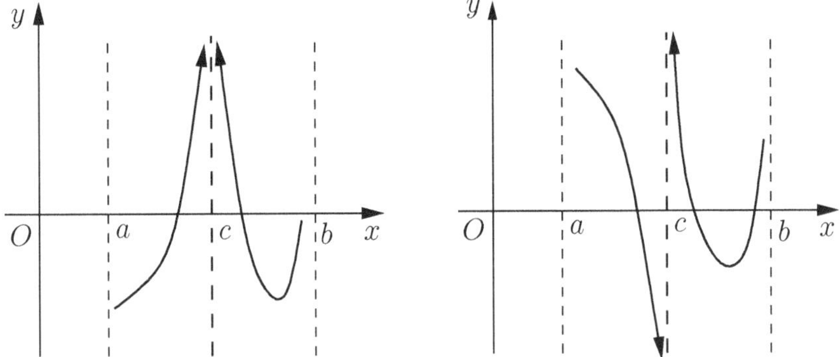

Fig. 7.7. Two functions that are unbounded at c.

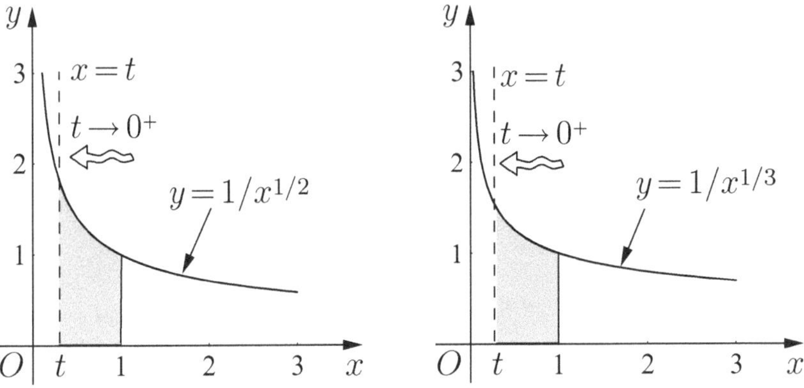

Fig. 7.8. Convergence of $\int_0^1 (1/x^{1/p})\,dx$ for $p = 1/2, 1/3$.

If we let $t \to 0+$, we see that $I(t) \to 2$. Hence it is natural to define

$$\int_0^1 \frac{dx}{\sqrt{x}} = \lim_{t\to 0+} \int_t^1 \frac{dx}{\sqrt{x}} = 2.$$

The same idea may be used if the integrand is unbounded at the right end-point. A general case is considered in Example 7.15.

Also, we adopt a similar procedure when functions have a discontinuity but not necessarily an infinite discontinuity. For example, consider

$$g(x) = x \log x, \quad 0 < x \le 1.$$

Then the function $g(x)$ is not defined at $x = 0$, although

$$\lim_{x\to 0+} x \log x = -\lim_{x\to 0+} \frac{\log(1/x)}{1/x} = -\lim_{x\to 0+} \frac{1/x}{-1/x^2} = 0.$$

In this case, we define the integral of $g(x)$ on $[0,1]$ by

$$\int_0^1 x\log x\,\mathrm{d}x = \lim_{t\to 0+}\int_t^1 \log x\,\mathrm{d}\left(x^2/2\right) = \lim_{t\to 0+}\left.\frac{x^2\log x}{2} - \frac{x^2}{4}\right|_t^1 = -\frac{1}{4}.$$

We remark that this does not fall in the improper integral category.

Here is the formal definition of improper integrals in cases in which the integrand tends to $\pm\infty$ at some point in the interval of integration.

Definition 7.10 (Improper integrals—second type). *Let a and b be finite real numbers.*

(a) *If $\int_t^b f(x)\,\mathrm{d}x$ exists for all t such that $a < t \leq b$, and if f is unbounded at a (i.e., $|f(x)| \to \infty$ as $x \to a+$), then we define*

$$\int_a^b f(x)\,\mathrm{d}x = \lim_{t\to a+}\int_t^b f(x)\,\mathrm{d}x := \lim_{\epsilon\to 0+}\int_{a+\epsilon}^b f(x)\,\mathrm{d}x,$$

provided the limit exists. If the limit exists (as a finite number), we say that the improper integral $\int_a^b f(x)\,\mathrm{d}x$ converges; otherwise, the improper integral diverges.

(b) *If $\int_a^t f(x)\,\mathrm{d}x$ exists for all t such that $a \leq t < b$ and f is unbounded at b (i.e., $|f(x)| \to \infty$ as $x \to b-$), then*

$$\int_a^b f(x)\,\mathrm{d}x = \lim_{t\to b-}\int_a^t f(x)\,\mathrm{d}x := \lim_{\epsilon\to 0+}\int_a^{b-\epsilon} f(x)\,\mathrm{d}x,$$

provided the limit exists. If the limit exists (as a finite number), we say that the improper integral $\int_a^b f(x)\,\mathrm{d}x$ converges; otherwise, the improper integral diverges.

(c) *If f is unbounded at an interior point c (i.e., f has a vertical asymptote at c), where $a < c < b$, and if the integrals $\int_a^{c-\epsilon} f(x)\,\mathrm{d}x$ and $\int_{c+\eta}^b f(x)\,\mathrm{d}x$ exist for $0 < \epsilon < c-a$ and $0 < \eta < b-c$, then*

$$\begin{aligned}\int_a^b f(x)\,\mathrm{d}x &=: \int_a^c f(x)\,\mathrm{d}x + \int_c^b f(x)\,\mathrm{d}x \\ &= \lim_{\epsilon\to 0+}\int_a^{c-\epsilon} f(x)\,\mathrm{d}x + \lim_{\eta\to 0+}\int_{c+\eta}^b f(x)\,\mathrm{d}x, \qquad (7.4)\end{aligned}$$

provided the limits exist. If both limits on the right exist (as a finite number), then we say that the improper integral $\int_a^b f(x)\,\mathrm{d}x$ converges; otherwise, the improper integral diverges. That is, the integral on the left diverges if either (or both) of the integrals on the right diverge. It is also possible for an integral to be improper because of an infinite discontinuity at a finite number of points in the interval $[a,b]$. In such cases, the same strategy is followed.

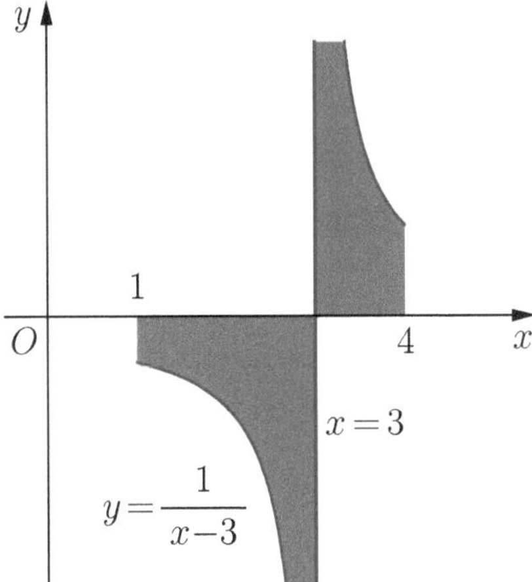

Fig. 7.9. Graph of $y = \dfrac{1}{x-3}$ on $[1, 4]$.

The absolute and conditional convergence of improper integrals of the second type may be defined similarly.

Remark 7.11. If f has a finite number of (infinite) discontinuities, say at $c_1, c_2, c_3, \dots, c_n$ $(c_1 < c_2 < \cdots < c_n)$, then we write

$$\int_{-\infty}^{\infty} f(x)\,\mathrm{d}x = \sum_{k=0}^{n} \int_{c_k}^{c_{k+1}} f(x)\,\mathrm{d}x \quad (c_0 = -\infty,\ c_{n+1} = \infty)$$

and say that the improper integral $\int_{-\infty}^{\infty} f(x)\,\mathrm{d}x$ is convergent, provided that each $\int_{c_k}^{c_{k+1}} f(x)\,\mathrm{d}x$ is convergent for $k = 0, 1, 2, \dots, n$ in the sense of Definitions 7.1 and 7.10. A similar procedure is applicable for integrals of the type $\int_a^{\infty} f(x)\,\mathrm{d}x$ and $\int_{-\infty}^{b} f(x)\,\mathrm{d}x$ when $f(x)$ satisfies a similar condition on $[a, \infty)$ and $(-\infty, b]$, respectively.

Remark 7.12. Sometimes we fail to notice an (infinite) discontinuity at an interior point. Find the faults in the following analysis(see Figure 7.9):

$$\int_1^4 \frac{\mathrm{d}x}{x-3} = \log|x-3|\,\Big|_1^4 = \log 1 - \log 2 = -\log 2.$$

Similarly, it is wrong to conclude that

$$\int_1^4 \frac{\mathrm{d}x}{(x-3)^2} = -\frac{1}{x-3}\Big|_1^4 = -\left(\frac{1}{2} + 1\right) = -\frac{3}{2},$$

which is clearly absurd, because $f(x) = (x-3)^{-2}$ is never negative. These examples show that mistakes such as these lead to the conclusion that the corresponding improper integrals converge, and so an improper integral should never be treated simply as an ordinary integral. Also, one must be cautious in using computer software with improper integrals, because it may not detect that the integral is improper. ●

Remark 7.13. Consider an improper integral $\int_a^b f(x)\,dx$, where f is unbounded at $c \in (a,b)$. It might happen that neither of the limits on the right of (7.4) exists as $\epsilon \to 0+$ and $\eta \to 0+$ independently, but that

$$\lim_{\epsilon\to 0+}\left[\int_a^{c-\epsilon} f(x)\,dx + \int_{c+\epsilon}^b f(x)\,dx\right] \tag{7.5}$$

exists. *Note that we do not define the improper integral of this kind by* (7.5). However, when the limit in (7.5) exists, it is called the *Cauchy principal value* of $\int_a^b f(x)\,dx$, and is denoted for brevity by $PV\ \int_a^b f(x)\,dx$. For example, $PV\ \int_{-1}^1 (1/x)\,dx = 0$, but $\int_{-1}^1 (1/x)\,dx$ does not exist. We note that if the limit exists in (7.4), then it also exists in the sense of (7.5), i.e., the Cauchy principal value of $\int_a^b f(x)\,dx$ exists and the two limits are equal. However, the converse is not true in general (see Example 7.14(**c**)). Also, for $f(x) \geq 0$ on $[a,b]$, the converse holds. ●

Example 7.14. Evaluate each of the following improper integrals if it exists.

(**a**) $\displaystyle\int_0^1 \frac{dx}{(x-1)^{2/3}}$. (**b**) $\displaystyle\int_1^2 \frac{dx}{(2-x)^2}$. (**c**) $\displaystyle\int_0^3 \frac{dx}{x-2}$.

Solution. (**a**) The function $f(x) = (x-1)^{-2/3}$ is unbounded at the right endpoint of the interval of integration (i.e., f has a vertical asymptote at the right end of the interval) and is continuous on $[0,\epsilon]$ for every ϵ with $0 < \epsilon < 1$. We find that (see Figure 7.10)

$$\lim_{\epsilon\to 1-}\int_0^\epsilon \frac{dx}{(x-1)^{2/3}} = \lim_{\epsilon\to 1-}\left[3(x-1)^{1/3}\right]\Big|_0^\epsilon = 3\lim_{\epsilon\to 1-}[(\epsilon-1)^{1/3}-(-1)] = 3.$$

That is, the given improper integral converges and has the value 3.

(**b**) Next we have (see Figure 7.10)

$$\int_1^2 \frac{dx}{(2-x)^2} = \lim_{\epsilon\to 0+}\int_1^{2-\epsilon}\frac{dx}{(2-x)^2} = \lim_{\epsilon\to 0+}\frac{1}{2-x}\Big|_1^{2-\epsilon} = \lim_{\epsilon\to 0+}\left(\frac{1}{\epsilon}-1\right),$$

which is ∞, and so the given integral diverges.

(**c**) The given integral is improper because the integrand is unbounded at the interior point $x=2$. We write

$$\int_0^3 \frac{dx}{x-2} = \int_0^2 \frac{dx}{x-2} + \int_2^3 \frac{dx}{x-2} = \lim_{\epsilon\to 2-}\int_0^\epsilon \frac{dx}{x-2} + \lim_{\epsilon\to 2+}\int_\epsilon^3 \frac{dx}{x-2}.$$

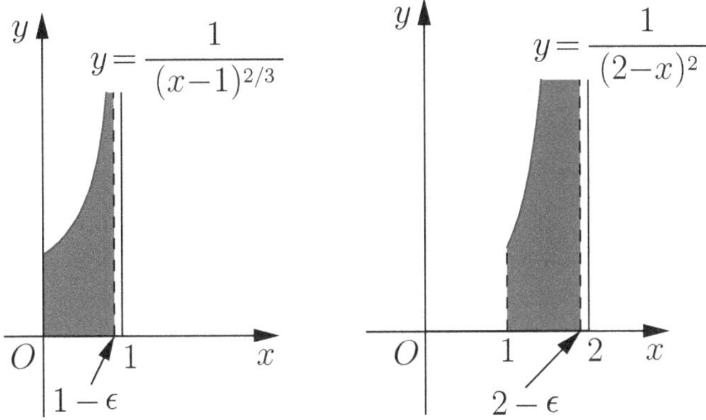

Fig. 7.10. Graphs of $y = \dfrac{1}{(x-1)^{2/3}}$ on $[0, 1)$, and $y = \dfrac{1}{(2-x)^2}$ on $[1, 2)$.

If either of these limits fails to exist, then the original integral diverges. However, because

$$\lim_{\epsilon \to 2-} \int_0^{\epsilon} \frac{dx}{x-2} = \lim_{\epsilon \to 2-} \log|x-2| \Big|_0^{\epsilon} = \lim_{\epsilon \to 2-} [\log|\epsilon - 2| - \log 2] = -\infty,$$

we find that the original integral diverges. On the other hand, we note that

$$\lim_{\epsilon \to 0+} \left[\int_0^{2-\epsilon} \frac{dx}{x-2} + \int_{2+\epsilon}^{3} \frac{dx}{x-2} \right] = \lim_{\epsilon \to 0+} [\log \epsilon - \log|-2| + \log 1 - \log \epsilon],$$

so that the Cauchy principal value of the given improper integral exists:

$$PV \int_0^3 \frac{dx}{x-2} = -\log 2. \qquad \bullet$$

Example 7.15. For what values of $p > 0$ is the improper integral $\displaystyle\int_0^1 \frac{dx}{x^p}$ convergent? To solve this, one can just use the change of variable $t = 1/x$ and apply Example 7.3. Alternatively, we simply compute

$$\begin{aligned}
\lim_{\epsilon \to 0+} \int_{\epsilon}^{1} \frac{dx}{x^p} &= \begin{cases} \displaystyle\lim_{\epsilon \to 0+} \frac{x^{-p+1}}{-p+1} \Big|_{\epsilon}^{1} & \text{if } p \neq 1, \\ \displaystyle\lim_{\epsilon \to 0+} \log x \Big|_{\epsilon}^{1} & \text{if } p = 1 \end{cases} \\
&= \begin{cases} \displaystyle\lim_{\epsilon \to 0+} \frac{1}{1-p} \left[1 - \epsilon^{1-p}\right] & \text{if } p \neq 1, \\ \displaystyle\lim_{\epsilon \to 0+} (-\log \epsilon) & \text{if } p = 1. \end{cases}
\end{aligned}$$

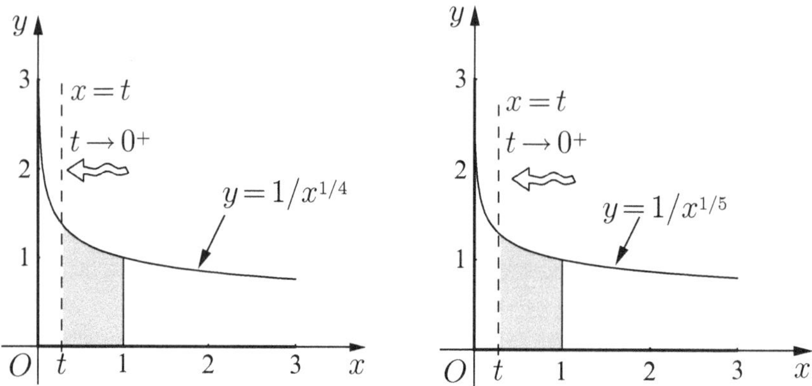

Fig. 7.11. Convergence of $\int_0^1 (1/x^{1/p})\,\mathrm{d}x$ for $p = 1/4, 1/5$.

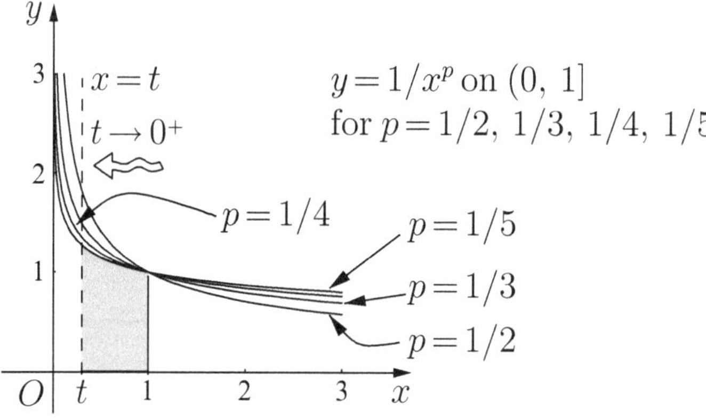

Fig. 7.12. Convergence of $\int_0^1 (1/x^{1/p})\,\mathrm{d}x$ for $p = 1/2, 1/3, 1/4, 1/5$.

Consequently (see Figures 7.11 and 7.12),

$$\int_0^1 \frac{\mathrm{d}x}{x^p} = \lim_{\epsilon \to 0+} \int_\epsilon^1 \frac{\mathrm{d}x}{x^p} = \begin{cases} \dfrac{1}{1-p} & \text{if } 0 < p < 1, \\ \infty & \text{if } p \geq 1. \end{cases}$$

Thus, the improper integral diverges for $p \geq 1$ and converges for $0 < p < 1$. Moreover, in view of the observation

$$\int_1^\infty \frac{\mathrm{d}y}{y^q} = \int_0^1 \frac{\mathrm{d}x}{x^p}, \quad p = 2 - q,$$

the desired conclusion follows quickly from Example 7.3.

More generally, it is straightforward to see that each of the improper integrals

$$\int_b^a \frac{\mathrm{d}x}{(x-b)^p} \quad \text{and} \quad \int_b^a \frac{\mathrm{d}x}{(a-x)^p}$$

converges if and only if $p < 1$. •

Example 7.16. Prove the following limits:

(a) $\lim_{n\to\infty} \frac{n}{(n!)^{1/n}} = \mathrm{e}$. (b) $\lim_{n\to\infty} \left[\left(1+\frac{1}{n}\right)\left(1+\frac{2}{n}\right)\cdots\left(1+\frac{4n}{n}\right)\right]^{1/n} = \frac{5^5}{\mathrm{e}^4}$.

Solution. **(a)** Set $S_n = (n^n/n!)^{1/n}$. Then (see also Example 2.61)

$$\log S_n = \frac{1}{n}\log\left(\frac{n}{n}\cdot\frac{n}{(n-1)}\cdot\frac{n}{(n-2)}\cdots\frac{n}{1}\right) = -\frac{1}{n}\sum_{k=1}^{n}\log\left(\frac{k}{n}\right).$$

Treating $\log S_n$ as a Riemann sum, we see that

$$\lim_{n\to\infty}\log S_n = -\int_0^1 \log x\,\mathrm{d}x := -\lim_{\epsilon\to 0+}\int_\epsilon^1 \log x\,\mathrm{d}x,$$

which is an improper integral. Note that $\log x$ is unbounded on $[0,1]$. Now we evaluate

$$\int_\epsilon^1 \log x\,\mathrm{d}x = x\log x\big|_\epsilon^1 - \int_\epsilon^1 x\cdot\frac{1}{x}\,\mathrm{d}x = -\epsilon\log\epsilon - (1-\epsilon),$$

and so by l'Hôpital's rule, we have

$$\lim_{\epsilon\to 0+}\int_\epsilon^1 \log x\,\mathrm{d}x = -1 + \lim_{\epsilon\to 0+}\frac{\log(1/\epsilon)}{1/\epsilon} = -1 + \lim_{\epsilon\to 0+}\frac{-1/\epsilon}{-1/\epsilon^2} = -1.$$

That is, $\log S_n \to 1$ as $n\to\infty$, and hence $S_n \to \mathrm{e}$ as $n\to\infty$.

(b) Set $S_n = \left[\left(1+\frac{1}{n}\right)\left(1+\frac{2}{n}\right)\left(1+\frac{3}{n}\right)\cdots\left(1+\frac{4n}{n}\right)\right]^{1/n}$. Then we see that

$$\log S_n = \frac{1}{n}\sum_{k=1}^{4n}\log\left(1+\frac{k}{n}\right) = 4\left[\frac{1}{4n}\sum_{k=1}^{4n}\log\left(1+4\left(\frac{k}{4n}\right)\right)\right],$$

and hence

$$\lim_{n\to\infty}\log S_n = 4\int_0^1 \log(1+4x)\,\mathrm{d}x = \int_0^4 \log(1+y)\,\mathrm{d}y.$$

Integration by parts gives the desired value. •

Now we state another useful result, which may be proved easily using the definition of continuity. However, in Section 5.2 (see Theorem 5.33), we have presented a similar comparison test for series.

Theorem 7.17 (Limit comparison test). *Suppose that f and g are two continuous functions on $[a,b)$, $a < b \leq \infty$, $g(x) > 0$, and $f(x) \geq 0$ on some subinterval $[a_1, b)$ $(a \leq a_1)$, and that*

$$\lim_{x \to b-} \frac{f(x)}{g(x)} = L. \tag{7.6}$$

Then we have the following:

(a) *If $0 < L < \infty$, then $\int_a^b f(x)\,dx$ and $\int_a^b g(x)\,dx$ both converge or diverge together.*

(b) *If $L = \infty$ and $\int_a^b g(x)\,dx = \infty$, then $\int_a^b f(x)\,dx = \infty$.*

(c) *If $L = 0$ and $\int_a^b g(x)\,dx < \infty$, then $\int_a^b f(x)\,dx < \infty$. In the last two cases, the two integrals do not necessarily converge or diverge together.*

Proof. **(a)** From the limit condition (7.6), there exists a point $a_2 \in [a_1, b)$ such that

$$\left| \frac{f(x)}{g(x)} - L \right| < \frac{L}{2} \quad \text{for } x \in [a_2, b).$$

This gives

$$\frac{L}{2} g(x) < f(x) < \frac{3L}{2} g(x) \quad \text{for } x \in [a_2, b).$$

By the direct comparison test, $\int_{a_2}^b f(x)\,dx$ and $\int_{a_2}^b g(x)\,dx$ both converge or diverge together, and in the latter case, they must diverge to ∞, since the integrands are nonnegative. We note that the integrals $\int_a^{a_2} f(x)\,dx$ and $\int_a^{a_2} g(x)\,dx$ exist, since f and g are continuous on $[a, a_2]$. The conclusion follows.

(b) If $L = \infty$, then by (7.6), there exists a point $a_2' \in [a_1, b)$ such that

$$f(x) \geq g(x) \quad \text{for } x \in [a_2', b).$$

Again the direct comparison test gives the desired conclusion.

(c) If $L = 0$, then there exists a point $a_2'' \in [a_1, b)$ such that

$$f(x) \leq g(x) \quad \text{for } x \in [a_2'', b),$$

and the conclusion is a consequence of the direct comparison test.

If $f(x) = 1/x$, $g(x) = 1/x^2$ with $a = 1$ and $b = \infty$, then

$$\lim_{x \to \infty} \frac{f(x)}{g(x)} = \infty \quad \text{and} \quad \lim_{x \to \infty} \frac{g(x)}{f(x)} = 0.$$

Note that $\int_1^\infty \frac{1}{x}\,dx = \infty$, whereas $\int_1^\infty \frac{1}{x^2}\,dx < \infty$. ∎

The limit comparison test as in Theorem 7.17 can be stated for improper integrals of other types, namely, when f and g are two positive continuous functions on $(a, b]$, where $-\infty \leq a < b$. We remark that the direct comparison test (Theorem 7.5) is a special case of the limit comparison test (Theorem 7.17). However, because of their independent interest, we formulate some special results that are often useful in solving our exercises.

Corollary 7.18. *Suppose that f and g are two positive continuous functions on $[a, N]$ for all $N > a$ such that*

$$\lim_{x\to\infty} \frac{f(x)}{g(x)} = L \quad (0 < L < \infty).$$

Then $\int_a^\infty f(x)\,dx$ is convergent if and only if $\int_a^\infty g(x)\,dx$ is convergent.

Example 7.19. The improper integral $\int_{-\infty}^{\infty}(1+x^8)^{-1/4}\,dx$ is convergent. Indeed, with $f(x) = 1/(1+x^8)^{1/4}$ and $g(x) = 1/(1+x^2)$, we have

$$\lim_{x\to\infty} \frac{f(x)}{g(x)} = \frac{1+x^2}{(1+x^8)^{1/4}} = 1.$$

Thus, by Corollary 7.18, the integral $\int_{-\infty}^{\infty}(1+x^8)^{-1/4}\,dx$ converges, because both $\int_{-\infty}^{0} g(x)\,dx$ and $\int_0^\infty g(x)\,dx$ converge. •

In particular, Corollary 7.18 gives the following simple result.

Corollary 7.20. *Suppose that $f(x)$ is a positive continuous function on $[a, \infty)$ such that*

$$\lim_{x\to\infty} x^p f(x) = L \quad (0 < L < \infty).$$

Then the improper integral $\int_0^\infty f(x)\,dx$ is convergent if and only if $p > 1$.

For example, to examine the convergence of $I = \int_1^\infty \frac{dx}{x^p(1+x^q)}$, it is natural to compare $f(x) = \frac{1}{x^p(1+x^q)}$ with $g(x) = \frac{1}{x^{p+q}}$ and observe that

$$\lim_{x\to\infty} \frac{f(x)}{g(x)} = \lim_{x\to\infty} \frac{x^{p+q}}{x^p(1+x^q)} = \lim_{x\to\infty} \frac{x^q}{1+x^q} = 1.$$

We also know that $\int_1^\infty g(x)\,dx$ converges if and only if $p+q > 1$. Consequently, the integral I converges if and only if $p+q > 1$.

Corollary 7.21. *Suppose that f and g are two positive continuous functions on $(a, b]$ (respectively $[a, b)$) such that*

$$\lim_{x\to a+} \frac{f(x)}{g(x)} = L \qquad \left(\lim_{x\to b-} \frac{f(x)}{g(x)} = L, \textit{ respectively}\right), \quad 0 < L < \infty.$$

Then $\int_a^b f(x)\,dx$ is convergent if and only if $\int_a^b g(x)\,dx$ is convergent.

For example, a comparison with $g(x) = (1-x)^{-1/2}$ shows that the integral $\int_0^1 (1-x^3)^{-1/2}\,dx$ is convergent. Indeed, it suffices to note that

$$\lim_{x\to 1-} \frac{f(x)}{g(x)} = \lim_{x\to 1-} \frac{(1-x)^{1/2}}{(1-x^3)^{1/2}} = \lim_{x\to 1-} \frac{1}{(1+x+x^2)^{1/2}} = \frac{1}{\sqrt{3}}.$$

Example 7.22. Let $f(x) = x^{-p}(1+x)^{-q}$. To see whether the improper integral $I = \int_0^\infty f(x)\,\mathrm{d}x$ converges, we decompose I into two improper integrals

$$I = \int_0^1 f(x)\,\mathrm{d}x + \int_1^\infty f(x)\,\mathrm{d}x =: I_1 + I_2.$$

Since

$$\lim_{x\to 0+} \frac{f(x)}{1/x^p} = \lim_{x\to 0+} \frac{1}{(1+x)^q} = 1$$

and $\int_0^1 \frac{\mathrm{d}x}{x^p}$ converges for $p < 1$ and diverges for $p \geq 1$, by Example 7.15, we conclude that $I_1 = \int_0^1 f(x)\,\mathrm{d}x$ converges for $p < 1$. Further, since

$$\lim_{x\to\infty} \frac{f(x)}{1/x^{p+q}} = \lim_{x\to\infty} \frac{x^q}{(1+x)^q} = 1$$

and $\int_1^\infty \frac{\mathrm{d}x}{x^{p+q}}$ converges for $p+q > 1$ and diverges for $p+q \leq 1$, by Example 7.3, it follows that $I_2 = \int_1^\infty f(x)\,\mathrm{d}x$ converges if and only if $p+q > 1$. Consequently, we conclude that I converges if and only if $p < 1$ and $p+q > 1$. ●

Example 7.23. Investigate the following improper integrals for convergence.

(a) $\displaystyle\int_0^{\pi/4} \frac{\mathrm{d}x}{\sin x}$. **(b)** $\displaystyle\int_1^\infty \frac{1+\sin^2 x}{x^p}\,\mathrm{d}x$. **(c)** $\displaystyle\int_0^1 \frac{3+\sin \pi x}{x^p}\,\mathrm{d}x$.

(d) $\displaystyle\int_0^{\pi/4} \frac{\mathrm{d}x}{\sin x^2}$. **(e)** $\displaystyle\int_1^\infty \frac{\mathrm{d}x}{\sqrt{x^2 - 1/2010}}$. **(f)** $\displaystyle\int_0^1 \left(\frac{1}{1+x} - \frac{1}{\mathrm{e}^x}\right)\frac{\mathrm{d}x}{x}$.

(g) $\displaystyle\int_1^\infty \frac{x^{-p}\mathrm{d}x}{(1+\mathrm{e}^x)}$. **(h)** $\displaystyle\int_0^\infty \left(\frac{1}{x} - \frac{1}{\sinh x}\right)\frac{\mathrm{d}x}{x}$. **(i)** $\displaystyle\int_0^1 \left(\frac{1}{1+x+\frac{x^2}{2}} - \frac{1}{\mathrm{e}^x}\right)\frac{\mathrm{d}x}{x^2}$.

Solution. For **(b)** and **(c)**, we observe that for $x \geq 1$,

$$\frac{1}{x^p} \leq \frac{1+\sin^2 x}{x^p} \leq \frac{2}{x^p} \quad \text{and} \quad \frac{2}{x^p} \leq \frac{3+\sin \pi x}{x^p} \leq \frac{4}{x^p}.$$

Consequently, by Theorem 7.5 and Example 7.15, the improper integral $\int_0^1 \frac{3+\sin \pi x}{x^p}\,\mathrm{d}x$ converges for $p < 1$ and diverges for $p \geq 1$.

By Theorem 7.5 and Example 7.3, the improper integral $\int_1^\infty \frac{1+\sin^2 x}{x^p}\,\mathrm{d}x$ converges for $p > 1$ and diverges for $p \leq 1$.

For **(a)** and **(d)**, we compare the integrands with $1/x$ and $1/x^2$, respectively. We observe that

$$\lim_{x\to 0+} \frac{1/x}{1/\sin x} = 1 \quad \text{and} \quad \lim_{x\to 0+} \frac{1/x^2}{1/(\sin x^2)} = 1.$$

Since $\int_0^{\pi/2} \frac{\mathrm{d}x}{x}$ diverges and $\int_0^{\pi/2} \frac{\mathrm{d}x}{x^2}$ converges, it follows that the integral in **(a)** diverges, while the integral in **(d)** converges.

(f) Since

$$f(x) = \left(\frac{1}{1+x} - \frac{1}{\mathrm{e}^x}\right)\frac{1}{x} = \frac{\mathrm{e}^x - (1+x)}{\mathrm{e}^x(1+x)x} > 0$$

and

$$\lim_{x\to 0+} \frac{f(x)}{1/x} = \lim_{x\to 0}\left(\frac{1}{(1+x)\mathrm{e}^x}\right)\left(\frac{\mathrm{e}^x - (1+x)}{x^2}\right) = \frac{1}{2!},$$

it is natural to set $g(x) = \frac{1}{x} > 0$ for $x \in (0,1]$. We see that $\int_0^1 f(x)\,\mathrm{d}x$ diverges, because $\int_0^1 g(x)\,\mathrm{d}x$ diverges. We leave the remaining integrals as exercises. ●

Theorem 7.24. *If f is continuous on $[0,\infty)$ such that*

$$\lim_{x\to 0+} f(x) = f_0 \quad \text{and} \quad \lim_{x\to\infty} f(x) = f_1,$$

then for $0 < a < b < \infty$, we have

$$\int_0^\infty \frac{f(ax) - f(bx)}{x}\,\mathrm{d}x = (f_0 - f_1)\log(b/a).$$

Proof. We have that

$$\begin{aligned}
\int_\epsilon^N \frac{f(ax) - f(bx)}{x}\,\mathrm{d}x &= \int_\epsilon^N \frac{f(ax)}{x}\,\mathrm{d}x - \int_\epsilon^N \frac{f(bx)}{x}\,\mathrm{d}x \\
&= \int_{a\epsilon}^{aN} \frac{f(y)}{y}\,\mathrm{d}y - \int_{b\epsilon}^{bN} \frac{f(y)}{y}\,\mathrm{d}y \quad (y = ax;\ y = bx) \\
&= \underbrace{\int_{a\epsilon}^{b\epsilon} \frac{f(y)}{y}\,\mathrm{d}y - \int_{aN}^{b\epsilon} \frac{f(y)}{y}\,\mathrm{d}y} - \int_{b\epsilon}^{bN} \frac{f(y)}{y}\,\mathrm{d}y \\
&= \int_{a\epsilon}^{b\epsilon} \frac{f(y)}{y}\,\mathrm{d}y - \int_{aN}^{bN} \frac{f(y)}{y}\,\mathrm{d}y.
\end{aligned}$$

Applying the generalized mean value theorem for integrals to the integrals on the right-hand side, we obtain that

$$\begin{aligned}
\int_\epsilon^N \frac{f(ax) - f(bx)}{x}\,\mathrm{d}x &= f(c)\int_{a\epsilon}^{b\epsilon} \frac{\mathrm{d}y}{y} - f(d)\int_{aN}^{bN} \frac{\mathrm{d}y}{y} \\
&= (f(c) - f(d))\log(b/a)
\end{aligned}$$

for some $c \in (a\epsilon, b\epsilon)$ and $d \in (aN, bN)$. The desired equality follows if we let $\epsilon \to 0$ and $N \to \infty$. ■

If we choose $f(x) = \arctan x$ and $f(x) = \mathrm{e}^{-x}$, it follows easily that

(a) $\displaystyle\int_0^\infty \frac{\arctan(ax) - \arctan(bx)}{x}\,\mathrm{d}x = -\frac{\pi}{2}\log\left(\frac{b}{a}\right);$

(b) $\displaystyle\int_0^\infty \frac{\mathrm{e}^{-ax} - \mathrm{e}^{-bx}}{x}\,\mathrm{d}x = \log\left(\frac{b}{a}\right);$

and by integration by parts, we easily have

$$\int_0^\infty \frac{\cos(ax) - \cos(bx)}{x^2}\,dx = \frac{\pi}{2}(b-a),$$

where $0 < a < b < \infty$.

7.1.3 The Gamma and Beta Functions

Next we consider the integral

$$\int_0^\infty x^{\alpha-1}e^{-x}\,dx \quad (\alpha > 0)$$

and show that the improper integral exists (meaning that it converges) for $\alpha > 0$. This integral is considered to be nonelementary because it cannot be evaluated in closed form in terms of so-called elementary functions. This integral, denoted by $\Gamma(\alpha)$, arises frequently in pure and applied mathematics and is referred to as Euler's *gamma function* defined on $(0,\infty)$. Also, we note that if $\alpha < 1$, then $f(x) = x^{\alpha-1}e^{-x} \to \infty$ as $x \to 0+$. So to discuss its convergence we need to decompose the integral as

$$\int_0^\infty x^{\alpha-1}e^{-x}\,dx = \int_0^c x^{\alpha-1}e^{-x}\,dx + \int_c^\infty x^{\alpha-1}e^{-x}\,dx =: I_1 + I_2,$$

where $c > 0$. The given integral converges if and only if each of the integrals on the right converges. For convenience, we may choose $c = 1$. The integral I_1 is proper if $\alpha \geq 1$, and is improper if $0 < \alpha < 1$, because the integrand has a point of infinite discontinuity at $x = 0$. In any case, the integral I_1 converges for $\alpha > 0$ because

$$0 < x^{\alpha-1}e^{-x} < x^{\alpha-1} \text{ for } x > 0, \text{ and } \int_0^1 x^{\alpha-1}\,dx = \frac{1}{\alpha},$$

so that by the direct comparison test, we conclude that $I_1 = \int_0^1 x^{\alpha-1}e^{-x}\,dx$ converges. Note also that since

$$\lim_{x\to 0} \frac{x^{\alpha-1}e^{-x}}{x^{\alpha-1}} = 1$$

(by Theorem 7.17), the convergence of I_1 follows from the fact that $\int_0^1 x^{\alpha-1}\,dx$ converges if and only if $\alpha > 0$. The second integral likewise converges. Indeed, for each fixed $\alpha > 0$, we know that

$$\lim_{x\to\infty} \frac{x^{\alpha-1}e^{-x}}{1/x^2} = \lim_{x\to\infty} x^{\alpha+1}e^{-x} = 0.$$

Since $\int_c^\infty \frac{dx}{x^2}$ is convergent, by the comparison test, I_2 converges for $\alpha > 0$. Consequently, it follows that

$$\Gamma(\alpha) = \int_0^\infty x^{\alpha-1} e^{-x}\, dx \quad (\alpha > 0)$$

is convergent. Although we cannot evaluate $\Gamma(\alpha)$ explicitly for most values of α, there are many interesting properties that may be stated here:

1. $\Gamma(1) = \int_0^\infty e^{-x}\, dx = \lim_{b\to\infty} (-e^{-b} + 1) = 1.$
2. Integration by parts easily gives

$$\Gamma(n+1) = \int_0^\infty x^n e^{-x}\, dx = -\, x^n e^{-x} \Big|_0^\infty + n \int_0^\infty x^{n-1} e^{-x}\, dx = 0 + n\Gamma(n),$$

for $n \in \mathbb{N}$, and more generally, $\Gamma(\alpha+1) = \alpha\Gamma(\alpha)$ holds for $\alpha > 0$ (prove this). In particular,

$$\Gamma(n+1) = n\Gamma(n) = n(n-1)\Gamma(n-1) = \cdots = n!\Gamma(1) = n!.$$

The gamma function thus gives us a way of extending the domain of the factorial function from the set of positive integers to the set of positive real numbers.
3. Also, $\Gamma(1/2) = \sqrt{\pi}$. Indeed, using the change of variable $x = y^2$, we obtain

$$\Gamma(1/2) = \int_0^\infty x^{-1/2} e^{-x}\, dx = 2 \int_0^\infty e^{-y^2}\, dy,$$

and those who are familiar with double integrals can easily see that

$$\begin{aligned}
\left(\Gamma\left(\frac{1}{2}\right)\right)^2 &= 4 \int_0^\infty e^{-u^2}\, du \int_0^\infty e^{-v^2}\, dv \\
&= 4 \int_0^\infty \int_0^\infty e^{-(u^2+v^2)}\, du\, dv \\
&= 4 \int_0^{\pi/2} \int_0^\infty e^{-r^2} r\, dr\, d\theta \quad (u = r\cos\theta,\ v = r\sin\theta) \\
&= 4 \left(\int_0^{\pi/2} d\theta\right) \left(-\frac{1}{2} \int_0^\infty e^{-r^2} (-2r)\, dr\right) \\
&= 4\left(\frac{\pi}{2}\right)\left(-\frac{1}{2}\right) \left(e^{-r^2}\Big|_0^\infty\right) = \pi.
\end{aligned}$$

Hence $\Gamma(1/2) = \sqrt{\pi}$.

We are now prepared to consider another important function, the beta function, which depends on two parameters a and b:

$$I = \int_0^1 x^{a-1} (1-x)^{b-1}\, dx, \quad a, b > 0.$$

This integral has been studied extensively. Clearly this integral is proper if $a \geq 1$ and $b \geq 1$. However, it is improper if $0 < a < 1$ or $0 < b < 1$ or both.

Indeed, the improper integral has points of infinite discontinuity at $x = 0$ if $a < 1$, and at $x = 1$ if $b < 1$.

To discuss the case that I is improper, we decompose I into the sum of two integrals $I =: I_1 + I_2$, where

$$I_1 = \int_0^{1/2} x^{a-1}(1-x)^{b-1}\,dx$$

and

$$I_2 = \int_{1/2}^1 x^{a-1}(1-x)^{b-1}\,dx = \int_0^{1/2} (1-y)^{a-1}y^{b-1}\,dy.$$

Therefore, it suffices to discuss the integral I_1. For any real b,

$$0 \le x \le 1/2 \Longleftrightarrow 1/2 \le 1-x \le 1 \Longrightarrow 0 < (1-x)^{b-1} \le M$$

for some constant $M > 0$. Using this observation, we find that

$$I_1 = \lim_{\epsilon\to 0+} \int_\epsilon^{1/2} x^{a-1}(1-x)^{b-1}dx \le M \lim_{\epsilon\to 0+} \int_\epsilon^{1/2} x^{a-1}\,dx = M\frac{(1/2)^a}{a}.$$

Thus, I_1 converges for $a > 0$. Alternatively, we see that $f(x) = x^{a-1}(1-x)^{b-1}$ and $g(x) = x^{a-1}$ are two positive continuous functions on $(0, 1/2]$ such that

$$\lim_{x\to 0+} \frac{f(x)}{g(x)} = \lim_{x\to 0+} \frac{x^{a-1}(1-x)^{b-1}}{x^{a-1}} = \lim_{x\to 0+} (1-x)^{b-1} = 1.$$

Since $\int_0^{1/2} g(x)\,dx$ converges if and only if $a > 0$ (by Corollary 7.21), we obtain that $I_1 = \int_0^{1/2} f(x)\,dx$ also converges whenever $a > 0$ and for any b. Interchanging the roles of a and b shows that I_2 converges for $b > 0$ and for any a.

The combination of the two situations shows that the integral I converges for $a > 0$ and $b > 0$. The integral I is called the *beta function* and is usually denoted by $B(a, b)$. Thus,

$$B(a,b) = \int_0^1 x^{a-1}(1-x)^{b-1}\,dx \quad (a, b > 0). \tag{7.7}$$

The beta function has many interesting properties. For example, we have the following:

(1) If we substitute $1 - x = t$, it follows that $B(a, b) = B(b, a)$.
(2) Next, if we substitute $x = t/(1 + t)$, then

$$1 - x = \frac{1}{1+t}, \quad dx = \frac{dt}{(1+t)^2}, \quad \text{and } 0 \le x < 1 \Longleftrightarrow 0 \le t < \infty,$$

so that (7.7) has an equivalent formulation

$$B(a,b) = \int_0^\infty \frac{t^{a-1}}{(1+t)^{a+b}}\,\mathrm{d}t.$$

The fact that $B(a,b) = B(b,a)$ gives

$$2B(a,b) = \int_0^\infty \frac{t^{a-1}+t^{b-1}}{(1+t)^{a+b}}\,\mathrm{d}t.$$

Also, by splitting this integral, we see that

$$\begin{aligned} 2B(a,b) &= \int_0^1 \frac{t^{a-1}+t^{b-1}}{(1+t)^{a+b}}\,\mathrm{d}t + \int_1^\infty \frac{t^{a-1}+t^{b-1}}{(1+t)^{a+b}}\,\mathrm{d}t \\ &= \int_0^1 \frac{t^{a-1}+t^{b-1}}{(1+t)^{a+b}}\,\mathrm{d}t + \int_0^1 \frac{x^{a-1}+x^{b-1}}{(1+x)^{a+b}}\,\mathrm{d}x \quad (t = 1/x), \end{aligned}$$

so that $B(a,b)$ has another equivalent form,

$$B(a,b) = \int_0^1 \frac{t^{a-1}+t^{b-1}}{(1+t)^{a+b}}\,\mathrm{d}t.$$

Since $B(a,b) = B(b,a)$, we have

$$\int_0^\infty \frac{t^{a-1}-t^{b-1}}{(1+t)^{a+b}}\,\mathrm{d}t = 0.$$

(3) Finally, we let $x = \sin^2\theta$, so that $1-x = \cos^2\theta$, and $\mathrm{d}x = 2\sin\theta\cos\theta\,\mathrm{d}\theta$. Thus, (7.7) becomes

$$B(a,b) = 2\int_0^{\pi/2} \sin^{2a-1}\theta\cos^{2b-1}\theta\,\mathrm{d}\theta \quad (a>0,\ b>0),$$

or equivalently,

$$B\left(\frac{p+1}{2}, \frac{q+1}{2}\right) = 2\int_0^{\pi/2} \sin^p\theta\cos^q\theta\,\mathrm{d}\theta \quad (p>-1,\ q>-1).$$

For example, we easily see that

- $\displaystyle\int_0^{\pi/2} \sqrt{\tan\theta}\,\mathrm{d}\theta = \int_0^{\pi/2} \sin^{1/2}\theta\cos^{-1/2}\theta\,\mathrm{d}\theta = \frac{1}{2}B\left(\frac{3}{2}, \frac{1}{2}\right)$,
- $\displaystyle\int_0^{\pi/2} \sin^p\theta\,\mathrm{d}\theta = \int_0^{\pi/2} \cos^p\theta\,\mathrm{d}\theta = \frac{1}{2}B\left(\frac{p+1}{2}, \frac{1}{2}\right)$, $p > -1$,
- $\displaystyle B\left(\frac{1}{2}, \frac{1}{2}\right) = 2\int_0^{\pi/2} \mathrm{d}\theta = \pi$.

(4) By the method of integration by parts, one can easily see that

$$B(a+1,b) = \frac{a}{a+b}B(a,b) \quad \text{for } a>0,\ b>0.$$

(5) Another important property, which is beyond the scope of this book, is

$$B(a,b)=\frac{\Gamma(a)\Gamma(b)}{\Gamma(a+b)},\quad \text{for } a>0,\ b>0.$$

(6) Setting $b=a$ in case (5) gives

$$\begin{aligned}
\frac{(\Gamma(a))^2}{\Gamma(2a)} &= B(a,a)=2\int_0^{\pi/2}\sin^{2a-1}\theta\cos^{2a-1}\theta\,\mathrm{d}\theta\\
&=\frac{2}{2^{2a-1}}\int_0^{\pi/2}\sin^{2a-1}(2\theta)\,\mathrm{d}\theta\\
&=\frac{1}{2^{2a-1}}\int_0^{\pi}\sin^{2a-1}\phi\,\mathrm{d}\phi\\
&=\frac{2}{2^{2a-1}}\int_0^{\pi/2}\sin^{2a-1}\phi\,\mathrm{d}\phi\\
&=\frac{1}{2^{2a-1}}B\Big(a,\frac{1}{2}\Big).
\end{aligned}$$

A simplification yields the Legendre duplication formula

$$\sqrt{\pi}\,\Gamma(2a)=2^{2a-1}\Gamma(a)\Gamma(a+1/2),\quad a>0,$$

or equivalently,

$$\Gamma(a)=\frac{2^{a-1}}{\sqrt{\pi}}\Gamma\Big(\frac{a}{2}\Big)\Gamma\Big(\frac{a+1}{2}\Big)\ \text{ for } a>0.$$

7.1.4 Wallis's Formula

We begin by evaluating

$$I_n=\int_0^{\pi/2}\sin^n x\,\mathrm{d}x,$$

where n is a positive integer. In order to do this, we use a different approach, namely integration by parts and the method of induction. If $n\geq 2$, we have

$$\begin{aligned}
I_n &= -\int_0^{\pi/2}\sin^{n-1}x\,\mathrm{d}(\cos x)\\
&= -\sin^{n-1}x\cos x\,\Big|_0^{\pi/2}+(n-1)\int_0^{\pi/2}\sin^{n-2}x\cos^2 x\,\mathrm{d}x\\
&=(n-1)\int_0^{\pi/2}\sin^{n-2}x(1-\sin^2 x)\,\mathrm{d}x,
\end{aligned}$$

and so

$$I_n=(n-1)I_{n-2}-(n-1)I_n,\quad \text{i.e.,}\quad I_n=\frac{n-1}{n}I_{n-2}.$$

It is easy to see that $I_0 = \pi/2$ and $I_1 = 1$. Hence, for $n = 2m$,

$$\begin{aligned} I_{2m} &= \frac{2m-1}{2m} I_{2(m-1)} = \cdots = \frac{2m-1}{2m}\,\frac{2m-3}{2m-1} \cdots \frac{3}{4}\,\frac{1}{2} I_0 \\ &= \frac{\pi}{2} \prod_{k=1}^{m} \frac{2k-1}{2k}, \end{aligned}$$

and for $n = 2m+1$,

$$\begin{aligned} I_{2m+1} &= \frac{2m}{2m+1} I_{2m-1} = \frac{2m}{2m+1}\,\frac{2m-2}{2m-1} \cdots \frac{4}{5}\,\frac{2}{3} I_1 \\ &= \prod_{k=1}^{m} \frac{2k}{2k+1}. \end{aligned}$$

We are done. The above two formulas for I_{2m} and I_{2m+1} have some interesting consequences, for example Wallis's formula, which expresses π in terms of an infinite product. From the last equations, we obtain

$$\frac{\pi}{2} = \frac{I_{2m}}{I_{2m+1}} \prod_{k=1}^{m} \frac{(2k)^2}{(2k+1)(2k-1)},$$

so that

$$\frac{\pi}{2} = \frac{I_{2m}}{I_{2m+1}} \left[\prod_{k=1}^{m} \left(\frac{2k}{2k-1} \right)^2 \right] \frac{1}{2m+1} \quad \text{for each } m \geq 1. \tag{7.8}$$

We shall first show that

$$\lim_{m\to\infty} \frac{I_{2m}}{I_{2m+1}} = 1. \tag{7.9}$$

In order to prove this, we recall that for $0 < x < \pi/2$,

$$0 < \sin^{2m+1} x \leq \sin^{2m} x \leq \sin^{2m-1} x,$$

and therefore by integrating this inequality from 0 to $\pi/2$, we obtain

$$I_{2m+1} \leq I_{2m} \leq I_{2m-1}, \text{ or } 1 \leq \frac{I_{2m}}{I_{2m+1}} = 1 + \frac{1}{2m}.$$

Thus (7.9) holds, and passing to the limit in (7.8), we have Wallis's product formula

$$\frac{\pi}{2} = \lim_{m\to\infty} \frac{1}{2m+1} \prod_{k=1}^{m} \left(\frac{2k}{2k-1} \right)^2.$$

Since

$$\frac{1}{2m+1} = \frac{1}{2m} \left(\frac{2m}{2m+1} \right) \quad \text{and} \quad \frac{2m}{2m+1} \to 1 \text{ as } m \to \infty,$$

this formula may be rewritten as

$$\frac{\pi}{2} = \lim_{m\to\infty} \frac{1}{2m} \prod_{k=1}^{m} \left(\frac{2k}{2k-1}\right)^2 \quad \text{or} \quad \sqrt{\frac{\pi}{2}} = \lim_{m\to\infty} \frac{1}{\sqrt{2m}} \prod_{k=1}^{m} \frac{2k}{2k-1},$$

or equivalently as

$$\begin{aligned}\sqrt{\frac{\pi}{2}} &= \lim_{m\to\infty} \frac{1}{\sqrt{2m}} \left[\frac{2\cdot 4\cdot 6\cdots 2(m-1)\cdot 2m}{3\cdot 5\cdot 7\cdots (2m-1)}\right] \\ &= \lim_{m\to\infty} \frac{1}{\sqrt{2m}} \left[\frac{2^2\cdot 4^2\cdots (2(m-1))^2\cdot (2m)^2}{(2m)!}\right] \\ &= \lim_{m\to\infty} \frac{1}{\sqrt{2m}} \left[\frac{2^m\cdot 2^m\cdot (m!)^2}{(2m)!}\right].\end{aligned}$$

Finally, we set

$$\sqrt{\pi} = \lim_{m\to\infty} \frac{2^{2m}(m!)^2}{\sqrt{m}(2m)!}.$$

7.1.5 The Integral Test

For series whose terms are nonnegative, Theorem 5.20 simplifies the problem of investigating convergence. Further, to establish the divergence of such series, it suffices to show that the sequence of partial sums has no upper bound. However, this theorem is often difficult to apply, since one is required to determine whether the sequence of partial sums has an upper bound. This task, in general, is not easy. We shall now discuss the integral test, which avoids this difficulty. For motivation, we begin our discussion with the harmonic series

$$\sum_{k=1}^{\infty} \frac{1}{k}, \quad S_n = \sum_{k=1}^{n} \frac{1}{k}.$$

We have already proved that the harmonic series diverges, by showing that the sequence of partial sums $\{S_n\}$ is unbounded. There is another way to estimate the partial sums. For instance, consider function $f(x) = 1/x$, which is clearly decreasing, continuous, and positive on $[1, \infty)$, and note that $f(k) = 1/k$ (Figure 7.13). Geometrically, it is evident that

$$\frac{1}{k+1} \le \int_k^{k+1} \frac{dx}{x} \le \frac{1}{k},$$

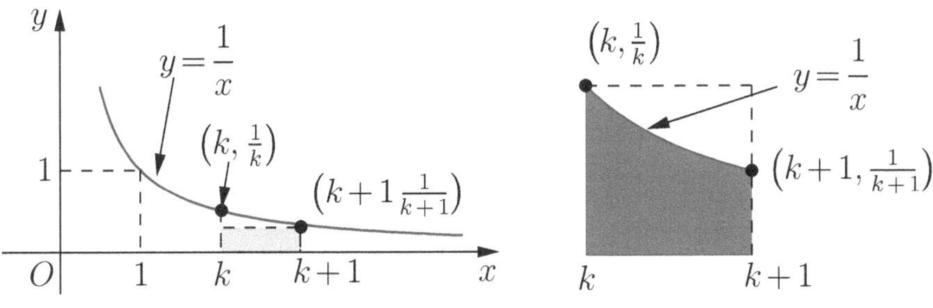

Fig. 7.13. Graph of $f(x) = \dfrac{1}{x}$ on $[1, \infty)$.

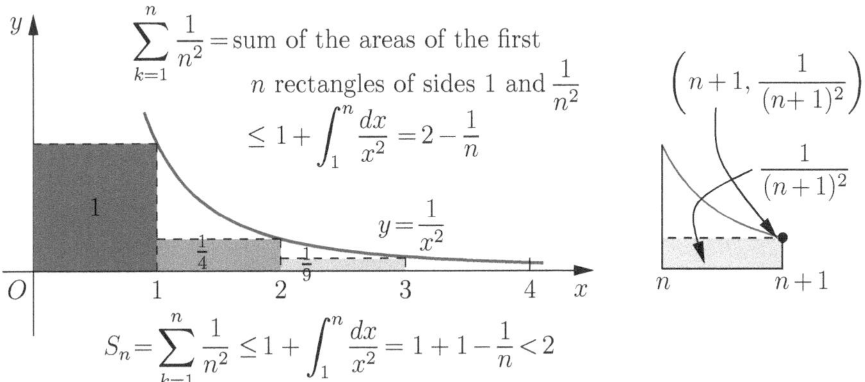

Fig. 7.14. Demonstration for integral test with $f(x) = 1/x^2$.

so that

$$S_{n+1} - 1 = \sum_{k=1}^{n} \frac{1}{k+1} \leq \sum_{k=1}^{n} \int_k^{k+1} \frac{dx}{x} = \int_1^{n+1} \frac{dx}{x} \leq \sum_{k=1}^{n} \frac{1}{k} = S_n,$$

which simplifies to

$$S_n + \frac{1}{n+1} - 1 \leq \log(n+1) \leq S_n, \quad \text{i.e.,} \quad \log(n+1) \leq S_n \leq \log(n+1) + \frac{n}{n+1}.$$

Hence $\{S_n\}$ diverges roughly at the same rate as that of the logarithm function.

The idea of the above example leads to what is called the *integral test*, which demonstrates a close relationship between the convergence of certain series and improper integrals. This test is extremely useful in determining the convergence or divergence of certain series.

Theorem 7.25 (The integral test). *Suppose that f is a nonnegative, continuous, and decreasing function of x for $x \geq 1$. Then either both*

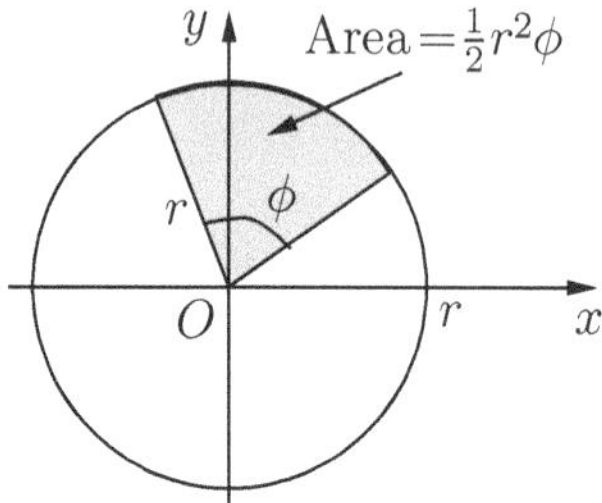

Fig. 7.15. Sketch for the integral test with $f(x) = 1/x^2$.

$$\sum_{k=1}^{\infty} f(k) \quad \textit{and} \quad \int_1^{\infty} f(x)\,\mathrm{d}x$$

converge or both diverge. Moreover, if the series converges, then

$$\int_1^{\infty} f(x)\,\mathrm{d}x \le \sum_{k=1}^{\infty} f(k) \le f(1) + \int_1^{\infty} f(x)\,\mathrm{d}x, \tag{7.10}$$

or equivalently,

$$\sum_{k=2}^{\infty} f(k) \le \int_1^{\infty} f(x)\,\mathrm{d}x \le \sum_{k=1}^{\infty} f(k).$$

Proof. Set $a_k = f(k)$ for $k = 1, 2, \dots$ and $S_n = \sum_{k=1}^{n} a_k$. Since f is nonnegative, continuous, and decreasing, the area under the curve $y = f(x)$ between $x = k$ and $x = k+1$ lies between the areas of the rectangles of unit width with height $f(k) = a_k$ and $f(k+1) = a_{k+1}$ respectively (see Figure 7.14 for $f(x) = 1/x^2$). In short,

$$a_{k+1} \le \int_k^{k+1} f(x)\,\mathrm{d}x \le a_k,$$

so that

$$\sum_{k=1}^{n} a_{k+1} \le \sum_{k=1}^{n} \int_k^{k+1} f(x)\,\mathrm{d}x = \int_1^{n+1} f(x)\,\mathrm{d}x \le \sum_{k=1}^{n} a_k. \tag{7.11}$$

Now suppose that $\int_1^{\infty} f(x)\,\mathrm{d}x$ converges as an improper integral and has the value I. Then by the left-hand side of (7.11), we have

$$S_{n+1} \le a_1 + \int_1^{n+1} f(x)\,\mathrm{d}x \le a_1 + I \quad \text{for each } n.$$

Since $a_k \ge 0$, it follows that the sequence of partial sums $\{S_n\}$ is increasing and bounded above by $a_1 + I$, and hence $\{S_n\}$ converges by BMCT (see Theorem 2.25). That is, the series $\sum_{k=1}^{\infty} a_k$ is convergent.

On the other hand, if the improper integral $\int_1^\infty f(x)\,dx$ diverges, then it must tend to infinity, because $f(x)$ is nonnegative. Then the right-hand side of (7.11) shows that $\{S_n\}$ has no upper bound, and so the series must diverge to infinity.

Finally, if the series converges and has the sum S, then (7.11) gives that

$$I_{n+1} = \int_1^{n+1} f(x)\,dx \le S_n \le S.$$

Since $f(x) \ge 0$ and $I_{n+1} \ge 0$ for all $n \ge 1$, it follows that the sequence $\{I_{n+1}\}$ is increasing and bounded above by S, and hence $\{I_{n+1}\}$ converges by BMCT; that is, $\lim_{n\to\infty} I_{n+1} = \int_1^\infty f(x)\,dx$ exists as an improper integral. Thus, the series and the improper integral either both converge or both diverge, as claimed. ∎

Remark 7.26. 1. From the proof of the integral test, it is clear that the sum and the integral could begin with any integer N. In other words, if f is a positive, continuous, and decreasing function of x for $x \ge N$, then

$$\sum_{k=N}^{\infty} f(k) \quad \text{and} \quad \int_N^\infty f(x)\,dx$$

either both converge or both diverge. Thus, it is not necessary to begin the series from $k = 1$ or the integral with the lower limit of integration with 1.

2. It is important to note that the sum of the series is not necessarily equal to the value of the integral. For instance, the series $\sum_{k=1}^\infty \frac{1}{k^2}$ is known to be $\pi^2/6$, which is definitely greater than $5/4$, whereas $\int_1^\infty \frac{dx}{x^2} = 1$ (see also Figure 7.14).
3. Note that $\lim_{k\to\infty} \sin(1/k) = \sin 0 = 0$. This does not mean that $\sum_{k=1}^\infty \sin(1/k)$ is convergent. On the other hand, it is easy to see that $\int_1^\infty \sin(1/x)\,dx$ diverges (see Exercise 7.31(**f**)). Thus, by the integral test, $\sum_{k=1}^\infty \sin(1/k)$ is divergent (see also Example 5.36(**d**)).
4. As noted before (see also (7.11) and Figure 7.13),

$$\sum_{k=1}^{n} \frac{1}{k+1} < \int_1^{n+1} \frac{dx}{x} = \log(n+1) < \sum_{k=1}^{n} \frac{1}{k} = \sum_{k=1}^{n+1} \frac{1}{k} - \frac{1}{n+1},$$

or equivalently,

$$\frac{1}{n+1} < \sum_{k=1}^{n+1} \frac{1}{k} - \log(n+1) < 1 \quad \text{for each } n \ge 1,$$

which may be rewritten as

$$\frac{1}{n} < \gamma_n < 1, \quad \gamma_n = \sum_{k=1}^{n} \frac{1}{k} - \log n \quad \text{for } n > 1.$$

We see that

$$\gamma_{n+1} - \gamma_n = \frac{1}{n+1} - \int_n^{n+1} \frac{dx}{x},$$

and by the graph of $y = 1/x$ on $[n, n+1]$, it follows that $\gamma_{n+1} - \gamma_n \leq 0$, showing that $\{\gamma_n\}_{n=1}^\infty$ is a decreasing bounded sequence of positive real numbers, and hence converges, say to γ. Thus, we have

$$\gamma := \lim_{n\to\infty} \left(\sum_{k=1}^n \frac{1}{k} - \log n \right).$$

Even though $\sum_{k=1}^\infty \frac{1}{k} = \infty$ and $\lim_{n\to\infty} \log n = \infty$, the limit γ exists and is finite. The number γ is called *Mascheroni's constant* or *Euler's constant*, after Leonhard Euler (1707–1783). Its value to six decimal places is 0.577216. In contrast to other familiar constants e and π, no other simple representation is known for Euler's constant. It is still unknown whether γ is rational or irrational.

5. Since

$$\frac{1}{k+1} \leq \int_k^{k+1} \frac{dx}{x} \leq \frac{1}{k} \quad \text{for } k \geq 1,$$

it follows that

$$\begin{aligned} \frac{1}{n+1} + \frac{1}{n+2} + \cdots + \frac{1}{n+n} &\leq \left(\int_n^{n+1} + \int_{n+1}^{n+2} + \cdots + \int_{2n-1}^{2n} \right) \frac{dx}{x} \\ &\leq \frac{1}{n} + \frac{1}{n+1} + \cdots + \frac{1}{2n-1}, \end{aligned}$$

which gives

$$\sum_{k=1}^n \frac{1}{n+k} \leq \int_n^{2n} \frac{dx}{x} = \log\left(\frac{2n}{n}\right) = \log 2 \leq \left(\frac{1}{n} - \frac{1}{2n}\right) + \sum_{k=1}^n \frac{1}{n+k},$$

or equivalently,

$$-\frac{1}{2n} + \log 2 \leq \sum_{k=1}^n \frac{1}{n+k} \leq \log 2.$$

Thus, we obtain that

$$\lim_{n\to\infty} \sum_{k=1}^n \frac{1}{n+k} = \log 2,$$

a fact that has been verified by another method (see Example 6.24(**a**)). Similarly, it is easy to see that

$$\lim_{n\to\infty} \sum_{k=1}^{pn} \frac{1}{n+k} = \log(p+1). \qquad \bullet$$

Example 7.27. There are many ways (see Examples 5.28 and 5.40(1), for instance) to show that the harmonic p-series defined by

$$\sum_{k=1}^{\infty} \frac{1}{k^p} = 1 + \frac{1}{2^p} + \frac{1}{3^p} + \cdots + \frac{1}{k^p} + \cdots$$

converges if $p > 1$ and diverges if $p \leq 1$.

We now illustrate the integral test by investigating the convergence of the harmonic p-series. For $p > 0$, consider the function $f(x) = 1/x^p$ for $x > 0$. Then f is positive, continuous, and decreasing for $x > 0$. Also, $f(k) = 1/k^p$ for all positive integers k. So the integral test is applicable. Therefore, $\sum_{k=1}^{\infty} \frac{1}{k^p}$ is convergent if and only if the improper integral $\int_1^{\infty} \frac{dx}{x^p}$ converges. Since the improper integral $\int_1^{\infty} \frac{dx}{x^p}$ converges to $1/(p-1)$ if and only if $p > 1$, we conclude that $\sum_{k=1}^{\infty} \frac{1}{k^p}$ converges if $p > 1$ and diverges if $p \leq 1$. ●

The integral test can also be used to obtain lower and upper bounds for the partial sum of the series as well as for the series itself.

Remark 7.28. 1. We have the estimate (apply Theorem 7.25 with $f(x) = 1/x^p$, so that $a_k = 1/k^p$)

$$\frac{1}{p-1} \leq \sum_{k=1}^{\infty} \frac{1}{k^p} \leq 1 + \frac{1}{p-1} \quad \text{for } p > 1.$$

For instance, with $p = 11/10$, we see that

$$10 \leq \sum_{k=1}^{\infty} \frac{1}{k^{1+1/10}} \leq 11,$$

whereas $\sum_{k=1}^{\infty}(1/k)$ diverges although $1/k^{1.1}$ is close to $1/k$. This clearly demonstrates that $\sum_{k=1}^{\infty}(1/k^{1.1})$ grows very slowly. Also, from Theorem 7.25, we obtain for instance

$$\frac{10}{11^{0.1}} = \int_{11}^{\infty} \frac{dx}{x^{1.1}} \leq \sum_{k=11}^{\infty} \frac{1}{k^{1.1}} \leq \int_{11}^{\infty} \frac{dx}{x^{1.1}} + a_{11} = \frac{1}{11^{1.1}} + \frac{10}{11^{0.1}} = 7.939\ldots,$$

so that $\sum_{k=11}^{\infty}(1/k^{1.1})$ lies between $10(11^{-0.1}) = 7.867\ldots$ and $7.939\ldots$. Moreover, by computing the value of $\sum_{k=1}^{10} k^{-1.1}$ ($\approx 2.690\ldots$), it can be seen that the value of

$$\sum_{k=1}^{\infty} \frac{1}{k^{1.1}} = \sum_{k=1}^{10} \frac{1}{k^{1.1}} + \sum_{k=11}^{\infty} \frac{1}{k^{1.1}}$$

lies between $10.548\ldots$ and $10.619\ldots$.

2. Although the graph of $y = 1/x^p$ $(p > 0)$ is decreasing for $x > 0$, we obtain that the series $\sum \frac{1}{k^p}$ converges only for $p > 1$. Why is this so? If $p > 1$, the curve $y = 1/x^p$ decreases fast enough to guarantee that the area under the curve for $p > 1$ is finite, whereas in the case of $p \leq 1$, the area under the curve is infinite (see Figure 7.5).
3. Under the hypothesis of Theorem 7.25 and from its proof, (7.10) may be rewritten as

$$S_n + \int_{n+1}^{\infty} f(x)\,dx \leq \sum_{k=1}^{\infty} f(k) \leq S_{n+1} + \int_{n+1}^{\infty} f(x)\,dx,$$

where $S_n = \sum_{k=1}^{n} f(k)$ and $\{S_n\}$ converges. With $f(x) = 1/x^2$, this gives

$$\sum_{k=1}^{n} \frac{1}{k^2} + \int_{n+1}^{\infty} \frac{dx}{x^2} \leq \sum_{k=1}^{\infty} \frac{1}{k^2} \leq \sum_{k=1}^{n+1} \frac{1}{k^2} + \int_{n+1}^{\infty} \frac{dx}{x^2}.$$

Because $\int_{n+1}^{\infty} \frac{dx}{x^2} = \frac{1}{n+1}$, the last inequality can be rewritten as

$$0 \leq \sum_{k=n+1}^{\infty} \frac{1}{k^2} - \frac{1}{n+1} \leq \frac{1}{(n+1)^2}.$$

In particular, this yields that if $n > 99$, then one has

$$\sum_{k=101}^{\infty} \frac{1}{k^2} < \frac{1}{100} + \frac{1}{10000}.$$

●

Example 7.29. Test each of the following series for convergence.

(a) $\sum_{k=1}^{\infty} \frac{k}{e^{k/7}}$. (b) $\sum_{k=1}^{\infty} \left(\frac{1}{e^k} - \frac{100}{k}\right)$. (c) $\sum_{k=2}^{\infty} \frac{1}{k(\log k)^p}$.

(d) $\sum_{k=2}^{\infty} \frac{1}{k \log k[\log(\log k)]^p}$ $(p \in \mathbb{R})$.

Solution. **(b)** The function $f(x) = xe^{-x/7}$ is positive and continuous for all $x > 0$. We find that

$$f'(x) = (1 - x/7)\,e^{-x/7}.$$

Since $e^x > 0$ on $\mathbb{R}$, we see that $f'(x) \leq 0$ for $x \geq 7$, so it follows that f is decreasing for $x \geq 7$. Thus, by the integral test, the given series and the improper integral $\int_7^{\infty} xe^{-x/7}\,dx$ either both converge or both diverge. It is a simple exercise to see that the improper integral $\int_7^{\infty} xe^{-x/7}\,dx$ converges. Consequently, the given series converges.

(b) We note that $\sum_{k=1}^{\infty} \frac{1}{e^k}$ converges, because it is a geometric series with $r = 1/e < 1$. Also, the harmonic series $\sum_{k=1}^{\infty} \frac{1}{k}$ is known to be divergent. Because one of the series in the difference converges and the other diverges, the given series must diverge (see Theorem 5.12).

(c) According to the integral test, it suffices to discuss the convergence of the improper integral $\int_a^\infty f(x)\,\mathrm{d}x$ for sufficiently large a, where $f(x) = \frac{1}{x(\log x)^p}$. Then

$$f'(x) = -\frac{(\log x)^p + px(\log x)^{p-1}\left(\frac{1}{x}\right)}{[x(\log x)^p]^2} = -\frac{(\log x)^{p-1}(\log x + p)}{x^2(\log x)^{2p}} \leq 0$$

for $x > 1$. Now

$$\int_a^N \frac{\mathrm{d}x}{x(\log x)^p} = \int_a^N \frac{1}{(\log x)^p}\,\mathrm{d}(\log x) = \begin{cases} \left.\dfrac{(\log x)^{1-p}}{1-p}\right|_a^N & \text{if } p \neq 1, \\ \log(\log x)|_a^N & \text{if } p = 1, \end{cases}$$

which implies that

$$\int_a^\infty \frac{\mathrm{d}x}{x(\log x)^p} = \lim_{N\to\infty} \int_a^N \frac{\mathrm{d}x}{x(\log x)^p} = \begin{cases} \infty & \text{if } p \leq 1, \\ \dfrac{(\log a)^{1-p}}{p-1} & \text{if } p > 1, \end{cases}$$

for any $a > 1$.

(d) We leave this as an exercise. •

7.1.6 Questions and Exercises

Questions 7.30.

1. What is an improper integral?
2. How many different types of improper integral are there?
3. Can we define the improper integral $\int_{-\infty}^\infty f(x)\,\mathrm{d}x$ by $\lim_{N\to\infty} \int_{-N}^N f(x)\,\mathrm{d}x$?
4. Suppose that f is continuous on $[1,\infty)$ and $\lim_{x\to\infty} xf(x) = \infty$. Must $\int_1^\infty f(x)\,\mathrm{d}x$ be divergent?
5. Suppose that f is continuous on $[1,\infty)$ such that $\lim_{x\to\infty} x^\alpha f(x) = \infty$ for some $\alpha \in (1,\infty)$. Must $\int_1^\infty f(x)\,\mathrm{d}x$ be divergent?
6. Suppose that f is continuous on $[a,b)$ and that there exists an $\alpha \in (0,1)$ such that $\lim_{x\to b-}(b-x)^\alpha f(x)$ exists as a finite number. Must $\int_a^b f(x)\,\mathrm{d}x$ be convergent absolutely?
7. Suppose that f is continuous on $[0,\infty)$. Does $\int_0^N f(x)\,\mathrm{d}x$ converge for each N with $N > 0$? Must $\int_0^\infty f(x)\,\mathrm{d}x$ be convergent?
8. Does $\int_0^\infty \sin x^2\,\mathrm{d}x$ converge? Does $\int_0^\infty \sin x^2\,\mathrm{d}x$ converge absolutely?
9. Does $\int_0^\infty x^{-1}\sin x\,\mathrm{d}x$ converge? Does it converge absolutely?
10. Does $\int_0^\infty \sin x\,\mathrm{d}x$ converge? Does $\int_{-\infty}^0 \sin x\,\mathrm{d}x$ converge? Does the Cauchy principal value of $\int_{-\infty}^\infty \sin x\,\mathrm{d}x$ exist?
11. Does $\int_0^\infty x\sin x\,\mathrm{d}x$ converge? Can $\int_0^\infty x^p\sin x\,\mathrm{d}x$ converge for each $p > 0$?
12. Does $\int_0^\infty \cos x\,\mathrm{d}x$ converge?
13. Does $\int_0^1 \log x\,\mathrm{d}x$ converge?

14. Is there something wrong in the following calculation?

$$\int_{-2}^{2} \frac{dx}{x^2} = -\frac{1}{x}\Big|_{-2}^{2} = -1.$$

15. Let $\alpha > 0$, $a > 0$, and let $\Gamma(\alpha)$ denote the gamma function.

    (a) Is it true that $\Gamma(\alpha) = 2\int_0^\infty t^{2\alpha-1}e^{-t^2}\,dt$?

    (b) Does $\Gamma(\alpha) = \int_0^1 (\log 1/t)^{\alpha-1}\,dt$?

    (c) Does $\Gamma(\alpha) = a^\alpha \int_0^\infty t^{\alpha-1}e^{-at}\,dt$?

16. Suppose that f is a nonnegative, continuous, and decreasing function of x for $x \geq N$. Set $a_k = f(k)$ for $k \geq N$. Does the $\sum_{k=1}^\infty a_k$ converge if and only if $\int_N^\infty f(x)\,dx$ exists?

Exercises 7.31.

1. Formulate the direct comparison test (see Theorem 7.5) for improper integrals of the second type.
2. Which of the following integrals are convergent? Which of them are divergent? In cases that the integrand is a function of p, find the range of p for which the corresponding integral converges.

    (a) $\displaystyle\int_1^\infty \frac{dx}{x^2-6x+8}$. (b) $\displaystyle\int_1^\infty \frac{dx}{\sqrt{1+x^3}}$. (c) $\displaystyle\int_1^\infty x\cos x\,dx$.

    (d) $\displaystyle\int_0^\infty e^{-x^3}\,dx$. (e) $\displaystyle\int_{-\infty}^\infty \frac{dx}{e^x+e^{-x}}$. (f) $\displaystyle\int_0^1 \frac{\sin(1/x)}{x^p}\,dx$.

    (g) $\displaystyle\int_0^\infty \frac{\cos x}{x^p}\,dx$. (h) $\displaystyle\int_1^\infty \frac{dx}{x\,(\log x)^p}$. (i) $\displaystyle\int_0^1 \frac{dx}{x^p\log x}$.

    (j) $\displaystyle\int_1^\infty \frac{x^{\frac{1}{2}}}{(1+x)^2}\,dx$. (k) $\displaystyle\int_{-1}^1 \frac{dx}{\sqrt{1-x^2}}$. (l) $\displaystyle\int_0^3 \frac{dx}{\sqrt{x}(2-x)^2}$.

    (m) $\displaystyle\int_3^\infty \frac{dx}{(2x-1)^p}$. (n) $\displaystyle\int_1^\infty \frac{x^2\,dx}{(x^3+2)^p}$. (o) $\displaystyle\int_0^\infty x^{p+3}e^{-x^p}\,dx$.

3. Prove the following:

    (a) $\displaystyle\int_0^\infty \frac{x}{(1+x)^3}\,dx = \frac{1}{2}\int_0^\infty \frac{dx}{(1+x)^2}$.

    (b) $\displaystyle\int_0^\infty xe^{-x^8}\,dx \times \int_0^\infty x^2e^{-x^4}\,dx = \frac{\pi}{16\sqrt{2}}$.

    (c) $\displaystyle\int_0^{\pi/2} \sqrt{\sin\theta}\,d\theta \times \int_0^{\pi/2} \frac{d\theta}{\sqrt{\sin\theta}} = \pi$.

    (d) $\displaystyle\int_0^{\pi/2} \sin^p\theta\,d\theta \times \int_0^{\pi/2} \sin^{p+1}\theta\,d\theta = \frac{\pi}{2(p+1)}$ $(p > -1)$.

(e) $\displaystyle\int_0^{\pi/2} \sin^n\theta\,\mathrm{d}\theta = \frac{\sqrt{\pi}\,\Gamma(\frac{n+1}{2})}{2\,\Gamma(\frac{n+2}{3})}$ $(n > -1)$.

(f) $\displaystyle\int_0^1 \frac{\mathrm{d}x}{\sqrt{1-x^n}} = \frac{\sqrt{\pi}\Gamma(\frac{1}{n})}{n\Gamma(\frac{1}{n}+\frac{1}{2})}$.

4. In the following integrals, either show that the improper integral converges and find its value, or show that it diverges.

(a) $\displaystyle\int_0^\infty x^6 \mathrm{e}^{-2x}\,\mathrm{d}x$. (b) $\displaystyle\int_0^2 \frac{x^2}{\sqrt{2-x}}\,\mathrm{d}x$. (c) $\displaystyle\int_0^\infty \frac{\mathrm{d}x}{\mathrm{e}^x+\mathrm{e}^{-x}}$.

(d) $\displaystyle\int_0^{\pi/3} \frac{\sec^2 x\,\mathrm{d}x}{1-\tan x}$. (e) $\displaystyle\int_0^6 \frac{x\,\mathrm{d}x}{x^2-4}$. (f) $\displaystyle\int_{-\infty}^\infty \frac{\mathrm{d}x}{(x-4)^3}$.

(g) $\displaystyle\int_0^6 \frac{x\,\mathrm{d}x}{(x^2-4)^{2/3}}$. (h) $\displaystyle\int_0^{\pi/2} \frac{\sin x\,\mathrm{d}x}{\sqrt[3]{1-2\cos x}}$. (i) $\displaystyle\int_0^{\pi/2} \frac{\mathrm{d}\theta}{\sqrt{\sin 2\theta}}$.

5. Find the area of the unbounded region between the x-axis and the curve $y = \frac{1}{(x-4)^3}$ for $x \geq 8$.
6. Complete Example 7.15, using Example 7.3 and the change of variable $x = 1/y$ with $a = 1$.
7. Discuss the following calculation: We have

$$\left(-2\sqrt{1-\sin x}\right)' = \sqrt{1+\sin x},$$

and so

$$\int_0^M \sqrt{1+\sin x}\,\mathrm{d}x = -2\sqrt{1-\sin x}\Big|_0^M = 2\left[1-\sqrt{1-\sin M}\right].$$

But as $M \to \infty$, the area under the curve $y = \sqrt{1+\sin x}$ between $x = 0$ and $x = M$ approaches infinity. On the other hand, the right-hand side of the last expression is never greater than 2. What is wrong, if anything, with this discussion?

8. Find $\int_{\pi/2}^{\pi} \sec x\,\mathrm{d}x$ if it exists. If it does not, explain why the improver integral diverges.
9. Let $a \neq 0$. Find the value α such that

$$\int_0^\infty \left(\frac{1}{(1+a^2x^2)^{1/2}} - \frac{\alpha}{x+1}\right)\mathrm{d}x$$

is convergent. Then evaluate that integral.

10. Show that the improper integral

$$\int_0^\infty \left(\frac{\alpha x}{1+x^2} - \frac{1}{1+2x}\right)\mathrm{d}x$$

is convergent if and only if $\alpha = 1/2$.

11. For what values of $\alpha > 0$ is the improper integral $\displaystyle\int_0^\infty \frac{x^{\alpha-1}}{1+x}\,dx$ convergent?
12. Formulate a statement of the limit comparison test with a proof when the functions involved satisfy the following conditions: Suppose that f and g are two positive continuous functions on $(a,b]$, $g(x) > 0$, and $f(x) \geq 0$ on some subinterval $(a,b_1]$ $(b_1 \leq b)$, and such that
$$\lim_{x\to a+} \frac{f(x)}{g(x)} = L \quad (0 \leq L \leq \infty),$$
where $-\infty \leq a < b$.
13. Evaluate $\int_0^1 x^p(1-x^q)^n\,dx$ $(p > -1,\ q > 0,\ n > -1)$. More generally, evaluate
$$\int_0^m x^p(m^q - x^q)^n\,dx.$$
14. Evaluate $\displaystyle\int_0^\infty x^n e^{-\alpha^2 x^2}\,dx$ $(\alpha > 0)$.
15. Give an example of a continuous nonmonotonic function $f(x)$ on $[1,\infty)$ such that $\sum_{k=1}^\infty f(k)$ converges but $\int_1^\infty f(x)\,dx$ does not.
16. Using the integral test, examine whether each of the following series converges or diverges.

(a) $\displaystyle\sum_{k=1}^\infty \frac{1}{(1+5k)^2}$. (b) $\displaystyle\sum_{k=2}^\infty \frac{1}{k\sqrt{k^2-1}}$. (c) $\displaystyle\sum_{k=1}^\infty (5+2k)^{-3/2}$.

(d) $\displaystyle\sum_{k=1}^\infty \frac{(\tan^{-1} 2k)^3}{1+4k^2}$. (e) $\displaystyle\sum_{k=1}^\infty \frac{1}{k(k+1)(k+2)}$. (f) $\displaystyle\sum_{k=2}^\infty \frac{1}{k^2+k}$.

17. Using the integral test, show that $\int_1^\infty \frac{x^{\alpha-1}}{1+x}\,dx$ converges if and only if $\alpha < 1$.

7.2 Applications of the Riemann Integral

We know that the area of a circle of radius r is $2\pi(r^2/2)$. Since the whole circle involves an angle of 2π at the center, the area of a circular sector of central angle ϕ in radian measure is $\frac{1}{2}r^2\phi$ (see Figure 7.16). In this section, we wish to find the area of a region bounded by arbitrary general curves that have path coordinates (r,θ).

7.2.1 Area in Polar Coordinates

Let a curve be given in polar coordinates by
$$r = f(\theta),$$
where $f(\theta)$ is a positive continuous function defined on $[\alpha,\beta]$, $0 \leq \alpha < \beta \leq 2\pi$. Sometimes it is convenient to fix α and β such that $-\pi \leq \alpha < \beta \leq \pi$, as we

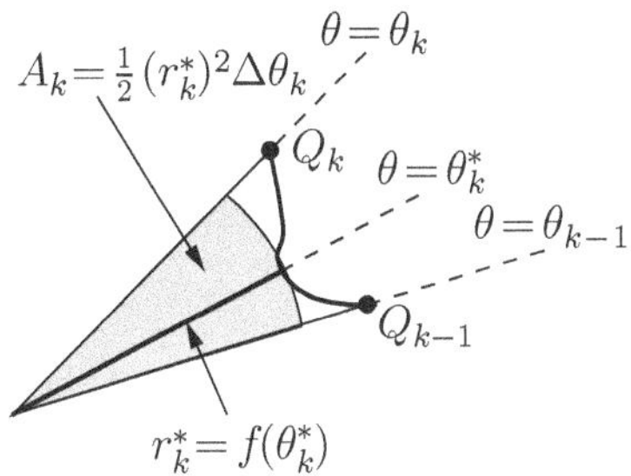

Fig. 7.16. Area of a circular sector.

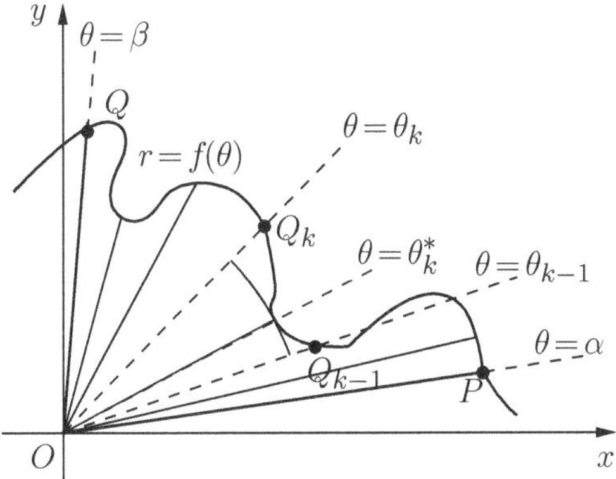

Fig. 7.17. Determining arc length of a polar curve.

shall see in a number of examples in this section. Consider the area A of the region (see POQ in Figure 7.16) bounded by the rays $\theta=\alpha$, $\theta=\beta$, and the curve $r=f(\theta)$, $\alpha\leq\theta\leq\beta$. Our experience with the Riemann integral suggests that we should divide the region into small sectors using radial lines, and a first guess is to approximate the sectorial area by the area of a circular arc. One of the general procedures is to consider a partition $P=\{\theta_0,\theta_1,\ldots,\theta_n\}$ of $[\alpha,\beta]$ given by

$$\alpha=\theta_0<\theta_1<\theta_2<\cdots<\theta_n=\beta.$$

That is, we divide the $\angle POQ$ into n parts. We draw the rays

$$\theta=\theta_k \quad \text{for} \quad k=0,1,2,\ldots,n.$$

These rays divide the area into n subregions A_k $(k=1,2,\ldots,n)$, as shown in Figure 7.16. Thus, the area A_k of the kth subregion will be bounded by the rays

$$\theta=\theta_{k-1}, \quad \theta=\theta_k,$$

and the portion of the curve $r=f(\theta)$ for which $\theta\in[\theta_{k-1},\theta_k]$. We first approximate the area A_k of the typical sector $Q_{k-1}OQ_k$ (see Figures 7.16 and 7.17).

Pick an arbitrary ray $\theta = \theta_k^*$, $\theta_{k-1} \leq \theta_k^* \leq \theta_k$. Denote by $r_k^* = f(\theta_k^*)$, the length of the radius vector corresponding to the angle θ_k^*. Then the area of the circular sector with radius r_k^* and central angle of radian measure $\Delta\theta_k = \theta_k - \theta_{k-1}$ is

$$\frac{1}{2}(r_k^*)^2\Delta\theta_k = \frac{1}{2}\left(f(\theta_k^*)\right)^2\Delta\theta_k,$$

which is an approximation to the area A_k. If we sum these areas, we get

$$\sum_{k=1}^{n}\frac{1}{2}(f(\theta_k^*))^2\Delta\theta_k,$$

which is the total area of the *steplike* circular sectors, and so the sum is a good approximation to A, the total area bounded by the given polar curve: $r = f(\theta)$, $\alpha \leq \theta \leq \beta$. Since f is a continuous function of θ for $\theta \in [\alpha, \beta]$, the above sum is clearly a Riemann sum of f on $[\alpha, \beta]$, and so we expect the approximation to improve for $\|P\| = \max\limits_{1\leq k\leq n} \Delta\theta_k \to 0$ as $n \to \infty$. Consequently, by the definition of the definite integral,

$$\lim_{\|P\|\to 0}\sum_{k=1}^{n}\frac{1}{2}(f(\theta_k^*))^2\Delta\theta_k = \int_\alpha^\beta \frac{1}{2}(f(\theta))^2\,\mathrm{d}\theta = \int_\alpha^\beta \frac{1}{2}r^2\,\mathrm{d}\theta = \int_\alpha^\beta \mathrm{d}A,$$

which is the required area A. Here

$$dA = \frac{1}{2}r^2\,\mathrm{d}\theta = \frac{1}{2}(f(\theta))^2\,\mathrm{d}\theta$$

is called the *differential element of area.* The above discussion gives the following result.

Theorem 7.32 (Area in polar coordinates). *Suppose that a polar curve is given by*

$$r = f(\theta),$$

where $f(\theta)$ is a positive continuous function defined on $[\alpha, \beta]$ $(0 \leq \alpha < \beta \leq 2\pi$ or $-\pi \leq \alpha < \beta \leq \pi)$. If A is the area of the region bounded by the curve $r = f(\theta)$ and the rays $\theta = \alpha$ and $\theta = \beta$, then the area A is given by

$$A = \int_\alpha^\beta \frac{1}{2}r^2\,\mathrm{d}\theta = \int_\alpha^\beta \frac{1}{2}(f(\theta))^2\,\mathrm{d}\theta.$$

More generally, suppose we wish to find the area A of the region that lies between two polar curves $r_2 = f_2(\theta)$, $r_1 = f_1(\theta)$ from $\theta = \alpha$ to $\theta = \beta$, where $f_2(\theta) \geq f_1(\theta) \geq 0$ for $\alpha \leq \theta \leq \beta$. For example, for the region shown in Figure 7.18, because

$$\text{Area of } (ABDCA) = \text{Area of } (OBAO) - \text{Area of } (ODCO),$$

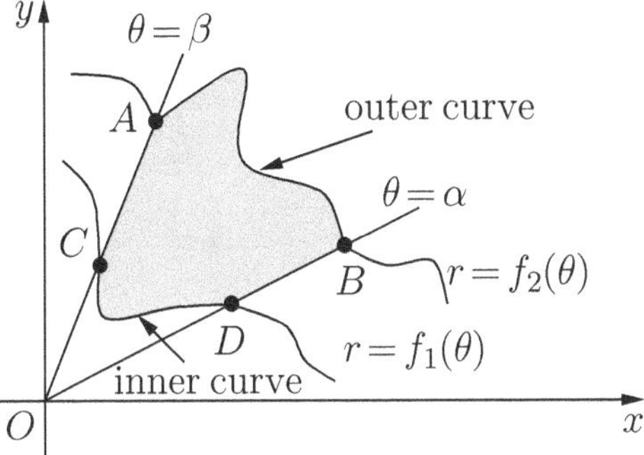

Fig. 7.18. Area between two polar curves.

we have the formula

$$A = \int_{\alpha}^{\beta} \frac{1}{2}(f_2(\theta))^2\, d\theta - \int_{\alpha}^{\beta} \frac{1}{2}(f_1(\theta))^2\, d\theta.$$

Sometimes, we may have to find the area A between two polar curves that have different intervals of integration (as in Figure 7.18 and Example 7.36**(a)**):

$$r_2 = f_2(\theta)\ (\alpha \le \theta \le \beta) \quad \text{and} \quad r_1 = f_1(\theta)\ (\alpha' \le \theta \le \beta'),$$

where $f_2(\theta) \ge f_1(\theta) \ge 0$. In that case, the corresponding formula for the area A is

$$A = \int_{\alpha}^{\beta} \frac{1}{2}(f_2(\theta))^2\, d\theta - \int_{\alpha'}^{\beta'} \frac{1}{2}(f_1(\theta))^2\, d\theta.$$

Just to indicate the *outer and the inner curves* in a convenient way to remember the formula, we may rewrite the above formula as

$$A = \int_{\alpha}^{\beta} \frac{1}{2} r_O^2\, d\theta - \int_{\alpha'}^{\beta'} \frac{1}{2} r_I^2\, d\theta,$$

where $r_O = f_2(\theta)$ and $r_I = f_1(\theta)$ represent the outer and inner curves, respectively.

In a given problem, the most difficult part of the problem is frequently to decide on the limits of integration. However, a decent sketch of the region should help in such problems.

Example 7.33. Find the area of the region bounded by following curves:

(a) $r = a(1 + \cos\theta)$, **(b)** $r = a(1 - \cos\theta)$,
(c) $r = a(1 - \sin\theta)$, **(d)** $r = a(1 + \sin\theta)$,

where $a > 0$. Each of these curves is called a *cardioid*, since it resembles a heart.

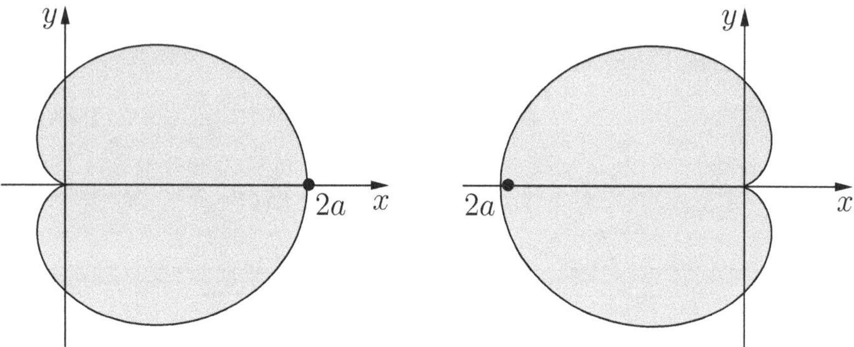

Fig. 7.19. Area enclosed by the curves $r = a(1 + \cos\theta)$ and $r = a(1 - \cos\theta)$.

Solution. **(a)** In this case we need to find the area of the entire region inside the curve, so we must let θ go from 0 to 2π (see Figure 7.19). According to the formula, the desired area A is given by

$$\begin{aligned} A &= \frac{1}{2}\int_0^{2\pi} a^2(1+\cos\theta)^2\, d\theta \\ &= \frac{a^2}{2}\int_0^{2\pi}\left(1 + 2\cos\theta + \frac{1+\cos 2\theta}{2}\right) d\theta \\ &= \frac{a^2}{2}\left[2\pi + 2\sin\theta\,\Big|_0^{2\pi} + \frac{1}{2}\left[2\pi + \frac{\sin 2\theta}{2}\Big|_0^{2\pi}\right]\right] \\ &= \frac{3}{2}a^2\pi. \end{aligned}$$

From Figures 7.19 and 7.20, it is clear that the areas of the regions bounded by the curves given by **(b)**–**(d)** remain the same. ●

Example 7.34. Find the area of the region that lies outside the cardioid $r = 2a(1 + \cos\theta)$ and inside the circle $r = 6a\cos\theta$, $a > 0$.

Solution. Recall that $r = 6a\cos\theta$ implies that

$$r^2 = 6a(r\cos\theta), \quad \text{i.e.,} \quad x^2 + y^2 = 6ax \quad \text{or} \quad (x - 3a)^2 + y^2 = (3a)^2,$$

which is a circle centered at $(3a, 0)$ with radius $3a$. Note that the full circle is described when θ runs from $-\pi/2$ to $\pi/2$. The points of intersection between the two curves are given by equating the two functions that represent these curves:

$$2a(1 + \cos\theta) = 6a\cos\theta \quad \text{or} \quad \cos\theta = \frac{1}{2}, \quad \text{i.e.,} \quad \theta = \pm\frac{\pi}{3}.$$

Therefore, $(3a, \pi/3)$ and $(3a, -\pi/3)$ are the intersection points of the given curves. The required area A corresponds to the region inside the circle and

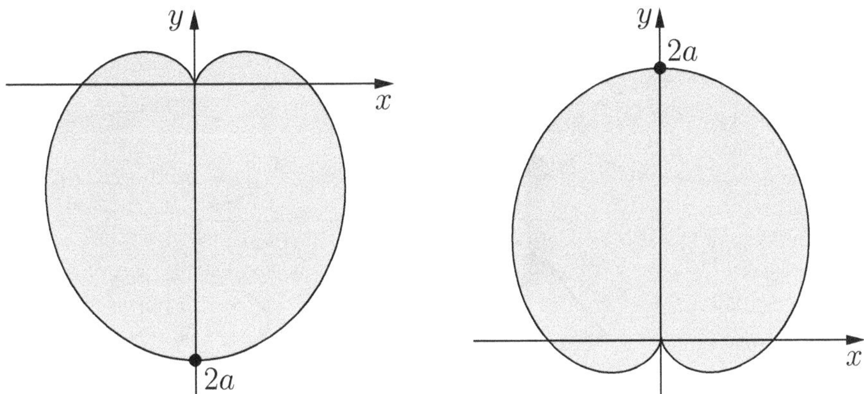

Fig. 7.20. Area enclosed by the curves $r = a(1 - \sin\theta)$ and $r = a(1 + \sin\theta)$.

outside the cardioid, provided θ lies between $-\pi/3$ and $\pi/3$. Thus (see Figure 7.21)

$$
\begin{aligned}
A &= \frac{1}{2}\int_{-\pi/3}^{\pi/3} \left\{(r_O(\theta))^2 - (r_I(\theta))^2\right\} d\theta \\
&= \frac{1}{2}\int_{-\pi/3}^{\pi/3} \left\{(6a\cos\theta)^2 - (2a(1+\cos\theta))^2\right\} d\theta \\
&= a^2 \int_0^{\pi/3} \left\{32\cos^2\theta - 8\cos\theta - 4\right\} d\theta \\
&= a^2 \int_0^{\pi/3} \{16(1+\cos 2\theta) - 8\cos\theta - 4\}\, d\theta \\
&= a^2 \left[12\left(\frac{\pi}{3} - 0\right) + \ 8\sin 2\theta\Big|_0^{\pi/3} \ - \ 8\sin\theta\Big|_0^{\pi/3}\right] \\
&= a^2\left[4\pi + 8\sin(2\pi/3) - 8\sin(\pi/3)\right],
\end{aligned}
$$

which by a simplification, gives $A = 4\pi a^2$. •

Remark 7.35. We note that in Example 7.34, we miss the origin among the points of intersection. Indeed, if we superimpose the two curves as in Figure 7.21, we find that these two curves also intersect at the origin, because $(0, \pi)$ and $(0, \pi/2)$ satisfy the cardioid and the circle, respectively. Such points will be missed in general when we solve the given pair of equations. •

Example 7.36. In each case find the area inside the first curve and outside the second one ($a > 0$):

(a) $r = a(1 + \cos\theta)$, $r = 2a\cos\theta$. **(b)** $r = 3a\cos\theta$, $r = a(1 + \cos\theta)$.

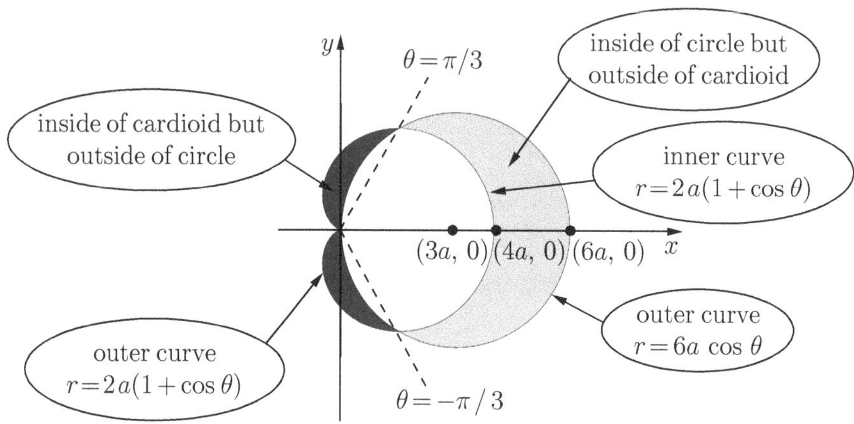

Fig. 7.21. Area outside $r = 2a(1 + \cos\theta)$ and inside $r = 6a\cos\theta$.

(c) $r = a\sin\theta$, $r = a(1 - \cos\theta)$. **(d)** $r = a$, $r = a(1 - \cos\theta)$.
(e) $r = a\cos\theta$, $r = a(1 - \cos\theta)$.

Solution. **(a)** Recall that $r = 2a\cos\theta$ is the equation of a circle centered at $(a, 0)$ and of radius a. The point of intersection of the two curves is given by

$$2a\cos\theta = a(1 + \cos\theta), \quad \text{i.e.,} \quad \cos\theta = 1 \quad \text{or} \quad \theta = 0.$$

We note that the origin is also a point of intersection, because although the coordinates $(0, \pi)$ and $(0, \pi/2)$ are different, both represent the origin, and the curves pass through the origin. The required area outside the circle and inside the cardioid is (see Figure 7.22)

$$\begin{aligned} A &= 2\left[\int_0^{\pi} \frac{1}{2} r_O^2 \, d\theta - \int_0^{\pi/2} \frac{1}{2} r_I^2 \, d\theta\right] \\ &= 2\left[\int_0^{\pi} \frac{1}{2}(a(1 + \cos\theta))^2 \, d\theta - \int_0^{\pi/2} \frac{1}{2}(2a\cos\theta)^2 \, d\theta\right] \\ &= a^2\left[\int_0^{\pi} (1 + \cos\theta)^2 \, d\theta - 4\int_0^{\pi/2} \cos^2\theta \, d\theta\right]. \end{aligned}$$

A computation shows that $A = \pi a^2/2$.

(b) This has been done in Example 7.34 except with the factor 2.
(c) The equation $r = a\sin\theta$ gives

$$r^2 = a(r\sin\theta), \quad \text{i.e.,} \quad x^2 + y^2 - ay = 0, \quad \text{or} \quad x^2 + (y - a/2)^2 = (a/2)^2,$$

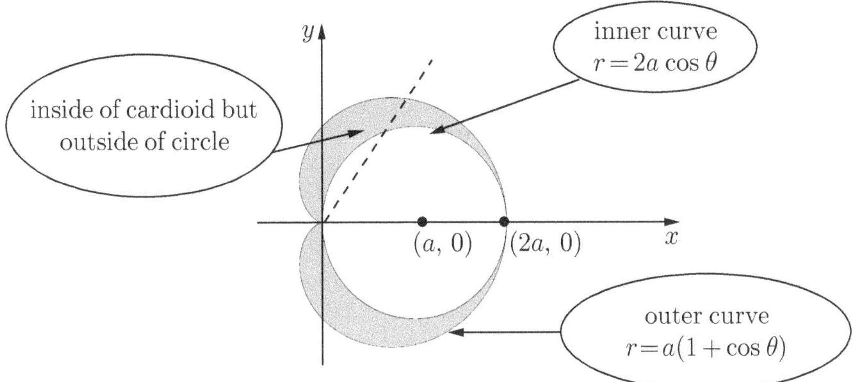

Fig. 7.22. Area outside of $r = 2a\cos\theta$ and inside of $r = a(1 + \cos\theta)$.

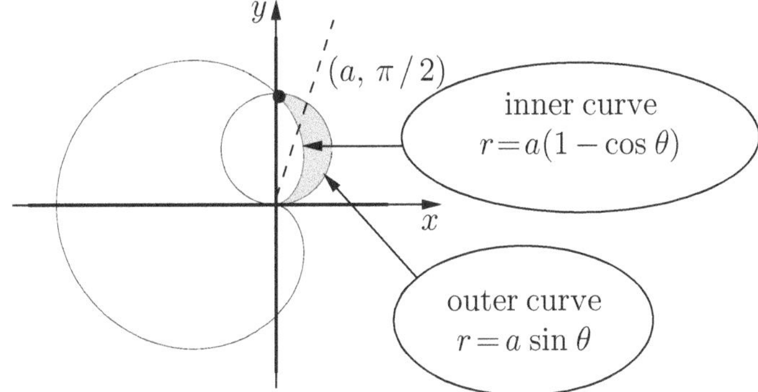

Fig. 7.23. Area outside of $r = a(1 - \cos\theta)$ and inside of $r = a\sin\theta$.

which is a circle. In polar coordinates, this is centered at $(a/2, \pi/2)$ and has radius $a/2$. The points of intersection are given by

$$a\sin\theta = a(1 - \cos\theta), \quad \text{i.e.,} \quad \sin\theta + \cos\theta = 1,$$

which gives $\theta = 0, \pi/2$. From the sketch (see Figure 7.23), the required area is

$$\begin{aligned} A &= \int_0^{\pi/2} \frac{1}{2} r_O^2 \, d\theta - \int_0^{\pi/2} \frac{1}{2} r_I^2 \, d\theta \\ &= \frac{1}{2} \int_0^{\pi/2} \left[a^2 \sin^2\theta - a^2(1 - \cos\theta)^2\right] d\theta. \end{aligned}$$

A computation gives $A = a^2(4 - \pi)/4$.

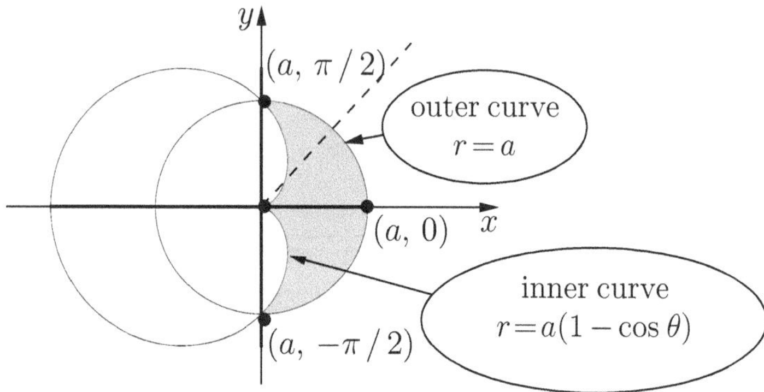

Fig. 7.24. Area inside of the circle $r = a$ and outside of the cardioid $r = a(1-\cos\theta)$.

(d) From the sketch (see Figure 7.24), the desired area is

$$\begin{aligned} A &= 2\left[\int_0^{\pi/2} \frac{1}{2}r_O^2\,\mathrm{d}\theta - \int_0^{\pi/2} \frac{1}{2}r_I^2\,\mathrm{d}\theta\right] \\ &= 2\left[\int_0^{\pi/2} \left(\frac{1}{2}a^2 - \frac{1}{2}a^2(1-\cos\theta)^2\right)\mathrm{d}\theta\right], \end{aligned}$$

and a computation shows that $A = a^2(2 - \pi/4)$.

(e) The curve $r = a\cos\theta$ describes the circle centered at $(a/2, 0)$ of radius $a/2$. The points of intersection of $r = a\cos\theta$ and $r = a(1-\cos\theta)$ are given by

$$a\cos\theta = a(1-\cos\theta), \quad \text{i.e.,} \quad \cos\theta = \frac{1}{2} \quad \text{or} \quad \theta = \pm\frac{\pi}{3},$$

and so the origin $(0, 0)$ and $(a, \pm\pi/3)$ are the points of intersection of the two curves. From the sketch (see Figure 7.25), the desired area is

$$\begin{aligned} A &= 2\left[\int_0^{\pi/3} \frac{1}{2}r_O^2\,\mathrm{d}\theta - \int_0^{\pi/3} \frac{1}{2}r_I^2\,\mathrm{d}\theta\right] \\ &= 2\left[\int_0^{\pi/3} \frac{1}{2}a^2\cos^2\theta\,\mathrm{d}\theta - \int_0^{\pi/3} \frac{1}{2}a^2(1-\cos\theta)^2]\,\mathrm{d}\theta\right] \\ &= a^2\int_0^{\pi/3} (2\cos\theta - 1)\,\mathrm{d}\theta, \end{aligned}$$

and a computation gives that $A = a^2(\sqrt{3} - \pi/3)$. ●

7.2.2 Arc Length of a Plane Curve

Physically speaking, the length of a curve (arc length) is quite a simple concept, although mathematically it is nontrivial. Suppose that we are given an

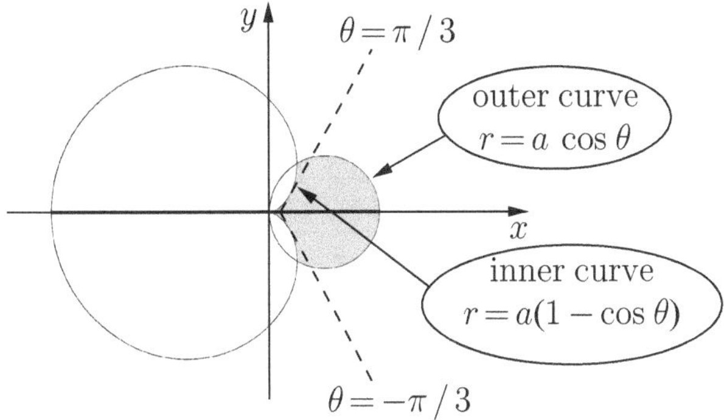

Fig. 7.25. Area outside of $r = a(1 - \cos\theta)$ and inside of $r = a\cos\theta$.

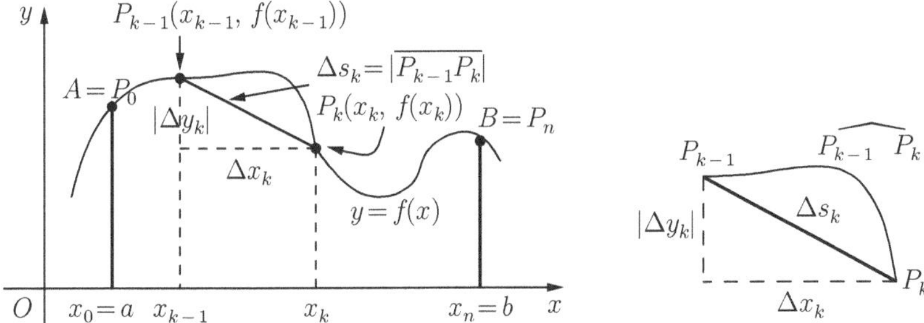

Fig. 7.26. Arc length in Cartesian coordinates.

arc of a curve $y = f(x)$ from A to B. The question is how to define the length of the arc $\widehat{AB}$ as a limit of Riemann sums, and hence as a definite integral. The aim of this subsection is to present a solid mathematical definition for arc length. Consider the equation

$$y = f(x), \quad x \in [a, b],$$

which describes the curve from A to B, denoted by $\widehat{AB}$, as shown in Figure 7.26. To find the length of the curve $\widehat{AB}$, we first try to approximate the length by computing the length of suitable polygonal paths that lie close to the curve. To do this, we consider a partition $P = \{x_0, x_1, \ldots, x_n\}$ of $[a, b]$. For $k = 1, 2, \ldots, n$, let $P_k = P_k(x_k, f(x_k))$ denote points on the curve $\widehat{AB}$, so that

$$P_0(x_0, f(x_0)) = P_0(a, f(a)) = A(a, f(a)) \text{ and } P_n(x_n, f(x_n)) = B(b, f(b)).$$

By joining points $P_0, P_1, \ldots, P_n$, we obtain polygonal paths (i.e., chords) whose lengths we shall denote by $\Delta s_1, \Delta s_2, \ldots, \Delta s_n$. The length of the kth chord is

$$\Delta s_k = |\overline{P_{k-1}P_k}| = \sqrt{\Delta x_k^2 + \Delta y_k^2} = \sqrt{1 + \left(\frac{\Delta y_k}{\Delta x_k}\right)^2}\,\Delta x_k,$$

or equivalently,

$$\Delta s_k = \sqrt{1 + \left(\frac{f(x_k) - f(x_{k-1})}{x_k - x_{k-1}}\right)^2}\,\Delta x_k. \tag{7.12}$$

The length of the polygonal path connecting the points $P_0, P_1, \dots, P_n$ on the graph of f is the sum

$$s_n = \sum_{k=1}^{n} \Delta s_k = \sum_{k=1}^{n} \sqrt{1 + \left(\frac{f(x_k) - f(x_{k-1})}{x_k - x_{k-1}}\right)^2}\,\Delta x_k,$$

which is an approximation to the length of the arc $\widehat{AB}$. When the number of division points is increased infinitely while $\|P\| = \max_{1\le k\le n} \Delta x_k \to 0$, we obtain the length s of the curve $y = f(x)$ between $x = a$ and $x = b$, defined as

$$s = \lim_{n\to\infty} s_n = \lim_{n\to\infty} \sum_{k=1}^{n} \sqrt{1 + \left(\frac{f(x_k) - f(x_{k-1})}{x_k - x_{k-1}}\right)^2}\,\Delta x_k, \tag{7.13}$$

provided the limit exists. First we note that the sum on the right-hand side of (7.13) is not in standard form, but it can be put into such a form as follows:

Assume that f is continuously differentiable on $[a, b]$. Then the mean value theorem is applicable to f on each subinterval $[x_{k-1}, x_k]$. This means that there exists a point x_k^* in (x_{k-1}, x_k) such that

$$f(x_k) - f(x_{k-1}) = f'(x_k^*)(x_k - x_{k-1}).$$

In view of this, the length of the kth chord (see (7.12)) is

$$\Delta s_k = \sqrt{1 + (f'(x_k^*))^2}\,\Delta x_k,$$

so that the arc length of the graph of $y = f(x)$ on $[a, b]$ may be approximated by the Riemann sum

$$s_n = \sum_{k=1}^{n} \sqrt{1 + (f'(x_k^*))^2}\,\Delta x_k.$$

Since $f'(x)$ is continuous on $[a, b]$, $\sqrt{1 + (f'(x))^2}$ is also continuous on $[a, b]$, and so $\lim_{\|P\|\to 0} s_n$ exists, which is in fact the definite integral

$$s = \int_a^b \sqrt{1 + (f'(x))^2}\, dx.$$

We have thus demonstrated the following theorem.

Theorem 7.37 (Arc length in Cartesian coordinates). *Let $f(x)$ be smooth (i.e., continuously differentiable) on $[a,b]$. Then the length s of the curve of $y = f(x)$ between $x = a$ and $y = b$ is given by*

$$s = \int_a^b \sqrt{1 + (f'(x))^2}\, \mathrm{d}x = \int_{x=a}^{x=b} \sqrt{1 + \left(\frac{\mathrm{d}y}{\mathrm{d}x}\right)^2}\, \mathrm{d}x.$$

This formula is valid even if f is not a positive function. Interchanging the roles of x and y, we can easily arrive at the following analogous result, which will sometimes be useful.

Theorem 7.38. *If $g(y)$ is smooth on $[c,d]$, then the length of the curve $x = g(y)$ between $y = c$ and $y = d$ is given by*

$$s = \int_c^d \sqrt{1 + (g'(y))^2}\, \mathrm{d}y = \int_{y=c}^{y=d} \sqrt{1 + \left(\frac{\mathrm{d}x}{\mathrm{d}y}\right)^2}\, \mathrm{d}y. \tag{7.14}$$

Proof. In the above discussion, we simply transform $\Delta s_k = \sqrt{\Delta x_k^2 + \Delta y_k^2}$ into

$$\Delta s_k = \sqrt{1 + \left(\frac{\Delta x_k}{\Delta y_k}\right)^2}\, \Delta y_k,$$

which gives the formula

$$s = \lim_{\|P\| \to 0} \sum_{k=1}^{n} \sqrt{1 + (g'(y_k^*))^2 \Delta y_k} = \int_c^d \sqrt{1 + (g'(y))^2}\, \mathrm{d}y.$$

The remaining details are easy to fill in, and so we omit them. ∎

The arc-length formula in Theorem 7.37 is often written with differentials instead of derivatives. If we use $\frac{\mathrm{d}y}{\mathrm{d}x}$ for $f'(x)$, then we may formally write

$$\sqrt{1 + (f'(x))^2}\, \mathrm{d}x = \sqrt{1 + \left(\frac{\mathrm{d}y}{\mathrm{d}x}\right)^2}\, \mathrm{d}x = \sqrt{(\mathrm{d}x)^2 + (\mathrm{d}y)^2} = \mathrm{d}s,$$

so that we end up with

$$s = \int_\alpha^\beta \sqrt{(\mathrm{d}x)^2 + (\mathrm{d}y)^2}. \tag{7.15}$$

Is this not reminiscent of the Pythagorean theorem? If we think of $\mathrm{d}x$ and $\mathrm{d}y$ as two sides of a small right triangle, then

$$\mathrm{d}s = \sqrt{(\mathrm{d}x)^2 + (\mathrm{d}y)^2}$$

represents the "hypotenuse" which is referred to as a differential of the arc length, while

$$s = \int \mathrm{d}s$$

is the differential formula for arc length.

Example 7.39. From basic calculus, we know that the circumference of the circle $x^2+y^2=a^2$ is $2\pi a$. Let us verify this using our definition. The circumference s of this circle is

$$s = 4\int_0^a \sqrt{1+(f'(x))^2}\,dx,$$

where $y=f(x)=\sqrt{a^2-x^2}$, $x\in[0,a]$, is the quarter-circle of radius a lying in the first quadrant. Note that

$$\sqrt{1+(f'(x))^2} = \sqrt{1+\left(\frac{-x}{\sqrt{a^2-x^2}}\right)^2} = \frac{a}{\sqrt{a^2-x^2}},$$

so that we end up with an improper integral for s:

$$s = 4a\int_0^a \frac{dx}{\sqrt{a^2-x^2}} = 4a\arcsin\left(\frac{x}{a}\right)\Big|_0^a = 4a\arcsin 1 = 2\pi a. \qquad \bullet$$

The value of s in this example may also be obtained using the parametric equation $x=a\cos t$ and $y=a\sin t$ for $0\le t\le 2\pi$. This formula is described later, in (7.16).

Example 7.40. Find the length of the curve

(a) $y=f(x)=\log\left(\dfrac{e^x-1}{e^x+1}\right)$ for $x\in[1,2]$,

(b) $y=f(x)=x^{2/3}$ for $x\in[-1,8]$.

Solution. **(a)** Clearly $f(x)$ is continuously differentiable on $[1,2]$ with

$$f'(x) = \frac{e^x}{e^x-1} - \frac{e^x}{e^x+1} = \frac{2e^x}{e^{2x}-1} = \frac{2}{e^x-e^{-x}},$$

so that

$$1+(f'(x))^2 = 1+\frac{4}{(e^x-e^{-x})^2} = \left(\frac{e^x+e^{-x}}{e^x-e^{-x}}\right)^2.$$

Thus

$$s = \int_1^2 \sqrt{1+(f'(x))^2}\,dx = \int_1^2 \frac{e^x+e^{-x}}{e^x-e^{-x}}\,dx = \log(e^x-e^{-x})\Big|_1^2,$$

which gives

$$s = \log\left(\frac{e^2-e^{-2}}{e-e^{-1}}\right) = \log(e+1/e).$$

(b) Let $f(x)=x^{2/3}$, $x\in[-1,8]$ (see Figure 7.27). Note that there is a vertical tangent at $x=0$, because

$$f'(x) = \frac{2}{3}\frac{1}{x^{1/3}} \to \infty \text{ as } x\to 0+ \quad \text{and } f'(x)\to -\infty \text{ as } x\to 0-,$$

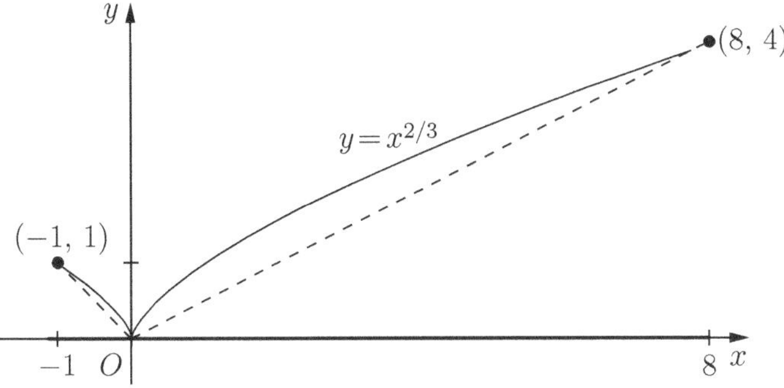

Fig. 7.27. The graph of $y = x^{2/3}$ on $[-1, 8]$.

and so the integrand in the arc-length formula in (7.13) approaches infinity as $x \to 0$. In this case, we may write

$$s = \int_{-1}^{0} \sqrt{1 + (f'(x))^2}\, dx + \int_{0}^{8} \sqrt{1 + (f'(x))^2}\, dx,$$

where both integrals on the right are improper. Thus,

$$s = \lim_{\epsilon \to 0-} \int_{-1}^{\epsilon} \sqrt{1 + \frac{4}{9x^{2/3}}}\, dx + \lim_{\eta \to 0+} \int_{\eta}^{8} \sqrt{1 + \frac{4}{9x^{2/3}}}\, dx,$$

and this can be computed. Alternatively, we may use the formula (7.14) to find the length of the curve. We need to find dx/dy, where

$$y = f(x) = x^{2/3} \quad \text{on} \quad [-1, 8] \Longleftrightarrow x = \pm y^{3/2} = g(y) \quad \text{on} \quad [0, 4].$$

Note that for $[-1, 0]$, we have

$$x = -y^{3/2}, \ y \in [0, 1],$$

and for $x \in [0, 8]$, we have

$$x = y^{3/2}, \ y \in [0, 4].$$

This observation leads to

$$s = \int_{0}^{1} \sqrt{1 + (g_1'(y))^2}\, dy + \int_{0}^{4} \sqrt{1 + (g_2'(y))^2}\, dy.$$

In either case, we have

$$\sqrt{1 + (g'(y))^2} = \sqrt{1 + \frac{9}{4}y},$$

and so

$$\begin{aligned} s &= \int_0^1 \sqrt{1+\frac{9}{4}y}\,\mathrm{d}y + \int_0^4 \sqrt{1+\frac{9}{4}y}\,\mathrm{d}y \\ &= \frac{4}{9}\left[\frac{(1+9y/4)^{3/2}}{3/2}\bigg|_0^1 + \frac{(1+9y/4)^{3/2}}{3/2}\bigg|_0^4\right] \\ &= \frac{1}{27}\left[13\sqrt{13}+80\sqrt{10}-16\right] \approx 10.5. \end{aligned}$$

Note that the sum of the lengths of the two inscribed chords is

$$\sqrt{2}+\sqrt{64+16} = \sqrt{2}+4\sqrt{5} \approx 10.4.$$

•

7.2.3 Arc Length for Parameterized Curves

Suppose that a smooth curve $y = f(x)$, $x \in [a, b]$, is given parametrically by

$$x = x(\theta),\ y = y(\theta) \quad (\alpha \le \theta \le \beta),$$

where $x(\theta)$ and $y(\theta)$ are continuously differentiable functions of θ, $\theta \in [\alpha, \beta]$, with $y'(\theta) \neq 0$ on $[\alpha, \beta]$, and the curve $y = f(x)$ does not intersect itself, except possibly for $\theta = \alpha$ and $\theta = \beta$. Here the points that corresponds to α and β are the endpoints of the curve. So we set $a = x(\alpha)$ and $b = y(\beta)$, and thus $b = f(a)$. Then

$$y = f(x) \Longleftrightarrow y(\theta) = f(x(\theta)); \quad \text{and} \quad y'(\theta) = f'(x(\theta))x'(\theta),$$

so that

$$\mathrm{d}s = \sqrt{1+(f'(x))^2}\,\mathrm{d}x = \sqrt{1+\left(\frac{y'(\theta)}{x'(\theta)}\right)^2}\,x'(\theta)\,\mathrm{d}\theta = \sqrt{x'(\theta)^2+y'(\theta)^2}\,\mathrm{d}\theta.$$

Then the Cartesian form of the arc length s takes the form

$$s = \int_a^b \sqrt{1+(f'(x))^2}\,\mathrm{d}x = \int_\alpha^\beta \sqrt{x'(\theta)^2+y'(\theta)^2}\,\mathrm{d}\theta,$$

and so the length s of the curve C (as described in (7.13)) becomes

$$s = \lim \sum_{k=1}^{n} \Delta s_k = \lim \sum_{k=1}^{n} \sqrt{x'(\theta_k^*)^2 + (y'(\theta_k^{**}))^2 \Delta\theta_k},$$

which is the definite integral

$$s = \int_\alpha^\beta \sqrt{x'(\theta)^2+y'(\theta)^2}\,\mathrm{d}\theta = \int_\alpha^\beta \sqrt{\left(\frac{\mathrm{d}x}{\mathrm{d}\theta}\right)^2 + \left(\frac{\mathrm{d}y}{\mathrm{d}\theta}\right)^2}\,\mathrm{d}\theta. \tag{7.16}$$

Now we shall consider examples for finding the arc length when the limits of integration are given.

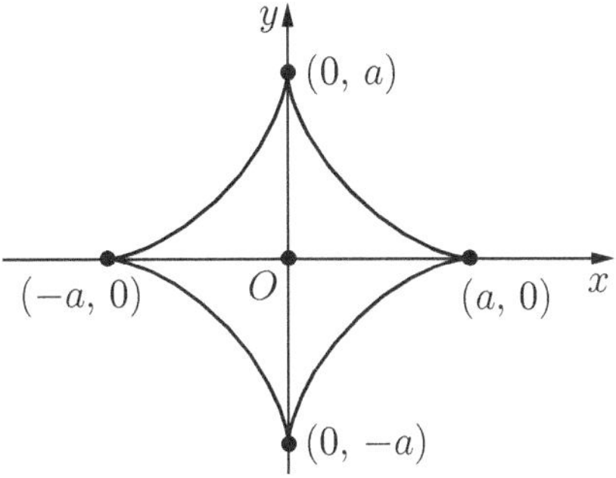

Fig. 7.28. The astroid $x^{2/3} + y^{2/3} = a^{2/3}$.

Example 7.41. Find the length of the curve parameterized by:

(a) $x = a\cos^3\theta$, $y = a\sin^3\theta$ for $0 \le \theta \le 2\pi$. This curve is known as an *astroid.*

(b) $x = a\cos^3\theta$, $y = b\sin^3\theta$ for $0 \le \theta \le 2\pi$.

(c) $x = e^\theta \sin\theta$, $y = e^\theta\cos\theta$ for $0 \le \theta \le \pi$.

(d) $x = a\cos\theta$, $y = b\sin\theta$ for $0 \le \theta \le 2\pi$. The curve is the general equation of an ellipse.

Solution. **(a)** Note that the Cartesian form of the given astroid is (Figure 7.28)

$$x^{2/3} + y^{2/3} = a^{2/3}, \quad \text{or} \quad \left(\frac{x}{a}\right)^{2/3} + \left(\frac{y}{a}\right)^{2/3} = 1.$$

From the symmetry of the astroid, it is apparent that the arc length of the curve in polar form is

$$s = 4\int_0^{\pi/2} \sqrt{x'(\theta)^2 + y'(\theta)^2}\, d\theta,$$

where $x(\theta) = a\cos^3\theta$ and $y(\theta) = a\sin^3\theta$. Since

$$x'(\theta)^2 + y'(\theta)^2 = [3a\cos^2\theta(-\sin\theta)]^2 + [3a\sin^2\theta(\cos\theta)]^2 = 9a^2\cos^2\theta\sin^2\theta,$$

we have

$$s = 4\int_0^{\pi/2} 3a\cos\theta\sin\theta\, d\theta = 12a\left(\frac{\sin^2\theta}{2}\right)\Big|_0^{\pi/2} = 6a.$$

(b) Proceeding exactly as in **(a)**, we see that (see Figure 7.29)

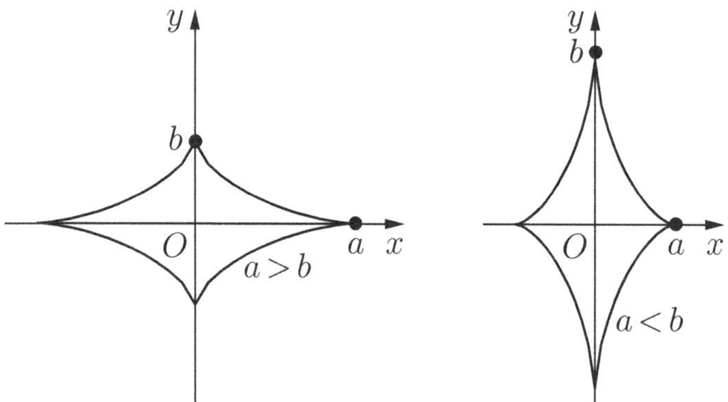

Fig. 7.29. The astroids $(x/a)^{2/3} + (y/b)^{2/3} = 1$ for $a > b$ and $a < b$ respectively.

$$s = 6\int_0^{\pi/2} \sqrt{a^2 + (b^2 - a^2)\sin^2\theta}\, \mathrm{d}(\sin^2\theta)$$

$$= \begin{cases} 6\left(\dfrac{[a^2 + (b^2 - a^2)\sin^2\theta]^{3/2}}{\frac{3}{2}(b^2 - a^2)}\right)\Bigg|_0^{\pi/2} & \text{if } b \neq a, \\ 6a\displaystyle\int_0^{\pi/2} \mathrm{d}(\sin^2\theta) & \text{if } b = a, \end{cases}$$

which may be simplified to obtain

$$s = \begin{cases} \dfrac{4(b^3 - a^3)}{b^2 - a^2} & \text{if } b \neq a, \\ 6a & \text{if } b = a. \end{cases}$$

(c) For the given curve, we easily see that (see Figure 7.30)

$$x'(\theta)^2 + y'(\theta)^2 = (\mathrm{e}^\theta(\cos\theta + \sin\theta)^2) + (\mathrm{e}^\theta(\cos\theta - \sin\theta))^2 = 2\mathrm{e}^{2\theta},$$

and so the desired arc length is

$$s = \int_0^\pi \sqrt{2}\mathrm{e}^\theta\, \mathrm{d}\theta = \sqrt{2}\mathrm{e}^\theta\Big|_0^\pi = \sqrt{2}(\mathrm{e}^\pi - 1).$$

(d) Without loss of generality, we assume that $a > b > 0$. In this case,

$$x'(\theta)^2 + y'(\theta)^2 = (a\sin\theta)^2 + (-b\sin\theta)^2 = a^2\left[1 - \left(1 - \frac{b^2}{a^2}\right)\cos^2\theta\right],$$

so that

$$s = 4\int_0^{\pi/2} a\sqrt{1 - k^2\cos^2\theta}\, \mathrm{d}\theta, \quad k = \frac{\sqrt{a^2 - b^2}}{a}.$$

We remind the reader that it is not possible to express this result in a simple form using elementary functions, although estimates for this integral are known. ●

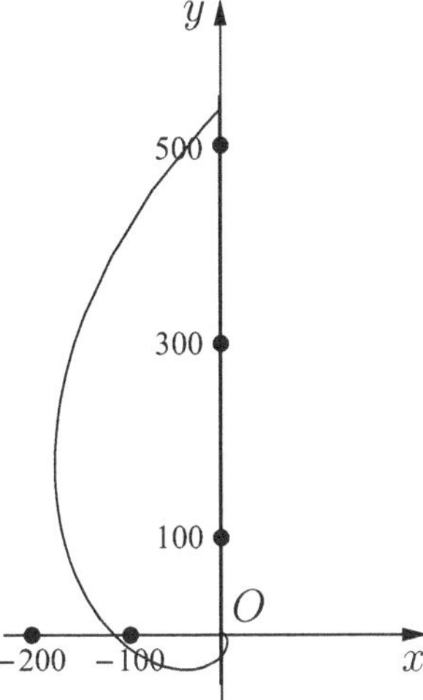

Fig. 7.30. The curve $x = \mathrm{e}^{\theta} \sin\theta$, $y = \mathrm{e}^{\theta}\cos\theta$ for $0 \le \theta \le 2\pi$.

7.2.4 Arc Length of Polar Curves

Now we turn our attention to finding the length of a curve C with polar equation

$$r = f(\theta) \quad (\alpha \le \theta \le \beta),$$

where r, θ, and $f(\theta)$ are respectively the radial vector, the polar angle, and a continuously differentiable function defined on $[\alpha, \beta]$ with $f(\theta) \ge 0$ on $[\alpha, \beta]$. We have the following parametric representation of C:

$$x = r\cos\theta = f(\theta)\cos\theta \quad \text{and} \quad y = r\sin\theta = f(\theta)\sin\theta.$$

We may regard these equations as the parametric equations of the curve. Again, by (7.14), the formula for computing the arc length of the polar curve $r = f(\theta)$ from $\theta = \alpha$ to $\theta = \beta$ is

$$s = \int_{\alpha}^{\beta} \mathrm{d}s = \int_{\alpha}^{\beta} \sqrt{(x'(\theta))^2 + (y'(\theta))^2}\, \mathrm{d}\theta.$$

Using the trigonometric identity $\sin^2\theta + \cos^2\theta = 1$, we see that

$$\begin{aligned}(x'(\theta))^2 + (y'(\theta))^2 &= (f'(\theta)\cos\theta - f(\theta)\sin\theta)^2 + (f'(\theta)\sin\theta + f(\theta)\cos\theta)^2 \\ &= (f(\theta))^2 + (f'(\theta))^2 = r^2 + \left(\frac{\mathrm{d}r}{\mathrm{d}\theta}\right)^2,\end{aligned}$$

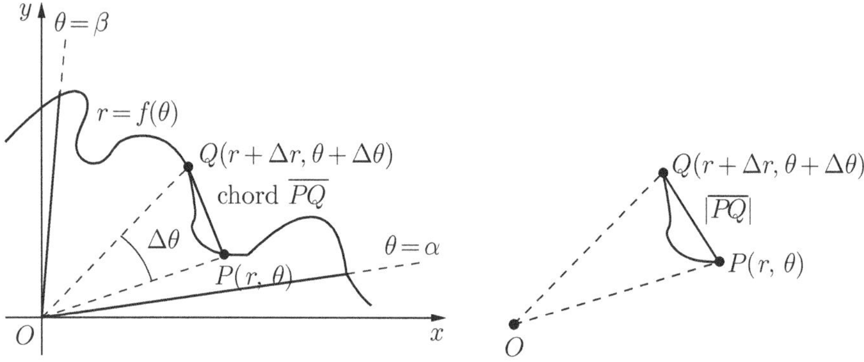

Fig. 7.31. Length of a polar curve.

and so we have the following formula for arc length s when the curve is specified in terms of polar coordinates:

$$s = \int_{\alpha}^{\beta} \sqrt{(f(\theta))^2 + (f'(\theta))^2}\,\mathrm{d}\theta, \quad \text{i.e.,} \quad s = \int_{\alpha}^{\beta} \sqrt{r^2 + \left(\frac{\mathrm{d}r}{\mathrm{d}\theta}\right)^2}\,\mathrm{d}\theta.$$

We now formulate the above discussion into a theorem.

Theorem 7.42 (Length of an arc of a polar curve). *Suppose that $r = f(\theta)$ is continuously differentiable for $\alpha \le \theta \le \beta$, $0 \le \beta - \alpha < 2\pi$. Then the length of the arc $r = f(\theta)$, $\alpha \le \theta \le \beta$, is*

$$s = \int_{\alpha}^{\beta} \mathrm{d}s, \quad \mathrm{d}s = \sqrt{r^2 + (r')^2}\,\mathrm{d}\theta. \tag{7.17}$$

Alternative proof of Theorem 7.42. Let $P(r, \theta)$ be an arbitrary point on the polar curve

$$r = f(\theta), \quad \alpha \le \theta \le \beta.$$

Consider a neighboring point Q on the polar curve such that the angle between the radial lines $\overline{OP}$ and $\overline{OQ}$ is $\Delta\theta$. Let $|\overline{PQ}|$ denote the length of the chord $\overline{PQ}$ (see Figure 7.31). By the law of cosines in trigonometry, it follows that

$$\begin{aligned} |\overline{PQ}|^2 &= r^2 + (r + \Delta r)^2 - 2r(r + \Delta r)\cos\Delta\theta \\ &= (2r^2 + 2r\Delta r)(1 - \cos\Delta\theta) + \Delta r^2 \\ &= \left[(2r^2 + 2r\Delta r)\frac{1 - \cos\Delta\theta}{\Delta\theta^2} + \left(\frac{\Delta r}{\Delta\theta}\right)^2\right](\Delta\theta)^2, \end{aligned}$$

so that the length of the chord $\overline{PQ}$ is

$$|\overline{PQ}| = \left[\sqrt{(2r^2 + 2r\Delta r)\frac{1 - \cos\Delta\theta}{\Delta\theta^2} + \left(\frac{\Delta r}{\Delta\theta}\right)^2}\right]\Delta\theta.$$

Since $f(\theta)$ is continuously differentiable, allowing $\Delta\theta \to 0$, we see that

$$\Delta r = f(\theta + \Delta\theta) - f(\theta) \to 0 \quad \text{as } \Delta\theta \to 0,$$

and so the term in parentheses approaches

$$\sqrt{r^2 + \left(\frac{dr}{d\theta}\right)^2}.$$

The remainder of the argument is left as an exercise. ∎

Example 7.43. Find the length of each of the following curves:

(a) $r = a(1 + \cos\theta)$. **(b)** $r^2 = a^2 \cos 2\theta$.
(c) $r = a\cos\theta$. **(d)** $r = e^{-a\theta}$, $0 \le \theta < \infty$, $a > 0$.

Show also that the upper half of the cardioid given by (a) is bisected by the line $\theta = \pi/3$.

Solution. **(a)** Set $r = f(\theta)$, where $f(\theta) = a(1 + \cos\theta)$. Using the arc-length formula (7.17) and the symmetry of the cardioid about the polar axis, we have (see Figure 7.32)

$$\begin{aligned} s &= 2\int_0^{\pi} \sqrt{r^2 + (r')^2}\, d\theta \\ &= 2\int_0^{\pi} \sqrt{a^2(1 + \cos\theta)^2 + (-a\sin\theta)^2}\, d\theta \\ &= 2a\int_0^{\pi} \sqrt{2(1 + \cos\theta)}\, d\theta \\ &= 2a\int_0^{\pi} 2\cos(\theta/2)\, d\theta \\ &= 4a\left(\frac{\sin(\theta/2)}{1/2}\right)\Bigg|_0^{\pi} = 8a. \end{aligned}$$

To show that the upper half of the cardioid is bisected by the line $\theta = \pi/3$, we need to show that the length of the arc s_1 of the cardioid between 0 and $\theta = \pi/3$ is $2a$. To see this, we compute

$$s_1 = \int_0^{\pi/3} \sqrt{r^2 + (r')^2}\, d\theta = 2a\left(\frac{\sin(\theta/2)}{1/2}\right)\Bigg|_0^{\pi/3} = 2a.$$

(b) Note that the Cartesian form of the given lemniscate is (see Figure 7.33)

$$(x^2 + y^2)^2 = a^2(x^2 - y^2).$$

Set $r = f(\theta)$, where $f(\theta) = \sqrt{a\cos 2\theta}$. The lemniscate is symmetric about both lines $\theta = 0$ and $\theta = \pi/2$. Hence, the total length of the lemniscate

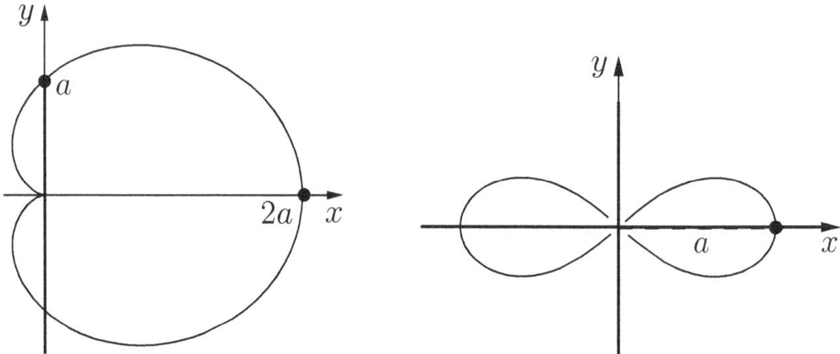

Fig. 7.32. $r = a(1 + \cos\theta)$ for $a = 2$.

Fig. 7.33. Graph of $r^2 = a^2 \cos 2\theta$.

is four times the arc length of the part that lies in the first quadrant between the rays $\theta = 0$ and $\theta = \pi/4$. That is,

$$s = 4 \int_0^{\pi/4} \sqrt{r^2 + (r')^2}\, d\theta.$$

Now on $[0, \pi/4)$, we have

$$\frac{f'(\theta)}{f(\theta)} = -\tan 2\theta, \quad \text{i.e.,} \quad r' = f'(\theta) = -r \tan 2\theta,$$

which gives $r^2 + (r')^2 = r^2 + r^2 \tan^2(2\theta) = r^2 \sec^2(2\theta) = a^2 \sec 2\theta$. Thus

$$\begin{aligned} s &= 4 \int_0^{\pi/4} \sqrt{a^2 \sec(2\theta)}\, d\theta \\ &= 4a \int_0^{\pi/4} \cos^{-\frac{1}{2}}(2\theta)\, d\theta \\ &= 2a \int_0^{\pi/2} \cos^{-\frac{1}{2}}(\phi)\, d\phi \\ &= aB(1/2, 1/4), \quad \text{where } B(a,b) = \frac{\Gamma(a)\Gamma(b)}{\Gamma(a+b)}. \end{aligned}$$

(c) Clearly, the given curve is a circle centered at $a/2$ with radius $a/2$. Since the length of the circle is not going to be altered by a change of center (since it depends only on the radius), the required length is π/a. In order to apply our formula, we set $r = f(\theta)$, where $f(\theta) = a\cos\theta$, $|\theta| \le \pi/2$. Using the formula and the symmetry of the circle about the polar axis, we have

$$s = 2 \int_0^{\pi/2} \sqrt{r^2 + (r')^2}\, d\theta = 2 \int_0^{\pi/2} \sqrt{a^2 \cos^2\theta + a^2 \sin^2\theta}\, d\theta = \pi a.$$

(d) Set $r = f(\theta)$, where $f(\theta) = \mathrm{e}^{-a\theta}$. Then $r' = -a\mathrm{e}^{-a\theta}$. We observe that as θ increases, the spiral winds around the pole O in the counterclockwise direction. The total length of the curve is given by the improper integral

$$s = \int_0^\infty \sqrt{r^2 + (r')^2}\, \mathrm{d}\theta = \sqrt{1+a^2} \int_0^\infty \mathrm{e}^{-a\theta}\, \mathrm{d}\theta = \frac{\sqrt{1+a^2}}{a},$$

since $\int_0^\infty \mathrm{e}^{-a\theta}\, \mathrm{d}\theta = \frac{1}{a}$. •

7.2.5 Questions and Exercises

Questions 7.44.

1. Let C be a smooth curve with parametric equations
$$x = x(\theta),\ y = y(\theta), \quad \alpha \le \theta \le \beta,$$
where $x'(\theta)$ and $y'(\theta)$ are bounded on $[\alpha, \beta]$. Must the length of C be finite?
2. Are there continuous curves defined on a bounded interval $[a, b]$ that have finite length?

Exercises 7.45.

1. Find the area of the region common to the circles $r = a\cos\theta$ and $r = a\sin\theta$.
2. Find the area of the common region included between the cardioids
$$r = a(1+\cos\theta) \quad \text{and} \quad r = a(1-\cos\theta).$$
3. Find the area of the common region included between the circle $r = a$ and the cardioid $r = a(1-\cos\theta)$.
4. Find the area of the region common to the parabola $2a = r(1+\cos\theta)$ and the cardioid $r = 2a(1+\cos\theta)$.
5. Find the area of the region within $r = a(1-2\sin\theta)$.
6. Find the area of the region bounded by one petal of $r = a\cos n\theta$ for $n = 2, 3, 4, 5$.
7. Find the area of the region inside the circle $r = a\cos\theta$ but outside $r = a\sin 2\theta$.
8. Find the area of the region common to the circle $r = a\cos\theta$ and the cardioid $r = a(1-\cos\theta)$.
9. Find the area of the region outside the circle $r = a$ but inside the lemniscate $r^2 = 2a^2\cos 2\theta$.
10. Find the area inside the lemniscate $r^2 = 2a^2\cos 2\theta$.
11. Compute the area enclosed by the loop of the folium of Descartes given by $x^3 + y^3 = 3axy$.
12. Find the length of the arc of the cardioid $r = a(1+\cos\theta)$ between the points whose vectorial angles are α and β.
13. Find the length of the cardioid $r = a(1-\cos\theta)$ between the points whose vectorial angles are α and β. Show that the arc of the upper half of the curve $r = a(1-\cos\theta)$ is bisected by $\theta = 2\pi/3$.

8

Power Series

In this chapter, we turn our attention to functional series of the form $\sum f_k(x)$ whose terms are functions of x rather than real numbers. In Section 8.1, we present two fundamental tests: the ratio and root tests for the convergence of numerical series. In Section 8.2, we shall begin our discussion with an important particular case in which $f_k(x) = a_k(x-a)^k$, and we shall be especially interested in deriving important properties of such series, which may be thought of as polynomials of infinite degree, although some of their properties are quite different from those of polynomials. In this section, we also discuss convergence of power series as well as term-by-term differentiation and integration of power series. The uniqueness of power series may be used in a number of ways. So whatever trick we use to find a convergent power series representing a function, it must be the Taylor series. In particular, we present some practical methods of computing the interval of convergence of a given power series.

8.1 The Ratio Test and the Root Test

The ratio and the root tests for series whose terms are real numbers are easy and useful consequences of the direct comparison test. Later, we shall obtain their analogues when the terms of the series are functions of x.

8.1.1 The Ratio Test

Theorem 8.1 (Ratio test). *Consider the series $\sum a_k$, where $a_k > 0$ for all $k \geq N_0$. Let*

$$L = \limsup_{k\to\infty} \frac{a_{k+1}}{a_k} \quad \text{and} \quad \ell = \liminf_{k\to\infty} \frac{a_{k+1}}{a_k}.$$

Then the series $\sum a_k$ converges if $L < 1$, and diverges if $\ell > 1$. This test offers no conclusion concerning the convergence of the series if $\ell \leq 1 \leq L$.

S. Ponnusamy, *Foundations of Mathematical Analysis*,
DOI 10.1007/978-0-8176-8292-7_8,

Proof. First we present a direct proof. The second proof is a simple consequence of the root test and Lemma 2.59.

Let $L < 1$. Choose any r such that $0 \leq L < r < 1$, e.g., $r = (1+L)/2$. Then by the definition of limit superior, there exists an $N > 0$ such that

$$\frac{a_{k+1}}{a_k} < r \quad \text{for all } k \geq N \quad (\geq N_0).$$

Thus,

$$a_{N+1} < a_N r,\ a_{N+2} < a_{N+1} r < a_N r^2, \dots,\ a_{N+k} < a_N r^k \quad \text{for } k \geq 1.$$

Then by the comparison test, the series $\sum a_k$ converges. Indeed, since $0 < r < 1$,

$$\sum_{k=0}^{\infty} a_{N+k} < a_N \sum_{k=0}^{\infty} r^k = \frac{a_N}{1-r},$$

which means that the series $\sum_{k=0}^{\infty} a_k$ is dominated by a convergent series

$$\sum_{k=0}^{N-1} a_k + a_N(1 + r + r^2 + \cdots) = \sum_{k=0}^{N-1} a_k + \frac{a_N}{1-r},$$

and so converges.

If $\ell > 1$, then we choose R, e.g., $R = (\ell+1)/2$, such that $\ell > R > 1$. Then there exists an $N\,(\geq N_0) > 0$ such that $a_{N+k} > a_N R^k$ for all $k \geq 1$. But $R > 1$, and so for $k \geq 1$,

$$a_{N+k} > a_N R^k > a_N,$$

and hence the general term cannot tend to zero. Thus by the divergence test, the series $\sum a_k$ is divergent.

To prove that the ratio test is inconclusive if $\ell \leq 1 \leq L$, we consider the harmonic p-series $\sum_{k=1}^{\infty} 1/k^p$ (with $p = 1, 2$), for which $\ell = L = 1$. ∎

Alternatively, one can quickly obtain a proof of the ratio test as a consequence of the root test (as demonstrated later in Section 8.1.2). However, since an absolutely convergent series is convergent, the ratio test is often stated in the following equivalent form.

Theorem 8.2 (Ratio test). *Given a series $\sum a_k$ of nonzero terms, let*

$$L = \limsup_{k\to\infty} \left|\frac{a_{k+1}}{a_k}\right| \quad \text{and} \quad \ell = \liminf_{k\to\infty} \left|\frac{a_{k+1}}{a_k}\right|.$$

Then the series $\sum a_k$ converges absolutely if $L < 1$ and diverges if $\ell > 1$. If $\ell \leq 1 \leq L$, then the series may or may not converge.

We recall that if $\lim_{k\to\infty} |a_{k+1}/a_k|$ exists, then it is equal to both ℓ and L, and hence the ratio test gives information unless, of course, $\lim_{k\to\infty} |a_{k+1}/a_k|$ equals 1. Thus, in this case the ratio test takes the following simple form, which is the familiar ratio test in calculus, especially when $a_k > 0$ for all k.

Corollary 8.3 (Simple form of the ratio test). *Given the series $\sum a_k$ of nonzero terms, let*

$$L = \lim_{k\to\infty} \left|\frac{a_{k+1}}{a_k}\right|.$$

Then the series $\sum a_k$ converges (absolutely) if $L < 1$ and diverges if $L > 1$. If $L = 1$, then the series may or may not converge.

Proof. Again, we present a direct proof because of its independent interest. If $L < 1$, choose r such that $0 \le L < r < 1$. Then for $\epsilon = r - L > 0$, there exists an N such that

$$\left|\frac{a_{k+1}}{a_k}\right| - L \le \left|\left|\frac{a_{k+1}}{a_k}\right| - L\right| < \epsilon \quad \text{for all } k \ge N.$$

In particular, $|a_{k+1}| < (L + \epsilon)|a_k| = |a_k| r$. Consequently,

$$|a_{N+k}| < |a_N| r^k \quad \text{for all } k \ge 1,$$

and thus by the comparison test, the series $\sum a_k$ converges absolutely.

The proof of the second part is similar. Indeed, we choose R such that $L > R > 1$. Then for $\epsilon = L - R > 0$, there exists an N such that

$$L - \left|\frac{a_{k+1}}{a_k}\right| \le \left|\left|\frac{a_{k+1}}{a_k}\right| - L\right| < L - R \quad \text{for all } k \ge N,$$

which implies that $|a_{N+k}| > |a_N| R^k$ for all $k \ge 1$, and so certainly $\{a_n\}$ does not converge to zero. The test gives no information, since $L = 1$ for the divergent series $\sum(1/k)$ and the convergent series $\sum(1/k^2)$. ∎

Here are some examples to illustrate the ratio test, namely Corollary 8.3. Moreover, the ratio test is most useful with series involving factorials or exponentials.

Example 8.4. Test the series $\sum_{k=1}^{\infty} a_k$ for convergence, where a_k equals:

(a) $\dfrac{5^k}{k!}$. (b) $\dfrac{k^k}{k!}$. (c) $\dfrac{1}{2k-3}$. (d) $\dfrac{k!}{k^k}$. (e) $\dfrac{2^k k!}{k^k}$. (f) $\dfrac{3^k k!}{k^k}$.

Solution. (a) Let $a_k = 5^k/k!$. Then we note that

$$L = \lim_{k\to\infty} \frac{a_{k+1}}{a_k} = \lim_{k\to\infty} \frac{5^{k+1} k!}{(k+1)! 5^k} = \lim_{k\to\infty} \frac{5}{k+1} = 0.$$

Thus $L < 1$, and the ratio test tells us that the given series converges.

(b) Let $a_k = k^k/k!$. Then the series $\sum a_k$ diverges, because

$$L = \lim_{k\to\infty} \frac{a_{k+1}}{a_k} = \lim_{k\to\infty} \frac{k!(k+1)^{k+1}}{k^k(k+1)!} = \lim_{k\to\infty} \left(1+\frac{1}{k}\right)^k = \mathrm{e} > 1.$$

Also note that $a_k \not\to 0$ as $k \to \infty$.

(c) Let $a_k = 1/(2k-3)$. Then $a_k > 0$ for all $k \geq 2$, and we find that

$$L = \lim_{k\to\infty} \frac{a_{k+1}}{a_k} = \lim_{k\to\infty} \frac{2k-3}{2k-1} = 1.$$

The ratio test is inconclusive. We can use the comparison test to determine convergence. Indeed, for $k \geq 2$,

$$a_k = \frac{1}{2k-3} > \frac{1}{2k} = c_k \quad \text{and} \quad \sum_{k=2}^{\infty} c_k \quad \text{diverges.}$$

Consequently, we conclude that the given series is divergent.

(d) From **(b)** we see that the corresponding limit value L in this case is $L = 1/\mathrm{e} < 1$, and so the series converges.

(e) Since

$$L = \lim_{k\to\infty} \frac{a_{k+1}}{a_k} = \frac{2}{\lim_{k\to\infty}\left(1+\frac{1}{k}\right)^k} = \frac{2}{\mathrm{e}} < 1,$$

the series $\sum_{k=1}^{\infty} a_k$ converges.

(f) In this case, we see that $L = (3/\mathrm{e}) > 1$, and so $\sum a_k$ diverges. •

8.1.2 The Root Test

On the one hand, it is easier to test the convergence of series such as $\sum_{k=1}^{\infty}(k!)^2/(2k!)$ by the ratio test. On the other hand, direct computation using the root test might lead to an unpleasant situation, although the structure of the test is quite similar to that of the ratio test. However, whenever the root test is applicable, it provides almost complete information, as demonstrated in a number of examples, The root test is particularly useful with a series involving a kth power in the kth term.

Theorem 8.5 (kth root test). *Suppose that $\{a_k\}$ is a sequence of nonnegative real numbers, and let*

$$L = \limsup_{k\to\infty} \sqrt[k]{a_k}.$$

Then the series $\sum a_k$ converges if $L < 1$, and diverges if $L > 1$. If $L = 1$, the root test is inconclusive. That is, if $L = 1$, the series may or may not converge.

Proof. Suppose that $L < 1$ and $a_k \geq 0$ for all k. To show that the series converges, it suffices to show that the sequence of partial sums is bounded above. Note that $L \geq 0$. Choose r such that $0 \leq L < r < 1$. Then by the definition of limit superior, for $\epsilon = r - L > 0$, there exists an N such that

$$0 \leq \sqrt[k]{a_k} < L + \epsilon = r, \quad \text{i.e., } 0 \leq a_k < r^k \text{ for all } k \geq N.$$

Since $\sum_{k=N}^{\infty} r^k = r^N/(1-r)$, the direct comparison test shows that the series $\sum_{k=N}^{\infty} a_k$ converges. Consequently, the series $\sum a_k$ also converges.

Suppose that $L > 1$. Then $a_k^{1/k} > 1$ for infinitely many values of k. But this implies that $a_k > 1$ for infinitely many k and thus $\{a_k\}$ does not converge to zero as $k \to \infty$. Therefore, the series diverges.

For the divergent series $\sum_{k=1}^{\infty}(1/k)$ and the convergent series $\sum_{k=1}^{\infty}(1/k^2)$, L turns out to equal 1, because

$$\left(\frac{1}{k}\right)^{1/k} = \frac{1}{k^{1/k}} \to 1 \quad \text{and} \quad \left(\frac{1}{k^2}\right)^{1/k} = \frac{1}{k^{2/k}} \to 1 \quad \text{as } k \to \infty.$$

Thus, the test is inconclusive if $L = 1$. ∎

Example 8.6. Consider the series

$$\frac{1}{3} + \frac{1}{4} + \frac{1}{3^2} + \frac{1}{4^2} + \frac{1}{3^3} + \frac{1}{4^3} + \cdots.$$

We may write the general term explicitly:

$$a_k = \begin{cases} \dfrac{1}{3^{(k+1)/2}} & \text{if } k \text{ is odd,} \\ \frac{1}{4^{k/2}} & \text{if } k \text{ is even.} \end{cases}$$

Then

$$\frac{a_{k+1}}{a_k} = \begin{cases} \left(\dfrac{3}{4}\right)^{(k+1)/2} & \text{if } k \text{ is odd,} \\ \dfrac{1}{3}\left(\dfrac{4}{3}\right)^{k/2} & \text{if } k \text{ is even,} \end{cases}$$

so that

$$\limsup_{k\to\infty} \frac{a_{k+1}}{a_k} = \infty \quad \text{and} \quad \liminf_{k\to\infty} \frac{a_{k+1}}{a_k} = 0.$$

The ratio test gives no information. On the other hand,

$$a_k^{1/k} = \begin{cases} \dfrac{1}{\sqrt{3}}\left(\dfrac{1}{3^{1/2k}}\right) & \text{if } k \text{ is odd,} \\ \dfrac{1}{2} & \text{if } k \text{ is even,} \end{cases}$$

so that $\limsup_{k\to\infty} a_k^{1/k} = 1/\sqrt{3} < 1$. Hence the series $\sum a_k$ converges. •

Finally, if $a_k = k$, then

$$\lim_{k\to\infty} a_k^{1/k} = 1 = \lim_{k\to\infty} \frac{a_{k+1}}{a_k},$$

and so both tests fail. However, $\sum k$ is divergent.

Since an absolutely convergent series is convergent, the root test is often stated in the following equivalent form.

Theorem 8.7 (Root test). *Suppose that $\{a_k\}$ is a sequence of real numbers, and let $L = \limsup_{k\to\infty} \sqrt[k]{|a_k|}$. Then the series $\sum a_k$ converges (absolutely) if $L < 1$, and diverges if $L > 1$. If $L = 1$ the series may or may not converge.*

If $\lim_{k\to\infty} \sqrt[k]{|a_k|}$ exists, then the root test gives information unless, of course, $\lim_{k\to\infty} \sqrt[k]{|a_k|} = 1$. Because of its independent interest, we include here a direct proof of it which is also referred to as a root test.

Corollary 8.8 (Simple form of the root test). *Suppose that $\{a_k\}$ is a sequence of real numbers, and let*

$$L = \lim_{k\to\infty} \sqrt[k]{|a_k|}.$$

Then the series $\sum a_k$ converges (absolutely) if $L < 1$, and diverges if $L > 1$. If $L = 1$, the series may or may not converge.

Proof. Suppose that $L < 1$. Then we note that $L \geq 0$. Choose any r such that $0 \leq L < r < 1$. Then for $\epsilon = r - L > 0$, there exists an N such that

$$\sqrt[k]{|a_k|} - L \leq \left| \sqrt[k]{|a_k|} - L \right| < \epsilon \quad \text{for all } k \geq N.$$

This gives

$$\sqrt[k]{|a_k|} < L + \epsilon = r \quad \text{or} \quad 0 \leq |a_k| < r^k \quad \text{for all } k \geq N,$$

which means that the series $\sum |a_k|$ is dominated by the convergent geometric series $\sum r^k$. Consequently, the series $\sum a_k$ converges absolutely by the comparison test.

Suppose that $L > 1$. Choose R such that $L > R > 1$. Then for $\epsilon = L - R > 0$, there exists an N such that

$$L - \sqrt[k]{|a_k|} \leq \left| \sqrt[k]{|a_k|} - L \right| < \epsilon \quad \text{for all } k \geq N,$$

which gives $\sqrt[k]{|a_k|} > R$, or $|a_k| > R^k$ for all $k \geq N$ where $R > 1$. Thus, the general term cannot tend to zero as $k \to \infty$. Therefore, the series diverges.

As remarked in the proof of Theorem 8.5, for the series $\sum_{k=1}^{\infty}(1/k)$ and $\sum_{k=1}^{\infty}(1/k^2)$, L turns out to be equal to 1, and therefore the test is inconclusive if $L = 1$. ∎

Remark 8.9. In the ratio and the root tests, the case $L > 1$ includes the case $L = \infty$. ●

Finally we outline the proof of the ratio test as a consequence of the root test.

Proof of Theorem 8.1. Let $\alpha = \limsup_{n\to\infty} a_n^{1/n}$, where L and ℓ are defined in Theorem 8.1. From Lemma 2.59, we obtain that $\ell \leq \alpha \leq L$. If $L < 1$, then the right-hand inequality give $\alpha < 1$, and so the series $\sum a_k$ converges by the root test. Similarly, if $\ell > 1$, then $\alpha > 1$, and so the series $\sum a_k$ diverges. The condition $\alpha = 1$ is equivalent to $\ell \leq 1 \leq L$, and so the test is inconclusive. ■

Example 8.10. Test the series $\sum_{k=2}^{\infty} a_k$ for convergence, where a_k equals

(a) $\dfrac{1}{(\log k)^{ck}}$ $(c > 0)$, **(b)** $(1 + 1/k)^{2k^2}$, **(c)** $\dfrac{k!}{1 \cdot 4 \cdot 7 \cdots (3k+1)}$.

Solution. Because a_k involves a power for cases **(a)** and **(b)**, it is appropriate to use the root test for those series.

(a) Let $a_k = 1/(\log k)^{ck}$, $c > 0$. Then $a_k > 0$ for all $k \geq 2$, and so

$$L = \lim_{k\to\infty} \sqrt[k]{a_k} = \lim_{k\to\infty} \frac{1}{(\log k)^c} = 0.$$

Because $L < 1$, the root test tells us that the given series converges.

(b) We note that

$$L = \lim_{k\to\infty} \left[\left(1 + \frac{1}{k}\right)^{2k^2}\right]^{1/k} = \lim_{k\to\infty} \left(1 + \frac{1}{k}\right)^{2k} = e^2 > 1.$$

Since $L > 1$, the series diverges.

(c) Because the corresponding a_k involves $k!$, we may try the ratio test. Now

$$\frac{a_{k+1}}{a_k} = \frac{[1 \cdot 4 \cdot 7 \cdots (3k+1)] \cdot (k+1)!}{k! \cdot [1 \cdot 4 \cdot 7 \cdots (3k+4)]} = \frac{k+1}{3k+4} \to \frac{1}{3} \quad \text{as } k \to \infty.$$

Thus $L = 1/3$, and by the ratio test, the series converges. ●

8.1.3 Questions and Exercises

Questions 8.11.

1. Which of the ratio and root tests is more appropriate on which occasions?
2. Which of the ratio and root tests implies the other? When do they produce the same conclusion? When do they not?
3. Does $\sum_{k=1}^{\infty} \left(3^k/(4 - (-1)^k)^k\right)$ converge?

Exercises 8.12.

1. For what values of $a \in \mathbb{R}$ can we use the ratio test to prove that $\sum_{k=1}^{\infty}(a^k k!/k^k)$ is a convergent series?
2. Test the convergence of $\sum_{k=1}^{\infty} a_k$, where
   (a) $a_{2k} = \dfrac{1}{2^k}, \quad a_{2k-1} = \dfrac{1}{2^{k+1}}$. (b) $a_{2k} = 3^k, \quad a_{2k-1} = 3^{-k}$.
3. Discuss the convergence of the series $\sum a_k$ by either the ratio test or the root test, whichever is applicable.
   (a) $a_k = 3^{(-1)^k - k}$. (b) $a_{2k} = 2^{-k}, \quad a_{2k-1} = 3^{-k}$.
4. Sum the series $\sum_{k=1}^{\infty} k3^{-k}$ and $\sum_{k=1}^{\infty}(k^2+k)5^{-k}$. Justify your method.
5. Which of the following series converge?
   (a) $\displaystyle\sum_{k=1}^{\infty} \frac{k^5}{5^k}$. (b) $\displaystyle\sum_{k=1}^{\infty} \left(1 + \frac{1}{\sqrt{k}}\right)^{-k^{3/2}}$. (c) $\displaystyle\sum_{k=1}^{\infty} \left(\frac{3k+1}{k+7}\right)^k$.
6. Sum the series $\sum_{k=1}^{\infty} a_k$, where a_k equals
   (a) $\dfrac{k}{3^{k-1}}$. (b) $\dfrac{k^2}{3^{k-1}}$. (c) $\dfrac{k^3}{3^{k-1}}$. (d) $\dfrac{k^2}{k!\,3^{k-1}}$.
7. Suppose that $\{a_n\}$ is a sequence of nonnegative real numbers, and
   $$L_1 = \limsup_{n\to\infty} \frac{a_{n+1}}{a_n} \quad \text{and} \quad L_2 = \limsup_{n\to\infty} \sqrt[n]{a_n}.$$
   Prove the following:
   (a) If either $0 \le L_1 < 1$ or $0 \le L_2 < 1$, then $a_n \to 0$ as $n \to \infty$.
   (b) If either $L_1 > 1$ or $L_2 > 1$, then $a_n \to \infty$ as $n \to \infty$.

8.2 Basic Issues around the Ratio and Root Tests

In analogy to numerical series whose terms are real numbers, we now consider *functional series*, which are series of functions of the form

$$\sum_{k=0}^{\infty} f_k(x),$$

where $f_k(x)$'s $(k \in \mathbb{N}_0)$ are functions defined on a subset of $\mathbb{R}$. Since the sequence $\{S_n(x)\}$ of partial sums given by

$$S_n(x) = \sum_{k=0}^{n} f_k(x)$$

is a function of x, we need to determine for what values of x the sequence $\{S_n(x)\}$ converges. The most interesting case is $f_k(x) = a_k(x-a)^k$, where a is a real constant. In this case, the functional series takes the form

$$\sum_{k=0}^{\infty} a_k(x-a)^k = a_0 + a_1(x-a) + a_2(x-a)^2 + \cdots,$$

which we call a *power series* with center at $x = a$ or a power series about a, or a Taylor series about $x = a$. The numbers $a_0, a_1, a_2, \dots$ are called the *coefficients* of the power series. If $a = 0$, then the series has the form

$$\sum_{k=0}^{\infty} a_k x^k = a_0 + a_1 x + a_2 x^2 + a_3 x^3 + \cdots$$

and is called a *Maclaurin series*. Clearly, a Maclaurin series is a Taylor series with $a = 0$. Moreover, these power series have a meaning only for those values of x for which the series converges. We also note that although each term of the power series is defined for all real x, it is not expected that the series will converge for all real x. The set of all values of x for which a given functional series $\sum f_k(x)$ (e.g., $f_k(x) = a_k(x-a)^k$) converges is called the *convergence set* or the *region of convergence* for $\sum f_k(x)$.

As a motivation, we shall begin our discussion with a number of examples by treating functional series as numerical series by fixing x and then applying the ratio test or root test to determine whether the series converges for that particular x. In addition, interesting conclusions will be drawn just by looking at a_k, the coefficients of the power series. We shall discuss this issue in detail later.

Example 8.13. Find all real values of x for which the series $\sum_{k=0}^{\infty} k^5 x^k$ converges.

Solution. Clearly, the series converges for $x = 0$. Keeping x fixed (that is, as a constant) nonzero real number, we apply the ratio test:

$$L = \lim_{k\to\infty} \left| \frac{(k+1)^5 x^{k+1}}{k^5 x^k} \right| = \lim_{k\to\infty} \left(\frac{k+1}{k} \right)^5 |x| = |x|.$$

Thus, according to the ratio test, the series converges if $|x| < 1$ and diverges if $|x| > 1$. The test fails if $|x| = 1$, i.e., if $x = 1$ or -1. When $x = 1$, the series becomes $\sum k^5$, which diverges by the divergence test. When $x = -1$, the series becomes $\sum (-1)^k k^5$, which diverges, because the general term does not tend to zero. Similarly, it is easy to see that the power series $\sum_{k=1}^{\infty} k x^k$ converges for $|x| < 1$ and diverges for $|x| \geq 1$. •

Example 8.14. Test the series $\sum_{k=1}^{\infty} (3 - \cos k\pi) x^k$ for convergence.

Solution. Set $a_k(x) = (3 - \cos k\pi) x^k$. Then $a_k(x) \neq 0$ for all $k \geq 1$ and for each fixed $x \neq 0$. In order to apply the root test, we need to compute

$$|a_k(x)|^{1/k} = (3 - \cos k\pi)^{1/k} |x| = \begin{cases} |x|(2^{1/k}) & \text{if } k \text{ is even,} \\ |x|(2^{2/k}) & \text{if } k \text{ is odd.} \end{cases}$$

Since $2^{1/k} = \exp((1/k)\log 2) \to \mathrm{e}^0 = 1$ (also $2^{2/k} = (2^{1/k})^2 \to 1$) as $k \to \infty$, we have

$$|a_k(x)|^{1/k} \to |x| \quad \text{as } k \to \infty,$$

and the root test (see Theorem 8.7 and Corollary 8.8) tell us that the series $\sum a_k(x)$ converges absolutely for all x with $|x| < 1$, and diverges for $|x| > 1$. For $|x| = 1$, the series clearly diverges, since the general term does not approach zero.

Notice that the ratio test as in Corollary 8.3 is not applicable, because

$$\frac{a_{k+1}(x)}{a_k(x)} = \begin{cases} 2|x| & \text{if } k \text{ is even,} \\ (1/2)|x| & \text{if } k \text{ is odd,} \end{cases}$$

showing that

$$\limsup_{k\to\infty} \left|\frac{a_{k+1}(x)}{a_k(x)}\right| = 2|x| \quad \text{and} \quad \liminf_{k\to\infty} \left|\frac{a_{k+1}(x)}{a_k(x)}\right| = \frac{|x|}{2}.$$

However, the ratio test as in Theorem 8.2 is applicable. According to this, $\sum a_k(x)$ converges absolutely for all x with $|x| < 1/2$ and diverges for all x with $|x| > 2$. Note that this test does not give information when $1/2 \leq |x| \leq 2$. However, the root test gives full information for $|x| < 1$ and for $|x| > 1$. ●

Example 8.15. Test the series $\sum_{k=0}^{\infty}(3+\sin(k\pi/2))(x-1)^k$ for convergence.

Solution. Set $a_k = (3 + \sin(k\pi/2))$. Then

$$a_k = 3 + \sin(k\pi/2) = \begin{cases} 3 + (-1)^{(k-1)/2} & \text{if } k \text{ is odd,} \\ 3 & \text{if } k \text{ is even,} \end{cases}$$

so that $\{a_k\}_{k\geq 0}$ is

$$\{3, 4, 3, 2, 3, 4, 3, 2, \ldots\}.$$

Therefore, if $a_k(x) = a_k(x-1)^k$, then $a_k(x) \neq 0$ for all $k \geq 0$ and for each fixed $x \neq 1$, so that the quotient $a_{k+1}(x)/a_k(x)$ assumes the values

$$\left\{\frac{4}{3}(x-1),\ \frac{3}{4}(x-1),\ \frac{2}{3}(x-1),\ \frac{3}{2}(x-1)\right\}$$

infinitely often. Fixing $x \neq 1$, we see that

$$\limsup_{k\to\infty} \left|\frac{a_{k+1}(x)}{a_k(x)}\right| = \frac{3}{2}|x-1| \quad \text{and} \quad \liminf_{k\to\infty} \left|\frac{a_{k+1}(x)}{a_k(x)}\right| = \frac{2}{3}|x-1|,$$

showing that (by Theorem 8.2) the series $\sum a_k(x)$ converges absolutely for all x with $|x-1| < 2/3$ and diverges for all x with $|x-1| > 3/2$. However, this test does not give information when $2/3 \leq |x-1| \leq 3/2$.

On the other hand, $|a_k(x)|^{1/k}$ assumes the values

$$\left\{3^{1/k}|x-1|,\ 4^{1/k}|x-1|,\ 2^{1/k}|x-1|\right\}$$

infinitely often. Thus, we have

$$\lim_{k\to\infty} |a_k(x)|^{1/k} = |x-1|,$$

and the root test (see Theorem 8.7 and Corollary 8.8) imply that the series $\sum a_k(x)$ converges absolutely for all x with $|x-1| < 1$, and diverges for $|x-1| > 1$. The root test does not tell us what happens when $|x-1| = 1$, but we by a direct verification that the series diverges (at $x = 0, 2$), since the general term does not approach zero. •

8.2.1 Convergence of Power Series

Power series of the form $\sum a_k(x-a)^k$ occur as representations for certain functions, such as

$$\frac{1}{1-x} = \sum_{k=0}^{\infty} x^k \quad (|x| < 1),$$

and in solutions of a large class of differential equations. Now we wish to continue our discussion on the convergence of power series. More precisely, we ask, *for what values of x does a given power series converge?* Theorem 8.17 answers this question for the case $a = 0$, and a corresponding theorem for a power series about a is a consequence of this result (see Theorem 8.22). We begin with a lemma.

Lemma 8.16. *For a power series*

$$\sum_{k=0}^{\infty} a_k x^k, \tag{8.1}$$

we have the following:

(1) *If the series converges at $x = x_0$ ($x_0 \neq 0$), then it converges absolutely for $|x| < |x_0|$.*
(2) *If the series diverges at x_1, then it diverges for $|x| > |x_1|$.*

Proof. Suppose that the series (8.1) converges at $x = x_0$ ($x_0 \neq 0$). Then $\sum a_k x_0^k$ converges, and so $a_k x_0^k \to 0$ as $k \to \infty$. In particular, since a convergent sequence is bounded, $|a_k x_0^k| \leq M$ for all $k \geq 0$. We claim that the series (8.1) converges absolutely for $|x| < |x_0|$. For each x with $|x| < |x_0|$, we can write

$$|a_k x^k| = |a_k x_0^k| \left|\frac{x}{x_0}\right|^k \leq Mr^k \quad \text{for all } k \geq 0 \quad (r = |x/x_0| < 1),$$

showing that the series $\sum |a_k x^k|$ is dominated by the convergent geometric series $M \sum r^k$. So the given series (8.1) is absolutely convergent for $|x| < |x_0|$, and hence must converge for $|x| < |x_0|$ (see Figure 8.1).

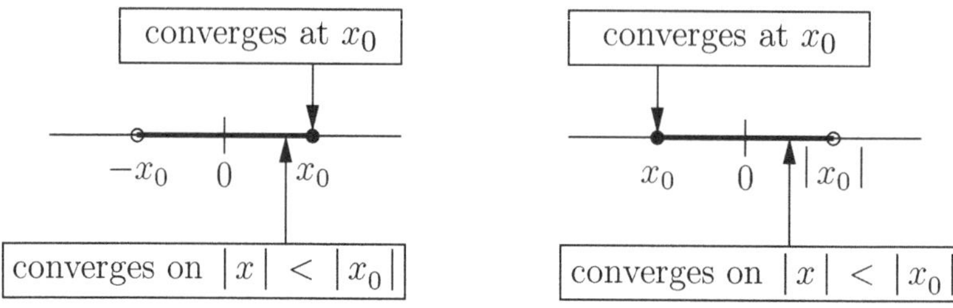

Fig. 8.1. Sketch for the convergence of $\sum a_k x^k$ at x_0.

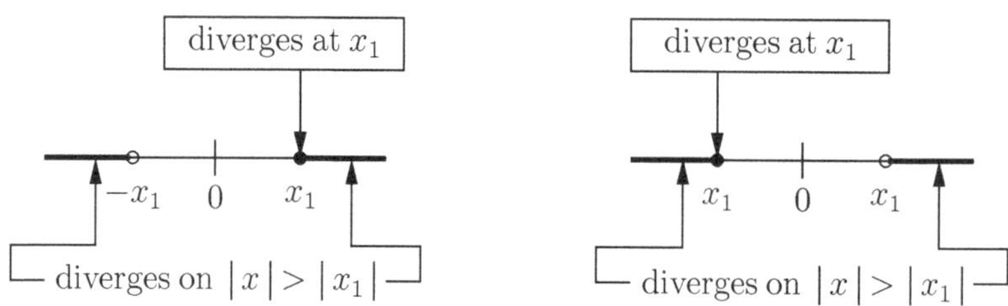

Fig. 8.2. Sketch for the divergence of $\sum a_k x^k$ at x_1.

Next suppose that the series $\sum a_k x_1^k$ diverges. If x is such that $|x| > |x_1|$ and $\sum a_k x^k$ converges, then by (1), $\sum a_k x_1^k$ converges absolutely, contrary to the assumption. Hence $\sum a_k x^k$ diverges for $|x| > |x_1|$ (see Figure 8.2). ∎

Theorem 8.17 (Convergence of Maclaurin series). *For a power series defined by* (8.1), *exactly one of the following is true:*

(1) *The series converges only for* $x = 0$.
(2) *The series converges for all* x.
(3) *Neither* (1) *nor* (2) *holds, and there exists an* $R > 0$ *such that the series converges absolutely for* $|x| < R$ *and diverges for* $|x| > R$.

Proof. The case (1) is self-explanatory, for the series (8.1) always converges when $x = 0$. If it diverges for all other values of x, then we set $R = 0$ by convention.

We can now suppose that the series $\sum a_k x^k$ converges at $x = x_0$ ($x_0 \neq 0$). By Lemma 8.16, the series $\sum a_k x^k$ must converge absolutely for all x with $|x| < |x_0|$.

Thus, if the series $\sum a_k x^k$ diverges for all x with $|x| > |x_0|$, then $|x_0|$ is the value of R mentioned in the statement of the theorem. Otherwise, there exists a number x_0' with $|x_0'| > |x_0|$ such that the series $\sum a_k x^k$ converges (absolutely) for all x with $|x| < |x_0'|$. Now let

$$I = \left\{ r > 0 : \sum |a_k x^k| \text{ converges for } |x| < r \right\}.$$

If I has no upper bound, then the series $\sum a_k x^k$ converges for all x. That is, it converges for $|x| < R$, $R = \infty$.

If I is bounded, then we let $R = \sup I$. We cannot say what happens at the endpoints $x = R, -R$. ∎

Examples 8.18. **(a)** By the alternating series test, the series $\sum_{k=1}^{\infty}(x^k/k)$ converges for $x = -1$, and so it must converge for all $|x| < 1$. Also, the series for $x = 1$ is actually the harmonic series, which is divergent. Hence the series must diverge for $|x| > 1$.

(b) The series $\sum_{k=1}^{\infty}(x^k/k^2)$ converges absolutely for $|x| \leq 1$, because $|x|^k/k^2 \leq 1/k^2$ and $\sum(1/k^2)$ converges.

(c) The series $\sum_{k=1}^{\infty}((-1)^{k-1}/k)x^k$ converges at $x = 1$ and diverges at $x = -1$. Therefore, it must converge for $|x| < 1$ and diverge for $|x| > 1$. ●

Note: When Case (3) in Theorem 8.17 occurs, then the series (8.1) could converge absolutely, converge conditionally, or diverge for $|x| = R$, that is, at the endpoints $x = R$ and $-R$.

In the case of a finite radius of convergence (defined below), Examples 8.18 illustrate why Theorem 8.17 makes no assertion about the behavior of a power series at the endpoints of the interval of convergence.

8.2.2 Radius of Convergence of Power Series

According to Theorem 8.17, the set of points at which the series $\sum_{k=0}^{\infty} a_k x^k$ converges is an interval about the origin. We call this interval the *interval of convergence* of the series; this interval is either $\{0\}$, the set of all real numbers, or an interval of positive finite length centered at $x = 0$ that may contain both, neither, or one of its endpoints; that is, the interval may be open, half-open, or closed. If this interval has length $2R$, then R is called the *radius of convergence* of $\sum_{k=0}^{\infty} a_k x^k$, as shown in Figure 8.3. We follow the convention that if the series about the origin converges only for $x = 0$, then we say that the series has radius of convergence $R = 0$, and if it converges for all x, we say that $R = \infty$, so that $\mathbb{R}$ is treated as an interval of infinite radius. Theorem 8.17 (with the above understanding in the extreme cases $R = 0, \infty$) may now be rephrased as follows:

Theorem 8.19. *For each power series about the origin, there is an R in $[0, \infty) \cup \{\infty\}$, the radius of convergence of the series, such that the series converges for $|x| < R$ and diverges for $|x| > R$.*

Note: If the radius of convergence R is a finite positive number, then as demonstrated in a number of examples (see also examples below), the behavior at the endpoints is unpredictable, and so the interval of convergence for the series $\sum_{k=0}^{\infty} a_k x^k$ is one of the four intervals

$$(-R, R), \ [-R, R), \ (-R, R], \ [-R, R].$$

The following examples show that each of the three possibilities stated in Theorem 8.17 occurs.

Example 8.20. Find the interval of convergence of the power series $\sum_{k=1}^{\infty} a_k x^k$, where a_k equals

$$\text{(a) } \frac{1}{k!}; \quad \text{(b) } k!; \quad \text{(c) } \frac{1}{\sqrt{k}}; \quad \text{(d) } \frac{5^k}{k}; \quad \text{(e) } \frac{5^k}{k!}; \quad \text{(f) } \left(\frac{k+1}{k}\right)^{2k^2}.$$

Solution. For convenience, we let $a_k(x) = a_k x^k$. If $x = 0$, then the series trivially converges.

(a) For $x \neq 0$, let $a_k(x) = x^k/k!$. Then $a_k(x) \neq 0$ for $x \neq 0$, and we use the ratio test to obtain

$$L = \lim_{k\to\infty} \left|\frac{a_{k+1}(x)}{a_k(x)}\right| = \lim_{k\to\infty} \left|\frac{x^{k+1}k!}{(k+1)!x^k}\right| = |x| \lim_{k\to\infty} \frac{1}{k+1} = 0.$$

Because $L = 0$ and thus $L < 1$, the series converges (absolutely) for all x. Thus $R = \infty$, and the interval of convergence is the entire real line.

(b) Let $a_k(x) = k!x^k$. Then for $x \neq 0$, we use the ratio test to obtain

$$L = \lim_{k\to\infty} \left|\frac{a_{k+1}(x)}{a_k(x)}\right| = \lim_{k\to\infty} \left|\frac{(k+1)!x^{k+1}}{k!x^k}\right| = |x| \lim_{k\to\infty} (k+1).$$

For any x other than 0, we have $L = \infty$. Hence, the power series converges only when $x = 0$. Thus the power series $\sum k!x^k$ has radius of convergence 0.

(c) For each fixed $x \neq 0$, let $a_k(x) = x^k/\sqrt{k}$. Then using the ratio test, we find that

$$L = \lim_{k\to\infty} \left|\frac{a_{k+1}(x)}{a_k(x)}\right| = |x| \lim_{k\to\infty} \frac{\sqrt{k}}{\sqrt{k+1}} = |x|.$$

The power series converges absolutely if $|x| < 1$, and diverges if $|x| > 1$. We must also check the convergence of the series at the endpoints of the interval $|x| < 1$, namely at $x = -1$ and 1:

- At $x = -1$:

$$\sum_{k=1}^{\infty} \frac{(-1)^k}{\sqrt{k}} \quad \textit{converges} \text{ by the alternating series test}$$

- At $x = 1$:

$$\sum_{k=1}^{\infty} \frac{1}{\sqrt{k}} \quad \textit{diverges} \ (p\text{-series with } p = \tfrac{1}{2} < 1).$$

Thus, the power series $\sum(x^k/\sqrt{k})$ converges for $-1 \leq x < 1$ and diverges otherwise. The interval of convergence is $[-1, 1)$, and the radius of convergence is $R = 1$.

(d) For $x \neq 0$, let $a_k(x) = 5^k x^k / k$. Then we apply the ratio test to obtain

$$L = \lim_{k\to\infty} \left| \frac{a_{k+1}(x)}{a_k(x)} \right| = \lim_{k\to\infty} \frac{5k}{k+1} |x| = 5\,|x|.$$

Thus, the series converges absolutely for $5\,|x| < 1$; that is, for $|x| < \frac{1}{5}$. The radius of convergence is $R = \frac{1}{5}$. At the endpoints, we have:

- At $x = \frac{1}{5}$,

$$\sum_{k=1}^{\infty} \frac{5^k}{k} \left(\frac{1}{5}\right)^k = \sum_{k=1}^{\infty} \frac{1}{k}, \quad \text{which is divergent.}$$

- At $x = -\frac{1}{5}$,

$$\sum_{k=1}^{\infty} \frac{5^k}{k} \left(-\frac{1}{5}\right)^k = \sum_{k=1}^{\infty} \frac{(-1)^k}{k}, \quad \text{which is convergent.}$$

The interval of convergence is $-\frac{1}{5} \leq x < \frac{1}{5}$.

(e) Set $a_k(x) = 5^k x^k / k!$, for $x \neq 0$. Applying the ratio test, we find that $L = 0$. Thus, the power series converges absolutely for all x, and so the radius of convergence is infinite, and interval of convergence is the entire real line.

(f) Using the root test, we find that

$$L = \lim_{k\to\infty} |a_k(x)|^{1/k} = \lim_{k\to\infty} \left(1 + \frac{1}{k}\right)^{2k} |x| = \mathrm{e}^2\,|x|.$$

Thus, the power series converges absolutely for $\mathrm{e}^2\,|x| < 1$, that is, for $|x| < \mathrm{e}^{-2}$. It follows that the radius of convergence is $R = \mathrm{e}^{-2}$. How about at the endpoints? ●

Example 8.21. The series $\sum_{k=0}^{\infty} x^k$ converges absolutely to $1/(1-x)$ for $|x| < 1$. It follows from Theorem 5.62 that the Cauchy product

$$\left(\sum_{k=0}^{\infty} x^k\right)^2 = \sum_{n=0}^{\infty} \left(\sum_{k=0}^{n} x^k x^{n-k}\right) = \sum_{n=0}^{\infty} (n+1)x^n$$

also converges absolutely and has sum $1/(1-x)^2$ for $|x| < 1$. ●

In some applications, we will encounter series about $x = a$. The procedure for determining the interval of convergence of such power series is exactly the same, and for the sake of completeness, it is illustrated in the following theorem and subsequent examples.

Theorem 8.22 (Convergence of a general Taylor series). *For a power series*

$$\sum_{k=0}^{\infty} a_k (x-a)^k, \tag{8.2}$$

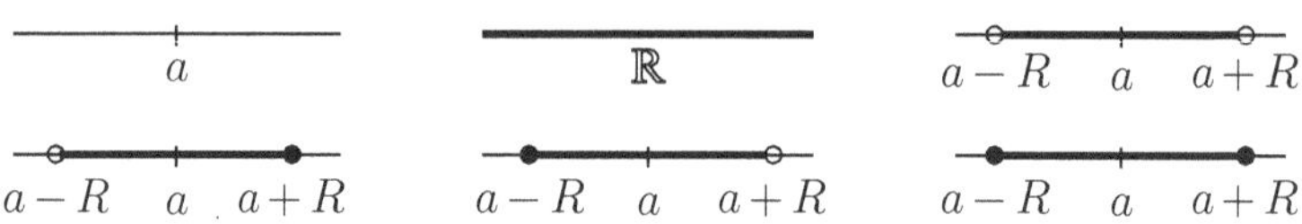

Fig. 8.3. Interval of convergence for a Taylor series.

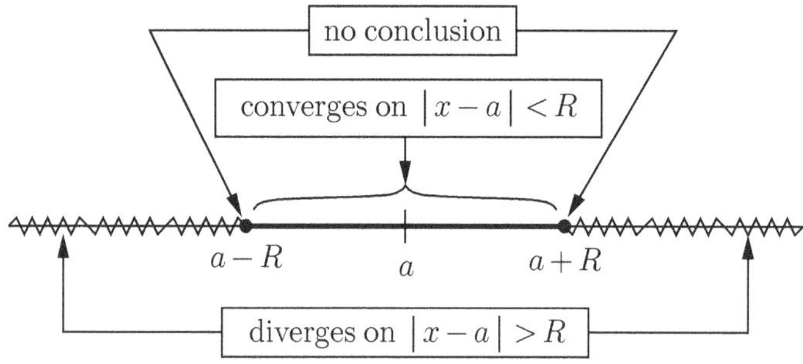

Fig. 8.4. Finite radius R of convergence of $\sum_{k\geq 1} a_k(x-a)^k$.

exactly one of the following is true:

(1) *The series converges only for $x = a$, i.e., $R = 0$.*
(2) *The series converges for all x, i.e., $R = \infty$.*
(3) *The series converges absolutely for $|x - a| < R$, i.e., for all x in the open interval $(a - R, a + R)$, and diverges for $|x - a| > R$, i.e., for all x in $(-\infty, a - R) \cup (a + R, \infty)$. It may converge absolutely, converge conditionally, or diverge at each of the endpoints of the interval, $x = a + R$ and $x = a - R$. Here R is called the radius of convergence of* (8.2).

Proof. Use the transformation $X = x - a$ and apply the proof of Theorem 8.17 to the power series in the new variable X. ∎

Figure 8.3 illustrates the various types of interval of convergence of the series (8.2). Figure 8.4 demonstrates the fact that no conclusion can be drawn about the convergence of the series the endpoints of the interval of convergence. A power series that converges for all x is called an *everywhere convergent power series.* A power series that converges only at $x = a$ is often referred to as a *nowhere convergent power series.*

Example 8.23. Consider the power series

$$\sum_{k=0}^{\infty} \frac{(x+1)^k}{3^k}.$$

With the introduction of a new variable $X = (x + 1)/3$, the given series becomes the geometric series $\sum X^k$, which converges for $|X| < 1$ and diverges for $|X| \geq 1$. Consequently, the given series converges absolutely for $|x+1| < 3$ and diverges for $|x+1| \geq 3$. Therefore, the interval of convergence of the given series is $(-4, 2)$. ●

8.2.3 Methods for Finding the Radius of Convergence

Applying the root test (Theorem 8.7), we have the following theorem.

Theorem 8.23 (Cauchy–Hadamard). *The power series $\sum_{k=0}^{\infty} a_k x^k$ has radius of convergence R, where*

$$\frac{1}{R} = \limsup_{n\to\infty} |a_n|^{1/n}.$$

Here we observe the conventions $1/0 = \infty$ and $1/\infty = 0$.

Proof. For a fixed $x \neq 0$, we have

$$L_x = \limsup_{n\to\infty} |a_n x^n|^{1/n} = |x| \limsup_{n\to\infty} |a_n|^{1/n} = \frac{1}{R}|x|.$$

Since $L_x < 1$ if and only if $|x| < R$, according to Theorem 8.7, the series $\sum_{k=0}^{\infty} a_k x^k$ converges absolutely when $|x| < R$, and diverges when $|x| > R$. In view of Theorem 8.19, the radius of convergence is R.

When $R = \infty$, the series converges everywhere; and when $R = 0$, the series converges only at $x = 0$. ∎

The following result generally suffices to explain the convergence of a Maclaurin or Taylor series.

Corollary 8.24. *The radius of convergence R of the power series $\sum_{k=0}^{\infty} a_k x^k$ is determined by*

(a) $\displaystyle \frac{1}{R} = \lim_{n\to\infty} |a_n|^{1/n}$, **(b)** $\displaystyle \frac{1}{R} = \lim_{n\to\infty} \left|\frac{a_{n+1}}{a_n}\right|$,

provided these limits exist. Again, we follow the conventions $1/0 = \infty$ and $1/\infty = 0$.

From Corollary 2.60 we recall that if $\lim_{n\to\infty} |a_{n+1}/a_n|$ exists (with the same limit), then $\lim_{n\to\infty} |a_n|^{1/n}$ exists, but the converse is not true. This observation gives the following corollary.

Corollary 8.25. *Suppose that $\lim_{n\to\infty} |a_{n+1}/a_n|$ exists. Then radius of convergence R of the power series $\sum_{k=0}^{\infty} a_k x^k$ is determined by*

$$\frac{1}{R} = \lim_{n\to\infty} |a_n|^{1/n} = \lim_{n\to\infty} \left|\frac{a_{n+1}}{a_n}\right|.$$

Example 8.28. Consider the power series

$$\sum_{k=0}^{\infty} 7^{-k} x^{5k}.$$

Note that in this power series the quotient a_{n+1}/a_n is undefined if $n \neq 5k$, $k \in \mathbb{N}_0$. Hence Corollary 8.24(**b**) is not applicable. However, we can either apply directly the ratio test for numerical series (by fixing x and ignoring the vanishing terms) or else introduce a change of variable. Thus, by the introduction of a new variable $X = x^5/7$, the given series becomes a geometric series $\sum X^k$, which converges for $|X| < 1$ and diverges for $|X| \geq 1$. Consequently, the given series converges absolutely for $|x| < \sqrt[5]{7}$ and diverges for $|x| \geq \sqrt[5]{7}$. Therefore, the radius of convergence of the given series is $R = \sqrt[5]{7}$. ●

Example 8.29. Discuss the convergence of $\sum_{n=1}^{\infty}(x^{5n}/n2^n)$.

Solution. For $x \neq 0$, let $a_n = x^{5n}/2^n n$. Then

$$\left|\frac{a_{n+1}}{a_n}\right| = \left|\frac{x^{5(n+1)}}{2^{n+1}(n+1)} \cdot \frac{2^n n}{x^{5n}}\right| = \frac{|x|^5}{2}\left(\frac{n}{n+1}\right) \to \frac{|x|^5}{2} \quad \text{as } n \to \infty,$$

so that by the ratio test, the series converges for $|x|^5 < 2$ (including the trivial case $x = 0$) and diverges for $|x|^5 > 2$. Let $x = 2^{1/5}$. Then $a_n = 1/n$, and the series is $\sum_{n=1}^{\infty}(1/n)$, which is divergent. If $x = -2^{1/5}$, then $a_n = (-1)^n/n$, which gives the convergent series $\sum_{n=1}^{\infty}((-1)^n/n)$. We conclude that the given series converges for $-2^{1/5} \leq x < 2^{1/5}$ and diverges for all other values of x. For each $-2^{1/5} \leq x < 2^{1/5}$, we readily obtain that (see, for instance, Examples 8.46)

$$\sum_{n=1}^{\infty} \frac{1}{n}\left(\frac{x^5}{2}\right)^n = -\log(1 - x^5/2).$$ ●

We wish to know whether a convergent power series $f(x) = \sum_{k=0}^{\infty} a_k x^k$ is differentiable. If term-by-term differentiation is allowed, then we get a new series $g(x) = \sum_{k=1}^{\infty} k a_k x^{k-1}$, which we call a *derived series*. It is natural to ask whether these two series have the same radius of convergence. Also, we ask whether f is differentiable, and if so, whether $g(x) = f'(x)$. We answer these questions below. Later, as an alternative proof, we also obtain our result as a special case of a more general result (see Corollary 9.44).

Lemma 8.30. *Suppose that $\{a_n\}_{n\geq 1}$ is a bounded sequence of real numbers. Then we have*

(a) *$\{n^{1/n}a_n\}$ is bounded;*
(b) $\limsup_{n\to\infty} a_n = \limsup_{n\to\infty} n^{1/n} a_n$.

Proof. **(a)** In Example 2.18, we have shown that $n^{1/n} \to 1$, and so **(a)** follows easily.

(b) Let $a = \limsup a_n$, $b = \limsup n^{1/n} a_n$, and let $\{a_{n_k}\}$ be a subsequence of $\{a_n\}$ converging to a. Using the properties of the limit superior, we easily have

$$n_k^{1/n_k} a_{n_k} \to a \quad \text{and} \quad a \leq b.$$

Now we suppose that (see Theorem 2.49) $\{m_k^{1/m_k} a_{m_k}\}$ converges to b. This implies that $\{a_{m_k}\}$ converges to b, because $m_k^{1/m_k} \to 1$ as $k \to \infty$. Consequently,

$$b \leq \limsup a_n, \quad \text{i.e., } b \leq a.$$

We obtain $a = b$, as desired. ∎

Lemma 8.31. *The two power series* $\sum_{k=0}^{\infty} a_k x^k$ *and* $\sum_{k=1}^{\infty} k a_k x^k$ *have the same radius of convergence.*

Proof. By Lemma 8.30, we have

$$\limsup_{n\to\infty} |na_n|^{1/n} = \lim_{n\to\infty} n^{1/n} \limsup_{n\to\infty} |a_n|^{1/n} = \limsup_{n\to\infty} |a_n|^{1/n}.$$

The result now follows from Theorem 8.23. ∎

It might be of interest to have a direct proof of Lemma 8.31 without using the definition of limit superior, which is needed in order to utilize Theorem 8.23 as well as the fact that $\lim n^{1/n} = 1$. Let R and R' be the radii of convergence of $\sum_{k=0}^{\infty} a_k x^k$ and $\sum_{k=1}^{\infty} k a_k x^{k-1}$, respectively. Fix an arbitrary x with $0 < |x| < R$. Choose x_0 such that $|x| < |x_0| < R$. Now, $\sum_{k=0}^{\infty} a_k x_0^k$ and $\sum_{k=1}^{\infty} a_k x_0^{k-1}$ both converge absolutely. Thus, $\{a_k x_0^{k-1}\}$ is bounded. This means that there exists an M such that $|a_k x_0^{k-1}| \leq M$ for all $k \geq 1$. Consequently,

$$\left|k a_k x^{k-1}\right| = k \left|a_k x_0^{k-1}\right| \left|\frac{x}{x_0}\right|^{k-1} \leq M k r^{k-1} \quad \text{for } k \geq 1 \quad (r = |x/x_0|).$$

Since $0 < r < 1$, the ratio test shows that $\sum_{k=1}^{\infty} k r^{k-1}$ converges. But then by the comparison test, $\sum_{k=1}^{\infty} k a_k x^{k-1}$ converges absolutely for $|x| < |x_0|$. Since x_0 is arbitrary, we conclude that the derived series converges absolutely for all $|x| < R$. Thus, $R \leq R'$.

To obtain the reverse inequality, we fix x with $|x| < R'$ and observe that

$$|a_k x^{k-1}| \leq |k a_k x^{k-1}| \quad \text{for all } k \geq 1,$$

so that if $\sum |k a_k x^k|$ converges at $x \neq 0$ ($|x| < R'$), then $\sum |a_k x^{k-1}|$, and hence $\sum |a_k x^k|$ converges at x. This shows that $R \geq R'$. Consequently, $R = R'$, and proof is complete.

Caution: Lemma 8.31 does not say that $\sum_{k=0}^{\infty} a_k x^k$ and $\sum_{k=1}^{\infty} k a_k x^{k-1}$ have the same *interval of convergence* (see Questions 8.50(8)), although they have the same radius of convergence.

We can repeat the differentiation process and obtain the following theorem.

Theorem 8.32. *A power series $\sum_{k\geq 0} a_k x^k$ and the n-fold derived series defined by $\sum_{k\geq n} k(k-1)\cdots(k-n+1)a_k x^{k-n}$ have the same radius of convergence.*

Next we present the following result.

Theorem 8.33 (Term-by-term differentiation in power series). *If $\sum_{k\geq 0} a_k x^k$ has radius of convergence $R>0$, then $f(x)=\sum_{k\geq 0} a_k x^k$ is differentiable in $|x|<R$ and*

$$f'(x)=\sum_{k\geq 1} k a_k x^{k-1} \quad (|x|<R). \tag{8.3}$$

Moreover, $f^{(n)}(x)$ exists for every $n\geq 1$ and every x with $|x|<R$, and

$$f^{(n)}(x)=\sum_{k=n}^{\infty} k(k-1)\cdots(k-n+1))a_k x^{k-n} \quad (|x|<R). \tag{8.4}$$

The coefficients a_n are uniquely determined, and $a_n=f^{(n)}(0)/n$.

Proof. Let $f(x)=\sum_{k\geq 0} a_k x^k$ have radius of convergence R. We have to prove the existence of $f'(x)$ in $|x|<R$ and that f' is of the stated form. By Theorem 8.32 with $k=1$, the derived series $\sum_{k\geq 1} k a_k x^{k-1}$ converges for $|x|<R$ and defines a function, say $g(x)$, in $|x|<R$. We show that

$$f'(x)=\lim_{h\to 0}\frac{f(x+h)-f(x)}{h}=g(x) \quad \text{for all } x\in(-R,R).$$

Let $x\in(-R,R)$ be fixed. Then choose a positive $r\,(<R)$ such that $|x|<r$, e.g., $r=(R+|x|)/2$. Also, let $h\in\mathbb{R}$ with $0<|h|<(R-|x|)/2$. We have

$$|x+h|\leq|x|+|h|<|x|+\frac{R-|x|}{2}=\frac{|x|+R}{2}=r,$$

and for all nonzero h such that $0<|h|<(R-|x|)/2$, we consider

$$\frac{f(x+h)-f(x)}{h}-g(x)=\sum_{k\geq 2} a_k\left(\frac{(x+h)^k-x^k}{h}-kx^{k-1}\right), \tag{8.5}$$

where x and $x+h$ are now such that $\max\{|x|,|x+h|\}\leq r<R$. As an application of Taylor's theorem (which we prove for convenience at a later stage) on the interval with endpoints x and $x+h$, we get

$$(x+h)^k=x^k+kx^{k-1}h+\frac{k(k-1)}{2}c_k^{k-2}h^2,$$

where c_k is some number between x and $x+h$. (This may also be verified directly.) Note also that $|c_k|\leq r$, and so

$$\left|\frac{(x+h)^k - x^k}{h} - kx^{k-1}\right| \le |h|\,\frac{k(k-1)}{2} r^{k-2}.$$

So we must show that as $h \to 0$,

$$\left|\frac{f(x+h)-f(x)}{h} - g(x)\right| = \left|\sum_{k\ge 2} a_k\left(\frac{(x+h)^k - x^k}{h} - kx^{k-1}\right)\right| \to 0.$$

Since the derived series $\sum_{k\ge 2} k(k-1)a_k x^{k-2}$ of $\sum_{k\ge 1} ka_k x^{k-1}$ is also convergent for $|x|<R$, we conclude that it is absolutely convergent for $|x| \le r\ (<R)$. Using this and the triangle inequality, we see that as $h \to 0$,

$$\left|\sum_{k\ge 2} a_k\left(\frac{(x+h)^k - x^k}{h} - kx^{k-1}\right)\right| \le \frac{|h|}{2}\sum_{k\ge 2} |a_k| k(k-1) r^{k-2} \to 0.$$

Consequently, by (8.5), it follows that $f'(x)$ exists and equals $g(x)$. Since x is arbitrary, this holds at any interior point in $|x|<R$.

A repeated application of this argument shows that all the derivatives $f', f'', \dots, f^{(n)}, \dots$ exist in $|x|<R$, and (8.3) holds. The substitution $x=0$ in (8.4) yields that $f^{(n)}(0) = n!a_n$, as required. ∎

The theorem just proved shows that inside the interval of convergence (not necessarily at the endpoints $x = R, -R$), every power series can be differentiated term by term, and the resulting derived series will converge to the derivative of the limit function of the original series.

For example, the geometric series $(1-x)^{-1} = \sum_{k\ge 0} x^k$, which converges for $|x|<1$, after n-fold differentiation yields

$$\frac{1}{(1-x)^{n+1}} = \sum_{k\ge n} \binom{k}{n} x^{k-n} = \sum_{m\ge 0} \frac{(m+n)!}{n!m!} x^m \quad \text{for } |x|<1.$$

In particular,

$$\frac{1}{(1-x)^2} = \sum_{k\ge 1} kx^{k-1} \quad \text{and} \quad \frac{2}{(1-x)^3} = \sum_{k=2}^{\infty} k(k-1)x^{k-2} \quad \text{for } |x|<1.$$

Consequently, expressions such as the one above may be used to evaluate sums such as

$$\sum_{k=1}^{\infty} (-1)^k \frac{k}{3^k}, \quad \sum_{k=1}^{\infty} \frac{k}{3^k}, \quad \sum_{k=2}^{\infty} \frac{k(k-1)(-1)^k}{3^k}.$$

Finally, we remark that the following corollary shows that there is one and only one Taylor series for a function f, meaning that whatever method one uses, one obtains the same Taylor coefficients.

Corollary 8.34 (Uniqueness of the coefficients). *Let $R > 0$ be the radius of convergence of $f(x) = \sum_{k=0}^{\infty} a_k x^k$ and let $g(x) = \sum_{k=0}^{\infty} b_k x^k$ be such that*

$$\sum_{k=0}^{\infty} a_k x^k = \sum_{k=0}^{\infty} b_k x^k \quad \text{for } |x| < R.$$

Then $a_k = b_k$ for each $k \geq 0$.

Proof. The proof follows easily from Theorem 8.33. ∎

The conclusion of Corollary 8.34 can be deduced from a weaker hypothesis (see Theorem 8.37), but a proof requires some preparation.

8.2.4 Uniqueness Theorem for Power Series

Suppose that $f(x) = \sum_{k=0}^{\infty} a_k x^k$ converges for $|x| < R$ and that there exists a sequence $\{x_n\}_{n\geq 1}$ of distinct points converging to zero and at each of these points

$$f(x_n) = \sum_{k=0}^{\infty} a_k x_n^k = 0 \quad \text{for each } n \geq 1.$$

Then by the continuity of $f(x)$ at the origin, $a_0 = f(0) = 0$. Thus f takes the form $f(x) = xg(x)$, where

$$g(x) = \sum_{k=1}^{\infty} a_k x^{k-1},$$

which is also a convergent series in $|x| < R$. Because $f(x_n) = 0$ for all $n \geq 1$, it follows that $g(x_n) = 0$. Continuity of g at the origin implies that $a_1 = 0$. Continuing this process, we obtain the following.

Lemma 8.35. *Suppose that the power series $f(x) = \sum_{k=0}^{\infty} a_k x^k$ converges for $|x| < R$. If there exists a sequence $\{x_n\}_{n\geq 1}$ of distinct points converging to zero such that $f(x_n) = 0$ for all $n \geq 1$, then $a_k = 0$ for all $k \geq 0$ and*

$$f(x) = \sum_{k=0}^{\infty} a_k x^k = 0 \quad \text{in } |x| < R.$$

In particular, this lemma implies that if there exists a neighborhood D_0 of zero in $|x| < R$ for which

$$f(x) = \sum_{k=0}^{\infty} a_k x^k = 0 \quad \text{in } D_0,$$

then $a_k = 0$ for all $k \geq 0$. It is natural to look for a similar result if the sequence $\{x_n\}$ converges to a point other than the center (origin). In order to solve this problem, we need to prove the following result.

Lemma 8.36. *Suppose that the series* $f(x) = \sum_{k=0}^{\infty} a_k x^k$ *converges for* $|x| < R$ *and* a *is a point such that* $|a| < R$. *Then the Taylor series expansion of* f *about* $x = a$ *is given by*

$$f(x) = \sum_{k=0}^{\infty} \frac{f^{(k)}(a)}{k!}(x-a)^k,$$

which converges at least for $|x-a| < R - |a|$.

Proof. We write

$$x^k = (x-a+a)^k = \sum_{m=0}^{k} \binom{k}{m} a^{k-m}(x-a)^m,$$

so that

$$\begin{aligned}\sum_{k=0}^{\infty} a_k x^k &= \sum_{k=0}^{\infty} a_k \sum_{m=0}^{k} \binom{k}{m} a^{k-m}(x-a)^m \\ &= \sum_{m=0}^{\infty} \left(\sum_{k=m}^{\infty} \binom{k}{m} a_k a^{k-m} \right)(x-a)^m.\end{aligned}$$

This implies that

$$f(x) = \sum_{m=0}^{\infty} \frac{f^{(m)}(a)}{m!}(x-a)^m, \tag{8.6}$$

as desired. We need justifications for two steps. The interchange of the order of summation is justified, since

$$\sum_{k=0}^{\infty} |a_k| \sum_{m=0}^{k} \binom{k}{m} |a|^{k-m}|x-a|^m = \sum_{k=0}^{\infty} |a_k|(|a| + |x-a|)^k,$$

and by hypothesis $\sum_{k=0}^{\infty} a_k x^k$ converges absolutely for $|x| < R$, so

$$\sum_{k=0}^{\infty} |a_k|(|x-a| + |a|)^k$$

converges at least for $|x-a| < R - |a|$. Also, by Theorem 8.33,

$$\frac{f^{(m)}(x)}{m!} = \sum_{k=m}^{\infty} \binom{k}{m} a_k x^{k-m} \quad \text{in } |x| < R.$$

In particular,

$$\frac{f^{(m)}(a)}{m!} = \sum_{k=m}^{\infty} \binom{k}{m} a_k a^{k-m},$$

which proves (8.6). ∎

Theorem 8.37 (Uniqueness/identity theorem for power series). *Suppose that the series $f(x)=\sum_{k=0}^{\infty} a_k x^k$ converges for $|x|<R$. Suppose that S, the set of all x for which $f(x)=0$, has a limit point in $|x|<R$. Then $a_k=0$ for all $k\geq 0$.*

Proof. Let $I=\{x: |x|<R\}$ and $S=\{x\in I: f(x)=0\}$. We spilt I into two sets:

$$A=\{x\in I: \ x \text{ is a limit point of } S\} \quad \text{and} \quad B=\{x\in I: x\notin A\}=I\backslash A.$$

Then $I=A\cup B$ and $A\cap B=\emptyset$. Clearly, B is open. Next we show that A is open. Since $a\in A$ by hypothesis, A is nonempty. Since $|a|<R$, by Lemma 8.36, $f(x)$ can be expanded in a power series about a:

$$f(x)=\sum_{k=0}^{\infty} c_k(x-a)^k \quad \text{for } |x-a|<R-|a|.$$

We claim that $c_k=0$ for all $k\geq 0$. Since a is a limit point of S, there exists a sequence of points x_n in S, $x_n\neq a$, $x_n\to a$, with $f(x_n)=0$ for all n so that

$$c_0=f(a)=\lim_{n\to\infty} f(x_n)=0$$

(by the continuity of f). Thus, it suffices to prove that $c_k=0$ for all $k>0$. Suppose not. Then there would be a smallest positive integer m such that $c_m\neq 0$. Thus, $f(x)$ has the form

$$f(x)=(x-a)^m g(x), \quad g(x)=\sum_{k=m}^{\infty} c_k(x-a)^{k-m} \quad \text{for } |x-a|<R-|a|.$$

The continuity of g at a and $g(a)=c_m\neq 0$ imply the existence of a $\delta>0$ such that

$$g(x)\neq 0 \quad \text{for } 0<|x-a|<\delta \quad (<R-|a|).$$

We conclude that $f(x)\neq 0$ in $0<|x-a|<\delta$. This contradicts the fact that a is a limit point of S. Consequently, $c_k=0$ for all $k\geq 0$, so that $f(x)=0$ in a neighborhood of a. Hence, A is open.

Since $A\cup B=I$ is connected and nonempty, it cannot be written as a union of two nonempty disjoint open sets. Hence we must have either $A=\emptyset$ or $B=\emptyset$. But by the hypothesis, $a\in A$, and therefore $B=\emptyset$. Thus $A=I$, which gives that $a_k=0$ for all $k\geq 0$ and $f(x)=\sum_{k=0}^{\infty} a_k x^k=0$ for all $|x|<R$, as desired. ∎

Corollary 8.38 (Identity/uniqueness theorem). *Suppose that $f(x)=\sum_{k=0}^{\infty} a_k x^k$ and $g(x)=\sum_{k=0}^{\infty} b_k x^k$ converge for $|x|<R$. Set*

$$S=\{x: |x|<R \text{ with } f(x)=g(x)\}.$$

If S has a limit point in $|x|<R$, then $a_n=b_n$ for all $n\geq 0$; i.e., $f(x)=g(x)$ for all $|x|<R$.

Proof. Apply Theorem 8.37 to $h(x)=f(x)-g(x)=\sum_{k=0}^{\infty}(a_k-b_k)x^k$ in $|x|<R$. ∎

8.2.5 Real Analytic Functions

Functions that are expressible as a convergent power series are of particular interest. For instance, consider $f(x) = 1/(1-x)$ on $\mathbb{R} \setminus \{1\}$. For each $a \neq 1$, $f(x)$ has a power series about a:

$$f(x) = \frac{1}{1-a}\left(\frac{1}{1-(x-a)/(1-a)}\right) = \sum_{k=0}^{\infty} \frac{(x-a)^k}{(1-a)^{k+1}} \quad \text{for } |x-a| < |1-a|.$$

This suggests the following definition of a real analytic function:

Definition 8.39. *Let $I \subset R$ be open, and $f\colon I \to \mathbb{R}$ and $a \in I$. We say that f is real analytic at a if f can be represented as a Taylor series about a valid in a neighborhood of a. We say that f is real analytic on I if it is real analytic at each $a \in I$. That is, for every $a \in I$, there exist a number $\delta_a > 0$ and a sequence $\{a_k\}$ of real numbers such that*

$$f(x) = \sum_{k=0}^{\infty} a_k (x-a)^k \quad \textit{for every } x \in I \textit{ with } |x-a| < \delta_a.$$

In view of Theorem 8.33, a_k in the above Taylor representation of f must be $f^{(k)}(a)/k!$. Thus, for f to be real analytic at $x = a$, it is necessary that $f^{(k)}(a)$ exist for each k. However, the converse is not necessarily true. In Example 8.48, we present an example of an infinitely differentiable function f on $\mathbb{R}$ such that $f^{(k)}(0)$ exists for $k \geq 0$, but the resulting Taylor series about $x = 0$ does not converge to f in any neighborhood of the origin. Thus, the function f in Example 8.48 is not real analytic.

Examples 8.40. 1. Every polynomial in x with real coefficients is real analytic in $\mathbb{R}$.
2. The exponential function e^x and the trigonometric functions $\sin x$ and $\cos x$ are all real analytic on $\mathbb{R}$.
3. The function $1/(1+x^2)$ is real analytic on $\mathbb{R}$, whereas $1/(1-x)$ is real analytic only on $\mathbb{R} \setminus \{1\}$.
4. If $f(x) = \sum_{k=0}^{\infty} a_k x^k$ has radius of convergence $R > 0$, then f is real analytic on $|x| < R$ (by Lemma 8.36 and Theorem 8.33).
5. The function $|x|$ is not real analytic at $x = 0$ because $f'(0)$ does not exist.
6. The function $|x|^3$ is not real analytic at $x = 0$ because, although $f'(0)$ and $f''(0)$ exist, $f'''(0)$ and other higher-order derivatives at the origin do not exist. ●

By Theorem 8.33, we have the following:

Theorem 8.41. *Real analytic functions are infinitely differentiable.*

8.2.6 The Exponential Function

In Example 2.33, we defined

$$e^x = \lim_{n\to\infty} T_n(x), \quad T_n(x) = \left(1+\frac{x}{n}\right)^n = \sum_{k=0}^{n} \binom{n}{k}\left(\frac{x}{n}\right)^k, \quad x > 0.$$

In Example 8.20, we showed that the series $\sum_{k=0}^{\infty}(x^k/k!)$ converges for all $x \in \mathbb{R}$. We now show that

$$e^x = \sum_{k=0}^{\infty} \frac{x^k}{k!} \quad \text{for all } x, \tag{8.7}$$

using the former definition of e^x. This can be easily done using the method of proof of Theorem 5.7.

Theorem 8.42. *For $x \in \mathbb{R}$,*

$$e^x = \sum_{k=0}^{\infty} \frac{x^k}{k!} = \lim_{n\to\infty}\left(1+\frac{x}{n}\right)^n.$$

Proof. Let us first supply a proof for $x > 0$. Let $S_n(x) = \sum_{k=0}^{n}(x^k/k!)$ and $T_n(x)$ be as above. The proof relies on the following simple observation:

$$\binom{n}{k}\frac{x^k}{n^k} = \frac{n!}{(n-k)!n^k}\frac{x^k}{k!} = \left(1-\frac{1}{n}\right)\left(1-\frac{2}{n}\right)\cdots\left(1-\frac{k-1}{n}\right)\frac{x^k}{k!} \le \frac{x^k}{k!},$$

which gives

$$0 < \left(1+\frac{x}{n}\right)^n \le S_n(x).$$

Thus, we have

$$\lim_{n\to\infty}\left(1+\frac{x}{n}\right)^n \le \lim_{n\to\infty} S_n(x), \quad \text{i.e., } e^x \le \lim_{n\to\infty} S_n(x). \tag{8.8}$$

Moreover, for $n \ge m$,

$$\begin{aligned} T_n(x) &\ge 1 + n\left(\frac{x}{n}\right) + \frac{n(n-1)}{2!}\left(\frac{x}{n}\right)^2 + \cdots + \frac{n(n-1)\cdots(n-(m-1))}{m!}\left(\frac{x}{n}\right)^m \\ &= 1 + x + \left(1-\frac{1}{n}\right)\frac{x^2}{2!} + \cdots + \left(1-\frac{1}{n}\right)\cdots\left(1-\frac{m-1}{n}\right)\frac{x^m}{m!}. \end{aligned}$$

Allow $n \to \infty$, keeping m fixed, and obtain

$$\lim_{n\to\infty} T_n(x) \ge S_m(x).$$

Because $\{S_m(x)\}$ is an increasing sequence for each fixed $x > 0$, allowing $m \to \infty$ in this inequality, we finally get

$$\lim_{n\to\infty} T_n(x) \geq \lim_{m\to\infty} S_m(x). \tag{8.9}$$

Equations (8.8) and (8.9) show that the theorem holds for $x > 0$. To present a proof for the case $x < 0$, we first claim that

$$\left(\sum_{k=0}^{\infty} \frac{x^k}{k!}\right)\left(\sum_{k=0}^{\infty} \frac{(-x)^k}{k!}\right) = 1.$$

By the Cauchy product rule for series, we can write the left-hand side of the last expression as $\sum_{k=0}^{\infty} c_n$, where $c_0 = 1$ and

$$c_n = \sum_{k=0}^{n} \frac{x^k}{k!}\frac{(-x)^{n-k}}{(n-k)!} = \frac{1}{n!}\sum_{k=0}^{n} \frac{n!}{k!(n-k)!}x^k(-x)^{n-k} = \frac{(x-x)^n}{n!} = 0$$

for each $n \geq 1$. The claim follows. Thus, for $x < 0$ (so that $-x > 0$), we have

$$\sum_{k=0}^{\infty} \frac{x^k}{k!} = \frac{1}{\sum_{k=0}^{\infty} \frac{(-x)^k}{k!}} = \frac{1}{\lim_{n\to\infty} T_n(-x)} = \frac{1}{\mathrm{e}^{-x}} = \mathrm{e}^x.$$

The proof of the theorem is complete. ∎

Let us now write down some basic properties of the exponential function.

(a) Because the series (8.7) that represents the exponential function e^x converges absolutely for all x, Theorem 5.62 on the Cauchy product is applicable with $a_k = x^k/k!$ and $b_k = y^k/k!$. This gives

$$c_n = \sum_{k=0}^{n} a_k b_{n-k} = \sum_{k=0}^{n} \frac{x^k}{k!}\frac{y^{n-k}}{(n-k)!} = \frac{1}{n!}\sum_{k=0}^{\infty} \binom{n}{k} x^k y^{n-k} = \frac{(x+y)^n}{n!},$$

and so we obtain the fundamental property of the exponential function—called the *addition formula*:

$$\mathrm{e}^{x+y} = \mathrm{e}^x \mathrm{e}^y \quad \text{for all } x, y.$$

(b) By Theorem 8.33, a power series can be differentiated term by term, and so we have

$$\frac{\mathrm{d}}{\mathrm{d}x}\mathrm{e}^x = \mathrm{e}^x \quad \text{for all } x.$$

In particular, the addition formula gives

$$\mathrm{e}^x \mathrm{e}^{-x} = \mathrm{e}^{x-x} = \mathrm{e}^0 = 1,$$

and so $\mathrm{e}^x \neq 0$ for all x.

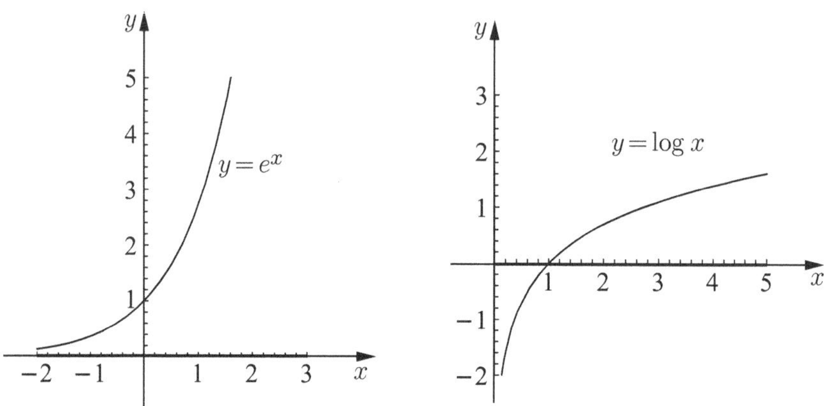

Fig. 8.5. Graphs of the exponential and logarithmic functions.

(c) By (8.7), $e^x > 0$ for $x \geq 0$, and the last relation in **(b)** shows that $e^x > 0$ for all real x. Also, (8.7) gives $e^x > 1 + x$ for $x > 0$, so that

$$e^x \to \infty \quad \text{as } x \to \infty \quad \text{and} \quad e^{-x} = \frac{1}{e^x} \to 0 \quad \text{as } x \to \infty.$$

Again, (8.7) gives

$$0 < x < y \Rightarrow e^x < e^y \ (\iff e^{-y} < e^{-x}),$$

or we may use the fact that $(e^x)' = e^x > 0$ for all $x \in \mathbb{R}$. Thus e^x is a strictly increasing continuous function on the whole real axis, and the image of $\mathbb{R}$ under e^x is $(0, \infty)$:

$$\exp(\mathbb{R}) = (0, \infty).$$

Thus, e^x is a bijection of $\mathbb{R}$ onto $(0, \infty)$.

These facts help us to obtain the graph of e^x, $x \in \mathbb{R}$ (see Figure 8.5).

8.2.7 Taylor's Theorem

We state and prove the single-variable version of Taylor's theorem for real-valued functions. We see that this is a generalization of the first mean value theorem. The several-variable version of Taylor's theorem will be proved in the author's book [7].

Theorem 8.43 (Taylor's theorem). *Suppose that $f : (\alpha, \beta) \subseteq \mathbb{R} \to \mathbb{R}$ is such that*

(i) *$f, f', \dots, f^{(n)}$ are all continuous on $[a, b] \subset (\alpha, \beta)$,*
(ii) *$f^{(n+1)}$ exists on (a, b).*

Then there exists a point c in (a,b) such that

$$f(b)=\sum_{k=0}^{n}\frac{f^{(k)}(a)}{k!}(b-a)^k+\frac{f^{(n+1)}(c)}{(n+1)!}(b-a)^{n+1}.$$

Proof. We want to show that

$$\left[f(a)+\sum_{k=1}^{n}\frac{f^{(k)}(a)}{k!}(b-a)^k+\frac{f^{(n+1)}(c)}{(n+1)!}(b-a)^{n+1}\right]-f(b)=0.$$

Fix the interval $[a,b]$ and introduce a new function ϕ by

$$\phi(x)=\left[f(x)+\sum_{k=1}^{n}\frac{f^{(k)}(x)}{k!}(b-x)^k+\frac{M}{(n+1)!}(b-x)^{n+1}\right]-f(b), \quad (8.10)$$

where M has been chosen in such a way that $\phi(a)=0$. Now

- ϕ is continuous on $[a,b]$, by **(i)**;
- ϕ is differentiable on (a,b), by **(i)** and **(ii)**;
- $\phi(a)=0=\phi(b)$.

By Rolle's theorem, it follows that there exists a number $c\in(a,b)$ such that $\phi'(c)=0$. Since

$$\begin{aligned}\phi'(x)&=f'(x)+\sum_{k=1}^{n}\left(\frac{f^{(k+1)}(x)}{k!}(b-x)^k-\frac{f^{(k)}(x)}{(k-1)!}(b-x)^{k-1}\right)-\frac{M}{n!}(b-x)^n\\&=\frac{f^{(n+1)}(x)}{n!}(b-x)^n-\frac{M}{n!}(b-x)^n,\end{aligned}$$

$\phi'(c)=0$ implies that $M=f^{(n+1)}(c)$. When this value is substituted into (8.10), the condition $\phi(a)=0$ yields the desired formula. ■

In particular, a small change in notation gives the following:

Theorem 8.44. *If f is $(n+1)$-times differentiable on an open interval containing $[a,x]$, then there exists some c between a and x such that*

$$f(x)=S_n(x)+R_n(x),$$

where

$$S_n(x)=\sum_{k=0}^{n}f^{(k)}(a)\frac{(x-a)^k}{k!}\quad\text{and}\quad R_n(x)=f^{(n+1)}(c)\frac{(x-a)^{n+1}}{(n+1)!}.$$

We recall the convention that $f^{(0)}(x)=f(x)$.

Here the polynomial $S_n(x)$ is called the nth-degree *Taylor polynomial* (or approximation) for f at a, while the remainder term $R_n(x)$ is usually called the *Lagrange form* of the remainder, or sometimes the *error term*. There are several other forms of the remainder term, which have some advantages in some situations, but the Lagrange form is the simplest. The integral form of the remainder term is stated in Exercise 8.51(20).

We remark that if $n = 1$, then Taylor's expansion is just the one-dimensional mean value theorem. Note that $S_n(x) \to f(x)$ if and only if $R_n(x) \to 0$. Moreover, Theorem 8.44 gives the following.

Corollary 8.45 (Taylor series). *If f has derivatives of all orders on an open interval containing points a and x, and if $R_n(x) \to 0$ as $n \to \infty$, then we have*

$$f(x) = \sum_{k=0}^{\infty} \frac{f^{(k)}(a)}{k!}(x-a)^k.$$

Unfortunately, it is not always the case that $R_n(x) \to 0$ as $n \to \infty$ even though f has derivatives of all orders (see Example 8.48). We call f the sum function of the corresponding power series, namely, the Taylor series. As remarked earlier, the series for the case $a = 0$ is often called a Maclaurin series. In a later section, we shall prove the converse of this corollary, namely that every convergent power series represents an infinitely differentiable function.

Examples 8.46. Using familiar differential properties, we can illustrate Taylor's theorem with some standard examples. Here is a list of a few well-known Maclaurin series expansions:

1. Consider $f(x) = \sin x$ for $x \in \mathbb{R}$. Then $f'(x) = \cos x = \sin(\pi/2 + x)$,

$$f''(x) = \cos(\pi/2+x) = \sin(\pi+x), \quad f'''(x) = \cos(\pi+x) = \sin(3\pi/2+x).$$

More generally, $f^{(n)}(x) = \sin(n\pi/2 + x)$, and so

$$f^{(n)}(0) = \sin(n\pi/2) = \begin{cases} 0 & \text{if } n \text{ is even,} \\ (-1)^{k-1} & \text{if } n \text{ is odd with } n = 2k-1. \end{cases}$$

Taylor's theorem applied for $|x| \leq r$ gives

$$|R_n(x)| = \left| \frac{\sin((n+1)\pi/2 + c)}{(n+1)!} x^{n+1} \right| \leq \frac{|x|^{n+1}}{(n+1)!} \leq \frac{r^{n+1}}{(n+1)!} =: a_{n+1}$$

for each n and for every $r > 0$ with $|x| \leq r$. Note that

$$\frac{a_{n+1}}{a_n} = \frac{r}{n+1} \to 0 \quad \text{as } n \to \infty,$$

and so $a_n \to 0$ as $n \to \infty$. Thus, $R_n(x) \to 0$ as $n \to \infty$, and for each x with $|x| < r$. This gives

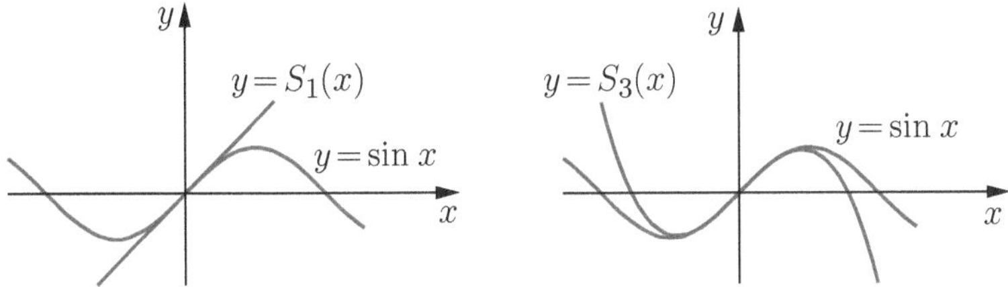

Fig. 8.6. The graphs of $y = \sin x$, $y = S_1(x)$; $y = \sin x$, $y = S_3(x)$ near $x = 0$.

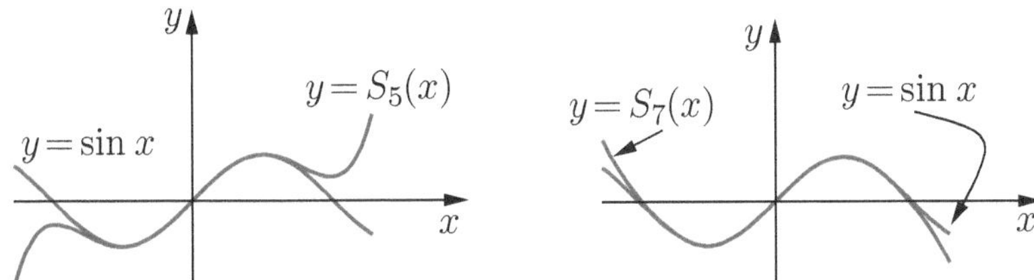

Fig. 8.7. The graphs of $y = \sin x$, $y = S_5(x)$; $y = \sin x$, $y = S_7(x)$ near $x = 0$.

$$\sin x = \sum_{k=0}^{\infty}(-1)^k \frac{x^{2k+1}}{(2k+1)!} \quad \text{for all } x \in \mathbb{R}.$$

Differentiating with respect to x, or directly, we see that

$$\cos x = \sum_{k=0}^{\infty}(-1)^k \frac{x^{2k}}{(2k)!} \quad \text{for all } x \in \mathbb{R}.$$

Finally, we remark that the Taylor polynomials for $f(x) = \sin x$ at 0 are

$$S_1(x) = S_2(x) = x,\ S_3(x) = S_4(x) = x - \frac{x^3}{3!},\ S_5(x) = S_6(x) = x - \frac{x^3}{3!} + \frac{x^5}{5!}$$

and

$$S_7(x) = S_8(x) = x - \frac{x^3}{3!} + \frac{x^5}{5!} - \frac{x^7}{7!}.$$

The graphs in Figures 8.6 and 8.7 illustrate how the approximation of $f(x) = \sin x$ by $S_n(x)$ gets better as n increases.

2. Consider $f(x) = \mathrm{e}^x$ for $x \in \mathbb{R}$. This function has the nice property that $f^{(n)}(x) = \mathrm{e}^x$ for all $n \in \mathbb{N}_0$, so that $f^{(n)}(0) = 1$ for all $n \in \mathbb{N}_0$. Then for each x, $|x| \leq r$, there exists c_n between 0 and x such that

$$\mathrm{e}^x = \sum_{k=0}^{n} \frac{x^k}{k!} + \frac{\mathrm{e}^{c_n}}{(n+1)!} x^{n+1} \quad \text{for all } |x| \leq r.$$

Since f is increasing on $\mathbb{R}$, $\mathrm{e}^{-|x|} \leq \mathrm{e}^c \leq \mathrm{e}^{|x|}$ for all c between 0 and x. Thus, as in the previous case, it follows that for $|x| \leq r$,

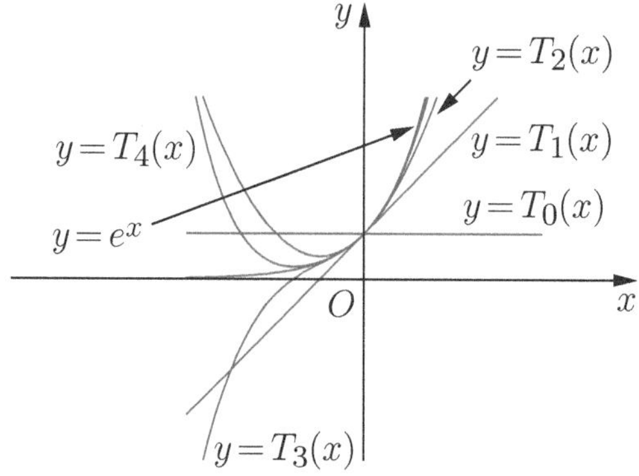

Fig. 8.8. The approximation of e^x.

$$|R_n(x)| = \left|\frac{e^{c_n}}{(n+1)!}x^{n+1}\right| \leq \frac{e^r r^{n+1}}{(n+1)!} \to 0 \quad \text{as } n \to \infty,$$

and since x was arbitrary, we obtain

$$e^x = \sum_{k=0}^{\infty} \frac{x^k}{k!} \quad \text{for all } x \in \mathbb{R}.$$

If one wishes to use Taylor's formula to approximate $e^{0.1}$ by a quadratic polynomial with an error estimate, we need to consider

$$\left|e^x - \left(1 + x + \frac{x^2}{2!}\right)\right| < \frac{e^{|c|}}{3!}|x|^3 \quad (|c| < 0.1),$$

so that for $x = 0.1$, this inequality gives

$$|e^{0.1} - (1.1 + 0.005)| < \frac{e^{0.1}}{6}\left(\frac{1}{10^3}\right), \quad \text{i.e., } |e^{0.1} - 1.105| < 0.000184,$$

which gives a fairly decent approximation of $e^{0.1}$. For large values of x, a good approximation for e^x requires a Taylor polynomial of degree n, where n is a large number. The graphs in Figure 8.8 illustrate how the approximation of $f(x) = e^x$ (for x near the origin) by the Taylor polynomials $T_n(x)$ gets better as n increases.

The last series expansion for e^x also gives

$$e^{-x} = \sum_{k=0}^{\infty} \frac{(-1)^k}{k!} x^k \quad \text{for all } x \in \mathbb{R},$$

so that

$$\sinh x = \frac{e^x - e^{-x}}{2} = x + \frac{x^3}{3!} + \frac{x^5}{5!} + \frac{x^7}{7!} + \cdots \quad \text{for all } x \in \mathbb{R}$$

and

$$\cosh x = \frac{e^x + e^{-x}}{2} = 1 + \frac{x^2}{2!} + \frac{x^4}{4!} + \frac{x^6}{6!} + \cdots \quad \text{for all } x \in \mathbb{R}.$$

3. Consider $f(x) = -\log(1-x)$ for $1 - x > 0$. Then $f(0) = 0$,

$$f'(x) = \frac{1}{1-x}, \quad f^{(k)}(x) = \frac{(k-1)!}{(1-x)^k} \quad \text{for } k \geq 1.$$

In particular, $f^{(k)}(0) = (k-1)!$. Thus, f has derivatives of all orders for $x < 1$. In particular, by Taylor's theorem applied to $(-R, 1)$, for any $R > 0$,

$$f(x) = \sum_{k=0}^{n} \frac{f^{(k)}(0)}{k!} x^k + R_n(x) = \sum_{k=1}^{n} \frac{x^k}{k} + R_n(x),$$

where

$$R_n(x) = \frac{f^{(n+1)}(c)}{(n+1)!} x^{n+1} = \frac{1}{n+1} \frac{x^{n+1}}{(1-c)^{n+1}} = \frac{1}{n+1} \left(\frac{x}{1-c} \right)^{n+1}$$

and c is a number in $(-R, 1)$ for any $R > 0$. We observe that $R_n(x) \to 0$ only when $|x| < 1$. Consequently,

$$-\log(1-x) = x + \frac{x^2}{2} + \frac{x^3}{3} + \frac{x^4}{4} + \cdots \quad \text{for all } |x| < 1.$$

Finally, by considering $\log(1+x) - \log(1-x)$, we see that

$$\frac{1}{2} \log\left(\frac{1+x}{1-x} \right) = \sum_{k=1}^{\infty} \frac{x^{2k-1}}{2k-1} \quad \text{if } |x| < 1.$$

•

Example 8.47. If we apply Taylor's theorem with

$$f(x) = \sin x \quad \text{for } x \in [-3, 3],$$

then we have (for $c \in (-3, 3)$)

$$\left| \sin x - x + \frac{x^3}{3!} \right| = \left| \sin(c) \frac{x^4}{4!} \right| < \frac{3^4}{4!} \quad \text{for } |x| < 3.$$

Similarly, we see that

$$\left| e^x - \left(1 + x + \frac{x^2}{2!} + \frac{x^3}{3!} \right) \right| = \left| e^c \frac{x^4}{4!} \right| < \frac{3^4 e^3}{4!} \quad \text{for } |x| < 3.$$

•

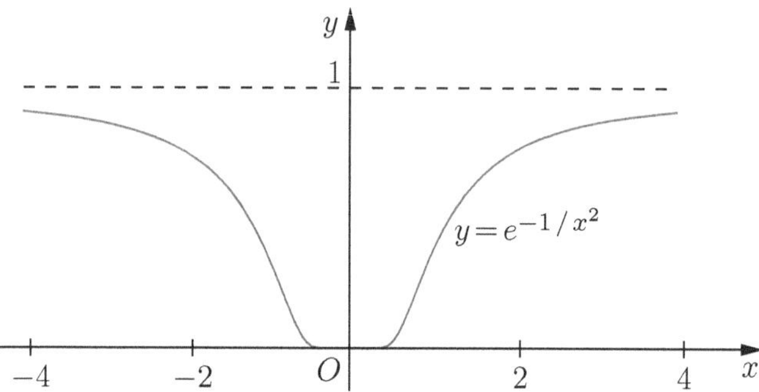

Fig. 8.9. The graph of $f(x) = \exp(-1/x^2)$, $f(0) = 0$.

Example 8.48 (Not all infinitely differentiable functions are analytic). Consider (see Figure 8.9)

$$f(x) = \begin{cases} e^{-1/x^2} & \text{if } x \neq 0, \\ 0 & \text{if } x = 0. \end{cases}$$

Note that this function is clearly continuous for all $x \in \mathbb{R}$. In fact,

$$\lim_{x\to 0} e^{-1/x^2} = \lim_{u\to +\infty} \frac{1}{e^{u^2}} = 0.$$

At this place, it might be important to remark that if $g(x) = e^{-1/x}$, $x \neq 0$, then it would not have worked in this way, since as $x \to 0-$, one would then get

$$\lim_{x\to 0-} e^{-1/x} = e^{+\infty} = \infty.$$

Thus, in this case, one could perhaps modify g to ϕ in the following form:

$$\phi(x) = \begin{cases} e^{-1/x} & \text{if } x > 0, \\ 0 & \text{if } x \leq 0. \end{cases}$$

However, we proceed to show that $f \in C^\infty(\mathbb{R})$, i.e., f is differentiable infinitely often in $(-\infty, \infty)$. Before we look at the properties of $f(x)$, we observe that for $x \neq 0$, $f'(x)$ exists by the chain rule, so that

$$f'(x) = 2x^{-3}e^{-1/x^2} \quad \text{for } x \neq 0.$$

Now,

$$\lim_{x\to 0} \frac{f(x) - f(0)}{x} = \lim_{x\to 0} \frac{1/x}{e^{1/x^2}},$$

so that, by l'Hôpital's rule,

$$\lim_{x\to 0+} \frac{1/x}{e^{1/x^2}} = \lim_{u\to +\infty} \frac{u}{e^{u^2}} = \lim_{u\to +\infty} \frac{1}{2ue^{u^2}} = 0$$

and

$$\lim_{x\to 0-} \frac{1/x}{e^{1/x^2}} = \lim_{u\to -\infty} \frac{u}{e^{u^2}} = -\lim_{u\to +\infty} \frac{u}{e^{u^2}} = 0.$$

Therefore, $f'(0)$ exists and equals zero. Next, we show that $f'(x)$ is continuous on $\mathbb{R}$. Obviously, f' is continuous for $x \neq 0$, and so we need to check the continuity only at 0. For this, we find that (again by l'Hôpital's rule)

$$\lim_{x\to 0} f'(x) = 2\lim_{x\to 0} \frac{1/x^3}{e^{1/x^2}} = 2\lim_{u\to \pm\infty} \frac{u^3}{e^{u^2}} = 0 = f'(0).$$

Note that because of the square term in the denominator, we need to consider $u \to +\infty$ and $u \to -\infty$ separately, and in both cases the limit value is seen to be 0. Thus, $f \in C^1(\mathbb{R})$. Our next aim is to show that $f \in C^\infty(\mathbb{R})$. For this, we observe that for $n = 0, 1, 2, \ldots,$

$$\lim_{x\to 0} \frac{e^{-1/x^2}}{x^n} = \lim_{u\to \pm\infty} \frac{u^n}{e^{u^2}}, \tag{8.11}$$

which is seen to be 0, by l'Hôpital's rule. The case $n = 0$ shows that f is continuous at 0, whereas the case $n = 1$ implies that f is continuously differentiable on $\mathbb{R}$. We have discussed these two cases above, and so we now consider the case $n \geq 2$. Note that the function $f(x)$ has the form $e^{-g(x)}$, where $g(x) = 1/x^2$. Clearly, for $x \neq 0$, the higher derivatives $f^{(n)}(x)$ are given by

$$f^{(n)}(x) = \frac{p_n(1/x)}{e^{1/x^2}},$$

where p_n is a polynomial of the form

$$p_n(x) = a_n x^n + a_{n-1}x^{n-1} + \cdots + a_1 x + a_0.$$

This fact can easily be proved by induction and the chain rule. By (8.11), we now have

$$\lim_{x\to 0} f^{(n)}(x) = \lim_{u\to \infty} \frac{p_n(u)}{e^{u^2}} = 0 \quad \text{for } n \geq 2,$$

since e^{u^2} goes to infinity *faster than any polynomial* (one can use l'Hôpital's rule). We have already shown that $f'(0) = 0$, and thus by induction it can be shown that $f^{(n)}(0) = 0$ for all n. In fact, if $f^{(k)}(0) = 0$ for all $k = 1, 2, \ldots, n$, then

$$\begin{aligned} f^{(n+1)}(0) &= \lim_{x\to 0} \frac{f^{(n)}(x) - f^{(n)}(0)}{x} \\ &= \lim_{x\to 0} \frac{(1/x)p_n(1/x)}{e^{1/x^2}} \\ &= \lim_{x\to 0} \frac{p_{n+1}(1/x)}{e^{1/x^2}} \\ &= \lim_{x\to 0} f^{(n+1)}(x) = 0, \end{aligned}$$

and hence $f^{(n+1)}(0)$ exists and is zero. Thus, f has derivatives of all orders at 0, and hence $f \in C^{\infty}(\mathbb{R})$. Therefore, Taylor's series certainly converges, but not to $f(x)$. •

Remark 8.49. From Example 8.48, we note that if $f(x) = e^{-1/x^2}$ had a Taylor series expansion about the origin, then we would have $f(x) = 0$ in a neighborhood of the origin, since $f^{(k)}(0) = 0$ for all k. Since the exponential function never vanishes, $f(x) = e^{-1/x^2}$ certainly cannot be identically zero and so does not admit a Taylor series expansion about the origin.

Is it correct to say that $f^{(n)}(0) = 0$ for all n implies $f(x) = 0$ for all x in a neighborhood of 0? (Compare with Example 8.48.) What is your answer in the case of a complex-valued function? Is it something to do with the uniqueness theorem for complex-valued analytic functions in some domain?[1] This is beyond the scope of this book, but an interested reader can compare it. •

8.2.8 Questions and Exercises

Questions 8.50.

1. Suppose that the series $\sum a_k x^k$ converges at all positive integer values of x. Must the series be convergent at all real values of x? What can be said about the radius of convergence of the power series?
2. Can a power series of the form $\sum_{k=0}^{\infty} a_k (x-2)^k$ be convergent at $x = 0$ and divergent at $x = 3$?
3. What can be said about the radius of convergence of the power series $\sum_{k=0}^{\infty} a_k (x-2)^k$ if it is convergent at $x = 1$ and divergent at $x = 5$?
4. Suppose that $\sum_{k=0}^{\infty} a_k (x-2)^k$ converges at $x = 5$ but diverges at $x = -1$. What can be said about the radius of converges of the given series?
5. Suppose that $f(x) = \sum_{k=0}^{\infty} a_k x^k$ on $(-R, R)$ $(R > 0)$ such that f is odd on $(-R, R)$. What can be said about f?
6. What can be said about the radius of convergence of the power series $\sum_{k=0}^{\infty} a_k x^k$ if there exists a real sequence $\{x_k\}$ converging to 0 and the series is divergent at each x_k?
7. If $\{b_n\}$ is a decreasing sequence converging to zero, must the functional series $\sum_{k=1}^{\infty} b_k \cos kx$ be convergent on $[0, 2\pi]$?
8. Let I_1, I_2, and I_3 denote the intervals of convergence of $\sum_{k=0}^{\infty} a_k x^k$, $\sum_{k=1}^{\infty} k a_k x^{k-1}$, and $\sum_{k=2}^{\infty} k(k-1) a_k x^{k-2}$, respectively. Must we have $I_1 = I_2$? Must we have $I_1 \subseteq I_2$? Must we have $I_2 \subseteq I_1$? Is there a simple relationship between the sets I_1, I_2, and I_3?
9. Suppose that $f(x) = \sum a_k x^k$ and $g(x) = \sum b_k x^k$ have positive radii of convergence R_1 and R_2, respectively.
   (a) Does the Cauchy product of the two power series converge for $x \neq 0$? If so, does it converge to $f(x)g(x)$ in $|x| < \min\{R_1, R_2\}$?

---

[1] Suppose that f is analytic in a domain D. If S, the set of zeros of f in D, has a limit point z^* in D, then $f(z) \equiv 0$ in D.

(b) If $|a_k| \leq |b_k|$ for all large values k, must there be a relationship between R_1 and R_2?

10. Are the sum and difference of two real analytic functions real analytic?
11. Is the product of two real analytic functions real analytic?
12. Is the reciprocal of a nowhere-vanishing real analytic function real analytic?
13. Are there infinitely differentiable functions that are not real analytic?
14. Suppose that f is real analytic on an open interval (a, b). Must f be infinitely differentiable on (a, b)?
15. Suppose that g is real analytic on an open set I and J is an open set such that $g(I) \subset J$ and f is analytic on J. Must the composition $f \circ g$ be real analytic on I?
16. Suppose that f is such that $f(0) = 0$ and $f(x) = \mathrm{e}^{-1/x^2}$ for $x \neq 0$. Must f be real analytic on $\mathbb{R} \setminus \{0\}$? Can f be real analytic on $\mathbb{R}$?
17. We have mentioned that $f(x) = 1/(1+x)^2$ is real analytic on $\mathbb{R}$ but the power series expansion of $f(x)$ about $x = 0$ converges only for $|x| < 1$. Why does the power series not converge at all values of $\mathbb{R}$?
18. Suppose that f is continuous on $\mathbb{R}$ and that both f^2 and f^3 are real analytic. Must f be real analytic?

Exercises 8.51.

1. Let $f(x)$ be as in Example 8.48, where we have shown that $f^{(n)}(0) = 0$ for all $n \geq 1$. Conclude that the remainder term in Taylor's theorem (with $a = 0$) does not converge to zero as $n \to \infty$ for $x \neq 0$.
2. Show that the power series $\sum_{k=1}^{\infty}((-1)^{k-1}/(2k-1))x^{2k-1}$ converges for $|x| \leq 1$.
3. If R is the radius of convergence of $\sum_{k=0}^{\infty} a_k x^k$, determine the radius of convergence of the following series:

   (a) $\displaystyle\sum_{k=0}^{\infty} a_k x^{pk}$, (b) $\displaystyle\sum_{k=0}^{\infty} a_{pk} x^k$, (c) $\displaystyle\sum_{k=0}^{\infty} a_k^p x^k$,

   where p is a positive integer.
4. Give an example of a power series that has:
   (a) radius of convergence zero.
   (b) radius of convergence ∞.
   (c) finite radius of convergence.
   (d) conditional convergence at both endpoints of the interval of convergence.
   (e) conditional convergence only at one of the endpoints.
   (f) absolute convergence at both endpoints of the interval of convergence.
5. Give an example of a power series $\sum_{k=0}^{\infty} a_k x^k$ (if it exists) that:
   (a) converges at $x = -1$ but diverges at $x = 2$.
   (b) converges at $x = 1$ but diverges at $x = -1$.
   (c) converges at $x = 2$ but diverges at $x = -1, 1$.
   (d) converges at $x = -1, 1$ but diverges at $x = 2$.

6. Determine the radius of convergence of each of the following power series $\sum_{k=1}^{\infty} a_k x^k$ when a_k equals:

$$\text{(a) } k^{\log k}. \quad \text{(b) } \frac{(3k)!}{(k!)^2}. \quad \text{(c) } k^{k^2}. \quad \text{(d) } \prod_{m=0}^{k} \frac{(a+m)(b+m)}{(c+m)(1+m)} \ (a,b,c>0).$$

$$\text{(e) } \frac{3^k}{k!}. \quad \text{(f) } k^{\sqrt{k}}. \quad \text{(g) } \frac{2^k}{k!}. \quad \text{(h) } r^{k^2} \ (|r|<1).$$

7. Determine the radius and the interval of convergence of each of the following power series:

$$\text{(a) } \sum_{k=1}^{\infty} k! x^{3k+3}. \qquad \text{(b) } \sum_{k=2}^{\infty} \frac{k^2+1}{k^2-1} x^k. \qquad \text{(c) } \sum_{k=0}^{\infty} \frac{k^3}{k!} \frac{x^{2k}}{2^k}.$$

$$\text{(d) } \sum_{k=1}^{\infty} x^{k^3}. \qquad \text{(e) } \sum_{k=1}^{\infty} \frac{x^{k^2}}{(k!)^k}. \qquad \text{(f) } \sum_{k=1}^{\infty} \left(1+\frac{2}{k}\right)^k (x-3)^k.$$

$$\text{(g) } \sum_{k=1}^{\infty} r^{k^3} (x+2)^k. \quad \text{(h) } \sum_{k=1}^{\infty} \frac{2^k}{k} (x+1)^{k+1}. \quad \text{(i) } \sum_{k=1}^{\infty} \frac{k(k-1)}{k^2+5} (x+2)^k.$$

8. Consider the power series $\sum_{k=0}^{\infty} a_k x^k$, where

$$a_k = \begin{cases} \dfrac{1}{5^k} & \text{if } k \text{ is odd,} \\ 3^k & \text{if } k \text{ is even,} \end{cases} \qquad k \in \mathbb{N}_0.$$

(a) Show that neither $\lim_{k\to\infty} \sqrt[k]{|a_k|}$ nor $\lim_{k\to\infty} |a_{k+1}/a_k|$ exists.
(b) Determine the radius of convergence of

$$\sum_{k=0}^{\infty} \frac{1}{5^{2k+1}} x^{2k+1} \quad \text{and} \quad \sum_{k=0}^{\infty} 3^{2k} x^{2k}.$$

(c) What is the radius of convergence of the original power series? Can this be obtained from (b)?
(d) What is the interval of convergence of the original power series?

9. For each $\alpha \in \mathbb{R}$, show that

$$(1+x)^{\alpha} = 1 + \sum_{k=1}^{\infty} \frac{\alpha(\alpha-1)\cdots(\alpha-k+1)}{k!} x^k \quad \text{for } |x|<1.$$

10. Show that

$$\sum_{k=0}^{\infty} (k+1)^2 x^k = \frac{1+x}{(1-x)^3} \quad \text{for } |x|<1.$$

11. Suppose that $1960 \le |a_n| \le 2008$ for all $n \ge 0$. Discuss the radius of convergence of the series $\sum_{k=0}^{\infty} a_k (x-2009)^k$, and justify your answer.

12. Find the set of values of x for which the power series

$$\sum_{k=1}^{\infty}(1+(-3)^{k-1})x^k$$

converges.

13. For what values of $x \in \mathbb{R}$ does the functional series $\sum_{k=0}^{\infty}((x-1)/(x+2))^k x^k$ converge absolutely.

14. Find all the values of $x \in \mathbb{R}$ for which the functional series $\sum_{k=0}^{\infty}(1-x^2)^n$ is absolutely convergent.

15. Suppose that

$$a_k = \begin{cases} 1 & \text{if } k = 3m, \\ \dfrac{(-1)^m}{m} & \text{if } k = 3m+1, \\ \dfrac{1}{m^2} & \text{if } k = 3m+2, \end{cases} \quad \text{and} \quad b_k = \begin{cases} 3^m & \text{if } k = 2m, \\ \dfrac{(-1)^m}{m} & \text{if } k = 2m+1, \end{cases}$$

for $m \geq 0$. Find the interval of convergence of $\sum a_k x^k$ and $\sum b_k x^k$, respectively.

16. Using the Cauchy product rule and the power series expansion of e^x and $1/(1-x)$, determine a_k for $k = 0, 1, 2, 3$ in

$$\frac{e^x}{1-x} = \sum_{k=0}^{\infty} a_k x^k.$$

17. Find the real solution greater than 1 of the equation

$$x = \sum_{n=1}^{\infty} \frac{n(n+1)}{x^n}.$$

18. Suppose the series $f(x) = \sum_{k=0}^{\infty} a_k x^k$ and $g(x) = \sum_{k=0}^{\infty} b_k x^k$ converge for $|x| < 1$ and $f(x) = g(x)$ for $x = 1/(n+1)$, $n \in \mathbb{N}$. Show that $a_k = b_k$ for all $k \geq 0$.

19. Let f be infinitely differentiable on $(-1, 1)$. Prove that f is real analytic in some neighborhood of the origin if and only if there exist two positive real numbers r and K such that

$$\left|\frac{f^{(n)}(x)}{n!}\right| \leq K \quad \text{for all } x \in (-1,1) \text{ with } |x| < r.$$

20. Let I be an open interval containing $[a, b]$ and $f \in C^n(I, \mathbb{R})$. Show that the Taylor formula can be put into the following form:

$$f(b) = \sum_{k=0}^{n} \frac{f^{(k)}(a)}{k!}(b-a)^k + R_n, \quad R_n = \frac{1}{n!}\int_a^b (b-t)^n f^{(n+1)}(t)\, dt,$$

where R_n is called the *integral form* of the remainder term.

9

Uniform Convergence of Sequences of Functions

In this chapter we consider sequences and series of real-valued functions and develop *uniform convergence tests*, which provide ways of determining quickly whether certain sequences and infinite series have limit functions. Our particular emphasis in Section 9.1 is to present the definitions and simple examples of pointwise and uniform convergence of sequences. In addition, we present characterizations for interchanging limit and integration signs in sequences of functions. In Section 9.2, we discuss a characterization for interchanging limit and integration signs, and interchange of limit and differentiation signs for uniform convergence of sequences and series of functions. At the end of the section, we also include some foundations for the study of summability of series, which is an attempt to attach a value to a series that may not converge, thereby generalizing the concept of the sum of a convergent series. Finally, we also discuss the Abel summability of series. At the end of Section 9.2, we state and prove an important result due to Weierstrass, which in a simple form states that "any continuous function on $[a, b]$ can be uniformly approximated by polynomials."

9.1 Pointwise and Uniform Convergence of Sequences

In Chapter 2, we considered numerical sequences and numerical series, whereas in Section 8.2 we discussed the convergence of power series. In this section, we consider sequences and series of real-valued functions. Of course, we did discuss some examples of this type in the earlier chapter. The theoretical importance of uniform convergence and its use will be discussed using a number of motivating examples.

S. Ponnusamy, *Foundations of Mathematical Analysis*,
DOI 10.1007/978-0-8176-8292-7_9,

9.1.1 Definitions and Examples

Let $\{f_n\}$ be sequence of real-valued functions defined on a set $E \subseteq \mathbb{R}$. To each $x_0 \in E$, $\{f_n\}$ gives rise to the sequence $\{f_n(x_0)\}$ of real numbers. Thus, we have the notion of pointwise and uniform convergence on E.

Definition 9.1 (Pointwise convergence). *We say that $\{f_n\}$ converges at $x \in E$ if the sequence $\{f_n(x)\}$ of real numbers is convergent. We say that $\{f_n\}$ converges pointwise on E if for each $x \in E$ the sequence $\{f_n(x)\}$ converges. If the sequence $\{f_n\}$ converges pointwise on a set E, then we can define $f : E \to \mathbb{R}$ by*

$$f(x) = \lim_{n\to\infty} f_n(x) \quad \textit{for each } x \in E.$$

It is a common practice to call f the (pointwise) limit of the sequence $\{f_n\}$. We write $f_n \to f$ pointwise on E or $\lim f_n = f$ pointwise on E. Throughout we simply use $f_n \to f$ on E or $\lim f_n = f$ on E to indicate the pointwise convergence on E.

Thus, in terms of ϵ-N notation, $\{f_n\}$ converges to f on E iff for each $x \in E$ and for an arbitrary $\epsilon > 0$, there exists an integer $N = N(\epsilon, x)$ such that

$$|f_n(x) - f(x)| < \epsilon \quad \text{whenever } n > N.$$

The integer N in the definition of pointwise convergence may, in general, depend on both $\epsilon > 0$ and $x \in E$. If, however, one integer can be found that works for all points in E, then the convergence is said to be *uniform*. That is, a sequence of functions $\{f_n\}$ *converges uniformly* to f on a set E if for each $\epsilon > 0$, there exists an integer $N = N(\epsilon)$ such that

$$|f_n(x) - f(x)| < \epsilon \quad \text{whenever } n > N(\epsilon) \text{ and for all } x \in E.$$

That is,

$$f(x) - \epsilon < f_n(x) < f(x) + \epsilon \quad \text{for all } x \in E \text{ and } n > N(\epsilon).$$

Often, we say that f is the uniform limit of the sequence $\{f_n\}$ on E and write $f_n \to f$ uniformly on E or $\lim f_n = f$ uniformly on E. Geometrically, this means that given $\epsilon > 0$, there exists an integer $N = N(\epsilon)$ such that for $n > N$, the graph of $y = f_n(x)$ on E must lie between the graphs of $y = f(x) - \epsilon$ and $y = f(x) + \epsilon$ on E. In Figure 9.1, the graphs of all functions $y = f_n(x)$ with $n > N$ would fit inside this band. This can happen if and only if

$$\sup_{x\in E} |f_n(x) - f(x)| < \epsilon \quad \text{for } n > N.$$

We emphasize that uniform convergence on a set implies (pointwise) convergence on that set. But the converse is not true, as we shall soon see in a number of examples. Thus, uniform convergence is a stronger form of convergence. Moreover, a formulation of the negation of uniform convergence will

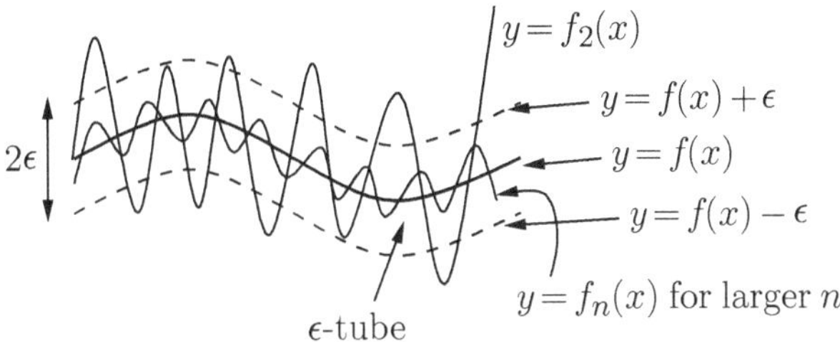

Fig. 9.1. Graphs of a neighborhood of f and an f_n.

be helpful for producing examples that show the converse to be false. Let $f_n \to f$ on a set E. Then the convergence of $\{f_n\}$ to f on E is not uniform if there exists an $\epsilon > 0$ such that to each integer N there corresponds an integer $n > N$ and a point $x_n \in E$ for which

$$|f_n(x_n) - f(x_n)| \geq \epsilon.$$

Finally, we remark that it is apparent that if a sequence of functions converges uniformly on a set E, then it converges on every compact subset of E.

We are now interested in knowing whether important properties of functions in the sequence will be "transferred" to the limit as well. In particular, the general theme of this section is to consider the following questions:

- Suppose that $f_n \to f$ pointwise on E, where $\{f_n\}$ is a sequence of continuous functions on E. Can f be continuous on E? If not, under what conditions can we assert that f is continuous on E?
- Suppose that $f_n \to f$ pointwise on E, where $\{f_n\}$ is a sequence of integrable functions on E. Can f be Riemann integrable on every interval of finite length in E? In case f is Riemann integrable on $[a, b] \subset E$, is it true that
$$\int_a^b f(t)\,dt = \lim_{n\to\infty} \int_a^b f_n(t)\,dt,$$
or in other words, are we allowed to interchange the limit and the integral sign? The answer is in general negative. Then under what conditions can we make the interchange?
- Suppose that $f_n \in C^1(E)$, i.e., continuously differentiable on E, and $f_n \to f$ pointwise on E. Can f be differentiable on E? If so, do we have $f_n' \to f'$ pointwise on E? If not, under what conditions can we assert that $f_n' \to f'$ pointwise on E?

We shall soon see that it will be more profitable to consider these questions with uniform convergence instead of pointwise convergence. For instance, we will show later that we may interchange the limit and the integral over $[a, b]$ if $\{f_n\}$ converges uniformly to some f on $[a, b]$.

We now recall the distinction between continuity and uniform continuity. A continuous function is uniformly continuous on a set if a single $\delta = \delta(\epsilon)$ can

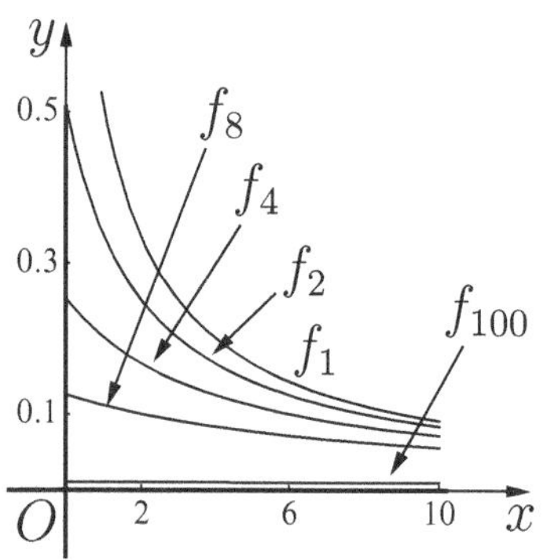

Fig. 9.2. $f_n(x) = 1/(n+x)$ on $[0,\infty)$.

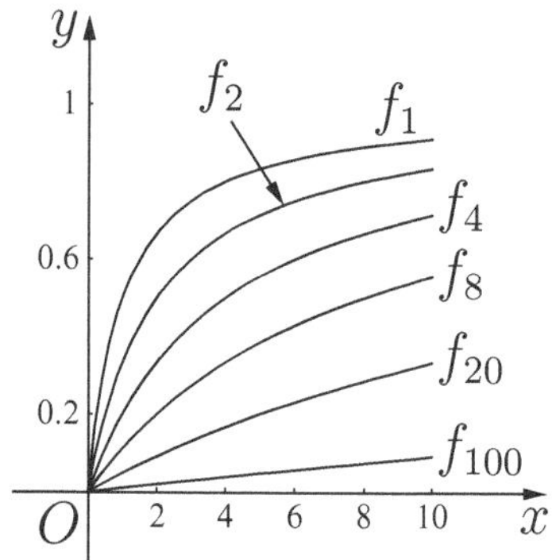

Fig. 9.3. $f_n(x) = x/(x+n)$ on $[0,\infty)$.

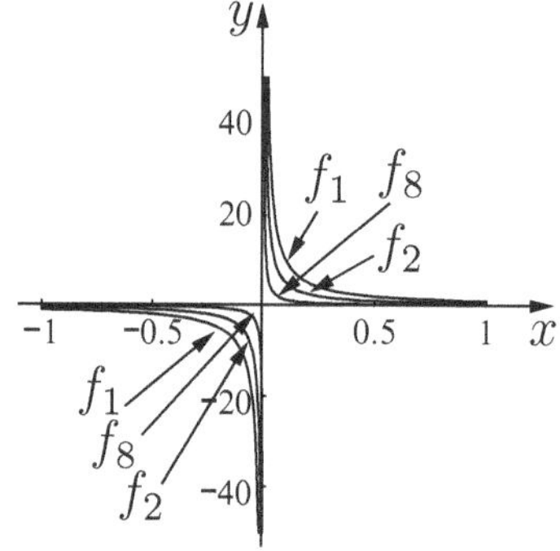

Fig. 9.4. Graph of $f_n(x) = 1/nx$ on $[-1,1] \setminus \{0\}$.

be found that works for all points in the set. Earlier, the function $f(x) = 1/x$ was shown to be continuous, but not uniformly, on the set $0 < |x| < 1$. The following example is an analogue for convergence.

Example 9.2. For $n \geq 1$, consider $f_n(x) = 1/(x+n)$ on $[0,\infty)$ (see Figure 9.2). For $\epsilon > 0$ we have

$$|f_n(x) - 0| = \frac{1}{n+x} \leq \frac{1}{n} < \epsilon \quad \text{for } n > \frac{1}{\epsilon},$$

showing that $f_n \to 0$ uniformly on $[0,\infty)$ (see also Figure 9.3). ●

Example 9.3 (Pointwise but not uniform convergence to a continuous function). Consider $f_n(x) = 1/(nx)$ on the set $0 < |x| < 1$; see Figure 9.4. Then we observe that for a given $\epsilon > 0$,

$$|f_n(x) - 0| = \left|\frac{1}{nx}\right| < \epsilon \Longleftrightarrow n > \frac{1}{\epsilon|x|}.$$

So the corresponding $N = N(x;\epsilon)$ is an integer greater than $1/(\epsilon|x|)$. Note that as $|x|$ decreases, the corresponding N increases without bound. Thus, the sequence $\{f_n(x)\}$ converges pointwise, but not uniformly, to the function $f(x) = 0$ on the set $0 < |x| < 1$.

Alternatively, we may argue as follow. If this convergence were uniform, then there would exist an integer N for which the inequality $|1/(Nx)| < \epsilon = 0.1$ would be valid for all x, $0 < |x| < 1$. But the inequality does not hold for $x = 1/N$. ●

Note that in the above example, we have shown that the convergence is not uniform because

$$\left| f_n\left(\frac{1}{n}\right) - f\left(\frac{1}{n}\right)\right| = 1 \quad \text{for all } n.$$

This observation suggests the following result, which is a precise formulation of nonuniform convergence.

Theorem 9.4 (Sufficiency for nonuniform convergence). *Suppose that f_n and f are functions defined on a set E. If there exists a sequence $\{x_n\}$ in E and a number $c \neq 0$ such that*

$$f_n(x_n) - f(x_n) \to c \quad \textit{as } n \to \infty,$$

then the sequence $\{f_n\}$ cannot converge uniformly to the function f on E.

Proof. If the convergence of $\{f_n\}$ were uniform, then for $\epsilon = |c|/2$ there would exist an integer $N = N(\epsilon)$ such that

$$|f_n(x) - f(x)| < |c|/2 \quad \text{for all } x \in E \text{ and } n > N.$$

In particular, this would yield

$$|f_n(x_n) - f(x_n)| < |c|/2 \quad \text{for all } n > N,$$

which is clearly a contradiction to $\lim_{n\to\infty} |f_n(x_n) - f(x_n)| = |c|$. ∎

By Theorem 9.4, we can easily conclude that (by taking $x_n = \pi/n$) $\{f_n\}$, $f_n(x) = \arctan(nx)$, does not converge uniformly on any interval containing the origin (see Figure 9.5). However, $\{f_n\}$ does converge pointwise to f, where

$$f(x) = \begin{cases} 0 & \text{if } x = 0, \\ \pi/2 & \text{if } x > 0, \\ -\pi/2 & \text{if } x < 0. \end{cases}$$

Example 9.5 (Pointwise convergence to a discontinuous function). For $n \geq 1$, consider $f_n(x) = 1/(1 + nx^2)$ on $\mathbb{R}$; see Figure 9.6 and 9.7. Then the sequence $\{f_n\}$ converges (pointwise) everywhere to the function

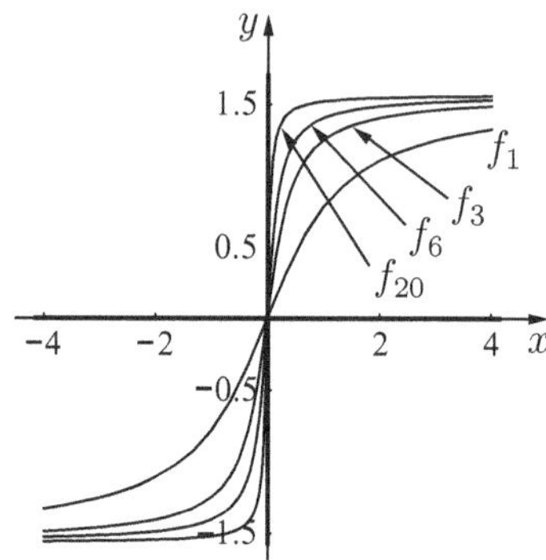

Fig. 9.5. Graph of $f_n(x) = \arctan(nx)$ on $[-4, 4]$ for $n = 1, 2, 6, 20$.

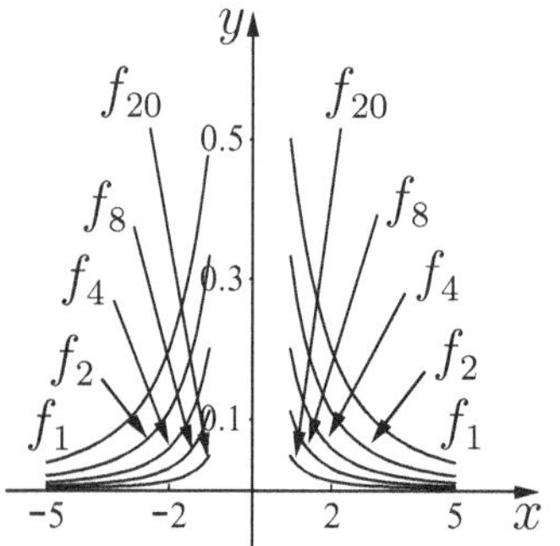

Fig. 9.6. $f_n(x) = 1/(1 + nx^2)$ on $[-5, -1] \cup [1, 5]$.

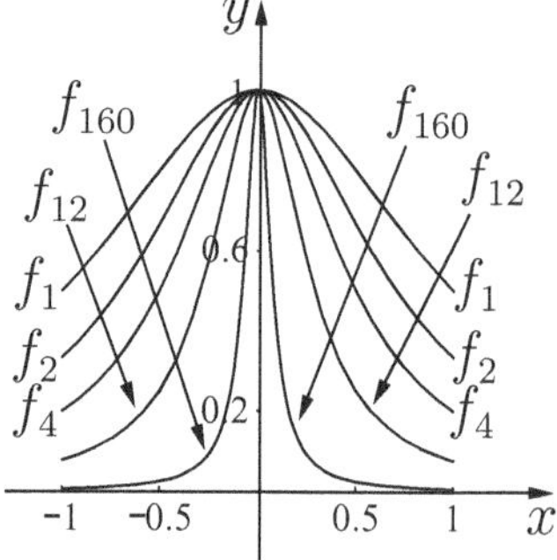

Fig. 9.7. $f_n(x) = 1/(1 + nx^2)$ on $(-1, 1)$.

$$f(x) = \begin{cases} 0 & \text{if } x \neq 0, \\ 1 & \text{if } x = 0. \end{cases}$$

If $|x| \geq c$ $(c > 0)$, then

$$|f_n(x)| = \frac{1}{1 + nx^2} \leq \frac{1}{nx^2} \leq \frac{1}{nc^2}.$$

Therefore, $|f_n(x)| < \epsilon$ whenever $n > 1/(\epsilon c^2)$, which proves uniform convergence for $|x| \geq c$ for $c > 0$. Consequently, the sequence $\{f_n\}$ converges uniformly for $|x| \geq c$, but does not converge uniformly on any interval containing the origin.

Alternatively, we can apply Theorem 9.4 and obtain the nonuniform convergence of the sequence. For instance, if the convergence were uniform for $|x| \leq 1$, then there would exist an integer N for which the inequality $|f_N(x) - f(x)| < \frac{1}{2}$ would be valid for all x with $|x| \leq 1$. But

$$\left| f_n\left(\frac{1}{\sqrt{n}}\right) - f\left(\frac{1}{\sqrt{n}}\right) \right| = \left| \frac{1}{1 + n \cdot (1/n)} - 0 \right| = \frac{1}{2} \quad \text{for every } n \geq 1. \quad \bullet$$

A reformulation of uniform convergence is the following:

Theorem 9.6 (Characterization of uniform convergence). *A sequence of functions $\{f_n\}$ defined on a set $E \subseteq \mathbb{R}$ converges uniformly to a function f on E if and only if*

$$\delta_n = \sup_{x \in E} |f_n(x) - f(x)| \to 0 \quad \text{as } n \to \infty.$$

Proof. The proof is straightforward, and so we leave it as an exercise. ∎

Since

$$|f_n(x) - f(x)| \leq \sup_{x \in E} |f_n(x) - f(x)|,$$

we see that if $\{f_n\}$ converges uniformly on E to f, then $\{f_n\}$ converges pointwise on E to f. The converse is not true, as shown, for instance, in Example 9.3. Thus, pointwise convergence is a necessary condition for uniform convergence, but it is not a sufficient condition. In order to test the uniform convergence of a sequence, our first step is to check whether it converges pointwise by trying to determine the pointwise limit. If it converges pointwise, then our second step is to check whether it also converges uniformly to the pointwise limit. Later, in Theorem 9.12, we show that if the pointwise limit of continuous functions is not continuous, then the convergence cannot be uniform.

Example 9.7. Consider $f_n(x) = x^n(1-x)$ on $[0,1]$; see Figure 9.8. Then $f_n(0) = 0 = f_n(1)$ for each $n \geq 1$, and it is clear that $f_n(x) \to 0$ for $|x| < 1$. In particular, $f_n \to 0$ pointwise on $[0,1]$. Next we compute, for $n \geq 1$,

$$f_n'(x) = x^{n-1}[n - (n+1)x] \quad \begin{cases} \geq 0 & \text{for } 0 \leq x \leq \frac{n}{n+1}, \\ \leq 0 & \text{for } \frac{n}{n+1} \leq x \leq 1. \end{cases}$$

Since f_n assumes the value 0 at both endpoints of the interval $[0,1]$, it follows that f_n has a maximum at $x = n/(n+1)$. Consequently,

$$\delta_n = \max_{x \in [0,1]} |f_n(x) - 0| = f_n\left(\frac{n}{n+1}\right) = \left(\frac{n}{n+1}\right)^n \frac{1}{n+1} < \frac{1}{n+1} \to 0$$

as $n \to \infty$. By Theorem 9.6, $f_n \to 0$ uniformly on $[0,1]$. Do we have $f_n'(x) \to f'(x) = 0$ on $[0,1]$?

One can consider $f_n(x) = n^2 x^n (1-x)$ on $[0,1]$ and obtain that this time, $\{f_n\}$ converges (but not uniformly) to a continuous function $f(x) = 0$ on $[0,1]$ (see Figure 9.9). ●

Example 9.8. Consider $f_n(x) = x^n$. Then from our earlier experience with sequences, it follows that $f_n(x)$ does not converge at $x = -1$, nor does it converge at any point x for which $|x| > 1$. Clearly if we consider f_n on $(-1,1]$, then we have

$$f_n(x) \to f(x) = \begin{cases} 0 & \text{for } x \in (-1,1), \\ 1 & \text{for } x = 1. \end{cases}$$

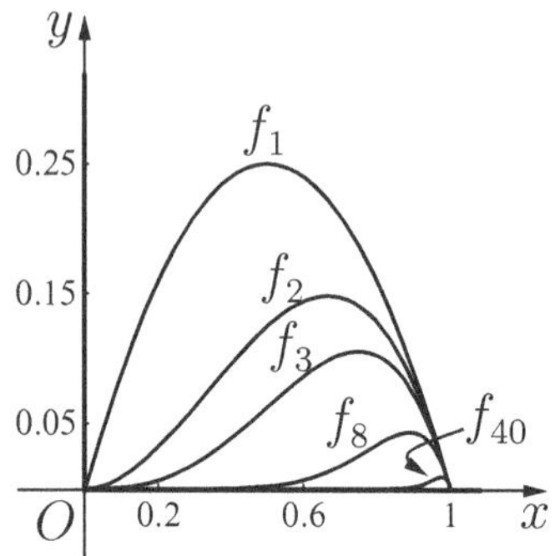

Fig. 9.8. $f_n(x) = x^n(1-x)$ on $[0,1]$.

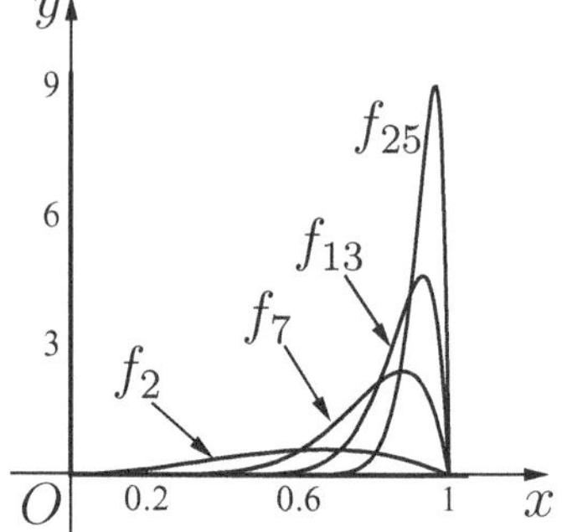

Fig. 9.9. $f_n(x) = n^2x^n(1-x)$ on $[0,1]$.

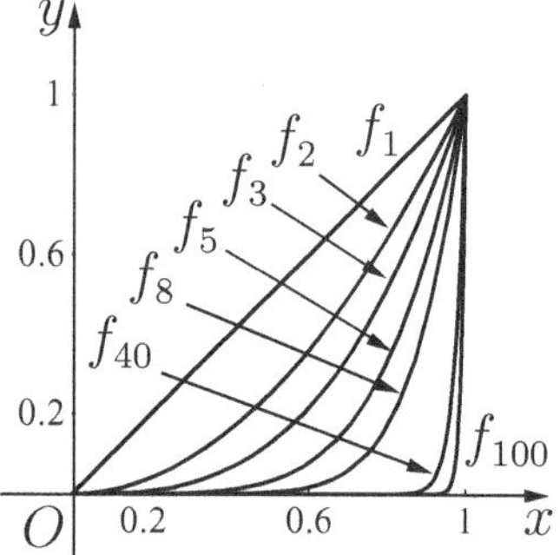

Fig. 9.10. $f_n(x) = x^n$ on $[0,1]$.

Thus, the limit function $f(x)$ is not continuous at $x = 1$ although each f_n is continuous everywhere on $\mathbb{R}$. Also, the limit function does not exist on $\mathbb{R}\backslash(-1,1]$. Thus, $\{f_n\}$ converges pointwise to f on $(-1,1]$ and *continuity is not preserved under convergence*. The graphs of

$$y_1 = f_1(x) = x,\ y_2 = f_2(x) = x^2, \ldots,\ y_n = f_n(x) = x^n$$

do not approach the graph of $y = f(x)$; see Figure 9.10. Pointwise convergence may be proved directly. Let $\epsilon > 0$ be given. Then for $0 < |x| < 1$,

$$|x^n| < \epsilon \Longleftrightarrow n > \frac{\log \epsilon}{\log |x|} = \frac{\log(1/\epsilon)}{\log(1/|x|)} = N(\epsilon, x).$$

Now for each fixed $0 < \epsilon < 1$, as $|x| \to 1-$, $N(\epsilon, x)$ is increased without bound because $N(\epsilon, x) \to \infty$. It follows that the convergence is not uniform for $|x| < 1$, although there is pointwise convergence on $(-1,1]$.

However, if $r < 1$ and $|x| \le r$, then $\log(1/|x|) \ge \log(1/r)$, and therefore since $r^n \to 0$ as $n \to \infty$, an integer $N = N(\epsilon)$ can be found for which $r^n < \epsilon$ $(n > N > (\log \epsilon)/\log r)$. But then

$$|x^n| \le r^n < \epsilon \quad (|x| \le r,\ n > N(\epsilon)).$$

Hence $\{f_n(x)\}$ converges uniformly to zero in every closed interval $|x| \le r$, $r < 1$. Recall that the choice of N with $N > (\log \epsilon)/\log r$ is possible for an arbitrary $\epsilon > 0$ and $0 < r < 1$. In particular, $\{f_n\}$ converges uniformly on $[0, 2/3]$ to $f(x) = 0$. In this case, the choice of N is such that

$$N > \frac{\log(1/\epsilon)}{\log(3/2)}.$$

Method 2: To see that $f_n(x)$ does not converge uniformly to $f(x)$ on $(-1, 1)$, we may also proceed as follows (or we can apply Theorem 9.4 straightaway). If the convergence of $\{x^n\}$ were uniform on the set $(-1, 1)$, then for sufficiently large n we would have

$$|x^n| < \epsilon \quad \text{for all } x \in (-1, 1).$$

Choosing $x_n = (1 - 1/n)^{1/n}$, we must then have

$$|f_n(x_n) - f(x_n)| = |x_n^n - 0| = 1 - \frac{1}{n} < \epsilon \quad \text{for large } n.$$

This is not possible for large n if we have chosen $\epsilon < 1$ and hence the convergence cannot be uniform on $|x| < 1$.

We could instead choose c_n with $c_n = 2^{-1/n}$ so that for large n,

$$|f_n(c_n) - f(c_n)| = c_n^n = \frac{1}{2},$$

which is not possible if we have chosen $\epsilon < 1/2$. Does Theorem 9.6 apply here?
Method 3: We also see that

$$\sup_{x \in [0,1]} |f_n(x) - f(x)| \ge \sup_{x \in [0,1]} |x^n| = 1,$$

and by Theorem 9.6, $\{f_n\}$ does not converge uniformly on $[0, 1]$. •

We see that the sequence $\{g_n(x)\}$, where

$$g_n(x) = \frac{x^n}{n} \quad \text{on } [-1, 1],$$

converges uniformly to $g(x) = 0$ on $[-1, 1]$.

By the Cauchy convergence criterion, we know that a sequence of real numbers $\{a_n\}$ converges if and only if it is a Cauchy sequence. We can now formulate a similar criterion for uniform convergence of sequences of functions. Since the proof is easy, we leave it as an exercise to the reader.

Theorem 9.9 (Cauchy criterion for uniform convergence of a sequence). *A sequence of real-valued functions $\{f_k\}_{k\geq 1}$ defined on a set E converges uniformly on E if and only if it is uniformly Cauchy on E; i.e., for an arbitrary $\epsilon > 0$ there is a number $N = N(\epsilon)$ such that*

$$|f_m(x) - f_n(x)| < \epsilon \quad \textit{whenever } m > n > N(\epsilon), \textit{ and for all } x \in E,$$

or equivalently,

$$\sup_{x\in E} |f_m(x) - f_n(x)| < \epsilon.$$

In Example 9.8, we showed that $x^n \to 0$ uniformly on $[-r, r] \subseteq (-1, 1)$, $0 < r < 1$. This can be proved using the Cauchy criterion. For $m > n$,

$$\sup_{x\in[-r,r]} |x^m - x^n| \leq 2r^n.$$

Since $r^n \to 0$ $(r < 1)$, an integer $N = N(\epsilon)$ can be found for which

$$2r^n < \epsilon \quad \text{for all } n > N(\epsilon).$$

Actually, any integer N greater than $(\log(\epsilon/2))/(\log r)$ will satisfy our requirement. Thus, for all $m > n > N$ and all $x \in [-r, r]$,

$$|x^m - x^n| \leq 2r^n < \epsilon.$$

We conclude that $x^n \to 0$ uniformly on any interval $[-r, r]$, $0 < r < 1$. Here we do not gain any computational advantage over the argument in Example 9.8. However, when we do not know the limit function, the Cauchy criterion is more useful than the definition. Now many more examples follow.

Example 9.10. The sequence $\{nx^n(1-x^n)\}$ does not converge uniformly on $[0, 1]$.

Solution. Set $f_n(x) = nx^n(1 - x^n)$. Then $f_n(1) = 0$ and observe that

$$\sum_{k=1}^{\infty} f_k(x) = \sum_{k=1}^{\infty} kx^k - \sum_{k=1}^{\infty} kx^{2k},$$

which converges for $|x| < 1$, and so the general term $f_n(x)$ approaches zero as $n \to \infty$. In particular, this shows that $f(x) = 0$, $x \in [0, 1]$, is the pointwise limit of $\{f_n(x)\}$ on $[0, 1]$.

Next we compute, for $n \geq 1$,

$$f_n'(x) = n^2x^{n-1}(1 - 2x^n) \begin{cases} \geq 0 & \text{for } 0 \leq x \leq 1/\sqrt[n]{2}, \\ < 0 & \text{for } 1/\sqrt[n]{2} < x \leq 1. \end{cases}$$

Since f_n assumes the value 0 at both endpoints of the interval $[0, 1]$, and $x_0 = 1/\sqrt[n]{2}$ is the only critical point in the interior,

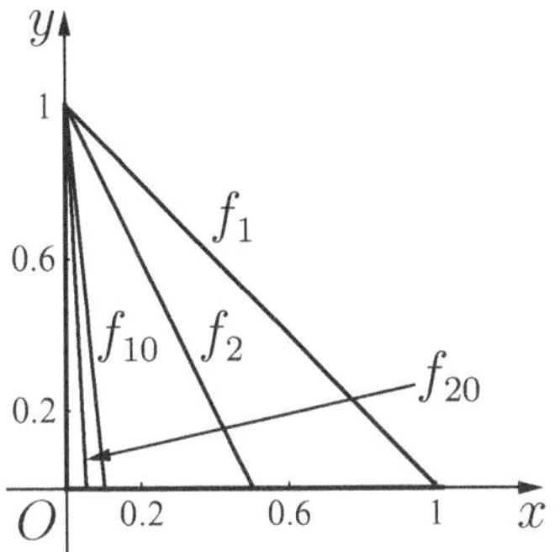

Fig. 9.11. $f_n(x) = 1 - nx$ for $0 \leq x \leq 1/n$ and 0 for $1/n < x \leq 1$.

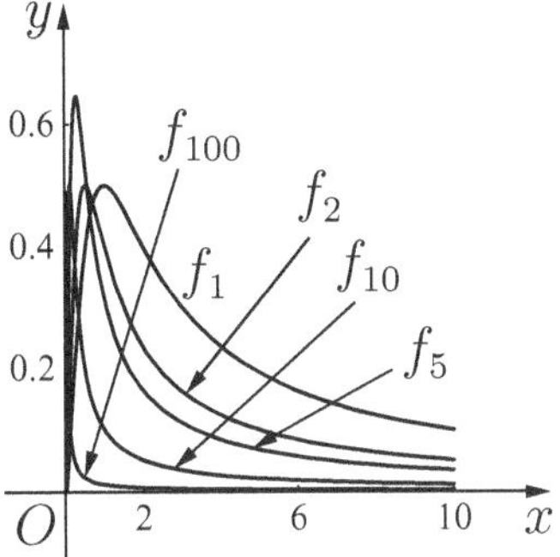

Fig. 9.12. $f_n(x) = nx/(1 + n^2x^2)$ on $\mathbb{R}$.

$$\delta_n = \max_{x\in[0,1]} |f_n(x) - 0| = f_n\left(\frac{1}{\sqrt[n]{2}}\right) = n\frac{1}{2}\left(1 - \frac{1}{2}\right) = \frac{n}{4} \to \infty \quad \text{as } n \to \infty.$$

Thus, by Theorem 9.4, $\{f_n(x)\}$ does not converge uniformly to $f(x) = 0$ on $[0, 1]$. ●

Examples 9.11. 1. For $n \geq 1$, define $f_n : [0, 1] \to \mathbb{R}$ by (see Figure 9.11)

$$f_n(x) = \begin{cases} 1 - nx & \text{if } 0 \leq x \leq 1/n, \\ 0 & \text{if } 1/n < x \leq 1. \end{cases}$$

Clearly, each $f_n(x)$ is continuous on $[0, 1]$ and $f_n \to f$ on $[0, 1]$, where

$$f(x) = \begin{cases} 1 & \text{if } x = 0, \\ 0 & \text{if } 0 < x \leq 1, \end{cases}$$

which is not continuous on $[0, 1]$.

2. Consider $\{f_n(x)\}_{n\geq 1}$, where $f_n(x) = nx/(1 + n^2x^2)$ for $x \in \mathbb{R}$. Clearly, $f_n(0) \to 0$, as $f_n(0) = 0$ for all $n \geq 1$. Let $x \neq 0$. For a given $\epsilon > 0$, we see that

$$|f_n(x) - 0| = \left|\frac{nx}{1 + n^2x^2}\right| \leq \frac{1}{n|x|} < \epsilon \quad \text{whenever } n > \frac{1}{\epsilon|x|}.$$

Note that $\sup_{x\in\mathbb{R}} 1/(\epsilon|x|) = \infty$. Thus, given $\epsilon > 0$, there exists an N (with $N \geq 1/(\epsilon|x|)$) such that $|f_n(x) - 0| < \epsilon$ for all $n \geq N$. Therefore, $\{f_n(x)\}$ converges everywhere to the function $f(x) = 0$ (see Figures 9.12–9.14). Observe that N depends on both x and ϵ. To test for uniform convergence, it is sufficient to note that for $x_n = 1/n$, we have

$$|f_n(x_n) - f(x_n)| = 1/2,$$

so that by Theorem 9.4, the sequence $\{f_n\}$ does not converge uniformly on $[0, 1]$. Does it converge uniformly on $|x| > c$, $c > 0$? Why?

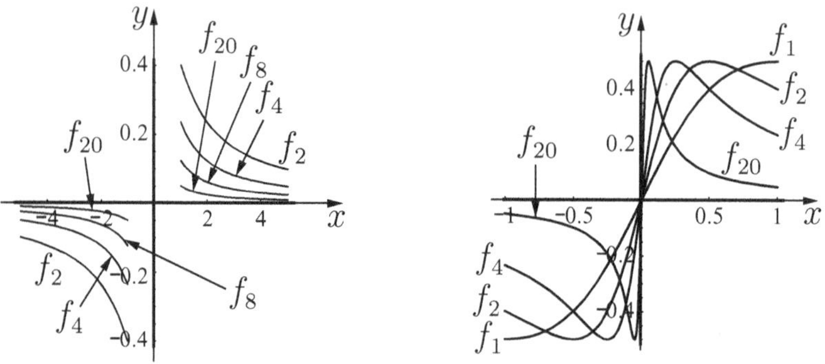

Fig. 9.13. $f_n(x) = nx/(1+n^2x^2)$ on $[-5,-1] \cup [1,5]$.

Fig. 9.14. $f_n(x) = nx/(1+n^2x^2)$ on $(-1,1)$.

3. Similarly, $\{x/n\}_{n\geq 1}$ converges to $f(x) = 0$ on $\mathbb{R}$ but does not converge uniformly on $\mathbb{R}$. Does it converge on any finite closed interval in $\mathbb{R}$? Does it converge uniformly on every bounded subset of $\mathbb{R}$? Does it converge uniformly on $[1,\infty)$?
4. Consider $f_n(x) = 1/(1+nx)$ on $[0,1]$. We see that

$$f_n(x) \to f(x) = \begin{cases} 1 & \text{for } x = 0, \\ 0 & \text{for } 0 < x \leq 1, \end{cases}$$

because for $0 < x \leq 1$,

$$|f_n(x) - 0| = \frac{1}{1+nx} < \epsilon \quad \text{whenever } n > \frac{1-\epsilon}{\epsilon x}.$$

That is, given $\epsilon > 0$, there exists an N (with $N > (1-\epsilon)/(\epsilon x)$) such that $|f_n(x)| < \epsilon$ for all $n \geq N$. Again, we note that N depends on both x and ϵ. Also, for $0 < \epsilon < 1$,

$$\sup_{x\in(0,1]} \frac{1-\epsilon}{\epsilon x} = \infty,$$

which shows that it is not possible to choose N independent of $x \in (0,1]$; see Figure 9.15. Note also that for $x_n = 1/n$, we have

$$|f_n(x_n) - f(x_n)| = 1/2,$$

so that by Theorem 9.4, the sequence $\{1/(1+nx)\}$ does not converge uniformly on $[0,1]$. The conclusion can also be drawn from Theorem 9.12 (see also Remark 9.13). How about uniform convergence on $x > c$, $c > 0$? (See Figure 9.16.) ●

9.1.2 Uniform Convergence and Continuity

Suppose that $\{f_n\}$ is a sequence of continuous functions on a set E such that $f_n \to f$ on E. Must f be continuous on E? Continuity of f at $a \in E$ demands that

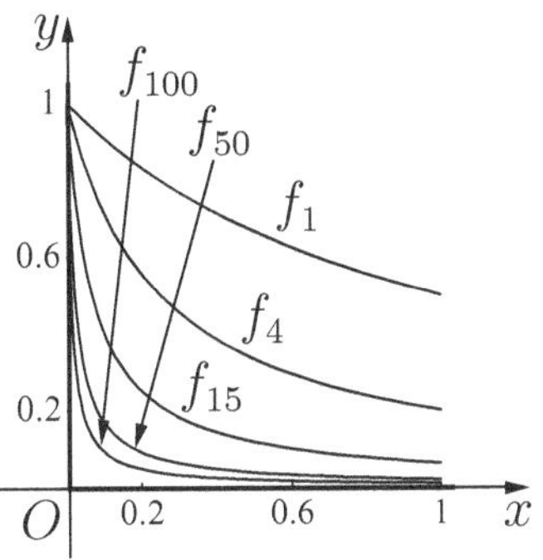

Fig. 9.15. $f_n(x) = 1/(1+nx)$ on $[0,1]$.

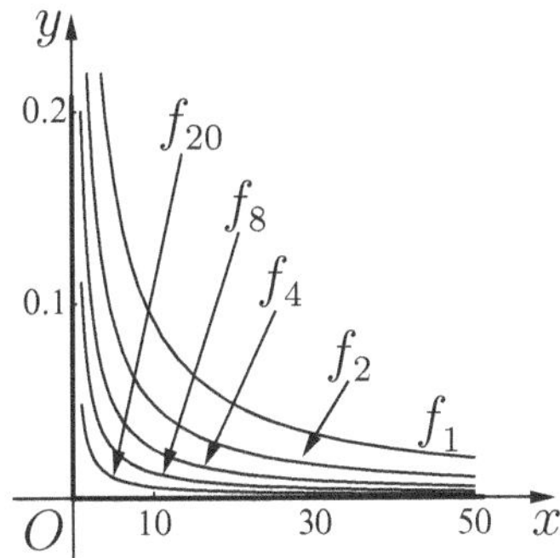

Fig. 9.16. $f_n(x) = 1/(1+nx)$ on $[1,\infty)$.

$$\lim_{x\to a} f(x) = f(a),$$

that is,

$$\lim_{x\to a}\left(\lim_{n\to\infty} f_n(x)\right) = \lim_{n\to\infty} f_n(a) = \lim_{n\to\infty}\left(\lim_{x\to a} f_n(x)\right).$$

Thus, to verify the continuity of f at a, we need to be assured that the order of passing to the limit is immaterial. Here the continuity at the endpoint refers to the right or the left limit, whichever is the case.

As observed in a number of examples, pointwise convergence does not allow the interchange of limit operations. For instance, for $f_n(x) = x^n$ on $(-1,1]$, we had

$$\lim_{x\to 1-}\left(\lim_{n\to\infty} x^n\right) \neq \lim_{n\to\infty}\left(\lim_{x\to 1-} x^n\right)$$

(see also Examples 9.11(1) and 9.5).

On the other hand, an important consequence of uniform convergence is that continuity is inherited by the limit function (Figure 9.17).

Theorem 9.12 (Necessary condition for the uniform convergence of a sequence). *The limit of a uniformly convergent sequence of continuous functions on E is continuous on E. That is, for each $a \in E$,*

$$\lim_{x\to a}\left(\lim_{n\to\infty} f_n(x)\right) = \lim_{n\to\infty}\left(\lim_{x\to a} f_n(x)\right).$$

Proof. Let each f_n be continuous on E and suppose that $f_n \to f$ uniformly on E. Let $\epsilon > 0$ be given. The uniform convergence of $\{f_n\}$ implies that there exists an integer N independent of x such that

$$|f(x) - f_N(x)| < \frac{\epsilon}{3} \quad (x \in E).$$

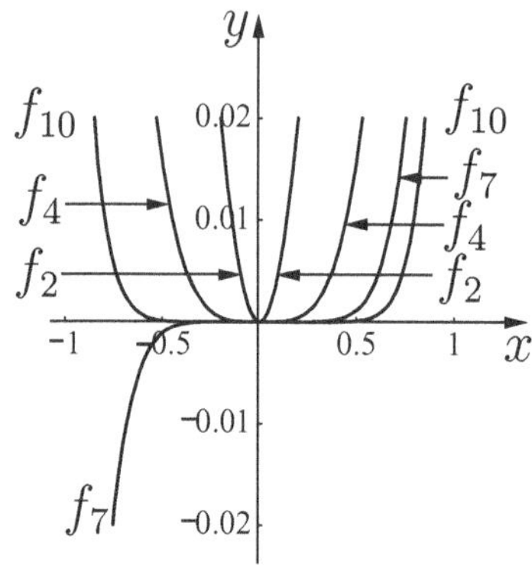

Fig. 9.17. $f_n(x) = x^n/n$ on $[-1, 1]$.

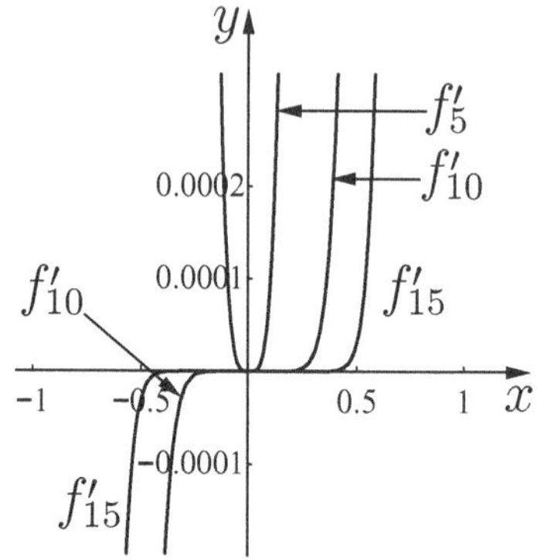

Fig. 9.18. $f_n(x) = x^n/n$ on $[0, 1]$.

Choose a point $a \in E$, and fix it. By the continuity of f_N at a, there exists a $\delta > 0$ such that

$$|f_N(x) - f_N(a)| < \frac{\epsilon}{3} \quad \text{for all } |x - a| < \delta, \; x \in E.$$

Further, for all x such that $|x - a| < \delta$ and $x \in E$, the triangle inequality gives

$$|f(x) - f(a)| \leq |f(x) - f_N(x)| + |f_N(x) - f_N(a)| + |f_N(a) - f(a)| < \epsilon.$$

Thus, f is continuous at a. Since a is arbitrary, the claim now follows. ∎

Remark 9.13. The sequence $\{x^n/n\}$ on $[0, 1]$ is a sequence of continuous functions converging uniformly to a continuous function (see Figure 9.18).

Next, consider

$$f_n(x) = \begin{cases} -1/n & \text{if } -1 \leq x < 0, \\ 1/n & \text{if } 0 \leq x \leq 1. \end{cases}$$

Then none of the functions $f_n(x)$ is continuous at the origin, but the limit function is $f(x) = 0$ on $[-1, 1]$, which is continuous on $[-1, 1]$. Thus, Theorem 9.12 gives a necessary, but not sufficient, condition for uniform convergence of a sequence of continuous functions.

Finally, from Example 9.8, we see that the sequence of continuous functions $\{x^n\}$ on $(-1, 1)$ converges to $f(x) = 0$ on $(-1, 1)$ but not uniformly on $(-1, 1)$. In the same example, $\{x^n\}$ on $(-1, 1]$ converges to a discontinuous function on $(-1, 1]$. The discontinuity of the limit function at $x = 1$ rules out uniform convergence for the sequence $\{x^n\}$ on $(-1, 1]$ (see Figure 9.19). ●

Examples 9.14. **(a)** For $n \geq 1$, consider $f_n(x) = (1 - x^2)^n$ for $x \in (-1, 1)$.

Then (see Figure 9.20) $f_n \to f$ pointwise on $(-1, 1)$, where

$$f(x) = \begin{cases} 1 & \text{if } x = 0, \\ 0 & \text{if } x \in (-1, 1) \setminus \{0\}. \end{cases}$$

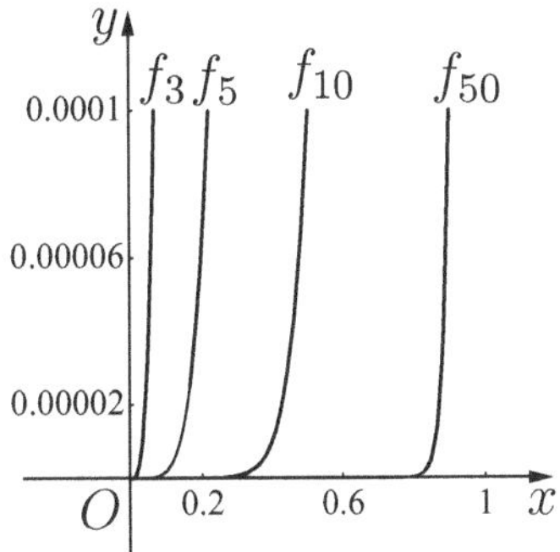

Fig. 9.19. $f_n'(x) = x^{n-1}$ on $(-1, 1)$.

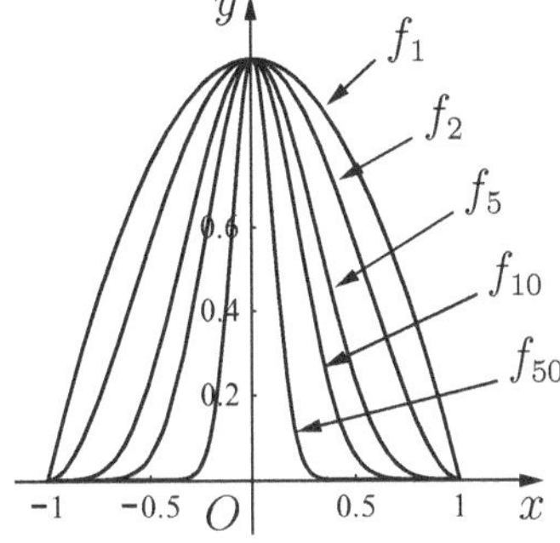

Fig. 9.20. $f_n(x) = (1 - x^2)^n$ on $[-1, 1]$.

In view of Theorem 9.12, $\{f_n\}$ does not converge uniformly on $(-1, 1)$ because the pointwise limit function f is not continuous on $(-1, 1)$. This fact can be also seen with the help of Theorem 9.4. If the sequence were uniformly convergent on $(-1, 1)$, then, for example for $\epsilon = 1/4$, there would exist an N such that

$$|f_n(x) - f(x)| < \frac{1}{4} \quad \text{for all } x \in (-1, 1) \text{ and } n \geq N.$$

But for $x_N^2 = 1 - 3^{-1/N}$, this would give

$$|f_N(x_N) - f(x_N)| = \frac{1}{3} > \frac{1}{4},$$

a contradiction. It follows that the sequence $\{f_n\}$ does not converge uniformly on $(-1, 1)$.

(b) For $n \geq 1$, define $f_n : [0, \pi/2] \to \mathbb{R}$ by

$$f_n(x) = \frac{4 \cos^n x}{3 + \cos^n x}, \quad x \in [0, \pi/2].$$

Then each f_n is continuous. For $x \in (0, \pi/2]$, $\cos^n x \to 0$ as $n \to \infty$, and so the sequence $\{f_n\}$ converges pointwise on $[0, \pi/2]$ to

$$f(x) = \begin{cases} 1 & \text{for } x = 0, \\ 0 & \text{for } x \in (0, \pi/2], \end{cases}$$

which is not continuous on $[0, \pi/2]$. By Theorem 9.12, the sequence $\{f_n\}$ cannot converge uniformly to $f(x)$ on $[0, \pi/2]$. How about if the number 3 in the definition of $f_n(x)$ is replaced by $3n$? ●

9.1.3 Interchange of Limit and Integration

We now examine some relationships between uniform convergence and integration. We begin with an example. For $n \geq 1$, define $g_n : [0, 1] \to \mathbb{R}$ by

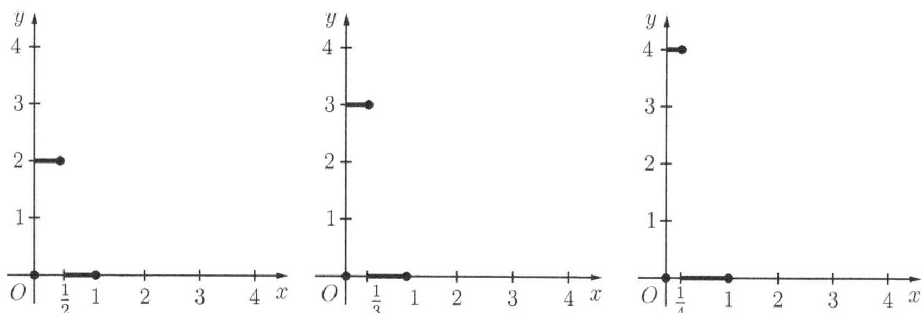

Fig. 9.21. Graph of $g_2(x)$. **Fig. 9.22.** Graph of $g_3(x)$. **Fig. 9.23.** Graph of $g_4(x)$.

$$g_n(x) = \begin{cases} 0 & \text{if } x = 0, \\ n & \text{if } 0 < x \leq 1/n, \\ 0 & \text{if } 1/n < x \leq 1. \end{cases}$$

Then $\{g_n(x)\}$ converges pointwise to $g(x) = 0$ on $[0,1]$. For instance, if $x = 1/N$ $(N \in \mathbb{N})$, then

$$g_n\left(\frac{1}{N}\right) = \begin{cases} N & \text{if } n \leq N, \\ 0 & \text{if } n > N. \end{cases}$$

The sequence $\{g_n(x)\}$ cannot converge uniformly to $g(x) = 0$ on $[0,1]$, although the limit function is continuous on $[0,1]$. Each g_n is Riemann integrable on $[0,1]$. For all $n \in \mathbb{N}$, we have

$$\sup_{x \in [0,1]} |g_n(x) - g(x)| = n \geq 1,$$

showing that $\{g_n\}$ does not converge uniformly on $[0,1]$ to g. Further,

$$\int_0^1 g_n(x)\,\mathrm{d}x = \left(\int_0^{1/n} n\,\mathrm{d}x + \int_{1/n}^1 0\,\mathrm{d}x\right) = 1 \quad \text{and} \quad \int_0^1 g(x)\,\mathrm{d}x = 0.$$

The last fact is easy to verify pictorially (see Figures 9.21–9.23). Thus,

$$\lim_{n\to\infty} \int_0^1 g_n(x)\,\mathrm{d}x \neq \int_0^1 \lim_{n\to\infty} g_n(x)\,\mathrm{d}x,$$

so that *the limit of a sequence of integrals of Riemann integrable functions on a set is not necessarily equal to the integral of the limit.* As another example, consider $f_n(x) = nx\mathrm{e}^{-nx^2}$ on $[0,1]$, which is continuous on $[0,1]$ for each $n \geq 1$ (see Figures 9.24 and 9.25). We see that $\{f_n(x)\}$ converges pointwise to $f(x) = 0$ on $[0,1]$. As n increases, the hump moves closer to $x = 0$ and becomes higher. So the convergence cannot be uniform, although the limit function is continuous on $[0,1]$ (see Exercise 9.20(12)). We see that

$$\int_0^1 f_n(x)\,\mathrm{d}x = \frac{1}{2}(1 - \mathrm{e}^{-n}) \to \frac{1}{2} \quad \text{as } n \to \infty \quad \text{and} \quad \int_0^1 f(x)\,\mathrm{d}x = 0.$$

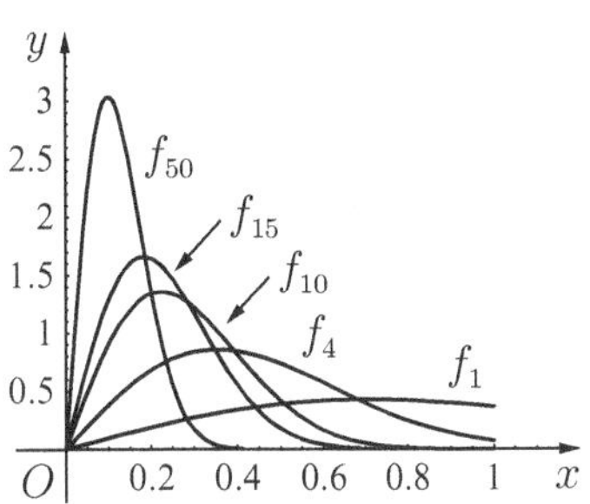

Fig. 9.24. Graphs of $f_n = nxe^{-nx^2}$ for $n = 1, 4, 10, 15, 50$.

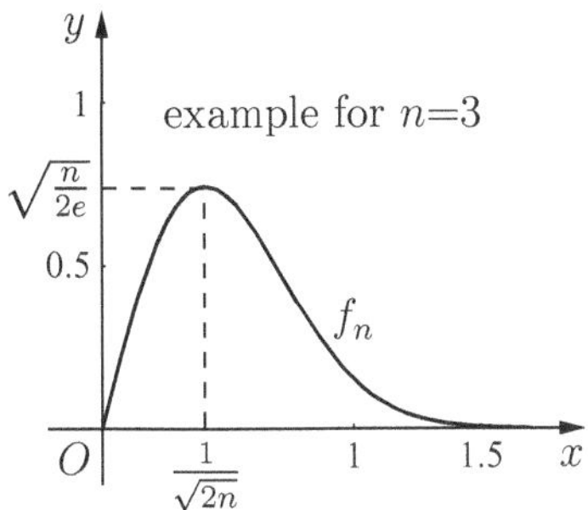

Fig. 9.25. Graph of $f_n = nxe^{-nx^2}$.

This observation, according to Theorem 9.15, helps to establish that $\{f_n\}$ does not converge uniformly on $[0, 1]$.

Moreover, the functions in Exercises 9.20(6) and 9.20(7) show that the integral of the limit of continuous functions is not necessarily equal to the limit of the integrals of those functions.

Theorem 9.15 (Interchange of limit and integration). *Suppose that $\{f_n(x)\}$ is a sequence of continuous functions on the interval $[a, b]$ and that $\{f_n(x)\}$ converges uniformly to $f(x)$ on $[a, b]$. Then*

$$\lim_{n\to\infty} \int_a^b f_n(x)\,dx = \int_a^b \lim_{n\to\infty} f_n(x)\,dx = \int_a^b f(x)\,dx.$$

Also, for each $t \in [a, b]$,

$$\lim_{n\to\infty} \int_a^t f_n(x)\,dx = \int_a^t f(x)\,dx,$$

and the convergence is uniform on $[a, b]$.

Proof. Note that by Theorem 9.12, f is continuous on $[a, b]$, so that $\int_a^b f(x)\,dx$ exists. Let $\epsilon > 0$ be given. Then since $f_n \to f$ uniformly on $[a, b]$, there is an integer $N = N(\epsilon)$ such that

$$|f_n(x) - f(x)| < \frac{\epsilon}{b-a} \quad \text{for all } n > N \text{ and all } x \text{ on } [a, b].$$

Again, since $f_n - f$ is continuous on $[a, b]$, it follows, for $n > N$, that

$$\begin{aligned} \left| \int_a^b f_n(x)\,dx - \int_a^b f(x)\,dx \right| &= \left| \int_a^b (f_n(x) - f(x))\,dx \right| \\ &\le \int_a^b |f_n(x) - f(x)|\,dx \\ &< \frac{\epsilon}{b-a}(b-a) = \epsilon. \end{aligned}$$

Since ϵ is arbitrary, the proof is complete. ∎

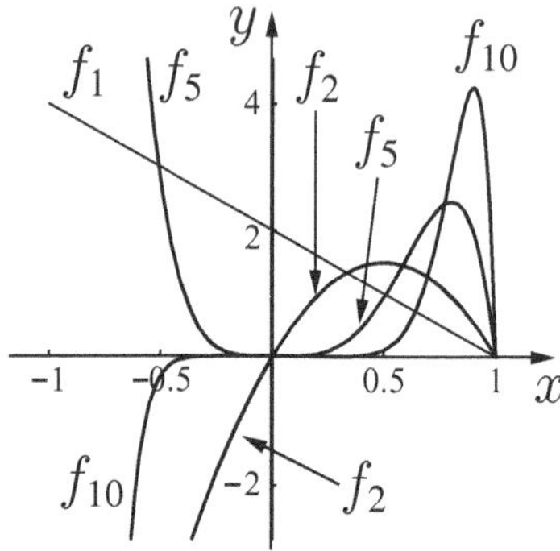

Fig. 9.26. $f_n(x) = n(n+1)x^{n-1}(1-x)$ on $(-1, 1]$.

The hypothesis of Theorem 9.15 will be sufficient for our purposes, but a stronger theorem holds (see Theorem 9.36). Theorem 9.15 may be used to show the nonuniform convergence of the sequence $\{f_n(x)\}$ on $[a, b]$. Also, it is important to point out that a direct analogue of Theorem 9.15 for derivatives is not true, but we do have a result for derivatives (see Theorem 9.40).

Example 9.16. We remark that the uniform convergence $\{f_n(x)\}$ is a sufficient condition, but not a necessary one, for the conclusion of Theorem 9.15. For example, we may recall that on the interval $[0, 1]$,

$$f_n(x) = x^n \to f(x) = \begin{cases} 0 & \text{for } x \in [0, 1), \\ 1 & \text{for } x = 1, \end{cases}$$

and so we obtain

$$\int_0^1 f_n(x)\, dx = \frac{1}{n+1} \to 0 = \int_0^1 f(x)\, dx.$$

Similarly, although $\{f_n\}$, $f_n(x) = 1/(1 + nx)$ on $[0, 1]$, does not converge uniformly on $[0, 1]$ to $f(x)$, we see that

$$\int_0^1 f_n(x)\, dx = \frac{\log(n+1)}{n} \to 0 = \int_0^1 0\, dx, \quad f(x) = \begin{cases} 0 & \text{for } x \in (0, 1], \\ 1 & \text{for } x = 0. \end{cases}$$

These two examples show that the conclusion of Theorem 9.15 holds without $\{f_n\}$ being convergent uniformly on $[0, 1]$. ●

Next, we consider $f_n(x) = n(n+1)x^{n-1}(1-x)$ for $x \in (-1, 1]$ (see Figure 9.26). Clearly, $f_n(x) = 0$ at $x = 0, 1$. For $0 < |x| < 1$, we have

$$\left| \frac{f_{n+1}(x)}{f_n(x)} \right| = \left(1 + \frac{2}{n}\right)|x| \to |x| \quad \text{as } n \to \infty,$$

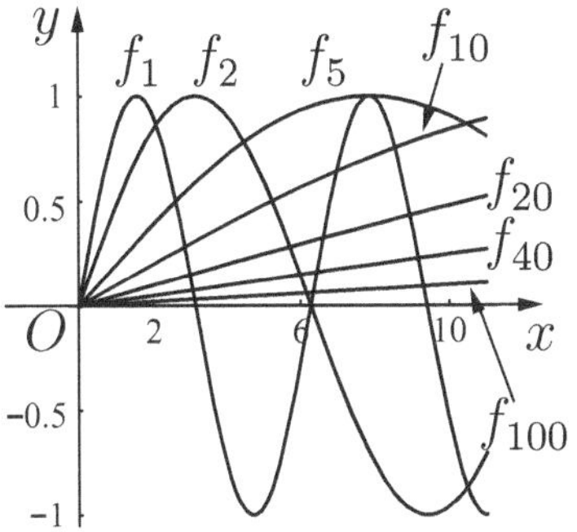

Fig. 9.27. $f_n(x) = \sin(x/n)$ for $x \in \mathbb{R}$.

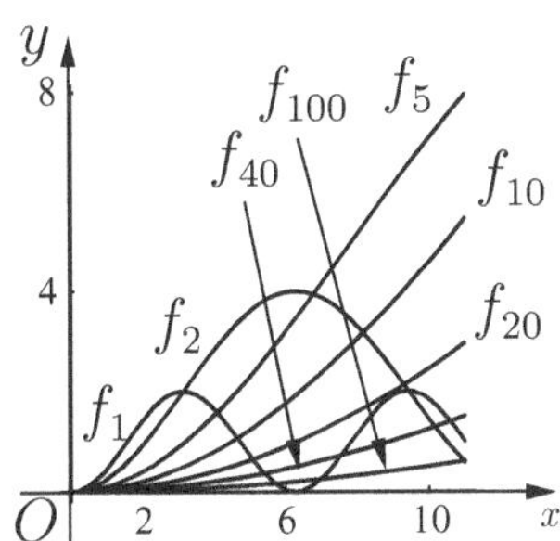

Fig. 9.28. $f_n(x) = n(1 - \cos(x/n))$ on $[0, 11]$.

showing that $\sum f_n(x)$ converges (whenever $|x| < 1$), and so $f_n(x) \to f(x) = 0$ pointwise for $-1 < x \leq 1$. Also, we see that

$$\int_0^1 f_n(x)\, dx = 1 \not\to \int_0^1 f(x)\, dx = 0.$$

Thus, by Theorem 9.15, the sequence $\{f_n\}$ cannot converge uniformly on $(-1, 1]$.

Thus, we conclude that *pointwise convergence does not preserve integrals.*

Example 9.17. Consider $f_n(x) = \sin(x/n)$ for $x \in \mathbb{R}$ (see Figure 9.27). Then each f_n is continuous on $\mathbb{R}$, and since $|\sin t| \leq |t|$ on $\mathbb{R}$, it follows that $f_n \to 0$ pointwise on $\mathbb{R}$, and $f_n \to 0$ uniformly on any finite interval $[a, b] \subseteq \mathbb{R}$. For instance, by Theorem 9.15, on $[0, t]$,

$$\int_0^t \sin(x/n)\, dx = n - n\cos(t/n) \to \int_0^t 0 \cdot dx = 0,$$

and so the sequence $\{n(1 - \cos(x/n))\}$ converges uniformly to 0 on $[0, t]$ (see Figure 9.28).

If we continue the process, it follows that

$$\int_0^t n(1 - \cos(x/n))\, dx = n(t - n\sin(t/n))$$

converges uniformly to zero on any finite interval. •

We remark that in Theorem 9.15, it is important to use the definite integral rather than the antiderivative, for an antiderivative of $f_n(x) = \sin(x/n)$ is $-n\cos(x/n)$, but $\{n\cos(x/n)\}$ does not converge for any x.

Example 9.18. Does $\{\int_1^3 \mathrm{e}^{-nx^2}\,dx\}$ converge on $[1,3]$? The convergence is easy to see, since

$$0 < \int_1^3 \mathrm{e}^{-nx^2}\,dx \le \mathrm{e}^{-n}\int_1^3 dx = 2\mathrm{e}^{-n} \to 0 \quad \text{as } n \to \infty.$$

Alternatively, set $f_n(x) = \mathrm{e}^{-nx^2}$ on $[1,3]$. Then for $n \ge 1$, we have

$$0 < f_n(x) \le \mathrm{e}^{-n},$$

from which we conclude that $f_n \to 0$ uniformly on $[1,3]$. By Theorem 9.15, we obtain

$$\lim_{n\to\infty}\int_1^3 \mathrm{e}^{-nx^2}\,dx = \int_1^3 0\cdot dx = 0.$$ •

9.1.4 Questions and Exercises

Questions 9.19.

1. What is the difference between the pointwise and the uniform convergence of a sequence of functions?
2. Suppose that a sequence $\{f_n\}$ does not converge pointwise on $[a,b]$. Can it converge uniformly to some f on $[a,b]$?
3. Suppose that $\{f_n\}$ converges uniformly on a set E. Does it converge uniformly on any subset of E?
4. If $\{f_n\}$ is pointwise convergent on $\mathbb{R}$, must it be uniformly convergent on every finite interval $[a,b]$?
5. Suppose that $f_n \to f$ (pointwise) on E. Is it possible that $f_n \to g$ uniformly on E with $f(x) \ne g(x)$ on E?
6. Suppose that $f_n \to f$ uniformly on E_1 as well as on E_2. Must $f_n \to f$ uniformly on $E_1 \cup E_2$? Must $f_n \to f$ uniformly on $E_1 \cap E_2$?
7. Suppose that $f_n \to f$ uniformly on each set of the infinite sequence $E_1, E_2, \ldots$. Must we have the convergence $f_n \to f$ uniformly on $E = \cup_{k=1}^{\infty} E_k$? on $E = \cap_{k=1}^{\infty} E_k$?
8. Suppose that E is a finite set of numbers, say $E = \{x_1, x_2, \ldots, x_n\}$. Is it true that $f_n \to f$ uniformly on E if and only if $f_n \to f$ pointwise on E?
9. Let $f : \mathbb{R} \to \mathbb{R}$ be continuous, and for each $n \ge 1$, define $f_n : \mathbb{R} \to \mathbb{R}$ by
$$f_n(x) = f\left(x + \frac{1}{n}\right), \quad x \in \mathbb{R}.$$
Must $\{f_n\}$ converge uniformly on $\mathbb{R}$?
10. Suppose that $g_n \to 0$ uniformly on E such that $|f_n(x)| \le g_n(x)$ for $n \ge 1$ and for $x \in E$. Must we have $f_n \to 0$ uniformly on E?
11. Suppose that $f_n \to f$ uniformly on E, and g is uniformly continuous on D, where D contains the range of f_n and f. Show that $g \circ f_n \to g \circ f$ uniformly on E. Is the uniform continuity of g necessary?

12. Suppose that $\{f_n\}$ and $\{g_n\}$ converge uniformly to f and g, respectively, on a set $E \subseteq \mathbb{R}$. Let α be a real constant. Must $\{f_n + g_n\}$ be uniformly convergent to $f+g$ on E? Must $\{\alpha f_n\}$ be uniformly convergent to αf on E? Must $f_n g_n \to fg$ be uniformly convergent on E if f and g are bounded functions on E? Must $\{f_n g_n\}$ be uniformly convergent to fg on E?
13. Does there exist a pointwise convergent sequence of continuous (respectively differentiable, Riemann integrable) functions whose limit function is not *continuous* (respectively differentiable, Riemann integrable)?
14. Does there exist a pointwise convergent sequence of differentiable functions whose limit is differentiable but the sequence of derivatives does not converge?
15. Does there exist a pointwise convergent sequence of Riemann integrable functions whose limit is Riemann integrable but

$$\lim_{n\to\infty} \int f_n(x)\, dx \neq \int \lim_{n\to\infty} f_n(x)\, dx?$$

16. Let $\{f_n\}$ be a sequence of continuous functions on $[0,1]$ converging pointwise to a continuous function f on $[0,1]$. If $f_n(x) \leq f_{n+1}(x)$ for all $x \in [0,1]$, must $\{f_n\}$ be uniformly convergent to f?
17. Suppose that $\{f_n\}$ is a sequence of continuous functions on E and that there exists a continuous function f on E such that $f_n \to f$ on E. If $f_n(x_n) \to f(x)$ for every sequence $\{x_n\}$ in E with $x_n \to x$, $x \in E$, must $f_n \to f$ uniformly on E?
18. Can the limit of a sequence of functions, each discontinuous at every point in $[0,1]$, be uniformly convergent on $[0,1]$?

Exercises 9.20.

1. Show by your own example that the limit of a sequence of continuous functions need not be continuous unless the convergence is uniform.
2. Does $\{x/n\}$ converge uniformly on $\mathbb{R}$? How about on $|x| < c$, $c > 0$? How about on $|x| > d$, $d > 0$?
3. Does $\{nx^n\}$ converge pointwise to $f(x) = 0$ on $[0,1)$? Does it converge uniformly on $[0,1)$?
4. Does the sequence $\{f_n\}$, $f_n(x) = nx/(n+x)$, converge uniformly on $[0,1]$?
5. Does $\{\arctan(nx)\}$ converge uniformly on $[0,b]$? Does it converge uniformly to $f(x) = \pi/2$ on $[a,b]$, $a > 0$? Does it converge uniformly on $[c,0]$? Does it converge uniformly to $f(x) = -\pi/2$ on $[c,a]$, $a < 0$?
6. Is each $f_n(x) = |x|^{1+1/n}$ differentiable on $[-1,1]$? Does $\{f_n(x)\}$ converge uniformly to $f(x) = |x|$ on $[-1,1]$? (See Figure A.12.)
7. For $n \geq 1$, let $f_n(x) = nx/(1+n^p x^2)$ for $x \in \mathbb{R}$, where $p > 1$. Show that $\{f_n\}$ converges pointwise on $\mathbb{R}$. Show also that $\{f_n\}$ converges uniformly on $\mathbb{R}$ if and only if $p > 2$.
8. Show that $\{f_n(x)\}$, $f_n(x) = 1/(x^{-n}+x^n)$, converges pointwise to $f(x) = 0$ on $(0,1)$, but not uniformly on $(0,1)$.

9. Show that $\{x^n(1-x^n)\}_{n\geq 1}$ does not converge uniformly on $[0,1]$.
10. Show that $\{f_n(x)\}$, $f_n(x) = x^n/(1+x^n)$, is not uniformly convergent on $[0,3]$.
11. Does $\{nxe^{-nx^2}\}$ converge pointwise on $\mathbb{R}$? Must $\{nxe^{-nx^2}\}$ be uniformly convergent on $[0,1]$? Does it converge uniformly on $\{x : |x| > r\}$, $r > 0$?
12. Let $f_n(x) = nxe^{-nx^2}$ for $0 \leq x \leq 1$ (see Figure 9.25). Show that $f_n \to 0$ on $[0,1]$ but not uniformly. Show also that

$$\lim_{n\to\infty} \int_0^1 f_n(x)\,dx \neq \int_0^1 \lim_{n\to\infty} f_n(x)\,dx.$$

13. Suppose that g is continuous on $[0,1]$. Show that $\{g(x)x^n\}$ converges uniformly on $[0,1]$ if and only if $g(1) = 0$.
14. Define f_n as below on the stated interval I:

(a) $\dfrac{n}{x+n}$, $x \geq 0$; and $0 \leq x \leq c$.

(b) $(x - 1/n)^2$, $x \in [0,1]$.

(c) $(1/n)\arctan(nx)$, $x \in \mathbb{R}$.

(d) $\dfrac{1-(-1)^n x^n}{1+x}$, $x \in [0,1]$.

(e) $\sin^n x$, $x \in [0,\pi]$.

(f) $n\sin(x/n)$, $x \in \mathbb{R}$.

(g) e^{-nx}, $x \geq 0$ and $c \leq x < \infty$, $c > 0$.

(h) $\dfrac{1}{1+nx}$, $0 \leq x \leq c$.

(i) $\dfrac{\sin^n x}{3n+\sin^n x}$, $x \in \left[-\dfrac{\pi}{2}, \dfrac{\pi}{2}\right]$.

(j) $\dfrac{\cos^n x}{2n - \cos^n x}$, $x \in [0,\pi]$.

(k) $\dfrac{xe^{-x^2}}{n}$, $x \in (-\infty,\infty)$.

(l) $1 - \dfrac{x^n}{n}$, $x \in [0,1]$.

Does the sequence $\{f_n(x)\}$ converge pointwise on I? If so, find the (pointwise) limit function $f(x)$ on I. Does it converge uniformly on I? Justify your answer.

15. Show that $\{(1/n)\sin nx\}_{n\geq 1}$ converges pointwise to $f(x) = 0$ on $\mathbb{R}$.
16. Show that $\{x^k e^{-nx}\}_{n\geq 1}$ ($k \in \mathbb{N}$ is fixed) converges pointwise to $f(x) = 0$ on $[0,\infty)$.
17. Show that $\{e^{x(n+1)/n}\}_{n\geq 1}$ converges uniformly to e^x on $[0,b]$, whereas $\{e^{-nx}\}_{n\geq 1}$ does not converge uniformly on $[0,b]$, $b > 0$.
18. If $f_n(x) = nx/(1+n^2x^2)$ on $[0,1]$, show that

$$\lim_{n\to\infty} \int_0^1 f_n(x)\,dx = \int_0^1 f(x)\,dx.$$

Can we conclude that $\{f_n\}$ converges uniformly on $[0,1]$? (Compare with Theorem 9.15.)

19. For each of the sequences $\{f_n\}_{n\geq 1}$ given by the functions $f_n : [0,1] \to \mathbb{R}$ below, find the pointwise limit of $\{f_n\}$ and determine whether the sequence converges uniformly on $[0,1]$ to its pointwise limit.

(a) $\dfrac{n}{x}\sin\left(\dfrac{x}{n}\right)$. (b) $\dfrac{1}{1+x^{2n}}$. (c) $\left(1+\dfrac{x}{n}\right)^n$. (d) $\dfrac{x^n}{1+x^n}$.

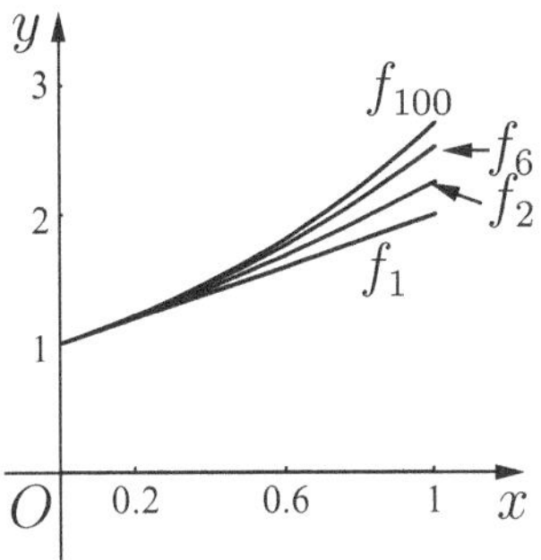

Fig. 9.29. $f_n(x) = (1 + x/n)^n$ on $[0, 1]$.

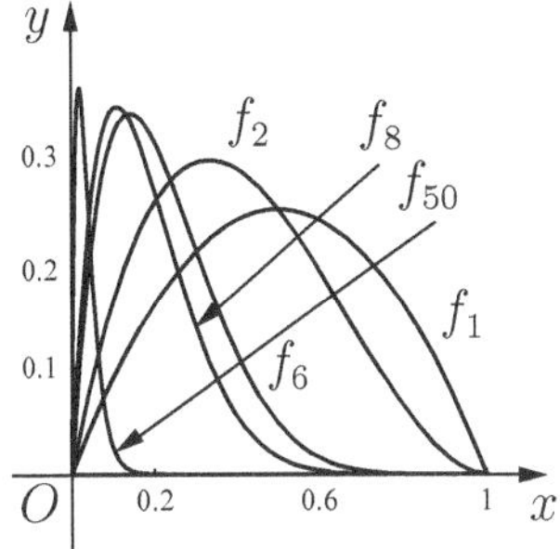

Fig. 9.30. $f_n(x) = nx(1 - x)^n$ on $[0, 1]$.

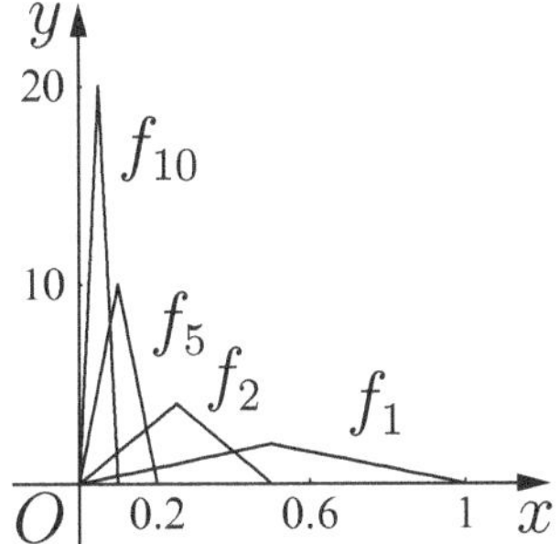

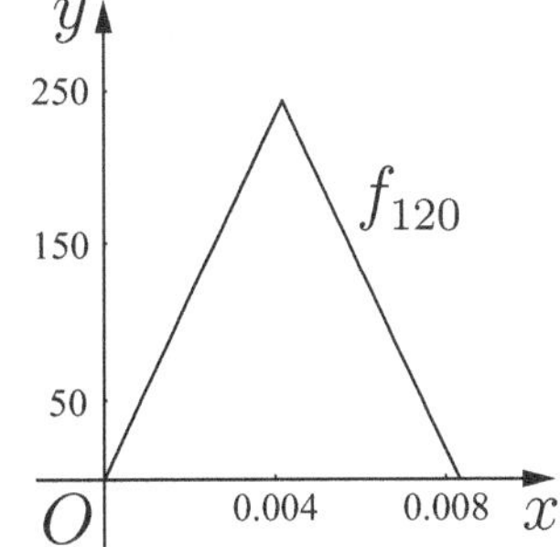

Fig. 9.31. Graph of $f_n(x)$ on $[0, 1/n]$ for some n, for Exercise 9.20(21).

For case **(a)**, assume that the function has the value 1 at $x = 0$. Verify whether

$$\lim_{n\to\infty} \int_0^1 f_n(x)\,\mathrm{d}x = \int_0^1 \left(\lim_{n\to\infty} f_n(x)\right)\,\mathrm{d}x.$$

20. Find the convergence set of each of the following sequences $\{f_n(x)\}$ of functions on $[0, 1]$. Also find sets on which these converge uniformly (see Figures 9.9, 9.29, and 9.30 for **(a)**, **(b)**, and **(c)**, respectively).
(a) $n^2x^n(1 - x)$. **(b)** $\left(1 + \dfrac{x}{n}\right)^n$. **(c)** $nx(1 - x)^n$.
21. For $n \geq 1$, define $\{f_n(x)\}$ on $[0, 1]$ (see Figure 9.31) by

$$f_n(x) = \begin{cases} 4n^2x & \text{for } 0 \leq x \leq 1/(2n), \\ -4n^2(x - 1/n) & \text{for } 1/(2n) \leq x \leq 1/n, \\ 0 & \text{for } 1/n \leq x \leq 1. \end{cases}$$

Show that $f_n \to 0$ pointwise on $[0, 1]$ and $\int_0^1 f_n(x)\,\mathrm{d}x = 1$.
Note: See also Examples 9.11.

22. Determine

(a) $\lim_{n\to\infty} \int_a^b \frac{\sin^n(x^2)}{x^3+n}\,dx, \quad 0<a<b.$ (b) $\lim_{n\to\infty} \int_0^1 \frac{n+\cos^n(e^x)}{4n+x^4}\,dx.$

9.2 Uniform Convergence of Series

Definition 9.1 suggests that we continue our discussion from *sequences* of real-valued functions to *series* of real-valued functions. Consider a sequence of functions $\{f_n(x)\}$ defined on a set E. Recall that the formal sum $\sum_{k=1}^{\infty} f_k$ is called a *series of functions*. Form a new sequence of partial sums of functions $\{S_n(x)\}$ defined by

$$S_n(x) = \sum_{k=1}^{n} f_k(x), \quad x \in E. \tag{9.1}$$

If the sequence $\{S_n(x)\}$ converges at a point $x \in E$, then we say that the series of functions $\sum_{k=1}^{\infty} f_k$ converges at x, and we write

$$\lim_{n\to\infty} S_n(x) = \sum_{k=1}^{\infty} f_k(x).$$

If the sequence $\{S_n(x)\}$ converges at all points of E, then we say that $\sum_{k=1}^{\infty} f_k$ *converges (pointwise)* on E and write the sum function as

$$f(x) := \lim_{n\to\infty} S_n(x) = \sum_{k=1}^{\infty} f_k(x).$$

Definition 9.21. *The series* $\sum_{k=1}^{\infty} f_k(x)$ *is said to be* uniformly convergent *to* $f(x)$ *on* E *if* $\{S_n(x)\}$ *converges uniformly to* $f(x)$ *on* E. *That is, the series converges uniformly to* $f(x)$ *on* E *if to each* $\epsilon > 0$, *there corresponds an integer* $N = N(\epsilon)$ *such that*

$$\left| \sum_{k=1}^{n} f_k(x) - f(x) \right| < \epsilon \quad \textit{whenever } n > N(\epsilon) \textit{ for all } x \in E.$$

In order to appreciate the importance of questions about convergence, we continue our investigation with a familiar sequence. Consider

$$S_n(x) = 1 + x + x^2 + \cdots + x^{n-1} = \frac{1-x^n}{1-x} \quad \text{for } x \neq 1.$$

We see that $\{S_n\}_{n=1}^{\infty}$ converges pointwise on $(-1,1)$ to $1/(1-x)$. Next, we show that $\{S_n\}$ converges uniformly on any closed subinterval of $(-1,1)$.

To see this, we consider a subinterval $[-1+r, 1-r]$, where $r < 1$ is an arbitrary positive number. In this interval $|x| < 1-r$, we have $|1-x| > r$. Consequently,

$$\left| S_n(x) - \frac{1}{1-x} \right| = \left| \frac{x^n}{1-x} \right| < \frac{(1-r)^n}{r} < \epsilon \quad \text{for all } n \geq N,$$

because

$$(1-r)^n < \epsilon r, \quad \text{i.e., } n \log(1-r) < \log(\epsilon r) \quad \text{or} \quad n > \frac{\log(\epsilon r)}{\log(1-r)},$$

and so N may be chosen as any positive integer greater than $\log(\epsilon r)/\log(1-r)$. As for the entire interval $(-1, 1)$, it contains points close to $x = 1$, and so

$$\lim_{x \to 1-} \left| \frac{-x^n}{1-x} \right| = \infty.$$

We may also note that $\sup_{r \in (0,1)} \log(\epsilon r)/\log(1-r) = \infty$ (with $\epsilon = 2$, for example) showing that $\{S_n\}$ does not converge uniformly on $(-1, 1)$. How about the convergence of $\{S_n\}$ on $|x| \geq 1$?

We may now rewrite Theorem 9.12 about sequences to get a necessary, but not sufficient, condition for the uniform convergence of series of functions.

Corollary 9.22 (Necessary condition for uniform convergence of series). *If f_n is continuous on E for each $n \geq 1$ and if $\sum_{k \geq 1} f_k(x)$ is uniformly convergent to $f(x)$ on E, then f must be continuous on E.*

A series $\sum_{k=1}^{\infty} f_k(x)$ is said to be *absolutely convergent* at x (respectively on E) if $\sum_{k=1}^{\infty} |f_k(x)|$ converges at x (respectively on E). Moreover, a necessary condition for $\sum_{k=1}^{\infty} f_k(x)$ to converge on E is that $f_k(x) \to 0$ for each $x \in E$. This fact is evident if we write

$$f_n = S_n - S_{n-1}$$

and allow $n \to \infty$. Also, rewording the Cauchy criterion, we have the following result, which may be used to test a series for uniform convergence without guessing what the limit function might be.

Theorem 9.23 (Cauchy criterion for functional series). *A series $\sum_{k=1}^{\infty} f_k(x)$ converges uniformly on a set E if and only if the sequence of partial sums is uniformly Cauchy on E, i.e., to each $\epsilon > 0$, there corresponds an integer $N = N(\epsilon)$ such that*

$$|S_m(x) - S_n(x)| = \left| \sum_{k=n+1}^{m} f_k(x) \right| < \epsilon \quad \textit{whenever } m > n > N(\epsilon) \textit{ for all } x \in E,$$

where $\{S_n\}$ is the sequence of partial sums defined by (9.1).

Proof. Apply Theorem 9.9. ∎

For instance, if $S_n(x) = 1 + x + x^2 + \cdots + x^{n-1}$, then for $m > n$ and all $|x| \leq r$, $0 < r < 1$,

$$|S_m(x) - S_n(x)| = \left|\frac{1-x^m}{1-x} - \frac{1-x^n}{1-x}\right| = \left|\frac{x^m - x^n}{1-x}\right| \leq \frac{2|x^n|}{1-|x|} \leq \frac{2r^n}{1-r},$$

so that

$$\sup_{x\in[-r,r]} |S_m(x) - S_n(x)| \leq \frac{2r^n}{1-r}.$$

Since $2r^n/(1-r) \to 0$ $(r < 1)$, an integer $N = N(\epsilon)$ can be found for which $2r^n/(1-r) < \epsilon$ for all $n > N(\epsilon)$. But then for all $m > n > N$ and all $x \in [-r, r]$,

$$|S_m(x) - S_n(x)| \leq \frac{2r^n}{1-r} < \epsilon,$$

showing that the series $\sum_{k=0}^{\infty} x^k$ converges uniformly on any interval $[-r, r]$, $0 < r < 1$. Note that here we do not bother about the limit function.

9.2.1 Two Tests for Uniform Convergence of Series

Theorem 9.23 enables us to establish an analogue of the comparison test, namely, a sufficient condition for the uniform convergence of a series.

Definition 9.24. *Let $\{M_n\}_{n\geq 1}$ be a sequence of nonnegative real numbers, and $\{f_n\}$ a sequence of functions defined on a set E such that $|f_n(x)| \leq M_n$ for all $x \in E$ and each $n \in \mathbb{N}$. Then the series $\sum_{k=1}^{\infty} f_k(x)$ is said to be a dominated series on E if $\sum_{k=1}^{\infty} M_k$ converges.*

Indeed, the following result is a simple and direct test for the uniform convergence of series.

Theorem 9.25 (Weierstrass M-test/dominated convergence test). *If $\sum_{k=1}^{\infty} f_k(x)$ is a dominated series on E, then it converges uniformly and absolutely on the set E.*

Proof. Suppose that $\sum_{k=1}^{\infty} f_k(x)$ is dominated by a convergent series $\sum_{k=1}^{\infty} M_k$. That the series $\sum_{k=1}^{\infty} |f_k(x)|$ converges on E follows immediately from the comparison test for real series. To verify the uniform convergence of $\sum_{k=1}^{\infty} f_k(x)$ on E, we invoke the Cauchy criterion for the series $\sum_{k=1}^{\infty} M_k$. Thus, given $\epsilon > 0$, there exists an integer $N = N(\epsilon)$ such that for $m > n > N$, we have

$$\sum_{k=n+1}^{m} M_k < \epsilon.$$

But for all $m > n > N$, we also have

$$|S_m(x) - S_n(x)| = \left|\sum_{k=n+1}^{m} f_k(x)\right| \leq \sum_{k=n+1}^{m} |f_k(x)| \leq \sum_{k=n+1}^{m} M_k < \epsilon.$$

Therefore, by the Cauchy criterion (see Theorem 9.23), the series $\sum_{k=1}^{\infty} f_k(x)$ converges uniformly and absolutely on E. ∎

As an illustration, consider the series $\sum_{k=0}^{\infty}(x^k/k!)$. For $|x| \leq r$ $(r \geq 0)$,

$$\left|\frac{x^k}{k!}\right| \leq \frac{r^k}{k!} =: M_k.$$

But the ratio test immediately implies that $\sum_{k=0}^{\infty} M_k$ converges. Consequently, by the Weierstrass M-test, the series $\sum_{k=0}^{\infty}(x^k/k!)$ converges uniformly for $|x| \leq r$. We remark that the series does not converge uniformly on $\mathbb{R}$, but since it converges uniformly on every bounded interval, we conclude that the limit is continuous on $\mathbb{R}$ (by Corollary 9.22). We may now formulate the following remarkable result, which is basic for many applications of power series in analysis.

Theorem 9.26. *Let $R > 0$ be the radius of convergence of the power series $\sum_{k=0}^{\infty} a_k x^k$ and let r be a positive number such that $r < R$. Then $\sum_{k=0}^{\infty} a_k x^k$ converges uniformly for $|x| \leq r$.*

Proof. Assume that $\sum_{k=0}^{\infty} a_k x^k$ converges for $|x| < R$. As usual, choose t such that $r < t < R$. Then there exists an $M > 0$ such that

$$|a_k t^k| \leq M \quad \text{for all } k \geq 0.$$

Also, for $|x| \leq r$ $(< t < R)$, we have

$$|a_k x^k| \leq |a_k| r^k = |a_k t^k| \left(\frac{r}{t}\right)^k \leq M\left(\frac{r}{t}\right)^k = M_k \quad \text{for all } k \geq 0.$$

We recall that $\sum_{k=0}^{\infty} M_k$ converges. Therefore, by the Weierstrass M-test, the series $\sum_{k=0}^{\infty} a_k x^k$ converges uniformly for $|x| \leq r$, for each $r < R$. ∎

For instance, each of the series $\sum_{k=0}^{\infty} x^k$ and $\sum_{k=1}^{\infty}(x^k/k)$ converges uniformly on every compact subset of $(-1, 1)$, because the radius of convergence of both series is 1.

Remark 9.27. Note that the Weierstrass M-test has a slight drawback, since it applies only to series that are also absolutely convergent. Thus, one needs to use a different method for testing the uniform convergence of nonabsolutely convergent series. Below, we present a uniformly convergent series that is not absolutely convergent. •

Examples 9.28. We begin by illustrating the Weierstrass M-test by a number of examples.

- For the functional series $\sum_{k=1}^{\infty}\left(1/(x^2+k^2)\right)$ and $\sum_{k=1}^{\infty}\left((\sin kx)/k^2\right)$, we have
$$\left\{\left|\frac{1}{x^2+k^2}\right|,\left|\frac{\sin kx}{k^2}\right|\right\}\leq\frac{1}{k^2}\quad\text{for all } x\in\mathbb{R}\text{ and } k\geq 1,$$
where $\sum_{k=1}^{\infty}(1/k^2)$ converges. Thus, both series converge uniformly on $\mathbb{R}$. By Corollary 9.22, each of the series represents a continuous function on $\mathbb{R}$.
- For the functional series $\sum_{k=1}^{\infty}\sin(x/k^p)$ $(p>1)$, for any x with $|x|\leq r$,
$$\left|\sin\left(\frac{x}{n^p}\right)\right|\leq\left|\frac{x}{n^p}\right|\leq\frac{r}{n^p}=M_n,$$
where $\sum M_k$ converges for $p>1$. Thus, the series converges uniformly (and absolutely) on every closed and bounded interval $[-r,r]$.
- As a consequence of the substitution $y=x/(1+x)$, the functional series $\sum_{k=1}^{\infty}\left(x/(1+x)\right)^k$ may be treated as a geometric series with the new variable y, and hence it converges pointwise for
$$|x/(1+x)|<1,\quad\text{i.e., for } x>-1/2,$$
and converges uniformly for
$$|x/(1+x)|\leq r\,(<1),\quad\text{i.e., for } x\in[\tfrac{-r}{1+r},\tfrac{r}{1-r}],\ r<1.$$
Thus, the series converges on any compact subset of $(-1/2,\infty)$.
- The functional series $\sum_{k=1}^{\infty}((\cos^k x)/k!)$ converges uniformly on $\mathbb{R}$, because for any real x,
$$\left|\frac{\cos^n x}{n!}\right|\leq\frac{1}{n!}=M_n,$$
and $\sum M_k$ converges.
- The functional series $\sum_{k=1}^{\infty}\left(x^k/(1+x^k)\right)$ converges uniformly for $|x|\leq 1/2$, because for $|x|\leq 1/2$,
$$|1+x^{-n}|\geq 2^n-1\geq 2^{n-1},\quad\text{i.e.,}\ \left|\frac{x^n}{1+x^n}\right|\leq\frac{1}{2^{n-1}}=M_n,$$
and $\sum M_k$ converges.
- The functional series $\sum_{k=1}^{\infty}[(1-k\cos(kx))/(k^{\alpha-\sin(kx)})]$ $(\alpha>3)$ converges uniformly on $[0,2\pi]$, because for real $x\in[0,2\pi]$,
$$\left|\frac{1-n\cos(nx)}{n^{\alpha-\sin(nx)}}\right|\leq\frac{1+n}{n^{\alpha-1}}\leq\frac{2n}{n^{\alpha-1}}=\frac{2}{n^{\alpha-2}}=M_n,$$
and $\sum M_k$ converges.

- Consider $\sum_{k=1}^{\infty}(-1)^{k-1}\left(x/(k+x^2)\right)$. Then for each fixed $x>0$, we have

$$a_n(x)=\frac{x}{n+x^2}\geq\frac{x}{n+1+x^2}=a_{n+1}(x)$$

  and $a_n(x)\to 0$ as $n\to\infty$, and so by the alternating series test, the series converges pointwise on $(0,\infty)$ (as well as $(-\infty,0)$) and hence on $\mathbb{R}$. Let the limit function be $f(x)$. Using the error estimate for alternating series (see Corollary 5.46), we see that the sequence of partial sums $\{S_n(x)\}$ satisfies

$$|S_n(x)-f(x)|\leq|a_{n+1}(x)|=\frac{|x|}{1+n+x^2}\leq\frac{|x|}{2n^{1/2}|x|}=\frac{1}{2n^{1/2}},$$

  showing that it converges uniformly on $\mathbb{R}$. In particular, the given series represents a continuous function on $\mathbb{R}$. Is the Weierstrass M-test applicable in this example? This result cannot be obtained from the Weierstrass M-test because $\sum_{k=1}^{\infty}(|x|/(k+x^2))$ diverges for all $x\neq 0$.
- As in the last item, we conclude that the series $\sum_{k=1}^{\infty}(-1)^{k-1}(1/(k+x))$ converges uniformly on $[0,\infty)$. ●

Theorem 9.29 (Dirichlet's test for uniform convergence). *Suppose that $\{b_n\}$ is a sequence of (nonnegative) functions on a set E such that $b_n(x)\geq b_{n+1}(x)$ and $b_n\to 0$ uniformly on E. If $\{a_n\}$ is a sequence of functions such that $|s_n(x)|\leq M$ if $n\geq 1$ and $x\in E$, where $s_n(x)=\sum_{k=1}^{n}a_k(x)$, then $\sum_{k=1}^{\infty}a_k(x)b_k(x)$ converges uniformly on E.*

Proof. It follows from Dirichlet's test (Corollary 5.55) that $\sum_{k=1}^{\infty}a_k(x)b_k(x)$ converges on E. Let $S_n(x)=\sum_{k=1}^{n}a_k(x)b_k(x)$. Then (see the proof of Theorem 5.54) from Abel's summation by parts, we have

$$\sum_{k=n+1}^{m}a_kb_k=s_mb_{m+1}-s_nb_{n+1}+\sum_{k=n+1}^{m}s_k(b_k-b_{k+1}).$$

Since $\{b_n(x)\}$ is decreasing and $\{s_n(x)\}$ is bounded, it follows that for $m>n$,

$$\begin{aligned}\left|\sum_{k=n+1}^{m}a_kb_k\right| &\leq Mb_{m+1}+Mb_{n+1}+M\sum_{k=n+1}^{m}(b_k-b_{k+1})\\ &=Mb_{m+1}+Mb_{n+1}+M(b_{n+1}-b_{m+1})\\ &=2Mb_{n+1},\end{aligned}$$

and so we conclude that if $m>n$, then

$$|S_m(x)-S_n(x)|\leq 2Mb_{n+1}(x)\quad\text{for all } x\in E.$$

This fact together with that fact that $b_n(x)\to 0$ uniformly on E implies that $\{S_n(x)\}$ converges uniformly on E. ∎

9.2.2 Interchange of Summation and Integration

Theorem 9.15 applied to a sequence of partial sums leads to the following result for series.

Corollary 9.30 (Interchange of summation and integration). *Suppose that $\{f_n(x)\}$ is a sequence of continuous functions on $[a,b]$ and that $\sum_{k=0}^{\infty} f_k(x)$ converges uniformly to a function $f(x)$ on $[a,b]$. Then*

$$\sum_{k=0}^{\infty}\left(\int_a^b f_k(x)\,dx\right) = \int_a^b f(x)\,dx.$$

Also, for each $t \in [a,b]$,

$$\sum_{k=0}^{\infty}\left(\int_a^t f_k(x)\,dx\right) = \int_a^t f(x)\,dx.$$

Example 9.31. We know that $\sum_{k=1}^{\infty}(1/k^p)\cos kx$ converges uniformly to some $f(x)$ on $\mathbb{R}$ for $p>1$ (by Weierstrass's M-test). In particular, the series is integrable on any closed interval in $\mathbb{R}$, and so

$$\int_0^{\pi/2} f(x)\,dx = \int_0^{\pi/2}\sum_{k=1}^{\infty}\frac{\cos kx}{k^p}\,dx = \sum_{k=1}^{\infty}\frac{1}{k^p}\int_0^{\pi/2}\cos kx\,dx = \sum_{k=1}^{\infty}\frac{\sin(k\pi/2)}{k^{p+1}},$$

which gives

$$\int_0^{\pi/2} f(x)\,dx = \sum_{k=1}^{\infty}\frac{(-1)^{k-1}}{(2k-1)^{p+1}}, \quad p>1.$$

Similarly, $\sum_{k=1}^{\infty}(1/k^p)\sin kx$ converges uniformly to some $g(x)$ on $\mathbb{R}$ for $p>1$, and therefore we easily see that

$$\int_0^{\pi} g(x)\,dx = \sum_{k=1}^{\infty}\frac{2}{(2k-1)^{p+1}}.$$

Again, consider the series $\sum_{k=1}^{\infty}(1/k^p)\cos kx$, where $0<p\leq 1$. Set $a_k(x)=\cos kx$ and $b_k=1/k^p$. Note that $\{b_k\}$ is convergent to 0. To evaluate $s_n(x)=\sum_{k=1}^{n}a_k(x)$, we may rewrite the partial sum (see Corollary 5.59(**ii**))

$$s_n(x) = \frac{1}{2\sin(x/2)}[\sin(n+1/2)x - \sin(x/2)]$$

for $x \neq 2m\pi$ ($m\in\mathbb{Z}$), and so

$$|s_n(x)| \leq \frac{1}{\sin(x/2)}$$

for any n and on any interval of the form $[\delta, 2\pi - \delta]$, where $\delta > 0$. Thus, $\{s_n(x)\}$ is uniformly bounded on $[\delta, 2\pi - \delta]$. So Dirichlet's test applies, and the series $\sum_{k=1}^{\infty}(1/k^p)\cos kx$ $(0 < p \leq 1)$ converges uniformly on $[\delta, 2\pi - \delta]$. In particular, $\sum_{k=1}^{\infty}(1/k^p)\cos kx$ converges for $0 < x < 2\pi$ (and hence for all x with $x \neq 2m\pi$, $m \in \mathbb{Z}$) and $0 < p \leq 1$.

Similarly, it can be shown that the series $\sum_{k=1}^{\infty}(1/k^p)\sin kx$ is uniformly convergent on $[\delta, 2\pi - \delta]$ for $0 < p \leq 1$. ●

Example 9.32. From the proof of Theorem 5.3 (or even directly), we see that

$$\frac{1}{1+r^2} = 1 - r^2 + r^4 - r^6 + \cdots + (-1)^{n-1}r^{2(n-1)} + (-1)^n \frac{r^{2n}}{1+r^2},$$

which by integration from 0 to 1 gives,

$$\tan^{-1} r\Big|_0^1 = \left(r - \frac{r^3}{3} + \frac{r^5}{5} - \frac{r^7}{7} + \cdots + (-1)^{n-1}\frac{r^{2n-1}}{2n-1}\right)\Bigg|_0^1 + R_n,$$

so that

$$\frac{\pi}{4} = 1 - \frac{1}{3} + \frac{1}{5} - \frac{1}{7} + \cdots + (-1)^{n-1}\frac{1}{2n-1} + R_n,$$

where

$$|R_n| = \left|(-1)^n \int_0^1 \frac{r^{2n}}{1+r^2}\,dr\right| = \int_0^1 \frac{r^{2n}}{1+r^2}\,dr < \int_0^1 r^{2n}\,dr = \frac{1}{2n+1},$$

which approaches zero as $n \to \infty$. Therefore, we have the Madhava-Leibniz-Gregory series (see also Examples 9.52 and 10.16)[1]

$$\frac{\pi}{4} = 1 - \frac{1}{3} + \frac{1}{5} - \frac{1}{7} + \cdots + (-1)^{n-1}\frac{1}{2n-1} + \cdots.$$

This example shows that if a series can be recognized as a variant of the geometric series (5.1), then it becomes easier to evaluate its limit. However, this formula is of limited value, because the rate of convergence is so slow in the representation of π. ●

Example 9.33. From our earlier knowledge, we know that

$$\sum_{k=0}^{\infty}(-1)^k x^{2k} = \lim_{n\to\infty} S_n(x) = \frac{1}{1+x^2} \quad \text{for } |x| < 1,$$

where $S_n(x) = \sum_{k=0}^{n-1}(-1)^k x^{2k}$, and the convergence is uniform on $[-r, r]$, for each $r < 1$. By Corollary 9.30, for $x \in [-r, r]$ $(r < 1)$,

$$\int_0^x S_n(t)\,dt = \sum_{k=0}^{n-1}\frac{(-1)^k}{2k+1}x^{2k+1} \to \int_0^x \frac{dt}{1+t^2} = \arctan x,$$

[1] This formula was first deduced by the Indian Mathematician Madhava of Sangramagrama (1350–1425) as a special case of the more general infinite series for $\arctan x$ discovered by him. The latter was rediscovered by Gregory during 1638–1675.

and the convergence is uniform on $[-r, r]$. Thus, we have a series expansion for $\arctan x$:

$$\arctan x = \sum_{k=0}^{\infty} \frac{(-1)^k}{2k+1} x^{2k+1} \quad \text{for } x \in [-r, r] \ (r < 1).$$

Observe that the radius of convergence of this series is 1. We remark that by the alternating series test, the above series representation of $\arctan x$ converges at both endpoints $x = \pm 1$, although we would not be able to conclude this from Corollary 9.30. ●

Remark 9.34. As an application to the last equation, we set $x = 1/2$ and $x = 1/3$ and add the resulting equations to obtain

$$\arctan \frac{1}{2} + \arctan \frac{1}{3} = \sum_{k=0}^{\infty} \frac{(-1)^k}{2k+1} \left(\frac{1}{2^{2k+1}} + \frac{1}{3^{2k+1}} \right).$$

The left-hand side is known to be $\pi/4$, because

$$\arctan \frac{1}{2} + \arctan \frac{1}{3} = \arctan \left(\frac{\frac{1}{2} + \frac{1}{3}}{1 - \frac{1}{2}\frac{1}{3}} \right) = \arctan 1 = \frac{\pi}{4}.$$

This gives

$$\pi = 4 \left[\arctan \frac{1}{2} + \arctan \frac{1}{3} \right],$$

so that

$$\pi = 4 \sum_{k=0}^{\infty} \frac{(-1)^k}{2k+1} \left(\frac{1}{2^{2k+1}} + \frac{1}{3^{2k+1}} \right) = 4 \left(\frac{5}{6} - \frac{35}{648} + \cdots \right), \tag{9.2}$$

which is a series representation of π.

To get another expansion of π, we set $x = 1/5$ and $\arctan(1/5) = A$, i.e., $\tan A = 1/5$. From the formula

$$\tan(2A) = \frac{2 \tan A}{1 - \tan^2 A},$$

it follows that $\tan(2A) = 5/12$. Again, using this formula once again, we get

$$\tan(4A) = \frac{2 \tan(2A)}{1 - \tan^2(2A)} = \frac{120}{119}.$$

Using this value of $\tan(4A)$, choose B such that

$$1 = \tan(4A - B) = \frac{\tan(4A) - \tan B}{1 + \tan(4A) \tan B} = \frac{(120/119) - \tan B}{1 + (120/119) \tan B}.$$

A computation gives $\tan B = 1/239$ and $4A - B = \pi/4$. That is,

$$\pi = 4\left[4\arctan\frac{1}{5} - \arctan\frac{1}{239}\right],$$

and therefore, in series form,

$$\pi = 4\sum_{k=0}^{\infty}\frac{(-1)^k}{2k+1}\left(\frac{4}{5^{2k+1}} - \frac{1}{239^{2k+1}}\right). \tag{9.3}$$

In order to compare the two expansions (9.2) and (9.3) of π, we consider the partial sums of these two series:

$$A_n = \sum_{k=0}^{n}\frac{(-1)^k}{2k+1}\left(\frac{4}{2^{2k+1}} + \frac{4}{3^{2k+1}}\right) \quad \text{and } B_n = \sum_{k=0}^{n}\frac{(-1)^k}{2k+1}\left(\frac{16}{5^{2k+1}} - \frac{4}{239^{2k+1}}\right).$$

A computation shows that the formula (9.3) can be used to approximate π up to 100 decimal places, whereas in the formula (9.2), the convergence is somewhat slow. For a comparison, we refer to Table 9.1. ●

n	A_n	B_n
0	3.33333333	3.18326360
1	3.11728395	3.14059703
2	3.14557613	3.14162103
3	3.14085056	3.14159177
4	3.14174120	3.14159268
5	3.14156159	3.14159265

Table 9.1. Iterated values of π.

Here is one more example. We have learned that the geometric series $\sum_{k=0}^{\infty} x^k$ converges to $1/(1-x)$ on $(-1,1)$ and converges uniformly on any closed interval $[a,b]$ contained in $(-1,1)$ (see Example 9.28). Since each $f_n(x) = x^n$ is continuous on $[a,b]$, by Corollary 9.30, term-by-term integration is permissible. Thus, we have for $-1 < t < 1$,

$$\sum_{k=0}^{\infty}\int_0^t x^k\,dx = \sum_{k=0}^{\infty}\frac{t^{k+1}}{k+1} \to \int_0^t \frac{dx}{1-x} = -\log(1-t) \quad \text{on } [a,b].$$

That is,

$$\sum_{k=0}^{\infty}\frac{t^{k+1}}{k+1} = -\log(1-t) \quad \text{on } (-1,1),$$

and the convergence is uniform on any closed subinterval of $(-1,1)$. It turns out that the new series also converges (by the alternating series test) at one of the endpoints, $t=-1$.

The following result is almost obvious and is in fact a straightforward application of Corollary 9.30.

Theorem 9.35 (Term-by-term integration in power series). *Suppose that $f(x)=\sum_{k=0}^{\infty} a_k x^k$ converges for $|x|<R$. Then for any closed interval $[c,x]$ contained in $(-R,R)$, the integral $\int_c^x f(t)\,dt$ exists and can be obtained by integrating the power series term by term. In particular, we have*

$$\int_0^x f(t)\,dt = \sum_{k=0}^{\infty} \frac{a_k}{k+1} x^{k+1} + \text{ constant for } |x|<R,$$

and the convergence is uniform on any closed subinterval of $(-R,R)$.

Theorem 9.35 tells us that inside the interval of convergence (not necessarily at the endpoints), a power series can be integrated term by term. Next, we see that term-by-term differentiation is also possible.

In Theorem 9.15, we needed for each f_n to be continuous on $[a,b]$. *Is it possible to weaken the hypothesis to permit integrable functions rather than just continuous functions?* Recall that if f and g are two Riemann integrable functions on $[a,b]$, then

$$\int_a^b f(x)\,dx \geq \int_a^b g(x)\,dx,$$

whenever $f(x)\geq g(x)$ on $[a,b]$.

Theorem 9.36 (Integration of sequences of integrable functions). *Suppose that $\{f_n\}$ is a sequence of Riemann integrable functions defined on an interval $[a,b]$. If $f_n\to f$ uniformly on $[a,b]$, then f is Riemann integrable on $[a,b]$, and*

$$\lim_{n\to\infty}\int_a^b f_n(x)\,dx = \int_a^b f(x)\,dx.$$

Also, for each $t\in[a,b]$,

$$\int_a^t f_n(x)\,dx \to \int_a^t f(x)\,dx \quad \text{uniformly on } [a,b].$$

Proof. Because of the argument in Theorem 9.15, we need only show that the limit function f is integrable on $[a,b]$. We see that the following statements hold:

- f_n is bounded, because each f_n is integrable on $[a,b]$.

- f is bounded, because

$$|f(x)| \le |f_n(x) - f(x)| + |f_n(x)| \le \delta_n + |f_n(x)|,$$

where $\delta_n = \sup_{x \in [a,b]} |f_n(x) - f(x)| \to 0$ (by Theorem 9.6).
- Because $f_n \to f$ uniformly on $[a, b]$, given $\epsilon > 0$, there exists an N such that

$$|f_n(x) - f(x)| < \frac{\epsilon}{3(b-a)} \quad \text{for all } x \in [a,b] \text{ and all } n \ge N. \tag{9.4}$$

- Since f_N is integrable, there exists a partition P of $[a, b]$ such that

$$U(P, f_N) - L(P, f_N) < \frac{\epsilon}{3}.$$

For each $x \in [a, b]$, (9.4) with $n = N$ implies that

$$f_N(x) - \frac{\epsilon}{3(b-a)} < f(x) < f_N(x) + \frac{\epsilon}{3(b-a)},$$

and therefore

$$L(P, f_N) - \frac{\epsilon}{3} < L(P, f) \le U(P, f) < U(P, f_N) + \frac{\epsilon}{3}.$$

Consequently,

$$U(P, f) - L(P, f) < U(P, f_N) - L(P, f_N) + \frac{2\epsilon}{3} < \frac{\epsilon}{3} + \frac{2\epsilon}{3} = \epsilon,$$

showing that f is integrable on $[a, b]$.

Finally, for $n \ge N$ and for each $t \in [a, b]$, (9.4) implies that

$$\begin{aligned} \left| \int_a^t f_n(x)\,dx - \int_a^t f(x)\,dx \right| &\le \int_a^t |f_n(x) - f(x)|\,dx \\ &\le \frac{\epsilon(t-a)}{3(b-a)} \le \frac{\epsilon(b-a)}{3(b-a)} = \frac{\epsilon}{3}, \end{aligned}$$

and the proof is complete. ■

The limit of a uniformly convergent series of integrable functions is integrable, and so term-by-term integration is permissible for such a series. More precisely, we have the following result concerning integration of a series of integrable functions.

Corollary 9.37 (Interchange of summation and integration). *Suppose that $\{f_n\}$ is a sequence of integrable functions on $[a, b]$. If $\sum f_k$ converges uniformly to f on $[a, b]$, then f is integrable on $[a, b]$, and*

$$\sum \int_a^b f_k(x)\,dx = \int_a^b \left(\sum f_k(x) \right) dx = \int_a^b f(x)\,dx.$$

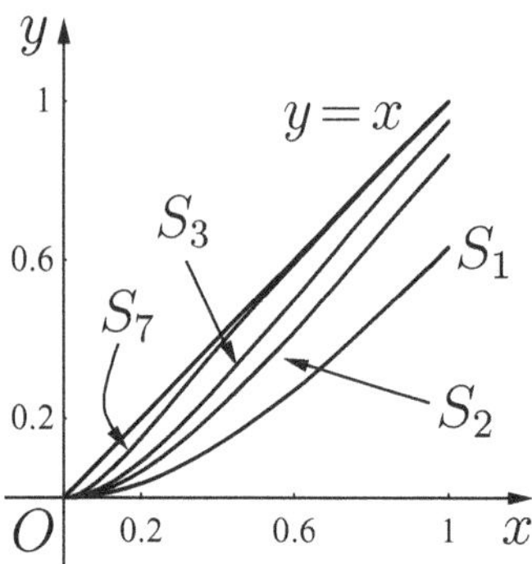

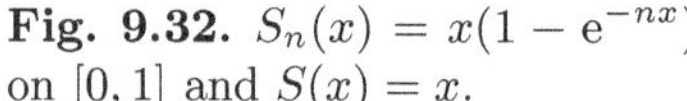
Fig. 9.32. $S_n(x) = x(1 - \mathrm{e}^{-nx})$ on $[0,1]$ and $S(x) = x$.

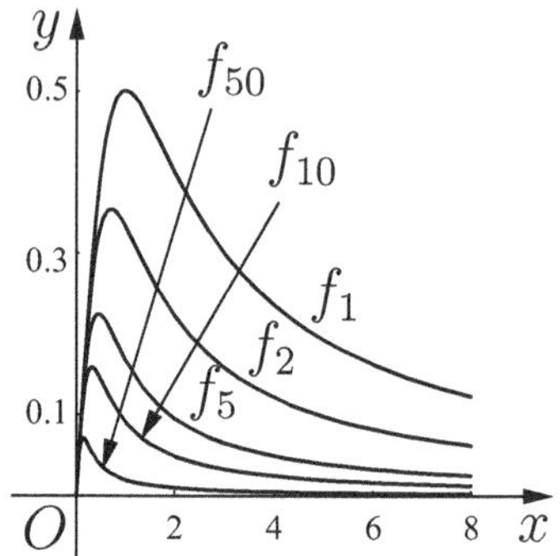

Fig. 9.33. $f_n(x) = x/(1+nx^2)$ on $\mathbb{R}$.

For instance, if $f_n(x) = x/(n(1+nx^2))$ on $[a,b] \subset \mathbb{R}$, Weierstrass's M-test shows that $\sum_{k=1}^{\infty} f_k$ converges uniformly on $[a,b]$. Consequently, term-by-term integration is permissible in this series.

Example 9.38. Evaluate $\sum_{k=1}^{\infty} \int_0^1 x(\mathrm{e}^x - 1)\mathrm{e}^{-kx}\,\mathrm{d}x$.

Solution. To evaluate this integral, we set $S_n(x) = \sum_{k=1}^{n} x(\mathrm{e}^x - 1)\mathrm{e}^{-kx}$. We observe that $S_n(0) = 0$ and for $x > 0$,

$$S_n(x) = x(\mathrm{e}^x - 1)\frac{\mathrm{e}^{-x}(1-\mathrm{e}^{-nx})}{1-\mathrm{e}^{-x}} = x(1-\mathrm{e}^{-nx}).$$

Thus, since $x\mathrm{e}^{-nx}$ attains its maximum at $x = 1/n$, we have

$$\delta_n = \sup_{x\geq 0} |S_n(x) - x| = \sup_{x \geq 0} |x\mathrm{e}^{-nx}| = \frac{1}{\mathrm{e}n} \to 0 \quad \text{as } n \to \infty,$$

so that $\{S_n(x)\}$ converges uniformly to $f(x) = x$ for $x \geq 0$ (see Theorem 9.6 and Figure 9.32). Thus, Corollary 9.37 is applicable with $f_k(x) = x(\mathrm{e}^x - 1)\mathrm{e}^{-kx}$, and we conclude that

$$\sum_{k=1}^{\infty} \int_0^1 f_k(x)\,\mathrm{d}x = \int_0^1 f(x)\,\mathrm{d}x = \int_0^1 x\,\mathrm{d}x = \frac{1}{2}. \qquad \bullet$$

Remark 9.39. Note that $T_n(x) = \sum_{k=0}^{n} x\mathrm{e}^{-kx}$ does not converge uniformly to a continuous function on $[0,1]$. •

9.2.3 Interchange of Limit and Differentiation

Consider $f_n(x) = x/(1+nx^2)$ on $\mathbb{R}$. Then

$$|f_n(x)| = \left|\frac{x}{1+nx^2}\right| \le \frac{|x|}{2\sqrt{n}|x|} = \frac{1}{2\sqrt{n}},$$

showing that $f_n(x) \to f(x) = 0$ uniformly on $\mathbb{R}$. Now we compute

$$f_n'(x) = \frac{1-nx^2}{(1+nx^2)^2} \to \begin{cases} 0 & \text{for } x \neq 0, \\ 1 & \text{for } x = 0, \end{cases}$$

and we see that $f_n'(0) = 1$. But $f'(x) = 0$ on $\mathbb{R}$. Thus, we have a sequence of differentiable functions $\{f_n\}$ on E such that (see Figure 9.33):

- $f_n \to f$ uniformly on E;
- f is differentiable on E;
- there exists $x \in E$ with $f'(x) \neq \lim_{n\to\infty} f_n'(x)$, because $f_n'(0) \to 1 \neq f'(0)$.

(See also Questions 9.19(5) and Exercises 9.20(6), in which the limit function is not differentiable although the convergence is uniform.) Thus, even if the limit of a uniformly convergent sequence (respectively series) of differentiable functions on E is differentiable on E, it may happen that the derivative of the limit is not the limit of the sequence (respectively sequence of partial sums) of derivatives of the differentiable functions.

Theorems 9.15 and 9.36 concern integration of sequences. In view of the fundamental theorem of calculus, it seems reasonable to expect a result about differentiation of sequences.

Theorem 9.40 (Differentiation of a sequence of functions). *Suppose that $\{f_n\}$ is a sequence of functions such that:*

(a) $f_n \in C^1[a,b]$;
(b) *there exists a point $x_0 \in [a,b]$ such that $\{f_n(x_0)\}$ converges;*
(c) $f_n' \to g$ *uniformly on $[a,b]$.*

Then $\{f_n\}$ converges uniformly to some f on $[a,b]$ such that $f'(x) = g(x)$ on $[a,b]$.

Proof. By **(c)**, $\{f_n'\}$ is uniformly convergent to g on any closed interval contained in $[a,b]$, say in an interval with endpoints x_0 and x, $x \in [a,b]$. Thus, for all $x \in [a,b]$, we have

$$\begin{aligned}\int_{x_0}^{x} g(t)\,dt &= \lim_{n\to\infty} \int_{x_0}^{x} f_n'(t)\,dt \quad \text{by Theorem 9.15,} \\ &= \lim_{n\to\infty} (f_n(x) - f_n(x_0)) \quad \text{by the fundamental theorem of calculus,}\end{aligned}$$

and the convergence is uniform on $[a,b]$. Since $\lim_{n\to\infty} f_n(x_0)$ exists by **(b)**, we can add this term to both sides and obtain

$$\lim_{n\to\infty} f_n(x) = \int_{x_0}^{x} g(t)\,dt + \lim_{n\to\infty} f_n(x_0) \quad \text{on } [a,b],$$

and the convergence is uniform on $[a,b]$. We may now set $f(x) = \lim_{n\to\infty} f_n(x)$, and the last equation then becomes

$$f(x) = \int_{x_0}^{x} g(t)\,dt + \lim_{n\to\infty} f_n(x_0).$$

Now, g, being the limit of a uniformly convergent sequence of continuous functions on $[a,b]$, is continuous on $[a,b]$, and so by the second fundamental theorem of calculus (Theorem 6.36), $G(x) = \int_{x_0}^{x} g(t)\,dt$ is differentiable and $G'(x) = g(x)$ on $[a,b]$. Therefore, the last inequality implies that

$$f'(x) = g(x), \quad \text{i.e., } f'(x) = \lim_{n\to\infty} f_n'(x) \text{ on } [a,b]. \qquad \blacksquare$$

We may now state a condition under which term-by-term differentiation of an infinite series is permissible.

Corollary 9.41 (Interchange of summation and differentiation). *Suppose that*

(a) $f_n \in C^1[a,b]$;
(b) *there exists a point* $x_0 \in [a,b]$ *such that* $\sum f_k(x_0)$ *converges;*
(c) $\sum f_k'(x)$ *converges uniformly on* $[a,b]$.

Then $\sum f_k$ *converges uniformly on* $[a,b]$ *to a differentiable function* F,

$$F'(x) = \frac{d}{dx}\left(\sum f_k(x)\right) = \sum \frac{d}{dx}(f_k(x)) \quad \text{on } (a,b),$$

and at the endpoints, we express the above equality as

$$F_+'(a) = \sum (f_k)_+'(a) \quad \text{and} \quad F_-'(b) = \sum (f_k)_-'(b).$$

Let us consider an easy example. Consider the series

$$\sum_{k=1}^{\infty} \frac{(-1)^{k-1}}{k} \cos\left(\frac{x}{k}\right), \tag{9.5}$$

which from the alternating series test converges at $x_0 = 0$. Clearly, by Weierstrass's M-test, the derived functional series

$$\sum_{k=1}^{\infty} \frac{(-1)^{k}}{k^2} \sin\left(\frac{x}{k}\right) \tag{9.6}$$

converges uniformly on $\mathbb{R}$. By Corollary 9.41, the given series (9.5) converges uniformly on every finite interval, and so the series (9.5) represents a differentiable function $F(x)$ on $\mathbb{R}$, and $F'(x)$ is the derived series given by (9.6). Because (9.5) converges uniformly on every finite interval, it follows easily that

$$\sum_{k=1}^{\infty} (-1)^{k-1} \sin\left(\frac{x}{k}\right)$$

converges uniformly on any finite interval.

Remark 9.42. The second condition in Corollary 9.41, namely the convergence of the given series at some point, is not superfluous. For instance, consider

$$\sum_{k=1}^{\infty} \cos\left(\frac{x}{k}\right). \tag{9.7}$$

Then the derived functional series

$$-\sum_{k=1}^{\infty} \frac{1}{k} \sin\left(\frac{x}{k}\right)$$

is known to be uniformly convergent on $[-r, r]$ for every $r > 0$, since

$$\left|\frac{1}{k} \sin\left(\frac{x}{k}\right)\right| \leq \frac{|x|}{k^2} \leq \frac{r^2}{k^2} \quad \text{for } |x| \leq r.$$

However, we cannot conclude that the given series (9.7) converges uniformly on $[-r, r]$. Indeed, it can be shown that (9.7) diverges for every x. •

Example 9.43. The series

$$\sum_{k=1}^{\infty} (-1)^{k-1} \frac{x^2 + k}{k^2}$$

converges on $\mathbb{R}$, by the alternating series test. The derived functional series converges uniformly on each $[-r, r]$, $r > 0$. By Corollary 9.41, the given series converges uniformly on every finite closed interval $[a, b]$, and it represents a differentiable function on $[a, b]$. Clearly, since

$$\left|(-1)^{k-1} \frac{x^2 + k}{k^2}\right| = \frac{x^2 + k}{k^2} > \frac{k}{k^2} = \frac{1}{k},$$

the given series does not converge absolutely. •

As a special case of Corollary 9.41, we have the following result, part of which has been proved in Theorem 8.32.

Corollary 9.44 (Term-by-term differentiation of power series). *Let $R > 0$ be the radius of convergence of $f(x) = \sum_{k=0}^{\infty} a_k x^k$. Then f is differentiable on $|x| < R$,*

$$f'(x) = \sum_{k=1}^{\infty} k a_k x^{k-1} \quad \textit{for } |x| < R,$$

and the convergence is uniform on any closed subinterval of $(-R, R)$.

Proof. We have already shown that $\sum_{k=0}^{\infty} a_k x^k$ and $\sum_{k=1}^{\infty} k a_k x^{k-1}$ have the same radius of convergence. Weierstrass's M-test (see Example 9.28) guarantees that the derived series $\sum_{k=1}^{\infty} k a_k x^{k-1}$ converges uniformly on $[-r, r]$ $(r < R)$. Theorem 9.26 implies that $\sum_{k=0}^{\infty} a_k x^k$ converges uniformly to f, where f is differentiable and

$$f'(x) = \sum_{k=1}^{\infty} k a_k x^{k-1} \quad \text{for all } x \in [-r, r]. \qquad \blacksquare$$

For instance, consider the geometric series

$$\frac{1}{1-x} = \sum_{k=0}^{\infty} x^k \quad \text{for } |x| < 1. \tag{9.8}$$

By Corollary 9.44, since the convergence in (9.8) is uniform on $|x| \leq r$, $0 < r < 1$ (see also Example 9.28), we may differentiate term by term and obtain

$$\frac{1}{(1-x)^2} = \sum_{k=1}^{\infty} k x^{k-1} \quad \text{for } |x| < 1.$$

Theorem 9.40 (and hence Corollary 9.41) continues to hold under a weaker hypothesis, namely by weakening the first assumption in Theorem 9.40. However, we cannot replace the third condition, namely, the uniform convergence of the sequence $\{f_n'\}$, with pointwise convergence, as the example $f_n(x) = x^n/n$ demonstrates. Now we state an improved version of Theorem 9.40 and leave its proof as an exercise (see Exercise 9.20(12)).

Theorem 9.45. *Suppose that $\{f_n(x)\}$ is a sequence of functions such that*

(a) *each f_n is differentiable on $[a, b]$;*
(b) *there exists a point $x_0 \in [a, b]$ such that $\{f_n(x_0)\}$ converges;*
(c) *$f_n' \to g$ uniformly on $[a, b]$.*

Then $\{f_n\}$ converges uniformly to some f on $[a, b]$ such that $f'(x) = g(x)$ on $[a, b]$.

Example 9.46. Consider $f_n(x) = (1/n)\log(1 + n^2x^2)$, $x \in [-a, a]$ $(a > 0)$. Then $\{f_n\}$ converges uniformly to $f(x) = 0$ on $[-a, a]$. Then (see Examples 9.11(2))

$$f_n'(x) = \frac{2nx}{1 + n^2x^2} \to g(x) = 0 \quad \text{pointwise on } [-a, a]$$

(see Figures 9.34 and 9.35). Observe that $f'(x) = g(x)$ on $[-a, a]$. •

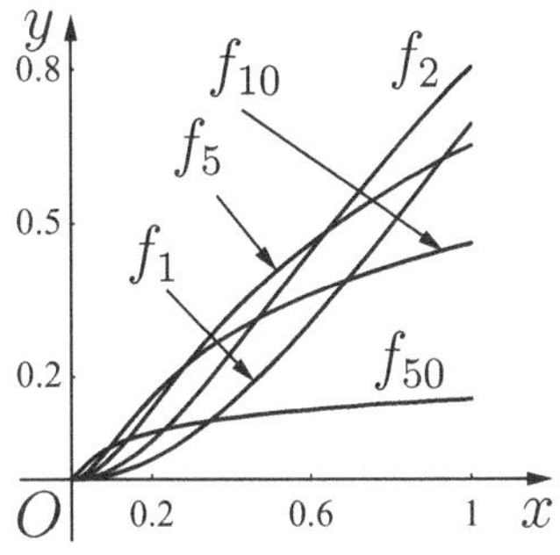

Fig. 9.34. $f_n(x) = (1/n)\log(1 + n^2x^2)$ on $[-1, 1]$.

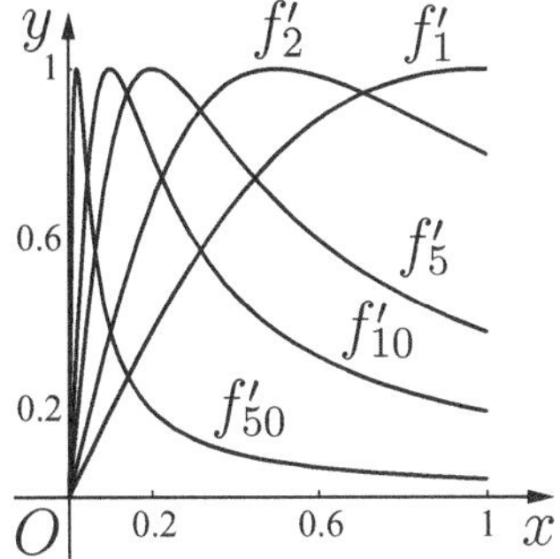

Fig. 9.35. $f'_n(x) = 2nx/(1 + n^2x^2)$ on $[-1, 1]$.

9.2.4 The Weierstrass Approximation Theorem

The proof of the Weierstrass approximation theorem is due to the Russian mathematician S.N. Bernstein, who in 1912 constructed for every continuous function f on $[0, 1]$, an explicit formula for a sequence of polynomials $B_n(f)$ converging to f. These are called the *Bernstein polynomials.*

Definition 9.47. *Let f be a function defined on the closed interval $[0, 1]$. The polynomial $(B_n(f))(x)$ defined by*

$$(B_n(f))(x) := B_n(x) = \sum_{k=0}^{n} f\left(\frac{k}{n}\right)\binom{n}{k}x^k(1-x)^{n-k}, \quad x \in [0, 1],$$

is called the Bernstein polynomial (associated to f) of degree at most n. Here $\binom{n}{k}$ denotes the usual binomial coefficient, defined by

$$\binom{n}{k} = \frac{n!}{k!(n-k)!}.$$

It is easy to see that

$$B_n(f + g) = B_n(f) + B_n(g) \quad \text{and} \quad B_n(\lambda f) = \lambda B_n(f) \ \ (\lambda \in \mathbb{R}).$$

Moreover, $B_n(f) \geq 0$ whenever $f \geq 0$, and therefore the map $f \mapsto B_n(f)$ is linear and positive.

We are now ready to state and prove the Weierstrass approximation theorem in the following form.

Theorem 9.48 (Weierstrass approximation theorem, 1885). *Suppose that $f : [0, 1] \to \mathbb{R}$ is continuous. Then there exists a sequence of polynomials, the Bernstein polynomials $B_n(f)$, such that $\{B_n(f)\}$ converges to f uniformly on $[0, 1]$.*

Before we present the proof of Theorem 9.48, it is essential to offer a few remarks and illustrations. A direct extension of Theorem 9.48 is can be stated thus: Every continuous function on a nonempty compact subset D of the complex plane can be approximated by complex polynomials in z and $\overline{z}$. Moreover, various other extensions of Theorem 9.48 can be found in advanced texts on this topic.

For a proof of the Weierstrass approximation theorem, we need to show that for each continuous function f on $[0,1]$ there exists a sequence of polynomials p_n such that $\lim_{n\to\infty}\left[\sup_{x\in[0,1]}|f(x)-p_n(x)|\right]=0$. Also, we remark that the underlying interval $[0,1]$ in Theorem 9.48 is of no consequence here. The intervals $[0,1]$ and $[-1,1]$ are popular choices, but it hardly matters which interval we choose. In fact, given a continuous function F on $[a,b]$ $(-\infty<a<b<\infty)$, the function f defined by

$$f(t)=F((b-a)t+a), \quad t\in[0,1],$$

is continuous on $[0,1]$. By Theorem 9.48, given $\epsilon>0$ there exists a polynomial p in $[0,1]$ such that

$$\sup_{t\in[0,1]}|f(t)-p(t)|<\epsilon,$$

which is equivalent to

$$\sup_{t\in[a,b]}|F(t)-P(t)|<\epsilon \quad \text{where } P(t)=p((t-a)/(b-a)).$$

Note that the space $\mathcal{P}[a,b]$ of all real polynomials

$$P(x)=a_0+a_1t+\cdots+a_nt^n, \quad n=0,1,2,\dots,$$

on $[a,b]$ is invariant under translation and is an infinite-dimensional subspace of the space of continuous function F on $[a,b]$. Thus, by Theorem 9.48, we can approximate a real-valued continuous function on $[a,b]$ arbitrarily closely in modulus by a real-valued polynomial in $[a,b]$.

Next, we prove a lemma.

Lemma 9.49. *For each $x\in\mathbb{R}$ and $n\in\mathbb{N}$, we have*

$$\sum_{k=0}^{n}\binom{n}{k}x^k(1-x)^{n-k}=1 \tag{9.9}$$

and

$$\sum_{k=0}^{n}\left[\frac{k}{n}-x\right]^2\binom{n}{k}x^k(1-x)^{n-k}\leq\frac{1}{4n}. \tag{9.10}$$

Proof. Let us start with the well-known binomial formula

$$(x+y)^n=\sum_{k=0}^{n}\binom{n}{k}x^ky^{n-k}.$$

Define $f_0 = 1$, $f_1 = x$, $f_2 = x^2$. We first show that

$$B_n f_0 = f_0, \quad B_n f_1 = f_1, \quad B_n f_2 - f_2 = \frac{f_1 - f_2}{n} = \frac{x(1-x)}{n}. \tag{9.11}$$

The binomial formula for $y = 1 - x$ gives (9.9), so that $B_n f_0 = f_0$. Differentiating the binomial formula partially with respect to x and then multiplying the resulting equation by x gives

$$\sum_{k=0}^{n} k \binom{n}{k} x^k y^{n-k} = nx(x+y)^{n-1},$$

and a similar operation on this equation yields

$$\sum_{k=0}^{n} k^2 \binom{n}{k} x^k y^{n-k} = n[(n-1)x(x+y)^{n-2} + (x+y)^{n-1}]x.$$

Substitution of $y = 1 - x$ in the last two identities and division by n and n^2, respectively, gives

$$\sum_{k=0}^{n} \left[\frac{k}{n}\right] \binom{n}{k} x^k (1-x)^{n-k} = x, \quad \text{i.e., } B_n f_1 = f_1, \tag{9.12}$$

and

$$\sum_{k=0}^{n} \left[\frac{k}{n}\right]^2 \binom{n}{k} x^k (1-x)^{n-k} = \left(1 - \frac{1}{n}\right) x^2 + \frac{1}{n} x, \tag{9.13}$$

so that

$$B_n f_2 = \left(1 - \frac{1}{n}\right) f_2 + \frac{1}{n} f_1, \quad \text{i.e., } B_n f_2 - f_2 = \frac{1}{n}(f_1 - f_2).$$

Thus, (9.11) follows. Finally, by (9.9), (9.12), and (9.13), it follows that

$$\sum_{k=0}^{n} \left[\frac{k}{n} - x\right]^2 \binom{n}{k} x^k (1-x)^{n-k} = \left(1 - \frac{1}{n}\right) x^2 + \frac{1}{n} x - 2x^2 + x^2 = \frac{x(1-x)}{n}.$$

Since $x(1-x)$ has the maximum value $1/4$ in $[0,1]$, this gives (9.10). ■

Now we are in a position to prove Theorem 9.48.

Proof. Let f be continuous on $[0,1]$ and $M = \sup_{t \in [0,1]} |f(t)|$. Suppose that $\epsilon > 0$ is given. Since f is uniformly continuous on $[0,1]$, there exists a $\delta > 0$ such that

$$|f(y) - f(x)| < \epsilon \quad \text{whenever } y, x \in [0,1] \text{ and } |y - x| < \delta.$$

Further, because of (9.9), we have

$$
\begin{aligned}
|B_n(f) - f(x)| &= \left| \sum_{k=0}^{n} \left[f\left(\frac{k}{n}\right) - f(x) \right] \binom{n}{k} x^k (1-x)^{n-k} \right| \\
&\leq \sum_{k=0}^{n} \left| f\left(\frac{k}{n}\right) - f(x) \right| \binom{n}{k} x^k (1-x)^{n-k}.
\end{aligned}
$$

Next, we observe that

$$
-\epsilon - \frac{2M}{\delta^2} \left| \frac{k}{n} - x \right|^2 \leq f\left(\frac{k}{n}\right) - f(x) \leq \epsilon + \frac{2M}{\delta^2} \left| \frac{k}{n} - x \right|^2 \tag{9.14}
$$

holds for all $x, k/n \in [0,1]$. Indeed, if $|(k/n) - x| < \delta$, then (9.14) trivially follows from

$$
-\epsilon \leq f\left(\frac{k}{n}\right) - f(x) \leq \epsilon.
$$

On the other hand, if $|(k/n) - x| \geq \delta$, then (9.14) follows from the inequalities

$$
-\frac{2M}{\delta^2} \left| \frac{k}{n} - x \right|^2 \leq -2M \leq f\left(\frac{k}{n}\right) - f(x) \leq 2M \leq \frac{2M}{\delta^2} \left| \frac{k}{n} - x \right|^2.
$$

Thus, for a given $\epsilon > 0$ and given x, we can split the sum into two parts:

(i) those k's in the set $K = \{0, 1, 2, \dots, n\}$ for which

$$
|(k/n) - x| < \delta;
$$

we shall call the subset of K that satisfies the last inequality K_1.

(ii) those k's in K for which

$$
|(k/n) - x| \geq \delta;
$$

we shall call this subset K_2.

Now $K = K_1 \cup K_2$, and we obtain

$$
\begin{aligned}
|B_n(f)(x) - f(x)| &\leq \sum_{k=0}^{n} \left| f\left(\frac{k}{n}\right) - f(x) \right| \binom{n}{k} x^k (1-x)^{n-k} \\
&\leq \epsilon \sum_{k \in K_1} \binom{n}{k} x^k (1-x)^{n-k} \\
&\quad + \frac{2M}{\delta^2} \sum_{k \in K_2} \left[\frac{k}{n} - x \right]^2 \binom{n}{k} x^k (1-x)^{n-k} \\
&< \epsilon \cdot 1 + \frac{2M}{\delta^2} \left(\frac{1}{4n} \right) \quad \text{by (9.9) and (9.10)} \\
&= \epsilon + \frac{M}{2n\delta^2} \leq 2\epsilon \quad \text{whenever } n \geq \frac{M}{2\epsilon\delta^2}.
\end{aligned}
$$

Thus, $\sup_{t\in[0,1]} |B_n f(t) - f(t)| < 2\epsilon$ must hold for all sufficiently large n. Hence,

$$\lim_{n\to\infty} B_n f(t) = f(t)$$

uniformly in $[0,1]$, and the proof is complete. ■

There are several other proofs and extensions (in various forms) of this theorem in the literature. We end the section with an example that illustrates Theorem 9.48.

Example 9.50. Consider the function $f : [0,1] \to \mathbb{R}$ defined by

$$f(x) = |x - c|, \quad c \in [0,1].$$

We provide a direct method of getting a polynomial approximation for f on $[0,1]$. First we let $c \in (0,1/2]$ and write

$$|x-c| = \left\{c^2 - [c^2 - (x-c)^2]\right\}^{1/2} = c(1-y)^{1/2} \quad \text{with } y = 1 - ((x-c)/c)^2,$$

so that the resulting series expansion for $|x-c|$ is given by the series

$$c\sum_{k=0}^{\infty} \frac{(-1/2,k)}{(1,k)} y^k.$$

Here $(a,0) = 1$ for $a \neq 0$, and (a,k) is the ascending factorial notation defined by

$$(a,k) = a(a+1)\cdots(a+k-1).$$

Therefore, we can rewrite the series expansion as

$$c\left[1 - \sum_{k=1}^{\infty} c_k y^k\right],$$

where $c_1 = 1/2$, and for $k \geq 2$,

$$-c_k = \frac{(-1/2,k)}{(1,k)} = -\frac{1}{2}\cdot\frac{1}{4}\cdot\frac{3}{6}\cdot\frac{2k-3}{2k}.$$

Note that $c_k > 0$ for all $k \geq 1$ and $c_k = a_{k-1} - a_k$, where $a_0 = 1$ and

$$a_k = \frac{1\cdot 3\cdot 5\cdots(2k-1)}{2\cdot 4\cdot 6\cdots(2k)},$$

so that

$$\sum_{k=1}^{n} c_k = a_0 - a_n = 1 - a_n < 1.$$

Thus, $\sum_{k=1}^{\infty} c_k < \infty$, and therefore the series

$$1 - \sum_{k=1}^{\infty} c_k y^k$$

converges absolutely and uniformly for $|y| \leq 1$. Equivalently, we say that the series

$$c\left[1 - \sum_{k=1}^{\infty} c_k \left(1 - ((x-c)/c)^2\right)^k\right]$$

converges uniformly to $|x - c|$ whenever

$$\left|1 - \left(\frac{x-c}{c}\right)^2\right| \leq 1,$$

and hence for $|x - c| \leq c$, or equivalently, $x \in [0, 2c]$. Thus, the sequence of polynomials of partial sums converges uniformly to $|x - c|$ on the interval $[0, 2c]$, and a fortiori on $[0, 1]$.

The conclusion for $c \in (1/2, 1)$ is similar if we replace c^2 by $1 - c^2$. ●

9.2.5 Abel's Limit Theorem

Suppose that R is the finite positive radius of convergence of the series $f(x) = \sum_{k=0}^{\infty} a_k x^k$, and r is any positive number such that $r < R$. Then we know that the series converges uniformly on $[-r, r]$, and hence by Corollary 9.22, the function $f(x)$ is continuous on $[-r, r]$. Also, we have seen from examples that the series may or may not converge at the endpoints $\pm R$. Suppose that the series converges at R. Does it follow that the limit function $f(x)$ is (left) continuous at R?

Concerning the behavior of the power series at the endpoints of the interval of convergence, we have the following result, which answers the above question.

Theorem 9.51 (Abel's continuity/limit theorem). *Suppose that R is the finite positive radius of convergence of the series $f(x) = \sum_{k=0}^{\infty} a_k x^k$. If the series converges at $x_0 = R$, then f is (left) continuous at $x_0 = R$ (Also, if it converges at $x_0 = -R$, then f is (right) continuous at $x_0 = -R$.) Moreover, the series converges uniformly on the closed interval with endpoints 0 and x_0.*

Proof. By the transformation $x \mapsto x_0 t$, the point $x_0 = R$ is transformed to $t = 1$, and $|x| < R$ is transformed to $|t| < 1$. Therefore without loss of generality, we may assume that $x_0 = 1$ and $R = 1$ from the beginning. In addition, we may assume that $f(1) = 0$; otherwise, replace a_0 by $a_0 - f(1)$.

Thus we assume that $f(x) = \sum_{k=0}^{\infty} a_k x^k$ converges for $-1 < x \leq 1$ and $f(1) = 0$. Now for convenience, we let

$$s_n(x) = \sum_{k=0}^{n} a_k x^k, \quad s_n(1) = s_n.$$

Since $f(1) = 0$, $s_n \to 0$ as $n \to \infty$. We have $s_0 = a_0$ and $s_k - s_{k-1} = a_k$ for $k \geq 1$. To prove that f is continuous at $x = 1$, we need to show that for any $\epsilon > 0$, there exists a $\delta > 0$ such that

$$|f(x) - f(1)| = |f(x)| < \epsilon \quad \text{if } 1 - \delta < x < 1$$

(we remark that we do not consider $x > 1$ because the domain of $f(x)$ is a subset of $[-1, 1]$). Now for $0 < x < 1$,

$$\begin{aligned} s_n(x) &= a_0 + \sum_{k=1}^{n} (s_k - s_{k-1}) x^k \\ &= \sum_{k=0}^{n-1} s_k x^k - \sum_{k=0}^{n-1} x s_k x^k + s_n x^n \\ &= s_n x^n + \sum_{k=0}^{n-1} (1 - x) s_k x^k. \end{aligned}$$

Since $\lim_{n\to\infty} s_n x^n = \left(\lim_{n\to\infty} s_n\right)\left(\lim_{n\to\infty} x^n\right) = 0$ for $0 < x < 1$, and $\lim_{n\to\infty} s_n(x) = f(x)$, we can allow $n \to \infty$ in the last equation, and we obtain

$$f(x) = (1 - x) \sum_{k=0}^{\infty} s_k x^k \quad \text{for } 0 < x < 1. \tag{9.15}$$

We now look at this identity more closely. Since $s_n \to 0$ as $n \to \infty$, given $\epsilon > 0$, there exists an N such that $|s_n| < \epsilon/2$ for all $n \geq N$. Set $M = \max_{0 \leq k \leq N-1} |s_k|$ and $F_N(x) = M \sum_{k=0}^{N-1} (1 - x) x^k$. From (9.15) we obtain

$$\begin{aligned} |f(x)| &\leq \sum_{k=0}^{N-1} |s_k| (1 - x) x^k + \sum_{k=N}^{\infty} |s_k| (1 - x) x^k \\ &\leq F_N(x) + \sum_{k=N}^{\infty} \frac{\epsilon}{2} (1 - x) x^k \\ &\leq F_N(x) + \frac{\epsilon}{2} \quad \text{for } 0 < x < 1. \end{aligned}$$

Since $F_N(x)$ is continuous and $F_N(1) = 0$, there exists a $\delta > 0$ such that

$$F_N(x) < \frac{\epsilon}{2} \quad \text{whenever } 1 - \delta < x < 1.$$

Thus, for $1-\delta < x < 1$, we have

$$|f(x)| \leq \frac{\epsilon}{2} + \frac{\epsilon}{2} = \epsilon,$$

showing that f is (left) continuous at $x = 1$. ∎

Example 9.52. We know that

$$\log(1+x) = \sum_{k=1}^{\infty} \frac{(-1)^{k-1}}{k} x^k, \quad |x| < 1,$$

and the series converges at $x = 1$. Therefore, by Abel's limit theorem,

$$\sum_{k=1}^{\infty} \frac{(-1)^{k-1}}{k} = \lim_{x \to 1-} \log(1+x) = \log 2.$$

A similar conclusion may be obtained for the series (see Example 9.33)

$$\arctan x = \sum_{k=0}^{\infty} \frac{(-1)^k}{2k+1} x^{2k+1} \quad \text{for } |x| < 1.$$

Because the series converges at $x = \pm 1$ by the alternating series test, Theorem 9.51 shows that the series converges uniformly for $|x| \leq 1$. Thus, allowing $x \to 1-$ or $x \to -1+$, we have

$$\frac{\pi}{4} = 1 - \frac{1}{3} + \frac{1}{5} - \frac{1}{7} + \cdots,$$

which is again a series representation of π, and the series here is referred to as Madhava-Leibniz-Gregory's series (see also Examples 9.32 and 10.16). ●

The most important case of Abel's limit theorem occurs when $R = 1$.

Corollary 9.53. *Suppose that the series $\sum_{k=0}^{\infty} a_k$ is convergent with sum A. Then $A = \lim_{x \to 1-} \sum_{k=0}^{\infty} a_k x^k$.*

Proof. By assumption, the series $\sum_{k=0}^{\infty} a_k x^k$ converges for $|x| < 1$. If the limit function is $f(x)$, then by Theorem 9.51, f is continuous at $x = 1$ and

$$A = \lim_{x \to 1-} f(x).$$ ∎

Abel's limit theorem has another interesting application.

Theorem 9.54 (Abel). *If $\sum_{k=0}^{\infty} a_k$ and $\sum_{k=0}^{\infty} b_k$ are convergent with sum A and B, respectively, and if the Cauchy product of these series is convergent, then its sum must be AB.*

Proof. Because the power series $\sum_{k=0}^{\infty} a_k x^k$ and $\sum_{k=0}^{\infty} b_k x^k$ converge at $x = 1$, it follows from Lemma 8.16 that they converge absolutely for $|x| < 1$. We know that if $\sum_{k=0}^{\infty} c_k$ is the Cauchy product of $\sum_{k=0}^{\infty} a_k$ and $\sum_{k=0}^{\infty} b_k$, then

$$\sum_{k=0}^{\infty} c_k x^k = \left(\sum_{k=0}^{\infty} a_k x^k\right)\left(\sum_{k=0}^{\infty} b_k x^k\right) \quad \text{for } |x| < 1.$$

Thus, if $\sum_{k=0}^{\infty} c_k$ converges, then by Abel's limit theorem we obtain the desired conclusion. ∎

9.2.6 Abel's Summability of Series and Tauber's First Theorem

Definition 9.55. *A series $\sum_{k=1}^{\infty} a_k$ is said to be* Abel summable *to A if the associated series $\sum_{k=0}^{\infty} a_k x^k$ converges for $|x| < 1$ to a function $f(x)$ and*

$$\lim_{x \to 1-} f(x) := \lim_{x \to 1-} \sum_{k=0}^{\infty} a_k x^k = A.$$

In this case, A is called the Abel sum *of the series $\sum_{k=0}^{\infty} a_k$, and we write*

$$\sum_{k=1}^{\infty} a_k = A \quad \text{(Abel)}.$$

Corollary 9.53 shows that ordinary convergence of a series implies Abel convergence. That is,

$$\sum_{k=1}^{\infty} a_k = A \quad \text{implies} \quad \sum_{k=1}^{\infty} a_k = A \quad \text{(Abel)}.$$

Example 9.56 (There are divergent series that are Abel summable).

(a) Consider $\sum_{k=0}^{\infty} (-1)^k$. Then the associated series is

$$\sum_{k=0}^{\infty} (-1)^k x^k,$$

which is convergent for $|x| < 1$ to the function f given by

$$f(x) = \frac{1}{1+x}.$$

Since $\lim_{x \to 1-} f(x) = 1/2$, we conclude that $\sum_{k=0}^{\infty} (-1)^k$ is Abel summable to $1/2$, but is not a convergent series. This example demonstrates that the converse of Abel's limit theorem is false in general.

(b) Consider the divergent series $\sum_{k=1}^{\infty}(-1)^k k$. Its associated series

$$\sum_{k=1}^{\infty}(-1)^k k x^k$$

is convergent for $|x| < 1$ to the function f given by

$$f(x) = \frac{-x}{(1+x)^2}.$$

Since $\lim_{x\to 1-} f(x) = -1/4$, we conclude that $\sum_{k=1}^{\infty}(-1)^k k$ is Abel summable to $-1/4$. ●

It is natural to ask under what conditions on a_n one can be certain of the convergence of the series $\sum_{k=0}^{\infty} a_k$. There are several results that deal with this question. In the next theorem, we place certain conditions on a_n to obtain a converse to Abel's limit theorem. There are many results of this type, and they are referred to as *Tauberian theorems.*

Theorem 9.57 (Tauber's first theorem). *Suppose that $\sum_{k=0}^{\infty} a_k$ is Abel summable to A, and $na_n \to 0$ as $n \to \infty$. Then $\sum_{k=0}^{\infty} a_k$ is convergent with sum A.*

Proof. Assume the hypotheses. Let $\epsilon > 0$ be given. Then there exists a function $f(x)$ such that

- $\sum_{k=0}^{\infty} a_k x^k = f(x)$ for $|x| < 1$,
- $\lim_{x\to 1-} f(x) = A$; i.e., $|f(1-1/n) - A| < \epsilon/3$ for large n,
- $na_n \to 0$ as $n \to 0$; i.e., $n|a_n| < \epsilon/3$ for large n, and so by Theorem 2.64, it follows that

$$\frac{1}{n}\sum_{k=1}^{n} k|a_k| < \frac{\epsilon}{3} \quad \text{for large } n.$$

We need to show that $S_n \to A$, $S_n = \sum_{k=0}^{n} a_k$. We write

$$\begin{aligned} S_n - A &= \sum_{k=0}^{n} a_k - A + f(x) - \sum_{k=0}^{\infty} a_k x^k \\ &= f(x) - A + \sum_{k=1}^{n} a_k(1-x^k) - \sum_{k=n+1}^{\infty} a_k x^k. \end{aligned}$$

Since $1 - x^k = (1-x)(1+x+x^2+\cdots+x^{k-1}) \le k(1-x)$ for $0 < x < 1$, and $n \le k$ for all $k \ge n+1$, it follows that for $x \in (0,1)$,

$$|S_n - A| \le |f(x) - A| + \sum_{k=1}^{n} k|a_k|(1-x) + \frac{1}{n}\sum_{k=1+n}^{\infty} k|a_k|x^k. \tag{9.16}$$

The assumptions imply that for a given $\epsilon > 0$, there exists an N such that

$$\left|f\left(1-\frac{1}{n}\right)-A\right| < \frac{\epsilon}{3}, \quad n|a_n| < \frac{\epsilon}{3}, \quad \text{and} \quad \sum_{k=1}^{n} k|a_k| < \frac{n\epsilon}{3} \quad \text{for all } n \geq N.$$

Thus for $n \geq N$ and $x = 1 - 1/n$, (9.16) becomes

$$|S_n - A| < \frac{\epsilon}{3} + \frac{1}{n}\left(\frac{n\epsilon}{3}\right) + \frac{1}{n}\left(\frac{\epsilon}{3}\right) \sum_{k=1+n}^{\infty} x^k = \epsilon$$

(where in the last step, we have used the fact that $\sum_{k=n+1}^{\infty} x^k \leq 1/(1-x) = n$). The conclusion follows. ∎

A counterpart of Theorem 9.57 for Cesàro summable series has already been discussed in Theorem 5.67.

9.2.7 (C, α) Summable Sequences

For $\alpha \geq 1$, we have

$$\frac{x}{(1-x)^{\alpha}} = \sum_{n=1}^{\infty} A_n^{(\alpha)} x^n,$$

where

$$A_n^{(\alpha)} := \frac{\alpha(\alpha+1)\cdots(\alpha+n-2)}{(n-1)!} \quad \text{for } n \geq 1,$$

which for $\alpha = m \in \mathbb{N}$, takes the form

$$A_n^{(m)} = \frac{(n+m-2)!}{(m-1)!(n-1)!} = \binom{n+m-2}{n-1}.$$

In particular, we have

$$A_n^{(1)} = 1, \quad A_n^{(2)} = n, \quad A_n^{(3)} = \frac{n(n+1)}{2}, \quad \text{and so on.}$$

Now we consider the identity

$$\frac{1}{1-x} \cdot \frac{x}{(1-x)^{\alpha}} = \frac{x}{(1-x)^{\alpha+1}},$$

which may be rewritten in terms of power series (use the definition of Cauchy product) as

$$\sum_{n=1}^{\infty} \left(\sum_{k=1}^{n} A_{n+1-k}^{(\alpha)} \right) x^n = \sum_{n=1}^{\infty} A_n^{(\alpha+1)} x^n.$$

Comparing the coefficients of x^n on both sides, we get

$$\frac{1}{A_n^{(\alpha+1)}}\sum_{k=1}^{n} A_{n+1-k}^{(\alpha)} = 1.$$

This basic property suggests that for a given sequence of real numbers $\{s_n\}_{n\geq 1}$, we can consider the mean

$$\sigma_n^{(\alpha)} := \frac{1}{A_n^{(\alpha+1)}}\sum_{k=1}^{n} A_{n+1-k}^{(\alpha)} s_k.$$

For $\alpha = m \in \mathbb{N}$, this formula takes the form

$$\sigma_n^{(m)} = \frac{1}{\binom{n+m-1}{n-1}}\sum_{k=1}^{n}\binom{n+m-1-k}{n-k} s_k.$$

The cases $m = 1, 2$ lead to

$$\sigma_n^{(1)} := \sigma_n = \frac{1}{n}\sum_{k=1}^{n} s_k \quad \text{and} \quad \sigma_n^{(2)} = \frac{2}{n(n+1)}\sum_{k=1}^{n}(n+1-k)s_k,$$

respectively. The above discussion helps to introduce the following.

Definition 9.58. *If $\{s_n\}_{n\geq 1}$ is a sequence of real numbers, then we say that $\{s_n\}_{n\geq 1}$ is (C,α) summable to L if the new sequence $\{\sigma_n^{(\alpha)}\}_{n\geq 1}$ converges to L. In this case, we write*

$$s_n \to L \quad (C,\alpha) \quad \text{or} \quad \lim_{n\to\infty} s_n = L \quad (C,\alpha).$$

We have shown that all convergent sequences are $(C,1)$ summable. Now we show that all $(C,1)$ summable sequences are $(C,2)$ summable. More precisely, we have the following.

Theorem 9.59. *If $s_n \to x$ $(C,1)$, then $s_n \to x$ $(C,2)$.*

Proof. First we observe that $n\sigma_n = \sum_{k=1}^{n} s_k$, so that

$$s_k = k\sigma_k - (k-1)\sigma_{k-1} \quad (\sigma_0 = 0). \tag{9.17}$$

Next, we note that

$$\begin{aligned}\frac{n(n+1)}{2}\sigma_n^{(2)} &= ns_1 + (n-1)s_2 + (n-2)s_3 + \cdots + s_n \\ &= s_1 + (s_1+s_2) + (s_1+s_2+s_3) + \cdots + (s_1+s_2+\cdots+s_n) \\ &= \sum_{k=1}^{n} k\sigma_k\end{aligned}$$

and

$$\frac{n(n+1)}{2}(\sigma_n^{(2)} - x) = \sum_{k=1}^{n} k(\sigma_k - x).$$

The last relation clearly implies that it suffices to prove the theorem for the case $x = 0$. So we let $s_n \to 0$ $(C, 1)$. Then $\sigma_n \to 0$ as $n \to \infty$. Therefore, given $\epsilon > 0$, there exists an $N \in \mathbb{N}$ such that $|\sigma_n| < \epsilon/2$ for all $n > N$. Now for $n > N$,

$$\begin{aligned}
|\sigma_n^{(2)}| &= \left| \frac{2}{n(n+1)} \sum_{k=1}^{n} k\sigma_k \right| \\
&\leq \frac{2}{n(n+1)} \left[\sum_{k=1}^{N} k|\sigma_k| + \sum_{k=N+1}^{n} k|\sigma_k| \right] \\
&= \frac{2}{n(n+1)} \left[M \sum_{k=1}^{N} k + \frac{\epsilon}{2} \sum_{k=N+1}^{n} k \right], \quad M = \max_{k=1,2,\ldots,N} |\sigma_k|, \\
&< \frac{2}{n(n+1)} \frac{MN(N+1)}{2} + \frac{\epsilon}{2}.
\end{aligned}$$

Note that M and N are independent of n, and $1/n \to 0$ as $n \to \infty$. Consequently, given $\epsilon > 0$, there exists an N_1 $(> N)$ such that

$$|\sigma_n^{(2)}| < \frac{\epsilon}{2} + \frac{\epsilon}{2} = \epsilon \quad \text{for all } \; n \geq N_1,$$

and so $\sigma_n^{(2)} \to 0$ whenever $\sigma_n \to 0$. ∎

Equation (9.17) gives the following simple result.

Corollary 9.60. *If* $\{s_n\}_{n\geq 1}$ *is a* $(C, 1)$ *summable sequence of real numbers, then* $\{s_n\}$ *converges to* 0.

Using this corollary, it follows that $1, -1, 2, -2, 3, -3, \ldots$ is not $(C, 1)$ summable, because

$$s_n = \begin{cases} \dfrac{n+1}{2} & \text{if } n \text{ is odd,} \\ -\dfrac{n}{2} & \text{if } n \text{ is even.} \end{cases}$$

9.2.8 Questions and Exercises

Questions 9.61.

1. Must the (pointwise) limit of a convergent sequence of integrable functions on E be integrable?

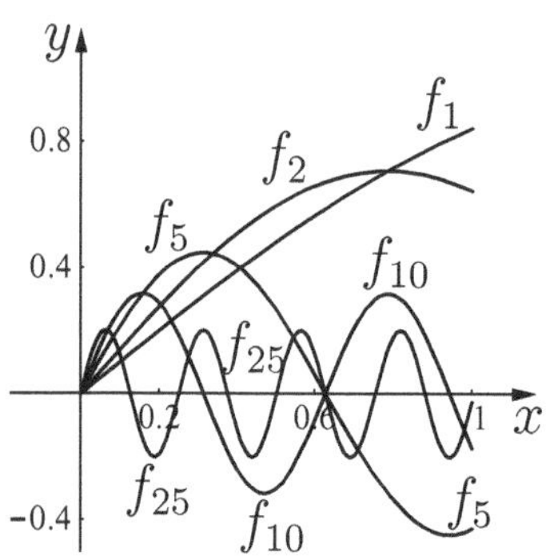

Fig. 9.36. Graph of $f_n(x) = (\sin nx)/\sqrt{n}$.

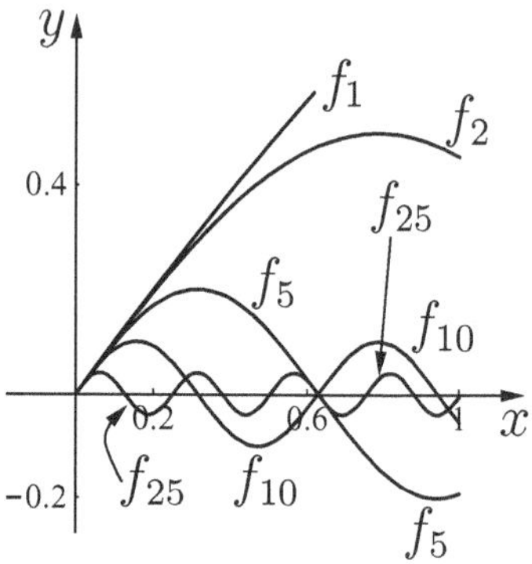

Fig. 9.37. Graph of $f_n(x) = (\sin nx)/n$.

2. Must the limit of a uniformly convergent sequence of integrable functions on $[a, b]$ be integrable?
3. Can the limit of a sequence of continuous functions on a set E be unbounded on E?
4. Must the (pointwise) limit of a convergent sequence of differentiable functions be continuous?
5. Does $\{x^n/n\}$ converge uniformly to $f(x) = 0$ on $-1 \leq x \leq 1$? Does $\{x^{n-1}\}$ converge to $f'(x) = 0$ on $-1 < x < 1$? Does $\{x^{n-1}\}$ converge uniformly to $f'(x) = 0$ on $-1 < x < 1$?
6. If $f_n(x) = (1/\sqrt{n}) \sin(nx)$, do we have $f_n \to 0$ uniformly on $\mathbb{R}$? What is the limit of the sequence $\{f_n'(0)\}$ if it exists? What is the limit of the sequence $\{f_n'(\pi)\}$ if it exists? What does this convey (see Figure 9.36)?
7. If $f_n(x) = (1/n) \sin(nx)$, do we have $f_n \to 0$ uniformly on $\mathbb{R}$? What is the limit of the sequence $\{f_n'(0)\}$? What is the limit of the sequence $\{f_n'(\pi)\}$ (see Figure 9.37)?
8. Can a sequence of Riemann integrable functions converge pointwise to a function that is not Riemann integrable? How about if the pointwise convergence is replaced by uniform convergence?
9. If $\sum_{k=1}^{\infty} f_k(x)$ converges uniformly on a set E, must the sequence $\{f_k\}$ be convergent on E? Must the sequence $\{f_k\}$ be uniformly convergent on E?
10. Suppose that $\sum_{k=1}^{\infty} f_k(x)$ converges to a function $f(x)$ on an interval I such that $\sup_{x \in I} |\sum_{k=1}^{n} f_k(x) - f(x)|$ does not approach to 0 as $n \to \infty$. Must the series be uniformly convergent on I?
11. If $\sum_{k=1}^{\infty} a_k$ converges absolutely, must $\sum_{k=1}^{\infty} a_k \cos kx$ and $\sum_{k=1}^{\infty} a_k \sin kx$ be convergent on $\mathbb{R}$?
12. Let $\{f_k\}$ be a sequence of constant functions. What does it mean in this case to say that the sequence $\{f_k\}$ converges uniformly on $[a, b]$? What does it mean in this case to say that the series $\sum f_k$ converges uniformly on $[a, b]$?

13. Suppose that $\{f_n\}$ is a sequence of bounded functions converging uniformly to f on E. Must f be bounded on E?
14. If $\sum_{k=1}^{\infty} |f_k|$ converges uniformly on E, must $\sum_{k=1}^{\infty} f_k$ be uniformly convergent on E? How about its converse?
15. If $\sum_{k=0}^{\infty} a_k$ is absolutely convergent, must we ahve

$$\int_0^1 \left(\sum_{k=0}^{\infty} a_k x^k\right) dx = \sum_{k=0}^{\infty} \frac{a_k}{k+1}?$$

16. Consider the series $\sum_{k=1}^{\infty}(-1)^{k-1}((x+k)/k^2)$. Does it converge on $\mathbb{R}$? If not, does it converge pointwise on $[0,\infty)$? Does it converge uniformly on $[0,R]$, $R>0$? Does it converge absolutely for some $x \in \mathbb{R}$?
17. Why can't we apply Abel's theorem (Theorem 9.51) for $\sum_{k=0}^{\infty} x^k$ at $x=1$ or $x=-1$?

Exercises 9.62.

1. Let f be continuous on $[0,1]$ and $g_n(x) = f(x^n)$ for $n \geq 1$. Verify whether $\{\int_0^1 g_n(x)\, dx\}_{n\geq 1}$ converges to $f(0)$.
2. Does the sequence $\{f_n\}$, $f_n(x) = nx/(1+n^2x^2)$, converge uniformly on every interval I containing 0? If f is the pointwise limit of $\{f_n\}$ on I, is f differentiable on I? If so, does $f_n'(x) \to f'(x)$ on E? Justify your answer.
3. For $x \in [0,1]$, consider $\sum_{k=1}^{\infty} f_k(x)$, where

$$f_n(x) = \frac{x^{2n-1}}{2n-1} - \frac{x^n}{2n}.$$

   (a) Does the sequence $\{S_n(x)\}$ of partial sums converge pointwise on $[0,1]$? If so, find its limit.
   (b) Does the series converge uniformly on $[0,1]$? Justify your answer.
4. Show that $\sum_{k=1}^{\infty}(x^k/k)$ does not converge uniformly on $(0,1)$.
5. Investigate the pointwise and uniform convergence of $\sum_{k=1}^{\infty} f_k(x)$ on $[0,1]$ if $f_k(x)$ equals
   (a) $\dfrac{x^k(1-x)}{k^2}$. (b) $\dfrac{x^k(1-x)}{k}$. (c) $\dfrac{\log(k+x)-\log k}{k}$.
6. Repeat Exercise 9.62(5) but for all real x when $f_k(x)$ equals
   (a) $\dfrac{1}{k^3+k^4x^2}$. (b) $\dfrac{x}{k+k^2x^2}$.
7. Show that the series

$$\sum_{k=1}^{\infty} \frac{x}{(1+(k-1)x)(1+kx)}$$

converges uniformly on $[r,R]$, for every finite $r>0$ and $R>r$, but does not converge uniformly on $[0,R]$. Does it converge pointwise on $[0,\infty)$?

8. If $f_k(x) = 1/(k^3(1+kx^3))$ on $[0,1]$, justify with a proof that
$$\frac{\mathrm{d}}{\mathrm{d}x}\Big(\sum_{k\geq 1} f_k(x)\Big) = \sum_{k\geq 1} f_k'(x) = -3x^2 \sum_{k\geq 1} \frac{1}{k^2(1+kx^3)^2} \quad \text{on } [0,1].$$

9. Examine the uniform convergence of the following series:

(i) $\displaystyle\sum_{n=1}^{\infty} \frac{x}{n(1+nx^2)},\ x \in \mathbb{R}.$ **(ii)** $\displaystyle\sum_{n=1}^{\infty} \frac{1}{n^3+n^4x^2},\ x \geq 0.$

(iii) $\displaystyle\sum_{n=1}^{\infty} \frac{x}{1+n^2x^2},\ 0 < c \leq x < \infty.$ **(iv)** $\displaystyle\sum_{n=0}^{\infty} \frac{1}{1+n^2+n^2x^2},\ x \geq 0.$

(v) $\displaystyle\sum_{n=1}^{\infty} \frac{1}{n^x},\ c \leq x < \infty,\ c > 1.$ **(vi)** $\displaystyle\sum_{n=1}^{\infty} \frac{1}{(n+x)^2},\ x \geq 0.$

(vii) $\displaystyle\sum_{n=1}^{\infty} \frac{\sin nx}{e^n},\ -\infty < x < \infty.$ **(viii)** $\displaystyle\sum_{n=1}^{\infty} \frac{e^{nx}}{5^n},\ -\infty < x \leq 0.$

(ix) $\displaystyle\sum_{n=1}^{\infty} \left(\frac{\log x}{x}\right)^n,\ 1 \leq x < \infty.$ **(x)** $\displaystyle\sum_{n=1}^{\infty} (x \log x)^n,\ 0 < x \leq 1.$

10. Find the interval of convergence of the following power series:

(i) $x + \frac{x^2}{3} + \frac{x^3}{5} + \frac{x^4}{7} + \cdots.$

(ii) $(x+1) + 4(x+1)^2 + 9(x+1)^3 + \cdots.$

(iii) $-\frac{1}{x} + \frac{1}{2x^2} - \frac{1}{3x^3} + \cdots.$

(iv) $\displaystyle\sum_{n=1}^{\infty} \frac{(-1)^n 3^n}{(4n-1)x^n}.$

(v) $\displaystyle\sum_{n=1}^{\infty} \frac{n(x+5)^n}{(2n+1)^3}.$

(vi) $\displaystyle\sum_{n=1}^{\infty} \frac{2^n(\sin x)^n}{n^2}.$

11. Show that the series $\sum_{k=1}^{\infty}(1/k^2)\sin(kx)$ is uniformly convergent on $\mathbb{R}$. If $f(x)$ is the sum of the series, then determine the value of $\int_0^{\pi} f(x)\,\mathrm{d}x$.
12. Prove Theorem 9.45.
13. Integrate suitable series to evaluate the sum

(a) $\displaystyle\sum_{k=1}^{\infty} \frac{1}{k(k+1)(k+2)}.$ **(b)** $\displaystyle\sum_{k=0}^{\infty} \frac{(-1)^k}{3k+1}.$

14. Differentiate suitable series to evaluate the sum

(a) $\displaystyle\sum_{k=0}^{\infty} \frac{(k+2)^2}{k!}.$ **(b)** $\displaystyle\sum_{k=1}^{\infty} \frac{k(k+1)(k+2)}{3^k}.$ **(c)** $\displaystyle\sum_{k=1}^{\infty} \frac{k(k+1)(k+2)}{k!}.$

15. Evaluate the integrals

(a) $\displaystyle\int_0^1 \frac{\log(1+x)}{x}\,\mathrm{d}x.$ **(b)** $\displaystyle\int_0^1 \log(1-x)\,\mathrm{d}x.$

16. If the power series $\sum_{k=0}^{\infty} a_k x^k$ converges to $f(x)$ for $|x| < 1$, then show that

$$\sum_{n=0}^{\infty} \left(\sum_{k=0}^{n} a_k \right) x^n = \frac{f(x)}{1-x} \quad \text{for } |x| < 1.$$

Deduce that

$$\frac{\log(1+x)}{1-x} = x + \left(1 - \frac{1}{2}\right)x^2 + \left(1 - \frac{1}{2} + \frac{1}{3}\right)x^3 + \cdots .$$

17. Show that

$$(\log(1+x))^2 = 2\sum_{k=2}^{\infty} (-1)^k \frac{s_{k-1}}{k} x^k \quad \text{for } |x| < 1,$$

where $s_n = \sum_{k=1}^{n} (1/k)$.

18. Show that a divergent series of positive numbers cannot be Abel summable.

10

Fourier Series and Applications

In a general course on functional analysis, one discusses a very general method of "Fourier analysis" in Hilbert space settings. Originally, the methods originated with the classical setting of real- or complex-valued periodic functions defined on the whole of $\mathbb{R}$. In this chapter we focus our attention mainly on describing the elementary theory of classical Fourier series (with the help of specific kernels) which have become indispensable tools in the study of periodic phenomena in physics and engineering. These kernels are mainly used to prove the convergence of Fourier series, and the study of Fourier series has led to many important problems and theories in the mathematical sciences. As a result of the introduction of Fourier series, much of the development of modern mathematics has been influenced by the theory of trigonometric series. We ask a number of questions concerning the nature of Fourier series and provide answers to these questions.

In Section 10.1 we shall discuss what a Fourier series is and present a number of examples to demonstrate the use of Fourier series, such as how a given function can be represented in terms of a series of sine and cosine functions. In Section 10.2, we discuss half-range series, which, as the name suggests, are defined over half of the normal range. Finally, we address the basic questions concerning the convergence of Fourier series.

10.1 A Basic Issue in Fourier Series

One of the fundamental methods of solving many problems in engineering fields such as mechanics, electronics, acoustics, and in most of the areas of applied mathematics such as ordinary and partial differential equations is to represent the *behavior* of a system by a combination of *simple behaviors*. Mathematically, this is related to representing a function $f(x)$ in the form of a functional series

$$f(x) = \sum_{k=1}^{\infty} c_k \phi_k(x).$$

S. Ponnusamy, *Foundations of Mathematical Analysis*,
DOI 10.1007/978-0-8176-8292-7_10,

Here the $\phi_k(x)$ are suitable *elementary functions*, also called the *base set* of functions, and the c_k are called the *coefficients of the expansion*. For instance, suppose one wants to write a program for a pocket calculator that has limited memory. Then, for example, for the function $f(x) = \cos x$, the calculator may just store n coefficients $c_1, c_2, \ldots, c_n$, and in this situation, one also needs to specify the best-suited $\phi_k(x)$ for the purpose, for example the Chebyshev polynomials. Other special functions such as Bessel functions, Legendre polynomials, and Hermite polynomials correspond to different coefficients $c_1, c_2, \ldots, c_n$. as another example, consider the familiar Taylor series expansion of $f(x)$ (e.g., e^x, $\sin x$, $\cos x$, $1/(1-x)$) of the form

$$f(x) = \sum_{k=0}^{\infty} c_k x^k, \quad |x| < R.$$

Here the set $\{1, x, \ldots, x^n, \ldots\}$ is considered a base set (or set of building blocks) for the Taylor series about the origin. If such an expansion is possible, then f must be infinitely differentiable on $|x| < R$, and the coefficients c_k are given by $f^{(k)}(0)/k!$, which shows that the Taylor series expansion can be used in the representation of only a rather small class of functions. Our main interest in this chapter is to discuss another type of expansion, the *Fourier series expansion*. The Fourier series is named after Jean Baptiste Joseph Fourier (1768–1830). Roughly speaking, a Fourier series expansion of a function is a representation of the function as a linear combination of sines and cosines, that is, the base set of the representation is $\{1, \cos nx, \sin nx\}_{n\geq 1}$ instead of $\{1, x, \ldots, x^n, \ldots\}$. An important difference between Fourier series and Taylor series is that Fourier series can be used to represent and approximate noncontinuous functions as well as continuous ones, whereas Taylor series are applicable only to infinitely differentiable functions.

Further, Fourier series are widely used to represent functions including solutions of partial differential equations, and also to approximate functions defined on an arbitrary (but finite) interval $[a, b]$. Usually, but not exclusively, we use Fourier series to represent periodic functions. We begin with some preliminary results about periodic functions.

10.1.1 Periodic Functions

A function $f : \Omega \subseteq \mathbb{R} \to \mathbb{R}$ is said to be *periodic* if there exists a nonzero real number ω such that

$$f(x) = f(x + \omega) \quad \text{for all } x \in \Omega. \tag{10.1}$$

The simplest examples of periodic functions from $\mathbb{R}$ into $\mathbb{R}$ include the well-known sine and cosine functions, since for each $k \in \mathbb{Z}\backslash\{0\}$,

$$\cos x = \cos(x + 2k\pi) \quad \text{and} \quad \sin x = \sin(x + 2k\pi) \quad \text{for every } x \in \mathbb{R}.$$

Moreover, for each integer n, the functions $\cos nx$ and $\sin nx$ are also periodic.

The complex-valued function $f : \mathbb{R} \to \mathbb{C}$ defined by $f(x) = e^{ix}$ is a periodic function, since

$$f(x) = f(x + 2k\pi) \quad \text{for each } k \in \mathbb{Z}\backslash\{0\} \text{ and for every } x \in \mathbb{R}.$$

In these examples, $\omega = 2k\pi$ for some $k \in \mathbb{Z}\backslash\{0\}$, and in all these cases, ω is a real quantity satisfying (10.1). However, if $f : \mathbb{C} \to \mathbb{C}$ is defined by $f(z) = e^z$, then from a basic result in complex analysis it follows that

$$f(z) = f(z + 2k\pi i) \quad \text{for each } k \in \mathbb{Z}\backslash\{0\} \text{ and for every } x \in \mathbb{R},$$

showing that $\omega = 2k\pi i$ for some $k \in \mathbb{Z}\backslash\{0\}$, which is a complex number. In such cases, one often says that the function has a *complex period.*

Observe that if $\omega = \omega_1$ and $\omega = \omega_2$ satisfy (10.1), then so does $\omega_1 \pm \omega_2$, since

$$f(x + (\omega_1 \pm \omega_2)) = f((x + \omega_1) \pm \omega_2) = f(x + \omega_1) = f(x) \quad \text{for every } x \in \mathbb{R}.$$

In most cases of interest, there is a smallest positive value ω of a periodic function f called the *primitive period* (or the basic period or the fundamental period) of $f(x)$. The reciprocal of the primitive period is called the *frequency* of the periodic function. Henceforth, in speaking about the *period* of a function, we mean the primitive period, and we consider only functions that have a real period. A periodic function f of period ω will be called an ω-periodic function. For example, $\sin x$ and $\cos x$ are 2π-periodic functions. The graphs of $\sin x$ and $\cos x$ are illustrated in Figure 10.1. The graphs of some nontrivial

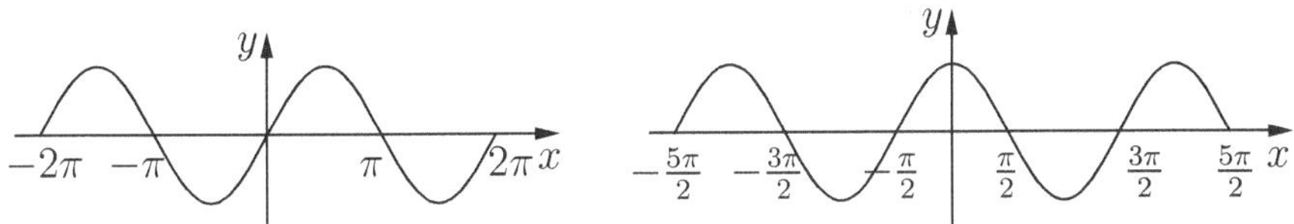

Fig. 10.1. Graphs of the periodic functions $\sin x$ and $\cos x$.

periodic functions are illustrated in Figures 10.2 and 10.3. We remark that

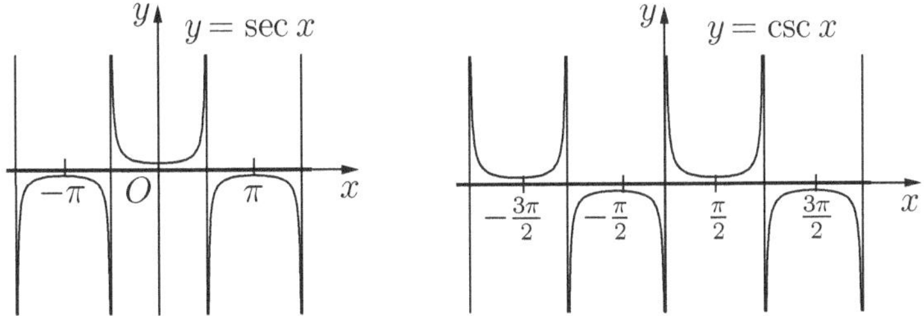

Fig. 10.2. Graphs of periodic functions $y = \sec x$ and $y = \csc x$.

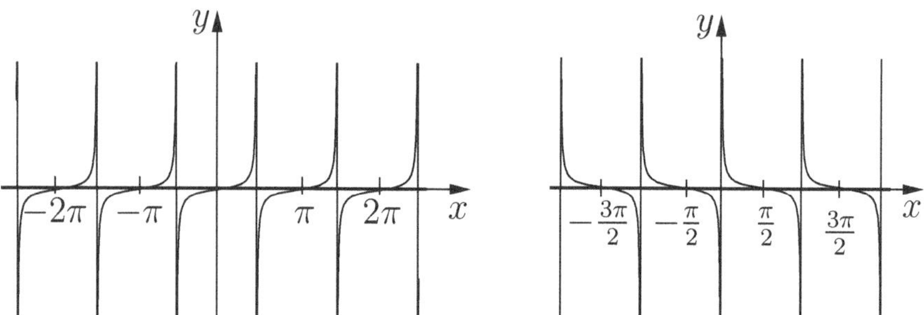

Fig. 10.3. Graphs of periodic functions $y = \tan x$ and $y = \cot x$.

if $f(x)$ is a constant, then every $\omega > 0$ satisfies (10.1) but has no smallest ω satisfying (10.1). Thus, constant functions do not have a primitive period.

The following simple lemmas and remarks will be useful in the sequel.

Lemma 10.1. *If $f : \mathbb{R} \to \mathbb{R}$ is a periodic function with period ω, then the period of $f(cx)$ is ω/c. If $f(x)$ and $g(x)$ are periodic with the same period ω, then $h(x) = af(x) + bg(x)$ is also is periodic with period ω. Here ω is not necessarily a primitive period.*

Proof. Let $\phi(x) = f(cx)$. Then

$$\phi(x) = f(cx) = f(cx + \omega) = f(c(x + \omega/c)) = \phi(x + \omega/c) \quad \text{for every } x \in \mathbb{R}.$$

This shows that ω/c is a period. For the proof of the second part, we simply note that

$$h(x + \omega) = af(x + \omega) + bg(x + \omega) = af(x) + bg(x) = h(x),$$

and thus $h(x)$ also has the period ω. ∎

For instance,

- $\sin(cx)$ and $\cos(cx)$ are periodic functions with period $2\pi/c$.
- $\sum_{k=1}^{\infty}(a_k \cos kx + b_k \sin kx)$ is a periodic function with period 2π, although individual functions, e.g., $\cos x$, $\cos 2x$, $\cos 3x, \ldots$, have period 2π, π, $2\pi/3$, $\ldots$, respectively.

Lemma 10.2. *If $f(x)$ is a periodic function with period ω, then*

$$\int_c^{c+\omega} f(x)\,\mathrm{d}x = \int_0^{\omega} f(x)\,\mathrm{d}x,$$

whenever f is integrable on $[0, \omega]$.

Proof. Geometrically, the proof is obvious (see Figure 10.4). Using the property

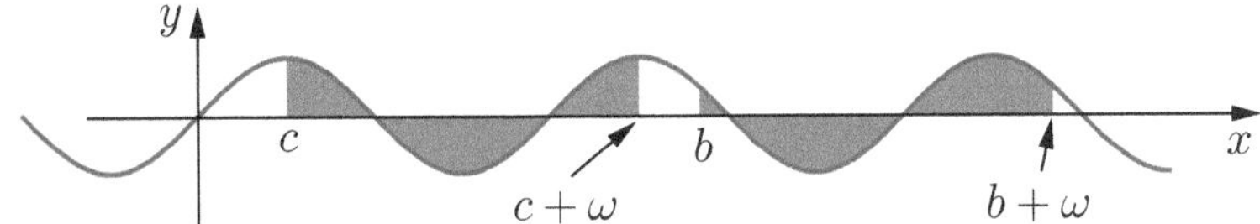

Fig. 10.4. Areas over one period are equal.

of the definite integral we obtain

$$\begin{aligned}\int_c^{c+\omega} f(x)\,\mathrm{d}x &= \int_c^0 f(x)\,\mathrm{d}x + \int_0^\omega f(x)\,\mathrm{d}x + \int_\omega^{c+\omega} f(x)\,\mathrm{d}x \\ &= -\int_0^c f(x)\,\mathrm{d}x + \int_0^\omega f(x)\,\mathrm{d}x + \int_0^c f(s)\,\mathrm{d}s \quad (x = s+\omega), \\ &= \int_0^\omega f(x)\,\mathrm{d}x,\end{aligned}$$

showing that the integral of a periodic function with period ω taken over an arbitrary interval of length ω always has the same value. ■

Definition 10.3 (Periodic extension). *Suppose that f is a function defined on $[a, a+\omega)$. Then the periodic extension of f over the infinite interval $(-\infty, \infty)$ is defined by the formula*

$$\tilde{f}(x) = \begin{cases} f(x) & \text{for } a \le x < a+\omega, \\ f(x - n\omega) & \text{for } a + n\omega \le x < a + (n+1)\omega, \end{cases}$$

where n is an integer.

Note that n in this definition is chosen such that $x - n\omega$ lies in the interval $[a, a+\omega)$. Examples of periodic extensions are shown in Figures 10.5 and 10.6.

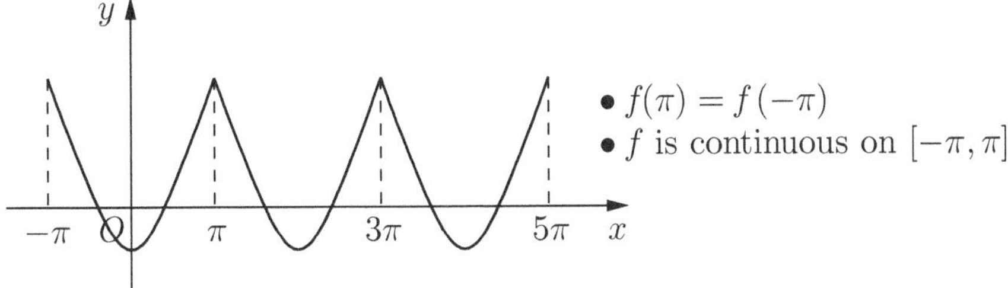

Fig. 10.5. Periodic extension of f from $[-\pi, \pi]$ with $f(\pi) = f(-\pi)$ to $\mathbb{R}$.

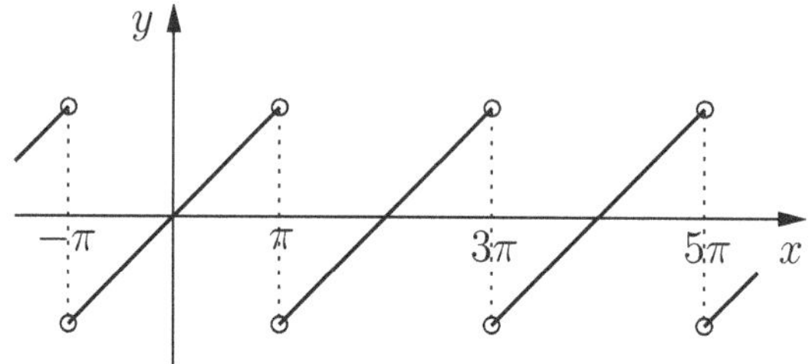

Fig. 10.6. Periodic extension of $f(x) = x$ on $(-\pi, \pi)$.

10.1.2 Trigonometric Polynomials

By a real trigonometric polynomial, we mean a (periodic) function of the form

$$s_n(x) = \frac{a_0}{2} + \sum_{k=1}^{n} (a_k \cos kx + b_k \sin kx) \quad \text{for every } x \in \mathbb{R}, \tag{10.2}$$

where a_k and b_k are some real constants. (There is a famous theorem due to Stone and Weierstrass stating that the trigonometric polynomials are dense in $C[a, b]$ for any closed interval $[a, b]$, provided that $b - a < 2\pi$.)

If the sequence $\{s_n\}$ given by (10.2) converges on a set E, then we may define a function $f\colon E \to \mathbb{R}$ by

$$f(x) = \lim_{n\to\infty} s_n(x) = \frac{a_0}{2} + \sum_{n=1}^{\infty} \big(a_n \cos nx + b_n \sin nx \big) \tag{10.3}$$

for all $x \in E$. The series on the right is called a *trigonometric series*. The constants a_0, a_k, b_k $(k \in \mathbb{N})$ are called *coefficients of the trigonometric series.* We have taken the constant term in (10.3) as $a_0/2$ rather than a_0 so that we can make $a_0/2$ fit in a general formula later. We observe that if the series on the right in (10.3) converges for all real $t \in [0, 2\pi]$, then the sum f must satisfy

$$f(x) = f(x + 2\pi) \quad \text{for all } x \in \mathbb{R}.$$

We ask a number of questions concerning the nature of f. For example, we ask the converse of the sequence $\{s_n(x)\}$ (see Problem 10.6).

10.1.3 The Space $\mathcal{E}$

Our main concern is to consider the space $\mathcal{E}$ of all $\mathbb{F}$-valued piecewise continuous functions f on the interval $[-\pi, \pi]$, where $\mathbb{F}$ is either $\mathbb{R}$ or $\mathbb{C}$ (see Definition 3.39). For $\mathbb{F} = \mathbb{C}$, every $f : [-\pi, \pi] \to \mathbb{C}$ in $\mathcal{E}$ can be written as $f = u + iv$, where u and v are real-valued piecewise continuous functions on $[-\pi, \pi]$. The restriction to this interval will be lifted later, but periodicity will always be assumed in our later discussion.

We assume that the reader is familiar with the notion of linear spaces and inner product spaces (see [6]). We now state a basic result that enables every function in $\mathcal{E}$ to be expressed as a Fourier series.

Lemma 10.4. *The space $\mathcal{E}$ is a linear space over $\mathbb{F}$. Moreover, $\mathcal{E}$ is an inner product space with respect to the inner product*

$$\langle f, g \rangle = \frac{1}{\pi} \int_{-\pi}^{\pi} f(x)\overline{g(x)}\, dx.$$

The set of functions (called a trigonometric system)

$$\Phi = \left\{ \frac{1}{\sqrt{2}}, \cos(nx), \sin(nx) : n \in \mathbb{N} \right\} \tag{10.4}$$

is an infinite orthonormal system in $\mathcal{E}$ with respect to the inner product defined in Lemma 10.4. *What is the analogous orthonormal system when the functions involved are considered in the interval $[-L, L]$?* The fact that $\mathcal{E}$ is linear is easy to verify, and so we leave it as an exercise, but the proof of the second part of Lemma 10.4 may be indicated quickly. We remark that it is possible to have $\langle f, f \rangle = 0$ without f being identically zero. For instance, if

$$f(t) = \begin{cases} 0 & \text{for } x \neq 0, \\ 1 & \text{for } x = 0, \end{cases}$$

then

$$\int_{-\pi}^{\pi} f(x)\, dx = 0 = \int_{-\pi}^{\pi} f^2(x)\, dx,$$

but f is not identically zero. It turns out that this is not a great difficulty. We merely regard two functions in $\mathcal{E}$ as being equivalent if they are equal at all but a finite number of points.

We shall consider only trigonometric systems in this chapter. There are many other orthogonal systems that are widely used, such as Legendre polynomials, Bessel functions, Hermite polynomials, and Jacobi polynomials.

Definition 10.5. *Let $\Phi = \{\phi_1, \phi_2, \ldots\}$ be an orthonormal basis of an infinite-dimensional inner product space $\mathcal{X}$, and let $f \in \mathcal{X}$. Then the infinite series*

$$\sum_{k=1}^{\infty} \langle f, \phi_k \rangle \phi_k(x) := \sum_{k=1}^{\infty} c_k \phi_k(x)$$

is called the Fourier series of f (relative to Φ), and the coefficients $c_k = \langle f, \phi_k \rangle$ are called the kth Fourier coefficients of f (relative to the orthonormal system Φ).

We could state a simpler form of Lemma 10.4 and Definition 10.5 (i.e. for the space over the real field $\mathbb{R}$). We retain the present form in order to have a better understanding of how it works in a general setting, and the modification to this effect is natural and easy. We introduce

$$\|f\|^2 = \langle f, f\rangle = \frac{1}{\pi}\int_{-\pi}^{\pi} |f(x)|^2\, dx,$$

and if f is real-valued, then $|f(x)|^2$ in this definition will be replaced by $(f(x))^2$.

10.1.4 Basic Results on Fourier Series

Suppose that we are given a trigonometric series of the form (10.3). Clearly, since each term of the series has period 2π, if it converges to a function $f(x)$, then $f(x)$ must be a periodic function with period 2π. Thus, only 2π-periodic functions are expected to have trigonometric series of the form (10.3). Although this condition is essential, the existence of such an expression is a very useful concept with a long history that has produced a rich theory for a variety of classes of functions.

Problem 10.6. *Suppose that f is a 2π-periodic function. Under what conditions does the function have a representation of the form* (10.3)*? When it does, what should be a_n, b_n?*

To answer these basic questions, it is convenient to assume for the moment that the series in (10.3) converges uniformly on $\mathbb{R}$ and that its sum is $f(t)$ (for example, this is the case if

$$\frac{|a_0|}{2} + \sum_{n=1}^{\infty}(|a_n| + |b_n|)$$

converges, so that the series (10.3) is dominated by a convergent series in $\mathbb{R}$. Then the series converges uniformly on $\mathbb{R}$ (in particular on the interval $[-\pi, \pi]$), and so term-by-term integration is permissible (see Corollary 9.30). Then we can find a formula for the coefficients a_0, a_n, b_n, $n \in \mathbb{N}$. First, we determine a_0:

$$\begin{aligned}
\frac{1}{\pi}\int_{-\pi}^{\pi} f(x)\, dx &= \frac{1}{\pi}\int_{-\pi}^{\pi}\left\{\frac{a_0}{2} + \sum_{n=1}^{\infty}(a_n\cos nx + b_n \sin nx)\right\} dx \\
&= \frac{a_0}{2}\left\{\frac{1}{\pi}\int_{-\pi}^{\pi} dx\right\} + \sum_{n=1}^{\infty}\left[\frac{a_n}{\pi}\int_{-\pi}^{\pi}\cos nx\, dx + \frac{b_n}{\pi}\int_{-\pi}^{\pi}\sin nx\, dx\right] \\
&= a_0 \quad \text{since} \int_{-\pi}^{\pi}\cos nx\, dx = 0 = \int_{-\pi}^{\pi}\sin nx\, dx.
\end{aligned}$$

Since f was assumed to be integrable on $[-\pi,\pi]$, the same is true for $f(x)\cos kx$ and $f(x)\sin kx$. Also, it is a simple exercise to see that for $n,k\in\mathbb{N}$,

$$\frac{1}{\pi}\int_{-\pi}^{\pi}\cos nx\cos kx\,dx=\delta_{nk}=\frac{1}{\pi}\int_{-\pi}^{\pi}\sin nx\sin kx\,dx,\quad \int_{-\pi}^{\pi}\cos nx\sin kx\,dx=0.$$

For a proof of these identities, one may require the following well-known identities:

$$\begin{aligned}\cos A\cos B &= \tfrac{1}{2}[\cos(A+B)+\cos(A-B)],\\ \sin A\sin B &= \tfrac{1}{2}[\cos(A-B)-\cos(A+B)],\\ \sin A\cos B &= \tfrac{1}{2}[\sin(A+B)+\sin(A-B)].\end{aligned}$$

Thus, if we multiply (10.3) by $\cos kt$, then

$$f(x)\cos kx=\frac{a_0}{2}\cos kx+\sum_{n=1}^{\infty}\left(a_n\cos nx\cos kx+b_n\sin nx\cos kx\right),$$

and so since the series for $f(x)\cos kx$ can be integrated term by term for each fixed k, we can determine a_k and b_k. Indeed,

$$\begin{aligned}\frac{1}{\pi}\int_{-\pi}^{\pi}f(x)\cos kx\,dx &= \frac{a_0}{2\pi}\int_{-\pi}^{\pi}\cos kx\,dx+\sum_{n=1}^{\infty}a_n\left\{\frac{1}{\pi}\int_{-\pi}^{\pi}\cos kx\cos nx\,dx\right\}\\ &\qquad +b_n\left\{\frac{1}{\pi}\int_{-\pi}^{\pi}\cos kx\sin nx\,dx\right\}\\ &= a_k \quad \text{for } k\in\mathbb{N},\end{aligned}$$

and similarly, by repeating the argument for $f(x)\sin kx$, we get the formula

$$\frac{1}{\pi}\int_{-\pi}^{\pi}f(x)\sin kx\,dx=b_k\quad\text{for } k\in\mathbb{N}.$$

Since the integrability of f does not depend on the value of the function f at a finite number of points in the interval of integration, f need not be defined at $x=\pm\pi$ nor at a finite number of discontinuities on $(-\pi,\pi)$. We shall now give a definition of Fourier series and present some examples.

Definition 10.7. *For any integrable function f on $[-\pi,\pi)$, the numbers a_k and b_k defined by*

$$\begin{cases} a_k=\dfrac{1}{\pi}\displaystyle\int_{-\pi}^{\pi}f(x)\cos kx\,dx & \text{for } k\ge 0,\\[2ex] b_k=\dfrac{1}{\pi}\displaystyle\int_{-\pi}^{\pi}f(x)\sin kx\,dx & \text{for } k\ge 1,\end{cases}\qquad (10.5)$$

are called the Fourier coefficients of f. The corresponding trigonometric series

$$\frac{a_0}{2} + \sum_{k=1}^{\infty} (a_k \cos kx + b_k \sin kx) \tag{10.6}$$

is called the Fourier series of f. We express this association by writing

$$f(x) \sim \frac{a_0}{2} + \sum_{k=1}^{\infty} (a_k \cos kx + b_k \sin kx), \tag{10.7}$$

to indicate that the Fourier series on the right may or may not converge to f at some point $t \in [-\pi, \pi]$.

We note that in the definition we use $\sim$ (or the symbol $\cong$ by some authors) and not $=$, to indicate thereby an association independent of any question of convergence of the Fourier series of f on the right. In fact, the reason for this notation will be clear soon. Also, we shall see that most functions are actually represented by their Fourier series. We remind the reader that if $\int_{-\pi}^{\pi} f(x)\, dx$ exists, then a_0, a_k, and b_k exist for all $k \geq 1$. Moreover, from the above discussion, we have the following result that justifies some of Fourier's original intuitions.

Theorem 10.8. *If the trigonometric series of the form* (10.6) *converges uniformly on $[-\pi, \pi]$, then it is the Fourier series of its sum. More precisely, if the trigonometric series* (10.6) *converges uniformly to f on $[-\pi, \pi]$, then the a_k and b_k are given by* (10.5).

Of course, we have no idea what happens if the series (10.6) does not converge uniformly. However, since

$$|a_k \cos kx + b_k \sin kx| \leq |a_k| + |b_k|,$$

Weierstrass's M-test (see Theorem 9.25) shows that the trigonometric series (10.6) converges absolutely and uniformly on every closed interval $[a, b]$ whenever $\sum_{k=1}^{\infty} (|a_k| + |b_k|)$ is convergent. Similar comments also apply to complex Fourier series, which will not be discussed in this book.

The complex Fourier series is much more convenient for theoretical purposes. In any case, once we prove convergence, we will be able to derive a large number of applications. To do this, we begin with a Fourier series of a given function f even if we do not have any idea whether it will converge uniformly to f or diverge.

There are special circumstances in which the calculation of the Fourier coefficients becomes a bit simpler than usual. For example, in the case of even and odd functions, certain coefficients are zero. The graph of an even function is symmetric about the y-axis, whereas the graph of an odd function is symmetric about the origin. Certain well-known even functions are $|x|$, x^2, $\cos x$, and well-known odd functions are x, $\sin x$, $\tan x$. Even and odd functions possess certain simple but useful properties:

- The product of two even (or odd) functions is an even function.
- The sum of two even (or odd) functions is an even (or odd) function.
- The product of an even and an odd function is an odd function.
- For a Riemann integrable function f defined on $[-c, c]$ $(c > 0)$, from the inspection of related figures with this assumption, it is evident that

$$\int_{-c}^{c} f(x)\,\mathrm{d}x = \begin{cases} 2\displaystyle\int_{0}^{c} f(x)\,\mathrm{d}x & \text{if } f \text{ is even,} \\ 0 & \text{if } f \text{ is odd.} \end{cases}$$

For instance,

$$\int_{-c}^{c} \sin kx\,\mathrm{d}x = 0 \quad \text{for each } k.$$

Suppose that $f(x)$ is a periodic function of period 2π. Let us further assume that f is even on $(-\pi, \pi)$, i.e., $f(x) = f(-x)$ for all $x \in (-\pi, \pi)$. Then the product function $f(x)\sin kx$ is odd, which means that $b_k = 0$ for all $k \geq 1$, and hence we have the *Fourier cosine series*

$$f(x) \sim \frac{a_0}{2} + \sum_{k=1}^{\infty} a_k \cos kx, \quad a_k = \frac{2}{\pi}\int_{0}^{\pi} f(x)\cos kx\,\mathrm{d}x.$$

If f is odd on $(-\pi, \pi)$, i.e., $f(x) = -f(-x)$ for all $x \in (-\pi, \pi)$, then the product function $f(x)\cos kx$ is odd, which means that $a_k = 0$ for all $k \geq 0$, and hence we have the *Fourier sine series*

$$f(x) \sim \sum_{k=1}^{\infty} b_k \sin kx, \quad b_k = \frac{2}{\pi}\int_{0}^{\pi} f(x)\sin kx\,\mathrm{d}x.$$

Example 10.9. Consider $f(x) = |x|$ on $[-\pi, \pi]$. Then f is even and continuous on $[-\pi, \pi]$. Also, $f'(x)$ exists on $(-\pi, \pi) \setminus \{0\}$. Obviously, $b_n = 0$ for all $n \geq 1$. Now because f is even and so is $f(x)\cos nx$, we have

$$a_n = \frac{2}{\pi}\int_{0}^{\pi} x\cos nx\,\mathrm{d}x.$$

The graph of f together with its periodic extension is shown in Figure 10.7. The extended function is continuous and piecewise smooth. Indeed, the Fourier series converges to $f(x) = |x|$ on $[-\pi, \pi]$. Clearly, $a_0 = \pi$. For $n \geq 1$, integration by parts yields

$$a_n = \frac{2}{\pi}\int_{0}^{\pi} x\,\mathrm{d}\left(\frac{\sin nx}{n}\right) = -\frac{2}{n\pi}\int_{0}^{\pi} \sin nx\,\mathrm{d}x = -\frac{2}{n\pi}\left(\frac{\cos nx}{-n}\right)\bigg|_{0}^{\pi},$$

so that

$$a_n = -\frac{2(1-(-1)^n)}{n^2\pi} = \begin{cases} -\dfrac{4}{\pi(2k+1)^2} & \text{if } n = 2k+1,\ k \geq 0, \\ 0 & \text{if } n \text{ is even.} \end{cases}$$

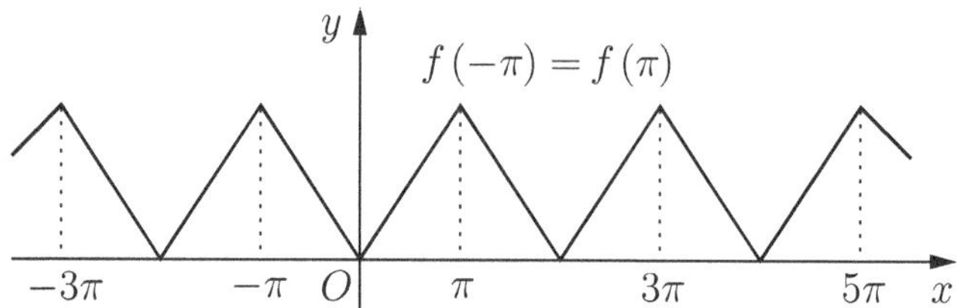

Fig. 10.7. Periodic (even) extension of the periodic function $f(x) = |x|$ on $[-\pi, \pi]$.

We have, for $x \in [-\pi, \pi]$,

$$|x| \sim \frac{\pi}{2} - \frac{4}{\pi} \sum_{k=0}^{\infty} \frac{\cos(2k+1)x}{(2k+1)^2}.$$

Note that the Fourier series here converges uniformly to $|x|$ on $[-\pi, \pi]$ but not on the whole interval $(-\infty, \infty)$, and so outside the interval $(-\pi, \pi)$, $f(x)$ is determined by the periodicity condition $f(x) = f(x + 2\pi)$. By Theorem 10.8, in the above expression, the symbol $\sim$ indicates equality. Thus, we can make use of this series to find the values of some numerical series. For instance, $x = 0$ yields

$$\frac{\pi^2}{8} = \sum_{k=1}^{\infty} \frac{1}{(2k+1)^2} \tag{10.8}$$

as a special case. The function to which the Fourier cosine series for $f(x) = |x|$,

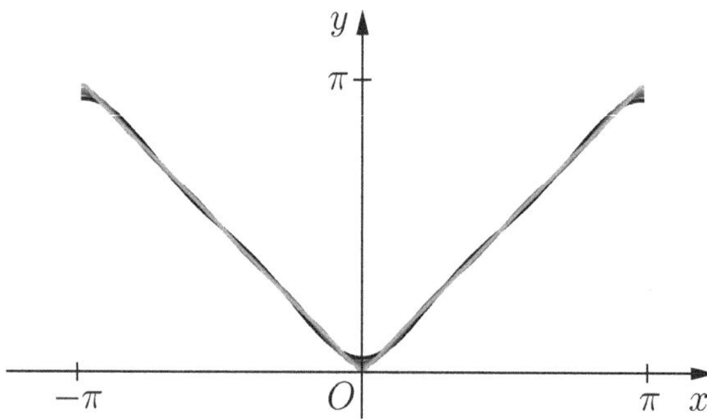

Fig. 10.8. Approximation of $|x|$ with partial sums of $\frac{\pi}{2} - \frac{4}{\pi} \sum_{k=0}^{\infty} \frac{\cos(2k+1)x}{(2k+1)^2}$.

converges is illustrated in Figure 10.8. •

10.1.5 Questions and Exercises

Questions 10.10.

1. When we say p-periodic, are we referring to a function's fundamental period p?
2. What is the major advantage of Fourier series over Taylor series?
3. If $f_1, f_2, \ldots, f_n$ are periodic functions of period p, must the sum $\sum_{k=1}^{n} a_k f_k$ be a periodic function of period p?
4. Is $\cos x + \cos \pi x$ periodic? If so, what is its period?
5. If f and g are periodic with period p, must the periods of fg and f/g $(g \neq 0)$ be p? What is the period of $\sin 2t = 2 \sin t \cos t$?
6. If f and g are ω-periodic, must $f + g$ be ω-periodic?
7. Suppose that f is ω-periodic. Must f' be ω-periodic? Must $\int_0^x f(t)\, dt$ be ω-periodic if and only if $\int_0^\omega f(t)\, dt = 0$?
8. Does the integrability of $\int_{-\pi}^{\pi} f(x)\, dx$ suffice for the existence of the Fourier series of f?
9. Suppose that the sequence of trigonometric polynomials $\{s_n(x)\}_{n\geq 0}$ converges uniformly to f on $[-\pi, \pi]$. Must the sequences $\{s_n(x) \cos kx\}_{n\geq 0}$ and $\{s_n(x) \sin kx\}_{n\geq 0}$ be uniformly convergent to the functions $f(x) \cos kx$ and $f(x) \sin kx$, respectively, for each $k \geq 1$, on $[-\pi, \pi]$? If so, does that imply that term-by-term integration of the Fourier series of f is permissible on $[-\pi, \pi]$?
10. Suppose that the Fourier series of f converges uniformly on $[-\pi, \pi]$. Must it be the Fourier series of exactly one continuous function?
11. Suppose that f is continuous on $[-\pi, \pi]$ such that $f(-\pi) \neq f(\pi)$. Must the periodic extension of f to $\mathbb{R}$ have discontinuities at $x = (2k - 1)\pi$, $k \in \mathbb{Z}$?

Exercises 10.11.

1. Define
   (a) $f(x) = x^2$ on $[-\pi, \pi)$; (b) $g(x) = x^2$ on $[0, 2\pi]$.
   Draw the graphs of f and g together with their periodic extensions.
2. Determine the Fourier series of a sawtooth function f defined by

$$f(x) = \begin{cases} (\pi - x)/2 & \text{for } 0 < x \leq 2\pi, \\ f(x + 2\pi) & \text{otherwise.} \end{cases}$$

3. Expand $f(x) = x^2$ on $[-\pi, \pi]$ in a Fourier series. Show that the Fourier series of f converges absolutely and uniformly to the periodic function extension of f. Also, deduce that

$$\frac{\pi^2}{6} = \sum_{n=1}^{\infty} \frac{1}{n^2}.$$

4. Plot the graph of each of the following functions for $L = 1, 2, \pi, 2\pi$:

   (a) $f(x) = \begin{cases} \cos\left(\frac{\pi}{L}x\right) & \text{if } 0 \le x \le L/2, \\ 0 & \text{if } L/2 < x \le L. \end{cases}$

   (b) $f(x) = \begin{cases} 0 & \text{if } 0 \le x \le L/2, \\ \sin\left(\frac{\pi}{L}x\right) & \text{if } L/2 < x \le L. \end{cases}$

   (c) $f(x) = \begin{cases} x & \text{if } 0 \le x \le L/2, \\ L - x & \text{if } L/2 \le x \le L. \end{cases}$

   Also graph each function's periodic extension onto the interval $[-L, 0]$ together with its period extension (with period $2L$) onto $\mathbb{R}$.

5. Find the Fourier series corresponding to the following function:

$$f(x) = \begin{cases} \sin x & \text{if } -\pi < x < 0, \\ \cos x & \text{if } 0 < x < \pi. \end{cases}$$

6. Consider $g(x) = 1 - (|x|/\pi)$ on $[-\pi, \pi]$ and extend it as a 2π-periodic function defined on $\mathbb{R}$. Show that the Fourier series of g is given by

$$g(x) = \frac{1}{2} + \frac{4}{\pi^2} \sum_{k=1}^{\infty} \frac{\cos(2k+1)x}{(2k+1)^2}.$$

7. Show that the Fourier series of the function $f(x) = x^2$ on $(0, 2\pi)$ is given by

$$\frac{4\pi^2}{6} + \sum_{n=1}^{\infty} 4\left(\frac{\cos nx}{n^2} - \frac{\pi \sin nx}{n}\right).$$

   Show that it converges uniformly on every closed interval $[a, b]$ in $(0, 2\pi)$, and hence converges on $(0, 2\pi)$.

8. Sketch the periodic function $f(x) = 1 - x^2$, $x \in [-\pi, \pi]$, and determine its Fourier sine and cosine coefficients. Do the same for the function $f(x) = \pi - x$ on $[0, \pi]$. Apply a convergence test to see whether the Fourier series corresponding to these two functions converge uniformly for $-\pi \le x \le \pi$.

9. Verify whether each of the following is true:

   (a) $x^2 = \frac{\pi^2}{3} + 4\sum_{k=1}^{\infty} \frac{(-1)^k \cos kx}{k^2}, \quad x \in [-\pi, \pi]$.

   (b) $\frac{\pi}{4} = \sum_{n=1}^{\infty} \frac{\sin(2n-1)x}{2n-1}, \quad x \in (0, \pi)$.

   (c) $x = 2\sum_{k=1}^{\infty} \frac{(-1)^{k-1}}{k} \sin kx, \quad x \in (-\pi, \pi)$.

   (d) $x = \pi - 2\sum_{k=1}^{\infty} \frac{\sin kx}{k}, \quad x \in (0, 2\pi)$.

   Does the identity **(b)** hold at the end points $x = 0$ or π? Give a valid reason to support your answer. How about the equality case at the endpoints in **(c)** and **(d)**?

10.2 Convergence of Fourier Series

A Fourier series is simply a trigonometric series considered formally with no claim of convergence, although the series is associated with a function f in the sense that coefficients have been obtained from f. Some natural questions arise:

(a) For what values of x does the Fourier series of f converge? Does it converge for all x in $[-\pi,\pi]$? If it converges on $[-\pi,\pi]$ but not to f, what will be its sum?
(b) If the Fourier series of f converges at x, does it converge to f?
(c) If the Fourier series of f converges to f on $[-\pi,\pi]$, does it converge uniformly to f on $[-\pi,\pi]$?

Compare these questions with the analogous questions about the Taylor series of a function.

The history of Fourier series is long and glorious and has been one of the most fruitful concepts in mathematics since its inception. However, continuity of f is not sufficient to guarantee convergence of the Fourier series of f on $[-\pi,\pi]$. In 1876, Paul du Bois-Reymond constructed a continuous function $f\colon [-\pi,\pi]\to\mathbb{R}$ whose Fourier series failed to converge to f at each point in a dense subset of $[-\pi,\pi]$.[1] Indeed, the following are true statements:

- There exists a continuous function whose Fourier series diverges at a point.
- There exists a continuous function whose Fourier series converges everywhere on $[-\pi,\pi]$, but not uniformly.
- There exists a continuous function whose Fourier series diverges for points in some set S and converges on $(-\pi,\pi)\backslash S$.

For proofs of these statements, see standard texts such as Z. Zygmond.[2]

10.2.1 Statement of Dirichlet's Theorem

It is important to establish simple criteria for determining when a Fourier series converges. Define

$$\begin{aligned}\mathcal{E}' = \Big\{ f\in\mathcal{E}:\ &\text{(i)} \lim_{h\to 0+}\frac{f(x+h)-f(x+)}{h} \quad \text{exists for each } x\in[-\pi,\pi)\\ &\text{(ii)} \lim_{h\to 0-}\frac{f(x+h)-f(x-)}{h} \quad \text{exists for } x\in(-\pi,\pi]\Big\}.\end{aligned}$$

Recall that $f:[a,b]\to\mathbb{R}$ is smooth if f and f' are continuous on $[a,b]$. For the definition of left- and right-hand limits, we refer to Section 3.1.4 and

---

[1] A subset $A\subseteq X$ is dense in X if $\overline{A}=X$; for example, $\overline{\mathbb{Q}}=\mathbb{R}$.

[2] Z. Zygmond, Trigonometric Series, 3rd Edition, Vols. I and II, Cambridge University Press, 2002.

the discussion following Theorem 3.13. For one-sided derivatives, we refer to Section 3.3.

Now we state two versions of the Dirichlet theorem, although the two relatively easy versions given in Theorems 10.33 and 10.35 are sufficient for our purposes.

Theorem 10.12 (Dirichlet's theorem). *Let $f \in \mathcal{E}'$. Then for each $x \in (-\pi, \pi)$, the Fourier series of $f(x)$ converges to the value*

$$\frac{f(x-) + f(x+)}{2}.$$

At the endpoints $x = \pm\pi$, the series converges to

$$\frac{f(\pi-) + f((-\pi)+)}{2}.$$

Remark 10.13. **(i)** If $f \in \mathcal{E}'$ is continuous at x, then $f(x-) = f(x+) = f(x)$, and so at such points,

$$\frac{f(x-) + f(x+)}{2} = f(x).$$

Thus, the Fourier series of f converges to $f(x)$ at the point x where it is continuous.

(ii) At the point of discontinuity x, the Fourier series of f assumes the mean of the one-sided limits of f. ●

Corollary 10.14. *If $f : [-\pi, \pi] \to \mathbb{R}$ is continuous, and if $f(-\pi) = f(\pi)$, $f'(x)$ exists and is piecewise continuous on $[-\pi, \pi]$, then the Fourier series of f converges to $f(x)$ at every point $x \in [-\pi, \pi]$.*

Theorem 10.15 (Dirichlet's theorem). *Suppose that $f : [-\pi, \pi] \to \mathbb{R}$ is piecewise continuous on $[-\pi, \pi]$ and piecewise monotone, i.e., there exists a partition $P = \{x_0, x_1, \dots, x_n\}$ of $[-\pi, \pi]$ such that the restriction $f|_{[x_{k-1}, x_k]}$, $k = 1, 2 \dots, n$, is either increasing or decreasing. Let $f(x)$ be defined for other values of x by the periodicity condition $f(x) = f(x + 2\pi)$. Then the Fourier series of f on $[-\pi, \pi]$ converges to*

(a) *$f(x)$ if f is continuous at $x \in (-\pi, \pi)$;*
(b) *$(f(x-) + f(x+))/2$ if f is discontinuous at x;*
(c) *$(f(\pi-) + f((-\pi)+))/2$ at the endpoints $x = \pm\pi$.*

The conditions imposed on $f(x)$ in Theorems 10.12 and 10.15 are called Dirichlet's conditions (see Figure 10.9).

Example 10.16. If $f(x) = x$ on $[-\pi, \pi)$ and $f(\pi) = -\pi$, graph the 2π-periodic extension of f to $\mathbb{R}$. Also, find the Fourier sine series of f.

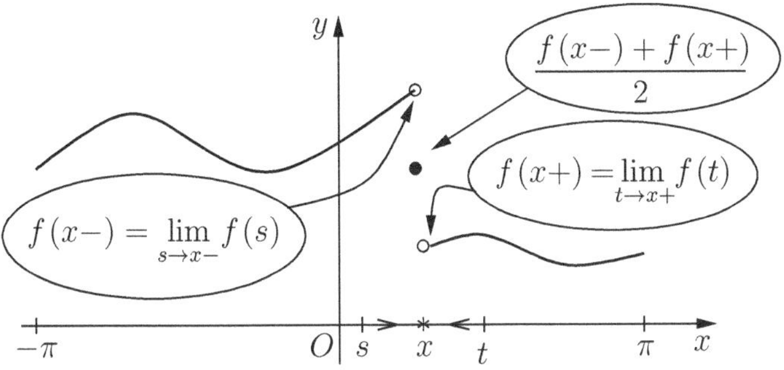

Fig. 10.9. Illustration for the convergence of Fourier series.

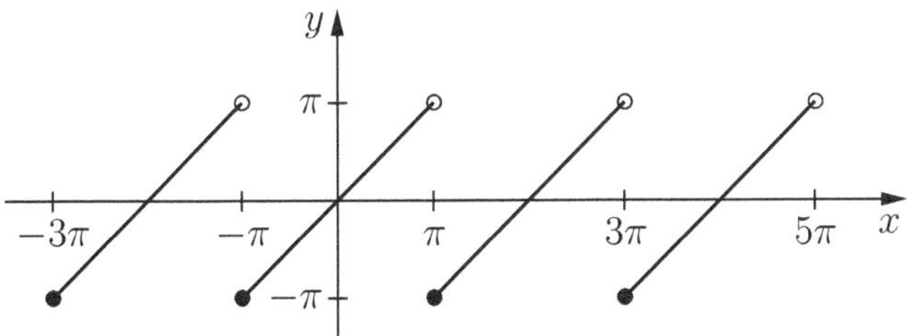

Fig. 10.10. Graphs of $f(x) = f(x + 2\pi)$, $f(x) = x$ on $[-\pi, \pi)$ with $f(\pi) = -\pi$.

Solution. Note that f is odd on $(-\pi, \pi)$ and $f(-\pi) = f(\pi)$ (Figure 10.10). The periodic extension of f to $\mathbb{R}$ may be given by

$$f(x + 2\pi) = f(x) \quad \text{for } x \in \mathbb{R}.$$

Note that $f((-\pi)+) = -\pi = -f(\pi+)$ and the 2π-periodic extension of f is not continuous. We see that $a_n = 0$ for $n \geq 0$ and

$$b_n = \frac{1}{\pi}\int_{-\pi}^{\pi} x \sin nx \, dx = \frac{2}{\pi}\int_0^{\pi} x \sin nx \, dx = \frac{2(-1)^{n-1}}{n}.$$

Consequently (see Figure 10.11),

$$x \sim 2\sum_{k=1}^{\infty} \frac{(-1)^{k-1}}{k} \sin kx.$$

Note that the Fourier series does not necessarily agree with $f(x) = x$ at every point in $[-\pi, \pi]$. In the present example, the Fourier series vanishes at both endpoints $x = \pm\pi$, whereas the function does not vanish at either endpoint. However, by Theorem 10.12, the series converges at every interior point of $(-\pi, \pi)$. For example, at $x = \pi/2$ the symbol $\sim$ can be replaced by an equal sign (why?), and so

$$\frac{\pi}{2} = 2\left[1 - \frac{0}{2} + \frac{(-1)}{3} - \frac{0}{4} + \frac{1}{5} + \cdots\right],$$

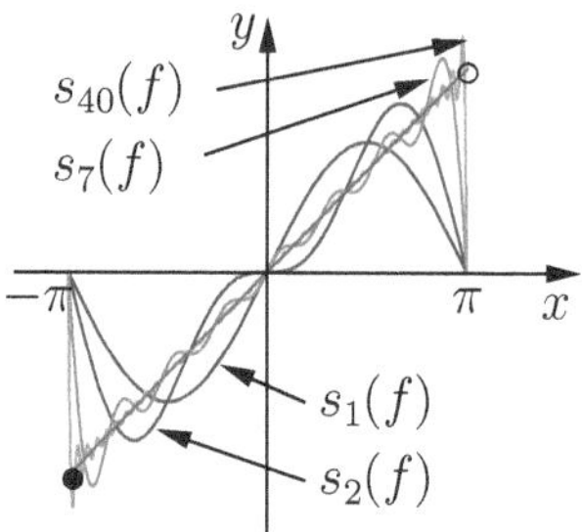

Fig. 10.11. The graphs of the nth partial sums of $2\sum_{k=1}^{\infty}((-1)^{k-1}/k)\sin(kx)$.

which gives the remarkable Madhava-Leibniz-Gregory formula (see also Example 9.32)

$$\frac{\pi}{4} = 1 - \frac{1}{3} + \frac{1}{5} - \frac{1}{7} + \cdots.$$

Finally, we remark that at the endpoints $x = \pm\pi$, the series converges to

$$\frac{f(\pi-) + f((-\pi)+)}{2} = \frac{\pi + (-\pi)}{2} = 0.$$

●

In the above example, we could also consider f as follows: $f(x) = x$ on $(-\pi, \pi)$ and $f(-\pi) = f(\pi) = 0$. The periodic extension is pictured in Figure 10.12.

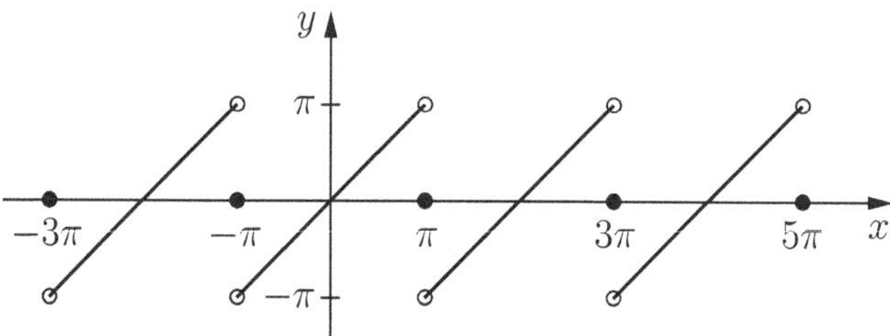

Fig. 10.12. Graphs of $f(x) = f(x + 2\pi)$, $f(x) = x$ on $(-\pi, \pi)$ with $f(\pi) = 0 = f(-\pi)$.

Example 10.17. If $f(x) = e^x$ on $[-\pi, \pi)$ and $f(x + 2\pi) = f(x)$ for $x \in \mathbb{R}$, determine the Fourier series of f.

Solution. It is convenient and simpler to use the following standard notation:

$$\frac{1}{a + \mathrm{i}b} = \frac{a}{a^2 + b^2} - \mathrm{i}\frac{b}{a^2 + b^2} \quad \text{and} \quad \mathrm{e}^{\mathrm{i}x} = \cos x + \mathrm{i}\sin x,$$

so that for $n \geq 0$, we have $e^{in\pi} = (-1)^n$ and

$$\int e^{inx}\, dx = \int \cos nx\, dx + i \int \sin nx\, dx.$$

According to this, the Fourier coefficients are easy to derive quickly by writing

$$\begin{aligned}
a_n - ib_n &= \frac{1}{\pi}\int_{-\pi}^{\pi} e^{-inx} e^x\, dx \\
&= \frac{1}{\pi} \frac{e^{(1-in)x}}{1-in}\bigg|_{-\pi}^{\pi} \\
&= \frac{1}{\pi}\left(\frac{e^{(1-in)\pi} - e^{-(1-in)\pi}}{1-in}\right) \\
&= \frac{(-1)^n(e^{\pi} - e^{-\pi})}{\pi(1-in)}, \quad \text{since } e^{\pm in\pi} = (-1)^n, \\
&= \frac{2(-1)^n \sinh \pi}{\pi(1+n^2)}(1+in),
\end{aligned}$$

and therefore, equating real and imaginary parts gives

$$a_n = \frac{2(-1)^n \sinh \pi}{\pi(1+n^2)} \quad \text{and} \quad b_n = \frac{2(-1)^{n-1} n \sinh \pi}{\pi(1+n^2)}.$$

Thus, we have

$$e^x \sim \frac{\sinh \pi}{\pi} + \frac{2\sinh \pi}{\pi}\sum_{n=1}^{\infty}\frac{(-1)^n}{1+n^2}\cos nx + \frac{2\sinh \pi}{\pi}\sum_{n=1}^{\infty}\frac{(-1)^{n-1}n}{1+n^2}\sin nx \tag{10.9}$$

for $x \in (-\pi, \pi)$. In particular, at the point of continuity $x = 0$, it follows that

$$1 = \frac{\sinh \pi}{\pi} + \frac{2\sinh \pi}{\pi}\sum_{n=1}^{\infty}\frac{(-1)^n}{1+n^2},$$

which reduces to

$$\frac{\pi \operatorname{csc} \pi - 1}{2} = \sum_{n=1}^{\infty}\frac{(-1)^n}{1+n^2}.$$

According to Dirichlet's theorem, at the endpoint $x = \pi$, the relation (10.9) yields

$$\frac{e^{\pi} + e^{-\pi}}{2} = \frac{\sinh \pi}{\pi} + \frac{2\sinh \pi}{\pi}\sum_{n=1}^{\infty}\frac{1}{1+n^2}, \quad \text{i.e. } \pi \coth \pi = 1 + 2\sum_{n=1}^{\infty}\frac{1}{1+n^2},$$

which reduces to

$$\frac{\pi \coth \pi - 1}{2} = \sum_{n=1}^{\infty}\frac{1}{1+n^2}.$$

●

10.2.2 Fourier Series of Functions with an Arbitrary Period

Having discussed periodic functions defined on $[-\pi,\pi]$ of period 2π, we are now ready for a discussion of a more general case, namely the development of a function in a Fourier series valid on an arbitrary interval. In this process, by letting the length of the interval increase indefinitely, we may obtain an expression valid for all x.

Suppose that f is a $2L$-periodic and Riemann integrable function. Then by Lemma 10.1, the function $f(at)$ has period $2L/a$. In particular, $f((L/\pi)t)$ is 2π-periodic, and so the Fourier series expansion has the following in terms of the variable t:

$$f\left(\frac{L}{\pi}t\right) \sim \frac{a_0}{2} + \sum_{k=1}^{\infty} a_k \cos kt + \sum_{k=1}^{\infty} b_k \sin kt, \quad t \in [-\pi,\pi], \tag{10.10}$$

where

$$a_k = \frac{1}{\pi}\int_{-\pi}^{\pi} f\left(\frac{L}{\pi}t\right)\cos kt \,\mathrm{d}t = \frac{1}{L}\int_{-L}^{L} f(x)\cos\left(\frac{k\pi}{L}x\right)\mathrm{d}x,$$

and similarly,

$$b_k = \frac{1}{L}\int_{-L}^{L} f(x)\sin\left(\frac{k\pi}{L}x\right)\mathrm{d}x.$$

By Lemma 10.2, we remark that the interval of integration in the last two formulas for the Fourier coefficients can be replaced with an arbitrary interval $[c, c+2L]$, of length $2L$. Changing the variable t, by setting $t = (\pi/L)x$, we can reformulate (10.10), and the above discussion as follows.

Theorem 10.18. *Let f be a periodic function with period $2L$. Then the Fourier expansion of f is given by*

$$f(x) \sim \frac{a_0}{2} + \sum_{k=1}^{\infty} a_k \cos\left(\frac{k\pi}{L}x\right) + \sum_{k=1}^{\infty} b_k \sin\left(\frac{k\pi}{L}x\right), \quad x \in [-L,L], \tag{10.11}$$

where

$$a_k = \frac{1}{L}\int_{-L}^{L} f(x)\cos\left(\frac{k\pi}{L}x\right)\mathrm{d}x \quad \textit{and} \quad b_k = \frac{1}{L}\int_{-L}^{L} f(x)\sin\left(\frac{k\pi}{L}x\right)\mathrm{d}x.$$

The interval of integration in the last formulas for the Fourier coefficients can be replaced with the interval $[c, c+2L]$, where c is any real number; we usually let $c = -L$. Most of the physically realizable periodic functions satisfy this theorem. Note that

$$\cos\left(\frac{k\pi}{L}(x+2L)\right) = \cos\left(\frac{k\pi}{L}x\right) \quad \text{and} \quad \sin\left(\frac{k\pi}{L}(x+2L)\right) = \sin\left(\frac{k\pi}{L}x\right).$$

Thus, the Fourier series in (10.11) has period $2L$, and therefore the sum should have period $2L$. That is, the sum cannot represent an arbitrary function on $\mathbb{R}$, but can represent a periodic function only. But a natural extension to $\mathbb{R}$ may be obtained by letting $L \to \infty$. This process actually leads to what is called the Fourier transform, which is outside the scope of this book.

As an illustration of Theorem 10.18, we consider

$$f(x) = \begin{cases} 0 & \text{for } -2 \leq x < 0, \\ 1 & \text{for } 0 \leq x < 2. \end{cases}$$

Then with $L = 2$, we apply Theorem 10.18 and obtain

$$a_0 = \frac{1}{2}, \quad a_n = 0 \quad \text{for } n \geq 1\,, \quad \text{and} \quad b_n = \frac{1 + (-1)^{n-1}}{n\pi} \qquad \text{for } n \geq 1.$$

Corollary 10.19. *If f is a periodic function with period $2L$, then*

$$f(x) \sim \sum_{k=0}^{\infty} d_k \cos\left(\frac{k\pi}{L}x + \phi_k\right),$$

where the coefficients d_k and the phases ϕ_k may be calculated from the coefficients a_k and b_k defined in Theorem 10.18.

Theorem 10.18 gives the following result, which is widely used in applications because many interesting functions that turn out to be even or odd periodic functions.

Corollary 10.20. *The Fourier series of an even function f with period $2L$ is a Fourier cosine series*

$$f(x) \sim \frac{a_0}{2} + \sum_{k=1}^{\infty} a_k \cos\left(\frac{k\pi}{L}x\right) \quad \textit{with} \quad a_k = \frac{1}{L}\int_c^{c+2L} f(x)\cos\left(\frac{k\pi}{L}x\right)dx,$$

and the Fourier series of an odd function f with period $2L$ is a Fourier sine series

$$f(x) \sim \sum_{k=1}^{\infty} b_k \sin\left(\frac{k\pi}{L}x\right) \quad \textit{with} \quad b_k = \frac{1}{L}\int_c^{c+2L} f(x)\sin\left(\frac{k\pi}{L}x\right)dx,$$

where c is any real number. In particular, a_k and b_k may be given with $c = 0$.

10.2.3 Change of Interval and Half-Range Series

Note that a given function is not necessarily even or odd (e.g., e^x and x^2e^x). Suppose that we wish to find the Fourier series of a function f defined only in $[0, \pi]$ instead of $[-\pi, \pi]$. We may define

$$F(x) = \begin{cases} g(x) & \text{for } -\pi \leq x < 0 \\ f(x) & \text{for } 0 \leq x \leq \pi, \end{cases}$$

where g is an arbitrary function on $[-\pi, 0)$. Since we are interested in f only on $[0, \pi]$, properties of the convergence of the series on $[-\pi, 0)$ are irrelevant (Figure 10.13). This means that we may set on $[-\pi, 0)$,

$$g(x) = f(-x) \quad \text{or} \quad g(x) = -f(-x) \quad \text{or} \quad g(x) = 0.$$

Assume that F satisfies the Dirichlet conditions.

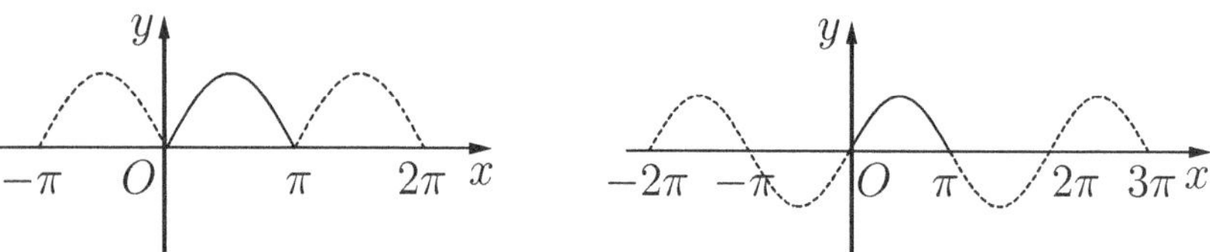

Fig. 10.13. Illustration for even and odd extensions.

Case (i) In the first case, we obtain an even function F on $[-\pi, \pi]$ (with $g(0) = 0$), so that Fourier series of $F(x)$ on $[-\pi, \pi]$ contains only cosine terms, and $F(x) = f(x)$ on $[0, \pi]$. Consequently, the cosine series of the even function $F(x) := f_{\text{even}}(x) = f(|x|)$ on $[-\pi, \pi]$ gives the cosine series of $f(x)$ on $[0, \pi]$. Thus, the Fourier cosine series of f on $[0, \pi]$ is given by

$$f(x) \sim \frac{a_0}{2} + \sum_{k=1}^{\infty} a_k \cos kx, \quad a_k = \frac{2}{\pi} \int_0^{\pi} f(x) \cos kx \, dx.$$

This is called the *half-range cosine series for* f defined on $[0, \pi]$. Thus, according to Theorem 10.15,

$$\frac{a_0}{2} + \sum_{k=1}^{\infty} a_k \cos kx = \begin{cases} f(x) & \text{for } x \in (0, \pi), \\ f(0) & \text{for } x = 0, \\ \dfrac{f(-\pi) + f(\pi)}{2} = f(\pi) & \text{for } x = \pi, \end{cases}$$

where in the last step we have used the fact that $f(\pi) = f(-\pi)$. We remark that if f is defined only on $[0, \pi)$, then we can define

$$f_{\text{even}}(-\pi) = \lim_{x \to \pi-} f(x),$$

provided the later limit exists.

Case (ii) In the second case, we consider $g(x) = -f(-x)$, and so F takes the form

$$F(x) := f_{\text{odd}}(x) = \begin{cases} f(x) & \text{for } 0 < x < \pi, \\ -f(-x) & \text{for } -\pi < x < 0, \\ 0 & \text{for } x = 0,\ x = \pm\pi, \end{cases}$$

where in the last step we have set $f(\pi-) = -f((-\pi)+)$ (provided $\lim_{x\to\pi-} f(x)$ exists), so that

$$\frac{f(\pi-) + f((-\pi)+)}{2} = 0,$$

and $F = f_{odd}$ is an odd function defined on $[-\pi,\pi]$ (Figure 10.14). Hence the Fourier series of F contains only sine terms, and $F(x) = f(x)$ on $[0,\pi]$. Consequently,

$$f(x) \sim \sum_{k=1}^{\infty} b_k \sin kx, \quad b_k = \frac{2}{\pi}\int_0^{\pi} f(x)\sin kx\, dx,$$

which is referred to as the *half-range sine series for f* on $[0,\pi]$.

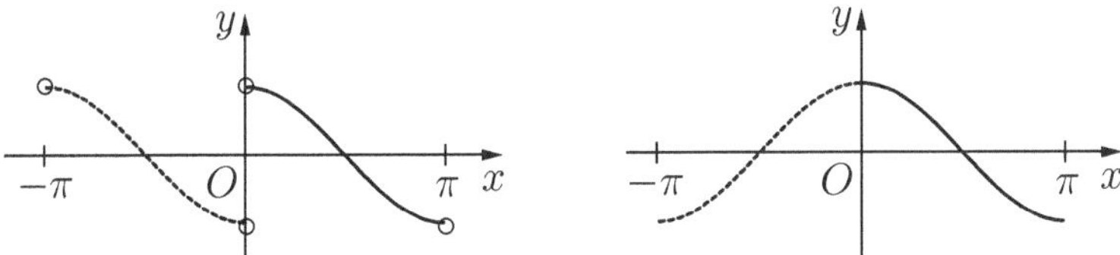

Fig. 10.14. Odd and even extensions of $f(x) = \cos x$ on $(0,\pi)$.

Odd and even extensions of f defined on $(0, L)$ may be defined similarly; see Figures 10.15 and 10.16.

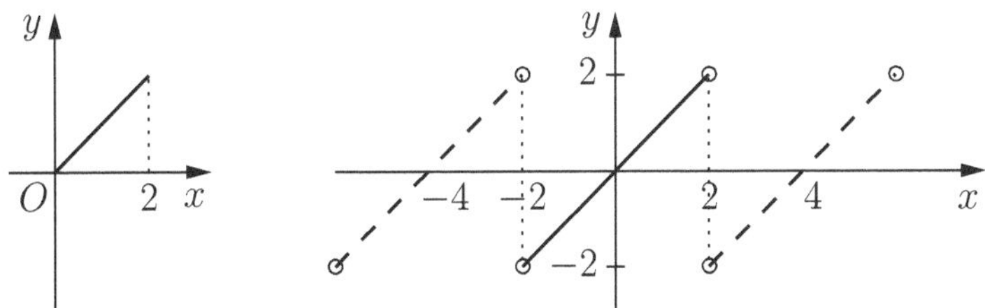

Fig. 10.15. Odd extension of $f(x) = x$ for $0 < x < 2$, period 4.

A change in scale can be made for the cosine and sine series, and we obtain the following:

Definition 10.21. *For an integrable function $f : [0, L] \to \mathbb{R}$, the Fourier cosine expansion of f is given by*

$$f(x) \sim \frac{a_0}{2} + \sum_{k=1}^{\infty} a_k \cos\left(\frac{k\pi x}{L}\right), \quad a_k = \frac{2}{L}\int_0^{L} f(x)\cos\left(\frac{k\pi x}{L}\right) dx.$$

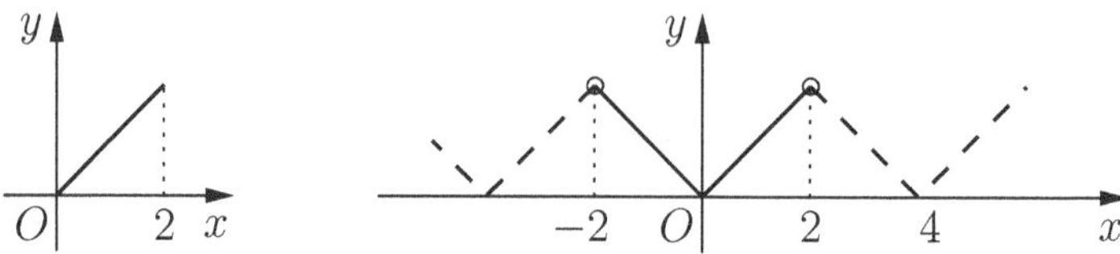

Fig. 10.16. Even extension of $f(x) = x$ for $0 < x < 2$, period 4.

The Fourier sine expansion of f on $[0, L]$ is given by

$$f(x) \sim \sum_{k=1}^{\infty} b_k \sin\left(\frac{k\pi x}{L}\right), \quad b_k = \frac{2}{L}\int_0^L f(x)\sin\left(\frac{k\pi x}{L}\right) dx.$$

Thus, given a function f defined on $[0, L]$, it is possible to obtain both cosine and sine series of f on $[0, L]$.

Example 10.22. Consider $f(x) = |\sin x|$ (see also Example 10.24). The function is defined for all x. We remark that $|\sin x|$ is π-periodic and $|\sin x| = \sin x$ on $[0, \pi]$ (see Figures 10.17 and 10.18). Clearly, f represents a continuous, piecewise smooth, even function of period π, and therefore it is everywhere equal to its Fourier series, consisting of cosine terms only. Now, for $k \geq 0$, Corollary 10.20 with $c = 0$ and $L = \pi/2$ yields that

$$\begin{aligned} a_k &= \frac{2}{\pi}\int_0^{\pi} f(x)\cos(2kx)\, dx \\ &= \frac{2}{\pi}\int_0^{\pi} \sin x \cos 2kx\, dx, \quad \text{since } |\sin x| = \sin x \text{ on } [0, \pi], \\ &= \frac{1}{\pi}\int_0^{\pi} [\sin(1 + 2k)x - \sin(2k - 1)x]\, dx \\ &= \frac{1}{\pi}\left(-\frac{\cos(2k+1)x}{2k+1} + \frac{\cos(2k-1)x}{2k-1}\right)\Bigg|_0^{\pi} \\ &= -\frac{1}{\pi}\left(\frac{(-1)^{2k+1} - 1}{2k+1} - \frac{(-1)^{2k-1} - 1}{2k-1}\right), \end{aligned}$$

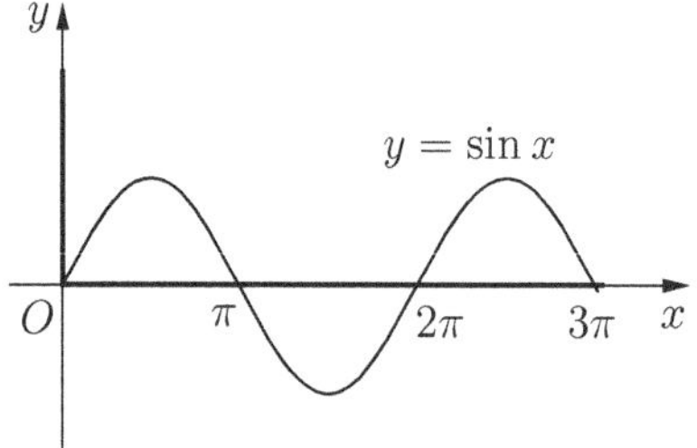

Fig. 10.17. Graph of $y = \sin x$.

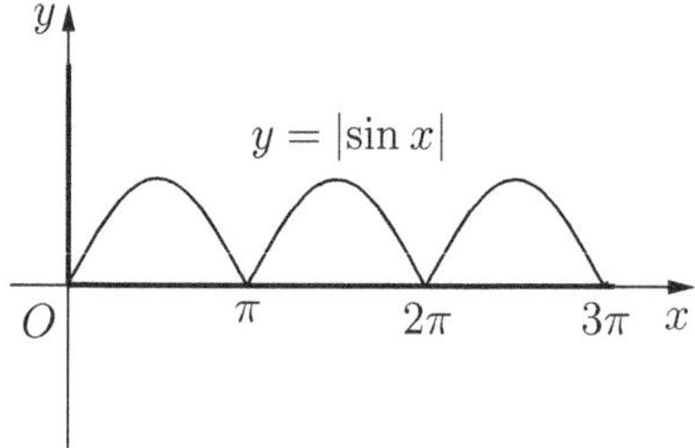

Fig. 10.18. Graph of $y = |\sin x|$.

which, after a simplification, shows that

$$a_k = -\frac{4}{\pi(4k^2-1)}, \quad k \geq 0.$$

In particular, $a_0 = 4/\pi$. Thus, the Fourier series expansion of the given function is (see Figure 10.19)

$$|\sin x| = \frac{2}{\pi} - \frac{4}{\pi}\sum_{k=1}^{\infty} \frac{\cos 2kx}{4k^2-1}, \quad x \in [-\pi, \pi].$$

Note that we have an equal sign rather than $\sim$, because the Fourier series converges absolutely and uniformly on $[-\pi, \pi]$, and hence on $\mathbb{R}$. ●

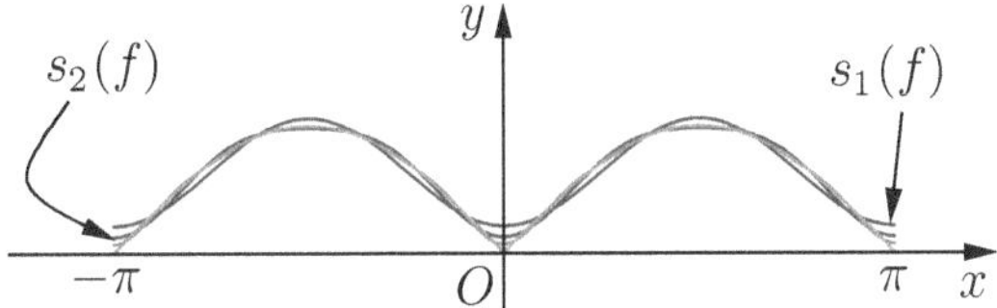

Fig. 10.19. The nth partial sums of $|\sin x| \sim \frac{2}{\pi} - \frac{4}{\pi}\sum_{k=1}^{\infty} \frac{\cos(2kx)}{(4k^2-1)}$, for $n = 1, 2, 4, 30$.

Example 10.23. **(a)** Expand $f(x) = \cos x$ as a half-range sine series on $[0, \pi]$.
(b) Does the series converge for each point in $[-\pi, \pi]$?
(c) Define $g(x) = \sum_{k=1}^{\infty} b_k \sin kx$ and sketch the graph of g on $[-\pi, \pi]$, where the b_k are the Fourier coefficients of f.
(d) Determine when $f(x) = g(x)$ on $[-\pi, \pi]$.

Solution. We have

$$\cos x \sim \sum_{k=1}^{\infty} b_k \sin kx, \quad b_k = \frac{2}{\pi}\int_0^{\pi} \cos x \sin kx\, dx.$$

Clearly, $b_1 = (1/\pi)\int_0^{\pi} \sin(2x)\, dx = 0$. For $k > 1$,

$$\begin{aligned} 2\int_0^{\pi} \cos x \sin kx\, dx &= \int_0^{\pi} (\sin(k+1)x + \sin(k-1)x)\, dx \\ &= -\left[\frac{\cos(k+1)x}{k+1} + \frac{\cos(k-1)x}{k-1}\right]\Bigg|_0^{\pi} \\ &= -\left[\frac{(-1)^{k+1}-1}{k+1} + \frac{(-1)^{k-1}-1}{k-1}\right] \\ &= -[(-1)^{k-1}-1]\frac{2k}{k^2-1} \\ &= \begin{cases} 0 & \text{if } k = 2n-1,\ n > 1, \\ \dfrac{8n}{4n^2-1} & \text{if } k = 2n,\ n \geq 1. \end{cases} \end{aligned}$$

Thus (see Figure 10.20),

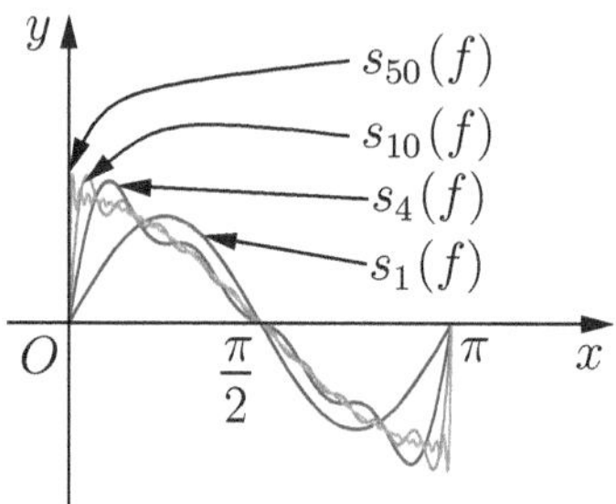

Fig. 10.20. $\cos x \sim \frac{8}{\pi} \sum_{n=1}^{\infty} \frac{n}{(4n^2-1)} \sin(2nx)$ on $(0, \pi)$.

$$\cos x \sim \frac{8}{\pi} \sum_{n=1}^{\infty} \frac{n}{4n^2-1} \sin(2nx) \quad \text{on } (0, \pi). \qquad \bullet$$

Example 10.24. Find the Fourier cosine series of $f(x) = \sin x$ on $[0, \pi)$.

Solution. Note that the even extension $F(x)$ of $f(x) = \sin x$ is given by $F(x) = |\sin x|$ on $(-\pi, \pi)$, and hence F may be treated as a π-periodic function as in Example 10.22. We could also use the above extension process of half-range cosine series. According to this,

$$\sin x \sim \frac{a_0}{2} + \sum_{k=1}^{\infty} a_k \cos kx, \quad a_k = \frac{2}{\pi} \int_0^{\pi} \sin x \cos kx \, dx.$$

We see that $a_0 = 4/\pi$ and $a_1 = 0$. For $k > 1$, we have

$$\begin{aligned} 2\int_0^{\pi} \sin x \cos kx \, dx &= \int_0^{\pi} [\sin(k+1)x - \sin(k-1)x] \, dx \\ &= -[(-1)^{k-1} - 1] \left[\frac{1}{k+1} - \frac{1}{k-1} \right] \\ &= [(-1)^{k-1} - 1] \frac{2}{k^2-1}. \end{aligned}$$

Thus on $(0, \pi)$,

$$\sin x \sim \frac{2}{\pi} + \frac{2}{\pi} \sum_{k=2}^{\infty} \frac{(-1)^{k-1} - 1}{k^2-1} \cos kx = \frac{2}{\pi} - \frac{4}{\pi} \sum_{n=1}^{\infty} \frac{\cos(2nx)}{4n^2-1}.$$

Since the sum of the series on the right is even, it converges to $|\sin x|$ rather than $\sin x$ when $-\pi \leq x \leq 0$. This implies, by periodicity, that the Fourier series indeed converges to $|\sin x|$ for all real x. Thus (see also Example 10.22)

$$|\sin x| = \frac{2}{\pi} - \frac{4}{\pi}\sum_{n=1}^{\infty}\frac{\cos(2nx)}{4n^2-1} \quad \text{for all } x \in \mathbb{R}.$$

●

Example 10.25. Consider $f(x) = x|x|$ for $-\pi \leq x < \pi$. Then f is an odd function with $L = \pi$. Therefore, $a_n = 0$ for $n \geq 0$. The periodic extension is given by

$$f(x) = f(x+2\pi), \quad f(\pi) = -\pi^2 = f(-\pi).$$

Also, we observe that $f(x)\sin nx$ is even, and so for all $n \geq 1$, we have

$$b_n = \frac{1}{\pi}\int_{-\pi}^{\pi} x|x|\sin nx\,dx = \frac{2}{\pi}\int_0^{\pi} x^2\sin nx\,dx,$$

so that

$$\begin{aligned}
b_n &= \frac{2}{\pi}\left[-\frac{x^2\cos nx}{n}\bigg|_0^{\pi} + \frac{2}{n}\int_0^{\pi} x\cos nx\,dx\right]\\
&= \frac{2}{\pi}\left[-\frac{\pi^2(-1)^n}{n} + \frac{2}{n}\left\{\frac{x\sin nx}{n}\bigg|_0^{\pi} - \frac{1}{n}\int_0^{\pi}\sin nx\,dx\right\}\right]\\
&= \frac{2}{\pi}\left[\frac{\pi^2(-1)^{n+1}}{n} + \frac{2}{n^2}\frac{\cos nx}{n}\bigg|_0^{\pi}\right]\\
&= \frac{2}{\pi}\left[\frac{\pi^2(-1)^{n+1}}{n} + \frac{2[(-1)^n-1]}{n^3}\right]\\
&= \frac{2}{\pi}\left[\frac{(-1)^{n-1}\pi^2n^2 + 2[(-1)^n-1]}{n^3}\right]
\end{aligned}$$

and a simplification gives

$$b_n = \begin{cases} \dfrac{2(\pi^2n^2-4)}{\pi n^3} & \text{if } n \text{ is odd,}\\ -\dfrac{2\pi}{n} & \text{if } n \text{ is even,} \quad n \geq 1. \end{cases}$$

The Fourier series of $f(x) = x|x|$ for $-\pi \leq x < \pi$ follows. ●

10.2.4 Issues Concerning Convergence

The following lemma, which is of central importance, tells us why Fourier coefficients are so important. Of course for our presentation, it suffices to consider Φ as in (10.4) (so that ϕ_k's, c_k's and d_k's are all real).

Lemma 10.26 (Best approximation). *Let $\Phi = \{\phi_1, \dots, \phi_n\}$ be an orthonormal set of functions in the inner product space $\mathcal{E}$, and let c_k be the Fourier coefficients of f relative to ϕ_k:*

$$c_k = \frac{1}{\pi}\int_{-\pi}^{\pi} f(x)\overline{\phi_k(x)}\,\mathrm{d}x := \langle f, \phi_k\rangle.$$

If $T_n(x)$ is an arbitrary Fourier polynomial relative to ϕ_k, that is, $T_n(x) = \sum_{k=1}^{n} d_k\phi_k(x)$ for some constants $d_1, \dots, d_n$, then we have

$$\left\| f - \sum_{k=1}^{n} c_k\phi_k(x)\right\|^2 \leq \|f - T_n\|^2,$$

with equality if and only if $c_k = d_k$ for each $k = 1, \dots, n$. Moreover,

$$\sum_{k=1}^{n} |c_k|^2 \leq \frac{1}{\pi}\int_{-\pi}^{\pi} |f(x)|^2\,\mathrm{d}x. \tag{10.12}$$

Proof. Set $S_n(x) = \sum_{k=1}^{n} c_k\phi_k(x)$. Then we have

$$\begin{aligned}
\|f - T_n\|^2 &= \frac{1}{\pi}\int_{-\pi}^{\pi} |f(x) - T_n(x)|^2\,\mathrm{d}x \\
&= \frac{1}{\pi}\int_{-\pi}^{\pi} |f(x)|^2\,\mathrm{d}x - 2\,\mathrm{Re}\sum_{k=1}^{n}\overline{d_k}\,\frac{1}{\pi}\int_{-\pi}^{\pi} f(x)\overline{\phi_k(x)}\,\mathrm{d}x \\
&\quad + \frac{1}{\pi}\int_{-\pi}^{\pi} |T_n(x)|^2\,\mathrm{d}x \\
&= \frac{1}{\pi}\int_{-\pi}^{\pi} |f(x)|^2\,\mathrm{d}x - 2\,\mathrm{Re}\sum_{k=1}^{n} c_k\overline{d_k} + \sum_{k=1}^{n} |d_k|^2 \\
&= \frac{1}{\pi}\int_{-\pi}^{\pi} |f(x)|^2\,\mathrm{d}x - \sum_{k=1}^{n} |c_k|^2 + \sum_{k=1}^{n} |c_k - d_k|^2 \\
&= \|f - S_n\|^2 + \sum_{k=1}^{n} |c_k - d_k|^2,
\end{aligned} \tag{10.13}$$

and therefore

$$\|f - T_n\|^2 \geq \|f - S_n\|^2,$$

with equality if and only if $c_k = d_k$ for each $k = 1, \dots, n$. Note that f and ϕ_k are fixed, while the d_k are allowed to vary. In particular, setting $d_k = c_k$ in (10.13) shows that the minimum value of $\|f - T_n\|^2$ is given by

$$\min_{T_n} \|f - T_n\|^2 = \frac{1}{\pi}\int_{-\pi}^{\pi} |f(x)|^2\,\mathrm{d}x - \sum_{k=1}^{n} |c_k|^2 = \|f\|^2 - \sum_{k=1}^{n} |c_k|^2,$$

which has to be nonnegative. This gives

$$\sum_{k=1}^{n} |c_k|^2 \leq \frac{1}{\pi} \int_{-\pi}^{\pi} |f(x)|^2 \, dx \quad \text{for all } n.$$ ∎

Upon letting $n \to \infty$, because the sequence $\left\{\sum_{k=1}^{n} |c_k|^2\right\}$ of partial sums of the series $\sum_{k=1}^{\infty} |c_k|^2$ is bounded above by $\|f\|^2$, we obtain the following result as a consequence of the monotone convergence theorem.

Corollary 10.27. *Suppose that* $\{\phi_k(x)\}_{k\geq 1}$ *is an orthonormal set in* $\mathcal{E}$ *and* $\sum_{k=1}^{\infty} c_k \phi_k(x)$ *is the Fourier series of* $f \in \mathcal{E}$ *relative to* $\{\phi_k\}_{k\geq 1}$. *Then*

(a) $\sum_{k=1}^{\infty} |c_k|^2 \leq (1/\pi) \int_{-\pi}^{\pi} |f(x)|^2 \, dx$, *where* $c_k = (1/\pi) \int_{-\pi}^{\pi} f(x) \overline{\phi_k(x)} \, dx$;
(b) $c_k \to 0$ *as* $k \to \infty$.

The inequality **(a)** is called Bessel's inequality. Part **(b)** of this corollary follows easily from the fact that the general term of a convergent series must approach zero.

Because the trigonometric system Φ defined by (10.4) forms an orthonormal basis with $\{a_0/\sqrt{2}, a_k, b_k\}$ as the corresponding Fourier coefficients associated with Φ, it follows that

$$\frac{a_0^2}{2} + \sum_{k=1}^{\infty} (a_k^2 + b_k^2) \leq \frac{1}{\pi} \int_{-\pi}^{\pi} f^2(x) \, dx, \tag{10.14}$$

which is the form of Bessel's inequality used for the basic trigonometric system Φ. Here the a_k and b_k are defined by (10.5). Note that the series on the left-hand side of this inequality is convergent whenever f is a square-integrable function. Equality holds in the inequality (10.14), although the inequality is sufficient for the convergence discussion. We may state (10.14) as the following theorem.

Theorem 10.28. *The sum of the squares of the Fourier coefficients of any square-integrable function always converges.*

Also, it is important to point out that there exists a function that is integrable (but not square-integrable) whose Fourier series diverges, but we shall not prove this fact in this book.

In particular, Bessel's inequality (10.14) gives that $a_k^2 \to 0$ and $b_k^2 \to 0$ as $k \to \infty$ and therefore,

$$\lim_{k\to\infty} a_k = 0 = \lim_{k\to\infty} b_k.$$

By the formula (10.5), it follows that

$$\lim_{k\to\infty} \int_{-\pi}^{\pi} f(t) \cos kt \, dt = 0 = \lim_{k\to\infty} \int_{-\pi}^{\pi} f(t) \sin kt \, dt. \tag{10.15}$$

Indeed, the following general result holds for an arbitrary interval $[a, b]$, although we do not include its proof here.

Lemma 10.29 (Riemann–Lebesgue lemma). *If f is piecewise continuous on $[a,b]$ and absolutely integrable on $[a,b]$, then*

$$\lim_{\alpha\to\pm\infty}\int_a^b f(t)\cos\alpha t\,dt = 0 = \lim_{\alpha\to\pm\infty}\int_a^b f(t)\sin\alpha t\,dt,$$

where $\alpha\in\mathbb{R}$.

10.2.5 Dirichlet's Kernel and Its Properties

In order to establish sufficient condition for the convergence of a Fourier series at a particular point, we pay some attention to representing the partial sums. To do this, we need a simple formula for calculating the sum of the "simplest" trigonometric polynomial

$$D_n(t) = \frac{1}{2} + \sum_{k=1}^{n}\cos kt. \tag{10.16}$$

We see that the formula for $D_n(t)$ is useful in examining the behavior of the partial sums of the Fourier series of a periodic function. First, we find that

$$D_n(t) = \begin{cases} \dfrac{\sin\left(n+\frac{1}{2}\right)t}{2\sin\frac{1}{2}t} & \text{if } t/2\pi\notin\mathbb{Z},\\[2mm] \dfrac{1}{2}+n & \text{if } t/2\pi\in\mathbb{Z}. \end{cases} \tag{10.17}$$

The function $D_n(t)$ is called *Dirichlet's kernel*, and it changes sign more rapidly as n increases. Indeed, if $t=2m\pi$ for some $m\in\mathbb{Z}$, then from (10.16), it follows that $D_n(2m\pi)$ equals $\frac{1}{2}+n$. If $\sin\frac{1}{2}t\neq 0$, then multiplying both sides of the equality (10.16) by $2\sin\frac{1}{2}t$, we have

$$\begin{aligned} 2D_n(t)\sin\frac{1}{2}t &= \sin\frac{1}{2}t + \sum_{k=1}^{n} 2\cos kt\sin\frac{1}{2}t\\ &= \sin\frac{1}{2}t + \sum_{k=1}^{n}\left\{\sin\left(k+\frac{1}{2}\right)t - \sin\left(k-\frac{1}{2}\right)t\right\}\\ &= \sin\left(n+\frac{1}{2}\right)t, \end{aligned}$$

and (10.17) follows.

We shall now formulate some preliminary properties of Dirichlet's kernel (see Figure 10.21 for the graph of $D_n(t)$). We observe that D_n is even. Recall that if we substitute $t=2m\pi$ in both sides of $D_n(t)$ in (10.16), we see that

$$D_n(2m\pi) = n+\tfrac{1}{2},\quad m\in\mathbb{Z}.$$

From the first expression on the right of $D_n(t)$ in (10.17), it follows that the points $t = 2m\pi$ ($m = 0, \pm 1, \ldots$) where both the numerator and denominator vanish are removable discontinuities:

$$\lim_{t\to 2m\pi} D_n(t) = \lim_{t\to 2m\pi} \frac{(n+\frac{1}{2})\cos(n+\frac{1}{2})t}{2\cdot\frac{1}{2}\cos\frac{1}{2}t} = \frac{(n+\frac{1}{2})(-1)^m}{(-1)^m} = n + \frac{1}{2}.$$

Thus, $D_n(t)$ is continuous on $\mathbb{R}$. Moreover, since $D_n(t)$ is even,

$$\frac{1}{\pi}\int_{-\pi}^{\pi} D_n(t)\,\mathrm{d}t = \frac{2}{\pi}\int_0^{\pi} D_n(t)\,\mathrm{d}t = \frac{2}{\pi}\left[\frac{1}{2}\int_0^{\pi}\mathrm{d}t + \sum_{k=1}^{n}\int_0^{\pi}\cos kt\,\mathrm{d}t\right] = 1,$$

for any n whatsoever. By the triangle inequality, for all $t \in \mathbb{R}$ and each $n \in \mathbb{N}$, we have

$$|D_n(t)| \le \frac{1}{2} + \sum_{k=1}^{n} |\cos kt| = \frac{1}{2} + n.$$

Finally, in view of the well-known Jensen's inequality, namely $\sin\theta \ge 2\theta/\pi$ for $\theta \in [0, \pi/2]$, we have

$$\frac{1}{2\sin(t/2)} \le \frac{\pi}{2t} \quad \text{for } t \in (0, \pi),$$

and therefore

$$|D_n(t)| \le \frac{\pi}{|t|} \quad \text{for } 0 < |t| < \pi.$$

Thus the basic properties of $D_n(t)$ may be summarized as follows.

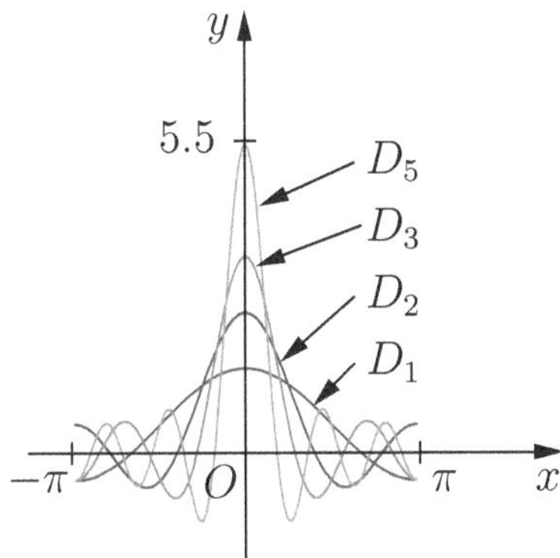

Fig. 10.21. The graph of $D_n(t) = \frac{1}{2} + \sum_{k=1}^{n}\cos(kt)$ for certain values of n.

Proposition 10.30. *The Dirichlet kernel $D_n(t)$ defined by* (10.16) *is even, and $|D_n(t)| \leq \frac{1}{2} + n$ for each $t \in \mathbb{R}$ and each $n \in \mathbb{N}$. Moreover,*

$$\frac{1}{\pi}\int_{-\pi}^{\pi} D_n(t)\,\mathrm{d}t = \frac{2}{\pi}\int_{0}^{\pi} D_n(t)\,\mathrm{d}t = 1 \quad \textit{for each } n. \tag{10.18}$$

In addition, for $0 < |t| < \pi$, $|tD_n(t)| \leq \pi$.

Next, we consider the partial sum $s_n(f)$ of the Fourier series of the 2π-periodic function f on $(-\pi, \pi)$ defined by

$$s_n(f)(t) = \frac{a_0}{2} + \sum_{k=1}^{n} (a_k \cos kt + b_k \sin kt),$$

where a_k and b_k are the Fourier coefficients of f defined by (10.5). Note that for each $n \in \mathbb{N}$,

- $s_n(f)(t)$ is continuous;
- $s_n(f)(t)$ is 2π-periodic, and $s_n(f)(\pi) = s_n(f)(-\pi)$.

Now we recall an important consequence of the uniform convergence theorem (see Theorem 9.12): if the Fourier series of f converges uniformly to f on $[-\pi, \pi]$, then f must be continuous on $[-\pi, \pi]$ with $f(\pi) = f(-\pi)$.

The Dirichlet kernel plays a key role in our next calculation, since it provides a convenient way of representing the nth partial sums $s_n(f)(t)$ in a more manageable form as an "integral transform" of f.

Proposition 10.31. *Let $f : \mathbb{R} \to \mathbb{R}$ be a 2π-periodic function that is integrable on $[-\pi, \pi]$. Then we have*

$$s_n(f)(x) = \frac{1}{\pi}\int_{0}^{\pi} (f(x+t) + f(x-t))D_n(t)\,\mathrm{d}t. \tag{10.19}$$

Proof. From the expression for the Fourier coefficients a_k and b_k in (10.5), we can rewrite our formula for $s_n(f)$ as follows: for each $x \in \mathbb{R}$,

$$\begin{aligned}
s_n(f)(x) &= \frac{a_0}{2} + \sum_{k=1}^{n} (a_k \cos kx + b_k \sin kx) \\
&= \frac{1}{\pi}\int_{-\pi}^{\pi} f(t)\left[\frac{1}{2} + \sum_{k=1}^{n} \cos kt \cos kx + \sin kt \sin kx\right] \mathrm{d}t \\
&= \frac{1}{\pi}\int_{-\pi}^{\pi} f(t)\left[\frac{1}{2} + \sum_{k=1}^{n} \cos k(t-x)\right] \mathrm{d}t \\
&= \frac{1}{\pi}\int_{-\pi}^{\pi} f(t)\, D_n(t-x)\,\mathrm{d}t, \quad \text{by (10.16)/(10.17)}, \\
&= \frac{1}{\pi}\int_{-\pi-x}^{\pi-x} f(x+u) D_n(u)\,\mathrm{d}u \quad (t-x=u) \\
&= \frac{1}{\pi}\int_{-\pi}^{\pi} f(x+t) D_n(t)\,\mathrm{d}t.
\end{aligned}$$

Here both $f(t)$ and $D_n(t)$ are periodic functions of period 2π, and therefore the last equality is a consequence (see Lemma 10.1) of the fact that for every 2π-periodic function g and for each real a, we have

$$\int_{-\pi-a}^{\pi-a} g(u)\,\mathrm{d}u = \int_{-\pi}^{\pi} g(u)\,\mathrm{d}u.$$

Since $D_n(t)$ is even, we have

$$\int_{-\pi}^{\pi} f(x+t)D_n(t)\,\mathrm{d}t = \int_{-\pi}^{\pi} f(x-t)D_n(t)\,\mathrm{d}t,$$

and therefore the last representation for $s_n(f)(x)$ can be rewritten as

$$s_n(f)(x) = \frac{1}{\pi}\int_{-\pi}^{\pi} f(x+t)D_n(t)\,\mathrm{d}t = \frac{1}{\pi}\int_{-\pi}^{\pi} f(x-t)D_n(t)\,\mathrm{d}t, \tag{10.20}$$

or equivalently

$$s_n(f)(x) = \frac{1}{2\pi}\int_{-\pi}^{\pi} (f(x+t)+f(x-t))D_n(t)\,\mathrm{d}t. \tag{10.21}$$

Since the integrand in the integral in (10.21) is even (in t), the desired representation (10.19) follows. ■

Multiplying both sides of (10.18) by $f(x)$ gives

$$f(x) = \frac{2}{\pi}\int_{0}^{\pi} f(x)D_n(t)\,\mathrm{d}t = \frac{1}{\pi}\int_{-\pi}^{\pi} f(x)D_n(t)\,\mathrm{d}t. \tag{10.22}$$

Subtracting this from (10.19) gives a fundamental relation

$$s_n(f)(x) - f(x) = \frac{2}{\pi}\int_{0}^{\pi} \left(\frac{f(x+t)+f(x-t)}{2} - f(x)\right) D_n(t)\,\mathrm{d}t.$$

Thus, we have the following result.

Proposition 10.32. *Assume the hypotheses of Proposition* 10.31. *Then the sequence $s_n(f)(x)$ converges to $f(x)$ if and only if*

$$\lim_{n\to\infty} \frac{2}{\pi}\int_{0}^{\pi} \left(\frac{f(x+t)+f(x-t)}{2} - f(x)\right) D_n(t)\,\mathrm{d}t = 0.$$

(Note that the integrand is even in t.)

10.2.6 Two Versions of Dirichlet's Theorem

First we show that the Fourier series of a piecewise continuous function converges to the function at points where the function is differentiable.

Theorem 10.33. *Let $f \in \mathcal{E}$ (i.e., piecewise continuous on $[-\pi, \pi]$) and 2π-periodic on $\mathbb{R}$. Suppose that f is differentiable at x_0. Then*

$$\lim_{n\to\infty} s_n(f)(x_0) = f(x_0).$$

Proof. Let x be a point at which f is differentiable. We have to prove that

$$\lim_{n\to\infty} s_n(f)(x) = f(x), \quad \text{i.e., } R_n(x) = f(x) - s_n(f)(x) \to 0 \text{ as } n \to \infty.$$

By (10.20) and (10.22), this is equivalent to proving that $R_n(x) \to 0$ as $n \to \infty$, where

$$\begin{aligned} R_n(x) &= \frac{1}{\pi}\int_{-\pi}^{\pi} (f(x) - f(x-t))D_n(t)\,dt \\ &= \frac{1}{\pi}\int_{-\pi}^{\pi} \frac{f(x) - f(x-t)}{2\sin\frac{1}{2}t} \sin\left(n + \frac{1}{2}\right)t\,dt \\ &= \frac{1}{\pi}\int_{-\pi}^{\pi} p(t)\cos\frac{1}{2}t\sin nt\,dt + \frac{1}{\pi}\int_{-\pi}^{\pi} \frac{f(x) - f(x-t)}{2}\cos nt\,dt, \end{aligned}$$

with

$$p(t) = \frac{f(x) - f(x-t)}{2\sin\frac{1}{2}t}.$$

By the Riemann–Lebesgue lemma (see (10.15)), the second integral in the last relation approaches zero as $n \to \infty$. Therefore, it suffices to prove that the first integral in the last integral relation approaches zero as $n \to \infty$. Since $f \in \mathcal{E}$ and $\sin(t/2)$ is continuous, the function $p(t)$ is piecewise continuous in t except possibly at $t = 0$. But since f is differentiable at x, it follows that

$$\lim_{t\to 0} p(t) = \lim_{t\to 0} \frac{f(x) - f(x-t)}{t}\,\frac{t/2}{\sin\frac{1}{2}t} = f'(x),$$

and therefore $p(t)\cos\frac{1}{2}t$ belongs to $\mathcal{E}$. Again, by the Riemann–Lebesgue lemma (see (10.15)), it follows that

$$\lim_{n\to\infty} \frac{1}{\pi}\int_{-\pi}^{\pi} p(t)\cos\frac{1}{2}t\sin nt\,dt = 0.$$

Thus, $R_n(x) \to 0$ as $n \to \infty$, i.e., $s_n(f)(x) \to f(x)$ as $n \to \infty$. ∎

Corollary 10.34. *If f is differentiable on $[-\pi, \pi]$ and 2π-periodic on $\mathbb{R}$, then the Fourier series of f converges pointwise to $f(x)$ at every point $x \in [-\pi, \pi]$.*

Next we shall present a sufficient condition for the convergence of a Fourier series of a function at a point of discontinuity.

Theorem 10.35. *Let f and f' be piecewise continuous on $[-\pi, \pi]$, and 2π-periodic on $\mathbb{R}$. Then*

$$\lim_{n\to\infty} s_n(f)(x) = \frac{f(x-) + f(x+)}{2}.$$

Proof. We need to prove that

$$s_n(f)(x) \to \frac{f(x-) + f(x+)}{2} \quad \text{as } n \to \infty.$$

By (10.19) and (10.22), this is equivalent to proving that

$$R_n(x) = s_n(f)(x) - \frac{f(x-) + f(x+)}{2} \to 0 \quad \text{as } n \to \infty,$$

where

$$R_n(x) = \frac{2}{\pi} \int_0^\pi \left(\frac{f(x+t) + f(x-t)}{2} - \frac{f(x-) + f(x+)}{2} \right) D_n(t)\, dt.$$

Using the expression for $D_n(t)$, we can rewrite $R_n(x)$ as

$$R_n(x) = \frac{1}{\pi} \int_0^\pi p(t) \sin\left(n + \frac{1}{2}\right) t\, dt + \frac{1}{\pi} \int_0^\pi q(t) \sin\left(n + \frac{1}{2}\right) t\, dt, \quad (10.23)$$

where

$$p(t) = \frac{f(x+t) - f(x+)}{2 \sin \frac{1}{2} t} \quad \text{and} \quad q(t) = \frac{f(x-t) - f(x-)}{2 \sin \frac{1}{2} t}.$$

We now show that the two integrals in (10.23) approach 0 as $n \to \infty$. We shall first deal with the first integral. Since f is a piecewise continuous function of t,

$$p(t) = \frac{f(x+t) - f(x+)}{t} \frac{t/2}{\sin \frac{1}{2} t}$$

is piecewise continuous on the half-open interval $(0, \pi]$, and

$$\lim_{t\to 0} p(t) = \lim_{t\to 0} \frac{f(x+t) - f(x+)}{t}.$$

Since f' is piecewise continuous, the limit on the right (right-hand derivative) exists. Thus, $p(t)$ is piecewise continuous on $[0, \pi]$. Similarly, $q(t)$ is piecewise continuous on $[0, \pi]$.

Next, expanding $\sin(n+\frac{1}{2})t$, the first integral in (10.23) becomes

$$\frac{1}{\pi}\int_0^\pi p(t)\sin\left(n+\frac{1}{2}\right)t\,dt = \frac{1}{\pi}\int_0^\pi p(t)\cos\frac{1}{2}t\sin nt\,dt + \frac{1}{\pi}\int_0^\pi \frac{f(x+t)-f(x+)}{2}\cos nt\,dt,$$

with each integral on the right approaching zero (by the Riemann–Lebesgue lemma) as $n\to\infty$. Thus,

$$\frac{1}{\pi}\int_0^\pi p(t)\sin\left(n+\frac{1}{2}\right)t\,dt \to 0 \quad \text{as } n\to\infty.$$

Similarly, the second integral in (10.23) approaches zero. Therefore, we conclude that $R_n(x)\to 0$ as $n\to\infty$, and the proof is complete. ∎

Corollary 10.36. *Let f and f' be piecewise continuous on $[-\pi,\pi]$ and 2π-periodic on $\mathbb{R}$. If f is continuous at x_0, then the Fourier series of f converges to $f(x_0)$.*

10.2.7 Questions and Exercises

Questions 10.37.

1. If f is even (respectively odd) and integrable on $[a,b]$, must $F(x) = \int_0^x f(t)\,dt$ be odd (respectively even) on $[a,b]$?
2. If f is even (respectively odd) and differentiable on $[a,b]$, must $F(x) = f'(x)$ be odd (respectively even) on $[a,b]$?
3. Suppose that f and g are either both even or both odd. Must fg be even?
4. If f is odd, must we have $f(0)=0$?
5. Suppose that f and g are even and odd, respectively. Must both the compositions $f\circ g$ and $g\circ f$ be even?
6. Suppose that f is a real-valued function defined on $[-c,c]$. Can f be expressed as the sum of even and odd functions?
7. Does there exist a function f on $[-\pi,\pi]$ such that its Fourier series is
$$\sin x + \sin 2x + \sin 3x + \cdots + \sin nx + \cdots?$$
8. Suppose that $f(x) = f(\pi - x)$ and that f is periodic with period 2π. What can be said about the Fourier coefficients of f?
9. Suppose that $f(x) = ax^2 + bx + c$ $(-\pi < x < \pi)$ for some real constants. What will be the graph of $f(x)$? Is the periodic extension of f a continuous function? What are its discontinuities?
10. Does there exist a periodic function that is not Riemann integrable?

Exercises 10.38.

1. Determine which of the following functions are even, which are odd, and which are neither.
   (a) $x \sin x$. **(b)** $x^5 \cos nx$. **(c)** e^{x^2-x}. **(d)** $xf(x^2)$. **(e)** $\log\left(\frac{1+x}{1-x}\right)$.
2. Suppose $f \in C^1([a,b])$ and $f(a) = f(b) = 0$. Using the Fourier series expansion of f, show that
$$\int_a^b |f(x)|^2 \,dt \le \left(\frac{b-a}{\pi}\right)^2 \int_a^b |f'(x)|^2 \,dx.$$
3. Assume that $f(x) = x^2$. Let
$$\frac{a_0}{2} + \sum_{n=1}^{\infty} (a_n \cos nx + b_n \sin nx) \quad \text{and} \quad \frac{A_0}{2} + \sum_{n=1}^{\infty} (A_n \cos nx + B_n \sin nx)$$
   be the Fourier series of f on $[-\pi, \pi]$ and of f on $[0, 2\pi]$, respectively. Define
$$h(x) = \frac{a_0 - A_0}{2} + \sum_{n=1}^{\infty} [(a_n - A_n) \cos nx + (b_n - B_n) \sin nx].$$
   Calculate h and give a careful sketch of the graph of h on $[-\pi, 2\pi]$.
4. Assume that $g(x) = x^2$ for $x \in [-\pi, \pi] \setminus \{\pi/2\}$ and $g(\pi/2) = 0$. How do the Fourier coefficients of g compare with those in Exercise 10.38(3) for $f(x)$? What can we conclude from these two problems?
5. Let $f : \mathbb{R} \to \mathbb{R}$ be a piecewise continuous function that is π-periodic. Let
$$\frac{a_0}{2} + \sum_{n=1}^{\infty} [a_n \cos 2nx + b_n \sin 2nx] \quad \text{and} \quad \frac{A_0}{2} + \sum_{n=1}^{\infty} (A_n \cos nx + B_n \sin nx)$$
   be the Fourier series of f on $[0, \pi]$ and of f on $[-\pi, \pi]$. Express A_n and B_n in terms of a_n and b_n.
6. Define
$$f(x) = \begin{cases} A \sin(2\pi L x) & \text{for } 0 < x < L/2, \\ 0 & \text{for } (L/2) \le x < L. \end{cases}$$
   Compute the Fourier series of f on the interval $[0, L]$.
7. Define $f(x) = x(\pi - x)$ for $x \in [0, \pi]$.
   (a) Find the sine and the cosine series of f.
   (b) Using the corresponding Fourier series, prove the following:
   (i) $\displaystyle\sum_{n=1}^{\infty} \frac{1}{n^6} = \frac{\pi^6}{945}$ and $\displaystyle\sum_{n=1}^{\infty} \frac{1}{n^4} = \frac{\pi^4}{90}$.
   (ii) $\displaystyle\sum_{\substack{n=1 \\ n-\text{odd}}}^{\infty} \frac{(-1)^{n-1}}{(2n-1)^3} = \frac{\pi^3}{32}$ and $\displaystyle\sum_{n=1}^{\infty} \frac{(-1)^{n-1}}{n^2} = \frac{\pi^2}{12}$.

8. Find the Fourier series representation for

$$\text{(a) } f(x) = \begin{cases} -1 & \text{if } x \in (-\pi, -\frac{\pi}{2}), \\ 1 & \text{if } x \in (-\frac{\pi}{2}, \frac{\pi}{2}), \\ -1 & \text{if } x \in (\frac{\pi}{2}, \pi). \end{cases} \qquad \text{(b) } f(x) = \begin{cases} -1 & \text{if } x \in (-\pi, -\frac{\pi}{2}), \\ 0 & \text{if } x \in (-\frac{\pi}{2}, \frac{\pi}{2}), \\ 1 & \text{if } x \in (\frac{\pi}{2}, \pi). \end{cases}$$

   How do the Fourier coefficients of f in **(a)** compare with those in **(b)**? What can we conclude from these two problems?

9. To obtain more practice with Fourier series, calculate the Fourier series for the five functions given by $|x|$, sign (x), x, x^2, and $|\sin x|$. Discuss the relationships among the Fourier series of these five functions.

10. Define $f(x) = 1 - x/2$ for $x \in (-\pi, \pi)$. Graph the 2π-periodic extension of f. Find the Fourier series expansion of f.

11. Define

$$f(x) = \begin{cases} 1 + x & \text{for } -1 < x < 0, \\ -1 & \text{for } 0 < x < 1. \end{cases}$$

   Compute the first five terms in the Fourier expansion of f. Plot f and $s_n(f)$ on $[-3/2, 3/2]$ on the same graph.

12. Does there exist a continuous function on $[-\pi, \pi]$ that generates the Fourier series

$$\sum_{n=1}^{\infty} \frac{(-1)^n}{n^3} \sin nx?$$

   If not, justify your answer. If yes, can we use Parseval's formula to prove that

$$\zeta(6) := \sum_{n=1}^{\infty} \frac{1}{n^6} = \frac{\pi^6}{945}?$$

13. Determine the Fourier series of

$$f(x) = 3x + 2 \quad \text{on } [-\pi, \pi), \quad f(\pi) = -3\pi + 2; \quad f(x) = f(x + 2\pi).$$

14. Show that

$$\cos(\alpha x) = \frac{\sin(\pi\alpha)}{\pi\alpha} + \frac{2\alpha \sin(\pi\alpha)}{\pi} \sum_{k=1}^{\infty} (-1)^{k-1} \frac{\cos(kx)}{k^2 - \alpha^2} \quad \text{for } x \in [-\pi, \pi],$$

   where $\alpha \notin \mathbb{Z}$. Deduce that

$$\pi \cot(\pi\alpha) = \frac{1}{\alpha} - \sum_{k=1}^{\infty} \frac{2\alpha}{k^2 - \alpha^2}.$$

15. Expand the function $f(x)$ defined by

$$f(x) = \begin{cases} x & \text{for } 0 \le x \le L/2, \\ L - x & \text{for } L/2 < x \le L, \end{cases}$$

   in a sine series.

16. Define

$$f(x) = \begin{cases} \pi & \text{for } 0 < x < \pi/2, \\ 0 & \text{for } x \in (-\pi, 0] \cup [\pi/2, \pi). \end{cases}$$

Find the Fourier series, Fourier cosine series, and Fourier sine series of f. In each case, sketch the graph of the sum of the series for x in the interval $[-4\pi, 4\pi]$.

17. Let f be a periodic function defined by

$$f(x) = \begin{cases} -\pi & \text{for } -\pi < x < 0, \\ x & \text{for } 0 < x < \pi, \end{cases} \quad \text{and} \quad f(x) = f(x + 2\pi).$$

Draw the graph of $f(x)$ and determine the Fourier series of f.

18. Define $f : [-\pi, \pi] \to \mathbb{R}$ by

$$f(x) = \begin{cases} 1 & \text{for } |x| \leq \pi/2, \\ 0 & \text{elsewhere.} \end{cases}$$

   (a) Draw the graph of $f(x)$ and extend to $\mathbb{R}$ as a 2π-periodic function.
   (b) Show that f is piecewise continuous, and determine the points of discontinuity of the periodic extension of f.
   (c) Show that f' is piecewise continuous.
   (d) Verify whether Theorem 10.35 is applicable.
   (e) Determine the Fourier series of f.

11

Functions of Bounded Variation and Riemann–Stieltjes Integrals

In Section 11.1, we introduce a special class of functions, namely, functions of bounded variation. In Section 11.1, we shall also discuss several nice properties of functions in the class $BV([a,b])$ of functions of bounded variation on $[a,b]$. Monotone functions on $[a,b]$ have nice properties. For example, they are integrable on $[a,b]$ and have only a countable number of jump discontinuities. In this section, we shall also show that every monotone function is a function of bounded variation, and hence the class $BV([a,b])$ contains the class of monotone functions on $[a,b]$. We shall show that increasing functions are in some sense the only functions of bounded variation. More precisely (see Theorem 11.19), every function of bounded variation is the difference of two increasing functions. As an application of functions of bounded variation, in Section 11.2 we shall consider important generalizations of the Darboux and Riemann integrals called the Darboux–Stieltjes and Riemann–Stieltjes integrals. The theory of Stieltjes integrals is almost identical to that of Riemann integrals, except that the notion of length of an integral is replaced by a more general concept of α-length. Stieltjes integrals are particularly useful in probability theory.

11.1 Functions of Bounded Variation

Let $f : [a,b] \to \mathbb{R}$ and let $P = \{x_0, x_1, \ldots, x_n\}$ be a partition of $[a,b]$. Introduce $V(P,f)$ by

$$V(P,f) = \sum_{k=1}^{n} |\Delta f_k|, \quad \Delta f_k := f(x_k) - f(x_{k-1}). \tag{11.1}$$

Note that $V(P,f) \geq 0$. We say that f is a function of *bounded variation* on $[a,b]$, or simply a BV function on $[a,b]$, if there exists a constant M such

S. Ponnusamy, *Foundations of Mathematical Analysis*,
DOI 10.1007/978-0-8176-8292-7_11,

that $V(P,f) \leq M$ for each partition P of $[a,b]$. We shall say simply that f is of bounded variation on $[a,b]$. The collection of all BV functions on $[a,b]$ is denoted by $BV([a,b])$. If

$$V_f[a,b] := \sup\{V(P,f) : P \in P[a,b]\},$$

then $V_f[a,b]$ is called the *total variation* of f on $[a,b]$. In general, $V_f[a,b]$ could be infinite. It follows that $V_f[a,b] = 0$ if and only if f is a constant on $[a,b]$. Also, it is clear that f is a function of bounded variation if and only if $V_f[a,b] < \infty$.

Examples 11.1. **(i)** If $f(x) = x$ on $[a,b]$, then by (11.1), it follows that

$$V(P,f) = \sum_{k=1}^{n} |x_k - x_{k-1}| = \sum_{k=1}^{n} (x_k - x_{k-1}) = x_n - x_0 = b - a,$$

and so $V_f[a,b] = b - a$.

(ii) Recall that $|\sin x| \leq |x|$ for $x \in \mathbb{R}$ and

$$|\sin x - \sin y| = \left|2\cos\left(\frac{x+y}{2}\right)\sin\left(\frac{x-y}{2}\right)\right| \leq 2\left|\frac{x-y}{2}\right| = |x-y|.$$

Alternatively, applying the mean value theorem on the interval $[x,y]$, it follows that

$$|\sin x - \sin y| = |\cos c|\,|y - x| \leq |x - y| \quad \text{for some } c \in (x,y).$$

Thus if $f(x) = \sin x$ on $[a,b]$, then by (11.1), we have

$$V(P,f) = \sum_{k=1}^{n} |\sin x_k - \sin x_{k-1}| \leq \sum_{k=1}^{n} |x_k - x_{k-1}| = b - a,$$

and so $V_f[a,b] \leq b - a$. •

11.1.1 Sufficient Conditions for Functions of Bounded Variation

There are several different classes of functions that are of bounded variation. For instance, we have these.

Theorem 11.2. *Suppose that $f : [a,b] \to \mathbb{R}$ satisfies any one of the following:*

(a) *f is monotone on $[a,b]$;*

(b) *f is Lipschitz on $[a,b]$, i.e., there exists an $M > 0$ such that*

$$|f(x) - f(y)| \leq M|x - y| \quad \textit{for each } x, y \in [a,b];$$

(c) *f is differentiable on $[a,b]$ such that $f'(x)$ is bounded on $[a,b]$.*

Then f is of bounded variation on $[a,b]$.

Proof. Suppose that $P = \{x_0, x_1, \ldots, x_n\}$ is a partition of $[a,b]$.

(a) Suppose that f is increasing on $[a,b]$. Then

$$V(P,f) = \sum_{k=1}^{n} |f(x_k) - f(x_{k-1})| = \sum_{k=1}^{n} (f(x_k) - f(x_{k-1})) = f(b) - f(a),$$

and if f is decreasing on $[a,b]$, then we see that

$$V(P,f) = f(a) - f(b).$$

Thus, if f is monotone on $[a,b]$, then $V_f[a,b] = |f(a) - f(b)|$.

(b) If f is Lipschitz on $[a,b]$, then

$$V(P,f) = \sum_{k=1}^{n} |f(x_k) - f(x_{k-1})| \leq M \sum_{k=1}^{n} (x_k - x_{k-1}) = M(b-a).$$

(c) In this case, the hypotheses imply (by the mean value theorem) that there exists a number c in (x,y) such that

$$f(x) - f(y) = f'(c)(y-x) \quad \text{for } [x,y] \subseteq [a,b],$$

and so because f' is bounded, there exists an M such that

$$|f(x) - f(y)| \leq M|x-y|, \quad \text{i.e., } f \text{ is Lipschitz on } [a,b].$$

As in **(b)**, it follows that $V(P,f) \leq M(b-a)$.

The conclusion of each case follows. ∎

Example 11.3 (A continuous function need not be of bounded variation). Consider $f : [0,1] \to \mathbb{R}$ defined by

$$f(x) = \begin{cases} x \sin\left(\frac{\pi}{2x}\right) & \text{for } 0 < x \leq 1, \\ 0 & \text{for } x = 0. \end{cases}$$

Clearly, f is continuous on $[0,1]$ and is bounded by 1. Since $\sin x = (-1)^m$ if and only if $x = (m + (1/2))\pi$ with $m \in \mathbb{Z}$, we consider the partition

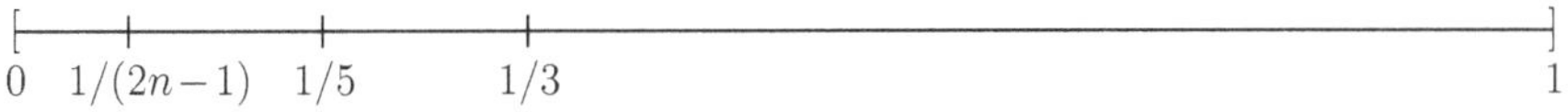

0 $1/(2n-1)$ $1/5$ $1/3$ 1

Fig. 11.1. Partition of $[0,1]$.

$$P = \left\{0, \frac{1}{2n-1}, \dots, \frac{1}{5}, \frac{1}{3}, 1\right\}, \quad \text{i.e. } x_0 = 0,\ x_k = \frac{1}{2(n-k)+1} \ (1 \le k \le n)$$

(see Figure 11.1), and find that

$$f(x_k) = \frac{1}{2(n-k)+1} \sin\left(\frac{\pi}{2}(2(n-k)+1)\right) = \frac{(-1)^{n-k}}{2(n-k)+1}.$$

Thus,

$$\begin{aligned} V(P,f) &= |f(x_1) - f(0)| + |f(x_2) - f(x_1)| + \cdots + |f(1) - f(x_{n-1})| \\ &= \frac{1}{2n-1} + \underbrace{\frac{1}{2n-3} + \frac{1}{2n-1}} + \underbrace{\frac{1}{2n-5} + \frac{1}{2n-3}} + \cdots + \underbrace{1 + \frac{1}{3}} \\ &> \sum_{k=1}^{n} \frac{1}{2k-1} \to \infty \quad \text{as } n \to \infty, \end{aligned}$$

and therefore $\{V(P,f) : P \in P[0,1]\}$ is not bounded above. Consequently, f is not a function of bounded variation on $[0,1]$. This is also an example of a uniformly continuous function that is not a function of bounded variation on $[0,1]$. ●

Example 11.4. For $\alpha \in \mathbb{R}$, consider $f_\alpha : [0,1] \to \mathbb{R}$ defined by

$$f_\alpha(x) = \begin{cases} x^2 \sin\left(\frac{\alpha}{x}\right) & \text{for } 0 < x \le 1, \\ 0 & \text{for } x = 0. \end{cases}$$

Then condition **(c)** of Theorem 11.2 is satisfied, and therefore f_α is a function of bounded variation on $[0,1]$. ●

Examples 11.5. Consider $f(x) = \sqrt{x}$ on $[0,1]$. Here f is increasing on $[0,1]$, and therefore f is a function of bounded variation on $[0,1]$. However, f is not Lipschitz, nor has it a bounded derivative on $[0,1]$.

Also, the function $f(x) = x^{1/3}$ is monotone on $[-1,1]$, but f' is not bounded on $[-1,1]$. Thus, *a function of bounded variation on $[a,b]$ need not have bounded derivative on (a,b)*. Note that f is not even differentiable at the origin. ●

Example 11.6. Consider $f(x) = x^3$ on $[-1,1]$. Then f is monotone, and hence it is a function of bounded variation. Also, for $x, y \in [-1,1]$,

$$|f(x) - f(y)| = |x^3 - y^3| = |(x-y)(x^2 + xy + y^2)| \le 3|x-y|,$$

and hence f is also Lipschitz. ●

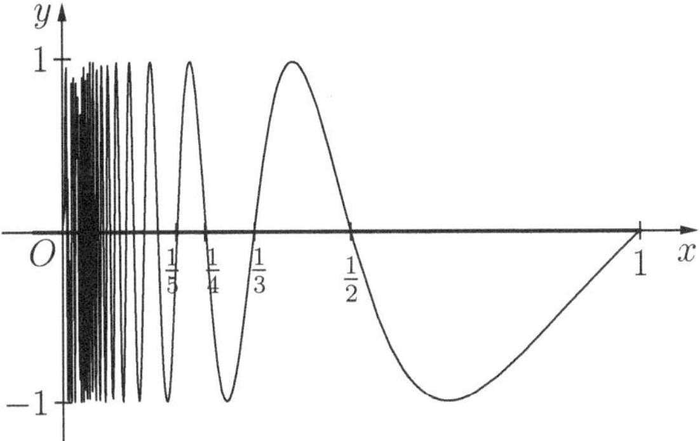

Fig. 11.2. The graph of $y = \sin(\pi/x)$ on $(0, 1]$.

Examples 11.7 (A bounded function need not be of bounded variation).

(a) Define f on $[0, 1]$ by

$$f(x) = \begin{cases} \sin(\pi/x) & \text{for } 0 < x \leq 1, \\ 0 & \text{for } x = 0. \end{cases}$$

Then f is not continuous at 0. We now show that $f \notin BV([0, 1])$. We observe that $\sin(\pi/x) = 0$ for $x = 1/n$, $n \in \mathbb{Z} \setminus \{0\}$, and

$$\sin(\pi/x) = (-1)^n \quad \text{for } x = 2/(2n + 1),\ n \in \mathbb{Z};$$

see Figure 11.2 for the graph of $y = \sin(\pi/x)$. This suggests that we choose a partition $P = \{0, 2/(2n + 1), \ldots, 2/5, 2/3, 1\}$ of $[0, 1]$. We find that $f \notin BV([0, 1])$, because

$$V(P, f) = \sum_{k=1}^{n} |f(x_k) - f(x_{k-1})| = 2n \to \infty \quad \text{as } n \to \infty.$$

(b) Define $f : [0, 1] \to \mathbb{R}$ by $f(x) = 0$ except at $x = n/(n + 1)$, where

$$f\left(\frac{n}{n+1}\right) = (-1)^n, \quad n \in \mathbb{N}.$$

For the partition $P = \{0, 1/2, 2/3, \ldots, n/(n + 1), 1\}$ of $[0, 1]$, it follows that

$$V(P, f) = \sum_{k=1}^{n} |f(x_k) - f(x_{k-1})| = 2(n - 1) \to \infty \quad \text{as } n \to \infty,$$

and hence f is not of bounded variation on $[0, 1]$. Note that f is bounded on $[0, 1]$. ●

11.1.2 Basic Properties of Functions of Bounded Variation

We see that f is of bounded variation on $[a,b]$ if f does not oscillate too wildly, and in particular, f must be bounded.

Theorem 11.8. *Let $f : [a,b] \to \mathbb{R}$ be a function of bounded variation on $[a,b]$. Then we have the following:*

(a) *f is bounded on $[a,b]$, but the converse is not necessarily true.*
(b) *$|f|$ is a function of bounded variation on $[a,b]$, but the converse is not always true.*
(c) *$1/f$ is a function of bounded variation on $[a,b]$ if $f(x) \geq m$ for some $m > 0$ and for all $x \in [a,b]$. In particular, $V_{1/f} \leq (1/m^2)V_f$.*
(d) *cf $(c \in \mathbb{R})$ is a function of bounded variation on $[a,b]$.*

Proof. Let f be a function of bounded variation on $[a,b]$.

(a) Let $x \in [a,b]$ be arbitrary. Consider the partition $P = \{a, x, b\}$ of $[a,b]$. For this partition P, there exists an $M > 0$ such that

$$V(P,f) = |f(x) - f(a)| + |f(b) - f(x)| \leq M,$$

which gives

$$|f(x)| \leq \frac{M + |f(a)| + |f(b)|}{2},$$

and so f is bounded on $[a,b]$.

(b) For any partition $P = \{x_0, x_1, x_2, \ldots, x_n\}$ of $[a,b]$, we have

$$\begin{aligned} V(P,|f|) &= \sum_{k=1}^{n} \big|\, |f(x_k)| - |f(x_{k-1})| \,\big| \\ &\leq \sum_{k=1}^{n} |f(x_k) - f(x_{k-1})| = V(P,f), \end{aligned}$$

and so $|f|$ is a function of bounded variation on $[a,b]$.

(c) Similarly, for any partition $P = \{x_0, x_1, x_2, \ldots, x_n\}$ of $[a,b]$, we have

$$\begin{aligned} V(P,1/f) &= \sum_{k=1}^{n} \left| \frac{f(x_{k-1}) - f(x_k)}{f(x_k)f(x_{k-1})} \right| \\ &\leq \frac{1}{m^2} \sum_{k=1}^{n} |f(x_k) - f(x_{k-1})| = \frac{1}{m^2} V(P,f). \end{aligned}$$

Since this is true for every partition P and $(1/m^2)V_f$ is an upper bound for every $V(P,1/f)$, $P \in P[a,b]$, this cannot be smaller than the least upper bound. Therefore, it follows that

$$V_{1/f} \leq \frac{1}{m^2} V_f.$$

(d) This case is trivial. ■

Theorem 11.9. *Let f and g be functions of bounded variation on $[a,b]$. Then so are their sum, difference, and product. In particular,*

$$V_{f\pm g} \leq V_f + V_g \quad \text{and} \quad V_{fg} \leq AV_g + BV_f,$$

where

$$A = \sup_{x\in[a,b]} |f(x)| \quad \text{and} \quad B = \sup_{x\in[a,b]} |g(x)|$$

*(note that by Theorem 11.8***(a)***, f and g must be bounded).*

Proof. Set $F(x) = f(x) + g(x)$ and $G(x) = f(x)g(x)$. Then for any partition $P = \{x_0, x_1, \ldots, x_n\}$ of $[a,b]$, we have

$$\begin{aligned} |F(x_k) - F(x_{k-1})| &= |(f(x_k) - f(x_{k-1})) + (g(x_k) - g(x_{k-1}))| \\ &\leq |f(x_k) - f(x_{k-1})| + |g(x_k) - g(x_{k-1})|, \end{aligned}$$

and so

$$V(P, f+g) \leq V(P, f) + V(P, g),$$

and therefore $f + g$ is of bounded variation on $[a,b]$. The case for $f - g$ is similar. Similarly,

$$\begin{aligned} |G(x_k) - G(x_{k-1})| &= |f(x_k)g(x_k) - f(x_{k-1})g(x_k) \\ &\quad + f(x_{k-1})g(x_k) - f(x_{k-1})g(x_{k-1})| \\ &\leq |g(x_k)|\,|f(x_k) - f(x_{k-1})| + |f(x_{k-1})|\,|g(x_k) - g(x_{k-1})| \\ &\leq B\,|f(x_k) - f(x_{k-1})| + A\,|g(x_k) - g(x_{k-1})|, \end{aligned}$$

and so

$$V(P, fg) \leq BV(P, f) + AV(P, g),$$

and therefore fg is of bounded variation on $[a,b]$. It follows that $V_{fg} \leq AV_g + BV_f$, where A and B are as in the statement. ■

Theorems 11.2**(a)** and 11.9 give the following corollary.

Corollary 11.10. *Every piecewise monotone function on $[a,b]$ is of bounded variation on $[a,b]$.*

If f and g are increasing on $[a,b]$, then so is their sum $f + g$, but the difference $f - g$ need not be. But the next corollary shows that $f - g$ is in the wider class of functions $BV([a,b])$.

Corollary 11.11. *Let f and g be increasing on $[a,b]$. Then $f - g$ is of bounded variation on $[a,b]$.*

Proof. This follows from Corollary 11.10 and Theorem 11.9. ■

This corollary has a converse that characterizes all functions in $BV([a,b])$. More precisely, we shall prove that every $f \in BV([a,b])$ can be expressed as the difference of two increasing functions on $[a,b]$. To accomplish this, we need some preparation.

Lemma 11.12. *Let f be a function of bounded variation on $[a,b]$, and let P and Q be two partitions of $[a,b]$. Then*

$$V(P,f) \leq V(Q,f) \quad \text{if } P \subseteq Q, \text{ i.e. if } Q \text{ is a refinement of } P.$$

Proof. Let $P = \{x_0, x_1, x_2, \ldots, x_n\}$ be a partition of $[a,b]$ and P_1 a new partition formed by adjoining one extra point, say $c \in (x_{k-1}, x_k)$, so that

$$P_1 = \{x_0, x_1, x_2, \ldots, x_{k-1}, c, x_k, \ldots, x_n\}.$$

Then since $|f(x_k) - f(x_{k-1})| \leq |f(x_k) - c| + |c - f(x_{k-1})|$, it follows easily that

$$V(P,f) \leq V(P_1,f).$$

Since Q is obtained by adjoining r $(r \geq 1)$ points not in P, after repeating the above argument a finite number of times, we have

$$V(P,f) \leq V(Q,f).$$

∎

Theorem 11.13 (Additivity property of total variation). *A function f is of bounded variation on $[a,b]$ if and only if f is of bounded variation on $[a,c]$ and on $[c,b]$, $c \in (a,b)$. In this case, we also have*

$$V_f[a,b] = V_f[a,c] + V_f[c,b].$$

Proof. Assume that $f \in BV([a,b])$ and $c \in (a,b)$. Let $P_1 = \{x_0, x_1, \ldots, x_p\}$ and $P_2 = \{x_p, x_{p+1}, \ldots, x_n\}$ be partitions of $[a,c]$ and $[c,b]$, respectively. Then

$$P = P_1 \cup P_2 = \{x_0, x_1, \ldots, x_p = c, x_{p+1}, \ldots, x_n\}$$

is a partition of $[a,b]$, and

$$\begin{aligned} V(P_1,f) + V(P_2,f) &= \sum_{k=1}^{p} |f(x_k) - f(x_{k-1})| + \sum_{k=p+1}^{n} |f(x_k) - f(x_{k-1})| \\ &= V(P,f) \leq V_f[a,b]. \end{aligned}$$

In particular,

$$V(P_1,f) \leq V_f[a,b] \quad \text{and} \quad V(P_2,f) \leq V_f[a,b],$$

and so f is of bounded variation on $[a,c]$ and on $[c,b]$. Moreover, from the basic properties of the supremum,

$$\begin{aligned} V_f[a,c]+V_f[c,b] &= \sup_{P_1\in P[a,c]} V(P_1,f)+\sup_{P_2\in P[c,b]} V(P_2,f) \\ &= \sup_{P=P_1\cup P_2} V(P,f) \\ &\le V_f[a,b] \quad \text{as } P_1\cup P_2\in P[a,b]. \end{aligned}$$

Next we prove the reverse inequality:

$$V_f[a,b]\le V_f[a,c]+V_f[c,b].$$

To do this, let $P=\{x_0,x_1,\dots,x_n\}$ be a partition of $[a,b]$ and $P_0=P\cup\{c\}$. Then $c\in[x_{k-1},x_k]$ for some k, say $k=p$, and so P_0 is a refinement of P if $c\notin P$ and $P_0=P$ if $c\in P$. Therefore, because

$$|f(x_k)-f(x_{k-1})|\le|f(x_k)-f(c)|+|f(c)-f(x_{k-1})|,$$

we have

$$V(P,f)\le V(P_0,f).$$

Next, we set

$$P_1=P_0\cap[a,c]=\{x_0,\dots,x_{p-1},c\} \quad\text{and}\quad P_2=P_0\cap[c,b]=\{c,x_{p+1},\dots,x_n\}.$$

Then P_1 and P_2 are partitions of $[a,c]$ and $[c,b]$, respectively. Also,

$$V(P,f)\le V(P_0,f)=V(P_1,f)+V(P_2,f)\le V_f[a,c]+V_f[c,b],$$

which holds for all partitions P of $[a,b]$. This means that $V_f[a,c]+V_f[c,b]$ is an upper bound for every $V(P,f)$, $P\in P[a,b]$. Thus,

$$V_f[a,b]\le V_f[a,c]+V_f[c,b],$$

and we have completed the proof.

We leave the converse as a simple exercise. ∎

For instance, to compute the total variation of $f(x)=|x|$ on $[-3,4]$, it suffices to observe that it is decreasing on $[-3,0]$ and increasing on $[0,4]$. Therefore, by Theorem 11.13,

$$V_f[-3,4]=V_f[-3,0]+V_f[0,4]=|f(-3)-f(0)|+|f(4)-f(0)|=7.$$

Example 11.14. Consider $f(x)=x+[x]$ on $[-1,2]$. Then

$$f(x)=\begin{cases} x-1 & \text{if } -1\le x<0,\\ x & \text{if } 0\le x<1,\\ x+1 & \text{if } 1\le x<2,\\ 4 & \text{if } x=2,\end{cases}$$

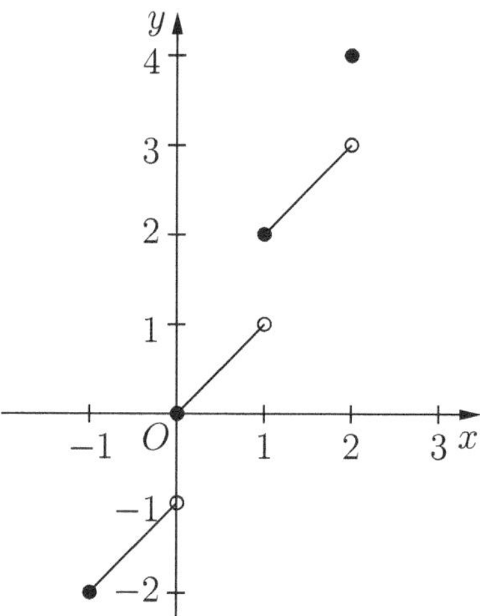

Fig. 11.3. The graph of $f(x) = x + [x]$ on $[-1, 2]$.

and we refer to Figure 11.3 for the graph of $f(x)$ on $[-1, 2]$. Note that f is piecewise monotone and bounded on $[-1, 2]$, and therefore $f \in BV([-1, 2])$. Now,

$$\begin{aligned} V_f[-1,0] &= \sup_{x\in[-1,0)} [V_f[-1,x) + |f(0) - f(x)|] \\ &= \sup_{x\in[-1,0)} [-f(-1) + f(x) + |f(0) - f(x)|] \\ &= \sup_{x\in[-1,0)} [x + 1 + |x - 1|] = 2. \end{aligned}$$

Similarly, $V_f[0, 1] = 1$ and $V_f[1, 2] = 4$. •

Theorem 11.15. *If $f \in BV([a, b])$ and g is bounded on $[a, b]$ such that $f(x) \neq g(x)$ only at a finite number of points on $[a, b]$, then $g \in BV([a, b])$.*

Proof. Let f be of bounded variation on $[a, b]$. Clearly, it suffices to prove the theorem when $f(x) = g(x)$ except at one point, say $c \in [a, b]$. Thus, if $f(x) \neq g(x)$ at $c \in [a, b]$, then we have

$$f(c) = g(c) + k$$

for some nonzero real constant k.

Case 1: If $c = a$, then we have (since $f(x_1) = g(x_1)$)

$$|g(x_1) - g(a)| = |f(x_1) - (f(a) - k)| \leq |f(x_1) - f(a)| + |k|.$$

Consequently, for any partition $P = \{x_0, x_1, \ldots, x_n\}$ of $[a, b]$ for which $c = a$, we have

$$V(P,g) \le V(P,f) + |k| \le V_f[a,b] + |k|,$$

and so g is of bounded variation on $[a,b]$. The proof is similar if $c = b$.

Case 2: If $c \in (a,b)$, then (because f is of bounded variation on $[a,b]$) f is of bounded variation on $[a,c]$ and $[c,b]$. Thus, f and g differ on $[a,c]$ (respectively, on $[c,b]$) only at one endpoint, namely at c. Therefore, by Case 1, g is of bounded variation on $[a,c]$ and on $[c,b]$. Hence g is of bounded variation on $[a,b]$. ∎

11.1.3 Characterization of Functions of Bounded Variation

Let $f \in BV([a,b])$. For $x \in [a,b]$, the *total variation* of f on $[a,x]$ defined by $V_f[a,x]$ is a function of x. Then, Theorem 11.13 helps us to introduce a function V_f on $[a,b]$ as follows:

$$V_f(x) = \begin{cases} V_f[a,x] & \text{if } x \in (a,b], \\ 0 & \text{if } x = a. \end{cases}$$

We call V_f, or simply V, the *variation function* of f on $[a,b]$.

Example 11.16. Consider $f(x) = 2x^3 - 3x^2 - 12x + 6$ on $[0,3]$. Then f, being a sum of monotone functions $f_1(x) = 2x^3$, $f_2(x) = -3x^2$, $f_3(x) = -12x$, and $f_4(x) = 6$, is of bounded variation on $[0,3]$. To determine the behavior of $f(x)$, we compute its derivative:

$$f'(x) = 6(x^2 - x - 2) = 6(x-2)(x+1).$$

Since $f'(x) > 0$ on $(2,3]$ and $f'(x) < 0$ on $[0,2)$, f is increasing on $[2,3]$ and decreasing on $[0,2]$. Therefore,

$$\begin{aligned} V_f[0,3] &= V_f[0,2] + V_f[2,3] \quad \text{by Theorem 11.13,} \\ &= (f(0) - f(2)) + (f(3) - f(2)) \quad \text{by Theorem 11.2(a),} \\ &= (6 + 14) + (-3 + 14), \end{aligned}$$

so that $V_f[0,3] = 31$. ●

Example 11.17. Let $f(x) = 3x^2 - x^3$ on $[-2,3]$. On $[-2,3)$ compute the following function (see Exercise 11.28(14)):

(a) total variation.
(b) positive variation.
(c) negative variation.

Solution. Clearly, $f \in BV([-2,3])$. To determine its behavior, we compute

$$f'(x) = 3x(2-x),$$

and observe that $f'(x) = 0$ at $x = 0, 2$. Further, $f'(x) \leq 0$ on $[-2, 0] \cup [2, 3]$ and $f'(x) \geq 0$ on $[0, 2]$. In particular, Theorem 11.2(**a**) gives

$$\begin{aligned} V_f[0,x] &= f(x) - f(0) \quad \text{for } 0 \leq x \leq 2 \\ V_f[-2,x] &= f(-2) - f(x) \quad \text{for } -2 \leq x \leq 0, \quad \text{and} \\ V_f[2,x] &= f(2) - f(x) \quad \text{for } 2 \leq x \leq 3. \end{aligned}$$

A computation gives

$$V_f[-2,x] = x^3 - 3x^2 + 20 \quad \text{for } -2 \leq x \leq 0.$$

In particular, $V_f[-2, 0] = 20$. Next, for $0 \leq x \leq 2$,

$$V_f[-2,x] = V_f[-2,0] + V_f[0,x] = 20 + (f(x) - f(0)) = -x^3 + 3x^2 + 20.$$

In particular, $V_f[-2, 2] = 24$. Finally, for $2 \leq x \leq 3$,

$$V_f[-2,x] = V_f[-2,2] + V_f[2,x] = 24 + f(2) - f(x) = x^3 - 3x^2 + 28.$$

Consequently, the total variation function V on $[-2, 3]$ is defined to be

$$V(x) = \begin{cases} x^3 - 3x^2 + 20 & \text{for } -2 \leq x \leq 0, \\ -x^3 + 3x^2 + 20 & \text{for } 0 \leq x \leq 2, \\ x^3 - 3x^2 + 28 & \text{for } 2 \leq x \leq 3. \end{cases}$$

The positive and negative variations can be obtained using the formulas in Exercise 11.28(14). ●

A restatement of the identity in Theorem 11.13 is given in the following lemma.

Lemma 11.18. *If $x, y \in [a, b]$ such that $a \leq x < y \leq b$, then we have*

$$V_f[x,y] = V(y) - V(x) \geq |f(y) - f(x)| \geq 0.$$

In particular, the function V is increasing on $[a, b]$ and

$$V(y) = V(x) + V_f[x,y].$$

Proof. We see that with $V_f[a, a] = 0$, Theorem 11.13 gives

$$V(y) - V(x) = V_f[a,y] - V_f[a,x] = V_f[x,y] \geq |f(y) - f(x)| \geq 0,$$

and so $V(y) \geq V(x)$, i.e., V is increasing on $[a, b]$. ■

Theorem 11.19 (Jordan decomposition of functions of bounded variation). *Every $f \in BV([a, b])$ can be written as $f = V - D$, where V and D are increasing functions on $[a, b]$. Conversely, if f can be expressed as the difference between two increasing functions on $[a, b]$, then $f \in BV([a, b])$.*

Proof. Let $x, y \in [a,b]$ such that $a \le x < y \le b$ and $D(x) = V(x) - f(x)$. Then

$$D(y) - D(x) = (V(y) - V(x)) - (f(y) - f(x)) = V_f[x,y] - (f(y) - f(x)) \geq 0,$$

and thus D is increasing on $[a,b]$. The converse is trivial. ■

For instance, $f(x) = [x] - x^2 \in BV([0,3])$, because both $f_1(x) = [x]$ and $f_2(x) = x^2$ are increasing on $[0,3]$.

The decomposition of $f \in BV([a,b])$ given in Theorem 11.19 is not unique.

Corollary 11.20. *A function of bounded variation f on $[a,b]$ has at most a countable number of points of discontinuity on $[a,b]$.*

Proof. By Theorem 11.19, $f = f_1 - f_2$, where f_1 and f_2 are increasing on $[a,b]$. Since monotone functions have at most a countable number of points of discontinuity, the corollary follows. ■

Theorem 11.21. *Suppose that f is of bounded variation on $[a,b]$. Then every point of continuity of f is also a point of continuity of the variation function V. The converse is also true.*

Proof. We begin the proof when $c \in (a,b)$.

Fig. 11.4. The partition P_1.

$\Rightarrow$: Let f be continuous at c. Then for a given $\epsilon > 0$, there exists a $\delta > 0$ such that

$$|f(x) - f(c)| < \frac{\epsilon}{2} \quad \text{whenever } 0 < |x - c| < \delta.$$

For the same ϵ there exists a partition $P_1 = \{c, x_1, x_2, \dots, b\}$ of $[c,b]$ such that

$$V(P_1, f) > V_f[c,b] - \frac{\epsilon}{2}.$$

Since $V(P_1, f)$ increases on adjoining more points to the partition P_1, we can assume that $0 < x_1 - c < \delta$ (see Figure 11.4). That is,

$$|f(x_1) - f(c)| < \frac{\epsilon}{2} \quad \text{whenever } 0 < x_1 - c < \delta.$$

Thus, we have

$$\begin{aligned} V_f[c,b] &< \frac{\epsilon}{2} + V(P_1, f) \\ &= \frac{\epsilon}{2} + |f(x_1) - f(c)| + \sum_{k=2}^{n} |f(x_k) - f(x_{k-1})| \\ &< \frac{\epsilon}{2} + \frac{\epsilon}{2} + V_f[x_1, b], \end{aligned}$$

so that

$$V_f[c,b] - V_f[x_1,b] < \epsilon \quad \text{whenever } 0 < x_1 - c < \delta.$$

Note that (see Lemma 11.18)

$$0 \leq V(x_1) - V(c) = V_f[c,x_1] = V_f[c,b] - V_f[x_1,b],$$

and so the last inequality becomes

$$0 \leq V(x_1) - V(c) < \epsilon \text{ whenever } 0 < x_1 - c < \delta,$$

showing that V is right continuous at c. A similar argument gives that V is left continuous at c. Proof of the continuity of V at the endpoints follows from a trivial modification of the proof.

$\Leftarrow$: Let V be continuous at $c \in (a,b)$. Since V is monotone, the right-hand and the left-hand limits $V(c+)$ and $V(c-)$ exist for each $c \in (a,b)$. By Theorem 11.19, the same is true for $f(c+)$ and $f(c-)$. If $a < c < x \leq b$, then Lemma 11.18 gives

$$0 \leq |f(x) - f(c)| \leq V(x) - V(c).$$

Letting $x \to c+$, we find that

$$0 \leq |f(c+) - f(c)| \leq V(c+) - V(c),$$

and similarly,

$$0 \leq |f(c) - f(c-)| \leq V(c) - V(c-).$$

These two inequalities show that f is continuous at c whenever V is continuous at c. ∎

As a consequence of Theorems 11.19 and 11.21, we have the following corollary.

Corollary 11.22. *Let f be continuous on $[a,b]$. Then $f \in BV([a,b])$ if and only if f can be expressed as the difference of two increasing continuous functions on $[a,b]$.*

Theorem 11.23. *If $f \in C^1([a,b])$, then f is of bounded variation and*

$$V_f[a,b] = \int_a^b |f'(t)|\,dt.$$

Proof. The first part is a consequence of Theorem 11.2**(c)**. For the proof of the second part, we apply the mean value theorem to f on each interval $[x_{k-1},x_k]$ of the partition $P = \{x_0, x_1, \ldots, x_n\}$ of $[a,b]$. According to this, there exists a $c_k \in (x_{k-1},x_k)$ for each $k = 1,2,\ldots,n$ such that

$$V(P,f) = \sum_{k=1}^{n} |f(x_k) - f(x_{k-1})| = \sum_{k=1}^{n} |f'(c_k)|(x_k - x_{k-1}).$$

Thus, by the definition of the Riemann integral, because $f'(x)$ is continuous on $[a,b]$, given $\epsilon > 0$, there exists a $\delta > 0$ such that

$$\left| V(P,f) - \int_a^b |f'(t)|\, \mathrm{d}t \right| < \frac{\epsilon}{2} \tag{11.2}$$

for every partition P of $[a,b]$ with $\|P\| < \delta$. Moreover, by the definition of bounded variation, it follows that for this $\epsilon > 0$ there exists a partition P_1 of $[a,b]$ such that

$$V_f[a,b] \geq V(P_1,f) > V_f[a,b] - \epsilon/2.$$

Set $Q = P \cup P_1$, the common refinement of P and P_1 of $[a,b]$. Then $\|Q\| < \delta$, and so

$$V_f[a,b] \geq V(Q,f) \geq V(P_1,f) > V_f[a,b] - \epsilon/2.$$

Again, because $\|Q\| < \delta$, (11.2) implies that

$$\left| V(Q,f) - \int_a^b |f'(t)|\, \mathrm{d}t \right| < \frac{\epsilon}{2}.$$

The last inequality gives

$$\left| V_f[a,b] - \int_a^b |f'(t)|\, \mathrm{d}t \right| < \epsilon.$$

Since $\epsilon > 0$ is arbitrary, the conclusion follows. ∎

11.1.4 Bounded Variation and Absolute Continuity

Definition 11.24. *A real-valued function f defined on $[a,b]$ is said to be* absolutely continuous *on $[a,b]$ if for every $\epsilon > 0$ there exists a $\delta > 0$ such that*

$$\sum_{k=1}^{n} |f(b_k) - f(a_k)| < \epsilon$$

for every n disjoint open subintervals (a_k, b_k) of $[a,b]$, $k = 1,2,\ldots,n$, with $\sum_{k=1}^{n}(b_k - a_k) < \delta$.

Clearly, we have the following:

- Every Lipschitz function is absolutely continuous.
- By choosing $n = 1$, we see that every absolutely continuous function is uniformly continuous (but the converse is not true), and hence continuous. Thus,

$$\text{Lipschitz} \Rightarrow \text{absolutely continuous} \Rightarrow \text{uniformly continuous} \Rightarrow \text{continuous}.$$

- There are continuous functions that are not Lipschitz.
- There are absolutely continuous functions that are not Lipschitz.
- Every function that has bounded derivative on $[a, b]$ is absolutely continuous on $[a, b]$.

Theorem 11.25. *Every absolutely continuous function on $[a, b]$ is of bounded variation.*

Proof. Let f be an absolutely continuous function on $[a, b]$. Then for $\epsilon = 1$, there exists a $\delta > 0$ such that

$$\sum_{k=1}^{n} |f(b_k) - f(a_k)| < 1$$

for every collection of n finite disjoint open subintervals (a_k, b_k) of $[a, b]$ satisfying the condition $\sum_{k=1}^{n}(b_k - a_k) < \delta$.

Let N be a positive integer such that $\delta > (b - a)/N$ and let $P_1 = \{y_1, y_2, \dots, y_N\}$ be the partition of $[a, b]$ with

$$y_k = a + \frac{k(b-a)}{N}, \quad k = 0, 1, 2, \dots, N.$$

Then $y_k - y_{k-1} = (b - a)/N < \delta$ for each $k = 1, 2, \dots, N$. Let $P_2 = \{x_0, x_1, \dots, x_n\}$ be a refinement of P_1. Then for each k, $[y_{k-1}, y_k]$ contains one or more subintervals $[x_{j-1}, x_j]$ of P_2 the sum of whose lengths is $y_k - y_{k-1} < \delta$. That is, if Q_k is a partition of $[y_{k-1}, y_k]$, then by the definition of absolute continuity,

$$\sum_{Q_k} |\Delta f| < 1 \quad \text{for each } k.$$

For any partition P of $[a, b]$, let P' be the refinement of P obtained by adjoining the points of P_1. Then $P' = \cup_{k=1}^{N} Q_k$, and therefore

$$\sum_{P} |\Delta f| \le \sum_{P'} |\Delta f| = \sum_{k=1}^{N} \left(\sum_{Q_k} |\Delta f| \right) < N.$$

This proves that f is of bounded variation. ∎

The converse of Theorem 11.25 is not true. Indeed, every continuous function that is not of bounded variation (for instance, see Example 11.3 and Exercise 11.28(3)) will serve as an example of a continuous function that is not absolutely continuous.

In view of Theorems 11.19 and 11.25, it follows that every absolutely continuous function can be written as the difference of two increasing functions:

$$f(x) = V(x) - (V(x) - f(x)),$$

where V is the variation function defined by $V(x) = V_f[a, x]$. We now show that V is also absolutely continuous, and hence Corollary 11.22 continues to hold if continuity is replaced by absolute continuity.

Theorem 11.26. *Let $f \in BV([a,b])$. Then f is absolutely continuous on $[a,b]$ if and only if the total variation function V, given by $V(x) = V_f[a,x]$, is absolutely continuous on $[a,b]$.*

Proof. Let f be absolutely continuous on $[a,b]$. Then corresponding to any given $\epsilon > 0$ there exists a $\delta > 0$ such that $\sum_{k=1}^n |f(b_k) - f(a_k)| < \epsilon$ for every finite system

$$S = \{(a_k, b_k) : 1 \leq k \leq n\}$$

of nonoverlapping subintervals (a_k, b_k) of $[a,b]$ with $\sum_{k=1}^n (b_k - a_k) < \delta$. For each k, let $P_k = \{a_k = a_{k_0}, a_{k_1}, \ldots, a_{k_{m_k}} = b_k\}$ be a partition of $[a_k, b_k]$. Then

$$\sum_{k=1}^n \sum_{j=1}^{m_k} (a_{k_j} - a_{k_{\underbrace{j-1}}}) = \sum_{k=1}^n (b_k - a_k) < \delta,$$

and therefore

$$\sum_{k=1}^n \sum_{j=1}^{m_k} (f(a_{k_j}) - f(a_{k_{j-1}})) < \epsilon.$$

Fixing the collection S and taking the supremum over all partitions P_k of $[a_k, b_k]$ for $k = 1, 2, \ldots, n$, we obtain

$$\sum_{k=1}^n V_f[a_k, b_k] = \sum_{k=1}^n (V(b_k) - V(a_k)) \leq \epsilon,$$

which shows that V is absolutely continuous. Since

$$|f(x_k) - f(x_{k-1})| \leq V(x_k) - V(x_{k-1})$$

for $a \leq x_{k-1} < x_k \leq b$, the converse is clear. ∎

11.1.5 Questions and Exercises

Questions 11.27.

1. Is every bounded function of bounded variation?
2. Must a function of bounded variation be monotone?
3. Must a function of bounded variation be continuous?
4. Must a function of bounded variation be integrable?
5. Can there exist a nondifferentiable function that is of bounded variation?
6. If f is a function of bounded variation, must the reciprocal $1/f$ be of bounded variation?
7. Suppose that $f \in BV([a,b])$ and $g \in BV([c,d])$, where $f([a,b]) \subset [c,d]$, so that $g \circ f$ is defined on $[a,b]$. Must $g \circ f \in BV([a,b])$?
8. Suppose that f is continuous and has a finite number of extrema on $[a,b]$. Must f be of bounded variation on $[a,b]$?

9. Must every polynomial function in x over a bounded interval $[a,b]$ be of bounded variation on $[a,b]$?
10. Is it true that $\sin x \in BV([a,b])$? Is it true that $\cos x \in BV([a,b])$?
11. If $f \in BV([a,b])$, must we have $-f \in BV([a,b])$? Must $V_f[a,b] = V_{-f}[a,b]$?
12. If $f \in BV([a,b])$, must we have $f \in BV([c,d])$ when $[c,d] \subset [a,b]$?
13. Suppose that the interval $[a,b]$ can be divided into a finite number of subintervals on each of which $f(x)$ is monotone. Must f be of bounded variation on $[a,b]$?
14. Suppose that f is of bounded variation on $[a,b]$ and V is the variation function of f on $[a,b]$. Must $V \pm f$ be increasing on $[a,b]$?
15. Must the representation of a function of bounded variation as a difference of two increasing functions be unique?
16. In the Jordan decomposition theorem, can we replace "increasing" by "strictly increasing"?
17. Does the set of absolutely continuous functions on $[a,b]$ form a vector space over $\mathbb{R}$?

Exercises 11.28.

1. Define $f : [0,1] \to \mathbb{R}$ by
$$f(x) = \begin{cases} 0 & \text{for } x \text{ rational in } [0,1], \\ 1 & \text{for } x \text{ irrational in } [0,1]. \end{cases}$$
Choosing a partition $P = \{x_0, x_1, \dots, x_{2n}\}$ where $x_{2k} \in \mathbb{Q}^c \cap [0,1]$ and $x_{2k+1} \in \mathbb{Q} \cap [0,1]$ $(k \geq 0)$, show that $f \notin BV([a,b])$.
2. Define $f : [0,1] \to \mathbb{R}$ by
$$f(x) = \begin{cases} -1 & \text{for } x \text{ rational in } [-1,1] \\ 1 & \text{for } x \text{ irrational in } [-1,1]. \end{cases}$$
Show that f is not a function of bounded variation. How about $|f(x)|$ on $[-1,1]$?
3. Define $f : [0,1] \to \mathbb{R}$ by
$$f(x) = \begin{cases} x\cos(\pi/2x) & \text{for } 0 < x \leq 1, \\ 0 & \text{for } x = 0. \end{cases}$$
By considering the partition $P = \{0, 1/2n, 1/(2n-1), \dots, 1/3, 1/2, 1\}$ of $[0,1]$, show that f is not a function of bounded variation on $[0,1]$.
4. Show that $f(x) = x^4(\cos x + x)$ on $[0,\pi/2]$ is of bounded variation on $[0,\pi/2]$.
5. Show that $BV([a,b])$ is a vector space over $\mathbb{R}$.
   (a) $BV([a,b])$ is closed with respect to addition and scalar multiplication.
   (b) $0 \in BV([a,b])$, $-f \in BV([a,b])$ if $f \in BV([a,b])$.

Note: If $\|f\| = |f(a)| + V_f[a,b]$, $f \in BV([a,b])$, then $BV([a,b])$ becomes a normed space.

6. Let $f(x) = \sin x$ on $[0, 2\pi]$.
   (a) Find the total variation of f on $[0, 2\pi]$.
   (b) Find a Jordan decomposition of f on $[0, 2\pi]$.
7. Suppose that
$$f(x) = \begin{cases} x\cos(\pi/x) & \text{for } x \in (0,1], \\ 0 & \text{for } x = 0. \end{cases}$$
If $P = \{0, 1/n, \dots, 1/3, 1/2, 1\}$ is a partition of $[0,1]$, determine $V(P,f)$. Conclude that $f \notin BV([0,1])$.
8. Define (see Figure 11.5)

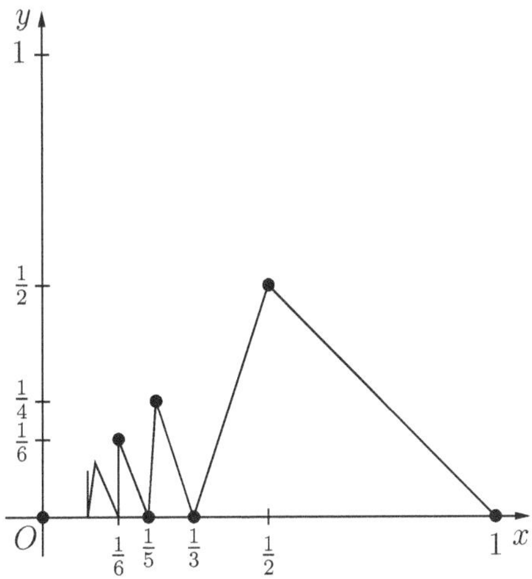

Fig. 11.5. Graph of $f(x)$ on $[0,1]$.

$$f(x) = \begin{cases} \dfrac{1}{2n} & \text{if } x = \dfrac{1}{2n}, \\ 0 & \text{if } x = \dfrac{1}{2n-1}, \qquad n \in \mathbb{N}. \\ 0 & \text{if } x = 0, \\ \text{linear} & \text{otherwise}, \end{cases}$$

Show that f is continuous but not of bounded variation on $[0,1]$.
9. If f is integrable on $[a,b]$ and $F(x) = \int_a^x f(t)\,dt$, then show that F is of bounded variation on $[a,b]$.
10. Determine all possible values of α and β for which the function f defined on $[0,1]$ by
$$f(x) = \begin{cases} x^\alpha \sin(\frac{1}{x^\beta}) & \text{for } x \in (0,1], \\ 0 & \text{for } x = 0, \end{cases}$$
is in $BV([0,1])$.

11. Given $f(x) = x^n e^{-x}$ on $[0, a]$, $a > n > 0$, find $V_f[0, a]$.
12. Define $f(x) = 2x^3 - 9x^2 + 12x$ on $[0, 4]$, and $g(x) = x - [x]$ on $[1, 3]$. Compute $V_f[0, 4]$ and $V_g[1, 3]$.
13. Suppose that f is continuous on $[a, b]$, $f \in BV([a, b])$, and $\{P_n\}$ is a sequence of partitions of $[a, b]$ such that $\|P_n\| \to 0$ as $n \to \infty$. Show that

$$V_f[a, b] = \lim_{n\to\infty} V(P_n, f).$$

14. For $f \in BV([a, b])$ and a partition $P = \{x_0, x_1, \ldots, x_n\}$ of $[a, b]$, define

$$A(P) = \{k : f(x_k) - f(x_{k-1}) > 0\} \text{ and } N(P) = \{k : f(x_k) - f(x_{k-1}) < 0\}.$$

Define the positive variation $p_f[a, b]$ and negative variation $n_f[a, b]$ of f as follows:

$$p_f[a, b] = \sup\left\{ \sum_{k\in A(P)} (f(x_k) - f(x_{k-1})) : P \in P[a, b] \right\}$$

and

$$n_f[a, b] = \inf\left\{ \sum_{k\in N(P)} (f(x_{k-1}) - f(x_k)) : P \in P[a, b] \right\}.$$

For each $x \in [a, b]$, let

$$p(x) = p_f[a, x], \quad n(x) = n_f[a, x], \quad V(x) = V_f[a, x]$$

and $p(a) = n(a) = V(a) = 0$. Show that

(a) $2p(x) - V(x) = f(x) - f(a)$ and $2n(x) - V(x) = f(a) - f(x)$.
(b) $p(x) - n(x) = f(x) - f(a)$.
(c) $V(x) = p(x) + n(x)$.
(d) p and n are increasing on $[a, b]$.
(e) $0 \le p(x) \le V(x)$ and $0 \le n(x) \le V(x)$.
(f) If f is continuous at $x = a$, then so are p and n at $x = a$.

11.2 Stieltjes Integrals

We have already shown that the Darboux and Riemann integrals are the same, and so we call them just the Riemann integral without distinguishing which approach we use to examine the integrability and possibly to compute the integral. In this section, we discuss Darboux–Stieltjes and Riemann–Stieltjes integrals, which are important generalizations of the Darboux and Riemann integrals. Also, we state and prove theorems to indicate when these two integrals are the same and when they are not. Since many results of this section are straightforward generalizations of results in Chapter 6, the development of these integrals requires only minor modifications to a more general setup, and therefore it is appropriate to omit certain parts or the entirety of proofs of some of the statements in this section.

As in the discussion of the Darboux integral in Section 6.1, we need to begin the discussion with standard notation: Let $P = \{x_0, x_1, \dots, x_n\}$ be a partition of $[a, b]$ and $x_k^* \in [x_{k-1}, x_k]$ $(1 \leq k \leq n)$ arbitrary. To motivate the definition of the Stieltjes integral, we consider the problem of finding the moment with respect to the y-axis of a distribution of mass over $[a, x]$. If $m(x)$ is the amount of mass over $[a, x]$, then the moment with respect to the y-axis is approximated by the sum

$$\sum_{k=1}^{n} x_k^* \Delta m_k,$$

where $\Delta m_k = m(x_k) - m(x_{k-1})$ is interpreted as the mass between x_{k-1} and x_k. In the same way, it is possible to approximate the moment of inertia of the mass distribution by the sum

$$\sum_{k=1}^{n} (x_k^*)^2 \Delta m_k.$$

Now we continue the discussion with a more general situation. For a real-valued function f defined and bounded on $[a, b]$ and for each k, $1 \leq k \leq n$, we use the following standard notation. Let

$$M_k = \sup_{x \in [x_{k-1}, x_k]} f(x), \quad m_k = \inf_{x \in [x_{k-1}, x_k]} f(x)$$

and similarly

$$M = \sup_{x \in [a,b]} f(x) \quad \text{and} \quad m = \inf_{x \in [a,b]} f(x),$$

so that

$$m \leq m_k \leq f(x_k^*) \leq M_k \leq M.$$

For an *increasing function* α defined on $[a, b]$, and for any partition P of $[a, b]$, define $\Delta \alpha_k = \alpha(x_k) - \alpha(x_{k-1})$ and the corresponding sums as follows:

$$U(P, f, \alpha) = \sum_{k=1}^{n} M_k \Delta \alpha_k, \quad L(P, f, \alpha) = \sum_{k=1}^{n} m_k \Delta \alpha_k,$$

and

$$\sigma(P, f, \alpha, x^*) = \sum_{k=1}^{n} f(x_k^*) \Delta \alpha_k.$$

We call $U(P, f, \alpha)$, $L(P, f, \alpha)$, and $\sigma(P, f, \alpha, x^*)$ the upper (Darboux–Stieltjes), the lower (Darboux–Stieltjes) and the Riemann (Stieltjes) sums of f with respect to α on $[a, b]$, respectively.

Since $m_k \leq f(x_k^*) \leq M_k$ and $\Delta\alpha_k$ is nonnegative, we easily obtain many key properties of these sums. For instance, for each partition P of $[a, b]$,

$$L(P, f, \alpha) \leq \sigma(P, f, \alpha, x^*) \leq U(P, f, \alpha), \tag{11.3}$$

i.e., the lower sum is always less than or equal to the Riemann–Stieltjes sum, which is in turn less than or equal to the upper sum. Moreover, two parts of Lemma 6.7 take the following form.

Lemma 11.29. *Let f be a bounded function on $[a, b]$, and let P and Q be two partitions of $[a, b]$. Then we have the following:*

(a) $L(P, f, \alpha) \leq L(Q, f, \alpha) \leq U(Q, f, \alpha) \leq U(P, f, \alpha)$ *if* $P \subseteq Q$.
(b) $m(\alpha(b) - \alpha(a)) \leq L(P, f, \alpha) \leq U(Q, f, \alpha) \leq M(\alpha(b) - \alpha(a))$ *for any P and Q.*

Proof. The proof of this lemma follows if we imitate the proof of Lemma 6.7 by replacing Δx_k by $\Delta\alpha_k$. ∎

As with the Darboux integral, the case **(a)** of this lemma shows that the upper sum is *decreasing* with respect to a refinement of the partition, while the lower sum is *increasing* with respect to a refinement of the partition. In particular, for any partitions P and Q of $[a, b]$, we have

$$L(P, f, \alpha) \leq L(P \cup Q, f, \alpha) \leq U(P \cup Q, f, \alpha) \leq U(Q, f, \alpha).$$

11.2.1 The Darboux–Stieltjes Integral

We now define the Darboux–Stieltjes integral.

Definition 11.30 (Darboux–Stieltjes integral). *Let f be a bounded function defined on the closed interval $[a, b]$ and let α be an* increasing function *on $[a, b]$. The upper (Darboux–Stieltjes) integral of f with respect to α over $[a, b]$ is defined by*

$$U_\alpha(f) := \overline{\int_a^b} f(x)\, \mathrm{d}\alpha(x) = \overline{\int_a^b} f\, \mathrm{d}\alpha = \inf\{U(P, f, \alpha) : P \in \mathcal{P}[a, b]\},$$

and the lower (Darboux–Stieltjes) integral of f with respect to α over $[a, b]$ is defined by

$$L_\alpha(f) := \underline{\int_a^b} f(x)\, \mathrm{d}\alpha(x) = \underline{\int_a^b} f\, \mathrm{d}\alpha = \sup\{L(P, f, \alpha) : P \in \mathcal{P}[a, b]\}.$$

We say that f is Darboux–Stieltjes integrable *with respect to α on $[a,b]$, or Stieltjes integrable with respect to α on $[a,b]$ in the sense of Darboux, if the upper and lower Darboux–Stieltjes integrals agree, i.e., if $U_\alpha(f) = L_\alpha(f)$. In this case, we denote their common value by*

$$\text{(DS)} \int_a^b f(x)\,\mathrm{d}\alpha(x) \quad \textit{or by} \quad \text{(DS)} \int_a^b f\,\mathrm{d}\alpha.$$

We call this integral the Darboux–Stieltjes integral of f with respect to α over $[a,b]$. Here the function f is called the integrand, *and the function α is called the* integrator. *We let $\mathcal{D}_\alpha[a,b]$ denote the set of all Darboux–Stieltjes integrable functions with respect to α on the interval $[a,b]$. If $U_\alpha(f) \neq L_\alpha(f)$, then we say that f is not Darboux–Stieltjes integrable.*

Throughout the section, unless otherwise stated, "f is Darboux–Stieltjes integrable" means that f is Darboux–Stieltjes integrable with respect to some α on $[a,b]$. A similar convention will be followed when we define what it means for a function to be Riemann–Stieltjes integrable with respect to some α on $[a,b]$.

Later, we simply denote the Darboux–Stieltjes integral by

$$\int_a^b f\,\mathrm{d}\alpha$$

for cases in which the Darboux–Stieltjes and Riemann–Stieltjes integrals are the same.

As with the Darboux integrals, the upper and lower (Darboux–Stieltjes) sums depend on the particular choice of the partition, while the upper and lower (Darboux–Stieltjes) integrals are independent of the partitions. Hence, a natural question is the following: *when do the two quantities, namely the upper and lower (Darboux–Stieltjes) integrals, coincide?*

If $\alpha(x) = x$, then the Darboux–Stieltjes integral reduces to the Darboux/Riemann integral of f over $[a,b]$. In this case, the upper and lower Darboux–Stieltjes sums are called the upper and lower (Riemann) sums, respectively.

Because many of the proofs of standard results are essentially the same as those for the Riemann upper and lower integrals, we shall not include their proofs. For example, the following corollary is a simple consequence of Lemma 11.29 and the definition of supremum and infimum.

Corollary 11.31. *We have*

$$m(\alpha(b) - \alpha(a)) \leq L_\alpha(f) = \underline{\int_a^b} f\,\mathrm{d}\alpha \leq U_\alpha(f) = \overline{\int_a^b} f\,\mathrm{d}\alpha \leq M(\alpha(b) - \alpha(a)).$$

Furthermore, if f is Darboux–Stieltjes integrable on $[a,b]$, then

$$m(\alpha(b) - \alpha(a)) \leq \int_a^b f\,\mathrm{d}\alpha \leq M(\alpha(b) - \alpha(a)).$$

Proof. The proof of the corollary follows along the lines of the proof of Lemma 6.7(**c**). ∎

Examples 11.32. (**a**) Let $\alpha(x) = k$ for all $x \in [a, b]$. Then for any bounded function f on $[a, b]$, the corresponding upper and lower sums of f are zero. Thus,

$$\underline{\int_a^b} f\, d\alpha = \overline{\int_a^b} f\, d\alpha = 0, \quad \text{i.e.,} \quad \int_a^b f\, d\alpha = 0,$$

and so $f \in \mathcal{D}_\alpha[a, b]$.

(**b**) Set $f(x) = k$ on $[a, b]$. Then for any increasing function α on $[a, b]$ and for each partition P of $[a, b]$,

$$U(P, f, \alpha) = k(\alpha(b) - \alpha(a)) = L(P, f, \alpha),$$

and so

$$f \in \mathcal{D}_\alpha[a, b] \quad \text{and} \quad \int_a^b f\, d\alpha = k(\alpha(b) - \alpha(a)).$$

(**c**) Let $f(x) = x$ on $[0, 1]$ and α be such that $\alpha(x) = 0$ on $[0, 1)$ and $\alpha(1) = 2$. Then for any partition P of $[a, b]$,

$$U(P, f, \alpha) = 2 \quad \text{and} \quad L(P, f, \alpha) = 2x_{k-1},$$

which shows that $\int_0^1 x\, d\alpha(x) = 2$.

(**d**) In the definition of Darboux–Stieltjes integral, α need not even be continuous. As a demonstration, let $f(x) = k$ on $[a, b]$ and

$$\alpha(x) = \begin{cases} 0 & \text{for } a \le x < c, \\ 1 & \text{for } c \le x \le b, \end{cases} \quad \text{for some } c \in (a, b).$$

Then for a partition $P = \{x_0, x_1, \ldots, x_n\}$ of $[a, b]$,

$$U(P, f, \alpha) - L(P, f, \alpha) = k \sum_{k=1}^{n} (\alpha(x_k) - \alpha(x_{k-1})),$$

and the situations described in Figure 11.6 occur for P. We see that the sum on the right is zero except for the subinterval that contains c, in which the sum is 1. Consequently, f is Darboux–Stieltjes integrable with

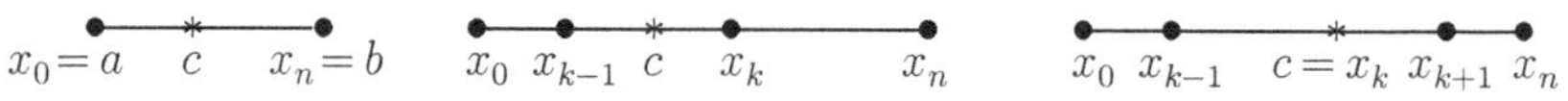

$x_0 = a$ c $x_n = b$ x_0 x_{k-1} c x_k x_n x_0 x_{k-1} $c = x_k$ x_{k+1} x_n

Fig. 11.6. The position of c in the partition P.

$$DS \int_a^b f\, d\alpha = k.$$

A similar situation is discussed in Example 11.36. ●

As with the Riemann integral (i.e., $\alpha(x) = x$), we seek a condition on f for which the difference $U(P, f, \alpha) - L(P, f, \alpha)$ can be made arbitrarily small.

Theorem 11.33 (Criterion for Darboux–Stieltjes integrability). *Suppose that f is a function that is bounded on $[a, b]$ and α is an* increasing *function on $[a, b]$. Then* (DS) $\int_a^b f\, d\alpha$ *exists if and only if for each $\epsilon > 0$ there is a partition P of $[a, b]$ such that*

$$U(P, f, \alpha) - L(P, f, \alpha) < \epsilon. \tag{11.4}$$

Proof. This follows by imitating the proof of Theorem 6.8. ∎

Theorem 11.33 provides a condition for Darboux–Stieltjes integrability, but it does not give any simple means for computing the integral. Recall that we encountered the same situation in our discussion on sequences and series.

Example 11.34. Consider

$$f(x) = \begin{cases} 5 & \text{for } 0 \leq x < 1, \\ 7 & \text{for } 1 \leq x \leq 2, \end{cases} \quad \text{and} \quad \alpha(x) = \begin{cases} 0 & \text{for } 0 \leq x \leq 1, \\ 2 & \text{for } 1 < x \leq 2. \end{cases}$$

Let $P = \{x_0, x_1, \dots, x_n\}$ be any partition of $[0, 2]$ that includes the number 1 (say $x_j = 1$). Then

$$U(P, f, \alpha) = M_{j+1}(\alpha(x_{j+1}) - \alpha(x_j)) = 7 \times 2 = 14$$

and

$$L(P, f, \alpha) = m_{j+1}(\alpha(x_{j+1}) - \alpha(x_j)) = 7 \times 2 = 14,$$

showing that $U(P, f, \alpha) - L(P, f, \alpha) = 0$, i.e., $f \in \mathcal{D}_\alpha[a, b]$. Moreover,

$$(\text{DS}) \int_0^2 f\, d\alpha = 14.$$

On the other hand, if $P' = \{x_0, x_1, \dots, x_n\}$ is a partition of $[0, 2]$ that does not include 1 as a partition point, then $x_{j-1} < 1 < x_j$ for some j. Then

$$\sigma(P', f, \alpha, x^*) = f(x_j^*)(\alpha(x_j) - \alpha(x_{j-1})) = 2f(x_j^*),$$

which in particular gives

$$\sigma(P', f, \alpha, x_{j-1}) = 10 \quad \text{and} \quad \sigma(P', f, \alpha, x_j) = 14.$$

Consequently the behavior of the upper and lower sums depends on the points in the partition rather than the norm of the partition. This difficulty is due to the fact that f and α have a common point of discontinuity. ●

There are some important classes of functions that one might expect to be Darboux–Stieltjes integrable. For example, the following theorem, which gives us a nice class of Darboux–Stieltjes integrable functions is important.

Theorem 11.35 (Darboux–Stieltjes integrability of continuous functions). *Let f be a function that is continuous on $[a, b]$. Then f is Darboux–Stieltjes integrable on $[a, b]$.*

Proof. Let α be an increasing function on $[a, b]$ and let f be continuous on $[a, b]$. There is nothing to prove if α is a constant on $[a, b]$ (see Example 11.32(**a**)), so we assume that $\alpha(a) < \alpha(b)$ and apply Theorem 11.33. Suppose that $\epsilon > 0$ is given. Since f is uniformly continuous on $[a, b]$, there is a $\delta > 0$ such that

$$|f(x) - f(y)| < \frac{\epsilon}{\alpha(b) - \alpha(a)} \quad \text{whenever } |x - y| < \delta \text{ and } x, y \in [a, b].$$

If we imitate the proof of Theorem 6.21, we see that

$$U(P, f, \alpha) - L(P, f, \alpha) = \sum_{k=1}^{n} (M_k - m_k)\Delta\alpha_k < \frac{\epsilon}{\alpha(b) - \alpha(a)} \sum_{k=1}^{n} \Delta\alpha_k = \epsilon.$$

Since $\epsilon > 0$ was arbitrary, by the Darboux–Stieltjes integrability criterion (Theorem 11.33), f is Darboux–Stieltjes integrable (with respect to any increasing function α on $[a, b]$). That is, $f \in \mathcal{D}_\alpha[a, b]$. ∎

For instance, let $f(x) = x$ and $\alpha(x) = x^2$ on $[0, 1]$. Then by Theorem 11.35, $f \in \mathcal{D}_\alpha[0, 1]$. How do we find the value of (DS) $\int_0^1 f \, d\alpha$?

Example 11.36. Let f be a function that is bounded on $[a, b]$ and continuous at $c \in (a, b)$. Define α on $[a, b]$ by

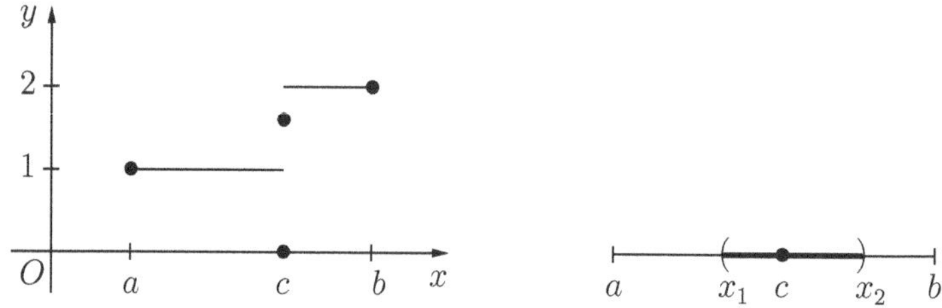

Fig. 11.7. Graph of $\alpha(x)$ on $[a, b]$ and a partition of $[a, b]$.

$$\alpha(x) = \begin{cases} 1 & \text{for } a \leq x < c, \\ 1.5 & \text{for } a \leq x < c, \\ 2 & \text{for } c < x \leq b. \end{cases}$$

The graph of $\alpha(x)$ is drawn in Figure 11.7. Then $f \in \mathcal{D}_\alpha[a, b]$ and therefore, we obtain DS $\int_a^b f d\alpha = f(c)$. •

Solution. Note that $\alpha(c) \in [1,2]$ is immaterial. Since f is continuous at c, given $\epsilon > 0$, there is a $\delta > 0$ such that

$$|f(x) - f(c)| < \frac{\epsilon}{2} \quad \text{whenever } |x - c| < \delta \text{ and } x \in [a,b].$$

In particular, there exist x_1 and x_2 such that

$$-\frac{\epsilon}{2} + f(c) < f(x) < f(c) + \frac{\epsilon}{2} \quad \text{for } x \in [x_1, x_2] \subset [a,b],$$

and therefore

$$-\frac{\epsilon}{2} + f(c) \le m_2 \le f(x) \le M_2 \le f(c) + \frac{\epsilon}{2}.$$

For the partition $P = \{a, x_1, x_2, b\}$, we see that

$$U(P, f, \alpha) = M_2(\alpha(x_2) - \alpha(x_1)) = M_2(2-1) = M_2,$$

and similarly, $L(P, f, \alpha) = m_2$. We thereby obtain

$$U(P, f, \alpha) - L(P, f, \alpha) = M_2 - m_2 \le f(c) + \frac{\epsilon}{2} - \left(f(c) - \frac{\epsilon}{2}\right) = \epsilon,$$

and so by Theorem 11.33, $f \in \mathcal{D}_\alpha[a,b]$. Also,

$$-\frac{\epsilon}{2} + f(c) \le L(P, f, \alpha) \le \int_a^b f \, \mathrm{d}\alpha \le U(P, f, \alpha) \le f(c) + \frac{\epsilon}{2}.$$

Since ϵ is arbitrary, it follows that $\int_a^b f \, \mathrm{d}\alpha = f(c)$. ●

Our next result provides us with a new class of Darboux–Stieltjes integrable functions.

Theorem 11.37 (Darboux–Stieltjes integrability of monotone functions). *Let f be a function that is monotone on $[a,b]$ and let α be increasing and continuous on $[a,b]$. Then $f \in \mathcal{D}_\alpha[a,b]$.*

Proof. If f is constant on $[a,b]$, then it is certainly Darboux–Stieltjes integrable on $[a,b]$ (see Examples 11.32**(b)**). So we shall assume that f is nonconstant, i.e., that $f(a) \neq f(b)$ in particular. Next, we assume that f is increasing on $[a,b]$ with $f(a) < f(b)$, since the proof for f decreasing is similar.

Let $\epsilon > 0$ be given. Then there exists a $K > 0$ such that

$$K(f(b) - f(a)) < \epsilon.$$

Since the function α is uniformly continuous on $[a,b]$, we consider a partition $P = \{x_0, x_1, \dots, x_n\}$ of $[a,b]$ with

$$\Delta\alpha_k = \alpha(x_k) - \alpha(x_{k-1}) < K \quad \text{for each } k \in \{1, 2, \dots, n\}.$$

Also, since $x_{k-1} < x_k$ and f is increasing, we have for each $k \in \{1, 2, \ldots, n\}$ that

$$M_k = \sup_{x \in [x_{k-1}, x_k]} f(x) = f(x_k) \quad \text{and} \quad m_k = \inf_{x \in [x_{k-1}, x_k]} f(x) = f(x_{k-1}).$$

As in Theorem 6.20, it follows easily that

$$U(P, f) - L(P, f) = \sum_{k=1}^{n} (f(x_k) - f(x_{k-1}))\, \Delta\alpha_k < K(f(b) - f(a) < \epsilon.$$

The integrability criterion (see Theorem 11.33) shows that $f \in \mathcal{D}_\alpha[a, b]$. ■

Corollary 11.38. *Suppose that f is bounded on $[a, b]$ and has only finitely many points of discontinuity on $[a, b]$. Let α be increasing and continuous at each point of discontinuity of f. Then $f \in \mathcal{D}_\alpha[a, b]$.*

Proof. We leave the proof as a simple exercise. ■

Theorem 11.39 (General properties of Darboux–Stieltjes integrals). *Suppose that $f, g \in \mathcal{D}_\alpha[a, b]$. Then we have the following:*

(a) *$c_1 f + c_2 g \in \mathcal{D}_\alpha[a, b]$ for constants c_1 and c_2. Also,*

$$\int_a^b [c_1 f + c_2 g]\, d\alpha = c_1 \int_a^b f\, d\alpha + c_2 \int_a^b g\, d\alpha.$$

This is called the linearity rule for Darboux–Stieltjes integrals.

(b) *If $f(x) \le g(x)$ on $[a, b]$, then*

$$\int_a^b f\, d\alpha \le \int_a^b g\, d\alpha.$$

This is called the dominance rule for Darboux–Stieltjes integrals.

(c) *If $m \le f(x) \le M$ for $x \in [a, b]$ and h is continuous on $[m, M]$, then $h \circ f \in \mathcal{D}_\alpha[a, b]$. In particular,*

$$m(\alpha(b) - \alpha(a)) \le \int_a^b f\, d\alpha \le M(\alpha(b) - \alpha(a)).$$

(d) *If $c \in (a, b)$, then $f \in \mathcal{D}_\alpha[a, c]$ and $f \in \mathcal{D}_\alpha[c, b]$. Moreover,*

$$\int_a^b f\, d\alpha = \int_a^c f\, d\alpha + \int_c^b f\, d\alpha.$$

(e) *$fg \in \mathcal{D}_\alpha[a, b]$.*

(f) $|f| \in \mathcal{D}_\alpha[a,b]$ *and*

$$\left|\int_a^b f\,\mathrm{d}\alpha\right| \le \int_a^b |f|\,\mathrm{d}\alpha.$$

In particular, if in addition we have $|f(x)| \le K$ *on* $[a,b]$*, then*

$$\left|\int_a^b f\,\mathrm{d}\alpha\right| \le K(\alpha(b)-\alpha(a)).$$

(g) *If in addition to* $f \in \mathcal{D}_\alpha[a,b]$*,* β *is increasing on* $[a,b]$ *and* $f \in \mathcal{D}_\beta[a,b]$*, then* $f \in \mathcal{D}_{\alpha+\beta}[a,b]$ *and* $f \in \mathcal{D}_{c\alpha}[a,b]$ *for* $c \ge 0$*. Also,*

$$\int_a^b f\,d(\alpha+\beta) = \int_a^b f\,\mathrm{d}\alpha + \int_a^b g\,\mathrm{d}\beta \quad \text{and} \quad \int_a^b f\,d(c\alpha) = c\int_a^b f\,\mathrm{d}\alpha.$$

There are cases in which one can interpret a Darboux–Stieltjes integral as a Riemann integral. In the following we show that if α is smooth, then the Darboux–Stieltjes integral of f with respect to α can be calculated using the ordinary Riemann integral. This result is useful for the computation of many Darboux–Stieltjes integrals.

Theorem 11.40. *Suppose* f *is Riemann integrable on* $[a,b]$ *and* α *is an increasing function that is differentiable on* $[a,b]$*. Then* $f \in \mathcal{D}_\alpha[a,b]$*, and* $f\alpha'$ *is Riemann integrable on* $[a,b]$*. Moreover,*

$$\int_a^b f\,\mathrm{d}\alpha = \int_a^b f(x)\alpha'(x)\,\mathrm{d}x \tag{11.5}$$

(the integral on the left is a Darboux–Stieltjes integral, whereas the integral on the right is a Riemann integral).

Proof. Let $\epsilon > 0$. By assumption, f and α' are Riemann integrable. Thus we have the following:

- $f\alpha'$ is Riemann integrable on $[a,b]$.
- α' is bounded on $[a,b]$ so that $\alpha'(x) \le K$ for some $K > 0$ and for all $x \in [a,b]$.
- There exists a partition $P = \{x_0, x_1, \ldots, x_n\}$ of $[a,b]$ such that

$$U(P,f) - L(P,f) < \epsilon/K.$$

- By the mean value theorem, there exists a $c_k \in (a,b)$ such that

$$\Delta\alpha_k = \alpha(x_k) - \alpha(x_{k-1}) = \alpha'(c_k)(x_k - x_{k-1}) = \alpha'(c_k)\Delta x_k < K\Delta x_k.$$

Using these observations, we deduce that

$$\begin{aligned} U(P,f,\alpha) - L(P,f,\alpha) &= \sum_{k=1}^{n} (M_k - m_k)\Delta\alpha_k \\ &< K\sum_{k=1}^{n} (M_k - m_k)\Delta x_k \\ &= K(U(P,f) - L(P,f)) < \epsilon, \end{aligned}$$

showing that $f \in \mathcal{D}_\alpha[a,b]$, by Theorem 11.33.

Next we prove the equality in (11.5), since the integrals on the left and the right exist. Call the integral on the right-hand side in (11.5) I.

Let $\epsilon > 0$ be given. Then by the Riemann integrability theorem, there exists a partition $P = \{x_0, x_1, \ldots, x_n\}$ of $[a,b]$ such that

$$I - \epsilon < \sigma(P, f\alpha', x^*) := \sum_{k=1}^{n} (f\alpha')(\xi_k)\,\Delta x_k < I + \epsilon$$

for any choice of $\xi_k \in [x_{k-1}, x_k]$. Now

$$\begin{aligned} U(P,f,\alpha) &= \sum_{k=1}^{n} M_k\,\Delta\alpha_k \\ &= \sum_{k=1}^{n} M_k\alpha'(c_k)\,\Delta x_k \\ &\geq \sum_{k=1}^{n} f(c_k)\alpha'(c_k)\,\Delta x_k \quad (\because \alpha'(x) \geq 0) \\ &> I - \epsilon, \end{aligned}$$

so that

$$\int_a^b f\,\mathrm{d}\alpha = \overline{\int_a^b} f\,\mathrm{d}\alpha \geq I. \tag{11.6}$$

Similarly,

$$L(P,f,\alpha) \leq \sum_{k=1}^{n} f(c_k)\alpha'(c_k)\,\Delta x_k < I + \epsilon,$$

so that

$$\int_a^b f\,\mathrm{d}\alpha = \underline{\int_a^b} f\,\mathrm{d}\alpha \leq I. \tag{11.7}$$

By (11.6) and (11.7), it follows that

$$\int_a^b f\,\mathrm{d}\alpha = I,$$

and the theorem follows. ∎

Our next result is a generalization of the customary integration by parts formula. This formula is useful especially in the computation of many Darboux–Stieltjes integrals (and hence Riemann–Stieltjes integrals; see Theorem 11.44).

Theorem 11.41 (Integration by parts). *Suppose that f and α are increasing on $[a,b]$. If $f \in \mathcal{D}_\alpha[a,b]$, then $\alpha \in \mathcal{D}_f[a,b]$ and*

$$\int_a^b f\,\mathrm{d}\alpha = f(b)\alpha(b) - f(a)\alpha(a) - \int_a^b \alpha\,\mathrm{d}f.$$

Proof. Let $P = \{x_0, x_1, \ldots, x_n\}$ be a partition of $[a,b]$. Then because f is increasing, we have $M_k(f) = f(x_k)$, and therefore

$$\begin{aligned} U(P,f,\alpha) &= \sum_{k=1}^n f(x_k)\,\Delta\alpha_k \\ &= \sum_{k=1}^n f(x_k)\alpha(x_k) - \sum_{k=1}^n f(x_k)\alpha(x_{k-1}) \\ &= f(x_n)\alpha(x_n) - f(x_0)\alpha(x_0) - \sum_{k=1}^n \alpha(x_{k-1})(f(x_k) - f(x_{k-1})), \end{aligned}$$

so that

$$U(P,f,\alpha) = f(b)\alpha(b) - f(a)\alpha(a) - L(P,\alpha,f), \tag{11.8}$$

because $m_k(\alpha) = \alpha(x_{k-1})$, since α is increasing. Similarly, interchanging the roles of α and f yields

$$U(P,\alpha,f) = f(b)\alpha(b) - f(a)\alpha(a) - L(P,f,\alpha). \tag{11.9}$$

Hence, subtracting (11.9) from (11.8) gives

$$U(P,f,\alpha) - L(P,f,\alpha) = U(P,\alpha,f) - L(P,\alpha,f).$$

By Theorem 11.33, $\alpha \in \mathcal{D}_f[a,b]$ iff $f \in \mathcal{D}_\alpha[a,b]$.

Now let $\epsilon > 0$ be given and $f \in \mathcal{D}_\alpha[a,b]$. Then by the integrability criterion, there exists a partition P of $[a,b]$ such that

$$U(P,f,\alpha) < \int_a^b f\,\mathrm{d}\alpha + \epsilon \quad \text{and} \quad L(P,f,\alpha) > \int_a^b f\,\mathrm{d}\alpha - \epsilon.$$

Thus,

$$
\begin{aligned}
f(b)\alpha(b) - f(a)\alpha(a) - \int_a^b f\,\mathrm{d}\alpha - \epsilon &< f(b)\alpha(b) - f(a)\alpha(a) - U(P,f,\alpha) \\
&= L(P,\alpha,f), \quad \text{by (11.8)}, \\
&\leq \underline{\int_a^b} \alpha\,\mathrm{d}f = \int_a^b \alpha\,\mathrm{d}f = \overline{\int_a^b} \alpha\,\mathrm{d}f \\
&\leq U(P,\alpha,f) \\
&= f(b)\alpha(b) - f(a)\alpha(a) - L(P,f,\alpha) \\
&\qquad \text{(by (11.9))} \\
&< f(b)\alpha(b) - f(a)\alpha(a) - \int_a^b f\,\mathrm{d}\alpha + \epsilon.
\end{aligned}
$$

Since $\epsilon > 0$ is arbitrary, it follows that

$$
\int_a^b \alpha\,\mathrm{d}f = f(b)\alpha(b) - f(a)\alpha(a) - \int_a^b f\,\mathrm{d}\alpha,
$$

and the theorem follows. ■

For instance

$$
\int_0^1 x\,d(x^2) = 1 - \int_0^1 x^2\,\mathrm{d}x = \frac{2}{3}.
$$

11.2.2 The Riemann–Stieltjes Integral

In the following, we take a more liberal point of view and introduce the definition of the Riemann–Stieltjes integral even though there is an alternative (but not equivalent) definition of the Riemann–Stieltjes integral. In the following we assume that α is not necessarily increasing.

Definition 11.42 (The Riemann–Stieltjes integral). *Let f be bounded on $[a,b]$. We say that f is Riemann–Stieltjes integrable with respect to α on $[a,b]$, or Stieltjes integrable with respect to α on $[a,b]$ in the sense of Riemann, and write $f \in \mathcal{R}_\alpha[a,b]$, if there exists a number I with the following property: For every $\epsilon > 0$, there corresponds a partition P_ϵ of $[a,b]$ such that for every partition P finer than P_ϵ we have*

$$
|\sigma(P,f,\alpha,x^*) - I| < \epsilon
$$

for every Riemann–Stieltjes sum $\sigma(P,f,\alpha,x^*)$ of f associated with the function α and the partition P. The value I is called the Riemann–Stieltjes integral of f with respect to α on $[a,b]$, and, in order to distinguish it from the Darboux–Stieltjes integral, it is temporarily denoted by

$$
\text{(RS)} \int_a^b f(x)\,\mathrm{d}\alpha(x), \quad \textit{or by} \quad \text{(RS)} \int_a^b f\,\mathrm{d}\alpha.
$$

Definition 11.43 (Riemann's condition). *A bounded function f defined on $[a,b]$ is said to satisfy Riemann's condition with respect to α on $[a,b]$ if for every $\epsilon > 0$, there exists a partition P_ϵ of $[a,b]$ such that*

$$U(P,f,\alpha) - L(P,f,\alpha) < \epsilon \quad \textit{whenever partition } P \textit{ is finer than } P_\epsilon.$$

The equivalence of Definitions 11.30 and 11.42 is well known when α is assumed to be increasing on $[a,b]$, in which case the Darboux–Stieltjes integral and the Riemann–Stieltjes integral of f are equal. Thus,

$$(\mathrm{DS}) \int_a^b f \, \mathrm{d}\alpha = (\mathrm{RS}) \int_a^b f \, \mathrm{d}\alpha,$$

and so we may write simply $\int_a^b f \, \mathrm{d}\alpha$.

Theorem 11.44. *Suppose that f is a function that is bounded on $[a,b]$ and α is an* increasing function *on $[a,b]$. Then the following are equivalent:*

(a) *$f \in \mathcal{D}_\alpha[a,b]$;*
(b) *$f \in \mathcal{R}_\alpha[a,b]$;*
(c) *f satisfies Riemann's condition.*

Proof. **(a)** $\Rightarrow$ **(b)**: Suppose that $f \in \mathcal{D}_\alpha[a,b]$, i.e., $U_\alpha(f) = L_\alpha(f)$, and $\beta =$ (DS) $\int_a^b f \, \mathrm{d}\alpha$, their common value. By Theorem 11.33, for each $\epsilon > 0$ there is a partition P of $[a,b]$ such that

$$U(P,f,\alpha) - L(P,f,\alpha) < \epsilon,$$

and by definition,

$$L(P,f,\alpha) \le L(Q,f,\alpha) \le \sigma(Q,f,\alpha,x^*) \le U(Q,f,\alpha) \le U(P,f,\alpha) < L(P,f,\alpha) + \epsilon$$

for all $P \subseteq Q$. Because $f \in \mathcal{D}_\alpha[a,b]$, it follows that

$$L(Q,f,\alpha) \le \beta \le U(Q,f,\alpha),$$

and so the last inequality implies that

$$|\sigma(Q,f,\alpha,x^*) - \beta| < \epsilon,$$

showing that $f \in \mathcal{R}_\alpha[a,b]$ with β as the Riemann–Stieltjes integral of f with respect to α on $[a,b]$.

(b) $\Rightarrow$ **(a)**: Suppose that $f \in \mathcal{R}_\alpha[a,b]$ in the sense of Definition 11.42. First we need to show that $f \in \mathcal{D}_\alpha[a,b]$. If $\alpha(b) = \alpha(a)$, then $\alpha(x)$ is constant on $[a,b]$, and for every partition,

$$U(P,f,\alpha) = L(P,f,\alpha) = \sigma(P,f,\alpha,x^*) = 0,$$

and so there is nothing to prove. So we may assume that $\alpha(b)-\alpha(a)>0$. By assumption, for each $\epsilon>0$, there exists a partition P_ϵ such that if $P_\epsilon\subseteq Q$, then

$$|\sigma(Q,f,\alpha,x^*)-I|<\epsilon$$

for some $I\in\mathbb{R}$. Then for $\epsilon>0$, there exists a partition $P=\{x_0,x_1,\ldots,x_n\}$ such that

$$|\sigma(P,f,\alpha,x^*)-I|<\epsilon/4,$$

regardless of the choice of $x_k^*\in[x_{k-1},x_k]$. By the definition of M_k and m_k, we have

$$M_k-\frac{\epsilon}{4(\alpha(b)-\alpha(a))}<f(x_k^*)\quad\text{for some } x_k^*\in[x_{k-1},x_k],$$

$$m_k+\frac{\epsilon}{4(\alpha(b)-\alpha(a))}>f(y_k^*)\quad\text{for some } y_k^*\in[x_{k-1},x_k],$$

so that

$$M_k-m_k<f(x_k^*)-f(y_k^*)+\frac{\epsilon}{2(\alpha(b)-\alpha(a))}.$$

Since $\sum_{k=1}^n\Delta\alpha_k=\alpha(b)-\alpha(a)$, it follows that

$$\begin{aligned}U(P,f,\alpha)-L(P,f,\alpha)&=\sum_{k=1}^n(M_k-m_k)\Delta\alpha_k\\&<\sum_{k=1}^n f(x_k^*)\Delta\alpha_k-\sum_{k=1}^n f(y_k^*)\Delta\alpha_k+\frac{\epsilon}{2}\\&=(\sigma(P,f,\alpha,x^*)-I)-(\sigma(P,f,\alpha,y^*)-I)+\frac{\epsilon}{2}\\&<\frac{\epsilon}{4}+\frac{\epsilon}{4}+\frac{\epsilon}{2}=\epsilon.\end{aligned}$$

Thus $f\in\mathcal{D}_\alpha[a,b]$, by Theorem 11.33.

It remains to show that I equals $\beta=\text{(DS)}\int_a^b f\,d\alpha$. Again, for a given $\epsilon>0$, there exists a partition P_1 of $[a,b]$ such that for every partition Q finer than P_1 we have

$$|\sigma(Q,f,\alpha,x^*)-I|<\epsilon/3.$$

Since $f\in\mathcal{D}_\alpha[a,b]$, there is a partition P_2 of $[a,b]$ such that

$$U(P_2,f,\alpha)-L(P_2,f,\alpha)<\epsilon/3.$$

Now we let $Q=P_1\cup P_2$. Then

$$\begin{aligned}|I-\beta|&\leq|I-\sigma(Q,f,\alpha,x^*)|+|\sigma(Q,f,\alpha,x^*)-L(Q,f,\alpha)|+|L(Q,f,\alpha)-\beta|\\&<(\epsilon/3)+(\epsilon/3)+(\epsilon/3)=\epsilon,\end{aligned}$$

because

$$L(Q,f,\alpha) \le \sigma(Q,f,\alpha,x^*) \le U(Q,f,\alpha) < L(Q,f,\alpha) + \epsilon/3$$

and

$$L(Q,f,\alpha) \le \beta \le U(Q,f,\alpha) < L(Q,f,\alpha) + \epsilon/3.$$

Since $\epsilon > 0$ is arbitrary, it follows that $I = \beta$.

We leave the implications **(b)** $\Longleftrightarrow$ **(c)** as an exercise. ■

Example 11.45. Compute the value of the integral $\displaystyle\int_0^2 x\, d[x]$.

Solution. Set $f(x) = x$ and $\alpha(x) = [x]$. By Theorem 11.44, f is both Darboux–Stieltjes and Riemann–Stieltjes integrable on $[0,2]$. Also, we note that

$$\alpha(x) = \begin{cases} 0 & \text{for } 0 \le x < 1, \\ 1 & \text{for } 1 \le x < 2, \\ 2 & \text{for } x = 2, \end{cases} \quad \text{and} \quad \Delta\alpha_i = \begin{cases} 0 & \text{for } 0 \le i < k-1, \\ 1 & \text{for } i = k, \\ 0 & \text{for } k+1 \le i < n-1, \\ 1 & \text{for } i = n. \end{cases}$$

Observe that f is continuous at $1, 2$, while α is discontinuous at $1, 2$. Let $P = \{x_0, x_1, \dots, x_n\}$ be any partition of $[0,2]$ with $x_k = 1$ for some k so that $1 \in P$. With $x_k^* \in [x_{k-1}, x_k]$ for $k = 1, \dots, n$, the Riemann–Stieltjes sum of f with respect to α takes the form

$$\sigma(P,f,\alpha,x^*) = \sum_{i=1}^{n} f(x_i^*)\Delta\alpha_i = f(x_k^*) + f(x_n^*) = x_k^* + x_n^*.$$

Now given $\epsilon > 0$, choose a partition P_ϵ such that $x_k - x_{k-1} < \epsilon/2$ for $k = 1, \dots, n$. Then for any refinement Q of P_ϵ,

$$|\sigma(Q,f,\alpha,x^*) - 3| \le |x_k^* - 1| + |x_n^* - 2| < (\epsilon/2) + (\epsilon/2) = \epsilon,$$

and therefore

$$\int_0^2 x\, d[x] = 3.$$ •

Since every function of bounded variation is the difference of two increasing functions (see Theorem 11.19), by Theorem 11.44, it follows that the Darboux–Stieltjes integral (and hence the Riemann–Stieltjes integral) exists when f is continuous and α is of bounded variation.

Corollary 11.46. *Suppose that f is continuous on $[a,b]$ and α is of bounded variation on $[a,b]$. Then $f \in \mathcal{R}_\alpha[a,b]$, i.e., f is Riemann–Stieltjes integrable on $[a,b]$.*

11.2.3 Questions and Exercises

Questions 11.47.

1. Why are the upper and lower integrals finite?
2. Suppose that f is bounded and continuous on $[1, n]$ and $\alpha(x) = [x]$. What is the value of $\int_1^n f\, d\alpha$? What happens when $f(x) = 1/x$?
3. Suppose that $f + g \in \mathcal{D}_\alpha[a, b]$. Are $f, g \in \mathcal{D}_\alpha[a, b]$?
4. Suppose that $f+g \in \mathcal{D}_\alpha[a, b]$ and $g \in \mathcal{D}_\alpha[a, b]$. Must we have $f \in \mathcal{D}_\alpha[a, b]$?
5. Is every Riemann–Stieltjes integrable function Darboux–Stieltjes integrable?
6. Suppose that f is Darboux–Stieltjes integrable with respect to a continuous function α. Is f Riemann–Stieltjes integrable with respect to α?
7. Suppose that f is bounded on $[a, b]$ and α is discontinuous at a point $c \in (a, b)$. Is f always Riemann–Stieltjes integrable with respect to α?
8. Suppose that α is bounded and increasing on $[a, b]$. Does there always exist a nonconstant function that is integrable with respect to α?
9. Suppose that f and α are increasing on $[a, b]$. Must $\int_a^b f\, d\alpha$ always exist? How about if $f(x) = \alpha(x) = [x]$ on $[0, 2]$?
10. If f is integrable with respect to itself (in the sense of Darboux) on $[a, b]$, then what is the value of $\int_a^b f\, d\alpha$? How about the value of the integral if f is integrable with respect to itself (in the sense of Riemann)?
11. In the definition of the Riemann–Stieltjes integral of f, why do we assume the boundedness of f? What will happen if we drop the boundedness condition on f?
12. If α is of bounded variation on $[a, b]$, what is the value of $\int_a^b d\alpha$?

Exercises 11.48.

1. Complete the proof of Theorem 11.33.
2. Complete the proof of Corollary 11.31.
3. Complete the proof of Corollary 11.38.
4. Consider

$$f(x) = \begin{cases} 0 & \text{for } 0 \le x < 1, \\ 1 & \text{for } 1 \le x \le 2, \end{cases} \quad \text{and} \quad \alpha(x) = \begin{cases} 0 & \text{for } 0 \le x < 1, \\ 2 & \text{for } 1 \le x \le 2. \end{cases}$$

   Show that (DS) $\int_0^2 f\, d\alpha = 0$.
5. Let f be continuous on $[a, b]$ and

$$\alpha(x) = \begin{cases} p & \text{for } a \le x < c, \\ q & \text{for } x = c, \\ r & \text{for } c < x \le b. \end{cases}$$

   Show that (DS) $\int_a^b f\, d\alpha = f(c)(r - p)$.

6. For $a < c < d < b$, we define
$$\alpha(x) = \begin{cases} 0 & \text{for } x < c, \\ 2 & \text{for } c \leq x \leq d, \\ 2 & \text{for } x \geq d. \end{cases}$$
Compute (DS) $\int_a^b x^2\, d\alpha$.
7. Let $f(x) = x^3$ and
$$\alpha(x) = \begin{cases} 3x^2 & \text{for } 0 \leq x \leq 1/2, \\ 3x^2 + 5 & \text{for } 1/2 < x \leq 1. \end{cases}$$
Determine whether f is Darboux–Stieltjes integrable. If it is, determine (DS) $\int_0^1 f\, d\alpha$.
8. Let f be continuous on $[0, 3]$ and
$$\alpha(x) = \begin{cases} x^2 & \text{for } x \in [0,3] \setminus (1,2), \\ 1 & \text{for } x \in (1,2). \end{cases}$$
Compute (DS) $\int_0^3 f\, d\alpha$ if f is Darboux–Stieltjes integrable.
9. If f is continuous and α is of bounded variation on $[a, b]$, then show that f is Darboux–Stieltjes integrable with respect to α.
10. Compute each of the following integrals:
(a) $\displaystyle\int_0^3 x\, d[x]$. (b) $\displaystyle\int_0^n x\, d[x]$. (c) $\displaystyle\int_0^{n/2} x\, d[x]$. (d) $\displaystyle\int_0^\pi x\, d(\sin x)$.

References for Further Reading

1. T. Apostol: *Mathematical Analysis*, Addison-Wesley, Reading, Massachusetts, 1967.
2. R. G. Bartle and D. R. Sherbert: *Introduction to Real Analysis*, John Wiley & Sons, 1982.
3. R. R. Goldberg: *Methods of Real Analysis*, 2nd ed., Wiley, 1976.
4. K. Plofker: *Mathematics in India*, Princeton University Press, 2008.
5. S. Ponnusamy: *Foundations of Complex Analysis*, Narosa Publishing House, India, 1995 (2005, Revised version).
6. S. Ponnusamy: *Foundations of Functional Analysis*, Narosa Publishing House, India, 2003.
7. S. Ponnusamy: *Foundations of Multivariable Calculus*, in preparation
8. S. Ponnusamy and H. Silverman: *Complex Variables with Applications*, Birkhäuser, Boston, 2006.
9. W. Rudin: *Principles of Mathematical Analysis*, 3nd ed., McGraw-Hill, New York, 1976.

S. Ponnusamy, *Foundations of Mathematical Analysis*,
DOI 10.1007/978-0-8176-8292-7,

Index of Notation

Symbol	Meaning
$\emptyset$	empty set
$a \in S$	a is an element of the set S
$a \notin S$	a is not an element of S
$\{x : \ldots\}$	the set of all elements with the property ...
$X \cup Y$	the set of all elements in X or Y; i.e., the union of the sets X and Y
$X \cap Y$	the set of all elements simultaneously in both X and Y; i.e., the intersection of the sets X and Y
$X \subset Y$	the set X is contained in the set Y; i.e., X is a subset of Y
$X \subsetneq Y$	$X \subset Y$ and $X \neq Y$; i.e., the set X is a proper subset of Y
$X \times Y$	the Cartesian product of the sets X and Y, $\{(x, y) : x \in X,\ y \in Y\}$
$X \setminus Y$ or $X - Y$	the set of all elements in X but not in Y
A^c	$\{x \in X : x \notin A\}$, the complement of $A \subset X$
$\Longrightarrow$	implies
$\Longleftrightarrow$	if and only if, or "iff"
$\longrightarrow$ or $\to$	converges to (approaches)
$\not\longrightarrow$ or $\not\to$	does not converge
$\not\Longrightarrow$	does not imply
$\mathbb{N}$	the set of all natural numbers, $\{1, 2, \ldots\}$
$\mathbb{N}_0$	$\mathbb{N} \cup \{0\} = \{0, 1, 2, \ldots\}$

S. Ponnusamy, *Foundations of Mathematical Analysis*,
DOI 10.1007/978-0-8176-8292-7,

$\mathbb{Z}$	the set of all integers (positive, negative, and zero)
$\mathbb{Q}$	the set of all rational numbers, $\{p/q : p,\ q \in \mathbb{Z},\ q \neq 0\}$
$\mathbb{R}$	the set of all real numbers, the real line
$\mathbb{R}_\infty$	$\mathbb{R} \cup \{-\infty, \infty\}$, the extended real line
$\mathbb{C}$	the set of all complex numbers, the complex plane
$\limsup x_n$	the upper limit of the real sequence $\{x_n\}$
$\liminf x_n$	the lower limit of the real sequence $\{x_n\}$
$\lim x_n$	the limit of the real sequence $\{x_n\}$
$\sup S$	the least upper bound, or supremum, of the set $S \subset \mathbb{R}_\infty$
$\inf S$	the greatest lower bound, or infimum, of the set $S \subset \mathbb{R}_\infty$
$\inf_{x\in S} f(x)$	the infimum of f in S
$\max S$	the the maximum of the set $S \subset \mathbb{R}$; the largest element in S
$\min S$	the minimum of the set $S \subset \mathbb{R}$; the smallest element in S
$f : A \to B$	f is a function from A into B
$f\big\rvert_A$	the restriction of f to A, $A \subset \text{domain}\,(f)$
f^{-1}	the inverse function of f
$f(x)$	the value of the function at x or the function of the variable x
$f(A)$	the direct image of a set A under f, i.e., set of all values $f(x)$ with $x \in A$; i.e., $y \in f(A) \Longleftrightarrow \exists\, x \in A$ such that $f(x) = y$
$f^{-1}(B)$	$\{x : f(x) \in B\}$, the inverse image of B under f
$f^{-1}(x)$	the preimage/inverse image of one element $\{x\}$
$f \circ g$	the composition mapping of f and g
$\text{dist}\,(x, A)$	the distance from the point x to the set A i.e., $\inf\{\lvert x - a\rvert : a \in A\}$
$\text{dist}\,(A, B)$	the distance between two sets A and B i.e., $\inf\{\lvert a - b\rvert : a \in A, b \in B\}$
$[x_1, x_2]$	the closed line segment connecting x_1 and x_2; $\{x = (1-t)x_1 + tx_2 : 0 \leq t \leq 1\}$
(x_1, x_2)	the open line segment connecting x_1 and x_2; $\{x = (1-t)x_1 + tx_2 : 0 < t < 1\}$

e^x or $\exp(x)$	$\lim_{n\to\infty}\left(1+\frac{x}{n}\right)^n = \lim_{n\to\infty}\left(1-\frac{x}{n}\right)^{-n} = \sum_{n\geq 0}\frac{x^n}{n!}$ the exponential function
$f'(a)$	the derivative of f evaluated at a
$f''(a), \ldots, f^{(n)}(a)$	second, ..., nth derivative of f at a
$f(x) = O(g(x))$ as $x \to a$	there exists a constant K such that $\lvert f(x)\rvert \leq K\lvert g(x)\rvert$ for all values of x near a
$f(x) = o(g(x))$ as $x \to a$	$\lim_{x\to a}\frac{f(x)}{g(x)} = 0$
$\lim_{n\to\infty} x_n = x$, or $x_n \to x$, or $d(x_n, x) \to 0$	the sequence $\{z_n\}$ converges to z with a metric d

Appendix A: Hints for Selected Questions and Exercises

Chapter 1: Questions 1.11

5. Since $x^3 - x - 7 = 0$ is equivalent to

$$x(x-1)(x+1) = 7,$$

this equation cannot be satisfied by any integer x, since the left-hand side is an even integer, whereas the right-hand side is not. Again, the equation cannot be satisfied by any rational number $x = m/n$, because for such an x the last equation would imply that

$$\frac{m^3}{n} = mn^2 + 7n^2,$$

a contradiction.

6. If $x = \sqrt{3+\sqrt{2}}$, then we see that

$$x^4 - 6x^2 + 7 = 0.$$

By Theorem 1.5, the only rational numbers that could possibly be solutions of this polynomial equation are $\pm 1, \pm 7$. Substituting them into the equation shows that they are not solutions. Thus, $\sqrt{3+\sqrt{2}}$ is irrational.

7. No. Suppose to the contrary that

$$\sqrt{p} = \frac{a}{b}, \quad \text{i.e., } pb^2 = a^2,$$

where a and b have no common factor greater than 1. Since the number a^2 is a multiple of a prime, it follows from the *unique factorization theorem* that a is a multiple of p. Thus, a has the form $a = pk$, and therefore

$$p^2k^2 = pb^2, \quad \text{i.e., } b^2 = pk^2,$$

which shows that b is multiple of p, a contradiction.

S. Ponnusamy, *Foundations of Mathematical Analysis*,
DOI 10.1007/978-0-8176-8292-7,

In the case of $p = 3$, our computation is easy to understand, and so we give the details here. Suppose to the contrary that there is a rational number $x = m/n$ such that $x^2 = 3$. Then

$$\left(\frac{m}{n}\right)^2 = 3, \quad \text{i.e., } m^2 = 3n^2,$$

where $m \in \mathbb{Z}$ and $n \in \mathbb{N}$ have no common factors other than 1. We observe that if m is divided by 3, then the remainder r will be either 0 or 1 or 2, and so we can write $m = 3k + r$, where $r \in \{0, 1, 2\}$. Thus,

$$m^2 = 3(3k^2 + 2kr) + r^2 = \begin{cases} 9k^2 & \text{if } r = 0, \\ 3(3k^2 + 2k) + 1 & \text{if } r = 1, \\ 3(3k^2 + 4k + 1) + 1 & \text{if } r = 2, \end{cases}$$

showing that m is a multiple of 3 iff m^2 is a multiple of 3. Thus, since $m^2 = 3n^2$, it follows that m is a multiple of 3, so that $m = 3a$ for some integer a. Substituting this in $m^2 = 3n^2$ implies that n is also a multiple of 3, which is a contradiction. Therefore, $\sqrt{3}$ is irrational.

Exercises 1.12:

4. **(b)** Let $n > 1$ and

$$\sqrt{n+1} + \sqrt{n-1} = \frac{p}{q}, \quad \text{i.e., } n - 1 = \frac{p^2}{q^2} + n + 1 - \frac{2p\sqrt{n+1}}{q}.$$

Then a simplification gives

$$\frac{2p\sqrt{n+1}}{q} = 2 + \frac{p^2}{q^2}, \quad \text{i.e., } n = \frac{p^2 + 4q^2}{4p^2q^2},$$

which is not an integer. This is a contradiction.

Questions 1.33:

1. No. Consider $A = \{-1/n : n \geq 1\}$.

Exercises 1.34:

1. We need to show that for every $x \in A$, there is an $a \in A$ with $x < a$. To do this, let $x > 0$ and $x^2 < 2$. Since $0 < h = 2 - x^2 < 2$, we can associate a number a:

$$a = x + \frac{h}{5}.$$

This gives

$$a^2 = x^2 + \frac{2xh}{5} + \frac{h^2}{25} < x^2 + \frac{4}{5}h + \frac{2}{25}h < x^2 + \frac{4}{5}h + \frac{1}{5}h = x^2 + h = 2,$$

and so $a \in A$. A similar argument can be made for B. In fact, if x is such that $x^2 > 2$ and $0 < x < 2$, then since $0 < h = x^2 - 2 < 2$, we can associate a number b:

$$b = x - \frac{h}{4}, \quad \text{i.e., } x = b + \frac{h}{4}.$$

This gives

$$b^2 = x^2 - \frac{xh}{2} + \frac{h^2}{16} > x^2 - h = 2,$$

showing that for each $x \in B$ there exists a $b \in B$ such that $x > b$. The conclusion follows.

8. Consider $f(x) = \tan x$.

Chapter 2: Questions 2.44

18. We recall that $\lim a_n = \infty$ if and only if given $\epsilon = 1/R > 0$, there exists an N such that $n \geq N$ implies that $a_n > R = 1/\epsilon$, which is equivalent to saying that given $\epsilon > 0$, we have

$$\left| \frac{1}{a_n} - 0 \right| < \epsilon \text{ whenever } n \geq N.$$

33. The proof follows from Theorem 2.8. Here is a direct proof. Let $a_n \to a$ as $n \to \infty$. Then, since every convergent sequence is bounded, there exists an $M > 0$ such that $|a_n + a| < M + |a|$ for all n. Again, for $\epsilon > 0$ there exists a positive integer N such that

$$|a_n - a| < \frac{\epsilon}{M + |a|} \quad \text{for all } n \geq N.$$

It follows that $a_n^2 \to a^2$, because

$$|a_n^2 - a^2| = |a_n + a|\,|a_n - a| < \epsilon \quad \text{for all } n \geq N.$$

We remark that this proof also yields the product rule in Theorem 2.8 if we use the identity

$$a_n b_n = (1/4)[(a_n + b_n)^2 - (a_n - b_n)^2]$$

and the linearity rule.

34. Let $\epsilon = |b|/2$. By the definition, there exists an N such that

$$|b| - |b_n| \leq |b_n - b| < |b|/2 \text{ for } n \geq N.$$

39. Consider $a_n = n$. As another example, let $s_n = \sum_{k=1}^{n} \frac{1}{k}$. Note that $s_{n+1} > s_n$, and we have already shown that $\{s_n\}_{n \geq 1}$ is unbounded. It is also easy to show by induction that $s_{2^n} > 1 + (n/2)$.
40. Consider $s_n = \sum_{k=1}^{n} 1/k$ or $s_n = \log n$.

43. Set $a_n = a^{1/2^n}$. Then $a_{n+1} = (a^{1/2})^{1/2^n} < a^{1/2^n} = a_n$, showing that $\{a_n\}$ is a bounded decreasing sequence and hence converges, say to L. Thus, we see that

$$L^2 = \lim_{n\to\infty} (a^{1/2^n} a^{1/2^n}) = \lim_{n\to\infty} a^{1/2^{n-1}} = L.$$

This gives either $L = 0$ or $L = 1$. But $L \neq 0$, and so $L = 1$.

44. Write $b_n = (1 + 1/n)^k (1 + 1/n)^n$ and apply Example 2.33 and the convergence property of the product of sequences.

Exercises 2.45:

3. Rewrite $(n^3 - 3)/(n + 2)$ as $(n^2 - 3/n)/(1 + 2/n)$, and use the same trick for the other case.

6. Set $a_n = a^{1/n}$ for $n \geq 1$, and $0 < a < 1$. Clearly, $0 < a_n < 1$ for all $n \geq 1$ and

$$a_{n+1} - a_n = a^{\frac{1}{n+1}} \left[1 - a^{\frac{1}{n(n+1)}}\right] > 0,$$

so that $\{a_n\}$ is increasing and bounded, and hence $\{a_n\}$ converges, say to L. Thus,

$$a_{2n} = a^{\frac{1}{2n}} \to L^{1/2}.$$

But $\{a_{2n}\}$, being a subsequence of $\{a_n\}$, must converge to the same limit L, and so

$$L = L^{1/2} \quad \text{or} \quad L(L - 1) = 0.$$

Since $a > 0$ and $0 < a_1 < a_2 \cdots$, we have $L \neq 0$, and therefore $L = 1$.

7. Use the fact that

$$a_{n-1} < a_n = \left(1 + \frac{1}{n}\right)^n < \left(1 + \frac{1}{n}\right)^n \left(1 + \frac{1}{n}\right).$$

8. First we observe that $a_1 < 2$ and $a_2 = \sqrt{2 + \sqrt{a_1}} > \sqrt{2} = a_1$. Next, we show by induction that $a_n < 2$ and $a_{n+1} > a_n$ for all $n \geq 1$. By the monotone convergence theorem, the sequence $\{a_n\}$ converges to L, $L > 1$. It follows that

$$L = \sqrt{2 + \sqrt{L}}, \quad \text{i.e.,} \quad (L^2 - 2)^2 = L \quad \text{or} \quad (L - 1)(L^3 + L^2 + 3L - 4) = 0.$$

Since $L > 1$, the limit must be a root of $x^3 + x^2 + 3x - 4 = 0$. Clearly, there is a unique root in $(1, 2)$.

9. First we observe that $a_1 < 2$ and $a_2 = \sqrt{2 + \sqrt{2}} > \sqrt{2} = a_1$. Next, we show by induction that $a_n < 2$ and $a_{n+1} > a_n$ for all $n \geq 1$.

10. **(a)** Use mathematical induction to show that $0 < a_n < 3$ and $a_{n+1} > a_n$ for each n. The monotone convergence theorem tells us that $\{a_n\}$ converges, say to L. Then $L > 1$ and $L = 1 + \sqrt{L}$, which yields that L is a solution of $L^2 - 3L + 1 = 0$. Thus, $L = (3 + \sqrt{5})/2$.

(c) By induction, show that $1 < a_n \leq L$ and $a_{n+1} = a_n/\sqrt{a_n} < a_n$. Hence, $\{a_n\}$ converges.
(g) If $\lim_{n\to\infty} a_n = L$, then we must have

$$L^2 = \frac{\alpha\beta^2 + L^2}{\alpha + 1}, \quad \text{i.e., } L = \beta,$$

because $L = -\beta$ is not possible. Also,

$$a_{n+1} = \beta\sqrt{\frac{\alpha + a_n^2/\beta^2}{\alpha + 1}},$$

so that $0 < a_n \leq \beta$ implies that $0 < a_{n+1} \leq \beta$. In view of this, since $a_1 = \alpha < \beta$, by induction we obtain $0 < a_n \leq \beta$ for all $n \geq 1$. Moreover,

$$a_{n+1}^2 - a_n^2 = \frac{\alpha(\beta^2 - a_n^2)}{\alpha + 1} > 0,$$

showing that $\{a_n\}$ is a bounded increasing sequence. Consequently, it converges to β.

14. Clearly, $\{a_n\}$ is bounded and decreasing. By BMCT, it is convergent.
15. **(e)** We have

$$\lim_{n\to\infty} \frac{3 - (\log n)/n^2}{1 + 3(1/n^{1/2})} = 3, \quad \text{because } \lim_{n\to\infty} \frac{\log n}{n^2} = 0.$$

(t) There exists an N such that

$$\frac{|a|}{n} < \frac{1}{2} \quad \text{for } n \geq N,$$

so that for $n \geq N$,

$$\frac{|a|^n}{n!} = \frac{|a|^N}{N!}\left(\frac{|a|}{N+1}\frac{|a|}{N+2}\cdots\frac{|a|}{n}\right) < \frac{|a|^N}{N!}\left(\frac{1}{2}\right)^{n-N},$$

which tends to 0 as $n \to \infty$.
(o) We have

$$\lim_{n\to\infty} a_n = \lim_{n\to\infty} \frac{1}{n}\sin\left(\frac{n\pi}{6}\right) + \lim_{n\to\infty}\left(\frac{5 + 1/n}{7 + 6/n}\right) = 0 + \frac{5}{7}.$$

Questions 2.67:

5. $a_n = n^{(-1)^n}$.
8. See Example 2.39(**d**).
9. $a_n = 1 + 1/n, 2 + 1/n^2, 3 + 2/n^3, 4 - 1/n^4$.

13. It is easy to see that $\{a_n\}$ is convergent. By Theorem 2.55, it is Cauchy. We next provide a direct proof. For $n > m$, we have

$$\begin{aligned}|a_n - a_m| &= \sum_{k=m+1}^{n} \frac{1}{k^2} < \sum_{k=m+1}^{n} \frac{1}{(k-1)k} = \sum_{k=m+1}^{n} \left[\frac{1}{k-1} - \frac{1}{k}\right] \\ &= \frac{1}{m} - \frac{1}{n} \\ &< \frac{1}{m} < \epsilon \quad \text{if } m > \frac{1}{\epsilon}.\end{aligned}$$

Thus, for any positive integer N greater than $1/\epsilon$, we have $|a_n - a_m| < \epsilon$ for $n > m > N$. That is, $\{a_n\}$ is Cauchy and hence $\{a_n\}$ is convergent.

Exercises 2.68:

1. Use the fact that if $\lim_{n\to\infty} |A_{n+1}/A_n| < 1$, then $A_n \to 0$. Alternatively, we may use the ratio/root test (which will be proved later) and show that $\sum_{k=1}^{\infty} k^p r^k$ converges for $|r| < 1$. Indeed, for $A_n = n^p r^n$,

$$\left|\frac{A_{n+1}}{A_n}\right| = \left(\frac{n+1}{n}\right)^p |r| \to |r|; \quad |A_n|^{1/n} = \left(n^{1/n}\right)^p |r| \to |r|,$$

showing that the series converges, and hence the general term of the series approaches zero.

2. For example,

$$a_n = \begin{cases} 1 & \text{if } n \text{ is odd}, \\ 1/n^2 & \text{if } n \text{ is even}, \end{cases} \qquad a_n = \begin{cases} 1 & \text{if } n = 3k-1, \\ 1/n & \text{if } n = 3k, \\ 1/n^2 & \text{if } n = 3k+1, \end{cases}$$

and

$$a_n = \begin{cases} n & \text{if } n \text{ is even}, \\ \dfrac{(-1)^{(n+1)/2}}{n} & \text{if } n \text{ is odd}. \end{cases}$$

5. Consider $a_n = (-1)^n$.
6. **(a)** Since $k! \geq 2^{k-1}$ for $k \geq 2$, we have for $n \geq m \geq 2$,

$$|a_n - a_m| \leq \sum_{k=m}^{n-1} \frac{1}{k!} \leq \sum_{k=m}^{n-1} \frac{1}{2^{k-1}} < \frac{2}{2^{m-1}}.$$

7. We remark that $\{a_{4n-3}\}$ and $\{a_{2n}\}$ are two subsequences converging to 1 and 0 respectively. Note that $\{a_{2n-1}\}$ is a divergent subsequence.
9. Apply Theorem 2.57, by noting that (since $|a_n| < 1/2$)

$$\begin{aligned}|a_{n+1} - a_{n+2}| &\leq (1/8)|a_{n+1} - a_n|\,|a_{n+1} + a_n| \\ &\leq (1/8)|a_{n+1} - a_n|\,(|a_{n+1}| + |a_n|) \\ &\leq (1/8)|a_{n+1} - a_n|.\end{aligned}$$

10. Note that the given sequence is

$$1,\ 1+\frac{1}{2},\ 1+\frac{1}{2+\frac{1}{2}},\ 1+\frac{1}{2+\frac{1}{2+\frac{1}{2}}},\ \ldots.$$

Clearly, $1 \leq a_n \leq 2$ for all n. For $n \geq 2$,

$$|a_{n+1}-a_n| = \left|\frac{1}{1+a_n} - \frac{1}{1+a_{n-1}}\right| = \frac{|a_n - a_{n-1}|}{(1+a_n)(1+a_{n-1})} \leq \frac{1}{2^2}|a_n - a_{n-1}|.$$

By Theorem 2.57, $\{a_n\}$ is Cauchy and hence converges, say to a. Then a must be given by

$$a = \lim_{n\to\infty} a_{n+1} = \lim_{n\to\infty}\left(1+\frac{1}{1+a_n}\right) = 1+\frac{1}{1+a}, \quad \text{i.e., } a=\sqrt{2}.$$

Again, we have a sequence of rational numbers converging to the irrational number $\sqrt{2}$ (see Remark 2.40).

11. Use the idea of Theorem 2.57, for example.
12. Set $a_n = x_{n+1} - x_n$. Then $a_n \to x$ and $\sum_{k=1}^{n} a_k = x_{n+1} - x_1$ for each n. So by Theorem 2.64, we have

$$\frac{x_{n+1}}{n+1} = \frac{x_1}{n+1} + \frac{n}{n+1}\left(\frac{1}{n}\sum_{k=1}^{n} a_k\right) \to x.$$

13. Apply the method of Example 2.61.

Chapter 3: Questions 3.24

9. No. Apply the algebra of limits for difference functions.
11. No.
22. $t = k\pi$, $k \in \mathbb{Z}$.
23. $t = 2k\pi$, $k \in \mathbb{Z}$.

Exercises 3.25:

2. Set $x_n = 1/n$ and $y_n = -1/n$. Then $x_n \to 0$ and $y_n \to 0$, whereas

$$f(x_n) = 3^n \to \infty \ \text{ and } \ f(y_n) = 3^{-n} \to 0 \ \text{ as } n \to \infty.$$

By Theorem 3.4, the conclusion follows.

3. Apply the idea of Example 3.7(**a**) to show that f has no limit as $x \to 0$; see Figure A.1.
6. (**a**) If $g(x) \leq f(x) \leq h(x)$ holds on (a, ∞) and

$$\lim_{x\to\infty} g(x) = \ell = \lim_{x\to\infty} h(x),$$

then $\lim_{x\to\infty} f(x) = \ell$.
(**b**) If $f(x) \geq g(x)$ on (a, ∞) and $g(x) \to \infty$ as $x \in \infty$, then $f(x) \to \infty$ as $x \to \infty$.
The squeeze rule for functions defined on $(-\infty, a)$ may be stated similarly.

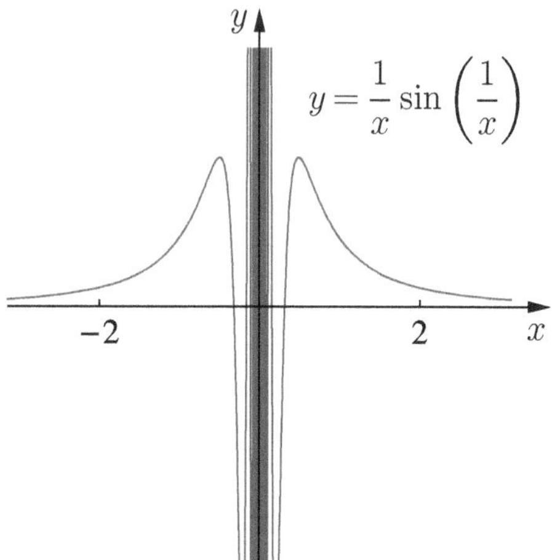

Fig. A.1. Graph of $\frac{1}{x}\sin\left(\frac{1}{x}\right)$ near the origin.

9. Use the Cauchy convergence criterion for sequences.

Questions 3.40:

7. Observe that
$$\big||f(x)| - |f(x_0)|\big| \leq |f(x) - f(x_0)|.$$
8. Observe that
$$\max\{f(x), g(x)\} = [(f(x) + g(x)) + |f(x) - g(x)|]/2$$
and
$$\min\{f(x), g(x)\} = -\max\{-f(x), -g(x)\} = [(f(x)+g(x))-|f(x)-g(x)|]/2.$$
11. Yes, $f(x) = 1$ on $\mathbb{R}$.
17. Yes.
18. Yes, because $\sin x$ is continuous and $\pi/n \to 0$.

Exercises 3.41:

1. Choose x_n such that $x_n = 1/\sqrt{(2n + 1/2)\pi}$ for $n \geq 1$. Then $f(x_n) \to \infty$, whereas $x_n \to 0$.
2. Observe that f is well defined for all $x > -2$. We must prove that given $\epsilon > 0$, there exists a $\delta > 0$ such that
$$|x - 2| < \delta \text{ implies that } |f(x) - f(2)| < \epsilon.$$
Now consider
$$|f(x) - f(2)| = \frac{|(\sqrt{x+2} - 2)(\sqrt{x+2} + 2)|}{\sqrt{x+2} + 2} = \frac{|x-2|}{\sqrt{x+2}+2} < \frac{|x-2|}{2}.$$
The conclusion follows if we choose $\delta = 2\epsilon$.

5. We have $(1+x)^2 = 1+2x+x^2 > 1+2x$, and so the inequality holds for $n=2$. Assume that the inequality is true for $n=k$. Then by the method of induction (see also Example 2.18),

$$\begin{aligned}(1+x)^{k+1} &= (1+x)^k(1+x)\\ &> (1+kx)(1+x) = 1+(k+1)x+kx^2 > 1+(k+1)x,\end{aligned}$$

which shows that the inequality holds for all n. Now we let $a_n = r^n$ and $0 < |r| < 1$. We may set $|r| = 1/(1+h)$, where $h > 0$. Then for every $n \geq 1$,

$$0 < |a_n| = \frac{1}{(1+h)^n} \leq \frac{1}{1+nh} < \frac{1}{nh} = \left(\frac{1-|r|}{|r|}\right)\frac{1}{n},$$

and so using the definition of convergence (or by the squeeze rule for sequences), it follows that $a_n \to 0$ as $n \to \infty$.
Next, for $r > 1$, we may let $r = 1+h$ with $h > 0$. Then $r^n > 1+nh$ for every $n \geq 1$. Again, by the squeeze rule, $\lim_{n\to\infty} r^n = \infty$. For $r=-1$, we obtain the sequence $\{(-1)^n\}$, which is bounded but oscillates between -1 and 1, and hence diverges. Finally, for $r < -1$, let $r = -t$, so that $t > 1$, and we have the oscillating sequence $\{(-1)^n t^n\}$, which has both large positive and negative terms. It follows that the sequence is unbounded, and hence $\{r^n\}$ diverges for $r < -1$.

8. Consider the three cases **(a)** $x \in \mathbb{Z}$, **(b)** $2k < x < 2k+1$, $k \in \mathbb{Z}$, **(c)** $2k-1 < x < 2k$, $k \in \mathbb{Z}$.

10. Because f is uniformly continuous on E, it follows that for a given $\epsilon > 0$, there exists a $\delta > 0$ such that

$$|f(x) - f(y)| < \epsilon \text{ whenever } |x-y| < \delta.$$

Since $\{x_n\}$ is Cauchy, for this $\delta > 0$ there exists an N such that

$$|x_n - x_m| < \delta \text{ for all } m, n > N.$$

In particular, this observation gives that

$$|f(x_n) - f(x_m)| < \epsilon \text{ whenever } n, m > N,$$

as desired. Set $f(x) = 1/x^2$ on $(0,1)$. Now, $\{1/n\}$ is a Cauchy sequence in $(0,1)$, whereas $\{f(1/n)\} = \{n^2\}$ is not Cauchy. Thus, $f(x) = 1/x^2$ cannot be uniformly continuous on $(0,1)$.

12. Let $\epsilon > 0$ be given. Then there exists an $R > 0$ such that

$$|f(x) - f(y)| \leq |f(x) - \ell| + |f(y) - \ell| < \epsilon \text{ for all } x, y \geq R.$$

The restriction of f to $[0, R]$ is uniformly continuous. Thus, there exists a $\delta > 0$ such that

$$|f(x) - f(y)| < \epsilon \quad \text{whenever } x, y \in [0, R] \text{ and } |x - y| < \delta.$$

Finally, if $x \geq R$ and $y < R$ with $|x - y| < \delta$, we have

$$|f(x) - f(y)| \leq |f(x) - f(R)| + |f(y) - f(R)| < \epsilon.$$

Thus, if $x, y \geq 0$ satisfy $|x - y| < \delta$, then we have $|f(x) - f(y)| < \epsilon$, as desired.

Questions 3.58:

10. We wish to show that $f'(x) = 0$ for each $x \in \mathbb{R}$. Now for $h \neq 0$,

$$\left|\frac{f(x+h) - f(x)}{h}\right| \leq |h|.$$

The squeeze rule shows that $f'(x)$ exists for each $x \in \mathbb{R}$ and $f'(x) = 0$ on $\mathbb{R}$. Thus, f is a constant function.

11. No. Consider $f(t) = \sin t$ on $[0, 1]$. Then the mean value theorem applied to $f(t)$ on $[x, y] \subseteq [0, 1]$ gives

$$|f(x) - f(y)| < |x - y| \quad \text{when } x \neq y.$$

If there exists an $M < 1$ such that

$$|f(x) - f(y)| < M|x - y| \quad \text{for all } x, y \in [0, 1],$$

then with $x = 0$ we have

$$|f(x) - f(y)| = |f(y) - 0| < M|y - 0| \quad \text{for } y \neq 0.$$

Allowing $y \to 0$, we see that $|f'(0)| \leq M$ (< 1), which is a contradiction.

14. We see that

$$\lim_{x \to a} \frac{g(x) - g(a)}{x - a} = \lim_{x \to a} \frac{|f(x)|}{x - a} = \lim_{x \to a} \frac{|f(x)|}{f(x)} \frac{f(x) - f(a)}{x - a}.$$

Exercises 3.59:

11. For $x \neq 0$,

$$\frac{f(x) - f(0)}{x - 0} = \frac{f(x)}{x} = \begin{cases} \dfrac{\sin x}{x} & \text{for } x \in \mathbb{Q}, \\ 1 & \text{for } x \in \mathbb{R} \setminus \mathbb{Q}, \end{cases}$$

and so by Example 4.24,

$$\cos x \leq \frac{f(x)}{x} \leq 1 \quad \text{for all } x \text{ with } 0 < x \leq \pi/2,$$

showing that $f'(0) = 1$.

12. Without loss of generality, we may assume $f(a) > 0$. Then $f(x) > 0$ in a neighborhood $B(a;\delta)$ of a. Therefore, for $x \in B(a;\delta)$,

$$\lim_{x\to a} \frac{g(x)-g(a)}{x-a} = \lim_{x\to a} \frac{f(x)-f(a)}{x-a} = f'(a) \text{ if } f(a) > 0.$$

Similarly, $g'(a) = -f'(a)$ if $f(a) < 0$.

14. (**b**) Set

$$\phi(x) = \begin{cases} x^2 \sin(1/x) & \text{for } x \neq 0, \\ 0 & \text{for } x = 0, \end{cases} \quad \text{and } g(x) = \sin x.$$

Then both ϕ and g are differentiable on $\mathbb{R}$. In particular, $h = \phi \circ g$ is differentiable at points x where $\sin x \neq 0$. Note that $\sin x = 0$ iff $x = n\pi$, $n \in \mathbb{Z}$, and on $(0,\pi)$,

$$h'(x) = \phi'(g(x))g'(x).$$

Chapter 4: Questions 4.9

4. No. Here is an example. On $[0,1]$, consider

$$f(x) = \frac{1}{4} - \left(x - \frac{1}{2}\right)^2 \text{ and } g(x) = \begin{cases} x & \text{for } x \in [0,1), \\ 3 & \text{for } x = 1. \end{cases}$$

5. No. Consider $f(x) = x$ and $g(x) = x - 1$ on $[0,1]$.

Exercises 4.10:

3. Consider $f\colon \mathbb{R} \to \mathbb{R}$ defined by

$$f(x) = \begin{cases} x & \text{for } x > 0, \\ x-1 & \text{for } x \leq 0. \end{cases}$$

We see that f is one-to-one but not onto. Also, it is not continuous at the origin.

6. By the logarithmic differentiation, we have

$$\begin{aligned} f'(x) &= f(x)\left[\log\left(1+\frac{1}{x}\right) + \frac{x}{1+\frac{1}{x}}\left(-\frac{1}{x^2}\right)\right] \\ &= f(x)\left(\log(1+y) - \frac{y}{1+y}\right), \quad y = \frac{1}{x} \\ &> 0 \text{ on } (0,\infty), \end{aligned}$$

because $\phi(y) = \log(1+y) - y/(1+y)$ is strictly increasing on $(0,\infty)$, and so $\phi(y) > \phi(0) = 0$.

7. Continuity of f at x_0 gives

$$f(x_0+h) = f(x_0) + \eta(h) = y_0 + \eta(h), \text{ where } \eta(h) \to 0 \text{ as } h \to 0,$$

and because f is one-to-one on I, $\eta(h) \neq 0$ for $h \neq 0$. Since

$$g(y) = x, \ y = f(x) \ \text{ for } x \in I \text{ and } y \in J,$$

we see that

$$g(y_0 + \eta(h)) - g'(y_0) = g(f(x_0 + h)) - g'(y_0) = (x_0 + h) - x_0 = h,$$

and so by the algebra of limits,

$$\frac{f(x_0 + h) - f(x_0)}{h} = \frac{\eta(h)}{g(y_0 + \eta(h)) - g(y_0)} \to \frac{1}{g'(y_0)} \quad \text{as } h \to 0,$$

and the conclusion follows.

Questions 4.39:

2. Consider $F(x) = f(x) - g(x)$. Then F is continuous on $[a, b]$ and $F(a) < 0 < F(b)$. Apply the intermediate value property.
4. No.
5. No.
6. No.
8. There is nothing to prove if $f(a) = a$ or $f(b) = b$. So we assume that $f(a) \neq a$ and $f(b) \neq b$. Define $g : [a, b] \to \mathbb{R}$ by

$$g(x) = f(x) - x, \quad x \in [a, b].$$

   Then g is continuous on $[a, b]$, $g(b) < 0 < g(a)$, since $f(a) \in (a, b]$ and $f(b) \in [a, b)$. By the intermediate value theorem, it follows that there exists a point $c \in (0, 1)$ such that $g(c) = 0$, i.e., $f(c) = c$.
10. Consider $f(x) = x^2 \sin(1/x)$ and $g(x) = x$.
16. Rolle's theorem applied to f on $[0, 1]$ shows that $f'(c_1) = 0$ for some $c_1 \in (0, 1)$. Again, apply Rolle's theorem to f' on $[0, c_1]$ and conclude that $f''(c_2) = 0$ for some $c_2 \in (0, c_1)$. Repeating the arguments implies that $f^{(n+1)}(c) = 0$ for some $c \in (0, 1)$.
17. Set $x' = (x + y)/2$. Apply the mean value theorem on $[x, x']$ and $[x', y]$, $x < y$.
18. Apply Corollary 4.21 to $F(x) = f(x)\mathrm{e}^{-cx}$. Then $F'(x) = 0$, and obtain $f(x) = f(0)\mathrm{e}^{-cx}$.
19. Given $\epsilon > 0$, there exists an $R > 0$ such that $|f'(x)| < \epsilon$ whenever $x > R$. Apply the mean value theorem for f on $[x, x+1]$, where $x > R$. This gives

$$f(x + 1) - f(x) = f'(c) \text{ for some } c \in (x, x + 1).$$

   Because $c > R$, this gives that for any given $\epsilon > 0$, there exists an $R > 0$ such that $|f(x + 1) - f(x)| < \epsilon$ whenever $x > R$.

Exercises 4.40:

5. **(a)** Here
$$\frac{f(1)-f(-1)}{1-(-1)} = -2 \text{ and } f'(c) = 3c^2 - 3,$$
so that $f'(c) = -2$, which gives $3c^2 = 1$, i.e., $c = \pm 1/\sqrt{3}$.
(b) Here
$$\frac{f(9/4)-f(1)}{(9/4)-1} = \frac{4}{9} \text{ and } f'(c) = \frac{1}{c^2},$$
so that $f'(c) = 4/9$, which gives $3c^2 = 1$, i.e., $c = \pm 3/2$.
8. Consider $h(x) = f(x)\exp(g(x))$ and apply Rolle's theorem.
13. Set $f(x) = \sin x - x^3 + x$ on $[\pi/4, \pi/2]$ and observe that
$$f(\pi/4) > 0 > f(\pi/2).$$
The desired conclusion follows from the intermediate value theorem.

Chapter 5: Questions 5.16

15. No. Choose $a_k = (-1)^k/\sqrt{k} = b_k$.
16. What is $\sum_{k=0}^{\infty}(a_k - b_k)^2$?
17. Observe that $T_n = \sum_{k=1}^{n}(a_k + a_{k+1}) = 2\sum_{k=1}^{n} a_k - a_1 + a_{n+1}$, showing that the sequence of partial sums $\{T_n\}$ converges to $2A - a_1$.
18. See Example 2.39(**a**). We have shown that $\sum_{k=1}^{\infty}(1/k^2)$ converges. If $S_n = \sum_{k=1}^{n}(1/k^2)$, then $a_n = S_{2n} - S_{n-1}$ converges to 0.

Exercises 5.17:

1. One is allowed to perform arithmetic operations only with convergent series.
2. Use the idea of the proof of Theorem 5.9.
6. **(a)** Write $k - 1 = 2k - (k+1)$, so that
$$a_k = \frac{k-1}{2^{k+1}} = \frac{k}{2^k} - \frac{k+1}{2^{k+1}}.$$
(b) Note that
$$a_k = \frac{\sqrt{k+1}-\sqrt{k}}{\sqrt{k}\sqrt{k+1}} = \frac{1}{\sqrt{k}} - \frac{1}{\sqrt{k+1}},$$
so that
$$\sum_{k=1}^{n} a_k = 1 - \frac{1}{\sqrt{n+1}} \to 1 \text{ as } n \to \infty.$$
(c) Use the properties of logarithms: $\log(xy) = \log(x) + \log(y)$, $x, y > 0$.

8. **(a)** Using the identity $2\sin x\cos y = \sin(x+y)+\sin(x-y)$, we have

$$\sum_{k=1}^{\infty} \sin\frac{1}{3^n}\cos\frac{2}{3^n} = \frac{1}{2}\lim_{n\to\infty}\sum_{k=1}^{n}\left(\sin\frac{1}{3^{k-1}} - \sin\frac{1}{3^k}\right) = \frac{1}{2}\sin 1.$$

9. If $s_n = \sum_{k=1}^n a_k$ and $S_n = \sum_{k=1}^n k(a_k - a_{k+1})$, then

$$S_n = \sum_{k=1}^{n}(ka_k - (k+1)a_{k+1}) + \sum_{k=1}^{n} a_{k+1} = s_{n+1} - (n+1)a_{n+1}.$$

10. What is the sum of the areas of the regions removed?
12. It suffices to note that

$$\frac{5^k \times 4^k}{(5^k - 4^k)(5^{k+1} - 4^{k+1})} = \frac{4^k}{5^k - 4^k} - \frac{4^{k+1}}{5^{k+1} - 4^{k+1}}.$$

13. Note: $0.\overline{12} = 0.12121212\ldots$.

Questions 5.42:

1. Set $S_n = \sum_{k=1}^n a_k$. Then the hypothesis implies that $\{S_n\}$ is Cauchy.
2. The converse fails, as the harmonic series shows. Note that if $a_k = 1/k$, then

$$0 < \frac{1}{n} + \frac{1}{n+1} + \cdots + \frac{1}{n+p} \le \frac{p+1}{n}.$$

3. No. For example, set

$$a_n = \begin{cases} \dfrac{1}{n} & \text{if } n \in S = \{1, 2^2, 3^2, 4^2, \ldots\}, \\[2ex] \dfrac{1}{n^2} & \text{otherwise.} \end{cases}$$

   Then (see Theorem 5.23),

$$\sum_{n\in S}^{\infty} a_n = \sum_{n\in S}^{\infty}\frac{1}{n^2},$$

   which is convergent. Also,

$$\sum_{n\notin S}^{\infty} a_n = \sum_{n\notin S}^{\infty}\frac{1}{n^2} \le \sum_{n=1}^{\infty}\frac{1}{n^2},$$

   so that $\sum_{n\notin S}^{\infty} a_n$ is convergent.
5. One may use the comparison test, because $a_k^2 < a_k$. Note that $\sum(1/k^2)$ converges although $\sum(1/k)$ does not.

7. Since $2xy \le x^2 + y^2$,
$$2|a_k|\,|b_k| \le a_k^2 + b_k^2 \ \text{ for all } k,$$
showing that $\sum_{k=1}^{\infty} a_k b_k$ converges absolutely.
10. Note that
$$\frac{|a_k b_k|}{a_k} = |b_k| \to 0 \ \text{ as } k \to \infty,$$
and so by the limit comparison test, it follows that $\sum_{k=1}^{\infty} a_k b_k$ converges absolutely.
11. Use the limit comparison test with $b_n = 1/n$ and $b_n = 1/n^2$.
16. Choose $a_k = 1/\log k$ for $k > 1$ and note that $a_k > 1/k$ for all $k > 1$.

Exercises 5.43:

1. Assume that $\sum |a_k|$ converges, and recall that
$$0 \le a_k + |a_k| \le 2|a_k| \text{ for all } k, \quad \text{and} \quad a_k = (a_k + |a_k|) - |a_k|.$$
Because the series $\sum |a_k|$ converges and both $\sum (a_k + |a_k|)$ and $\sum |a_k|$ are series of nonnegative terms, by the direct comparison test, we obtain that $\sum (a_k + |a_k|)$ also converges. By the linearity rule, the second relation above implies that $\sum a_k$ also converges, and the proof of Theorem 5.14 is complete.
2. Since
$$\frac{\sqrt{n+1} - \sqrt{n}}{n^a} = \frac{1}{(\sqrt{n+1} + \sqrt{n})n^a},$$
compare with $1/n^{a+1}$ and apply the limit comparison test.
4. $\Rightarrow$: Since $b_k < a_k$, the convergence of $\sum_{k=1}^{\infty} a_k$ implies that $\sum_{k=1}^{\infty} b_k$ converges.
$\Leftarrow$: Suppose that $\sum_{k=1}^{\infty} b_k$ converges. Note that $0 < b_k < 1$ and $a_k = b_k/(1 - b_k)$. Convergence of $\sum_{k=1}^{\infty} b_k$ implies that $b_k \to 0$ as $k \to \infty$. Thus, there exists an N such that
$$0 < b_k < \epsilon = 1/2, \ \text{ i.e., } 1 - b_k > 1/2, \quad \text{for all } k \ge N,$$
and so $a_k < 2b_k$. Thus, $\sum_{k=1}^{\infty} a_k$ converges by the comparison test.
6. Set $S = \sum_{k=1}^{\infty} a_k$ and observe that
$$S_n = \sum_{k=1}^{n} ((k+1)/k) a_k^2 \le 2 \sum_{k=1}^{n} a_k^2 \le 2 \left(\sum_{k=1}^{n} a_k \right)^2 \le 2S^2.$$
Thus, $\{S_n\}$ is an increasing sequence of nonnegative real numbers that is bounded, and so it converges.
7. Use the convergence of $\{b_n\}$ and the limit comparison test.

10. Since the given condition implies that the sequence $\{a_n/b_n\}$ is decreasing and bounded above by a_1/b_1, we have

$$a_n \leq \left(\frac{a_1}{b_1}\right) b_n \quad \text{for all } n \geq 1,$$

and the desired conclusion follows from the comparison test.

Questions 5.68:

12. Consider $a_n = (-1)^n$ and $b_n = 1 + (1/n)$.
13. For the first part, it suffices to observe that

$$\sum_{k=1}^{n} |a_{k+1} - a_k| \leq \sum_{k=1}^{n} (|a_{k+1}| + |a_k|).$$

For the second part, just consider $a_n = (-1)^{n-1}(1/n)$ and observe that $\{a_n\}$ is not of bounded variation but $\sum a_k$ is convergent.
14. Yes. Consider $s_n = \sum_{k=1}^{n}(b_k - b_{k+1}) = 1 - b_{n+1}$, and so $\{s_n\}$ is convergent iff $\{b_n\}$ is convergent.
15. Yes. Since $|b_k - b_{k+1}| = b_k - b_{k+1}$ or $b_{k+1} - b_k$, $\sum_{k=1}^{n} |b_k - b_{k+1}|$ is either $b_1 - b_{n+1}$ or $b_{n+1} - b_1$. Now apply BMCT.
16. Set $a_k = k(-1)^{k-1}$ and note that

$$\sigma_n = \frac{1}{n}\sum_{k=1}^{n} s_n = \begin{cases} 0 & \text{if } n \text{ is even,} \\ \dfrac{1}{2}\left(\dfrac{n+1}{n}\right) & \text{if } n \text{ is odd,} \end{cases}$$

showing that $\{\sigma_n\}$ does not converge, and so the given series is not $(C, 1)$ summable.

Exercises 5.69:

2. The given series is

$$1 + \frac{1}{2} - \frac{1}{3} - \frac{1}{4} + \frac{1}{5} + \frac{1}{6} - \frac{1}{7} - \frac{1}{8} + (-1)^k\left(\frac{1}{2k+1} + \frac{1}{2k+2}\right) + \cdots,$$

which is clearly not an alternating series. However, pairing the terms suitably leads to an alternating series

$$\frac{3}{2} - \frac{7}{12} + \frac{11}{30} - \frac{15}{56} + \cdots + (-1)^k \frac{4k+3}{(2k+1)(2k+2)} + \cdots,$$

which satisfies the conditions of the alternating series test, and so the rearranged series converges. Consequently, the given series converges (since the general term of the original series tends to zero as $k \to \infty$).
4. Let n be the required number. From our observation in Corollary 5.46, we must have $|S - S_n| \leq a_{n+1}$. If we let $\frac{1}{2n+1} = a_{n+1} \leq 0.0001$, then $n \geq \frac{10000-1}{2} = 4999.5$. Consequently, if we take the first 5000 terms or more, the error will be less than 0.0001.

8. When the parentheses are removed, compute the even sequence $\{S_{2n}\}$ and the odd sequence $\{S_{2n-1}\}$, and obtain that they converge to a different limit. Note that

$$S_{2n+1} = S_{2n} + 1.$$

9(**ii**). Recall that

$$\tan^{-1} x = x - \frac{x^3}{3} + \frac{x^5}{5} - \cdots \quad |x| \leq 1,$$

so that

$$\frac{\pi}{4} = 1 - \frac{1}{3} + \frac{1}{5} - \frac{1}{7} + \cdots.$$

Group the terms as

$$\left(1 - \frac{1}{3}\right) + \left(\frac{1}{5} - \frac{1}{7}\right) + \cdots$$

to obtain the series

$$\frac{2}{1 \cdot 3} + \frac{2}{5 \cdot 7} + \frac{2}{9 \cdot 11} + \cdots.$$

The conclusion follows from Riemann's rearrangement theorem. The second part follows similarly.

11. Apply Corollary 5.59.
13. A computation gives

$$c_n = \sum_{k=0}^{n} a_k a_{n-k} = \frac{2(-1)^n}{n+2} \sum_{k=0}^{n} \frac{1}{k+1} =: 2(-1)^n d_n.$$

Since (by Theorem 2.64)

$$d_n = \frac{2}{n+2} \sum_{k=0}^{n} \frac{1}{k+1} = 2\left(\frac{n+1}{n+2}\right)\left(\frac{1}{n+1} \sum_{k=0}^{n} \frac{1}{k+1}\right) \to 2 \cdot 0 = 0$$

as $n \to \infty$ and

$$\begin{aligned} d_{n+1} - d_n &= 2 \sum_{k=0}^{n} \frac{1}{k+1} \left[\frac{1}{n+3} - \frac{1}{n+2}\right] + \frac{2}{(n+3)(n+1)} \\ &= -\frac{2}{(n+2)(n+3)} \sum_{k=1}^{n} \frac{1}{k+1} < 0, \end{aligned}$$

the alternating series test shows that $\sum_{n=0}^{\infty} c_n = \sum_{n=0}^{\infty} (-1)^n d_n$ is convergent.

14. **(a)** $b_n = (-1)^n$ **(b)** $b_n = \begin{cases} 1 & \text{if } n \text{ is even,} \\ \frac{n}{n+1} & \text{if } n \text{ is odd.} \end{cases}$ **(c)** $b_n = \dfrac{(-1)^{n-1}}{n^2}$.

16. Suppose that $\sum_{k=1}^{\infty} a_k$ diverges, where $a_k > 0$. Then the sequence $\{s_n\}$ of the partial sums must be unbounded above. Thus, given any $M > 0$, there exists an N such that

$$s_n > M \quad \text{for all } n > N.$$

Now for $n > N$, we have

$$\begin{aligned}\sigma_n &= \left(\frac{1}{n}\sum_{k=1}^{N} s_k\right) + \frac{1}{n}\sum_{k=N+1}^{n} s_k \\ &> \left(\frac{1}{n}\sum_{k=1}^{N} s_k\right) + \frac{(n-N)M}{n} \\ &= \left(\frac{1}{n}\sum_{k=1}^{N} s_k\right) + M\left(1-\frac{N}{n}\right),\end{aligned}$$

so that for any $M > 0$, we see that $\limsup_{n\to\infty} \sigma_n \geq M$.

17. Compare with Theorem 5.18 concerning a necessary condition for convergent series. See also Corollary 9.60.

Chapter 6: Questions 6.31

6. **(a)** Yes, the partition norm decreases under refinement. Indeed, since Q has additional points beyond points in P, the maximal length of any subinterval determined from Q is less than or equal to the maximal length of any subinterval determined from P. Thus, $\|P\| \geq \|Q\|$.

14. The first half holds for $c < 0$, while the second half holds for $c > 0$, because
    - $L(P, cf) = c\,U(P, f)$ and $U(P, cf) = c\,L(P, f)$ whenever $c < 0$;
    - $L(P, cf) = c\,L(P, f)$ and $U(P, cf) = c\,U(P, f)$ whenever $c > 0$.

17. No. Let f be as in Example 6.12 and

$$g(x) = \begin{cases} 1 & \text{for } x = 0, \\ 0 & \text{for } x \in (0, 1]. \end{cases}$$

Then $g \circ f$ defined by

$$g(f(x)) = \begin{cases} 1 & \text{for } x \in \mathbb{Q}^c \cap [0, 1], \\ 0 & \text{for } x \in \mathbb{Q} \cap [0, 1], \end{cases}$$

is not integrable.

19. Choose the partition $P = \{x_0, x_1, \ldots, x_n\}$ such that $x_k^* = (x_{k-1} + x_k)/2$. Then the corresponding Riemann sum has the constant value $S_n = (b^2 - a^2)/2$. The conclusion follows from Theorem 6.21 (see also Example 6.6).

20. Note that

$$f(x_k^*)\Delta x_k = (x_k^*)^2 \Delta x_k = \frac{x_k^3 - x_{k-1}^3}{3}.$$

22. $\Longleftarrow$: Assume that the identity holds. The left-hand-side integral can be written as

$$\int_{-x}^{x} f(t)\,dt = \int_{-x}^{0} f(t)\,dt + \int_{0}^{x} f(t)\,dt = \int_{0}^{x} f(-t)\,dt + \int_{0}^{x} f(t)\,dt,$$

which equals the right-hand side if and only if

$$\int_{0}^{x} f(t)\,dt = \int_{0}^{x} f(-t)\,dt. \tag{A.1}$$

If we let $G(x) = \int_0^x f(t)\,dt$, then G is continuous on $\mathbb{R}$ and $G'(x) = f(x)$ on $\mathbb{R}$. Because of (A.1), we also have $G'(x) = f(-x)$ on $\mathbb{R}$. That is, $f(x) = f(-x)$ on $\mathbb{R}$. The converse is trivial (see Figures A.2 and A.3), because if f is even on $\mathbb{R}$, then we have

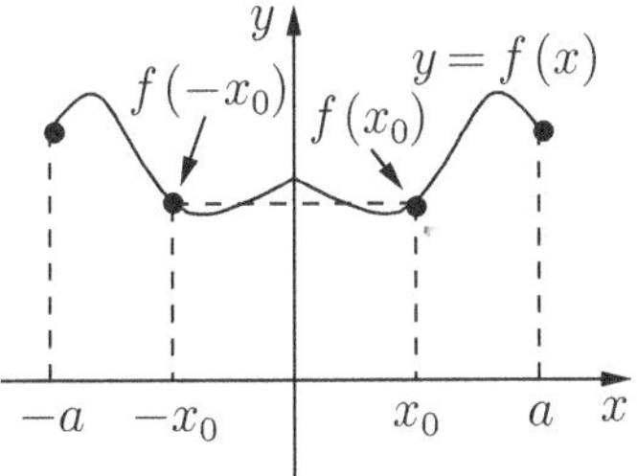

Fig. A.2. Graph of an even function.

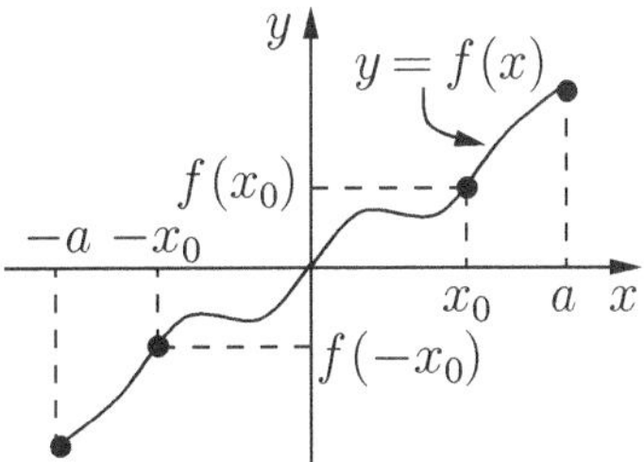

Fig. A.3. Graph of an odd function.

$$\int_{-x}^{x} f(t)\,dt = \int_{-x}^{0} f(t)\,dt + \int_{0}^{x} f(t)\,dt = \int_{0}^{x} f(-t)\,dt + \int_{0}^{x} f(t)\,dt,$$

and the result follows.

Exercises 6.32:

4. Use Example 6.17 with $\|P\| < \dfrac{\epsilon}{nb^{n-1}(b-a)}$.
5. Set $x_k^* = x_k = ah^{k-1}$ for $k \geq 0$. Then $\Delta x_k = a(h-1)h^{k-1}$, and proceed exactly as in Example 6.18. As a demonstration, we refer to Example 6.23.
6. Note that $m_k \leq (f(x_{k-1}) + f(x_k))/2 \leq M_k$, and so

$$L(P,f) \leq T_n(P,f) \leq U(P,f).$$

10. (**a**) Taking the logarithm, we have

$$\log a_n = \frac{1}{n}\sum_{k=1}^{n} \log\left(1+\frac{k}{n}\right) \to \int_0^1 \log(1+x)\,dx = \log 4 - 1 \text{ as } n \to \infty,$$

because

$$\int_0^1 \log(1+x)\,dx = x\log(1+x)\Big|_0^1 - \int_0^1 \frac{x\,dx}{1+x}$$

$$= \log 2 - (x - \log(1+x))\Big|_0^1 = 2\log 2 - 1.$$

Thus $\lim_{n\to\infty} a_n = e^{\log 4 - 1} = 4/e$.

(**b**) The convergence of this has been discussed in Section 2.1 (see also Remark 7.26(5)). But using the Riemann idea, we have (see Figure A.4)

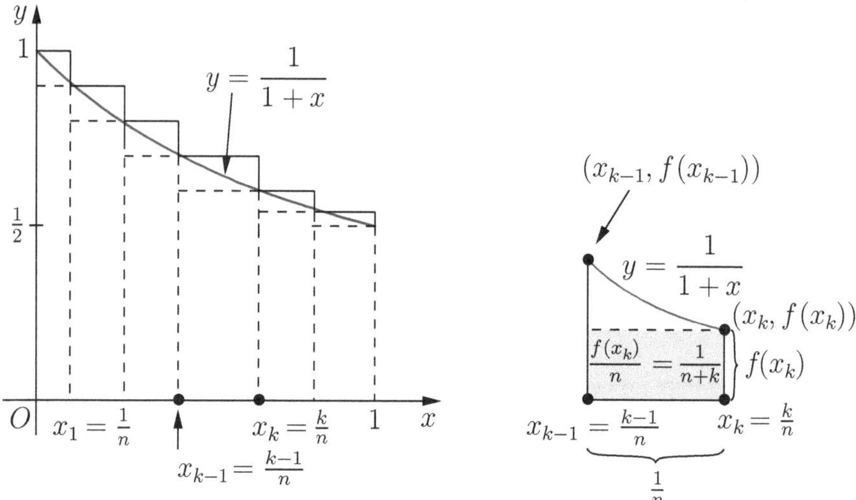

Fig. A.4. Riemann sum for $\int_0^1 \frac{dx}{1+x}$.

$$a_n = \frac{1}{n}\sum_{k=1}^{n} \frac{1}{1+k/n} \to \int_0^1 \frac{dx}{1+x} = \log 2.$$

(**c**) We may rewrite

$$a_n = \frac{1}{n} + \frac{1}{n}\sum_{k=1}^{n} \frac{n^3}{(n+k)^3} = \frac{1}{n} + \frac{1}{n}\sum_{k=1}^{n} \frac{1}{(1+k/n)^3},$$

so that

$$a_n \to 0 + \int_0^1 \frac{dx}{(1+x)^3} = \frac{3}{8}.$$

(**d**) Taking the logarithm, we have

$$\log a_n = \frac{1}{n}\sum_{k=1}^{n} \log\left(1+\frac{k^2}{n^2}\right) \to \int_0^1 \log(1+x^2)\,dx \quad \text{as } n\to\infty,$$

which, by integration by parts, gives $\lim_{n\to\infty} \log a_n = \log 2 - 2 + \pi/2$. Thus, $\lim_{n\to\infty} a_n = 2e^{(\pi/2)-2}$.

(**e**) See also Remark 7.26(5). Indeed,

$$\begin{aligned} a_n &= \frac{1}{p-1}\sum_{k=1}^{(p-1)n} \frac{1}{(k/(p-1)) + n/(p-1)} \\ &= \frac{1}{n(p-1)}\sum_{k=1}^{(p-1)n} \frac{1}{(k/(n(p-1))) + 1/(p-1)} \\ &= \frac{1}{N}\sum_{k=1}^{N} \frac{1}{(k/N) + 1/(p-1)}, \quad N=(p-1)n, \\ &\to \int_0^1 \frac{dx}{x+1/(p-1)} = \log p. \end{aligned}$$

(**f**) We have $a_n = \frac{1}{n}\sum_{k=1}^{n} \frac{1}{\sqrt{2(k/n)-(k/n)^2}} \to \int_0^1 \frac{dx}{\sqrt{2x-x^2}}$ as $n\to\infty$.

(**g**) We have $a_n = \frac{1}{n}\sum_{k=1}^{n} \frac{(k/n)^2}{1+(k/n)^3} \to \int_0^1 \frac{x^2}{1+x^3}\,dx$ as $n\to\infty$.

(**h**) $\log a_k = \sum_{k=1}^{n} \frac{1}{k}\log\left(1+\frac{k}{n}\right) = \frac{1}{n}\sum_{k=1}^{n}\frac{n}{k}\log\left(1+\frac{k}{n}\right) \to \int_0^1 \frac{\log(1+x)}{x}dx.$

For further details, we refer to Example 6.24(e).

13. (**a**) $f(x) = [x]$, the greatest integer function, on $x\in[0,8]$

(**b**) $f(x) = \begin{cases} 2 & \text{for } -1\le x<0, \\ 3 & \text{for } 0<x\le 1. \end{cases}$

16. Consider the partition $P=\{x_0,x_1,\dots,x_n\}$, where $x_k = k\pi/(4n)$, $k=0,1,2,\dots,n$. Then $\Delta x_k = \pi/(4n)$. Further, f is bounded, and for $x\in[a,b]\subset[0,\pi/4]$, $\sin x$ is increasing on $[a,b]$. Consequently,

$$\inf_{x\in[a,b]} f(x) = \sin a, \quad \sup_{x\in[a,b]} f(x) = \cos a,$$

and so

$$m_k = \inf_{x\in[x_{k-1},x_k]} f(x) = \sin(k-1)\frac{\pi}{4n}$$

and

$$M_k = \sup_{x\in[x_{k-1},x_k]} f(x) = \cos(k-1)\frac{\pi}{4n}.$$

Finally, it is easy to see that $\lim_{\|P\|\to 0} U(P,f) \neq \lim_{\|P\|\to 0} L(P,f)$.

17. **(d)** Consider $[2/\pi, 1] \subsetneq (0, 1]$ and P a partition on $[2/\pi, 1]$. For $x \in [2/\pi, 1]$, $(1/x) \in [1, \pi/2]$, and so $\sin(1/x) \geq \sin 1$ on $[2/\pi, 1]$. Consequently, $M_k \geq \sin 1$ and $m_k = 0$ for all k. This gives

$$U(P, f) \geq (\sin 1)(1 - 2/\pi) \quad \text{and} \quad L(P, f) = 0.$$

Consequently, $U(f) \geq (\sin 1)(1 - 2/\pi)$ and $L(f) = 0$. Thus, f is not integrable.

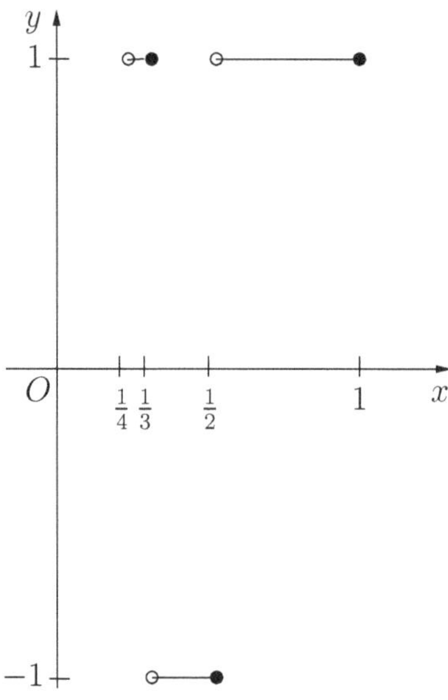

Fig. A.5. Graph of $f(x)$ on $[0, 1]$.

18. Note that (see Figure A.5)

$$\int_{1/(k+1)}^{1/k} f(x)\,dx = (-1)^{k-1}\left(\frac{1}{k} - \frac{1}{k+1}\right)$$
$$= \frac{2(-1)^{k-1}}{k} + (-1)^k\left(\frac{1}{k} + \frac{1}{k+1}\right),$$

and so

$$\int_{1/(n+1)}^{1} f(x)\,dx = \sum_{k=1}^{n} \int_{1/(k+1)}^{1/k} f(x)\,dx = 2\sum_{k=1}^{n} \frac{(-1)^{k-1}}{k} - 1 - \frac{(-1)^n}{n+1}$$
$$\to 2\log 2 - 1 \quad \text{as } n \to \infty.$$

Since $-1 \leq f(x) \leq 1$ on $[0, 1]$ and since for f on $[0, 1/(n+1)]$,

$$-\frac{1}{n+1} \leq L(f) \leq U(f) \leq \frac{1}{n+1} \quad \text{on } [0, 1/(n+1)],$$

it follows that $\int_0^{1/(n+1)} f(x)\,dx \to 0$ as $n \to \infty$. Consequently,

$$\int_0^1 f(x)\,dx = 2\log 2 - 1.$$

19. (**a**) Set $M = \sup_{x\in[a,b]} f(x)$, $m = \inf_{x\in[a,b]}$ and $c \in (a,b)$. Let $\{c_n\}$ be a sequence of points in (a,b) such that $c_n \to c$. Then for every $\epsilon > 0$, there exists an N such that

$$|c_n - c| < \frac{\epsilon}{2(M-m)} = \delta \ \text{ for all } n \geq N.$$

By assumption 1, f is integrable on $[c_n, b]$ for all $n \geq 1$. Consequently, there exists a partition P of $[c_n, b]$ such that

$$U(P,f) - L(P,f) < \frac{\epsilon}{2}.$$

Set $P' = \{a\} \cup P$. Then since $P \subset P'$, for all $n \geq N$,

$$U(P',f) - L(P',f) < (M-m)(b-c_n) + U(P,f) - L(P,f) < \frac{\epsilon}{2} + \frac{\epsilon}{2} = \epsilon,$$

which implies that f is integrable on $[a,b]$, since $c \in (a,b)$ is arbitrary.
(**b**). Follows similarly.
(**c**). Apply (**a**) and (**b**) to $[c,(a+b)/2]$ and $[(a+b)/2, d]$, respectively.

Questions 6.56:

2. For instance, $f : [0,1] \to \mathbb{R}$ defined by

$$f(x) = \begin{cases} -1 & \text{for } x \in [0,1/2), \\ 1 & \text{for } x \in [1/2,1], \end{cases}$$

is integrable but does not have a primitive on $[0,1]$.
3. If $f(c) > 0$ for some point c in (a,b), then by the continuity of f at c (with $\epsilon = f(c)/2$) we obtain $f(x) > 0$ for all x in some closed neighborhood of c, namely $[c-\delta, c+\delta] \subset [a,b]$. Thus, $\int_a^b f(x)\,dx > 0$, which is a contradiction. Alternatively, we observe that for each $x \in [a,b]$,

$$0 \leq F(x) := \int_a^x f(x)\,dx \leq \int_a^b f(x)\,dx = 0,$$

showing that $F(x) = 0$ and $F'(x) = f(x) = 0$, and the proof is complete.
5. Observe that $G(x) = \int_a^x (f(t) - g(t))\,dt$ satisfies the hypothesis of Rolle's theorem. It follows that there exists a point $c \in (a,b)$ such that $f(c) = g(c)$.

6. Continuity of f at c implies that there exists a closed interval $I = [a_1, b_1]$ ($a_1 < b_1$) such that $c \in I$ and $I \subseteq [a, b]$ with $f(x) \geq m$ on I for some $m > 0$. But then

$$\begin{aligned}\int_a^b f(x)\,dx &= \int_a^{a_1} f(x)\,dx + \int_{a_1}^{b_1} f(x)\,dx + \int_{b_1}^{b} f(x)\,dx \\ &\geq 0 + \int_{a_1}^{b_1} f(x)\,dx + 0 \geq m(b_1 - a_1) > 0.\end{aligned}$$

7. Yes. By hypothesis, $|f(x)| \leq M$ on $[a, b]$ for some $M > 0$. In particular,

$$|f(x)| = \left|\int_a^x f(t)\,dt\right| \leq M(x - a) \quad \text{for all } x \in [a, b].$$

The process may be continued to obtain

$$|f(x)| \leq \frac{M(x-a)^n}{n!} \quad \text{for all } x \in [a, b] \text{ and each } n \geq 1.$$

Allow $n \to \infty$ to get the result.

10. No; for instance, if $f(x) = 1$ on $[a, b]$, then

$$\frac{1}{b-a}\int_a^b f(x)\,dx = 1 = f(c)$$

for all $c \in [a, b]$. We have already demonstrated the nonuniqueness of c in Example 6.46.

12. Set $f(x) = \sin x - x + x^3/6$. Then $f''(x) = 1 - \cos x > 0$ for $x > 0$, and since

$$f(x) = \int_0^x f'(t)\,dt, \quad f'(x) = \int_0^x f''(t)\,dt, \quad f''(x) = \int_0^x f'''(t)\,dt,$$

it follows that $f(x) > 0$ for all $x > 0$.

13. Set $g(x) = x^n$ and apply the generalized mean value theorem (see Corollary 6.48).

14. Use the change of variable $t = \pi - x$ for the first part.

15. Consider $f(x) = x^3 3^{-x}$ for $x > 0$. Then

$$f'(x) = x^2 3^{-x}[3 - x\log 3] \leq 0 \quad \text{for } x \geq \tfrac{3}{\log 3},$$

so that f is decreasing on $[3/\log 3, \infty)$. In particular, since $\pi > 3 > 3/\log 3$, we have $f(3) > f(\pi)$, which is equivalent to $1 > \pi^3 3^{-\pi}$, i.e., $3^\pi > \pi^3$.

Exercises 6.57:

1. Clearly, f is continuous on $\mathbb{R}$ except at $x = 0$. On the other hand, $f(x)$ admits an antiderivative $F(x)$ given by

$$F(x) = \begin{cases} x^2 \sin(1/x) & \text{for } x \neq 0, \\ 0 & \text{for } x = 0. \end{cases}$$

Thus, $\int_0^1 f(x)\,\mathrm{d}x = F(1) - F(0) = \sin 1$.

5. Since f is continuous on $[-a, a]$ except at the origin and is bounded on $[-a, a]$, f is integrable on $[-a, a]$ for each $a > 0$. Clearly, f admits an antiderivative F given by

$$F(x) = \begin{cases} x^3 \cos(c/x^2) & \text{for } x \neq 0, \\ 0 & \text{for } x = 0. \end{cases}$$

Thus $\int_{-a}^{a} f(t)\,\mathrm{d}t = F(a) - F(-a) = 2a^3 \cos(c/a^2)$.

7. As in Example 6.38, we compute

$$G(x) = \begin{cases} x^2 & \text{if } 0 \leq x \leq 1, \\ x^2 - 2x + 2 & \text{if } 1 \leq x \leq 2, \end{cases}$$

which is continuous on $[0, 2]$.

10. Note that f is integrable on $[a, b]$. Define

$$g(x) = f(a)(x - a) + f(b)(b - x).$$

Clearly, g is continuous, $g(a) = f(b)(b - a)$, and $g(b) = f(a)(b - a)$. Because f is monotonic on $[a, b]$, the number $\mu = \int_a^b f(x)\,\mathrm{d}x$ lies between $g(a)$ and $g(b)$. Moreover, because g is continuous, there exists a c such that $g(c) = \mu$, which gives the desired conclusion.

11. First we note that the bracketed term in the limit can be recognized as a Riemann sum associated with the function $f(x) = \mathrm{e}^{\sqrt{x}}$. Thus, the limit is $\int_0^1 \mathrm{e}^{\sqrt{x}}\,\mathrm{d}x$. Now apply the mean value theorem for integrals to get the desired result.

12. We see that

$$\frac{\cos a - \cos b}{b} = \frac{1}{b}\int_a^b \sin x\,\mathrm{d}x \leq \int_a^b \frac{\sin x}{x}\,\mathrm{d}x \leq \frac{1}{a}\int_a^b \sin x\,\mathrm{d}x = \frac{\cos a - \cos b}{a},$$

which gives

$$\frac{-2}{a} \leq \frac{-2}{b} \leq \frac{\cos a - \cos b}{b} \leq \int_a^b \frac{\sin x}{x}\,\mathrm{d}x \leq \frac{\cos a - \cos b}{a} \leq \frac{2}{a},$$

which proves the inequality.

13. (a) Since $1 + x < \mathrm{e}^x < 1/(1-x)$ for $x > 0$, it follows that

$$\int_0^{1/3} (1+x^2)\,\mathrm{d}x < \int_0^{1/3} \mathrm{e}^{x^2}\,\mathrm{d}x < \int_0^{1/3} \frac{\mathrm{d}x}{1-x^2},$$

and observe that on the interval $[0, 1/3]$,

$$1 < \frac{1}{1-x^2} < \frac{9}{8}.$$

This gives

$$\frac{1}{3}\left(1 + \frac{1}{27}\right) < \int_0^{1/3} \mathrm{e}^{x^2}\,\mathrm{d}x < \frac{1}{3}\left(\frac{9}{8}\right) = \frac{3}{8}.$$

(b) Set $g(x) = \sqrt{4 - x^2 + x^\alpha}$. Then for $x \in (0,1)$, we have

$$\sqrt{4-x^2} < g(x) < \sqrt{4} = 2, \text{ i.e., } \frac{1}{2} < \frac{1}{g(x)} < \frac{1}{\sqrt{4-x^2}}.$$

The result follows as in the previous case.

(c) Set $G(x) = \int_0^{x^2} \mathrm{e}^{\sqrt{1+t^2}}\,\mathrm{d}t$. Then $G'(x) = 2x\mathrm{e}^{\sqrt{1+x^4}}$. By l'Hôpital's rule,

$$\lim_{x\to 0} \frac{G(x)}{x^2} = \lim_{x\to 0} \frac{G'(x)}{2x} = \lim_{x\to 0} \mathrm{e}^{\sqrt{1+x^4}} = \mathrm{e}.$$

(d) Since $\sin x$ is strictly increasing on $(0, \pi/2)$,

$$\frac{1}{2} = \sin\frac{\pi}{6} < \sin x < \sin\frac{\pi}{2} = 1, \text{ i.e., } 1 < \frac{1}{\sin x} < 2 \text{ for } x \in \left(\frac{\pi}{6}, \frac{\pi}{2}\right),$$

so that

$$\int_{\pi/6}^{\pi/2} x\,\mathrm{d}x < \int_{\pi/6}^{\pi/2} \frac{x}{\sin x}\,\mathrm{d}x < 2\int_{\pi/6}^{\pi/2} x\,\mathrm{d}x,$$

and a simplification gives the desired inequalities.

14. Apply l'Hôpital's rule and the second fundamental theorem of calculus to obtain

$$\begin{aligned}
\lim_{x\to 0} \frac{\int_{-2x}^{2x} f(t)\,\mathrm{d}t}{\int_0^{3x} f(t+2)\,\mathrm{d}t} &= \lim_{x\to 0} \frac{\frac{\mathrm{d}}{\mathrm{d}x}\left(-\int_0^{-2x} f(t)\,\mathrm{d}t + \int_0^{2x} f(t)\,\mathrm{d}t\right)}{\frac{\mathrm{d}}{\mathrm{d}x}\left(\int_0^{3x} f(t+2)\,\mathrm{d}t\right)} \\
&= \lim_{x\to 0} \frac{2f(-2x) + f(2x)}{3f(3x+2)} \\
&= \frac{f(0)}{f(2)}.
\end{aligned}$$

16. Apply the generalized mean value theorem (Corollary 6.48) with $g(x) = x^3$ on $[0, 1]$.

17. Decompose
$$\int_a^b f(x)\,\mathrm{d}x = \int_a^c f(x)\,\mathrm{d}x + \int_c^b f(x)\,\mathrm{d}x.$$
By the mean value theorem for integrals, there exists a $\lambda \in [a, c]$ such that
$$\frac{1}{c-a}\int_a^c f(x)\,\mathrm{d}x = f(\lambda).$$
Also, since f is monotonically increasing, $f(a) \le f(\lambda) \le f(c)$, so that
$$f(a) \le \frac{1}{c-a}\int_a^c f(x)\,\mathrm{d}x \le f(c).$$
Similarly, applying the mean value theorem for the second integral yields
$$f(c) \le \frac{1}{b-c}\int_c^b f(x)\,\mathrm{d}x \le f(b).$$
Adding the last two inequalities yields the desired result.
22. It suffices to show that $L(2) < 1 = L(0)$ and $L(3) > L(\mathrm{e}) = 1$.

Chapter 7: Questions 7.30

8. By the change of variable $t = x^2$, we have
$$\int_0^N \sin x^2\,\mathrm{d}x = \int_0^{N^2} t^{-1/2}\sin t\,\mathrm{d}t,$$
which converges conditionally, by Example 7.8.
10. The function $f(x) = \sin x$ is integrable an $[0, N)$, and
$$\lim_{N\to\infty}\int_0^N \sin x\,\mathrm{d}x = \lim_{N\to\infty}[-\cos x]\Big|_0^N = 1 - \lim_{N\to\infty}\cos N.$$
Since $\cos n\pi = (-1)^n$, $\cos N$ oscillates between -1 and 1 in a neighborhood of ∞, and therefore the limit on the right does not exist. We might even say that the integral $\int_0^\infty \sin x\,\mathrm{d}x$ "diverges by oscillation."
11. We have
$$\int_0^N x\mathrm{d}(-\cos x) = (-x\cos x + \sin x)|_0^N = -N\cos N + \sin N,$$
and $\lim_{N\to\infty}(-N\cos N + \sin N)$ does not exist, since it oscillates between $-\infty$ and ∞.
12. Indeed, $\int_0^N \cos x\,\mathrm{d}x = \sin N$, and $\lim_{N\to\infty}\sin N$ does not exist. Note that if $x_n = n\pi$ and $y_n = 2n\pi + \pi/2$, then $\sin x_n = 0$ and $\sin y_n = 1$ for all $n \in \mathbb{Z}$.

13. Clearly

$$\int_\epsilon^1 \log x\,dx = x\log x\big|_\epsilon^1 - \int_\epsilon^1 dx = -\epsilon\log\epsilon - (1-\epsilon),$$

so that

$$\lim_{\epsilon\to 0+}\int_\epsilon^1 \log x\,dx = -1 + \lim_{\epsilon\to 0+}\frac{\log(1/\epsilon)}{1/\epsilon} = -1 + \lim_{\epsilon\to 0+}\left(\frac{-1/\epsilon}{-1/\epsilon^2}\right) = -1.$$

Exercises 7.31:

2. **(f)** Note that $\lim_{x\to 0+}\sin(1/x)$ does not exist. However, using the substitution $y = 1/t$, we obtain

$$I(c) := \int_c^1 \frac{\sin(1/x)}{x^p}\,dx = \int_1^{1/c}\frac{\sin t}{t^{2-p}}\,dx \quad (c>0),$$

showing that (see Example 7.8(**c**)) $\lim_{c\to 0+} I(c)$ exists for $p<2$, and so $\int_0^1 \frac{\sin(1/x)}{x^p}\,dx$ converges for $p<2$.

7. At $x = 2n\pi + \pi/2$, the function $f(x) = -2\sqrt{1-\sin x}$ doesn't have a derivative.
8. Because $\sec x$ is unbounded at the left endpoint $\frac{\pi}{2}$ of the interval of integration and is continuous on $[t,\pi]$ for any t with $\frac{\pi}{2} < t \le \pi$, we find that the integral diverges, because

$$\int_{\pi/2}^{\pi}\sec x\,dx = \lim_{t\to(\pi/2)+}\log|\sec x + \tan x|\Big|_t^{\pi} = -\infty.$$

9. Consider

$$\begin{aligned} I(N) &= \int_0^N \left(\frac{1}{a(a^{-2}+x^2)^{1/2}} - \frac{\alpha}{x+1}\right)dx \\ &= \left[\frac{1}{a}\log\left(x + \left(a^{-2}+x^2\right)^{1/2}\right) - \alpha\log(x+1)\right]\Bigg|_0^N \\ &= \log\left[\frac{\left[N+(a^{-2}+N^2)^{1/2}\right]^{1/a}}{(N+1)^\alpha}\right] - \frac{1}{a}\log\left(\frac{1}{a}\right). \end{aligned}$$

We note that

$$\lim_{N\to\infty}\frac{\left[N+(a^{-2}+N^2)^{1/2}\right]^{1/a}}{(N+1)^\alpha} = \lim_{N\to\infty}\frac{N^{(1/a)-\alpha}\left(1+(\frac{1}{a^2N^2}+1)^{1/2}\right)^{1/a}}{(1+1/N)^\alpha},$$

which exists and is zero if $\frac{1}{a}-\alpha<0$, whereas the limit is $2^{1/a}$ if $\alpha = 1/a$. The limit fails to exist if $\frac{1}{a}-\alpha>0$.

11. Use Example 7.22 with $q=1$ and $p = 1-\alpha$. Then, according to Example 7.22, the given improper integral converges if and only if $0<\alpha<1$.

13. Set $x^q = t$. Then $qx^{q-1}\mathrm{d}x = \mathrm{d}t$, so that $\mathrm{d}x = \frac{1}{q}t^{\frac{1-q}{q}}\,\mathrm{d}t$, and

$$\int_0^1 x^p(1-x^q)^n\,\mathrm{d}x = \frac{1}{q}\int_0^1 t^{\frac{p}{q}+\frac{1}{q}-1}(1-t)^n\,\mathrm{d}t = \frac{1}{q}B\left(\frac{p+1}{q}, n+1\right).$$

Next, we may rewrite the given integral as

$$\begin{aligned}\int_0^m x^p(m^q-x^q)^n\,\mathrm{d}x &= m^{qn+p+1}\int_0^m \left(\frac{x}{m}\right)^p\left(1-\left(\frac{x}{m}\right)^q\right)^n \mathrm{d}\left(\frac{q}{m}\right)\\ &= \frac{m^{qn+p+1}}{q}B\left(\frac{p+1}{q}, n+1\right).\end{aligned}$$

14. Set $\alpha^2x^2 = t$. Then $2\alpha^2x\mathrm{d}x = \mathrm{d}t$, $\mathrm{d}x = \frac{\mathrm{d}t}{2\alpha\sqrt{t}}$, so that for $\alpha > 0$,

$$\int_0^\infty x^n \mathrm{e}^{-\alpha^2x^2}\mathrm{d}x = \frac{1}{2\alpha^{n+1}}\int_0^\infty t^{\frac{n}{2}-\frac{1}{2}}\mathrm{e}^{-t}\,\mathrm{d}t = \frac{1}{2\alpha^{n+1}}\Gamma\left(\frac{n+1}{2}\right).$$

16. **(b)** We observe that

$$\int_2^N \frac{\mathrm{d}x}{x\sqrt{x^2-1}} = \sec^{-1}x\,\Big|_2^N \to \frac{\pi}{2} - \sec^{-1}2 \text{ as } N\to\infty.$$

17. Set $f(x) = \frac{x^{\alpha-1}}{1+x}$ and observe that f is positive, continuous, and decreasing on $[1,\infty)$, and

$$\sum_{k=1}^\infty f(k) = \sum_{k=1}^\infty \frac{1}{k^{1-\alpha}(1+k)},$$

which is convergent for $1-\alpha > 0$.

Questions 7.44:

1. By hypothesis, $|x'(\theta)| \le M_1$ and $|y'(\theta)| \le M_2$ for some constants M_1 and M_2, and so

$$s = \int_\alpha^\beta \sqrt{(x'(\theta))^2 + (y'(\theta))^2}\,\mathrm{d}\theta \le \sqrt{M_1^2+M_2^2}\int_\alpha^\beta \mathrm{d}\theta \le M(\beta-\alpha).$$

2. Define $f : [-\pi/2, \pi/2] \to \mathbb{R}$ by

$$f(x) = \begin{cases} x\sin(1/x) & \text{for } x \ne 0,\\ 0 & \text{for } x = 0,\end{cases} \quad x \in [-\pi/2, \pi/2].$$

Exercises 7.45:

1. The points of intersection of the given circles are the origin (i.e., at $(0,\pi/2)$, $(0,0)$) and $(a/\sqrt{2}, \pi/4)$ (i.e., on the radial line $\theta = \pi/4$); see Figure A.6. The required area is

$$A = \int_0^{\pi/4}\frac{1}{2}(a\sin\theta)^2\,\mathrm{d}\theta + \int_{\pi/4}^{\pi/2}\frac{1}{2}(a\cos\theta)^2\,\mathrm{d}\theta,$$

and a computation gives $A = a^2(\pi-2)/8$.

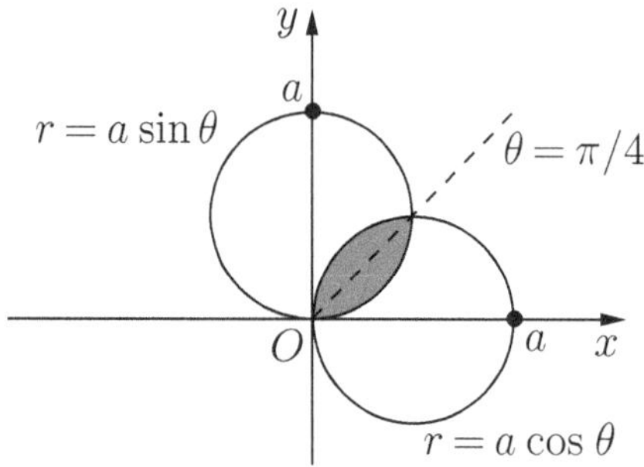

Fig. A.6. Area of the region common to $r = a\cos\theta$ and $r = a\sin\theta$.

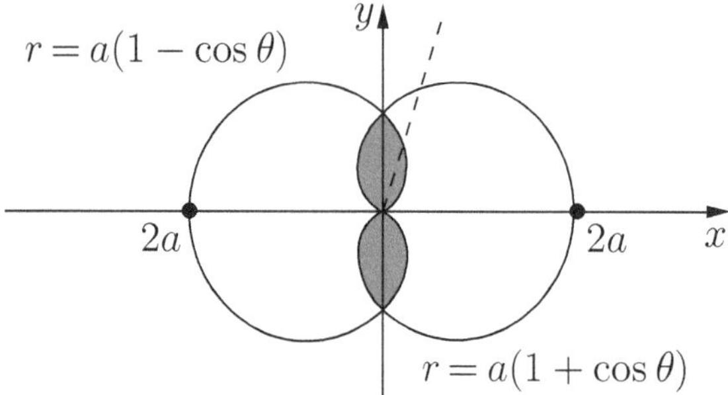

Fig. A.7. Area of the region common to $r = a(1+\cos\theta)$ and $r = a(1-\cos\theta)$.

2. The points of intersection are given by

$$a(1+\cos\theta) = a(1-\cos\theta), \quad \text{i.e., } \cos\theta = 0, \text{ i.e., } \theta = \pm\frac{\pi}{2}.$$

   In view of the symmetry (see Figure A.7),

$$A = 4\left[\int_0^{\pi/2} \frac{1}{2}\, a^2(1-\cos\theta)^2 \,\mathrm{d}\theta\right],$$

   which gives $A = (a^2/2)(3\pi - 8)$.
3. The points of intersection are given by $a = a(1-\cos\theta)$, i.e., $\theta = \pm\pi/2$. In view of the symmetry, the required area is

$$A = 2\left[\int_0^{\pi/2} \frac{1}{2}\, a^2(1-\cos\theta)^2 \,\mathrm{d}\theta + \int_{\pi/2}^{\pi} \frac{1}{2}\, a^2 \,\mathrm{d}\theta\right] = a^2\left(\frac{5\pi - 8}{4}\right).$$

4. Note that $2a = r(1+\cos\theta)$ in Cartesian form is

$$2a = \sqrt{x^2+y^2} + x, \quad \text{or} \quad y^2 = 4a(a-x).$$

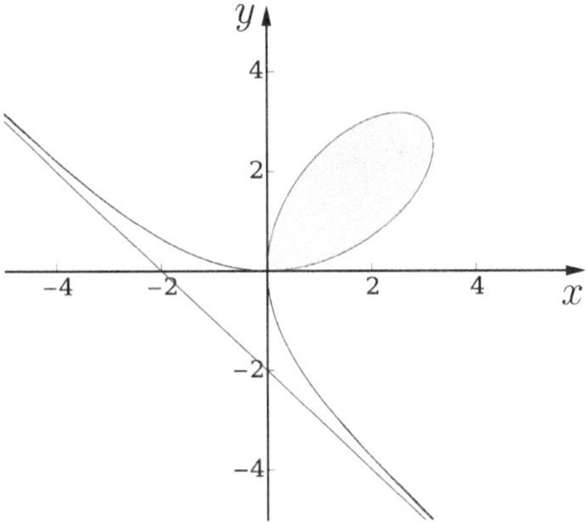

Fig. A.8. Loop of the folium of Descartes $x^3 + y^3 = 3axy$.

The intersection points of these two curves are $\theta = \pm\pi/2$, and both the curves are symmetric about $x = 0$. The desired area is

$$A = 2\left[\int_0^{\pi/2} \frac{1}{2}\left(\frac{2a}{1+\cos\theta}\right)^2 \mathrm{d}\theta + \int_{\pi/2}^{\pi} \frac{1}{2}(2a(1+\cos\theta))^2 \,\mathrm{d}\theta\right].$$

A computation gives $A = a^2(9\pi - 16)/3$.

9. The required area A is given by

$$A = 4\left[\frac{a^2}{2}\int_0^{\pi/6} (2\cos 2\theta - 1)\mathrm{d}\theta\right],$$

which gives $A = a^2(3\sqrt{3} - \pi)/3$.

11. Here it is not possible to express y in terms of x explicitly, and the curve has a loop at the origin. In order to evaluate the area within the loop, we use polar coordinates. The polar form (see Figure A.8) of the given curve is obtained by substituting $x = r\cos\theta$ and $y = r\sin\theta$:

$$r = \frac{3a\cos\theta\sin\theta}{\sin^3\theta + \cos^3\theta} = f(\theta).$$

The required area is

$$A = \frac{1}{2}\int_0^{\pi/2} r^2 \,\mathrm{d}\theta = \frac{9a^2}{2}\int_0^{\pi/2} \frac{\sin^2\theta\cos^2\theta}{(\sin^3\theta + \cos^3\theta)^2}\,\mathrm{d}\theta.$$

A computation gives $A = \frac{3}{2}a^2$ (use the substitution $t = \tan\theta$).

12. Following Example 7.43(**a**), if $r = f(\theta)$, $f(\theta) = a(1+\cos\theta)$, then the required length is given by

$$s = \int_\alpha^\beta \sqrt{r^2 + (r')^2}\,\mathrm{d}\theta = 4a\sin(\theta/2)\Big|_\alpha^\beta = 4a(\sin(\beta/2) - \sin(\alpha/2)).$$

13. See Exercise 7.45(12) and Example 7.43(**a**). Indeed, rotate the graph of $r = a(1 + \cos\theta)$ through an angle π to get the resulting answer.

Chapter 8: Questions 8.11

3. Let a_k be the general term. Then we have

$$a_k = \begin{cases} 1 & \text{if } k \text{ is even },\\ \dfrac{3^k}{5^k} & \text{if } k \text{ is odd,} \end{cases} \quad \text{and} \quad a_k^{1/k} = \begin{cases} 1 & \text{if } k \text{ is even },\\ \dfrac{3}{5} & \text{if } k \text{ is odd,} \end{cases}$$

so that $\limsup_{k\to\infty} a_k^{1/k} = 1$. So the root test is not applicable. Recall that $a_k = 1$ for all k even, and so $\{a_k\}$ does not converge to zero. Thus the series diverges by the divergence test.

Exercises 8.12:

1. We have shown that for $a = 2$, the series converges, and for $a = 3$, the series diverges.
3. (**a**) Since $a_k \leq 3^{1/(k-1)}$ for all k, it is appropriate to apply the comparison test. Alternatively, compute

$$a_k = \begin{cases} \dfrac{3}{3^k} & \text{if } k \text{ is even,}\\ \dfrac{3^{-1}}{3^k} & \text{if } k \text{ is odd,} \end{cases} \quad \text{and} \quad a_k^{1/k} = \begin{cases} \dfrac{3^{1/k}}{3} & \text{if } k \text{ is even,}\\ \dfrac{3^{-1/k}}{3} & \text{if } k \text{ is odd,} \end{cases}$$

and

$$\frac{a_{k+1}}{a_k} = \begin{cases} \dfrac{1}{27} & \text{if } k \text{ is even,}\\ 3 & \text{if } k \text{ is odd.} \end{cases}$$

It follows that

$$\lim_{k\to\infty} a_k^{1/k} = \frac{1}{3} \quad \text{and} \quad \limsup_{k\to\infty} \frac{a_{k+1}}{a_k} = 3 > 1 > \liminf_{k\to\infty} \frac{a_{k+1}}{a_k} = \frac{1}{27}.$$

Thus the series converges by the root test, whereas the ratio test gives no conclusion.

Questions 8.50:

2. No. If it converge at $x = 0$, then it should converge for $|x-2| < |0-2| = 2$, i.e., for $0 < x < 4$. But $x = 3$ lies inside the interval $(0, 4)$, so the series would converge there.
4. $R = 3$.
7. Yes, on $(0, 2\pi)$. But at the points 0 and 2π, the functional series reduces to the numerical series $\sum_{k=1}^{\infty} b_k$, and so the convergence depends on the choice of the b_k's, e.g., $b_k = 1/k,\ 1/k^2$.
8. Consider $\sum \dfrac{x^k}{k^2}, \sum \dfrac{x^k}{k}, \sum x^k$.

16. See Example 8.48.
18. Yes.

Exercises 8.51:

2. Set $a_k(x) = (-1)^{k-1}x^{2k-1}/(2k-1)$ for $x \neq 0$. Then, using the ratio test, we find that

$$\left|\frac{a_{k+1}(x)}{a_k(x)}\right| = |x|^2 \left|\frac{2k-1}{2k+1}\right| \to |x|^2 \quad \text{as } k \to \infty\ ,$$

so that the power series converges absolutely if $|x| < 1$, and diverges if $|x| > 1$. At the endpoints:

- At $x = 1$: $\displaystyle\sum_{k=1}^{\infty} \frac{(-1)^{k-1}}{2k-1}$ *converges* by the alternating series test.
- At $x = -1$: $\displaystyle-\sum_{k=1}^{\infty} \frac{(-1)^{k-1}}{2k-1}$ *converges* by the alternating series test.

3. We have

$$\limsup_{n\to\infty} |a_n x^{pn}|^{1/n} = |x|^p \limsup_{n\to\infty} |a_n|^{1/n} = \frac{1}{R}|x|^p$$

and

$$\limsup_{n\to\infty} |a_{pn} x^n|^{1/n} = |x| \limsup_{n\to\infty} \left(|a_{pn}|^{1/pn}\right)^p = \frac{1}{R^p}|x|,$$

so that the series in **(a)** and **(b)** have radii of convergence $\sqrt[p]{R}$ and R^p, respectively.

10. Apply Theorem 8.33 suitably to geometric series.
17. The series on the right converges for $|x| > 1$. Since

$$\sum_{n=0}^{\infty} x^{n+1} = \frac{x}{1-x} \quad \text{for } |x| < 1,$$

we have

$$\sum_{n=1}^{\infty} (n+1)nx^n = x\left(\frac{x}{1-x}\right)'' = \frac{2x}{(1-x)^3} \quad \text{for } |x| < 1,$$

so that

$$\sum_{n=1}^{\infty} \frac{(n+1)n}{x^n} = \frac{2(1/x)}{(1-1/x)^3} = \frac{2x^2}{(x-1)^3} \quad \text{for } |x| > 1.$$

Thus the given equation is equivalent (with $|x| > 1$) to

$$x = \frac{2x^2}{(x-1)^3} \quad \text{or} \quad (x-1)^3 = 2x \quad \text{or} \quad x^3 - 3x^2 + x - 1 = 0.$$

19. Assume that f is real analytic on $(-\delta, \delta)$. Choose r such that $2r < \delta$, i.e., $(-2r, 2r) \subset (-\delta, \delta)$. By assumption,

$$f(x) = \sum_{k=0}^{\infty} \frac{f^{(k)}(0)}{k!} x^k \quad \text{for } |x| < \delta,$$

and so it converges when $x = 2r$. In particular,

$$\frac{f^{(k)}(0)}{k!}(2r)^k \to 0 \quad \text{as } k \to \infty,$$

and so there exists an $N \in \mathbb{N}$ such that

$$\left| \frac{f^{(k)}(0)}{k!} \right| \leq \frac{1}{(2r)^k} \quad \text{for all } k \geq N.$$

Moreover,

$$f^{(n)}(x) = n! \sum_{k=n}^{\infty} \binom{k}{k-n} \frac{f^{(k)}(0)}{k!} x^{k-n} \quad \text{for } |x| < \delta,$$

and thus for $|x| < r$, we get

$$\begin{aligned} |f^{(n)}x| &\leq n! \sum_{k=n}^{\infty} \binom{k}{k-n} \frac{1}{(2r)^k} r^{k-n} \\ &= \frac{n!}{(2r)^n} \sum_{k=n}^{\infty} \binom{k}{k-n} \frac{1}{2^{k-n}} \\ &< \frac{n!}{(2r)^n} \left[\frac{1}{(1-\frac{1}{2})^{n+1}} \right] < n! \left(\frac{2}{r^n} \right). \end{aligned}$$

The result follows. We leave the converse as an exercise.

20. By integration by parts, we get

$$\begin{aligned} R_n &= \frac{1}{n!} \int_a^b (b-t)^n \, \mathrm{d}\left(f^{(n)}(t) \right) \\ &= \frac{1}{n!} \left((b-t)^n f^{(n)}(t) \Big|_a^b + n \int_a^b (b-t)^{n-1} f^{(n)}(t) \, \mathrm{d}t \right) \\ &= -\frac{1}{n!}(b-a)^n f^{(n)}(a) + \frac{1}{(n-1)!} \int_a^b (b-t)^{n-1} f^{(n)}(t) \, \mathrm{d}t. \end{aligned}$$

The desired formula follows if we continue the integration by parts in this way.

Alternatively, one could also prove the desired form using the fundamental theorem of calculus, which gives

$$f(b) = f(a) + \int_a^b f'(t)\,\mathrm{d}t.$$

According to this, one has

$$\begin{aligned}\int_a^b f'(t)\,\mathrm{d}t &= \int_0^{b-a} f'(b-u)\,\mathrm{d}u \quad (t = b-u)\\ &= uf'(b-u)\big|_0^{b-a} + \int_0^{b-a} uf''(b-u)\,\mathrm{d}u\\ &= (b-a)f'(a) + \int_0^{b-a} f''(b-u)\,\mathrm{d}(u^2/2)\\ &= (b-a)f'(a) + \frac{(b-a)^2}{2}f''(a) + \frac{1}{2}\int_0^{b-a} u^2 f'''(b-u)\,\mathrm{d}u\\ &= (b-a)f'(a) + \frac{(b-a)^2}{2}f''(a) + \frac{1}{2}\int_0^{b} (b-t)^2 f'''(t)\,\mathrm{d}t.\end{aligned}$$

More generally, we have

$$\int_a^b f'(t)\,\mathrm{d}t = (b-a)f'(a) + \cdots + \frac{(b-a)^n}{n!}f^{(n)}(a) + \frac{1}{n!}\int_a^b (b-t)^n f^{(n+1)}(t)\,\mathrm{d}t,$$

and the desired formula follows.

Chapter 9: Questions 9.19

7. No.
11. Define $f_n(x) = x + \frac{1}{n}$, $g(x) = x^2$ and $h_n(x) = g(f_n(x))$ for $x \in \mathbb{R}$. Then $f_n \to x$ uniformly on $\mathbb{R}$, and g is not uniformly continuous on $\mathbb{R}$. Now $f(x) = x$ and $g(f(x)) = x^2$,

$$g(f_n(x)) = \left(x + \frac{1}{n}\right)^2 \to x^2 \quad \text{as } n \to \infty \text{ on } \mathbb{R}.$$

Therefore,

$$g(f_n(x)) - g(f(x)) = \frac{1}{n^2} + \frac{2}{n}x,$$

so that at $x = n \in \mathbb{N}$, we have

$$|g(f_n(x)) - g(f(x))| \geq \frac{1}{n^2} + 2 > 2,$$

showing that $g \circ f_n$ does not converge uniformly to $g \circ f$ on $\mathbb{R}$.

12. Yes, for the first two cases (use the definition). The third case can be proved easily. The final case is false, and so uniform convergence does not carry over to product functions. Our argument below works for any unbounded function $f(x)$ on E instead of $f(x) = 1/x$ on $E = (0,1)$. Consider

$$f_n(x) = \frac{1}{x} + \frac{1}{n} \quad \text{and} \quad g_n(x) = f_n(x) \quad \text{on } E = (0,1).$$

Then $f_n(x) \to f(x) = 1/x$ uniformly on $(0,1)$. Now

$$f_n^2(x) - f^2(x) = \left(\frac{1}{x} + \frac{1}{n}\right)^2 - \frac{1}{x^2} = \frac{1}{n^2} + \frac{2}{n}.\frac{1}{x},$$

which for $x_0 = 1/n$ or $1/n^2$ in $(0,1)$ shows that

$$f_n^2(x_0) - f^2(x_0) \geq 2 \quad \text{for all } n \geq 1.$$

Hence $\{f_n^2\}$ does not converge uniformly on $(0,1)$.
The reader may also try with different pairs of sequences. For instance (see Figure 9.8),
(a) $f_n(x) = x$ and $g_n(x) = 1/n$ for $x \in \mathbb{R}$.
(b) $f_n(x) = x^n(1-x)$ and $g_n(x) = 1/(1-x)$ for $x \in (0,1)$.

17. No. Consider $f_n(x) = x/(x+n)$ for $x \in (0,\infty)$.
18. Yes. For $n \geq 1$ define

$$f_n(x) = \begin{cases} 1/n & \text{if } x \text{ is rational,} \\ 0 & \text{if } x \text{ is irrational.} \end{cases}$$

Exercises 9.20:

1. How about $f_n(x) = \sin^n x$ on $[0,\pi]$? (See Figure A.9.)
2. Note that $\{x/n\}$ converges pointwise to $f(x) = 0$ on $\mathbb{R}$. On the other hand, for $x \neq 0$,

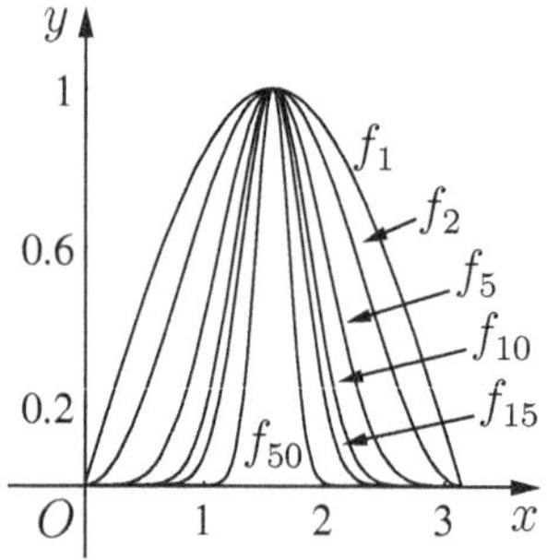

Fig. A.9. $f_n(x) = \sin^n x$ on $[0,\pi]$.

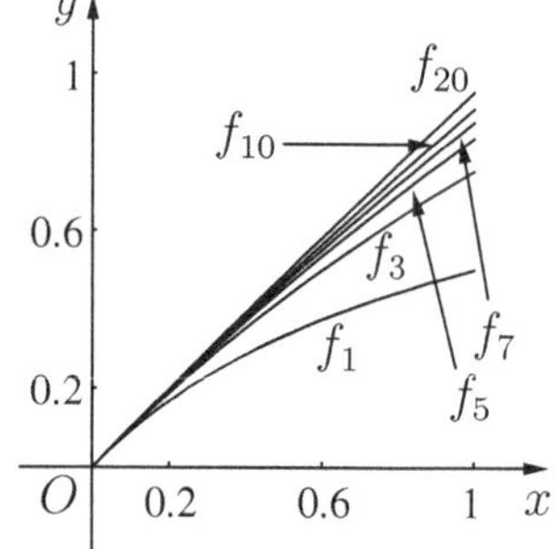

Fig. A.10. $f_n(x) = \dfrac{nx}{n+x}$ on $[0,1]$.

$$\left|\frac{x}{n} - 0\right| = \left|\frac{x}{n}\right| < \epsilon \ \text{ if and only if } \ n > \frac{\epsilon}{|x|},$$

and $\sup_{x\in\mathbb{R}\{0\}} \epsilon/|x| = \infty$. So the convergence is not uniform on $\mathbb{R}$.

3. For $x \in (0,1)$, $|nx^n - 0| < \epsilon$ if and only if $x^n < \epsilon/n$ $(\leq \epsilon)$, which gives $n > \frac{\log \epsilon}{\log x} = \frac{\log(1/\epsilon)}{\log(1/x)}$. Thus $\{nx^n\}$ converges to 0 pointwise on $[0,1)$. If it were convergent uniformly on $[0,1)$, then there would exist an N such that

$$|nx^n - 0| = nx^n < 1/2 = \epsilon \ \text{ for all } x \in [0,1) \text{ and } n \geq N.$$

In particular, $Nx^N < 1/2$ for all $x \in [0,1)$. But then, for $x = (1/N)^{1/N}$, this fails to satisfy. Thus, the convergence is not uniform. Note: Pointwise convergence of $\{nx^n\}$ is easy to obtain, because it is the general term of a convergent power series $\sum nx^n$ for $|x| < 1$.

4. Clearly, $\{f_n\}$ converges pointwise to $f(x) = x$ on $[0,1]$, see Figure A.10. Now,

$$|f_n(x) - x| = \frac{x^2}{n+x} \leq \frac{1}{n+1} \to 0 \quad \text{as } n \to \infty,$$

and so the convergence is uniform on $[0,1]$.

5. See Figures A.11 and A.12.

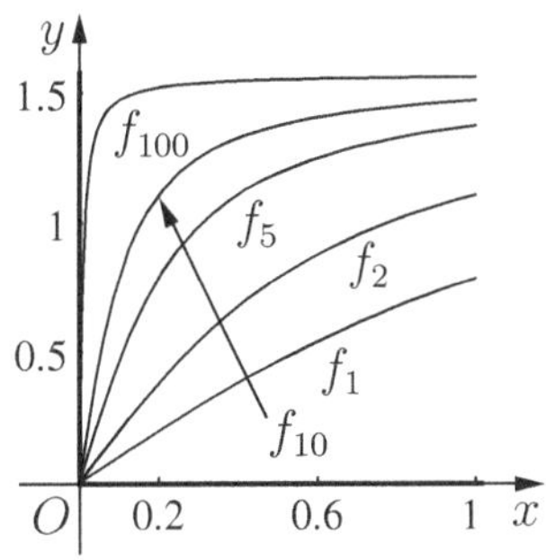

Fig. A.11. $f_n(x) = \arctan(nx)$ on $[0,1]$.

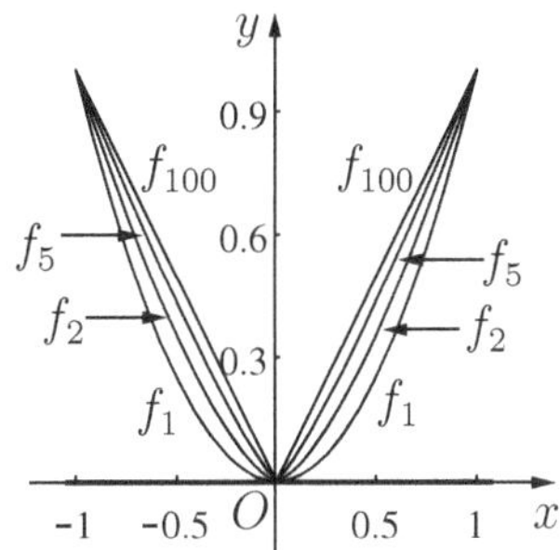

Fig. A.12. $f_n(x) = |x|^{1+\frac{1}{n}}$ on $[-1,1]$.

7. We note that $f_n(0) = 0$, and for $x \neq 0$,

$$f_n(x) = \frac{x}{\frac{1}{n} + n^{p-1}x^2} = \frac{n^{1-p}x}{n^{-p} + x^2} \to 0 \quad \text{as } n \to \infty$$

whenever $p - 1 > 0$. Thus, $\{f_n\}$ converges pointwise to $f(x) = 0$ on $\mathbb{R}$. Further,

$$|f_n(x) - 0| = \frac{|x|}{\frac{1}{n} + n^{p-1}x^2} \leq \frac{|x|}{2(1/\sqrt{n})n^{(p-1)/2}|x|} = \frac{1}{2n^{(p/2)-1}},$$

and so the convergence is uniform if $p > 2$. On the other hand, if $1 < p \leq 2$, we have

$$|f_n(1/n) - 0| = \frac{1}{1 + n^{p-2}} \geq \frac{1}{1+1} = \frac{1}{2},$$

which means that the convergence is not uniform whenever $1 < p \leq 2$.

8. For $x \in (0, 1)$, we have

$$\sup_{x \in (0,1)} |f_n(x) - 0| = \sup_{x \in (0,1)} \frac{x^n}{1 + x^{2n}} = \frac{1}{2} \text{ for all } x \in (0, 1).$$

Using the definition of uniform convergence with $\epsilon = \frac{1}{4}$ for x near 1, we conclude that the convergence is not uniform (see Figure A.13).

9. Use Theorem 9.15 (and see also Example 9.16 and Figure A.14).
10. Note that

$$f_n(x) \to \begin{cases} 0 & \text{for } 0 \leq x < 1, \\ \frac{1}{2} & \text{for } x = 1, \\ 1 & \text{for } 1 < x \leq 3. \end{cases}$$

11. We see that $f_n \to 0$ on $\mathbb{R}$. Also

$$\left| f_n\left(\frac{1}{\sqrt{n}}\right) - 0 \right| = \frac{\sqrt{n}}{\mathrm{e}} \to \infty \text{ as } n \to \infty.$$

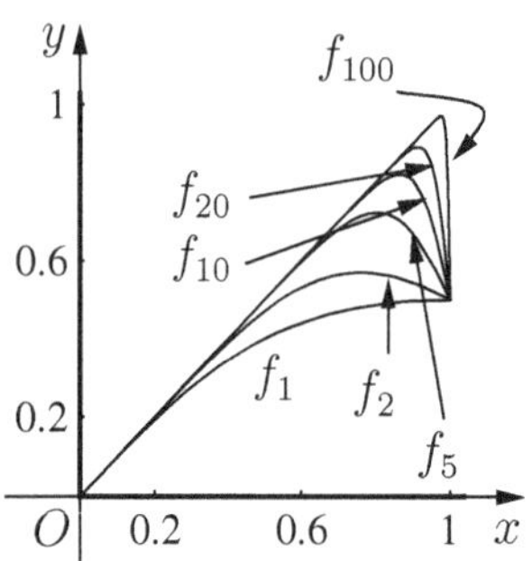

Fig. A.13. $f_n(x) = \dfrac{x^n}{1 + x^{2n}}$ on $(0, 1)$.

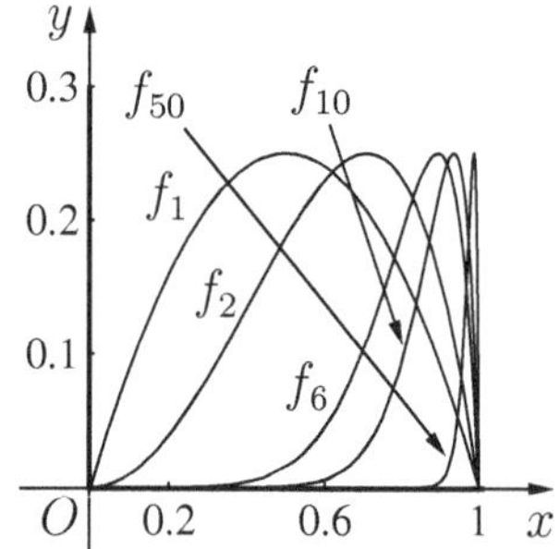

Fig. A.14. $f_n(x) = x^n(1 - x^n)$ on $[0, 1]$.

12. One may use l'Hôpital's rule to show that $f_n \to 0$ on $(0, 1]$. Also,

$$\max_{x \in [0,1]} |f_n(x) - f(x)| = \max_{x \in [0,1]} f_n(x) = f_n\left(\frac{1}{\sqrt{2n}}\right) = \sqrt{\frac{n}{2\mathrm{e}}} \to \infty,$$

and so the convergence is not uniform. We may also use Theorem 9.4 by choosing $x_n = 1/\sqrt{n}$. Finally, we see that

$$\int_0^1 f_n(x)\,\mathrm{d}x = \frac{1 - \mathrm{e}^{-n}}{2} \to \frac{1}{2} \quad \text{as } n \to \infty,$$

and hence by Theorem 9.15, the convergence cannot be uniform on $[0, 1]$.

13. Set $f_n(x) = g(x)x^n$. If $f_n \to f$ uniformly on $[0,1]$, then f must be continuous on $[0,1]$. Since

$$f(x) = \lim_{n\to\infty} f_n(x) = \begin{cases} 0 & \text{for } 0 \le x < 1, \\ g(1) & \text{for } x = 1, \end{cases}$$

and $f(1) = \lim_{x\to 1-} f(x) = 0$, by the continuity of f, $f(1) = g(1) = 0$. Conversely, let $g(1) = 0$. Since g is continuous on $[0,1]$, g is bounded on $[0,1]$. Indeed, given $\epsilon > 0$, there exists a $\delta > 0$ such that

$$|g(x) - g(1)| = |g(x)| < \epsilon \ \text{ whenever } 1 - \delta < x \le 1$$

and $|g(x)| \le M$ on $[0,1]$. Consequently,

$$f(x) = \lim_{n\to\infty} f_n(x) = \lim_{n\to\infty} g(x)x^n = 0 \ \text{ on } \ [0,1],$$

and so

$$|f_n(x)| = |g(x)x^n| \le \begin{cases} M(1-\delta)^n & \text{for } 0 < x \le 1-\delta, \\ \epsilon & \text{for } 1-\delta < x \le 1. \end{cases}$$

Therefore, $f_n \to 0$ uniformly on $[0,1]$, since $(1-\delta)^n \to 0$ as $n \to \infty$.

14. **(a)** First, $f_n \to 1$ for $x \ge 0$. Next, for $x \ge 0$,

$$|f_n(x) - 1| = \frac{x}{n+x} < \epsilon \iff n \ge \frac{x(1-\epsilon)}{\epsilon},$$

where for $0 < \epsilon < 1$, $\sup_{x\in[0,\infty)} \frac{x(1-\epsilon)}{\epsilon} = \infty$. Thus, the convergence is not uniform for $x \ge 0$. However, for $0 \le x \le c$,

$$\frac{x}{n+x} \le \frac{c}{n+x} \le \frac{c}{n} < \epsilon \quad \text{for } n > \tfrac{c}{\epsilon} = N(\epsilon),$$

and so the convergence is uniform on $[0,c]$.

(i) Pointwise convergence to $f(x) = 0$ on $[-\pi/2, \pi/2]$ is obvious, because

$$\left|\frac{\sin^n x}{3n + \sin^n x}\right| \le \frac{1}{3n-1} \to 0 \ \text{ as } n \to \infty.$$

Uniform convergence follows from Theorem 9.6 and

$$0 \le \sup_{x\in[-\pi/2,\pi/2]} \left|\frac{\sin^n x}{3n + \sin^n x} - 0\right| = \frac{1}{3n-1} \to 0 \ \text{ as } n \to \infty.$$

(j) Uniformly convergent to $f(x) = 0$ on $[0,\pi]$.

(k) Uniformly convergent to $f(x) = 0$ on $[-\infty, \infty]$.

15. Observe that

$$0 \le \left|\frac{\sin nx}{n}\right| \le \frac{1}{n}.$$

16. For $x \geq 0$, we have $e^{nx} > n^k x^k / k!$, and so

$$0 \leq x^k e^{-nx} < \frac{k!}{n^k}.$$

17. We observe that

$$\sup_{x \in [0,b]} |e^{x(n+1)/n} - e^x| = \sup_{x \in [0,b]} |e^x (e^{x/n} - 1)| = e^b (e^{b/n} - 1) \to 0 \text{ as } n \to \infty$$

and

$$e^{-nx} \to \begin{cases} 1 & \text{if } x = 0, \\ 0 & \text{if } x \in (0, b]. \end{cases}$$

20. **(a)** As in Example 9.7, if $f_n(x) = n^2 x^n (1-x)$, then $f_n \to 0$ pointwise on $[0,1]$. On the other hand, as in Example 9.7, we have

$$\delta_n = \max_{x \in [0,1]} |f_n(x) - 0| = f_n\left(\frac{n}{n+1}\right) = n\left(\frac{n}{n+1}\right)^{n+1} \sim \frac{n}{e},$$

so that δ_n does not approach 0 as $n \to \infty$. Thus, $\{f_n\}$ cannot converge uniformly on $[0,1]$. Also,

$$\int_0^1 f_n(x)\, dx = \frac{n^2}{(n+1)(n+2)} \to 1 \text{ and } \int_0^1 f(x)\, dx = 0.$$

Again, by Theorem 9.15, $\{f_n\}$ cannot converge uniformly to $f(x) = 0$ on $[0,1]$.

22. **(a)** Denote the integrand by $f_n(x)$. Then each f_n is continuous on $[a,b]$, and so the integral exists. Moreover,

$$|f_n(x)| \leq \frac{1}{a^3 + n} \to 0 \text{ as } n \to \infty.$$

Also, since

$$\lim_{n \to \infty} \sup_{x \in [a,b]} |f_n(x)| \leq \lim_{n \to \infty} \frac{1}{a^3 + n} = 0,$$

it follows that $\{f_n\}$ converges uniformly to $f(x) = 0$ on $[a,b]$, by Theorem 9.6. The desired limit is zero, by Theorem 9.15.

(b) Note that for $x \in [0,1]$,

$$f_n(x) = \frac{n + \cos^n(e^x)}{4n + x^4} = \frac{1 + \frac{1}{n}\cos^n(e^x)}{4 + \frac{1}{n}x^4} \to \frac{1}{4} \text{ as } n \to \infty,$$

so that

$$0 \leq \left| f_n(x) - \frac{1}{4} \right| = \left| \frac{\cos^n(e^x) - x^4}{4n + x^4} \right| \leq \frac{2}{4n + x^4} \leq \frac{1}{2n},$$

which implies that $f_n(x) \to \frac{1}{4}$ uniformly on $[0,1]$. Consequently, by Theorem 9.15,

$$\lim_{n \to \infty} \int_0^1 f_n(x)\, dx = \int_0^1 \frac{1}{4}\, dx = \frac{1}{4}.$$

Questions 9.61:

3. Yes. For $n \geq 1$ define (see Figure A.15)

$$f_n(x) = \begin{cases} n^2x & \text{for } 0 \leq x \leq 1/n, \\ 1/x & \text{for } 1/n \leq x \leq 1. \end{cases}$$

Then, we see that $f_n \to f$ pointwise on $[0,1]$, where

$$f(x) = \begin{cases} 1/x & \text{for } 0 < x \leq 1, \\ 0 & \text{for } x = 0, \end{cases}$$

which is unbounded. Note also that each f_n is continuous on $[0,1]$ and so integrable on $[0,1]$. But f is not integrable on $[0,1]$. Also, for $n \geq 1$,

$$\int_0^1 f_n(x)\,dx = \int_0^{1/n} f_n(x)\,dx + \int_{1/n}^1 f_n(x)\,dx = \frac{1}{2} + \log n \to \infty \text{ as } n \to \infty,$$

showing that the sequence of integrals $\left\{\int_0^1 f_n(x)\,dx\right\}$ is an unbounded sequence.

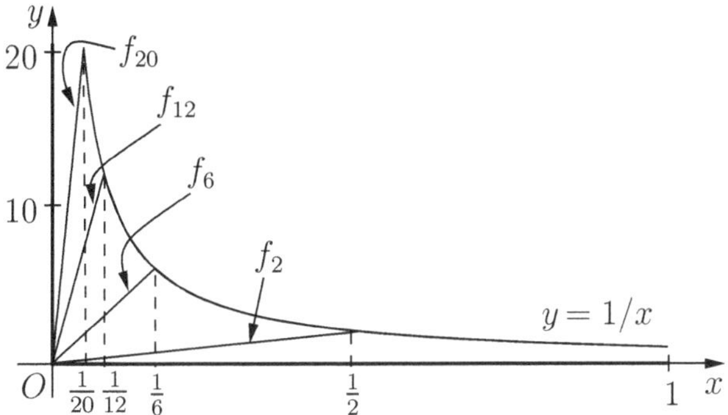

Fig. A.15. $f_n(x) = n^2x$ for $0 \leq x \leq \frac{1}{n}$, and $\frac{1}{x}$ for $\frac{1}{n} \leq x \leq 1$.

5. Set $f_n(x) = x^n/n$. Then the uniform convergence on $[-1,1]$ is clear (use the method of Example 9.8), and

$$f_n'(x) = x^{n-1} \to \begin{cases} 0 & \text{for } -1 < x < 1, \\ 1 & \text{for } x = 1, \\ \text{doesn't exist} & \text{for } x = -1. \end{cases}$$

Observe that even if $f_n(x)$ is differentiable on E for all n and $f_n \to f$ uniformly on E, then $\{f_n'\}$ need not converge to f' on E. It is true that $f_n'(x)$ is continuous on $[-1,1]$ for all $n \geq 1$. Note that $f_n'(1) = 1$ and $f'(1) = 0$, and so at $x = 1$,

$$f'(x) \neq \lim_{n\to\infty} f_n'(x).$$

7. We see that $f_n \to 0$ uniformly on $\mathbb{R}$, but $f_n'(x) = \cos(nx)$ converges only at integer multiples of 2π.
8. Define $f_m(x) = \lim_{n\to\infty} (\cos(m!\pi x))^{2n}$, for $m \in \mathbb{N}$. Clearly

$$f_m(x) = \begin{cases} 1 & \text{if } m!x \text{ is an integer,} \\ 0 & \text{otherwise.} \end{cases}$$

If x is irrational, then $f_m(x) = 0$ for every $m \in \mathbb{N}$. For rational x, say $x = p/q$, where p and q are integers, we see that $m!x$ is an integer if $m \geq q$. Hence

$$f(x) = \lim_{m\to\infty} f_m(x) = \begin{cases} 0 & \text{if } x \text{ irrational,} \\ 1 & \text{if } x \text{ rational,} \end{cases}$$

which is everywhere discontinuous and not Riemann integrable.
11. Apply Weierstrass's M-test.
13. The definition of uniform convergence implies that there is an N such that

$$|f_n(x) - f(x)| < \epsilon = 1 \quad \text{for all } x \in E \text{ and all } n \geq N.$$

Since $\{f_n\}$ is a bounded sequence, there is an $M > 0$ such that $|f_n(x)| \leq M$ for all n and for all $x \in E$. In particular,

$$|f(x)| \leq |f(x) - f_N(x)| + |f_N(x)| \leq 1 + M.$$

14. For the sequence of partial sums $\{S_n(x)\}$, we have for all $m > n > N$,

$$|S_m(x) - S_n(x)| = \left|\sum_{k=n+1}^{m} f_k(x)\right| \leq \sum_{k=n+1}^{m} |f_k(x)|,$$

showing that the series $\sum_{k=1}^{\infty} f_k(x)$ converges uniformly on E whenever $\sum_{k=1}^{\infty} |f_k|$ converges uniformly on E. The converse is not true in general.
15. Yes.

Exercises 9.62:

2. Note that $f_n(x) \to 0$ pointwise on $\mathbb{R}$ (see Example 9.11(2)). Clearly, $f_n(0) = 0$, but $f_n(1/n) = 1/2$ for all n. On the other hand, it converges uniformly on $\{x : |x| \geq k\}$, $k > 0$, because

$$|f_n(x) - 0| = \frac{n|x|}{1 + n^2x^2} < \frac{1}{n|x|} \leq \frac{1}{nk} < \epsilon \quad \text{for } n > \frac{1}{k\epsilon}.$$

Because $f_n(1/n) = 1/2$, the sequence $\{f_n(x)\}$ cannot converge uniformly to $f(x) = 0$ on any interval containing zero.

3. For $x = 1$, our previous experience with series implies that $\sum f_k(1)$ converges with sum $\log 2$. Also, for $0 \leq x < 1$,

$$S_n(x) \to \frac{1}{2}\log\left(\frac{1+x}{1-x}\right) + \frac{1}{2}\log(1-x) = \frac{1}{2}\log(1+x).$$

Consequently,

$$S_n(x) \to f(x) = \begin{cases} \frac{1}{2}\log(1+x) & \text{if } 0 \leq x < 1, \\ \log 2 & \text{if } x = 1, \end{cases}$$

pointwise on $[0,1]$. By Corollary 9.22, the series does not converge uniformly on $[0,1]$. Note: We observe that f is integrable on $[0,1]$. Does this imply that the uniform convergence of the series $\sum f_n$ to f on $[a,b]$ in Corollary 9.30 is sufficient but not necessary for the integrability of the limit function f?

4. We present a direct proof. If this were true, then for $\epsilon > 0$ there would exist an N such that

$$|S_m(x) - S_n(x)| = \sum_{k=n+1}^{m} \frac{x^k}{k} < \epsilon \quad \text{for all } x \in (0,1) \text{ and } m > n \geq N,$$

which in particular, holds for $n = N$. But then, for x close to 1, we have $x^k > \frac{1}{2}$ for some $k > N$, so that

$$\sum_{k=N+1}^{m} \frac{x^k}{k} > \frac{1}{2}\left[\frac{1}{N+1} + \frac{1}{N+2} + \cdots + \frac{1}{m}\right] \to \infty \quad \text{as } m \to \infty.$$

Therefore, the convergence cannot be uniform on $(0,1)$.

7. To prove the pointwise convergence on $[0,\infty)$, we may write the general term of the series as

$$f_k(x) = \frac{1}{1+(k-1)x} - \frac{1}{1+kx},$$

so that

$$s_n(x) = \sum_{k=1}^{n} f_k(x) = 1 - \frac{1}{1+nx} \to f(x) = \begin{cases} 0 & \text{if } x = 0, \\ 1 & \text{if } x > 0. \end{cases}$$

On the interval $[r, R]$,

$$\sup_{x\in[r,R]} |s_n(x) - s(x)| = \sup_{x\in[r,R]} \frac{1}{1+nx} = \frac{1}{1+nr} \to 0 \text{ as } n \to \infty,$$

so that $s_n(x) \to 1$ uniformly on $[r, R]$, but not on $[0, R]$, because

$$s_n\left(\frac{1}{n}\right) - s\left(\frac{1}{n}\right) = \frac{1}{2} \quad \text{for each } n \geq 1.$$

Observe that $\lim_{n\to\infty} \int_0^R s_n(x)\,dx = \int_0^R \lim_{n\to\infty} s_n(x)\,dx$ although $\{s_n\}$ does not converge uniformly on $[0, R]$.

8. Apply Corollary 9.41 and the Weierstrass M-test.
9. Assume that S_n denotes the nth partial sum of the given series of functions.

   (a) $\left|\dfrac{x}{n(1+nx^2)}\right| < \dfrac{1}{n^{1/3}}$ (since $1+nx^2 > 2\sqrt{n}x$).
   (b) $\left|\dfrac{1}{n^3+n^4x^2}\right| < \dfrac{1}{n^3}$ for $x \in \mathbb{R}$.
   (d) $\left|\dfrac{1}{1+n^2+n^2x^2}\right| \le \dfrac{1}{n^2}$ for $x \in \mathbb{R}$.
   (e) $\dfrac{1}{n^x} < \dfrac{1}{n^c}$ for $c > 1$.
   (f) $\dfrac{1}{(n+x)^2} \le \dfrac{1}{n^2}$ for $x \ge 0$.
   (g) $\left|\dfrac{\sin nx}{e^n}\right| \le \left(\dfrac{1}{e}\right)^n$ for $x \in \mathbb{R}$.
   (h) $\left|\dfrac{e^{nx}}{5^n}\right| \le \left(\dfrac{1}{5}\right)^n$ for $x \le 0$.
   (i) $\left|\dfrac{\log x}{x}\right| < 1$.
   (j) $|x \log x| \le x^2$.

10. (a) $-1 \le x < 1$. (b) $-2 < x < 0$. (c) $x < -1,\ x \ge 1$.
    (d) $x < -3,\ x \ge 3$. (e) $-6 \le x \le -4$. (f) $-\frac{\pi}{6} \le x \le \frac{\pi}{6}$.
11. The series converges uniformly by Weierstrass's M-test. By Corollary 9.22, f is continuous on $[0,\pi]$ and
$$\int_0^\pi f(x)\,dx = \sum_{k=1}^\infty \int_0^\pi \frac{\sin(kx)}{k^2} = \sum_{k=1}^\infty \frac{1}{k^3}(1-\cos k\pi) = \sum_{k=1}^\infty \frac{2}{(2k-1)^3}.$$
12. (See Theorem 9.40). By conditions (**b**) and (**c**) for Theorem 9.45, there exists an N such that for all $m,n \ge N$, we have
$$|f_n(x_0)-f_m(x_0)| < \frac{\epsilon}{2} \quad \text{and} \quad |f_n'(t)-f_m'(t)| < \frac{\epsilon}{2(b-a)} \quad \text{for all } t \in [a,b].$$
Because $h = f_n - f_m$ is differentiable on $[a,b]$ (for each pair of integers m and n), the mean value theorem implies that for $u,v \in [a,b],\ u \ne v$,
$$h(u)-h(v) = h'(\theta)(u-v), \quad \text{i.e.,} \quad |h(u)-h(v)| \le \frac{\epsilon}{2}|u-v|,$$
for some θ lying in the open interval with end points u and v. In particular, for any $x \in [a,b]$, $x \ne x_0$, on the interval $I \cup \partial I = [x_0, x]$ (or on the interval $[x, x_0]$ if $x < x_0$), and for $m,n \ge N$,
$$\begin{aligned}|h(x)| &\le |h(x)-h(x_0)| + |h(x_0)| \\ &\le \frac{\epsilon}{2}\frac{|x-x_0|}{b-a} + \frac{\epsilon}{2} \le \frac{\epsilon}{2} + \frac{\epsilon}{2} = \epsilon.\end{aligned}$$

Since N depends only on ϵ, the assertion is true for all $x \in [a, b]$. Consequently, by Cauchy's convergence criterion, $\{f_n\}$ converges uniformly to a continuous function $f(x)$ on $[a, b]$. So we define f and F on $[a, b]$, since

$$f(x) = \lim_{n\to\infty} f_n(x) \quad \text{and} \quad F(y) = \begin{cases} \dfrac{f(x) - f(y)}{x - y} & \text{for } y \in [a, b] \setminus \{x\}, \\ g(x) & \text{for } y = x. \end{cases}$$

Now for a fixed x, define $F_n : [a, b] \to \mathbb{R}$ by

$$F_n(y) = \begin{cases} \dfrac{f_n(x) - f_n(y)}{x - y} & \text{for } y \in [a, b] \setminus \{x\}, \\ f_n'(x) & \text{for } y = x. \end{cases}$$

For each n, the differentiability of f_n implies that F_n is continuous at x. Moreover, for all $m, n \geq N$,

$$\begin{aligned} |F_n(y) - F_m(y)| &= \left| \frac{f_n(x) - f_n(y) - (f_m(x) - f_m(y))}{x - y} \right| \\ &\leq \frac{1}{|x - y|} \frac{\epsilon |x - y|}{2} = \frac{\epsilon}{2} \quad \text{for all } y \in [a, b] \setminus \{x\}, \end{aligned}$$

and for $y = x$,

$$|F_n(y) - F_m(y)| = |f_n'(x) - f_m'(x)| < \frac{\epsilon}{2}.$$

Consequently, $\{F_n\}$ converges uniformly on $[a, b]$, and clearly, $\{F_n\}$ converges to F.

15. We know that

$$\phi(x) := \frac{\log(1 + x)}{x} = \sum_{k=1}^{\infty} \frac{(-1)^{k-1}}{k} x^{k-1} \quad \text{for } -1 < x < 1,$$

where $\phi(0) = 1$. The series on the right converges at $x = 1$, whereas it diverges at $x = -1$. By Abel's limit theorem,

$$\lim_{x\to 1-} \phi(x) = \log 2,$$

and the series on the right is uniformly convergent on $[0, 1]$. Thus term-by-term integration is permissible on $[0, 1]$. This gives

$$\int_0^1 \frac{\log(1 + x)}{x} \, \mathrm{d}x = \sum_{k=1}^{\infty} \frac{(-1)^{k-1}}{k^2}.$$

The same idea may be used to show that

$$\int_0^1 (-\log(1 - x)) \, \mathrm{d}x = \sum_{k=1}^{\infty} \frac{1}{k(k + 1)} = 1.$$

16. What is the Cauchy product of $\sum_{k=0}^{\infty} a_k x^k$ and $\sum_{k=0}^{\infty} x^k$? For the second part, choose

$$f(x) = \frac{\log(1+x)}{x} = \sum_{k=0}^{\infty} \frac{(-1)^k}{k+1} x^k.$$

Chapter 10: Questions 10.10

11. Yes. See Figure 10.6.

Exercises 10.11:

3. We have $b_n = 0$ for $n \geq 1$,

$$a_0 = \frac{2\pi^2}{3}, \quad \text{and} \quad a_n = \frac{4(-1)^n}{n^2}.$$

6. Example 10.9 and Exercise 10.11(6) may be obtained each from the other using the linearity property, for if $f(x) = |x|$, then

$$1 - \frac{|x|}{\pi} = m|x| + c$$

with $c = 1$, $m = -1/\pi$.

Questions 10.37:

2. f is even on $(-c, c)$ if and only if $2f(x) = f(x) + f(-x)$.
6. On $|x| < c$, define

$$g(x) = (f(x) + f(-x))/2 \quad \text{and} \quad h(x) = (f(x) - f(-x))/2.$$

Then g and h are even and odd, respectively.
8. We see that $a_k = 0$ for odd values of k, and $b_k = 0$ for even values of k. Also, for $k \geq 0$,

$$a_{2k} = \frac{2}{\pi} \int_{-\pi/2}^{\pi/2} f(x) \cos 2kx \, dx \quad \text{and} \quad b_{2k+1} = \frac{2}{\pi} \int_{-\pi/2}^{\pi/2} f(x) \sin(2k+1)x \, dx.$$

9. Continuity of the periodic extension will depend on the choices of a, b, c.
10. No. For example, the periodic extension of the characteristic function of the set of all rational numbers between 0 and 1 is not Riemann integrable.

Exercises 10.38:

3. We have

$$x^2 = \frac{\pi^2}{3} + 4 \sum_{n=1}^{\infty} \frac{(-1)^n}{n^2} \cos nx, \qquad x \in [-\pi, \pi],$$

and in particular, for $x = 0$, we have

$$\frac{\pi^2}{12} = \sum_{n=1}^{\infty} \frac{(-1)^{n-1}}{n^2}.$$

Also,

$$x^2 = \frac{4\pi^2}{3} + 4\sum_{n=1}^{\infty} \frac{\cos nx}{n^2} - 4\pi \sum_{n=1}^{\infty} \frac{\sin nx}{n}, \qquad x \in (0, 2\pi),$$

and at $x = 0, 2\pi$, the series converges to the value $2\pi^2$.

13. It suffices to consider $f(x) = x$, so that by Example 10.15,

$$3x + 2 \sim 2 + 6\sum_{k=1}^{\infty} \frac{(-1)^{k-1}}{k} \sin kx, \quad x \in [-\pi, \pi].$$

14. Use the method of Example 10.17.
15. Consider the odd extension of f to $[-L, 0]$, and then the periodic extension (for the period $2L$) onto $\mathbb{R}$; see Figure A.16.
17. See Figure A.17. It is a simple exercise to see that $a_0 = -\pi/2$,

$$a_n = \frac{(-1)^n - 1}{n^2\pi} \quad \text{and} \quad b_n = \frac{1 - 2(-1)^n}{n}.$$

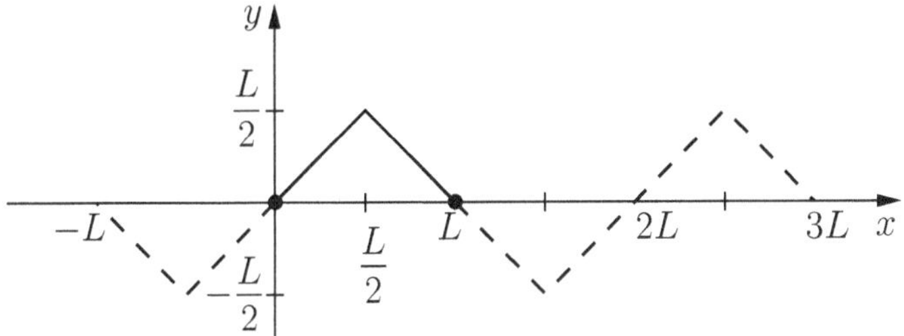

Fig. A.16. Odd extension of f to $[-L, 0]$, $2L$-periodic extension to $\mathbb{R}$.

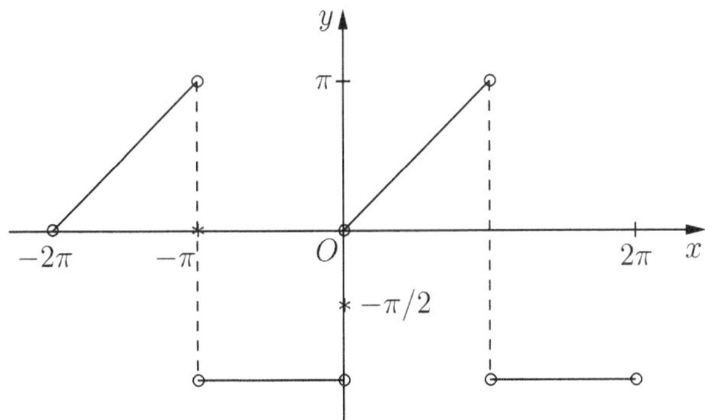

Fig. A.17. Graph of the given function.

By Dirichlet's theorem, equality in (10.7) holds at all points of continuity, since f has been defined to be periodic. At $x=0$ and $x=\pi$, the Fourier series with the above Fourier coefficients converge to

$$\frac{f(0+)+f(0-)}{2}=-\frac{\pi}{2} \text{ and } \frac{f((-\pi)+)+f(\pi-)}{2}=\frac{-\pi+\pi}{2}=0,$$

respectively. In either case, we can obtain (10.8) as a special case.

Chapter 11: Questions 11.27

1. No.
2. No; for instance, $f(x)=\sin x$ on $[0,\pi]$ and $f(x)=x^2$ on $[-1,1]$.
3. No; for instance, consider $f(x)=[x]$ on $[0,2]$.
4. By Theorem 11.8, it is bounded.
6. No. Consider $f(x)=x$ on $[0,1]$. Indeed if f is a function such that $f(x)\to 0$ as $x\to x_0$ for some $x_0\in(a,b)$, then $1/f$ cannot be bounded on any interval containing x_0, and hence $1/f$ cannot be a function of bounded variation on such an interval.
9. Yes.
13. Yes. Apply Theorem 11.13.
14. Yes. Set $G_{\pm}=V\pm f$. For $a\leq x<y\leq b$,

$$\begin{aligned} G_{\pm}(y)-G_{\pm}(x) &= (V(y)-V(x))\pm(f(y)-f(x)) \\ &= V_f[x,y]\pm(f(y)-f(x)), \quad \text{by Lemma 11.18,} \\ &\geq 0, \quad \text{since } V_f[x,y]\geq|f(y)-f(x)|. \end{aligned}$$

   Thus, both $V+f$ and $V-f$ are increasing on $[a,b]$. Note that

$$f=(V+f)-V=V-(V-f)=(V+g)-(V-f+g)$$

   for an arbitrary increasing function g on $[a,b]$.
15. No. Suppose that $f=f_1-f_2$, where f_1 and f_2 are increasing. Then for any arbitrary increasing function g, we could write

$$f=(f_1+g)-(f_2+g)=F_1-F_2,$$

   where F_1 and F_2 are again increasing.
16. Yes. In the above hint choose g to be strictly increasing. Then Theorem 11.19 continues to hold if "increasing" is replaced by "strictly increasing."
17. If f and g are absolutely continuous on $[a,b]$, then

$$|(f+g)(b_k)-(f+g)(a_k)|\leq|f(b_k)-f(a_k)|+|g(b_k)-g(a_k)|,$$

   and for $\alpha\in\mathbb{R}$,

$$|(\alpha f)(b_k)-(\alpha f)(a_k)|=|\alpha|\,|f(b_k)-f(a_k)|,$$

   showing that $f+g$ and αf are absolutely continuous on $[a,b]$.

Exercises 11.28:

1. We see that $V(P,f) = 1 + 2(n-1) = 2n-1$.
3. $V(P,f) = \sum_{k=1}^{n} \frac{1}{k}$.
4. Both $\cos x$ and x are in $BV([0,\pi/2])$, and so is their sum. Finally, apply the product rule, since x^4 is also in $BV([0,\pi/2])$.
5. Follows Theorem 11.9.
7. Note that $x_0 = 0$, $f(0) = 0$, $x_k = 1/(n+1-k)$ for $k \geq 1$ and $\cos(\pi/x_k) = (-1)^{n+1-k}$. Thus, we find that

$$\begin{aligned} V(P,f) &= |f(x) - f(0)| + \sum_{k=2}^{n} |f(x_k) - f(x_{k-1})| \\ &= \frac{1}{n} + \sum_{k=2}^{n} \left(\frac{1}{n+1-k} + \frac{1}{n+2-k} \right) \\ &= 1 + 2\left(\sum_{k=2}^{n} \frac{1}{k} \right) \to \infty \quad \text{as } n \to \infty. \end{aligned}$$

8. Consider the partition $P = \{0, \frac{1}{n}, \ldots, \frac{1}{3}, \frac{1}{2}, 1\}$. We observe that f is piecewise monotone and

$$\sum_{k=1}^{n} |f(x_k) - f(x_{k-1})| > 1 + \frac{1}{2} + \frac{1}{3} + \cdots + \frac{1}{n-1}.$$

9. For a partition $P = \{x_0, x_1, x_2, \ldots, x_n\}$ of $[a,b]$, we have

$$F(x_k) - F(x_{k-1}) = \int_a^{x_k} f(t)\,\mathrm{d}t + \int_{x_{k-1}}^{a} f(t)\,\mathrm{d}t = \int_{x_{k-1}}^{x_k} f(t)\,\mathrm{d}t,$$

and $|f(t)|$ is also integrable on $[a,b]$. Thus,

$$V(P,F) = \sum_{k=1}^{n} \left| \int_{x_{k-1}}^{x_k} f(t)\,\mathrm{d}t \right| \leq \sum_{k=1}^{n} \int_{x_{k-1}}^{x_k} |f(t)|\,\mathrm{d}t = \int_a^b |f(t)|\,\mathrm{d}t < \infty.$$

10. We show that $f \in BV([0,1])$ if and only if $\alpha > \beta$. The given function f is clearly differentiable on $(0,1]$, and for $\alpha > 0$, $\beta > 0$,

$$|f'(x)| \leq \alpha x^{\alpha-1} + \beta x^{\alpha-\beta-1}.$$

For the partition $P = \{x_0, x_1, \ldots, x_n\}$ of $[\epsilon, 1]$, we have

$$|f(x_k) - f(x_{k-1})| = \left| \int_{x_{k-1}}^{x_k} f'(t)\,\mathrm{d}t \right| \leq \int_{x_{k-1}}^{x_k} |f'(t)|\,\mathrm{d}t,$$

and so

$$\sum_{k=1}^{n} |f(x_k) - f(x_{k-1})| \le \sum_{k=1}^{n} \int_{x_{k-1}}^{x_k} |f'(t)|\, dt = \int_{\epsilon}^{1} |f'(t)|\, dt,$$

and $\int_\epsilon^1 |f'(t)|\, dt$ exists for all $\epsilon > 0$ and for $\alpha > 0$ and $\alpha > \beta \geq 0$ and

$$\int_0^1 |f'(t)|\, dt = \lim_{\epsilon \to 0} \int_\epsilon^1 |f'(t)|\, dt = \alpha\left(\frac{1}{\alpha}\right) + \frac{\beta}{\alpha - \beta} = \frac{\alpha}{\alpha - \beta}.$$

Thus, $f \in BV([0,1])$ if $\alpha > \beta > 0$.
For $\alpha = 0$, it is easy to see that $f \notin BV([0,1])$.
For $0 < \alpha \leq \beta$, by choosing a suitable partition, it can be shown that $f \notin BV([0,1])$. Because of the appearance of $\sin(1/x^\beta)$, we may choose the partition points from those x for which

$$\frac{1}{x^\beta} = \frac{(2m-1)\pi}{2}, \quad m \in \mathbb{Z},$$

so that $\sin(1/x^\beta) = (-1)^m$. Thus for $\beta > 0$, we consider the partition P of $[0,1]$ as

$$P = \left\{0, \left(\frac{2}{\pi(2n-1)}\right)^{1/\beta}, \left(\frac{2}{\pi(2n-3)}\right)^{1/\beta}, \left(\frac{2}{3\pi}\right)^{1/\beta}, \left(\frac{2}{\pi}\right)^{1/\beta}, 1\right\},$$

and obtain

$$V(P,f) \to \infty \ \text{ as } n \to \infty.$$

11. Use Theorem 11.23.
13. Let $\epsilon > 0$ be given. Then for any partition P of $[a,b]$, we have

$$V(P,f) > V_f[a,b] - \frac{\epsilon}{2}.$$

Since f is continuous, there exists a $\delta > 0$ such that

$$|f(x) - f(y)| < \frac{\epsilon}{4n} \ \text{ whenever } x, y \in [a,b],\ |x - y| < \delta.$$

It can be easily shown that for every partition Q of $[a,b]$ with $\|Q\| < \delta$, we have

$$V(Q,f) > V(P,f) - \frac{\epsilon}{2}.$$

This gives

$$V_f[a,b] > V(Q,f) > V_f[a,b] - \epsilon,$$

and the result follows.

14. Let P be a partition of $[a, x]$. Then

$$\begin{aligned} V(P, f) &= \sum_{k \in A(P)} (f(x_k) - f(x_{k-1})) - \sum_{k \in N(P)} (f(x_k) - f(x_{k-1})) \\ &= S^+(P) + S^-(P), \quad \text{say.} \end{aligned}$$

Then $S^+(P) - S^-(P) = f(x) - f(a)$. Adding and subtracting the last two equations gives

$$2S^+(P) = V(P, f) + f(x) - f(a) \;\text{ and }\; 2S^-(P) = V(P, f) - (f(x) - f(a)),$$

respectively. Taking the supremum as P varies over $P[a, x]$ gives (**a**). Adding and subtracting the two equations in (**a**) gives (**b**) and (**c**). The proofs for (**d**) and (**e**) follow as a consequence of (**c**) and the fact that $p(x) \geq 0$ and $n(x) \geq 0$. Finally, for $a \leq x < y \leq b$, (**a**) gives

$$\begin{aligned} 2[p(y) - p(x)] &= (V(y) - V(x)) + (f(y) - f(x)) \\ &= V_f[x, y] + (f(y) - f(x)) \geq 0, \end{aligned}$$

and similarly,

$$2[n(y) - n(x)] = V_f[x, y] - (f(y) - f(x)) \geq 0.$$

Thus, (**f**) follows.

Questions 11.47:

10. By Theorem 11.41, the value of the integral is $(f^2(b) - f^2(a))/2$.

Exercises 11.48:

4. Let $P = \{x_0, x_1, \ldots, x_n\}$ be any partition of $[0, 2]$ with $x_j = 1$. It follows easily that

$$U(P, f, \alpha) = 0 \;\text{ and }\; L(P, f, \alpha) = 0.$$

The conclusion follows from Theorem 11.33. On the other hand, if $P' = \{x_0, x_1, \ldots, x_n\}$ is a partition of $[0, 2]$ that does not include 1 as a partition point, then $x_{j-1} < 1 < x_j$, and therefore we see that

$$U(P', f, \alpha) = 1 \;\text{ and }\; L(P', f, \alpha) = 0,$$

no matter how small the norm of the partition.

Index

S. Ponnusamy, *Foundations of Mathematical Analysis*,
DOI 10.1007/978-0-8176-8292-7,

Zeitfracht Medien GmbH
Ferdinand-Jühlke-Straße 7
99095 Erfurt, Deutschland
produktsicherheit@kolibri360.de